Encyclopedia of
Structural Health Monitoring

Encyclopedia of Structural Health Monitoring

Volume 3

Editors-in-Chief

Christian Boller
Saarland University & Fraunhofer Institute for Non-Destructive Testing, Saarbrücken, Germany
(and formerly of The University of Sheffield, Sheffield, UK)

Fu-Kuo Chang
Stanford University, Stanford, CA, USA

Yozo Fujino
University of Tokyo, Tokyo, Japan

A John Wiley and Sons, Ltd., Publication

This edition first published 2009
© John Wiley & Sons Ltd

Registered office

John Wiley & Sons Ltd, The Atrium, Southern Gate, Chichester, West Sussex, PO19 8SQ, United Kingdom

For details of our global editorial offices, for customer services and for information about how to apply for permission to reuse the copyright material in this book please see our website at www.wiley.com.

All rights reserved. No part of this publication may be reproduced, stored in a retrieval system, or transmitted, in any form or by any means, electronic, mechanical, photocopying, recording or otherwise, except as permitted by the UK Copyright, Designs and Patents Act 1988, without the prior permission of the publisher.

Wiley also publishes its books in a variety of electronic formats. Some content that appears in print may not be available in electronic books.

Designations used by companies to distinguish their products are often claimed as trademarks. All brand names and product names used in this book are trade names, service marks, trademarks or registered trademarks of their respective owners. The publisher is not associated with any product or vendor mentioned in this book. This publication is designed to provide accurate and authoritative information in regard to the subject matter covered. It is sold on the understanding that the publisher is not engaged in rendering professional services. If professional advice or other expert assistance is required, the services of a competent professional should be sought.

Library of Congress Cataloging-in-Publication Data

Encyclopedia of structural health monitoring / editors-in-chief, Christian Boller, Fu-Kuo Chang, Yozo Fujino.
     p. cm.
   Includes bibliographical references and index.
   ISBN 978-0-470-05822-0 (cloth : alk. paper)
  1. Structural analysis (Engineering)—Encyclopedias. 2. Buildings—Inspection—Encyclopedias. 3. Structural frames—Inspection—Encyclopedias. 4. Structural failures—Risk assessment—Encyclopedias. 5. Detectors—Encyclopedias. I. Boller, C. (Christian) II. Chang, Fu-Kuo. III. Fujino, Yozo, 1949-
   TA656.5.E53 2009
   624.1'7—dc22

                                                                                         2008049836

A catalogue record for this book is available from the British Library.

ISBN-13: 978-0-470-05822-0

Set in 9.5/11.5 pt Times by Laserwords Private Limited, Chennai, India
Printed and bound by Grafos S.A., Barcelona, Spain

Editorial Board

Editors-in-Chief

Christian Boller
Saarland University & Fraunhofer Institute for Non-Destructive Testing
Saarbrücken
Germany

(and formerly of The University of Sheffield, Sheffield, UK)

Fu-Kuo Chang
Stanford University
Stanford, CA
USA

Yozo Fujino
University of Tokyo
Tokyo
Japan

Subject Editors

PHYSICAL MONITORING PRINCIPLES
Douglas E. Adams
Purdue University
West Lafayette, IN
USA

SIGNAL PROCESSING
Wieslaw J. Staszewski
and Keith Worden
University of Sheffield
Sheffield
UK

SIMULATION
Srinivasan Gopalakrishnan
Indian Institute of Science
Bangalore
India

SENSORS
Victor Giurgiutiu
University of South Carolina
Columbia, SC
USA

SYSTEMS AND SYSTEM DESIGN
Daniel L. Balageas
ONERA (The French Aerospace Lab)
Châtillon
France

PRINCIPLES OF SHM-BASED STRUCTURAL MONITORING, DESIGN AND MAINTENANCE
Michael K. Bannister
CRC-ACS
Melbourne, VIC
Australia

AEROSPACE APPLICATIONS
Stephen C. Galea
Defence Science and Technology Organisation
Melbourne, VIC
Australia

CIVIL ENGINEERING APPLICATIONS
Helmut Wenzel
VCE Holding GmbH
Vienna
Austria

OTHER APPLICATIONS
Daniel J. Inman
Virginia Tech
Blacksburg, VA
USA

Contents

VOLUME 1

Contributors xiii

Preface xxxiii

Abbreviations and Acronyms xxxvii

Part 1: Introduction 1

1. Structural Health Monitoring—An Introduction and Definitions 3
   Christian Boller

Part 2: Physical Monitoring Principles 27

Section 1: Physics-based and Data-driven Modeling of Structural Component Behavior 29

2. Free and Forced Vibration Models 31
   Muhammad Haroon
3. Fundamentals of Guided Elastic Waves in Solids 59
   Carlos E. S. Cesnik, Ajay Raghavan
4. Acoustic Emission 79
   Michael R. Gorman
5. Electromechanical Impedance Modeling 103
   Andrei N. Zagrai, Victor Giurgiutiu
6. Thermomechanical Models 123
   Minh P. Luong

Section 2: Modeling of Load and Damage Mechanisms in SHM 141

7. Civil Infrastructure Load Models 143
   Udo Peil
8. Static Damage Phenomena and Models 171
   Shankar Sundararaman
9. Damage Evolution Phenomena and Models 203
   Alten F. Grandt Jr.
10. Failure Modes of Aerospace Materials 219
    Kumar V. Jata, Ajit Roy, Triplicane A. Parthasarathy

Section 3: Passive and Active Measurements with Data Analysis Approaches for Load and Damage Identification 237

11. Modal–Vibration-based Damage Identification 239
    Keith Worden, Michael I. Friswell
12. Data Interrogation Approaches with Strain and Load Gauge Sensor Arrays 277
    Michael D. Todd
13. Applications of Acoustic Emission for SHM: A Review 289
    Martine Wevers, Kasper Lambrighs
14. Ultrasonic Methods 303
    Wolfgang Hillger
15. Nonlinear Acoustic Methods 321
    Dimitri M. Donskoy
16. Guided-wave Array Methods 333
    Paul D. Wilcox
17. Lamb Wave-based SHM for Laminated Composite Structures 353
    Constantinos Soutis
18. Piezoelectric Impedance Methods for Damage Detection and Sensor Validation 365
    Gyuhae Park, Charles R. Farrar
19. Thermal Imaging Methods 379
    Daniel L. Balageas
20. Eddy-current Methods 393
    Neil Goldfine, Andrew Washabaugh, Yanko Sheiretov, Mark Windoloski

VOLUME 2

Part 3: Signal Processing — 413

21 Signal Processing for Damage Detection — 415
   Wieslaw J. Staszewski, Keith Worden
22 Data Preprocessing for Damage Detection — 423
   Andrew Halfpenny
23 Statistical Time Series Methods for SHM — 443
   Spilios D. Fassois, John S. Sakellariou
24 Cepstral Methods of Operational Modal Analysis — 473
   Robert B. Randall
25 Hilbert Transform, Envelope, Instantaneous Phase, and Frequency — 487
   Michael Feldman
26 Time–frequency Analysis — 503
   Rosario Ceravolo
27 Wavelet Analysis — 525
   Amy N. Robertson, Biswajit Basu
28 Damage Detection Using the Hilbert–Huang Transform — 541
   Darryll J. Pines, Liming W. Salvino
29 Higher Order Statistical Signal Processing — 567
   Paul White
30 Statistical Pattern Recognition — 579
   Hoon Sohn, Chang Kook Oh
31 Machine Learning Techniques — 597
   Fulei Chu, Shengfa Yuan, Zhike Peng
32 Artificial Neural Networks — 611
   Steve Reed
33 Dimensionality Reduction Using Linear and Nonlinear Transformation — 625
   Gaëtan Kerschen, Jean-Claude Golinval
34 Nonlinear Features for SHM Applications — 639
   Jonathan M. Nichols, Michael D. Todd
35 Novelty Detection — 653
   Lionel Tarassenko, David A. Clifton, Peter R. Bannister, Steve King, Dennis King
36 Model-based Statistical Signal Processing for Change and Damage Detection — 677
   Michèle Basseville
37 Data Fusion of Multiple Signals from the Sensor Network — 697
   Zhongqing Su, Xiaoming Wang, Lin Ye
38 Optimization Techniques for Damage Detection — 709
   Keith Worden, Wieslaw Staszewski, Graeme Manson, Aldo Ruotulo, Cecilia Surace
39 Uncertainty Analysis — 731
   Graeme Manson, Keith Worden, S. Gareth Pierce, Thierry Denoeux

Part 4: Simulation — 753

Section 1: Constitutive Modeling of Smart Materials — 755
40 Piezoceramic Materials—Phenomena and Modeling — 757
   Jayabal Kaliappan, Srinivasan M. Sivakumar
41 Constitutive Modeling of Magnetostrictive Materials — 773
   Srinivasan Gopalakrishnan

Section 2: Modeling Methods for SHM — 789
42 Modeling Aspects in Finite Elements — 791
   Srinivasan Gopalakrishnan
43 Finite Elements: Modeling of Piezoceramic and Magnetostrictive Sensors and Actuators — 811
   Srinivasan Gopalakrishnan
44 A Simplified Damage Model for SHM Metallic and Composite Structures — 833
   N. Hu, D. R. Mahapatra, Srinivasan Gopalakrishnan
45 Modeling for Detection of Degraded Zones in Metallic and Composite Structures — 851
   Wiesław Ostachowicz, Marek Krawczuk
46 Fatigue Life Assessment of Structures — 867
   Thomas Bruder

Section 3: Damage Detection Algorithms and Techniques — 887
47 Damage Detection Using Piezoceramic and Magnetostrictive Sensors and Actuators — 889
   Srinivasan Gopalakrishnan
48 Damage Measures — 907
   Massimo Ruzzene
49 Modeling of Lamb Waves in Composite Structures — 923
   Abir Chakraborty

Section 4: Probabilistic Approaches — 941
50 Probabilistic Approaches to Sensor Layout Design, Data Processing, and Damage Detection — 943
   Sankaran Mahadevan, Xiaomo Jiang, Robert F. Guratzsch
51 Development of Fuzzy Rules for Damage Detection and Location — 963
   Ranjan Ganguli

VOLUME 3

Part 5: Sensors 979

52 Piezoelectricity Principles and Materials 981
    Victor Giurgiutiu
53 Operational Loads Sensors 993
    William F. Ranson, Reginald I. Vachon
54 Nondestructive Evaluation/Nondestructive Testing/Nondestructive Inspection (NDE/NDT/NDI) Sensors—Eddy Current, Ultrasonic, and Acoustic Emission Sensors 1003
    Mark Blodgett, Eric Lindgren, Shamachary Sathish, Kumar V. Jata
55 Piezoelectric Wafer Active Sensors 1013
    Lingyu Yu, Victor Giurgiutiu
56 Damage Presence/Growth Monitoring Sensors 1029
    Hua Gu, Ming L. Wang
57 Piezoelectric Paint Sensors for Ultrasonics-based Damage Detection 1037
    Yunfeng Zhang
58 Eddy-current in situ Sensors for SHM 1051
    Neil Goldfine, Vladimir Zilberstein, Darrell Schlicker, Dave Grundy
59 Fiber-optic Sensor Principles 1065
    Kara Peters
60 Intensity-, Interferometric-, and Scattering-based Optical-fiber Sensors 1083
    Kara Peters
61 Fiber Bragg Grating Sensors 1097
    Kara Peters
62 Novel Fiber-optic Sensors 1113
    Kara Peters
63 Electric and Electromagnetic Properties Sensing 1125
    Michel B. Lemistre
64 Directed Energy Sensors/Actuators 1137
    James L. Blackshire
65 Full-field Sensing: Three-dimensional Computer Vision and Digital Image Correlation for Noncontacting Shape and Deformation Measurements 1153
    Michael A. Sutton
66 Global Navigation Satellite Systems (GNSSs) for Monitoring Long Suspension Bridges 1169
    Xiaolin Meng, Wei Huang
67 Nanoengineering of Sensory Materials 1189
    Inpil Kang, Gunjan Maheshwari, YeoHeung Yun, Vesselin Shanov, Sachit Chopra, Jandro Abot, Gyeongrak Choi, Mark Schulz
68 Miniaturized Sensors Employing Micro- and Nanotechnologies 1211
    Kenneth J. Loh, Jerome P. Lynch

Part 6: Systems and System Design 1225

Section 1: Sensor/Actuator Network Configuration 1227

69 Wireless Sensor Network Platforms 1229
    Reinhard Bischoff, Jonas Meyer, Glauco Feltrin
70 Sensor Placement Optimization 1239
    Robert J. Barthorpe, Keith Worden
71 Sensor Network Paradigms 1251
    Charles R. Farrar, Gyuhae Park, Kevin M. Farinholt
72 Nondestructive Evaluation of Cooperative Structures (NDECS) 1269
    Daniel L. Balageas

Section 2: Information Fusion and Data Management 1281

73 Web-based SHM 1283
    Vistasp M. Karbhari, Hong Guan
74 Design of Active Sensor Network and Multilevel Data Fusion 1301
    Xiaoming Wang, Zhongqing Su

Section 3: Autonomous Sensing 1315

75 Energy Harvesting and Wireless Energy Transmission for SHM Sensor Nodes 1317
    Kevin M. Farinholt, Gyuhae Park, Charles R. Farrar
76 On the Way to Autonomy: the Wireless-interrogated and Self-powered "Smart Patch" System 1329
    Stephen C. Galea, Stephen Van der Velden, Scott Moss, Ian Powlesland
77 Energy Harvesting using Thermoelectric Materials 1351
    Daniel J. Inman, Henry A. Sodano

Section 4: Examples of Systems 1361

78 Stanford Multiactuator–Receiver Transduction (SMART) Layer Technology and Its Applications 1363
    Xinlin P. Qing, Shawn J. Beard, Amrita Kumar, Irene Li, Mark Lin, Fu-Kuo Chang
79 Hybrid PZT/FBG Sensor System 1389
    Zhanjun Wu, Xinlin P. Qing, Fu-Kuo Chang
80 The HELP-Layer® System 1403
    Michel B. Lemistre

81	Microelectromechanical Systems (MEMS)	1413
	Jonas Meyer, Reinhard Bischoff, Glauco Feltrin	

VOLUME 4

Part 7: Principles of SHM-based Structural Monitoring, Design and Maintenance 1423

Section 1: Design and Assessment 1425

82	Principles of Structural Degradation Monitoring	1427
	Charles R. Farrar, Keith Worden, Janice Dulieu-Barton	
83	Design Principles for Aerospace Structures	1449
	Jens Telgkamp	
84	Design Principles for Civil Structures	1467
	Vistasp M. Karbhari	

Section 2: Loads and Environment Monitoring 1477

85	Loads Monitoring in Aerospace Structures	1479
	Steve Reed	
86	Risk Monitoring of Aircraft Fatigue Damage Evolution at Critical Locations	1497
	Michael Shiao	
87	Risk Monitoring of Civil Structures	1511
	Narito Kurata	
88	Environmental Monitoring of Aircraft	1523
	Nicholas C. Bellinger, Marcias Martinez	

Section 3: Maintenance 1531

89	Maintenance Principles for Civil Structures	1533
	Dan M. Frangopol, Thomas B. Messervey	
90	Military Aircraft	1563
	Mark M. Derriso, Steven E. Olson, Martin P. DeSimio	
91	Use of Leave-in-place Sensors and SHM Methods to Improve Assessments of Aging Structures	1579
	Dennis Roach	

Section 4: Management 1619

92	Usage Management of Military Aircraft Structures	1621
	Rolf H. Neunaber	
93	Usage Management of Civil Structures	1635
	Ayaho Miyamoto	

94	Value Assessment Approaches for Structural Life Management	1673
	Enrique A. Medina, John C. Aldrin	

Part 8: Aerospace Applications 1691

Section 1: Full-Scale Aerospace Vehicles 1693

95	Commercial Fixed-wing Aircraft	1695
	Grant A. Gordon, Christian Boller	
96	History of SHM for Commercial Transport Aircraft	1711
	Roy Ikegami, Christian Boller	
97	Fatigue Monitoring in Military Fixed-wing Aircraft	1723
	Matthias Buderath	
98	Agile Military Aircraft	1743
	Loris Molent, Jason Agius	
99	Flight Demonstration of a SHM System on a USAF Fighter Airplane	1761
	Matthew C. Malkin	
100	Operational Loads Monitoring in Military Transport Aircraft and Military Derivatives of Civil Aircraft	1767
	Len Meadows, Steve Reed, Mike Duffield	
101	Unmanned Aerial Vehicles	1787
	Matthias Buderath	
102	Health and Usage Monitoring Systems (HUM Systems) for Helicopters: Architecture and Performance	1805
	Kenneth Pipe	
103	Experience with Health and Usage Monitoring Systems in Helicopters	1819
	Dy Dinh Le	
104	Thermal Protection System Monitoring of Space Structures	1827
	William H. Prosser, Eric I. Madaras, George F. Studor, Michael R. Gorman	

Section 2: Technologies 1837

105	Validation of SHM Sensors in Airbus A380 Full-scale Fatigue Test	1839
	Christophe Paget, Holger Speckmann, Thomas Krichel, Frank Eichelbaum	
106	Comparative Vacuum Monitoring (CVM™)	1849
	Duncan P. Barton	
107	Development of an Active Smart Patch for Aircraft Repair	1867
	Nik Rajic	
108	Aerospace Applications of SMART Layer Technology	1881
	Xinlin P. Qing, Shawn J. Beard, Roy Ikegami, Fu-Kuo Chang, Christian Boller	

109	Fiber-optic Sensors *Peter Foote*	1897	123	Modular Architecture of SHM System for Cable-supported Bridges 2089 *Kai-Yuen Wong, Yi-Qing Ni*

Section 3: Aircraft Structural Design   **1913**

110   Design Benefits in Aeronautics Resulting from SHM   1915
    Hans-Juergen Schmidt, Bianka Schmidt-Brandecker

111   Design, Analysis, and SHM of Bonded Composite Repair and Substructure   1923
    Constantinos Soutis, Jeong-Beom Ihn

112   Aircraft Structural Diagnostic and Prognostic Health Monitoring for Corrosion Prevention and Control   1941
    Stephen C. Galea, Tony Trueman, Len Davidson, Peter Trathen, Bruce Hinton, Alan Wilson, Tim Muster, Ivan Cole, Penny Corrigan, Don Price

Section 4: Aircraft Systems   **1969**

113   Video Landing Parameter Surveys   1971
    Thomas DeFiore, Richard P. Micklos

114   Landing Gear   1983
    R. Kyle Schmidt, Pia Sartor

115   Monitoring of Aircraft Engines   1995
    Visakan Kadirkamanathan, Peter Fleming

116   Monitoring of Solid Rocket Motors   2011
    Gregory A. Ruderman

117   Health Monitoring, Diagnostics, and Prognostics of Avionic Systems   2021
    Michael Pecht, Yan-Cheong Chan

VOLUME 5

Part 9: Civil Engineering Applications   2029

118   The Character of SHM in Civil Engineering   2031
    Helmut Wenzel

119   Ambient Vibration Monitoring   2039
    Helmut Wenzel

120   The Influence of Environmental Factors   2057
    Helmut Wenzel

121   Long-term Monitoring of Dynamic Loads on the Brandenburg Gate   2067
    Werner Rücker

122   Development of a Monitoring System for a Long-span Cantilever Truss Bridge   2077
    F. Necati Catbas, A. Emin Aktan

123   Modular Architecture of SHM System for Cable-supported Bridges   2089
    Kai-Yuen Wong, Yi-Qing Ni

124   Monitoring of Bridges in Korea   2107
    Hyun-Moo Koh, Hae-Sung Lee, Sungkon Kim, Jinkyo F. Choo

125   Bridge Monitoring in Japan   2131
    Masato Abe, Yozo Fujino

126   Continuous Vibration Monitoring and Progressive Damage Testing on the Z24 Bridge   2149
    Edwin Reynders, Guido De Roeck

127   Continuous Monitoring of the Øresund Bridge: Data Acquisition and Operational Modal Analysis   2159
    Bart Peeters

128   Condition Compensation in Frequency Analyses —a Basis for Damage Detection   2175
    Robert Veit-Egerer

129   Modal Testing of the Vasco da Gama Bridge, Portugal   2183
    Elsa Caetano, Álvaro Cunha

130   Multiple-model Structural Identification   2199
    Ian F. C. Smith

131   Construction Process Monitoring at the New Berlin Main Station   2209
    Rosemarie Helmerich

132   SHM Actions on the Holy Shroud Chapel in Torino   2223
    Alessandro De Stefano

133   SHM of a Tall Building   2233
    James M. W. Brownjohn, Tso-Chien Pan

134   Dynamic Response of Buildings of the Cultural Heritage   2243
    Paolo Clemente, Giacomo Buffarini

135   Suspended Roof of Braga Sports Stadium, Portugal   2253
    Álvaro Cunha, Filipe Magalhães, Elsa Caetano

136   Dams   2265
    Reto Cantieni

137   Condamine Floating Dock, Monaco   2289
    Luis M. Ortega, Manuel A. Floriano

138   Soil–Structure Interaction and Seismic Effects   2305
    Günther Achs

139   System Identification for Soil–structure Interaction   2319
    Erdal Safak

140   Loads and Temperature Effects on a Bridge   2327
    Ming L. Wang

141   Environmental Factors Derived from Satellite Data of Java, Indonesia   2343
    Barbara Teilen-Willige, Farah Mulyasari Sule, Helmut Wenzel

Part 10: Other Applications — 2355

142 Monitoring Marine Structures — 2357
Liming W. Salvino, Matthew D. Collette

143 Diagnosing Offshore Machines and Power Plants Using Vibration Methods — 2373
Andrzej Grzadziela

144 Ship and Offshore Structures — 2387
Myung Hyun Kim, Do Hyung Kim

145 Noncontact Rail Monitoring by Ultrasonic Guided Waves — 2397
Piervincenzo Rizzo, Stefano Coccia, Ivan Bartoli, Francesco Lanza di Scalea

146 Sensor Technologies for Direct Health Monitoring of Tires — 2411
Ronald D. Moffitt, Scott M. Bland, Mohammad R. Sunny, Rakesh K. Kapania

147 Wind Turbines — 2419
Goutham R. Kirikera, Mannur Sundaresan, Francis Nkrumah, Gangadhararao Grandhi, Bashir Ali, Sai L. Mullapudi, Vesselin Shanov, Mark Schulz

148 Large Rotating Machines — 2443
Tomasz Gałka

149 Gas Turbine Engines — 2457
Michael J. Roemer

150 Prognostics and Health Management of Electronics — 2473
Michael Pecht

151 Multiwire Strands — 2487
Francesco Lanza di Scalea, Ivan Bartoli, Piervincenzo Rizzo, Alessandro Marzani, Elisa Sorrivi, Erasmo Viola

152 SHM and Lifetime Management of Industrial Piping Systems — 2505
Frank Schubert, Bernd Frankenstein, Thomas Klesse, Klaus Kerkhof, Xaver Schuler, Herbert Friedmann, Fritz-Otto Henkel, Helmut Wenzel

153 Fatigue Monitoring in Nuclear Power Plants — 2525
Wilhelm Kleinöder, Christian Pöckl

154 Landfills — 2539
Kai Münnich, Jan Bauer, Klaus Fricke

Part 11: Specifications and Standardization — 2549

155 Reliable Use of Fiber-optic Sensors — 2551
Wolfgang R. Habel

156 Integrated Sensor Durability and Reliability — 2565
James L. Blackshire, Kumar V. Jata

157 Open Systems Architecture for Condition-based Maintenance — 2581
Robert L. Walter IV, David Boylan, Daniel Gilbertson

Glossary — 2591

Index — 2637

Contributors

Masato Abe
BMC Corporation, Mihama-Ku, Chiba, Japan
Chapter 125: Bridge Monitoring in Japan

Jandro Abot
Aerospace Engineering, University of Cincinnati, Cincinnati, OH, USA
Chapter 67: Nanoengineering of Sensory Materials

Günther Achs
VCE – Vienna Consulting Engineers, Vienna, Austria
Chapter 138: Soil–Structure Interaction and Seismic Effects

Jason Agius
Directorate General Technical Airworthiness, RAAF Williams, VIC, Australia
Chapter 98: Agile Military Aircraft

A. Emin Aktan
Drexel Intelligent Infrastructure Institute, Drexel University, Philadelphia, PA, USA
Chapter 122: Development of a Monitoring System for a Long-span Cantilever Truss Bridge

John C. Aldrin
Computational Tools, Gurnee, IL, USA
Chapter 94: Value Assessment Approaches for Structural Life Management

Bashir Ali
Department of Mechanical and Chemical Engineering, North Carolina A&T State University, Greensboro, NC, USA
Chapter 147: Wind Turbines

Daniel L. Balageas
Structure and Damage Mechanics Department, ONERA (The French Aerospace Lab), Châtillon, France
Chapter 19: Thermal Imaging Methods
Chapter 72: Nondestructive Evaluation of Cooperative Structures (NDECS)

Peter R. Bannister
Department of Engineering Science, University of Oxford, Oxford, UK
Chapter 35: Novelty Detection

Robert J. Barthorpe
Department of Mechanical Engineering, University of Sheffield, Sheffield, UK
Chapter 70: Sensor Placement Optimization

Ivan Bartoli
Department of Structural Engineering, University of California, San Diego, CA, USA
Chapter 145: Noncontact Rail Monitoring by Ultrasonic Guided Waves
Chapter 151: Multiwire Strands

Duncan P. Barton	*Structural Monitoring Systems Ltd., Perth, WA, Australia* Chapter 106: Comparative Vacuum Monitoring (CVM™)
Michèle Basseville	*IRISA, Campus de Beaulieu, Rennes, France* Chapter 36: Model-based Statistical Signal Processing for Change and Damage Detection
Biswajit Basu	*Department of Civil, Structural and Environmental Engineering, Trinity College Dublin, Dublin, Ireland* Chapter 27: Wavelet Analysis
Jan Bauer	*Leichtweiss-Institute, Department of Waste and Resource Management, Technical University of Braunschweig, Braunschweig, Germany* Chapter 154: Landfills
Shawn J. Beard	*Acellent Technologies, Inc., Sunnyvale, CA, USA* Chapter 78: Stanford Multiactuator–Receiver Transduction (SMART) Layer Technology and Its Applications Chapter 108: Aerospace Applications of SMART Layer Technology
Nicholas C. Bellinger	*Institute for Aerospace Research, National Research Council Canada, Ottawa, Ontario, Canada* Chapter 88: Environmental Monitoring of Aircraft
Reinhard Bischoff	*Structural Engineering Research Laboratory, Empa, Swiss Federal Laboratories for Materials Testing and Research, Dübendorf, Switzerland* Chapter 69: Wireless Sensor Network Platforms Chapter 81: Microelectromechanical Systems (MEMS)
James L. Blackshire	*Air Force Research Laboratory, Wright Patterson Air Force Base, Dayton, OH, USA* Chapter 64: Directed Energy Sensors/Actuators Chapter 156: Integrated Sensor Durability and Reliability
Scott M. Bland	*Aerospace and Ocean Engineering, Virginia Polytechnic Institute and State University, Blacksburg, VA, USA* Chapter 146: Sensor Technologies for Direct Health Monitoring of Tires
Mark Blodgett	*Air Force Research Laboratory, Wright Patterson Air Force Base, Dayton, OH, USA* Chapter 54: Nondestructive Evaluation/Nondestructive Testing/Nondestructive Inspection (NDE/NDT/NDI) Sensors—Eddy Current, Ultrasonic, and Acoustic Emission Sensors
Christian Boller	*Saarland University & Fraunhofer Institute for Non-Destructive Testing, Saarbrücken, Germany (and formerly of The University of Sheffield, Sheffield, UK)* Chapter 1: Structural Health Monitoring—An Introduction and Definitions Chapter 95: Commercial Fixed-wing Aircraft Chapter 96: History of SHM for Commercial Transport Aircraft Chapter 108: Aerospace Applications of SMART Layer Technology

David Boylan	*Applied Research Laboratory, Pennsylvania State University, University Park, PA, USA* Chapter 157: Open Systems Architecture for Condition-based Maintenance
James M. W. Brownjohn	*Department of Civil and Structural Engineering, University of Sheffield, Sheffield, UK* Chapter 133: SHM of a Tall Building
Thomas Bruder	*Fraunhofer Institute for Structural Durability and System Reliability (LBF), Darmstadt, Germany* Chapter 46: Fatigue Life Assessment of Structures
Matthias Buderath	*Product Support, EADS Military Air Systems, Ottobrunn, Germany* Chapter 97: Fatigue Monitoring in Military Fixed-wing Aircraft Chapter 101: Unmanned Aerial Vehicles
Giacomo Buffarini	*ENEA, Casaccia Research Centre, Rome, Italy* Chapter 134: Dynamic Response of Buildings of the Cultural Heritage
Elsa Caetano	*Faculty of Engineering, University of Porto, Porto, Portugal* Chapter 129: Modal Testing of the Vasco da Gama Bridge, Portugal Chapter 135: Suspended Roof of Braga Sports Stadium, Portugal
Reto Cantieni	*rci dynamics, Duebendorf, Switzerland* Chapter 136: Dams
F. Necati Catbas	*Civil and Environmental Engineering Department, University of Central Florida, Orlando, FL, USA* Chapter 122: Development of a Monitoring System for a Long-span Cantilever Truss Bridge
Rosario Ceravolo	*Dipartimento di Ingegneria Strutturale e Geotecnica, Politecnico di Torino, Torino, Italy* Chapter 26: Time–frequency Analysis
Carlos E. S. Cesnik	*Department of Aerospace Engineering, University of Michigan, Ann Arbor, MI, USA* Chapter 3: Fundamentals of Guided Elastic Waves in Solids
Abir Chakraborty	*India Science Laboratory, General Motors R & D, Bangalore, India* Chapter 49: Modeling of Lamb Waves in Composite Structures
Fu-Kuo Chang	*Department of Aeronautics and Astronautics, Stanford University, Stanford, CA, USA* Chapter 78: Stanford Multiactuator–Receiver Transduction (SMART) Layer Technology and Its Applications Chapter 79: Hybrid PZT/FBG Sensor System Chapter 108: Aerospace Applications of SMART Layer Technology
Yan-Cheong Chan	*Electrical Engineering Department, City University, Kowloon, Hong Kong, China* Chapter 117: Health Monitoring, Diagnostics, and Prognostics of Avionic Systems
Gyeongrak Choi	*Korean Institute of Industrial Technology, Chonan-Si, South Korea* Chapter 67: Nanoengineering of Sensory Materials

Jinkyo F. Choo	*Department of Civil and Environmental Engineering, Seoul National University, Seoul, Korea* Chapter 124: Monitoring of Bridges in Korea
Sachit Chopra	*Department of Chemical and Materials Engineering, University of Cincinnati, Cincinnati, OH, USA* Chapter 67: Nanoengineering of Sensory Materials
Fulei Chu	*Department of Precision Instruments and Mechanology, Tsinghua University, Beijing, China* Chapter 31: Machine Learning Techniques
Paolo Clemente	*ENEA, Casaccia Research Centre, Rome, Italy* Chapter 134: Dynamic Response of Buildings of the Cultural Heritage
David A. Clifton	*Department of Engineering Science, University of Oxford, Oxford, UK and Oxford BioSignals Ltd, Abingdon, UK* Chapter 35: Novelty Detection
Stefano Coccia	*Department of Structural Engineering, University of California, San Diego, CA, USA* Chapter 145: Noncontact Rail Monitoring by Ultrasonic Guided Waves
Ivan Cole	*Commonwealth Scientific and Industrial Research Organisation (CSIRO) Materials Science and Engineering, Clayton, VIC, Australia* Chapter 112: Aircraft Structural Diagnostic and Prognostic Health Monitoring for Corrosion Prevention and Control
Matthew D. Collette	*SAIC Advanced Systems and Technology Division, Bowie, MD, USA* Chapter 142: Monitoring Marine Structures
Penny Corrigan	*Commonwealth Scientific and Industrial Research Organisation (CSIRO) Materials Science and Engineering, Clayton, VIC, Australia* Chapter 112: Aircraft Structural Diagnostic and Prognostic Health Monitoring for Corrosion Prevention and Control
Álvaro Cunha	*Faculty of Engineering, University of Porto, Porto, Portugal* Chapter 129: Modal Testing of the Vasco da Gama Bridge, Portugal Chapter 135: Suspended Roof of Braga Sports Stadium, Portugal
Len Davidson	*Air Vehicles Division, Defence Science and Technology Organisation (DSTO), Melbourne, VIC, Australia* Chapter 112: Aircraft Structural Diagnostic and Prognostic Health Monitoring for Corrosion Prevention and Control
Guido De Roeck	*Department of Civil Engineering, Katholieke Universiteit Leuven, Leuven, Belgium* Chapter 126: Continuous Vibration Monitoring and Progressive Damage Testing on the Z24 Bridge
Alessandro De Stefano	*Politecnico di Torino, Torino, Italy* Chapter 132: SHM Actions on the Holy Shroud Chapel in Torino

Thomas DeFiore	*US Federal Aviation Administration, Atlantic City International Airport, NJ, USA* Chapter 113: Video Landing Parameter Surveys
Thierry Denoeux	*Université de Technologie de Compiègne, Compiègne, France* Chapter 39: Uncertainty Analysis
Mark M. Derriso	*US Air Force Research Laboratory, Wright Patterson Air Force Base, OH, USA* Chapter 90: Military Aircraft
Martin P. DeSimio	*University of Dayton Research Institute, Dayton, OH, USA* Chapter 90: Military Aircraft
Dimitri M. Donskoy	*Stevens Institute of Technology, Hoboken, NJ, USA* Chapter 15: Nonlinear Acoustic Methods
Mike Duffield	*QinetiQ, Farnborough, UK* Chapter 100: Operational Loads Monitoring in Military Transport Aircraft and Military Derivatives of Civil Aircraft
Janice Dulieu-Barton	*School of Engineering Sciences, University of Southampton, Southampton, UK* Chapter 82: Principles of Structural Degradation Monitoring
Frank Eichelbaum	*IABG, Dresden, Germany* Chapter 105: Validation of SHM Sensors in Airbus A380 Full-scale Fatigue Test
Kevin M. Farinholt	*Engineering Institute, Los Alamos National Laboratory, Los Alamos, NM, USA* Chapter 71: Sensor Network Paradigms Chapter 75: Energy Harvesting and Wireless Energy Transmission for SHM Sensor Nodes
Charles R. Farrar	*Engineering Institute, Los Alamos National Laboratory, Los Alamos, NM, USA* Chapter 18: Piezoelectric Impedance Methods for Damage Detection and Sensor Validation Chapter 71: Sensor Network Paradigms Chapter 75: Energy Harvesting and Wireless Energy Transmission for SHM Sensor Nodes Chapter 82: Principles of Structural Degradation Monitoring
Spilios D. Fassois	*Department of Mechanical and Aeronautical Engineering, University of Patras, Patras, Greece* Chapter 23: Statistical Time Series Methods for SHM
Michael Feldman	*Technion–Israel Institute of Technology, Haifa, Israel* Chapter 25: Hilbert Transform, Envelope, Instantaneous Phase, and Frequency
Glauco Feltrin	*Structural Engineering Research Laboratory, Empa, Swiss Federal Laboratories for Materials Testing and Research, Dübendorf, Switzerland* Chapter 69: Wireless Sensor Network Platforms Chapter 81: Microelectromechanical Systems (MEMS)

Peter Fleming	*Rolls-Royce University Technology Centre, Department of Automatic Control and Systems Engineering, University of Sheffield, Sheffield, UK* Chapter 115: Monitoring of Aircraft Engines
Manuel A. Floriano	*GEOCISA, Madrid, Spain* Chapter 137: Condamine Floating Dock, Monaco
Peter Foote	*BAE Systems, Advanced Technology Centre, Bristol, UK* Chapter 109: Fiber-optic Sensors
Dan M. Frangopol	*Department of Civil and Environmental Engineering, Lehigh University, Bethlehem, PA, USA* Chapter 89: Maintenance Principles for Civil Structures
Bernd Frankenstein	*Fraunhofer Institute for Nondestructive Testing (IZFP-D), Dresden, Germany* Chapter 152: SHM and Lifetime Management of Industrial Piping Systems
Klaus Fricke	*Leichtweiss-Institute, Department of Waste and Resource Management, Technical University of Braunschweig, Braunschweig, Germany* Chapter 154: Landfills
Herbert Friedmann	*Wölfel Beratende Ingenieure, Höchberg, Germany* Chapter 152: SHM and Lifetime Management of Industrial Piping Systems
Michael I. Friswell	*Department of Aerospace Engineering, University of Bristol, Bristol, UK* Chapter 11: Modal–Vibration-based Damage Identification
Yozo Fujino	*Department of Civil Engineering, University of Tokyo, Tokyo, Japan* Chapter 125: Bridge Monitoring in Japan
Stephen C. Galea	*Air Vehicles Division, Defence Science and Technology Organisation (DSTO), Melbourne, VIC, Australia* Chapter 76: On the Way to Autonomy: the Wireless-interrogated and Self-powered "Smart Patch" System Chapter 112: Aircraft Structural Diagnostic and Prognostic Health Monitoring for Corrosion Prevention and Control
Ranjan Ganguli	*Department of Aerospace Engineering, Indian Institute of Science, Bangalore, India* Chapter 51: Development of Fuzzy Rules for Damage Detection and Location
Tomasz Gałka	*Institute of Power Engineering, Warsaw, Poland* Chapter 148: Large Rotating Machines
Daniel Gilbertson	*Boeing Phantom Works, St. Louis, MO, USA* Chapter 157: Open Systems Architecture for Condition-based Maintenance
Victor Giurgiutiu	*Mechanical Engineering Department, University of South Carolina, Columbia, SC, USA* Chapter 5: Electromechanical Impedance Modeling Chapter 52: Piezoelectricity Principles and Materials Chapter 55: Piezoelectric Wafer Active Sensors

Neil Goldfine	JENTEK Sensors, Inc., Waltham, MA, USA
	    Chapter 20: Eddy-current Methods
	    Chapter 58: Eddy-current *in situ* Sensors for SHM

Jean-Claude Golinval	Aerospace and Mechanical Engineering Department (LTAS), University of Liège, Liège, Belgium Chapter 33: Dimensionality Reduction Using Linear and Nonlinear Transformation

Srinivasan Gopalakrishnan	Department of Aerospace Engineering, Indian Institute of Science, Bangalore, India
	    Chapter 41: Constitutive Modeling of Magnetostrictive Materials
	    Chapter 42: Modeling Aspects in Finite Elements
	    Chapter 43: Finite Elements: Modeling of Piezoceramic and Magnetostrictive Sensors and Actuators
	    Chapter 44: A Simplified Damage Model for SHM Metallic and Composite Structures
	    Chapter 47: Damage Detection Using Piezoceramic and Magnetostrictive Sensors and Actuators

Grant A. Gordon	Research Technology Centre, Honeywell Inc., Phoenix, AZ, USA Chapter 95: Commercial Fixed-wing Aircraft

Michael R. Gorman	Digital Wave Corporation, Englewood, CO, USA
	    Chapter 4: Acoustic Emission
	    Chapter 104: Thermal Protection System Monitoring of Space Structures

Gangadhararao Grandhi	Department of Mechanical and Chemical Engineering, North Carolina A&T State University, Greensboro, NC, USA Chapter 147: Wind Turbines
Alten F. Grandt Jr.	School of Aeronautics and Astronautics, Purdue University, West Lafayette, IN, USA Chapter 9: Damage Evolution Phenomena and Models
Dave Grundy	JENTEK Sensors, Inc., Waltham, MA, USA Chapter 58: Eddy-current *in situ* Sensors for SHM
Andrzej Grzadziela	Faculty of Mechanical and Electrical Engineering, Naval University of Gdynia, Gdynia, Poland Chapter 143: Diagnosing Offshore Machines and Power Plants Using Vibration Methods
Hong Guan	HDR, Los Angeles, CA, USA Chapter 73: Web-based SHM
Robert F. Guratzsch	Vanderbilt University, Nashville, TN, USA Chapter 50: Probabilistic Approaches to Sensor Layout Design, Data Processing, and Damage Detection
Hua Gu	Advanced Structures Group, Caterpillar Production System Division, Caterpillar Inc., Peoria, IL, USA Chapter 56: Damage Presence/Growth Monitoring Sensors

Wolfgang R. Habel	*Division VIII.1 Measurement and Testing Technology: Sensors, BAM Federal Institute for Materials Research and Testing, Berlin, Germany* Chapter 155: Reliable Use of Fiber-optic Sensors
Andrew Halfpenny	*nCode International Ltd., Innovation Technology Centre, Rotherham, UK* Chapter 22: Data Preprocessing for Damage Detection
Muhammad Haroon	*Purdue University, West Lafayette, IN, USA* Chapter 2: Free and Forced Vibration Models
Rosemarie Helmerich	*Division VIII.2 Non-destructive Damage Assessment and Environmental Measurement Methods, BAM Federal Institute for Materials Research and Testing, Berlin, Germany* Chapter 131: Construction Process Monitoring at the New Berlin Main Station
Fritz-Otto Henkel	*Wölfel Beratende Ingenieure, Höchberg, Germany* Chapter 152: SHM and Lifetime Management of Industrial Piping Systems
Wolfgang Hillger	*German Aerospace Center (DLR), Braunschweig, Germany* Chapter 14: Ultrasonic Methods
Bruce Hinton	*Air Vehicles Division, Defence Science and Technology Organisation (DSTO), Melbourne, VIC, Australia* Chapter 112: Aircraft Structural Diagnostic and Prognostic Health Monitoring for Corrosion Prevention and Control
Wei Huang	*Intelligent Transportation Systems Research Centre, Southeast University, Nanjing, China* Chapter 66: Global Navigation Satellite Systems (GNSSs) for Monitoring Long Suspension Bridges
N. Hu	*Department of Aerospace Engineering, Tohoku University, Sendai, Japan* Chapter 44: A Simplified Damage Model for SHM Metallic and Composite Structures
Jeong-Beom Ihn	*The Boeing Company, Advanced Structures Technology, Phantom Works, Seattle, WA, USA* Chapter 111: Design, Analysis, and SHM of Bonded Composite Repair and Substructure
Roy Ikegami	*Acellent Technologies, Inc., Sunnyvale, CA, USA* Chapter 96: History of SHM for Commercial Transport Aircraft Chapter 108: Aerospace Applications of SMART Layer Technology
Daniel J. Inman	*Center for Intelligent Material Systems and Structures, Virginia Polytechnic Institute and State University, Blacksburg, VA, USA* Chapter 77: Energy Harvesting using Thermoelectric Materials
Kumar V. Jata	*Air Force Research Laboratory, Wright Patterson Air Force Base, Dayton, OH, USA* Chapter 10: Failure Modes of Aerospace Materials

	Chapter 54: Nondestructive Evaluation/Nondestructive Testing/Nondestructive Inspection (NDE/NDT/NDI) Sensors—Eddy Current, Ultrasonic, and Acoustic Emission Sensors Chapter 156: Integrated Sensor Durability and Reliability
Xiaomo Jiang	*Vanderbilt University, Nashville, TN, USA* Chapter 50: Probabilistic Approaches to Sensor Layout Design, Data Processing, and Damage Detection
Visakan Kadirkamanathan	*Rolls-Royce University Technology Centre, Department of Automatic Control and Systems Engineering, University of Sheffield, Sheffield, UK* Chapter 115: Monitoring of Aircraft Engines
Jayabal Kaliappan	*Department of Applied Mechanics, Indian Institute of Technology, Chennai, India* Chapter 40: Piezoceramic Materials—Phenomena and Modeling
Inpil Kang	*Division of Mechanical Engineering, Pukyong National University, Busan, South Korea* Chapter 67: Nanoengineering of Sensory Materials
Rakesh K. Kapania	*Aerospace and Ocean Engineering, Virginia Polytechnic Institute and State University, Blacksburg, VA, USA* Chapter 146: Sensor Technologies for Direct Health Monitoring of Tires
Vistasp M. Karbhari	*University of Alabama in Huntsville, Huntsville, AL, USA* Chapter 73: Web-based SHM Chapter 84: Design Principles for Civil Structures
Klaus Kerkhof	*MPA Universität Stuttgart, Stuttgart, Germany* Chapter 152: SHM and Lifetime Management of Industrial Piping Systems
Gaëtan Kerschen	*Aerospace and Mechanical Engineering Department (LTAS), University of Liège, Liège, Belgium* Chapter 33: Dimensionality Reduction Using Linear and Nonlinear Transformation
Do Hyung Kim	*Research and Development, Lloyd's Register Asia, Busan, Korea* Chapter 144: Ship and Offshore Structures
Myung Hyun Kim	*Department of Naval Architecture and Ocean Engineering, Pusan National University, Busan, Korea* Chapter 144: Ship and Offshore Structures
Sungkon Kim	*Department of Structural Engineering, Seoul National University of Technology, Seoul, Korea* Chapter 124: Monitoring of Bridges in Korea
Dennis King	*Rolls-Royce Civil Aero-Engines, Derby, UK* Chapter 35: Novelty Detection
Steve King	*Rolls-Royce Civil Aero-Engines, Derby, UK* Chapter 35: Novelty Detection

Goutham R. Kirikera	*Center for Quality Engineering and Failure Prevention, Northwestern University, Evanston, IL, USA* Chapter 147: Wind Turbines
Wilhelm Kleinöder	*AREVA NP GmbH, Erlangen, Germany* Chapter 153: Fatigue Monitoring in Nuclear Power Plants
Thomas Klesse	*Fraunhofer Institute for Nondestructive Testing (IZFP-D), Dresden, Germany* Chapter 152: SHM and Lifetime Management of Industrial Piping Systems
Hyun-Moo Koh	*Department of Civil and Environmental Engineering, Seoul National University, Seoul, Korea* Chapter 124: Monitoring of Bridges in Korea
Marek Krawczuk	*Institute of Fluid Flow Machinery, PAS, Gdansk, Poland and Faculty of Electrical and Control Engineering, Gdansk University of Technology, Gdansk, Poland* Chapter 45: Modeling for Detection of Degraded Zones in Metallic and Composite Structures
Thomas Krichel	*Airbus, Hamburg, Germany* Chapter 105: Validation of SHM Sensors in Airbus A380 Full-scale Fatigue Test
Amrita Kumar	*Acellent Technologies, Inc., Sunnyvale, CA, USA* Chapter 78: Stanford Multiactuator–Receiver Transduction (SMART) Layer Technology and Its Applications
Narito Kurata	*Kobori Research Complex, Kajima Corporation, Tokyo, Japan* Chapter 87: Risk Monitoring of Civil Structures
Kasper Lambrighs	*Department of Metallurgy and Materials Engineering, Katholieke Universiteit Leuven, Leuven, Belgium* Chapter 13: Applications of Acoustic Emission for SHM: A Review
Francesco Lanza di Scalea	*Department of Structural Engineering, University of California, San Diego, CA, USA* Chapter 145: Noncontact Rail Monitoring by Ultrasonic Guided Waves Chapter 151: Multiwire Strands
Hae-Sung Lee	*Department of Civil and Environmental Engineering, Seoul National University, Seoul, Korea* Chapter 124: Monitoring of Bridges in Korea
Michel B. Lemistre	*Laboratoire SATIE/CNRS, Ecole Normale Supérieure de Cachan, Cachan, France* Chapter 63: Electric and Electromagnetic Properties Sensing Chapter 80: The HELP-Layer® System
Dy Dinh Le	*Federal Aviation Administration, Air Traffic Organization, William J. Hughes Technical Center, Atlantic City International Airport, NJ, USA* Chapter 103: Experience with Health and Usage Monitoring Systems in Helicopters

Irene Li	*Acellent Technologies, Inc., Sunnyvale, CA, USA* Chapter 78: Stanford Multiactuator–Receiver Transduction (SMART) Layer Technology and Its Applications
Mark Lin	*Department of Aeronautics and Astronautics, Stanford University, Stanford, CA, USA* Chapter 78: Stanford Multiactuator–Receiver Transduction (SMART) Layer Technology and Its Applications
Eric Lindgren	*Air Force Research Laboratory, Wright Patterson Air Force Base, Dayton, OH, USA* Chapter 54: Nondestructive Evaluation/Nondestructive Testing/Nondestructive Inspection (NDE/NDT/NDI) Sensors—Eddy Current, Ultrasonic, and Acoustic Emission Sensors
Kenneth J. Loh	*Department of Civil and Environmental Engineering, University of Michigan, Ann Arbor, MI, USA* Chapter 68: Miniaturized Sensors Employing Micro- and Nanotechnologies
Minh P. Luong	*LMS CNRS UMR7649, Ecole Polytechnique, Palaiseau, France* Chapter 6: Thermomechanical Models
Jerome P. Lynch	*Department of Civil and Environmental Engineering, University of Michigan, Ann Arbor, MI, USA* Chapter 68: Miniaturized Sensors Employing Micro- and Nanotechnologies
Eric I. Madaras	*NASA Langley Research Center, Hampton, VA, USA* Chapter 104: Thermal Protection System Monitoring of Space Structures
Filipe Magalhães	*Faculty of Engineering, University of Porto, Porto, Portugal* Chapter 135: Suspended Roof of Braga Sports Stadium, Portugal
Sankaran Mahadevan	*Vanderbilt University, Nashville, TN, USA* Chapter 50: Probabilistic Approaches to Sensor Layout Design, Data Processing, and Damage Detection
D. R. Mahapatra	*Department of Aerospace Engineering, Indian Institute of Science, Bangalore, India* Chapter 44: A Simplified Damage Model for SHM Metallic and Composite Structures
Gunjan Maheshwari	*Department of Mechanical Engineering, University of Cincinnati, Cincinnati, OH, USA* Chapter 67: Nanoengineering of Sensory Materials
Matthew C. Malkin	*Phantom Works, Boeing, Seattle, WA, USA* Chapter 99: Flight Demonstration of a SHM System on a USAF Fighter Airplane
Graeme Manson	*Department of Mechanical Engineering, University of Sheffield, Sheffield, UK* Chapter 38: Optimization Techniques for Damage Detection Chapter 39: Uncertainty Analysis
Marcias Martinez	*Institute for Aerospace Research, National Research Council Canada, Ottawa, Ontario, Canada* Chapter 88: Environmental Monitoring of Aircraft

Alessandro Marzani	DISTART, University of Bologna, Bologna, Italy Chapter 151: Multiwire Strands
Len Meadows	Transport and Surveillance Aircraft, Air Vehicles Division, Defence Science and Technology Organisation (DSTO), Fisherman's Bend, VIC, Australia Chapter 100: Operational Loads Monitoring in Military Transport Aircraft and Military Derivatives of Civil Aircraft
Enrique A. Medina	Radiance Technologies Inc., Dayton, OH, USA Chapter 94: Value Assessment Approaches for Structural Life Management
Xiaolin Meng	Institute of Engineering Surveying and Space Geodesy, University of Nottingham, Nottingham, UK Chapter 66: Global Navigation Satellite Systems (GNSSs) for Monitoring Long Suspension Bridges
Thomas B. Messervey	Department of Mathematical Sciences, United States Military Academy, West Point, NY, USA Chapter 89: Maintenance Principles for Civil Structures
Jonas Meyer	Structural Engineering Research Laboratory, Empa, Swiss Federal Laboratories for Materials Testing and Research, Dübendorf, Switzerland Chapter 69: Wireless Sensor Network Platforms Chapter 81: Microelectromechanical Systems (MEMS)
Richard P. Micklos	PE, Warminster, PA, USA Chapter 113: Video Landing Parameter Surveys
Ayaho Miyamoto	Graduate School of Science and Engineering, Yamaguchi University, Ube, Japan Chapter 93: Usage Management of Civil Structures
M. R. Mofakhami	Department of Mechanical Engineering, The University of Sheffield, Sheffield, UK Glossary
Ronald D. Moffitt	Institute for Advanced Learning and Research (IALR), Virginia Polytechnic Institute and State University, Danville, VA, USA Chapter 146: Sensor Technologies for Direct Health Monitoring of Tires
Loris Molent	Air Vehicles Division, Defence Science and Technology Organisation (DSTO), Melbourne, VIC, Australia Chapter 98: Agile Military Aircraft
Scott Moss	Air Vehicles Division, Defence Science and Technology Organisation (DSTO), Melbourne, VIC, Australia Chapter 76: On the Way to Autonomy: the Wireless-interrogated and Self-powered "Smart Patch" System
Sai L. Mullapudi	Department of Mechanical Engineering, University of Cincinnati, Cincinnati, OH, USA Chapter 147: Wind Turbines
Farah Mulyasari Sule	German–Indonesian Technical Cooperation, Environmental Geological Centre, Bandung, Indonesia and Institut Teknologi Bandung, Center for Disaster Mitigation, Bandung, Indonesia Chapter 141: Environmental Factors Derived from Satellite Data of Java, Indonesia

Kai Münnich	*Leichtweiss-Institute, Department of Waste and Resource Management, Technical University of Braunschweig, Braunschweig, Germany* Chapter 154: Landfills
Tim Muster	*Commonwealth Scientific and Industrial Research Organisation (CSIRO) Materials Science and Engineering, Clayton, VIC, Australia* Chapter 112: Aircraft Structural Diagnostic and Prognostic Health Monitoring for Corrosion Prevention and Control
Rolf H. Neunaber	*Industrieanlagen Betriebsgesellschaft mbH, Ottobrunn, Germany* Chapter 92: Usage Management of Military Aircraft Structures
Jonathan M. Nichols	*US Naval Research Laboratory, Washington, DC, USA* Chapter 34: Nonlinear Features for SHM Applications
Yi-Qing Ni	*Department of Civil and Structural Engineering, Hong Kong Polytechnic University, Kowloon, Hong Kong, China* Chapter 123: Modular Architecture of SHM System for Cable-supported Bridges
Francis Nkrumah	*Department of Mechanical and Chemical Engineering, North Carolina A&T State University, Greensboro, NC, USA* Chapter 147: Wind Turbines
Chang Kook Oh	*Department of Civil Engineering, California Institute of Technology, Pasadena, CA, USA* Chapter 30: Statistical Pattern Recognition
Steven E. Olson	*University of Dayton Research Institute, Dayton, OH, USA* Chapter 90: Military Aircraft
Luis M. Ortega	*GEOCISA, Madrid, Spain* Chapter 137: Condamine Floating Dock, Monaco
Wiesław Ostachowicz	*Institute of Fluid Flow Machinery, PAS, Gdansk, Poland and Faculty of Navigation, Gdynia Maritime University, Gdynia, Poland* Chapter 45: Modeling for Detection of Degraded Zones in Metallic and Composite Structures
Christophe Paget	*Airbus, Filton, UK* Chapter 105: Validation of SHM Sensors in Airbus A380 Full-scale Fatigue Test
Tso-Chien Pan	*Nanyang Technological University, Singapore* Chapter 133: SHM of a Tall Building
Gyuhae Park	*Engineering Institute, Los Alamos National Laboratory, Los Alamos, NM, USA* Chapter 18: Piezoelectric Impedance Methods for Damage Detection and Sensor Validation Chapter 71: Sensor Network Paradigms Chapter 75: Energy Harvesting and Wireless Energy Transmission for SHM Sensor Nodes

Triplicane A. Parthasarathy	*Air Force Research Laboratory, Wright Patterson Air Force Base, Dayton, OH, USA and UES Incorporated, Air Force Research Laboratory, Wright Patterson Air Force Base, Dayton, OH, USA* Chapter 10: Failure Modes of Aerospace Materials
Michael Pecht	*Center for Advanced Life Cycle Engineering (CALCE), University of Maryland, College Park, MD, USA* Chapter 150: Prognostics and Health Management of Electronics Chapter 117: Health Monitoring, Diagnostics, and Prognostics of Avionic Systems
Bart Peeters	*LMS International, Leuven, Belgium* Chapter 127: Continuous Monitoring of the Øresund Bridge: Data Acquisition and Operational Modal Analysis
Udo Peil	*Institute for Steel Structures, University of Technology Carolo-Wilhelmina at Braunschweig, Braunschweig, Germany* Chapter 7: Civil Infrastructure Load Models
Zhike Peng	*Department of Precision Instruments and Mechanology, Tsinghua University, Beijing, China* Chapter 31: Machine Learning Techniques
Kara Peters	*Department of Mechanical and Aerospace Engineering, North Carolina State University, Raleigh, NC, USA* Chapter 59: Fiber-optic Sensor Principles Chapter 60: Intensity-, Interferometric-, and Scattering-based Optical-fiber Sensors Chapter 61: Fiber Bragg Grating Sensors Chapter 62: Novel Fiber-optic Sensors
S. Gareth Pierce	*Department of Electronic and Electrical Engineering, University of Strathclyde, Glasgow, UK* Chapter 39: Uncertainty Analysis
Darryll J. Pines	*Department of Aerospace Engineering, University of Maryland, College Park, MD, USA* Chapter 28: Damage Detection Using the Hilbert–Huang Transform
Kenneth Pipe	*Humaware, Petersfield, UK* Chapter 102: Health and Usage Monitoring Systems (HUM Systems) for Helicopters: Architecture and Performance
Christian Pöckl	*AREVA NP GmbH, Erlangen, Germany* Chapter 153: Fatigue Monitoring in Nuclear Power Plants
Ian Powlesland	*Air Vehicles Division, Defence Science and Technology Organisation (DSTO), Melbourne, VIC, Australia* Chapter 76: On the Way to Autonomy: the Wireless-interrogated and Self-powered "Smart Patch" System
Don Price	*Commonwealth Scientific and Industrial Research Organisation (CSIRO) Materials Science and Engineering, Lindfield, NSW, Australia* Chapter 112: Aircraft Structural Diagnostic and Prognostic Health Monitoring for Corrosion Prevention and Control

William H. Prosser	*NASA Langley Research Center, Hampton, VA, USA* Chapter 104: Thermal Protection System Monitoring of Space Structures
Xinlin P. Qing	*Acellent Technologies, Inc., Sunnyvale, CA, USA* Chapter 78: Stanford Multiactuator–Receiver Transduction (SMART) Layer Technology and Its Applications Chapter 79: Hybrid PZT/FBG Sensor System Chapter 108: Aerospace Applications of SMART Layer Technology
Ajay Raghavan	*Department of Aerospace Engineering, University of Michigan, Ann Arbor, MI, USA* Chapter 3: Fundamentals of Guided Elastic Waves in Solids
Nik Rajic	*Defence Science and Technology Organisation (DSTO), Fisherman's Bend, VIC, Australia* Chapter 107: Development of an Active Smart Patch for Aircraft Repair
Robert B. Randall	*School of Mechanical and Manufacturing Engineering, University of New South Wales, Sydney, NSW, Australia* Chapter 24: Cepstral Methods of Operational Modal Analysis
William F. Ranson	*Direct Measurements, Inc., Atlanta, GA, USA* Chapter 53: Operational Loads Sensors
Steve Reed	*QinetiQ, Farnborough, UK* Chapter 32: Artificial Neural Networks Chapter 85: Loads Monitoring in Aerospace Structures Chapter 100: Operational Loads Monitoring in Military Transport Aircraft and Military Derivatives of Civil Aircraft
Edwin Reynders	*Department of Civil Engineering, Katholieke Universiteit Leuven, Leuven, Belgium* Chapter 126: Continuous Vibration Monitoring and Progressive Damage Testing on the Z24 Bridge
Piervincenzo Rizzo	*Department of Civil and Environmental Engineering, University of Pittsburgh, Pittsburgh, PA, USA* Chapter 145: Noncontact Rail Monitoring by Ultrasonic Guided Waves Chapter 151: Multiwire Strands
Dennis Roach	*Sandia National Laboratories, Albuquerque, NM, USA* Chapter 91: Use of Leave-in-place Sensors and SHM Methods to Improve Assessments of Aging Structures
Amy N. Robertson	*IMTECH, Boulder, CO, USA* Chapter 27: Wavelet Analysis
Michael J. Roemer	*Impact Technologies, Rochester, NY, USA* Chapter 149: Gas Turbine Engines

Ajit Roy	Air Force Research Laboratory, Wright Patterson Air Force Base, Dayton, OH, USA Chapter 10: Failure Modes of Aerospace Materials
Gregory A. Ruderman	Air Force Research Laboratory, AFRL/RZSB, Edwards AFB, CA, USA Chapter 116: Monitoring of Solid Rocket Motors
Aldo Ruotulo	Department of Geotechnical and Structural Engineering, Politecnico di Torino, Torino, Italy Chapter 38: Optimization Techniques for Damage Detection
Massimo Ruzzene	School of Aerospace Engineering, Georgia Institute of Technology, Atlanta, GA, USA Chapter 48: Damage Measures
Werner Rücker	Division VII.2 Buildings and Structures, BAM Federal Institute for Materials Research and Testing, Berlin, Germany Chapter 121: Long-term Monitoring of Dynamic Loads on the Brandenburg Gate
Erdal Safak	Department of Earthquake Engineering, Kandilli Observatory and Earthquake Research Institute, Bogazici University, Istanbul, Turkey Chapter 139: System Identification for Soil–structure Interaction
John S. Sakellariou	Department of Mechanical and Aeronautical Engineering, University of Patras, Patras, Greece Chapter 23: Statistical Time Series Methods for SHM
Liming W. Salvino	Structures and Composites, Carderock Division, Naval Surface Warfare Center, West Bethesda, MD, USA Chapter 28: Damage Detection Using the Hilbert–Huang Transform Chapter 142: Monitoring Marine Structures
Pia Sartor	Department of Mechanical Engineering, University of Sheffield, Sheffield, UK Chapter 114: Landing Gear
Shamachary Sathish	Air Force Research Laboratory, Wright Patterson Air Force Base, Dayton, OH, USA Chapter 54: Nondestructive Evaluation/Nondestructive Testing/Nondestructive Inspection (NDE/NDT/NDI) Sensors—Eddy Current, Ultrasonic, and Acoustic Emission Sensors
Darrell Schlicker	JENTEK Sensors, Inc., Waltham, MA, USA Chapter 58: Eddy-current *in situ* Sensors for SHM
Bianka Schmidt-Brandecker	AeroStruc—Aeronautical Engineering, Buxtehude, Germany Chapter 110: Design Benefits in Aeronautics Resulting from SHM
Hans-Juergen Schmidt	AeroStruc—Aeronautical Engineering, Buxtehude, Germany Chapter 110: Design Benefits in Aeronautics Resulting from SHM
R. Kyle Schmidt	Messier-Dowty SA, Vélizy-Villacoublay, France Chapter 114: Landing Gear
Frank Schubert	Fraunhofer Institute for Nondestructive Testing (IZFP-D), Dresden, Germany Chapter 152: SHM and Lifetime Management of Industrial Piping Systems

Xaver Schuler	*MPA Universität Stuttgart, Stuttgart, Germany* Chapter 152: SHM and Lifetime Management of Industrial Piping Systems
Mark Schulz	*Department of Mechanical Engineering, University of Cincinnati, Cincinnati, OH, USA* Chapter 67: Nanoengineering of Sensory Materials Chapter 147: Wind Turbines
Vesselin Shanov	*Department of Chemical and Materials Engineering, University of Cincinnati, Cincinnati, OH, USA* Chapter 67: Nanoengineering of Sensory Materials Chapter 147: Wind Turbines
Yanko Sheiretov	*JENTEK Sensors, Inc., Waltham, MA, USA* Chapter 20: Eddy-current Methods
Michael Shiao	*Airport and Aircraft Safety Group, Aviation Research & Development Office, Federal Aviation Administration, Atlantic City International Airport, NJ, USA* Chapter 86: Risk Monitoring of Aircraft Fatigue Damage Evolution at Critical Locations
Srinivasan M. Sivakumar	*Department of Applied Mechanics, Indian Institute of Technology, Chennai, India* Chapter 40: Piezoceramic Materials—Phenomena and Modeling
Ian F. C. Smith	*Institute of Structural Engineering, Ecole Polytechnique Fédérale de Lausanne (EPFL), Lausanne, Switzerland* Chapter 130: Multiple-model Structural Identification
Henry A. Sodano	*Department of Mechanical and Aerospace Engineering, Arizona State University, Tempe, AZ, USA* Chapter 77: Energy Harvesting using Thermoelectric Materials
Hoon Sohn	*Department of Civil and Environmental Engineering, Korea Advanced Institute of Science and Technology, Daejeon, Republic of Korea* Chapter 30: Statistical Pattern Recognition
Elisa Sorrivi	*DISTART, University of Bologna, Bologna, Italy* Chapter 151: Multiwire Strands
Constantinos Soutis	*Aerospace Engineering, University of Sheffield, Sheffield, UK* Chapter 17: Lamb Wave-based SHM for Laminated Composite Structures Chapter 111: Design, Analysis, and SHM of Bonded Composite Repair and Substructure
Holger Speckmann	*Airbus, Bremen, Germany* Chapter 105: Validation of SHM Sensors in Airbus A380 Full-scale Fatigue Test
Wieslaw J. Staszewski	*Department of Mechanical Engineering, University of Sheffield, Sheffield, UK* Chapter 21: Signal Processing for Damage Detection Chapter 38: Optimization Techniques for Damage Detection
George F. Studor	*NASA Johnson Space Center, Houston, TX, USA* Chapter 104: Thermal Protection System Monitoring of Space Structures

Shankar Sundararaman	*Ray W. Herrick Laboratories, Purdue University, West Lafayette, IN, USA* Chapter 8: Static Damage Phenomena and Models
Mannur Sundaresan	*Department of Mechanical and Chemical Engineering, North Carolina A&T State University, Greensboro, NC, USA* Chapter 147: Wind Turbines
Mohammad R. Sunny	*Aerospace and Ocean Engineering, Virginia Polytechnic Institute and State University, Blacksburg, VA, USA* Chapter 146: Sensor Technologies for Direct Health Monitoring of Tires
Cecilia Surace	*Department of Geotechnical and Structural Engineering, Politecnico di Torino, Torino, Italy* Chapter 38: Optimization Techniques for Damage Detection
Michael A. Sutton	*Department of Mechanical Engineering, University of South Carolina, Columbia, SC, USA* Chapter 65: Full-field Sensing: Three-dimensional Computer Vision and Digital Image Correlation for Noncontacting Shape and Deformation Measurements
Zhongqing Su	*Department of Mechanical Engineering, Hong Kong Polytechnic University, Kowloon, Hong Kong, China* Chapter 37: Data Fusion of Multiple Signals from the Sensor Network Chapter 74: Design of Active Sensor Network and Multilevel Data Fusion
Lionel Tarassenko	*Department of Engineering Science, University of Oxford, Oxford, UK* Chapter 35: Novelty Detection
Barbara Teilen-Willige	*Department of Hydrogeology and Bureau of Applied Geoscientific Remote Sensing (BAGF), Technical University of Berlin, Stockach, Germany* Chapter 141: Environmental Factors Derived from Satellite Data of Java, Indonesia
Jens Telgkamp	*Structure Design Principles, Airbus, Hamburg, Germany* Chapter 83: Design Principles for Aerospace Structures
Michael D. Todd	*Department of Structural Engineering, University of California, San Diego, CA, USA* Chapter 12: Data Interrogation Approaches with Strain and Load Gauge Sensor Arrays Chapter 34: Nonlinear Features for SHM Applications
Peter Trathen	*Air Vehicles Division, Defence Science and Technology Organisation (DSTO), Melbourne, VIC, Australia* Chapter 112: Aircraft Structural Diagnostic and Prognostic Health Monitoring for Corrosion Prevention and Control
Tony Trueman	*Air Vehicles Division, Defence Science and Technology Organisation (DSTO), Melbourne, VIC, Australia* Chapter 112: Aircraft Structural Diagnostic and Prognostic Health Monitoring for Corrosion Prevention and Control

Reginald I. Vachon	*Direct Measurements, Inc., Atlanta, GA, USA* Chapter 53: Operational Loads Sensors
Stephen Van der Velden	*Air Vehicles Division, Defence Science and Technology Organisation (DSTO), Melbourne, VIC, Australia* Chapter 76: On the Way to Autonomy: the Wireless-interrogated and Self-powered "Smart Patch" System
Robert Veit-Egerer	*VCE – Vienna Consulting Engineers, Vienna, Austria* Chapter 128: Condition Compensation in Frequency Analyses—a Basis for Damage Detection
Erasmo Viola	*DISTART, University of Bologna, Bologna, Italy* Chapter 151: Multiwire Strands
Robert L. Walter IV	*Applied Research Laboratory, Pennsylvania State University, University Park, PA, USA* Chapter 157: Open Systems Architecture for Condition-based Maintenance
Ming L. Wang	*Department of Civil and Materials Engineering, University of Illinois, Chicago, IL, USA* Chapter 56: Damage Presence/Growth Monitoring Sensors Chapter 140: Loads and Temperature Effects on a Bridge
Xiaoming Wang	*CSIRO Sustainable Ecosystems, Commonwealth Scientific and Industrial Research Organisation, Melbourne, VIC, Australia* Chapter 37: Data Fusion of Multiple Signals from the Sensor Network Chapter 74: Design of Active Sensor Network and Multilevel Data Fusion
Andrew Washabaugh	*JENTEK Sensors, Inc., Waltham, MA, USA* Chapter 20: Eddy-current Methods
Helmut Wenzel	*VCE Holding GmbH, Vienna, Austria* Chapter 118: The Character of SHM in Civil Engineering Chapter 119: Ambient Vibration Monitoring Chapter 120: The Influence of Environmental Factors Chapter 141: Environmental Factors Derived from Satellite Data of Java, Indonesia Chapter 152: SHM and Lifetime Management of Industrial Piping Systems
Martine Wevers	*Department of Metallurgy and Materials Engineering, Katholieke Universiteit Leuven, Leuven, Belgium* Chapter 13: Applications of Acoustic Emission for SHM: A Review
Paul White	*Institute of Sound and Vibration Research, University of Southampton, Southampton, UK* Chapter 29: Higher Order Statistical Signal Processing
Paul D. Wilcox	*Department of Mechanical Engineering, University of Bristol, Bristol, UK* Chapter 16: Guided-wave Array Methods

Alan Wilson	Air Vehicles Division, Defence Science and Technology Organisation (DSTO), Melbourne, VIC, Australia Chapter 112: Aircraft Structural Diagnostic and Prognostic Health Monitoring for Corrosion Prevention and Control
Mark Windoloski	JENTEK Sensors, Inc., Waltham, MA, USA Chapter 20: Eddy-current Methods
Kai-Yuen Wong	Highways Department, Government of Hong Kong, China Chapter 123: Modular Architecture of SHM System for Cable-supported Bridges
Keith Worden	Department of Mechanical Engineering, University of Sheffield, Sheffield, UK Chapter 11: Modal–Vibration-based Damage Identification Chapter 21: Signal Processing for Damage Detection Chapter 38: Optimization Techniques for Damage Detection Chapter 39: Uncertainty Analysis Chapter 70: Sensor Placement Optimization Chapter 82: Principles of Structural Degradation Monitoring
Zhanjun Wu	Department of Aeronautics and Astronautics, Stanford University, Stanford, CA, USA Chapter 79: Hybrid PZT/FBG Sensor System
Lin Ye	School of Aerospace, Mechanical and Mechatronic Engineering, University of Sydney, Sydney, NSW, Australia Chapter 37: Data Fusion of Multiple Signals from the Sensor Network
Shengfa Yuan	Department of Precision Instruments and Mechanology, Tsinghua University, Beijing, China Chapter 31: Machine Learning Techniques
YeoHeung Yun	Department of Mechanical Engineering, University of Cincinnati, Cincinnati, OH, USA Chapter 67: Nanoengineering of Sensory Materials
Lingyu Yu	University of South Carolina, Columbia, SC, USA Chapter 55: Piezoelectric Wafer Active Sensors
Andrei N. Zagrai	New Mexico Institute of Mining and Technology, Socorro, NM, USA Chapter 5: Electromechanical Impedance Modeling
Yunfeng Zhang	Department of Civil and Environmental Engineering, University of Maryland, College Park, MD, USA Chapter 57: Piezoelectric Paint Sensors for Ultrasonics-based Damage Detection
Vladimir Zilberstein	JENTEK Sensors, Inc., Waltham, MA, USA Chapter 58: Eddy-current *in situ* Sensors for SHM

Preface

Structural Health Monitoring (SHM) is an expression that has been in existence for less than 20 years. It has emerged from the field of "smart structures", a field which embodies the idea of making sensing and actuation devices an integral part of a structure or of the structural material itself. Although inspired by nature, a smart structure is far from being a simple recreation of nature in the engineering world. Indeed, if that were the case, it might be logical to start with the simplest structure first and then gradually move onto the more complex one. This step is achieved with smart structures if sensing becomes the major point to be addressed, which is the case with SHM. SHM might therefore be the first approach when someone intends to realize a smart structure. Perhaps this explains why SHM has gathered so much recent attention and why an international workshop dedicated to it alone is held at least once every year, with many further conferences, symposia, and seminars also devoting much time to the topic. A number of scientific journals deal wholly or in large part with SHM and a variety of books have been published by different publishers, primarily John Wiley & Sons.

SHM has been made possible as a consequence of sensors and sensor technology becoming extremely miniaturized and low cost. It has been further enhanced by significant progress in materials manufacturing and processing technology, in the context of composite and functional materials, which allow sensing devices to be integrated and/or attached to structures of various kinds. Advances in materials technologies are gradually enabling the spraying or printing of sensitive materials onto structures, while a further trend sees the integration of sensitive elements into different materials to be used as additives in structural components. There is also increasing interest in e-textiles—another logical development in materials processing and design.

The remarkable growth in computational power has allowed sensor signals of increasing complexity to be handled next to the structure and even at the source of damage, while further processing is carried out at remote locations. In addition, smart materials have made possible energy harvesting in situ, generating new ideas related to autonomous monitoring with further data transmitted to remote data acquisition units via wireless communication. Computation technology combined with different means of high-speed communications has allowed the consideration of different techniques of data processing, from artificial neural networks to genetic algorithms and multi-agent systems, to name just a few.

Last but not least, SHM makes use of a variety of physical principles associated with what is classically known as *nondestructive testing*. These include principles based on strain, vibrations, acoustics, electromagnetics, temperature, and so on. One of the ambitions of SHM is to develop these principles into an integral part of structural materials so that structures made of these materials have the capability of self-sensing. Self-sensing, however, is not the end of the story. The signals provided by the self-sensing mechanism have to be "translated" into the language of the technically interested human being, meaning the signals received have to be digitized and processed so that they can be visualized in the form of a time-domain signal. However, this may not be sufficient since an interpretation of the signals may be required in terms of the structure's condition. This requires a reference signal, a damage index, and a scaling factor which then allow the damage condition to be characterized. The information related to the damage condition may still not be sufficient as, in many cases, the residual life of a structure may be the major issue of interest. Therefore, the damage information provided needs to be fed into a prognostic algorithm so that the residual fatigue life of the structure can be predicted. Finally, much consideration should also be placed on the graphical user interface (GUI), the man–machine interface to whoever might be involved with SHM.

Allowing SHM to become a part of structures has also led to a change in structural design principles. An obvious example is the enhanced ability of a structure to be inspected and thus maintained, as inspection is possible at virtually any time at the press of a button, removing the need for laborious dismantling and reassembly of components in the way of the area to be inspected. This option of continuous inspection therefore allows damages of a larger extent to exist in structures since uncertainties in prognostics are minimized. Damage in structures with SHM can be tolerated to the maximum allowable level, which in the nature of damage tolerance design allows structures to become lighter in weight. In consideration of the increased number of transducers required and signals received, signal processing again plays a major role.

SHM is an area of research and development that clearly involves a variety of disciplines. Starting from the physical principles of a material's deterioration and monitoring, it evolves across areas such as signal processing, sensors, simulation, and complete sensing systems. Once the sensing or SHM system is achieved, SHM-based structural design principles that consider structural strength as well as inspection, have to be established.

The diversity of disciplines involved in SHM is perhaps one of the reasons why it has become a relatively highly published field. Each of these disciplines feels "at home" in terms of finding an interesting field of application. Some of these disciplines may even claim that SHM emerged specifically from their own. The origin of SHM may possibly be worth a study in its own right. The term itself came into use in the late 1980s and early 1990s, though aviation people might date it to the 1950s. Strain analysts will date it to the invention of the first strain measurements by Lord Kelvin in the 1850s and ceramics producers might even date condition monitoring to the early inventions of pottery, some millenia ago.

Many different books have been written about SHM, many of which have been published by John Wiley & Sons. All of these books take different directions and it was Wiley's idea in early 2005 to find out if there might be a market for an encyclopedia in this still very young field. After initial reluctance, a plan was drawn up to identify those topics which would be required in such a major reference work. An international group of Subject Editors was identified to cover the various parts of the world where SHM is considered to be well established in terms of research and development. The acceptance of invitations to join the Editorial Board then went at a remarkable pace. A structure for the encyclopedia was soon finalized within the Editorial Board and names for Subject Editors were quickly proposed. This work finally resulted in the list of sections given below, with the names of the Subject Editors mentioned in parentheses:

1. Introduction and Definitions (*Christian Boller, Saarland University & Fraunhofer Institute for Non-Destructive Testing, Saarbrücken, Germany; and formerly of The University of Sheffield, Sheffield, UK*)
2. Physical Monitoring Principles (*Douglas E. Adams, Purdue University, West Lafayette, IN, USA*)
3. Signal Processing (*Wieslaw J. Staszewski and Keith Worden, University of Sheffield, Sheffield, UK*)
4. Simulation (*Srinivasan Gopalakrishnan, Indian Institute of Science, Bangalore, India*)
5. Sensors (*Victor Giurgiutiu, University of South Carolina, Columbia, SC, USA*)
6. Systems and System Design (*Daniel L. Balageas, ONERA (The French Aerospace Lab), Châtillon, France*)
7. Principles of SHM-based Structural Monitoring, Design and Maintenance (*Michael K. Bannister, CRC-ACS, Melbourne, VIC, Australia*)
8. Aerospace Applications (*Stephen C. Galea, Defence Science and Technology Organisation, Melbourne, VIC, Australia*)
9. Civil Engineering Applications (*Helmut Wenzel, VCE Holding GmbH, Vienna, Austria*)
10. Other Applications (*Daniel J. Inman, Virginia Tech, Blacksburg, VA, USA*)
11. Standardization and Specification (*Christian Boller, Saarland University & Fraunhofer Institute for Non-Destructive Testing, Saarbrücken, Germany; and formerly of The University of Sheffield, Sheffield, UK*).

Within each section, a structure was then established and a variety of potential authors were contacted. Each section was configured in detail by the Subject Editor(s) on the basis of a general template. The response to invitations for article submission was very positive and with the exception of Part 1, which provides an introduction and Part 11, which looks into the very new field of SHM standardization and specification, where little work has been done so far, the number of articles per section can be considered to reflect the current activities in the respective field. The articles in the other sections have been placed in a well-defined order. In summary, these articles make up the largest collection of work ever to have been presented in a complete set of volumes. The encyclopedia will hopefully become a major reference work for scientists, engineers, students, and whoever might be interested in the subject of SHM, allowing the intentions of the Publisher and the Editors to be met.

Trying to achieve absolute completeness in an encyclopedia is almost impossible, especially in a field where innovations can be observed on a near-daily basis. Although every effort has been made to achieve 100% completeness, this was a challenge within our set time frame. It was decided not to compromise the needs of the publication market, as parties interested in the reference work might have had to wait much longer, which may have been detrimental in such an area of emerging science.

In the sense of laterally integrating different areas of science, a remarkable effort was made to cross-reference between the different articles. This might again be far from exhaustive but will certainly help to establish the links between the various disciplines and will specifically help those parties who are not so involved with the subject of SHM.

An option to overcome all the missing bits and pieces to some degree and to allow a major reference work to become a living document is provided through the Internet. The Publisher wisely considered this option when configuring this work, and an online version of the encyclopedia will be established a few months after publication of the printed version. This will allow missing articles and cross-references to be included at a later stage, and also for new articles to be added. This continuous updating may result in a new printed edition in the future, at a standard which might not have been possible in the past.

Books and, nowadays, websites, are sources of organized information that can be obtained at an extremely low cost. This is only possible because of the large number of people contributing to these pieces of work on a mainly voluntary basis. We are most grateful to the Subject Editors of this encyclopedia, who agreed to take on an editorial role and did an outstanding job in getting their sections organized in terms of structuring, recruiting suitable authors, and reviewing the articles. Different meetings were held at various conferences where the "SHM community" meets and the directions were discussed and decided, resulting in the final encyclopedia. After the first batch of invitations were sent out to the various authors, the hugely positive response made the initiative highly remarkable and the project encouraging. It is with great recognition that we want to thank all these authors who may have spent their leisure time writing their articles and bringing them to a level that met the standards requested by the Editorial Board. Many of those authors do not even get this effort recognized by their profession, either because they work in industry or the academic environment requests articles with science of unique novelty. It is therefore with even higher appreciation that we recognize these contributions that provide valuable information from engineering applications to engineering science on the one hand and an educational compendium on the other, for those readers who want to learn what SHM is all about.

Books, both printed and electronic, cannot be published at an adequate quality without a publisher. Indeed, the very idea of the encyclopedia came from our publisher John Wiley & Sons. As Development Editor, Tony Carwardine was part of this project from the very beginning until the very end. This project would possibly not have survived in the time frame and form it has without Tony's eagerness, patience, and diplomacy. He was always there, the "spirit" behind us, tracking by e-mail, on the phone, at conferences, and even on-site, without ever becoming burdensome. Tony was joined early on by Liz Grainge who as Project Editor worked hard to keep track of all the articles in order to keep the project on schedule. In these tasks Liz was ably supported by Debbie Allen, Madeleine Baldrich and Anne Hunt. Another most valuable partner has been Laserwords in India, represented by Sangeetha, who took care of the production and editing. It should be remembered that many of the contributors to this work are not native English speakers. Nevertheless they have shown the skill to communicate in another language, and the courage to do so despite the extra effort which this entails. Laserwords, and specifically all the people working with them, have been our facilitators and we appreciate all their hard work in the background to make us authors understandable in the foreground.

As we have said before, the encyclopedia as it stands represents only the first step in a longer story. This first step has been a hard one to take but we hope that it has got the ball rolling. The work is due to be published online shortly; we hope that this will allow continued updating of the work and will further improve its quality. Our hope is that it becomes the leading reference work in the field of SHM. We also hope that the community which has been established will continue to grow in the future. Thank you again to all who have contributed to this work so far and who will contribute further into the future.

Christian Boller
Saarbrücken, Germany

Fu-Kuo Chang
Stanford, CA, USA

Yozo Fujino
Tokyo, Japan

January, 2009

Abbreviations and Acronyms

2D-DIC	Two-Dimensional Digital Image Correlation		AIC	Akaike Information Criterion
3D-DIC	Three-Dimensional Digital Image Correlation		AIMDS	Aircraft Integrated Monitoring and Diagnostic System
3G	Third-Generation		AISG	Aerospace Industry Steering Group
AANN	Autoassociative Neural Networks		AIST	National Institute of Advanced Industrial Science and Technology
Ab	Abutment		ALAVR	Average Ratio of Local Attractor Variance
ABDS	Adhesive Bond Degradation Sensors		ALM	Adaptive Life Management
ABS	American Bureau of Shipping		AM	Amplitude-Modulated
AC	Advisory Circular		AMC	Aircraft Mission Computer
AC	Alternating Current		ANCPSD	Average Normalized Cross Power Spectrum Density
A/C	Aircraft		ANDES	Automated Nondestructive Evaluation System
ACAMS	Aircraft Condition Analysis and Management System		ANNs	Artificial Neural Networks
acf	Autocovariance Function		ANPSD	Averaged Normalized Power Spectral Density
ACI	Analytical Condition Inspection		AOSG	Aircraft Operation Support Group
ACM	Advanced Composite Materials		AP	Ammonium Perchlorate
ACMS	Aircraft Condition Monitoring Systems		APB	Antiphase Boundary
AC-UT	Air-Coupled Ultrasonic Testing		APCB	A-Shaped Printed Circuit Board
ACS	Aircraft Control System		APU	Auxiliary Power Unit
AD	Accidental Damage		AR	Autoregressive
A/D	Analog-to-Digital		ARDU	Aircraft Research and Development Unit
ADC	Analog-to-Digital Conversion Module		AREMA	American Railway Engineering and Maintenance-of-way Association
ADC	Analog-to-Digital Converter		ARIMA	Autoregressive Integrated Moving Average
ADF	Australian Defence Force		ARINC	Aeronautical Radio Inc
ADPR	Average Driving-Point Residue		ARL	Applied Research Laboratory
ADRs	Accident Data Recorders		ARMA	Autoregressive Moving Average
ADT	Average Daily Traffic		ARMAX	Autoregressive Moving Average with Exogenous Excitation
ADTT	Average Daily Truck Traffic		ARMAX	Autoregressive Moving Average with Exogenous Input
AEs	Acoustic Emissions		ARV	Autoregressive Volatility
AET	Acoustic Emission Technique		ARX	Auto-Regressive eXogenous Input
AF	Ambiguity Function			
AFB	Air Force Base			
AFM	Atomic Force Microscope			
AFTS	Flight Test Squadron			
AG	Advisory Generation			
AHB	Antimetric Horizontal Bending			
AI	Artificial Intelligence			

ARX	Autoregressive Exogenous	BPSP	Battery-Powered Smart Patch
ARX	Autoregressive Models with Exogenous Inputs	BRE	British Research Establishment
		BSS	Blind Source Separation
ARX	Autoregressive with Exogenous Excitation	BVID	Barely Visible Impact Damage
		BVP	Boundary Value Problem
ARX	Autoregressive with Exogenous Inputs	bw	Broken Wire
ASCII	American Standard Code for Information Interchange	CA	Constant-Amplitude
ASHM	Atlanta Short Haul Mission	CAA	Civil Aviation Authority
ASI	Aircraft Structural Integrity	CAD	Computer Aided Design
ASIC	Application-Specific Integrated Circuit	CAE	Computer Aided Engineering
ASIMP	Aircraft Structural Integrity Management Plan	CALCE	Center for Advanced Life Cycle Engineering
ASIMS	Autonomous Structural Integrity Monitoring System	CAM	Computer Aided Manufacturing
		CAN	Controller Area Network
ASIP	Aircraft Structural Integrity Program	CASA	Civil Aviation Safety Authority
ASME	American Society of Mechanical Engineers	CBA	Cost–Benefit Analysis
		CBB	Commodore John Barry Bridge
ASNT	American Society for Nondestructive Testing	C-BIT	Continuous Bit
		CBLC	Cycle-Based Load Counting
ASP	Active Smart Patch	CBM	Condition-Based Maintenance
ASTM	American Society for Testing and Materials	CBM+	Condition-Based Maintenance Plus
		CBR	Composite Bonded Repair
ATA	Air Transport Association	CCD	Charge Coupled Devices
ATLSS	Advanced Technology for Large Structural Systems	ccf	Cross Covariance Function
		CCNS	Commercial Cabling Network System
AU	Acousto-Ultrasonics	CCS	Camera Coordinate System
AUSAM	Acousto-Ultrasonic Structural Health Monitoring Array Module	CCTV	Closed Circuit Television
		CCU	Command and Control Unit
AVS	Ambient Vibration Survey	CD	Crack Detection Sensors
AVT	Ambient Vibration Testing	CDF	Cumulative Density Function
AW	Axle Weight	CDF	Cumulative Distribution Function
AWG	Arrayed Wave Guide Grating	CDP	Compressor Delivery Pressure
		CDPC	Compressor Delivery Pressure
BAES ATC	Bae Systems Advanced Technology Centre	CEA	Cost–Effectiveness Analysis
		CEFIT	Elastodynamic Finite Integration Technique
BARF	Burn Anomaly Rate Factor		
BBA	Basic Belief Assignment	CEM	Conditional Expectation Method
BCS	Bridge Coordinate System	CES	Corrosive Environment Sensors
BEM	Boundary Element Method	CFD	Computational Fluid Dynamics
b/ep	Boron/epoxy	CFL	Courant–Friedrich–Lewy
BHM	Buckling Health Monitoring	CFRP	Carbon Fiber Reinforced Plastic
BI	Bayesian Inference	CFRP	Carbon Fiber Reinforced Polymer
BIC	Bayesian Information Criterion	CFV	Critical Failure Volume
BITE	Built-in Test Equipment	CG	Center of Gravity
BITs	Built-in Tests	CGPS	Continuous Global Positioning System
BMS	Bridge Management System	CIBrE	Center for Innovative Bridge Engineering
BP	Back Propagation		
BPA	Basic Probability Assignment	CIMSS	Center for Intelligent Material Systems

CM	Condition Monitoring	D2D	Data to Decision
CM	Correlation Measure	DA	Data Acquisition
CMC	Central Maintenance Computer	D/A	Digital-to-Analog
CMCs	Ceramic Matrix Composites	DAC	Digital-to-Analog Converter
CMDDB	Computational Modeling Data Database	DAC	Distance Amplitude Control Unit
		DAL	Development Assurance Level
CMFD	Condition Monitoring and Fault Diagnosis	DAME	Distributed Aircraft Maintenance Environment
CMOS	Complementary Metal Oxide Semiconductor	DAQ	Data Acquisition
		DARPA	Defense Advanced Research Projects Agency
CMS	Control and Monitoring System		
cMUT	Capacitive Micromachined Ultrasonic Transducers	DAS	Data Acquisition System
		DATS	Data Acquisition and Transmission System
CND	Cannot Duplicate		
CNF	Carbon Nanofibers	DAU	Data Acquisition Unit
CNSCs	Carbon Nanosphere Chains	DB	Database
CNTs	Carbon Nanotubes	DBST	Dual Bond Stress Temperature
COD	Crack Opening Displacement	dc	Direct Current
COMAC	Combination of Modal Assurance Criterion	DCB	Double Cantilever Beam
		DCO	Digitally Controlled Oscillator
COMAC	Coordinate Modal Assurance Criterion	DCS	Direct Corrosion Sensors
COTS	Commercial off the Shelf	DDFs	Digital Damage Fingerprints
COTS	Component off the Shelf	DE	Differential Equation
COV	Coefficient of Variation	DE	Differential Evolution
CPB	Constant Percentage Bandwidth	DEM	Digital Elevation Model
CPM	Corrosion-Prediction Model	DERA	Defence Evaluation Research Agency
CPU	Central Processing Unit	DF	Detail Fracture
CRC-ACS	Cooperative Research Center	DfR	Design for Reliability
CSA	Clonal Selection Algorithm	DFS	Digitally Filtered, Shuffled
CSD	Cross Spectral Density	DFT	Discrete Fourier Transformation
CSI/ref	reference-based combined deterministic-stochastic subspace identification	DGEBPA	Diglycidyl Ether of Bisphenol A
		DGPS	Differential Global Positioning System
		DI	Damage Index
CSIRO	Commonwealth Scientific and Industrial Research Organisation	DIC	Deming Inspection Criterion
		DIV	Damage-Induced Voltage
CSLDVs	Continuously Scanning Laser Doppler Vibrometers	DLIRWF	Damage Location Indicator Based on the Residual Wavelet Force
CSMUs	Crash Survivable Memory Units	DLL	Design Limit Load
CSSC	Cyclic Stress–Strain Curve	DLVs	Damage Locating Vectors
CT	Computed Tomography	DM	Data Manipulation
CUA	Cost–Utility Analysis	DM	Deeper Maintenance
CUPT	Coordinates Update	DMA	Direct Memory Access
CUSUM	Cumulative Sum	DMF	N,N-Dimethylformamide
CVD	Chemical Vapour Deposition	DMP	Data Management Platform
CVM	Comparative Vacuum Method	DNM	Duplicate Node Method
CVM	Comparative Vacuum Monitoring	DNV	Det Norske Veritas
CW	Continuous Wave	DOC	Direct Operating Cost
CWT	Choi–Williams Transform	DoD	Department of Defence
CWT	Continuous Wavelet Transform	DoFs	Degrees of Freedom

DPCS	Data Processing and Control System	EIFS	Equivalent Initial Flaw Size
DPSM	Distributed Point Source Method	EIS	Electrical Impedance Spectroscopy
DRIE	Deep Reactive Ion Etching	EIS	Electrochemical Impedance Spectroscopy
DS	Dempster–Shafer		
DSN	Distributed Sensor Networks	EIS	Entry into Service
DSP	Digital Signal Processing	EKF	Extended Kalman Filter
DSS	Decision Support System	EL	Endurance Limit
DSTO	Defence Science and Technology Organisation	ELIMA	Environmental Life-Cycle Information Management and Acquisition
DT	Damage Tolerance	EM	Electromagnetic
DVC	Digital Video Converter	EM	Electromechanical
DWC	Digital Wave Corporation	E/M	Electromechanical
DWPT	Discrete Wavelet Packet Transform	EMA	Experimental Modal Analysis
DWT	Discrete Wavelet Transform	EMAT	Electromagnetic Acoustic Transducer
		EMD	Empirical Mode Decomposition
EAC	Environmentally Assisted Cracking	emf	Electromotive Force
EAP	Electroactive Polymer	EMI	Electromagnetic Interference
EAS	Equivalent Airspeed	EMI	Electromechanical Impedance
EASA	European Aviation Safety Agency	EN	East–North
EBF	Event-Based Fatigue	EN	Emergency Node
EBT	Euler–Bernoulli Theory	EoE	Envelope of the Envelope
EC	Eddy Current	EP	Energy Processor
ECEF	Earth-Centered Earth-Fixed	EPDM	Ethylene Propylene Diene Monomer
ECFS	Eddy-Current Fail Sensors	EPFM	Elastic–Plastic Fracture Mechanics
ECOMAC	Enhanced Coordinate Modal Assurance Criterion	EPR	European Pressurized Water Reactor
		ERA	Eigensystem Realization Algorithm
EDM	Electrical Discharge Machining	ES	East–South
EDM	Electronic Distance Meter	ESD	Electrostatic Discharge
EEEU	End Effector Electronics Unit	ESPI	Electronic Speckle Pattern Interferometry
EEG	Europe Études Gecti		
EELS	Electron Emission Loss Spectroscopy	ET	Eddy-Current Testing
EEMD	Ensemble Empirical Mode Decomposition	ETFS	Eddy-Current Foil Sensors
		ETM	Enhanced Thematic Mapper
EEPROM	Electrically Erasable Programmable Read-Only Memory	ETR	Energy Transfer Ratio
		EU	European Union
EFDD	Enhanced Frequency-Domain Decomposition	EUSR	Embedded Ultrasonic Structural Radar
		EVP	Eigenvalue Vector Product
EFI-DPR	Effective Independence Driving-Point Residue	EVS	Extreme Value Statistics
		EVT	Extreme Value Theory
EFPI	Extrinsic Fabry–Perot Interferometer		
EFPI	Extrinsic Fabry–Perot Interferometric Fiber Optic	FAA	Federal Aviation Administration
		FAA	Federal Aviation Authority
EGNOS	European Geostationary Navigation Service	FAD	Failure Assessment Diagram
		FADEC	Full Authority Digital Engine Control
EGT	Exhaust Gas Temperature	FAMOS	Fatigue Monitoring System
EI	Effective Independence	FAR	Federal Aviation Regulations
EI-DPR	Effective Independence Driving-Point Residue	FAS	Fourier Amplitude Spectrum
		FBDD	Frequency-Based Damage Detection

FBG	Fiber-Optic Sensors Based on Bragg Gratings	FP	Fixpoints
FBGs	Fiber Bragg Gratings	FP	Functionally Pooled
FCAs	Fatigue Critical Areas	FPA	Focal Plane Array
FCC	Federal Communications Commission	FP-ARX	Functionally Pooled Autoregressive with Exogenous Excitation
FCS	Flight Control System	FPGA	Field-Programmable Gate Array
FDA	Frequency-Domain Analysis	FPI	Fabry–Perot Interferometer
FDD	Frequency Domain Decomposition	FPS	Full Parameter Set
FDM	Finite Difference Method	FPSO	Floating Production Storage and Offload
FDR	Flight Data Recorder	FRA	Federal Railroad Administration
FDS	Fatigue Detecting System	FRAM	Ferroelectric Nonvolatile Random Access Memory
FE	Final Elements	FREEVIB	Free Vibration
FE	Finite Element	FRFs	Frequency Response Functions
FEA	Finite Element Analyses	FRP	Fiber-Reinforced Polymers
FEISS	Finite Element Interfacing Software System	FRP	Fiberglass-Reinforced Plastic
FEM	Finite Element Model	FRPs	Fiber Reinforced Plastics
FEM	Finite Elements Method	FS	Factor of Safety
FEMs/BEMs	Finite Element Or Boundary Element Methods	FSDT	First-Order Shear Deformation Theory
FET	Field-Effect Transistor	FTs	Fourier Transforms
FFP-TF	Fiber Fabry–Perot Tunable Filter	FUG	Fuel Gauging System
FFT	Fast Fourier Transformation	FVT	Forced Vibration Testing
FFT	Forward Fourier Transformation	FVT	Free Vibration Test
FFTs	Fast Fourier Transforms		
FH	Fight Hours	GA	Genetic Algorithms
FHWA	Federal Highway Administration	GAGAN	Gps and Geo Augmented Navigation
FIM	Fisher Information Matrix	GAs	Genetic Algorithms
FIR	Finite Impulse Response	GBR	Ground-Based Reasoner
FIs	Fatigue Indices	GBS	Ground-Based Station
FJSP	Flexible Job Shop Problem	GCM	Gulf Coast Mission
FKM	Forschungskuratorium Maschinenbau	GCNS	Global Cabling Network System
FLI	Fleet Leader Inspection	GCS	Ground Control Station
FLM	Flight Loads Measurement	GDP	Gross Domestic Product
FLS	Flight Loads Survey	GDR	German Democratic Republic
FLS	Fuzzy Logic System	GE	General Electric
FM	Functional Model	GEP	Generalized Evidence Processing
FMEA	Failure Modes Effects Analysis	GESA	Geometrically Exact Structural Analysis
FMECA	Failure Modes Effects and Criticality Analysis	GESA	German Society for Experimental Stress Analysis
FMMEA	Failure Modes, Mechanisms, and Effects Analysis	GEV	Generalized Extreme Value Distribution
FMS	Foreign Military Services	GFRP	Glass Fiber Reinforced Plastics
FO	Fiber-Optic	GFRP	Glass Fiber-Reinforced Polymer
FOD	Foreign Object Damage	GIS	Geographic Information Systems
FONS	Fiber-Optic Nervous System	GLASS	Graphical Linking and Assembly of Syntax Structure
FOOM	Frequency of Occurrence Matrix		
FORCEVIB	Forced Vibration		
FOS	Fiber-Optic Sensor	GLR	Generalized Likelihood Ratio

GM	Geometric Method	HPWREN	High Performance Wireless Research and Education Network
GPA	Gas Path Analysis		
GPIB	General Purpose Interface Bus (IEEE488)	HSMS	Hull Stress Monitoring System
		HSWC	Hong Kong-Shenzhen Western Corridor
GPR	Ground Penetrating Radar		
GPRS	General Packet Radio Service	HT	Hilbert Transform
GPS	Global Positioning System	HTPB	Hydroxyl-Terminated Polybutadiene
GRAS	Ground-Based Regional Augmentation System	HUMSs	Health and Usage Monitoring Systems
		HVAC	Heating Ventilation and Air Conditioning
GRIN	Graded Index		
GSM	Global System for Mobile Communications	HVD	Hilbert Vibration Decomposition
		HyPFO	Hybrid Piezoelectric/fiber-optic
GSS	Ground Support System	I^2C	Inter-Integrated Circuit
GTAC	Ground-to-Air Cycle	IAAFT	Iterative Amplitude-Adjusted Fourier Transform
GTE	Gas Turbine Engine		
GTG	Gas Turbine Generators	IABMAS	International Association for Bridge Maintenance and Safety
GTP	Guyed Tower Platform		
GUWs	Guided Ultrasonic Waves	IAC	Intellegent Automation Inc
GVI	General Visual Inspection	IACS	International Annealed Copper Standard
GVW	Gross Vehicular Weight		
GW	Guided-Wave	IACS	International Association of Classification Societies
GWUT	Guided-Wave Ultrasonic Testing		
HA	Head Area	IAF	Israeli Air Force
HA	Health Assessment	IALCCE	International Association for Life-Cycle Civil Engineering
HARP	Helicopter Airworthiness Review Panel		
		IAT	Individual Aircraft Tracking
HBRRP	Highway Bridge Replacement and Rehabilitation Program	I-BIT	Interruptive Bit
		IC	Image Correlation
HCF	High Cycle Fatigue	IC	Integrated Circuits
HDPE	High Density Polyethylene	ICA	Independent Components Analysis
HEDDB	Health Evaluation Data Database	ICE	Integrated Cost Estimation
HELP layer	Hybrid Electromagnetic Performing Layer	ID	Inner Diameter
		ID	Internal Defect
HEPC	Hanshin Expressway Public Corporation	IDEs	Interdigital Electrodes
		IDS	Initial Discontinuity States
HFEC	High-Frequency Eddy-Current	IDTs	Interdigital Transducers
HFRT	High-Frequency Resonance Techniques	IETP	Interactive Electronic Technical Publication
HHT	Hilbert–Huang Transform	IF	Instantaneous Frequency
HMI	Human Machine Interface	IFFT	Inverse Fast Fourier Transformation
HMMs	Hidden Markov Models	i.i.d.	Identically Independently Distributed
HMS	Hull Monitoring Systems	i.i.d.	Independent, Identically Distributed
HNDT	Holographic Nondestructive Testing	IIR	Infinite Impulse Response
HOLSIP	Holistic Structural Integrity Process	IM	Impact Modulation
HOS	Higher Order Statistical	IMA	Integrated Modular Avionics
HP	High-Pressure	IMAC	International Modal Analysis Conference
HPC	High-Pressure Compressor		
HPT	High-Pressure Turbine	IMFs	Intrinsic Mode Functions

IMO	International Maritime Organization	LANL	Los Alamos National Laboratories
IMRS	Integrated Monitoring and Recording System	LBC	Link-Bearing Connection
		LbL	Layer-by-Layer
IMS	Inspection and Maintenance System	LBU	Laser-Based Ultrasonics
INS	Inertial Navigation System	LCCs	Life-Cycle Costs
IP	Intermediate-Pressure	LCF	Low Cycle Fatigue
IP	Internet Protocol	LCM	Life-Cycle Maintenance Management
IR	Infrared	LCM	Life-Cycle Management
IRF	Impulse Response Function	LCNS	Local Cabling Network System
IRR	Internal Rate of Return	LDV	Laser Doppler Vibrometry
IS	Importance Sampling	LE	Lyapunov Exponent
ISC	International Standard Classification	LEDs	Light Emitting Diodes
ISDB	In-Service Database	LEFM	Linear Elastic Fracture Mechanics
ISG	Integral Strain Gauge	LEO	Low Earth Orbit
ISHM	Integrated Systems Health Management	LESS	Leading-Edge Support Structure
		LHS	Latin Hypercube Sampling
ISHM	International Society for Structural Health Monitoring	LISA	Local Interaction Simulation Approach
		LIST	Laboratory for Intelligent Structure Technology
ISIS	Intelligent Sensing for Innovative Structures		
		LM-LS	Levenberg–Marquardt Least Squares
ISO	International Standards Organization	LN_2	Liquid Nitrogen
ISR	Intelligence, Surveillance, and Reconnaissance	LNG	Liquefied Natural Gas
		LP	Linearly Polarized
IT	Information Technology	LP	Low-Pressure
IT	Inner Triangle	L–P	Littlewood–Paley
ITD	Ibrahim Time Domain	LPC	Low-Pressure Compressor
IU	Imaging Ultrasonic	LPG	Long Period Grating
IVHM	Integrated Vehicle Health Management	LPGs	Long Period Fiber Bragg Gratings
IVHM	Integrated Vehicle Health Monitoring	LPI	Liquid Penetrant Inspection
IVM	Inspection Value Method	LPR	Linear Polarization Resistance
I&C	Instrumentation & Control	LPT	Low-Pressure Turbine
		LR	Lloyd's Register of Shipping
JAR	Joint Aviation Regulations	LRFD	Load and Resistance Factor Design
JMA	Japan Meteorological Agency	LRM	Line Replaceable Module
JNR	Japanese National Railway	LRUs	Line Replaceable Units
JSF	Joint Strike Fighter	LS	Least Squares
KAMs	Knowledge-Acquisition Methods	LTMS	Lifetime Monitoring System
KBM	Kinematics-based Method	LVDT	Linear Variable Differential Transducer
KE	Kinetic Energy	LVDTs	Linear Variable Displacement Transducers
KEM	Kinetic Energy Method		
KICT	Korean Institute of Construction Technology	LVM	Location Vector Method
		LVQ	Learning Vector Quantization
KLD	Karhunen–Loève Decomposition	LWS	Lower Wing Skin
KOBMS	Korean Bridge Management System		
KOH	Potassium Hydroxide	MA	Moving Average
KSMB	Kap Shui Mun Bridge	MAC	Modal Assurance Criteria
		MAE	Modal Acoustic Emission
LADAR	Laser Detection and Ranging	MAFT	Major Airframe Fatigue Test
LAN	Local Area Network	MAPD	Mean Absolute Percentage Deviation

MARSE	Measured Area under the Rectified Signal Envelope	MPI	Magnetic Particle Inspection
MBDD	Mode-Shape-Based Damage Detection	MPRS	Mission Planning and Restitution System
MCDS	Mechanical Components Diagnostics System	MR	Maintenance Recorder
		MRA	Multiresolution Analysis
MCL	Main Coolant Lines	MRBR	Maintenance Review Board Report
MCP	Magnetic Carpet Probes	MRO	Maintenance, Repair and Overhaul
MCS	Monte Carlo Simulation	MSAS	Mtsat Satellite-Based Augmentation System
MCU	Microcontroller Unit		
MDC	Mean Differential Cepstrum	MSC	Mode-Shape Curvature
MDLAC	Multiple Damage Location Assurance Criterion	MSD	Mahalanobis Squared Distance
		MSD	Multiple Site Damage
MDOF	Multidegree of Freedom	MSD	Multiple Site Fatigue Damage
MDP	Maintenance Data Panel	MSD	Multisite Damage
MDS	Maintenance Data System	MSDRS	Maintenance Signal Data Recording System
MDS	Measured Damage Signature		
MEL	Minimum Equipment List	MSE	Mean Square Error
MEMS	Microelectromechanical Sensors	MSEI	Modal Strain-Energy Index
MEMSs	Microelectromechanical Systems	MSET	Multivariate State Estimation Technique
MESA	Multi Element Scanning Array		
MFC	Macro Fiber Composite	MSG	Maintenance Steering Group
MFCS	Multifunctional Corrosion Sensors	MSI	Medium Scale Integration
MFI	Modal Flexibility Index	MSP	Mode Superposition
MFL	Magnetic Flux Leakage	MSR	Mechanical Strain Recorders
MFOP	Maintenance-Free Operating Period	MsS	Magnetostrictive Sensors
MFOPS	Maintenance-Free Operating Period Survivability	MSS	Multispectral Scanner
		MSW	Municipal Solid Waste
MI	Modulation Index	MTFB	Mean Time between Failures
MIF	Modal Indicator Function	MUMPs	Multiuser Mems Processes
MIMO	Multiple Input, Multiple Output	MUSE	Mobile Ultrasonic Equipment
MIMOSA	Machinery Information Management Open Systems Alliance	MVP	Mixed Variable Programming
		MWCNTs	Multiwall Carbon Nanotubes
MIR	Micropower Impulse Radar	MWM	Meandering Winding Magnetometer
MISO	Multi-Input Single-Output		
MIT	Massachusetts Institute of Technology	NAALDAS	Naval Aircraft Approach and Landing Data-Acquisition System
MKE	Modal Kinetic Energy		
ML	Maximum Likelihood	NACA	National Advisory Committee on Aeronautics
MLG	Main Landing Gear		
MLP	Multilayer Perceptron	NADDB	Numerical Analyzed Data Database
MMOD	Micrometeoroid and Orbital Debris	NARMAX	Nonlinear Autoregressive Moving Average with Exogenous Excitation
MMS	Mission Management System		
MOCT	Korean Ministry of Construction and Transportation	NARMAX	Nonlinear Autoregressive Moving Average with Exogenous Inputs
MoD	Ministry of Defense	NASA	National Aeronautics and Space Administration
MOSFET	Metal-Oxide Semiconductor Field-Effect Transistor		
		NAT	Nonlinear Acoustic Technique
MPD	Maintenance Planning Data	NAVAIR	The Naval Air Systems Command
MPD	Maintenance Planning Document	NBI	National Bridge Inventory
MPDDB	Measured and Processed Data Database	NCPSD	Power Spectrum Densities

NDE	Nondestructive Evaluation	OLM	Operational Loads Monitoring or Measurement
NDECS	Nondestructive Evaluation of Cooperative Structures	OLMOS	Onboard Life Monitoring System
NDERM	Nondestructive Evaluation Ready Material	OLMS	Operational Loads Monitoring System
NDE/NDT	Nondestructive Evaluation and Testing	OMA	Operational Modal Analysis
NDI	Nondestructive Inspection	OMCs	Organic Matrix Composites
NDT	Nondestructive Techniques	ONERA	Office National d'Etudes et de Recherches Aérospatiales
NDT	Nondestructive Testing	ONS	Optimal Number of Sensors
NDT&E	Nondestructive Testing and Evaluation	OOR	Ordered Overall Range
NDTs	Nondestructive Evaluation Techniques	OPMS	Oleo Pressure Monitoring System
NDVI	Normalized Difference Vegetation Index	ORA	Operational Risk Assessment
NETD	Noise Equivalent Temperature Difference	OSA	Optical Spectrum Analyzer
NFF	No-Fault-Found	OSA-CBM	Open Systems Architecture for Condition-Based Maintenance
NFI	No Fault Indicated	OSA-EAI	Open Systems Architecture for Enterprise Application Integration
NI	Novelty Index	OSP	Optimal Sensor Placement
NIST	National Institute of Standards and Technology	OT	Outer Triangle
NLLS	Nonlinear Least Squares	OTDR	Optical Time-Domain Reflectometry
NLPCA	Nonlinear Principal Component Analysis	OTF	On-the-Fly
NN	Neural Network	PA	Prognostics Assessment
NOFRFs	Nonlinear Output Frequency Response Functions	PAN/PITCH	Polyacrylonitrile/Pitch Resin
NPPs	Nuclear Power Plants	PBAN	Polybutadiene Acrilonitrile
NPV	Net Present Value	PC	Personal Computer
NRC	National Research Council Canada	PCA	Principal Component Analysis
NREL	National Renewable Energy Laboratory	PCB	Printed Circuit Boards
NS	North-South direction	PCI	Peripheral Components Interconnect
NSF	National Science Foundation	PCs	Principal Components
NSS	Neutral Salt Spray	PDA	Personal Digital Assistant
NTF	No Trouble Found	PDEs	Partial Differential Equations
NURBS	Non-Uniform Rational B-Spline	PDF	Probability Distribution Function
NWs	Nanowires	PDFs	Probability Density Functions
NZ	Normal Accelerometer	PDM	Periodic Depot Maintenance
		PDS	Predicted Damage Signature
OC	Oblique Cut	PDSD	Phase Difference Standard Deviation
OCV	Open Circuit Voltage	PE	Prediction Error
OD	Outer Diameter	PEEK	Polyetheretherketone
ODEs	Ordinary Differential Equations	PEO	Poly-Ethylene Oxide
ODSs	Operational Deflection Shapes	PFCs	Piezo Fiber Composites
OE	Output Error	PFEM	Perturbation Theory Enhanced Finite-Element Modal
OEMs	Original Equipment Manufacturers	PFM	Proportional Flexibility Matrix
OIDA	Optoelectronics Industry Development Association	PGA	Peak Ground Acceleration
		PGSL	Probabilistic Global Search Lausanne
		PHM	Prognostics and Health Management
		PHM	Prognostics and Health Monitoring
		PITS	Point-in-the-Sky

PLD	Programmable Logic Device	QP	Quadratic Programming
PLI	Propellant–Liner–Insulator	QZSS	Quasi-Zenith Satellite System
PLZT	Lead Lanthanum Zirconate Titanate		
PM	Polarization Maintaining	RAAF	Royal Australian Air Force
PMAC	Partial Modal Assurance Criterion	RAF	Royal Air Force
PMAT	Portable Multipurpose Access Terminal	RAN	Royal Australian Navy
PMCs	Polymer Matrix Composites	RASU	Representative Active Sensing Unit
PMDS	Portable Maintenance Data Store	RBF	Radial Basis Function
PMMA	Poly-Methyl-Meth-Acrylate	RC	Reinforced Concrete
PMN	$Pb(Mg_{1/3}Nb_{2/3})O_3$	RCC	Reinforced Carbon–Carbon
PMT	Parameterized Modeling Technique	RDAU	Remote Data Acquisition Unit
PNN	Probabilistic Neural Network	RDC	Remote Data Concentrator
PoD	Probability of Detection	RF	Radio Frequency
POD	Proper Orthogonal Decomposition	RFEC	Remote-Field Eddy Current
PoF	Physics-of-Failure	RFEC-MCP	Remote-Field Eddy Current Magnetic Carpet Probes
POF	Polymer Optical Fiber	RFID	Radio Frequency Identification
POF	Probability of Failure	RGM	Residual Gust and Maneuver
POMs	Proper Orthogonal Modes	RH	Relative Humidity
POVs	Proper Orthogonal Values	RI	Refractive Index
PPS	Pilot Parameter Set	RID	Risk Information Delivery
PR	Pattern Recognition	RIE	Reactive Ion Etching
PRA	Probabilistic Risk Assessment	RIF	Reliability Importance Factor
ProDTA	Probabilistic Damage Tolerance Analysis	RIMUs	Remote Inertial Measurement Units
PSD	Position Sensitive Diode	RINEX	Receiver Independent Exchange
PSD	Power Spectral Densities	RIU	Remote Interface Unit
PSD	Power Spectrum Densities	RMS	Root Mean Square
PSF	Point Spread Function	RMSD	Root Mean Square Deviation
PSS	poly(sodium 4-styrene sulfonate)	RMSE	Root Mean Square Error
PTCa/PEKK	Lead Titanate/poly(Ether Ketone Ketone)	RNN	Recurrent Neural Network
PTZ	Pan-Tilt-Zoom	RPI	Recursive Probability Integration
PU	Polyurethane	RPS	Recorder Parameter Set
PUFEM	Partition of Unity Finite Element Method	rps	Revolutions per Second
		RPV	Reactor Pressure Vessel
PVA	Poly-Vinyl Alcohol	rrms	Relative Root Mean Square
PVC	Polyvinyl Chloride	RSRM	Reusable Solid Rocket Motor
PVDF	Poly-Vinyl-Dene Fluoride	RSSI	Receive Signal Strength Indicator
PWAS	Piezoelectric Wafer Active Sensors	RTCM	The Radio Technical Commission for Maritime Services
PWASs	Piezo Wafer Active Sensors	RTD	Resistance Temperature Detector
PWT	Partial Wave Technique	RTK	Real-Time Kinematic
PY	Pylon	RTK GPS	Real-Time Kinematic Global Positioning System
PZT	Lead Zirconate Titanate		
PZT	Lead Zirconium Titanate	RTM	Resin Transfer Molding
PZT	Piezoelectric	RTOK	Retest Ok
PZT	Piezoelectric Lead Zirconate Titanate	RUL	Remaining Usable Life
PZT	Piezoelectric Transducers	RVDT	Rotary Variable Differential Transformer
QA/QC	Quality Assurance/quality Control	RVE	Representative Volume Element

RVM	Relevance Vector Machine	SISTeC	Smart Infra-Structure Technology Center
RWF	Residual Wavelet Force	SLDs	Superluminescent Diodes
R&D	Research and Development	SLDV	Scanning Laser Doppler Vibrometer
SA	Simulated Annealing	SLE	Spectral Layer Element
SADDB	Statistical Analyzed Data Database	SMA	Shape Memory Alloy
SAFE	Semianalytical Finite Element	SMART	Stanford Multiactuator Receiver Transduction
SAM	Scanning Acoustic Microscope		
SAMCO	Structural Assessment Monitoring and Control	SMIP	Strong Motion Instrumentation Program
SAP	Stress-Applying Portion	SMS	Structural Monitoring Systems
SAT	Selected Aircraft Tracking	SN	Stationary Node
SAW	Surface Acoustic Wave	S/N	Signal-to-Noise
SBAS	Structural/Ballistic Analysis System	S/N	Stress Vs. Number Of Cycles
SBBMS	Sensor-Based Bridge Monitoring System	SNP	Self-Nulling Probe
		SNR	Signal-to-Noise Ratio
SC	Surface Cut	SNS	Structural Neural System
SCAL	Scalogram	SNSAP	Structural Neural System Analog Processor
SCB	Stonecutters Bridge		
SCC	Stress Corrosion Cracking	SoC	System-on-a-Chip
SCE	Saturated Calomel Electrode	SOFM	Self-Organizing Feature Map
SCF	Stress Concentration Factors	SOM	Self-Organizing Map
SCS	Sensor Coordinate System	SONAR	Sound Navigation and Ranging
SD	State Detection	SOV	Space Operations Vehicle
SDI	Special Detailed Inspection	SPATE	Stress Pattern Analysis by Thermal Emissions
SDOF	Single Degree of Freedom		
SEA	Statistical Energy Analysis	SPC	Statistical Process Control
SEAD	Suppression of Enemy Air Defenses	SPCs	Sortie Profile Codes
SEM	Scanning Electron Microscope	SPEC	Spectrogram
SEM	Spectral Element Method	SPI	Serial Peripheral Interface
SFA	Standard Fatigue Axle	SPO	Sensor Placement Optimization
SFEM	Spectral Finite Element Method	SPR	Statistical Pattern Recognition
SFV	Standard Fatigue Vehicle	SPRT	Sequential Probability Ratio Test
SH	Shear Horizontal	SPS	Smart Patch System
SHAUNN	Structural Health and Usage Neural Network	SPSP	Self-Powered (Wireless) Smart Patch
		SRB	Solid Rocket Booster
SHB	Symmetric Horizontal Bending	SRIM	System Realization Using Information Matrix
SHDMS	Structural Health Data Management System		
		SRM	Stiffness Reduction Method
SHM	Structural Health Management	SRM	Structural Repair Manual
SHM	Structural Health Monitoring	SRMs	Solid Rocket Motors
SHM-NDE	Structural Health Monitoring—Nondestructive Evaluation	SRMS	Space Shuttle Remote Manipulator System
SHMSs	Structural Health Monitoring Systems	SRTM	Shuttle Radar Topography Mission
SI	System Identification	SSBF	Simplified Stress-Based Fatigue
SIM	Sharp Interface Method	SSC	Systems, Structures and Components
SIMO	Single Input, Multiple Output	SSE	Sum of Squared Error
SIMS	Structural Integrity Monitoring System	SSI	Soil–Structure Interaction
SISO	Single Input Single Output	SSI	Stochastic Subspace Identification

SSI	Structure Significant Item	TM	Transverse Magnetic
SSI-COV	Covariance-Driven Stochastic Subspace Identification	TMB	Tsing Ma Bridge
		ToF	Time-of-Flight
SSTL	Smart Structures Technology Laboratory	TOW	Time of Wetness
		TPIS	Tire Pressure Indication System
STF	System Transfer Function	TPMSs	Tire Pressure Monitoring Systems
STFT	Short Time Fourier Transforms	TPS	Thermal Protection Systems
STSA	Symbolic Time Series Analysis	TPSs	Thermal Protective Systems
SV	Shear Vertical	TRD	Time-Rate-of-Decay
SVD	Singular Value Decomposition	TRL	Technology Readiness Level
SVMs	Support Vector Machines	TSA	Time Synchronous Averaging
SWBM	Still Water Bending Moment	TSK	Wave Height Meter
SWCNTs	Single-Wall Carbon Nanotubes	TSP	Travelling Salesperson Problem
SwRI	Southwest Research Institute	TTT	Through-the-Thickness
		TWIST	Transport Wing Standard
TAG	Touch and Go		
TANGO	Technology Application to the Near-Term Business Goals and Objectives of the Aerospace Industry	UART	Universal Asynchronous Receiver/Transmitter
		UAST	Uniaxial Strain Transducer
		UAV	Unmanned Aerial Vehicles
TARX	Time-Varying Autoregressive with Exogenous Excitation	UAV	Unmanned Air Vehicles
		UCAVs	Unmanned Combat Air Vehicles
TAT	Temporary Aircraft Tracking	UCL	Upper Control Limit
TATEM	Technologies and Techniques for New Maintenance Concepts	UCSD	University of California at San Diego
		UHF	Ultra High Frequency
T/BEST	Technology Benefit Estimator	UIs	Usage Indices
TBT	Timoshenko Beam Theory	UML	Unified Modeling Language
TC-OFS	Technical Committee "Fiber-Optic Sensors"	UML	Uniform Material Law
		UMMC	Utility Mission in Morgan City
TCP/IP	Transmission Control Protocol/Internet Protocol	UMTS	Universal Mobile Telecommunications System
TCU	Terminal Collection Unit	UPV	Ultrasonic Pulse Velocity
TDA	Time-Domain Analysis	UQP	Uncertainty Quantification and Propagation
TDD	Time Domain Decomposition		
TDDB	Time-Dependent Dielectric Breakdown	US	Ultrasound
TDM	Time Division Multiplexing	USAF	United States Air Force
TDNN	Time Delayed Neural Network	USM	Ubiquitous Structural Monitoring
TE	Transverse Electric	USN	United States Navy
TEGs	Thermoelectric Generators	UT	Ultrasonic Testing
TERSA	Thermal Evaluation for Residual Stress Analysis	UT	Ultrasonically Excited Infrared Thermography
TF	Transverse Fissure	U.T.L.	Ultimate Load
TFIE	Time–Frequency Instantaneous Estimators	UTS	Ultimate Tensile Strength
		UV	Ultraviolet
TFR	Time–Frequency Representation	UWB	Ultra Wideband
TFs	Transmittance Functions		
TFTs	Time Frequency Transforms	VA	Variable Amplitude
TKB	Ting Kau Bridge	VB	Vertical Bending
TLR	Top Level Requirement	VC	Vapnik–Chervonenkis
TM	Thematic Mapper		

VDD	Vibration-Based Damage Detection	WIM	Weigh-in-Motion
VHF	Very High Frequency	WLAN	Wireless Local Area Network
VIMS	Vehicle Intelligent Monitoring System	WLEIDS	Wing Leading-Edge Impact-Detection System
VM	Variance Method		
VM	Vibromodulation	WNN	Wavelet Neural Network
VQPCA	Vector Quantization Principal Component Analysis	WOW	Weight on Wheels
		WPERI	Wavelet Packet Energy Rate Index
VSI	Vacuum Sensor Integrated	WPT	Wavelet Packet Transform
		WRBM	Wing Root Bending Moment
WAAS	Wide Area Augmentation Systems	WSNs	Wireless Sensor Networks
WASHMS	Wind and Structural Health Monitoring System	WT	Wavelet Transformation
		WT	Wind Turbine
WCS	World Coordinate System	WTB	Wind Turbine Blade
WD	Wigner Distribution	WVD	Wigner–Ville Distributions
WDM	Wavelength Division Multiplexing		
WFD	Widespread Fatigue Damage	XML	Extensible Markup Language
WG	Working Group		
WGS84	World Geodetic System of 1984	YAG	Yttrium Aluminum Garnet
WIBM	Wave-Induced Bending Moment	ZUPT	Zero Velocity Update
WID	Wireless Impedance Device		

PART 5
Sensors

Chapter 52
Piezoelectricity Principles and Materials

Victor Giurgiutiu
Department of Mechanical Engineering, University of South Carolina, Columbia, SC, USA

1 Introduction	981
2 Basic Equations	981
3 Ferroelectric Perovskites	983
4 Typical Electroactive Ceramics	985
5 Summary and Conclusions	989
Related Articles	990
Further Reading	991

1 INTRODUCTION

Piezoelectricity (discovered in 1880 by Jacques and Pierre Curie) describes the phenomenon of generating an electric field when the material is subjected to a mechanical stress (direct effect), or, conversely, generating a mechanical strain in response to an applied electric field. The *direct piezoelectric effect* predicts how much electric field is generated by a given mechanical stress. This *sensing effect* is utilized in receiving piezoelectric sensors. The *converse piezoelectric effect* predicts how much mechanical strain is generated by a given electric field. This *actuation effect* is utilized in transmitting piezoelectric sensors. Piezoelectric active sensors act as both transmitters and receivers of elastic waves.

Piezoelectric properties occur naturally in some crystalline materials, e.g., quartz crystals (SiO_2) and Rochelle salt. The latter is a natural ferroelectric material, possessing an orientable domain structure that aligns under an external electric field and thus enhances its piezoelectric response. Piezoelectric response can also be induced by electrical poling in certain polycrystalline materials, such as piezoceramics. In recent years, affordable high-performance piezoceramics have become commercially available at affordable prices.

Piezoelectric materials are some of the major building blocks of the structural health monitoring (SHM) sensors. The intrinsic active behavior of these materials, which change dimensions in response to electric fields, and produce electricity in response to mechanical strain and stress, makes them ideal for actuation and sensing as required in SHM applications.

2 BASIC EQUATIONS

2.1 Piezoelectric equations

Electroactive materials can be distinguished into piezoelectric and electrostrictive. *Linear piezoelectric materials* obey the constitutive relations between the mechanical and electrical variables, which can be

Encyclopedia of Structural Health Monitoring. Edited by Christian Boller, Fu-Kuo Chang and Yozo Fujino © 2009 John Wiley & Sons, Ltd. ISBN: 978-0-470-05822-0.

written in tensor notations as

$$S_{ij} = s^E_{ijkl} T_{kl} + d_{ijk} E_k + \delta_{ij} \alpha^E_i \theta \quad (1)$$

$$D_j = d_{jkl} T_{kl} + \varepsilon^T_{jk} E_k + \tilde{D}_j \theta \quad (2)$$

where S_{ij} is the mechanical strain, T_{kl}, mechanical stress, E_k, electrical field, and D_j, electrical displacement. The variable s^E_{ijkl} is the mechanical compliance of the material measured at zero electric field ($E = 0$), ε^T_{jk} is the dielectric permittivity measured at zero mechanical stress ($T = 0$), and d_{kij} is the piezoelectric coupling between the electrical and mechanical variables. The variable θ is the temperature, and α^E_i is the coefficient of thermal expansion under constant electric field. The coefficient $\tilde{D}_j$ is the coefficient that connects electric displacement with temperature. The stress and strain variables are second-order tensors, while the electric field and the electric displacement are first-order tensors. Since thermal effects only influence the diagonal terms, the respective coefficients, α_i and $\tilde{D}_j$, have single subscripts. The term δ_{ij} is the Kroneker delta ($\delta_{ij} = 1$ if $i = j$; zero otherwise).

Compressed matrix notations (Voigt notations) are often used in engineering practice in lieu of the tensorial equations (1) and (2). The stress and strain tensors are arranged as six-component vectors, with the first three components representing *direct* stress and strain, and the last three representing *shear* stress and strain. When written in compact form, equations (1) and (2) become

$$S_p = s^E_{pq} T_q + d_{kp} E_k + \delta_{pq} \alpha^E_q \theta, \quad p, q = 1, \ldots, 6$$
$$k = 1, 2, 3 \quad (3)$$

$$D_i = d_{iq} T_q + \varepsilon^T_{ik} E_k + \tilde{D}_i \theta, \quad q = 1, \ldots, 6$$
$$i, k = 1, 2, 3 \quad (4)$$

Please note that the piezoelectric matrix in equation (3) is the transpose of the piezoelectric matrix in equation (4). Equations (3) and (4) can also be written in matrix format, i.e.,

$$\{S\} = [s]\{T\} + [d]^t\{E\} + \{\alpha\}\theta \quad (5)$$

$$\{D\} = [d]\{T\} + [\varepsilon]\{E\} + \{\tilde{D}\}\theta \quad (6)$$

Electromechanical coupling coefficient is the square root of the ratio between the mechanical energy stored and the electrical energy applied to a piezoelectric material, i.e.,

$$k^2 = \frac{\text{Mechanical energy stored}}{\text{Electrical energy applied}} \quad (7)$$

For direct actuation, $k^2_{33} = d^2_{33}/s_{33}\varepsilon_{33}$, where Voigt matrix notations are used. For transverse actuation, $k^2_{31} = d^2_{31}/s_{11}\varepsilon_{33}$; and for shear actuation, $k^2_{15} = d^2_{15}/s_{55}\varepsilon_{11}$. For uniform in-plane actuation, one uses the planar coupling coefficient, $\kappa_p = \kappa_{13}\sqrt{2/(1-\nu)}$, where ν is the Poisson ratio. The piezoelectric response in axial and shear directions is depicted in Figure 1.

2.2 Electrostrictive equations

In contrast to linear piezoelectricity, the *electrostrictive response* is quadratic in electric field. Hence, the direction of electrostriction does not change as the polarity of the electric field is reversed. The general constitutive equations incorporate both piezoelectric and electrostrictive terms as follows:

$$S_{ij} = s^E_{klij} T_{kl} + d_{kij} E_k + M_{klij} E_k E_l \quad (8)$$

$$D_m = d_{mkl} T_{kl} + \varepsilon^T_{mn} E_n + 2M_{mnij} E_n T_{ij} \quad (9)$$

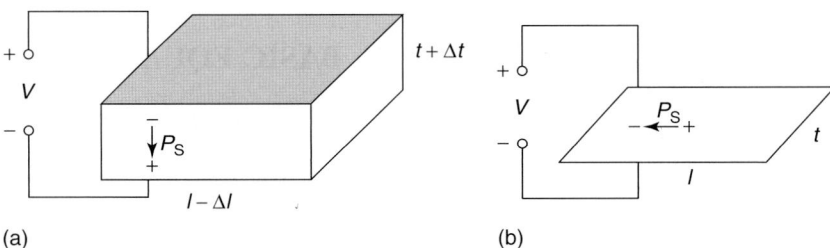

Figure 1. Induced-strain responses of piezoelectric materials: (a) axial strain and (b) shear strain.

Note that the first two terms in each equation are similar to the linear piezoelectric behavior. However, in electrostrictive materials, the linear piezoelectric response is much weaker than in piezoelectric materials. The third term represents the strong nonlinear electrostrictive behavior. The coefficients M_{klij} are the electrostrictive coefficients.

2.3 Magnetostrictive equations

Most magnetoactive materials are based on the magnetostrictive effect. The *magnetostrictive constitutive equations* contain both linear and quadratic terms as follows:

$$S_{ij} = s^E_{ijkl} T_{kl} + d_{kij} H_k + M_{klij} H_k H_l \quad (10)$$

$$B_m = d_{mkl} T_{kl} + \mu^T_{mk} H_k + 2 M_{mnij} E_n T_{ij} \quad (11)$$

where, in addition to the already defined variables, H_k is the magnetic field intensity, B_j is the magnetic flux density, and μ^T_{jk} is the magnetic permeability under constant stress. The coefficients d_{kij} and M_{klij} are defined in terms of magnetic units. The magnetic field intensity in a rod surrounded by a coil with n turns per unit length depends on the coil current, I, i.e.,

$$H = nI \quad (12)$$

Magnetostrictive material response is quadratic in the magnetic field, i.e., the magnetostrictive response does not change sign when the magnetic field is reversed. However, the nonlinear magnetostrictive behavior can be linearized about an operating point through the application of a bias magnetic field. In this case, *pseudo-piezomagnetic* behavior, in which response reversal accompanies field reversal, can be obtained. In Voigt matrix notations, the equations of linear piezomagnetism are as follows:

$$S_i = s^H_{ij} T_j + d_{ki} H_k, \quad i,j = 1, \ldots, 6;$$
$$k = 1, 2, 3 \quad (13)$$

$$B_m = d_{mj} T_j + \mu^T_{mk} H_k, \quad j = 1, \ldots, 6;$$
$$k, m = 1, 2, 3 \quad (14)$$

where, S_i is the mechanical strain, T_j is the mechanical stress, H_k is the magnetic field intensity, B_m is the magnetic flux density, and μ^T_{mk} is the magnetic permeability under constant stress. The coefficient s^H_{ij} is the mechanical compliance measured at zero magnetic field ($M = 0$). The coefficient μ^T_{mk} is the magnetic permeability measured at zero mechanical stress ($T = 0$). The coefficient d_{ki} is the *piezomagnetic constant*, which couples the magnetic and mechanical variables and expresses how much strain is obtained per unit applied magnetic field.

3 FERROELECTRIC PEROVSKITES

The popular piezoceramic materials owe their piezoelectric behavior to the ferroelectric phenomenon observed in a class of crystalline materials called *ferroelectric perovskites*.

3.1 Ferroelectricity

Ferroelectricity is the property of having permanent electric polarization, and of being able to alter it by the application of an external electric field. The term *ferroelectricity* was derived by analogy with the term *ferromagnetism*, which describes how permanent magnetization is altered by the application of an external magnetic field. As the electric field applied to a ferroelectric material is increased beyond a critical value called *coercive field*, E_c, the polarization suddenly increases to a high value. This value is more or less maintained when the electric field is decreased, such that at zero electric field the ferroelectric material retains a permanent spontaneous polarization P_S. When a negative electric field is applied beyond the negative value $-E_c$, the polarization suddenly switches to a large negative value, which is roughly maintained as the electric field is decreased. At zero electric field, the permanent spontaneous polarization is $-P_S$. As the electric field is again increased into the positive range, the polarization is again switched to a positive value, as the field increases beyond E_c. Characteristic of this behavior is the high hysteresis loop traveled during a cycle. The ferroelectric behavior can be explained through the existence of aligned internal dipoles that have their direction switched when the electric field is sufficiently strong.

3.2 Perovskite structure

Perovskites are a large family of crystalline oxides with the metal-to-oxygen ratio 2:3. Perovskites derive their name from a specific mineral, perovskite, first described in 1839 by the geologist Gustav Rose, who named it after the famous Russian mineralogist Count Lev Aleksevich von Perovski (1792–1856). The simplest perovskite lattice has the expression, $X_m Y_n$, in which the X atoms are rectangular close-packed and the Y atoms occupy the octahedral interstices. The rectangular close-packed X atoms may be a combination of various species, X^1, X^2, X^3, etc. For example, in the barium titanate perovskite, $BaTiO_3$, we have $X^1 = Ba^{2+}$ and $X^2 = Ti^{4+}$, while $Y = O^{2-}$ (Figure 2). In the lattice structure, the Ba^{2+} divalent cations are at the corners, the Ti^{4+} tetravalent cation is in the center, while the O^{2-} anions are on the faces. The Ba^{2+} cations are larger, while the Ti^{4+} cations are smaller. The size of the Ba^{2+} cation affects the overall size of the lattice structure. Perovskite arrangements like in $BaTiO_3$ are generically designated ABO_3. Their commonality is that they have a small, tetravalent metal ion, e.g., titanium or zirconium, in a lattice of larger, divalent metal ions, e.g., lead or barium and oxygen ions (Figure 2). Under conditions that confer tetragonal or rhombohedral symmetry, each crystal has a dipole moment.

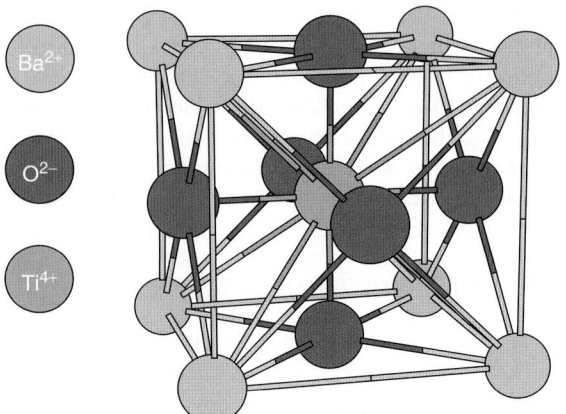

Figure 2. Crystal structure of a typical perovskite, $BaTiO_3$: the Ba^{2+} cations are at the cube corners, the Ti^{4+} cation is in the cube center, and the O^{2-} anions on the cube faces.

3.3 Spontaneous strain and spontaneous polarization of ferroelectric perovskites

At elevated temperatures, the primitive perovskite arrangement is symmetric face-centered cubic (fcc) and does not display electric polarity (Figure 3a). This symmetric lattice arrangement forms the *paraelectric phase* of the perovskite, which exists at

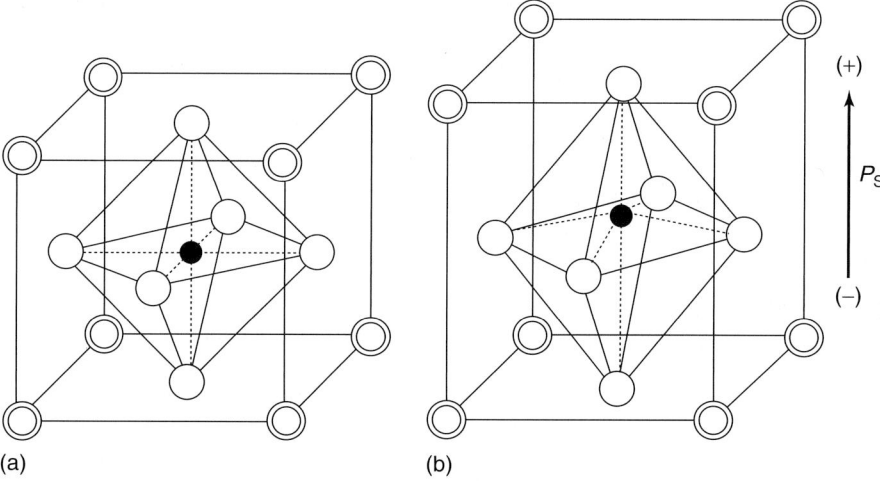

Figure 3. Spontaneous strain and polarization in a perovskite structure: (a) above the Curie point, the crystal has a cubic lattice, displaying a symmetric arrangement of positive and negative charges and no polarization (paraelectric phase) and (b) below the Curie point, the crystal has a tetragonal lattice, with an asymmetrically placed central atom, thus displaying polarization (ferroelectric phase).

elevated temperatures. As the temperature decreases, the lattice shrinks and the symmetric arrangement is no longer stable. For example, in barium titanate, the Ti^{4+} cation snaps from the cube center to other minimum-energy locations situated off center. This is accompanied by the corresponding motion of the O^{2-} anions. Shifting of the Ti^{4+} and O^{2-} ions causes the structure to be altered, creating strain and electric dipoles. The crystal lattice becomes distorted and slightly elongated in one direction, i.e., tetragonal (Figure 3b). In barium titanate, the distortion ratio is $c/a = 1.01$, corresponding to 1% strain in the c direction with respect to the a direction. This change in dimensions along the c axis is called *spontaneous strain*, S_S. The orthorhombic tetragonal structure has polarity because the centers of the positive and negative charges no longer coincide, yielding a net electric dipole, i.e., spontaneous polarization P_S. This polar lattice arrangement forms the *ferroelectric phase* of the perovskite, which exists at lower temperatures. The transition from one phase into the other takes place at the phase transition temperature, commonly called the *Curie temperature*, T_C. In barium titanate, $BaTiO_3$, the phase transition temperature is around $130\,°C$. As the perovskite is cooled below the transition temperature, T_C, the paraelectric phase changes into the ferroelectric phase and the material displays spontaneous strain, S_S, and spontaneous polarization, P_S; vice versa, when the perovskite is heated above the transition temperature, the ferroelectric phase changes into the paraelectric phase, and the spontaneous strain and spontaneous polarization are no longer present.

4 TYPICAL ELECTROACTIVE CERAMICS

Ceramics are hard, brittle, heat resistant, and corrosion-resistant materials made by shaping and then firing a nonmetallic mineral, such as clay, at a high temperature. Ceramics are polycrystalline materials. Commonly, ceramics are electrical and thermal insulators.

Electroactive ceramics are a class of ceramics that display very strong piezoelectric and/or electrostrictive response. Electroactive ceramics consist of polycrystalline structures of ferroelectric perovskites with strong piezoelectric and/or electrostrictive properties. The electroactive ceramics are synthetic compounds that can be fabricated with engineered properties tailored to meet specific operational requirements. On a macroscale, ferroelectric ceramics are given single-crystal symmetry by poling. Poled ferroelectric ceramics are commonly called *piezoelectric ceramics* (ANSI/IEEE 176). Most piezoelectrics are crystalline solids. Some piezoelectric materials are single crystals, either natural or synthetic. Others are polycrystalline materials that are given macroscale single-crystal-like symmetry through poling.

Electroactive ceramics display significant mechanical response under applied electric field, and electrical response under applied mechanical action. Typical strain response of commercially available electroactive ceramics is around 0.1% in the quasi-linear range. Higher strain response can be obtained by taking the electroactive ceramics in the strongly nonlinear range at high electric fields. Nonlinear strains of up to 0.2% have been reported in certain electroactive ceramics. The operation in the high nonlinear range is usually associated with a marked increase in hysteresis. This results in significant internal heating under high-frequency operation. Operation in the high nonlinear domain also results in a marked decrease in fatigue life.

The ceramic perovskites display both piezoelectric and electrostrictive behavior. One or the other of these two properties is usually enhanced through chemical formulation and processing. Some of these electroactive ceramics can display, according to detailed formulation and processing, either a predominantly piezoelectric or a predominantly electrostrictive behavior.

4.1 Fabrication and poling of electroactive ceramics

Conventional fabrication of ferroelectric ceramics is done in several stages as follows:

1. synthesis of the ferroelectric perovskite powders;
2. sintering and compaction of the perovskite powders into ferroelectric ceramics;
3. electric poling of the ferroelectric ceramics.

More novel methods for fabrication of ferroelectric ceramics include the following:

1. coprecipitation
2. sol–gel (alkoxide hydrolysis)
3. thin-film fabrication
4. single-crystal growth.

During cooling, the lead zirconate titanate (PZT) ceramic undergoes phase transformation from paraelectric state to ferroelectric state. This transformation takes place as the material cools below the Curie temperature, T_C. The resulting ferroelectric ceramic has a polycrystalline structure (grains) with randomly oriented ferroelectric domains (Figure 4a). If the grains are large, ferroelectric domains can exist even inside each grain. Owing to the random orientation of the electric domains, individual polarizations cancel each other, and the net polarization of the virgin ferroelectric ceramic is zero.

This random orientation can be transformed into a preferred orientation through *poling*. Poling aligns the dipole domains and gives a net polarization to the piezoceramic material. A poled ferroelectric ceramics behaves more or less like a single crystal. Poling of piezoceramics is attained at elevated temperatures in the presence of a high electric field. The application of a high electric field at elevated temperatures results in the alignment of the crystalline domains. This alignment is locked in place when the piezoceramic is cooled with the poling electric field still applied, thus resulting in permanent polarization. During poling, the orientation of the piezoelectric domains also produces a mechanical deformation. When the piezoceramic is cooled, this deformation is locked in place (permanent strain). Poling is performed in silicon oil bath at elevated temperature under a dc electric field of $1-3\,\text{kV}\,\text{mm}^{-1}$.

4.2 PZT piezoceramics

PZT is a ferroelectric perovskite consisting of a solid solution of $\text{Pb}(\text{Zr}_{1-x}\text{Ti}_x)\text{O}_3$. In the PZT perovskite unit cell, lead, Pb^{2+}, occupies the corners, oxygen, O^{2-}, the faces, and titanium/zirconium, $\text{Zr}^{4+}/\text{Ti}^{4+}$, the octahedral voids. To date, many PZT formulations exist, the main differentiation being between "soft" (e.g., PZT 5-H) and "hard" (e.g., PZT 8). PZT attains the highest piezoelectric coupling and the maximum electric permittivity near the morphotropic phase boundary. This corresponds to the change in the crystal structure from the tetragonal phase to the rhombohedral phase, which happens when the Zr/Ti ratio is approximately 53/47. The explanation for this phenomenon is as follows: above the Curie temperature, PZT has a cubic lattice and is paraelectric. The Curie temperature varies with the alloying proportion, from $\sim 250\,^\circ\text{C}$ for pure PbZrO_3 to $\sim 500\,^\circ\text{C}$ for pure PbTiO_3. Below the Curie temperature, PZT is ferroelectric; but its lattice can be either tetragonal or rhombohedral, according to the alloying proportion.

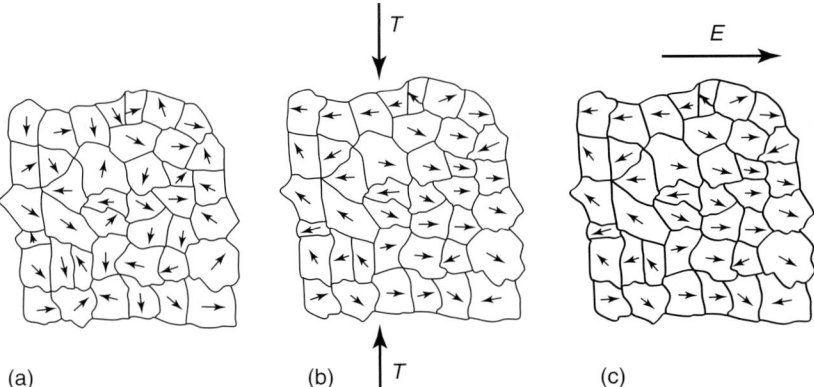

Figure 4. Piezoelectric effect in polycrystalline perovskite ceramics: (a) in the absence of stress and electric field, the electric domains are randomly oriented; (b) application of stress produces orientation of the electric domains perpendicular to the loading direction. The oriented electric domains yield a net polarization; and (c) application of an electric field orients the electric domains along the field lines and produces induced strain.

On the phase diagram, the line separating the two phases is called the *morphotropic phase boundary*. At room temperature, this boundary is placed around the 47/53 alloying ration.

Within the linear range, PZT-like piezoelectric ceramics produce strains that are more or less proportional to the applied electric field or voltage. Induced strains in excess of $1000\,\mu\varepsilon$ (0.1%) are encountered (Figure 5).

4.3 Electrostrictive ceramics—relaxor ferroelectrics

Electrostrictive ceramics are perovskite materials in which the electrostrictive response is dominant. Perovskites that display a large electrostrictive response are the disordered complex perovskites, which have high electrostrictive coefficient, M, with respect to the electric field and a diffuse transition temperature. Such ferroelectric ceramics are also called *relaxor ferroelectrics*, because they display large dielectric relaxation, i.e., frequency dependence of the dielectric permittivity. In a relaxor ferroelectric, the permittivity decreases as the test frequency increases. In addition, the value of temperature at which the permittivity peaks shifts upward. This behavior is in contrast to that of conventional ferroelectrics, for which the temperature at which the permittivity peaks hardly changes with frequency. The dielectric relaxation phenomenon can be attributed to the presence of microdomains in the crystal structure.

In relaxor materials, the transition between piezoelectric behavior and loss of piezoelectric capability does not occur at a specific temperature (Curie point), but instead occurs over a temperature range (Curie range). Thus, electrostrictive ceramics have a rather diffused phase transition that spans a temperature range around the transition temperature. Hence, the temperature dependence of electrostrictive ceramics around the transition temperature is markedly less than that of normal perovskite solid solutions. Sometimes, their transition temperature range is lower than the room temperature, which is beneficial for stable operation at elevated temperatures.

Lead magnesium niobate/lanthanum formulations, and lead nickel niobate currently are among the most studied relaxor electrostrictive ferroelectrics. They have very high dielectric permittivity and polarization. The coercive field of electrostrictive ceramics is much smaller than that of piezoelectric ceramics.

A common electrostrictive ceramics is lead magnesium niobate, $Pb(Mg_{1/3}Nb_{2/3})O_3$, also known as *PMN*. Another commonly used electrostrictive ceramic is lead titanate, $PbTiO_3$, also known as *PT*. Combination of these two formulations are also common, under the designation PMN–PT. Another electrostrictive ceramic is $(Pb, La)(Zr, Ti)O_3$, also known as *PLZT*. Other ferroelectric ceramic systems that have been found to display strong electrostrictive behavior include lead barium zirconate titanate, $(Pb, Ba)(Zr, Ti)O_3$, and barium stannate titanate, $Ba(Sn, Ti)O_3$. To obtain large electrostriction, it is essential that ferroelectric microdomains in the ceramic structure are generated. Various methods are used in this property, such as the doping with ions of a different valence or ionic radius, or the creation of vacancies, which introduce microscopic spatial inhomogeneity.

The strain versus electric field curves of electrostrictive ceramics display the typical quadratic behavior (Figure 6a). On such curves, a positive mechanical strain is obtained for both positive and negative electric fields. However, the strain field curve is strongly nonlinear, as appropriate to quadratic behavior. What is remarkable about electrostrictive ceramics is their very low hysteresis. Figure 6(a) shows that the increasing and decreasing curves superpose almost everywhere, with only a small exception in a limited region at relatively low fields.

Commercially available PMN formulations are internally biased and optimized to give quasi-linear behavior. In this situation, they display much less

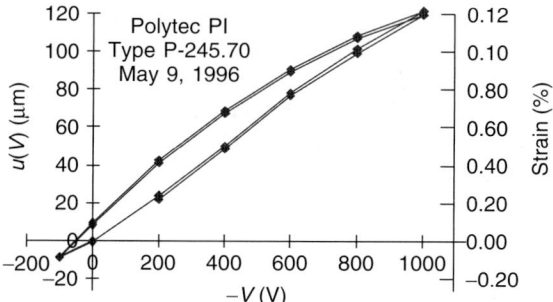

Figure 5. Induced-strain displacement versus applied voltage for Polytec PI model P-245.70 PZT actuator of 100-mm nominal length; induced strain of around 0.12% is observed.

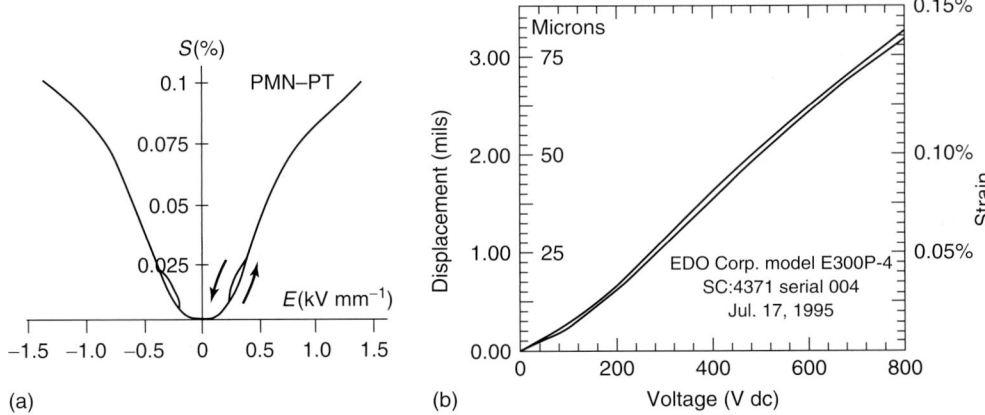

Figure 6. Electrostrictive ceramic: (a) field induced strain in 90–10 PMN–PT and (b) induced strain and displacement versus applied voltage for a 57-mm-long stack of EDO Corporation EC-98 electrostrictive PMN ceramic.

nonlinearity than the conventional quadratic electrostriction, and resemble more the conventional linear piezoelectricity (Figure 6b). Linearized electrostrictive ceramics retain the very low hysteresis of quadratic electrostrictive ceramics. Thus, from this standpoint, they are superior to conventional piezoelectric ceramics. However, linearized electrostrictive ceramics do not accept field reversal. After linearization, the constitutive equations of electrostrictive ceramics resemble those of conventional piezoceramics, i.e.,

$$S_{ij} = s^E_{ijkl}T_{kl} + \tilde{d}_{ijk}E_k \quad (15)$$

$$D_m = \tilde{d}_{mkl}T_{kl} + \varepsilon^T_{mn}E_n \quad (16)$$

The symbol $\sim$ indicates that the piezoelectric constants, $\tilde{d}_{ijk}$, of equations (15) and (16) are different from the corresponding constants d_{ijk} in the original equations (8) and (9). This is due to the linearization process. In equations (8) and (9), the d_{ijk} constants were quite small, since the main effect was due to the quadratic effects represented by the m_{klij} constants. In equations (15) and (16), the $\tilde{d}_{ijk}$ constants are quite significant, since they represent the effect of the linearization of equations (8) and (9).

4.4 Piezopolymers

Piezoelectric polymers are polymers that display piezoelectric properties similar to those of quartz and piezoceramics. Piezoelectric polymers are supplied in the form of thin films. They are flexible and show large compliance. Piezoelectric polymers are cheaper and easier to fabricate than piezoceramics. The flexibility of the polyvinylidene fluoride (PVDF) overcomes some of the drawbacks associated with the brittle nature of piezoelectric ceramics. Its applicability has been proven in keyboards, headphones, speakers, and high-frequency ultrasonic transducers. A typical piezoelectric polymer is the PVDF or PVF_2. This polymer has strong piezoelectric and pyroelectric properties. Its chemical formulation is $(-CH_2-CF_2-)_n$. This polymer displays a crystallinity of 40–50%. The PVDF crystal is dimorphic, the two types being designated I (or β) and II (or α). In the β phase (i.e., type I), PVDF is polar and piezoelectric. In the α phase, PVDF is not polar and is commonly used as electrical insulator. To impart piezoelectric properties, the α phase is converted into β phase and then polarized. Stretching α-phase material produces the β phase. The symmetry of PVDF is $mm2$. Remarkable progress has recently been made in developing piezoelectric polymeric materials through the use of copolymers. The copolymer VDF/TrFE consists mostly of piezoelectric β phase with high crystallinity ($\sim$90%). The film is then cut into various sizes to form piezopolymer sensors.

The piezopolymer surface is metallized to produce the surface electrodes. Silver ink can be screen-printed in patterns onto clear PVDF film. Leads are attached according to customers' specifications. Crimp or eyelet lead attachments are used. After surface metallization, polarization is obtained through the application of a strong electric field. Improved polarization is obtained through *field cooling*, i.e., cooling with the electric field applied to the sample. To achieve this, the PVDF specimen is to be held at 90–130 °C for 15–120 min under a field of 500–1000 kV cm^{-1}. It is then cooled to room temperature while maintaining the same electric field level.

One important advantage of piezopolymer films is their low acoustic impedance ($Z = \rho c$). The acoustic impedance values of piezopolymers are closer to those of water, biotissue, and other organic materials than the acoustic impedance of piezoceramics. For example, the acoustic impedance of PVDF film is only 2.6 times larger than that of water, whereas the acoustic impedance of piezoceramics is typically 11 times that of water. When the acoustic impedance of two media has similar values, the transmission between the two media is enhanced, and reflection at the interface is reduced.

4.5 Magnetostrictive materials

In simple terms, *magnetostriction* is the material property that causes certain ferromagnetic materials to change shape when an external magnetic field is applied. Magnetostrictive materials expand in the presence of a magnetic field, as their magnetic domains align with the field lines. Magnetostriction was initially observed in nickel, cobalt, iron, and their alloys but the values were small ($<50\,\mu\varepsilon$). Large strains ($\sim 10\,000\,\mu\varepsilon$) were observed in the rare-earth elements terbium (Tb) and dysprosium (Dy) at cryogenic temperatures (i.e., below 180 K). Large room-temperature magnetostriction exists in terbium–iron alloy $TbFe_2$. The binary alloy Terfenol-D ($Tb_{0.3}Dy_{0.7}Fe_{1.9}$), developed at Ames Laboratory and the Naval Ordnance Laboratory (now Naval Surface Weapons Center), displays magnetostriction of up to 2000 $\mu\varepsilon$ at room temperature and up to 80 °C and higher. Current Terfenol-D binary alloy formulations are of the form $Tb_{1-x}Dy_xFe_{1.9-2}$ where x is the relative proportion of dysprosium, while the proportion of iron can vary between 1.9 and 2.

5 SUMMARY AND CONCLUSIONS

This article has reviewed the equations of piezoelectricity and piezomagnetism, given a brief physical explanation of the phenomenon origin, and briefly discussed basic types of electroactive and magnetoactive materials.

Electroactive and magnetoactive materials are materials that modify their shape in response to electric or magnetic stimuli. Such behavior is essential in SHM sensors applications. On the one hand, elastic wave sensing with electroactive and magnetoactive materials creates direct conversion of mechanical energy into electric and magnetic energy. Under dynamic conditions, strong and clear voltage signals are obtained directly from the piezosensor without the need for intermediate gauge bridges, signal conditioners, and signal amplifiers. This direct sensing effect is especially significant at high frequencies when the rapid alternation of polarity prevents significant charge leaking. On the other hand, elastic wave actuation is achieved with active materials that display dimensional changes when energized by electric or magnetic fields.

Piezoelectric, electrostrictive, and magnetostrictive materials have been presented and analyzed. Of these, piezoelectric (PZT), electrostrictive (PMN), and magnetostrictive (Terfenol-D) materials have been shown to have excellent frequency response and good induced-strain capabilities ($\sim 0.1\%$). Figure 7 compares induced-strain response of some commercially available piezoelectric, electrostrictive, and magnetostrictive materials. It can be seen that the electrostrictive materials have less hysteresis, but more nonlinearity. The little hysteresis of electrostrictive ceramics can be an important addition in certain applications, especially at high frequencies. However, one should be aware that this low hysteresis is strongly temperature dependent. As the temperature decreases, the hysteresis of electrostrictive ceramics increases, such that, below a certain temperature, the hysteresis of electrostrictive ceramics may exceed that of piezoelectric ceramics. In general, since the beneficial behavior of the electrostrictive ceramics is related to the diffuse phase transition in the relaxor range, their properties degrade as the operation temperature gets outside the relaxor phase transition range.

In summary, one can conclude that the potential of active materials for sensing and actuation SHM applications has been demonstrated in several successful applications. However, this field is still in its infancy

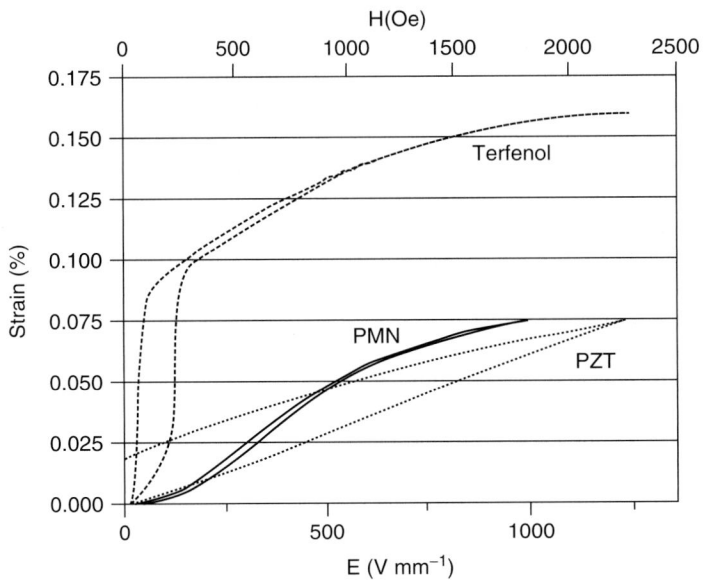

Figure 7. Strain versus electric-field behavior of currently available induced-strain materials.

and further research and development is being undertaken to establish these active materials as reliable, durable, and cost-effective options for large-scale engineering applications.

RELATED ARTICLES

Chapter 5: Electromechanical Impedance Modeling

Chapter 17: Lamb Wave-based SHM for Laminated Composite Structures

Chapter 18: Piezoelectric Impedance Methods for Damage Detection and Sensor Validation

Chapter 40: Piezoceramic Materials—Phenomena and Modeling

Chapter 41: Constitutive Modeling of Magnetostrictive Materials

Chapter 43: Finite Elements: Modeling of Piezoceramic and Magnetostrictive Sensors and Actuators

Chapter 47: Damage Detection Using Piezoceramic and Magnetostrictive Sensors and Actuators

Chapter 50: Probabilistic Approaches to Sensor Layout Design, Data Processing, and Damage Detection

Chapter 54: Nondestructive Evaluation/Nondestructive Testing/Nondestructive Inspection (NDE/NDT/NDI) Sensors—Eddy Current, Ultrasonic, and Acoustic Emission Sensors

Chapter 55: Piezoelectric Wafer Active Sensors

Chapter 57: Piezoelectric Paint Sensors for Ultrasonics-based Damage Detection

Chapter 68: Miniaturized Sensors Employing Micro- and Nanotechnologies

Chapter 74: Design of Active Sensor Network and Multilevel Data Fusion

Chapter 75: Energy Harvesting and Wireless Energy Transmission for SHM Sensor Nodes

Chapter 78: Stanford Multiactuator–Receiver Transduction (SMART) Layer Technology and Its Applications

Chapter 104: Thermal Protection System Monitoring of Space Structures

Chapter 107: Development of an Active Smart Patch for Aircraft Repair

Chapter 111: Design, Analysis, and SHM of Bonded Composite Repair and Substructure

Chapter 112: Aircraft Structural Diagnostic and Prognostic Health Monitoring for Corrosion Prevention and Control

Chapter 147: Wind Turbines

Chapter 156: Integrated Sensor Durability and Reliability

FURTHER READING

ANSI/IEEE Std 176, *IEEE Standard on Piezoelectricity*. The Institute of Electrical and Electronics Engineers: New York, 1987.

Engdahl G. *Handbook of Giant Magnetostrictive Materials*. Academic Press, 2000.

Giurgiutiu V. *Structural Health Monitoring with Piezoelectric Wafer Active Sensors*. Elsevier Academic Press: New York, 2008.

Ikeda T. *Fundamentals of Piezoelectricity*. Oxford University Press, 1996.

Lines ME, Glass AM. *Principles and Applications of Ferroelectrics and Related Materials*. Clarendon Press: Oxford, 2001.

MIL-STD-1376B(SH), *Military Standard Piezoelectric Ceramic Materials and Measurements Guidelines for Sonar Transducers*. Defense Quality and Standardization Office: Falls Church, VA, February 1995.

du Tremolet de Lacheisserie E. *Magnetostriction: Theory and Applications of Magnetoelasticity*. CRC Press, 1993.

Uchino K. *Piezoelectric Actuators and Ultrasonic Motors*. Kluwer Academic Publishers, 1997.

Chapter 53
Operational Loads Sensors

William F. Ranson and Reginald I. Vachon
Direct Measurements, Inc., Atlanta, GA, USA

1 Introduction	993
2 Accelerometers	993
3 Strain Gauges and Strain Measurement	996
4 Concluding Remarks	1000
References	1001

1 INTRODUCTION

Load sensors for structural health monitoring (SHM) measure force, torque, pressure, and strain. The data are translated into information to determine the effect of loads and/or vibration on operating dynamic systems and static infrastructures. Sensors used to make these measurements are accelerometers, frequency devices, or strain gauges. The theory of operation of each is fundamental. Nonetheless, there are a number of manifestations of theory for each by commercial companies and research laboratories providing accelerometer, frequency, and strain gauge devices. Some of these devices are described and the general theory for the major types of accelerometers and strain gauges are presented.

Encyclopedia of Structural Health Monitoring. Edited by Christian Boller, Fu-Kuo Chang and Yozo Fujino © 2009 John Wiley & Sons, Ltd. ISBN: 978-0-470-05822-0.

The transition from usage monitoring to individual component damage tracking is the challenge for these types of sensors. The areas of regime recognition, loads, strain prediction, and component serialization need to improve for enhanced structural health assessment. Of these three requirements, loads and strain prediction along with serialization represent the direct application for SHM.

The current developments of sensors of this type are to overcome the limitations of complex instrumentation, long wiring lengths, and in rotating components of the elimination of slip rings. Strain sensor development is focused on wireless technology, serialization, and energy harvesting for power. The accelerometers used in health usage monitoring systems (HUMS) such as the Navy V-22 and the Army Apache and Blackhawk helicopters are incorporating regime recognition in the onboard data acquisition systems.

2 ACCELEROMETERS

2.1 Overview

Accelerometers are one of the basically three vibration measuring instruments, which measure displacements, vibrations, and accelerations as a result of harmonic motion of a base relative to a mass system. The most common types are the displacement and acceleration measurement instruments. They consist

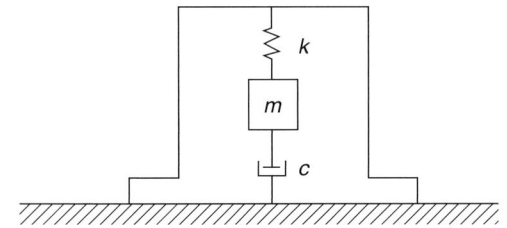

Figure 1. Schematic of an accelerometer.

of a case containing a spring mass damper system of the type shown in Figure 1 and a device measuring the displacement of the mass relative to the case. The mass is usually constrained to move along a given axis. Measurement of the mass relative to the case is measured electrically. When the frequency of the base is low relative to the spring mass damper of the system, the instrument is proportional to the acceleration of the case.

2.2 Theory

A single-degree-of-freedom mechanical model with viscous damping as shown in Figure 1 can be used to describe the mechanical behavior of instruments to measure force, pressure, and motion. Accelerometers are a special case of this type of instrument where a component with frequencies far below the natural frequency of the transducer, the relative motion between the base and the transducer, is proportional to the base acceleration. Figure 1 is a schematic that represents the vibration of a base by $y(t) = Y \sin \omega t$. From Newton's second law of motion [1], the equation of motion for the seismic mass is obtained. Thus

$$m\frac{d^2z}{dt^2} + c\frac{dz}{dt} + kz = m\omega^2 Y \sin \omega t \qquad (1)$$

where m is the mass of the seismic body, k is the linear spring constant, and c is the viscous damping constant, $z = x - y$ is the relative displacement between seismic mass and the base, and $z, dz/dt, d^2z/dt^2$ are the relative displacement, velocity, and acceleration.

A steady-state solution is assumed in the form $z = Z \sin(\omega t - \phi)$ and $Z = m\omega^2 Y$.

The solution yields

$$Z = \frac{Y(\omega/\omega_n)^2}{\sqrt{\left(1 - (\omega/\omega_n)^2\right)^2 - \left(2\zeta(\omega/\omega_n)\right)^2}} \qquad (2)$$

$$\tan \phi = \frac{2\zeta(\omega/\omega_n)}{1 - (\omega/\omega_n)} \qquad (3)$$

where $\zeta = c/c_c$ is the viscous damping ratio and ω_n is the seismic mass natural frequency.

Accelerometers have a small frequency ratio $\omega/\omega_n \ll 1$ and $\zeta \ll 1$ (small damping); then

$$Z = \frac{\omega^2 Y}{\omega_m^2} = \frac{\text{Base acceleration}}{\omega_n^2} \qquad (4)$$

The crystal accelerometer behaves like an underdamped, spring mass system with a single degree of freedom. Equation (1), a classical second-order differential equation, can be used to describe the behavior of the electromechanical crystal accelerometer system.

Piezoelectric crystals subjected to a force exhibit an electrical charge. Slicing a quartz crystal, for example, relative to an xyz axis results in a sensor producing a charge based on the direction of the applied force. This fact is used to design accelerometers used in dynamic systems as well as for quasi-static measuring systems using a quartz crystal with special signal conditioning. Examples can be given for quasi-static measurements over periods of minutes and even hours.

Piezoelectric crystal sensors are of high impedance or low impedance. Low-impedance sensors have a built-in charge-to-voltage converter and require an external power source to power the device and to separate the output voltage from the bias dc voltage. These low-impedance systems are tailored to a particular application. High-impedance units require a charge amplifier or external impedance converter for charge-to-voltage conversion. The high-impedance sensors are more versatile than the low-impedance devices.

Time constant and drift are two terms associated with the use of charge amplifiers with piezoelectric devices. The time constant is the discharge time of an ac-coupled circuit. One time constant is the period of time the input decays by 63–37% of its original value.

The product of the time constant resistor R_t and range capacitor C_t yields the time constant. Figure 2 depicts a high-gain inverting voltage amplifier typically used with crystal sensors.

The amplifier yields an output voltage that can be expressed by

$$V = -\frac{Q}{C_r}\left[\frac{1}{\left[1 + \frac{1}{A}C_r(C_t + C_r + C_c)\right]}\right] \quad (5)$$

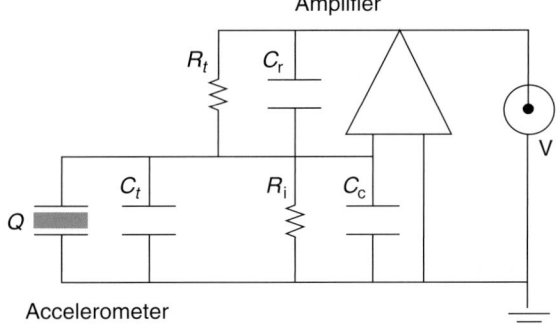

Figure 2. Typical accelerometer circuit.

where C_t is the sensor capacitance, R_i is the cable and sensor resistance, C_c is the cable capacitance, C_r is the range capacitor, R_t is the time constant, A is the open-loop gain, and Q is the charge generated.

When the open-loop gain is high, the output voltage is a function of the input charge and range capacitance.

$$V = -\frac{Q}{C_r} \quad (6)$$

High-impedance crystals coupled with an amplifier with a high open-loop gain produce a usable output voltage. Quartz-based piezoelectric sensors are widely used because they operate up to temperatures of 498.9 °C, and stresses of up to approximately 137 895 Pa, ultrahigh insulation resistance of 1014 Ω, which permits low frequency (1 Hz), negligible hysteresis, high rigidity, and high linearity.

There are a variety of accelerometers. Listing includes capacitive spring mass based; electromechanical servo; null balance; resonance; magnetic induction; optical; surface acoustic wave; laser; bulk micromachined piezoresistive; bulk micromachined capacitive; capacitive spring mass based; piezofilm; piezoelectric; shear mode accelerometer; thermal

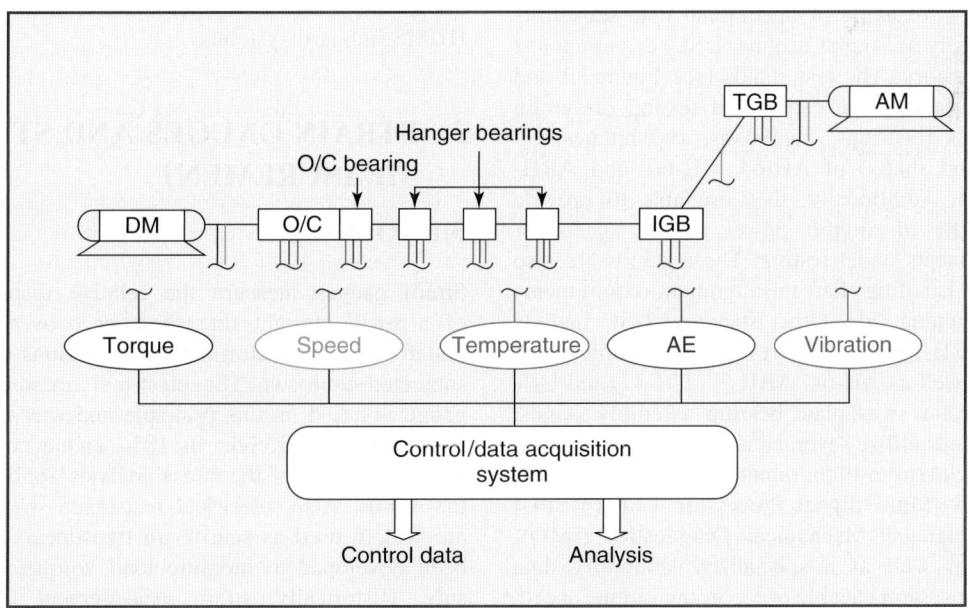

Figure 3. Schematic of the test stand (courtesy of Dr Abdel Bayoumi, Department of Mechanical Engineering, University of South Carolina). [Reproduced with permission from Dr. Abdel Bayoumi.]

Figure 4. Photograph of the test stand (courtesy of Dr Abdel Bayoumi, Department of Mechanical Engineering, University of South Carolina). [Reproduced with permission from © Dr. Abdel Bayoumi.]

(submicrometer CMOS process); and modally tuned impact hammers.

2.3 Example of operational load sensors for SHM

An excellent example of operational load sensors is the University of South Carolina (USC) drivetrain test stand laboratory. The test stands (see Figures 3 and 4) in this facility are capable of testing drivetrain components (bearings, gearboxes, swashplates, oil coolers, and shafts) of AH-64, UH-60, and ARH-70 aircraft. Additionally, they are able to provide up to 150% of aircraft power at full speed for the components under testing. These stands are also capable of handling shaft misalignment requirements while remaining safe. Other stands at USC include AH-64, ARH-70, CH-47, and UH-60 hydraulic pump stands, as well as AH-64, ARH-70, CH-47, and UH-60 main rotor swashplate bearing assembly stands. All test stands utilize several data acquisition systems, including current in-flight monitoring systems such as the HUMS (Multi Signal Processing Unit (MSPU) and/or Integrated Mechanical Diagnostics (IMD)-HUMS), as well as a specialized laboratory data acquisition system capable of recording torque, speed, temperature, vibration, and acoustic emissions. All test stands are controlled on the basis of monitored data measures of torque, speed, and temperature, which are collected every 2 s. All vibration and acoustic emissions data are collected every 2 min. Operational load sensors of this type have been developed for rotary wing aircraft. References 2–5 describe applications in maneuver regime recognition and SHM using drivetrain, gearbox, engine, rotor track, and balance accelerometers as part of the HUMS onboard system.

3 STRAIN GAUGES AND STRAIN MEASUREMENT

3.1 Overview

Strain gauges measure the relative displacements of a small straight line segment between usually undeformed and deformed configurations of a body subjected to loads. The electrical resistance strain gauge is based on this principle and was first established by Lord Kelvin in 1856 and accounted for more than 80% of the stress analysis applications in the 1980s. Also, electrical resistance strain gauges are widely used as sensors in transducers that have been developed to measure load, torque, and pressure. Historically, strain measurement for operational SHM has relied on electrical resistance strain gauge applications. In recent years, optical-based techniques have been developed to measure strains. Some of these very useful laboratory technologies [6, 7] include digital image correlation, interferometry, photoelasticity, holographic interferometry, and fiber-optic strain gauges.

The concentration here is the electrical resistance strain gauge and one optical technique suitable for in-the-field measurement and verification of engineered residual strains intended to enhance fatigue life for operating systems such as fastener holes in aircraft structures. The theory of each is discussed and examples are given.

3.2 Electrical resistance strain gauge

The characteristics for electrical resistance gauges are listed in [6, 8] based on the premise that they are usually the most economical. Some of the gauge characteristics are as follows:

1. The gauge constant should be stable with respect to both temperature and time.
2. Strain measurement accuracy of $\pm 1\,\mu\text{m}\,\text{m}^{-1}$ (μin./in.) over a range of $\pm 5\%$ strain.
3. The gauge length and width should be small.
4. The inertia of the gauge should be minimal for the measurement of dynamic strains.
5. The response or output of the gauge should be linear over the entire strain range of the gauge.

These characteristics are not only criteria but also constraints predicated on the limitations of the electrical resistance gauge. They also provide a basis for comparison and contrasting of other strain gauges such as inductance, capacitance, and optical gauges.

3.3 Theory of electrical resistance strain gauge

The electrical resistance strain gauge uses the analog of the change in resistance to resistance ratio to translate the change in resistance of the wire gauge as it undergoes strain into strain readings. The strain sensitivity (S_A) is termed the *gauge factor*. This is expressed by

$$S_A = \frac{dR/R}{dl/l} \qquad (7)$$

where dR is the change in resistance, R is the initial resistance, dl is the length change of a small line segment, and l is the initial length of a small line segment.

Equation (7) can be now expressed in terms of the strain:

$$\varepsilon_g = S_A \frac{dR}{R} \qquad (8)$$

where $\varepsilon_g = dl/l$ is the gauge strain from the measured resistance change.

Theoretically, the change in the resistance of the gauge is a function of surface deformations. Real considerations that affect the change in resistance include the adhesive used to affix the gauge to the surface, stability of the surface, temperature, and the material. Also, axial strain measurement is not sufficient to achieve complete analysis. A combination of gauges is required to yield orthogonal strains and associated shear strains. The electrical resistance strain gauge has wide use and utility. There are a variety of gauge configurations and wire filaments used to achieve stability, temperature sensitivity, and strain characteristics of the sensor. Many strain gauge filament materials degrade over time and are temperature sensitive. Thus, long-term continuous applications require attention to drift and temperature compensation. Selecting a gauge is based on the gauge characteristics that include gauge factor, thermal coefficient of resistively, resistance, and temperature coefficient of gauge factor. There are several manufacturers of gauges and the wide variety of gauges can be appreciated by reviewing available products.

3.4 Example of operational strain gauge application

The need exists to improve the accuracy of load/strain prediction by directly measuring data in flight. Microstrain has demonstrated a sensor that relies on piezoelectric energy harvesters capable of measuring strain, acceleration, and temperature. This type of sensor is a powerful combination of real-time sampling rates up to 4 KHz and onboard data storage. Microstrain sensors are configured to be efficient when using energy to sample, record, and communicate over a bidirectional radio link. Microstrain has

Figure 5. Wireless strain gauge on a Bell pitch link.

successfully demonstrated this technology to measure operational loads/strains on a Bell 412 helicopter pitch link as shown in Figure 5 [9].

3.5 Theory of optical strain gauge

The theoretical basis for the optical strain gauge is shown in Figure 6. Straight line segments are recorded in the undeformed and deformed configurations with the end points $MNOP$ mapped to the endpoints $M^*N^*O^*P^*$ [10]. The strains along these orthogonal directions are calculated from the following formulas:

$$E_{MN} = \frac{M^*N^* - MN}{MN} \tag{9}$$

$$E_{NO} = \frac{N^*O^* - NO}{NO} \tag{10}$$

$$E_{OP} = \frac{O^*P^* - OP}{OP} \tag{11}$$

$$E_{PM} = \frac{P^*M^* - PM}{PM} \tag{12}$$

This type of strain gauge [11] has further properties of serialization as shown in Figure 7. The encoded serialization is shown in the boundaries of the gauge similar to a bar code. The target gauge can be in the form of a thin polymer with the gauge design laser machined into the multilayer polymer to achieve the light and dark patterns or the gauge can be laser bonded onto the surface. The polymer gauge can be affixed to the surface with M-Bond 200, which is the same process used for thin-film electrical resistance gauges. The other alternative application is to laser bond the pattern onto the surface.

This is accomplished by spraying the surface with fine metallic particles suspended in a vehicle of water or alcohol in a normal hand-held aerosol spray can. When the surface has dried, a low-power laser is used to fuse the particles to the surface to create the pattern. The surface is then cleaned with a cloth and water. A third application is to paint the surface with a specialized paint that discolors in response to exposure to a low-power laser. The result is that the gauge is integral to the paint layer. Examples of the

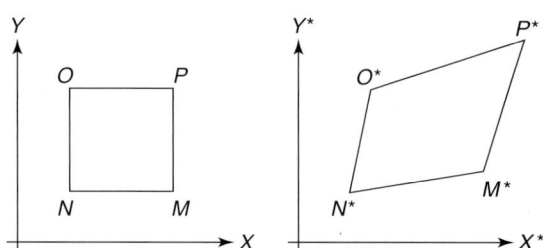

Figure 6. Optical strain gauge in undeformed and deformed configurations.

Figure 7. Gauge serialization.

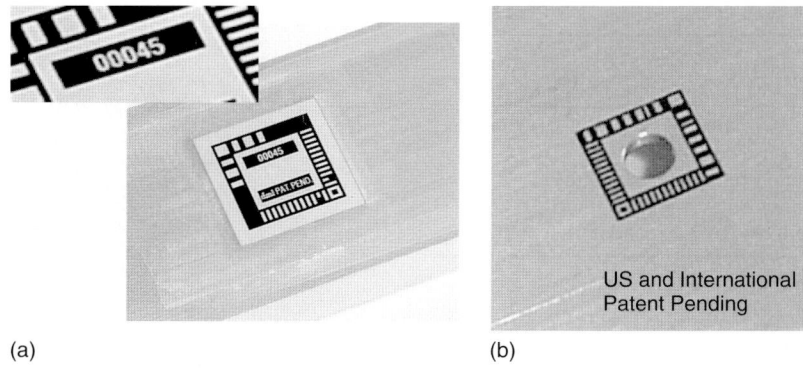

Figure 8. (a) Polymer gauge with serialization and (b) laser-bonded gauge.

polymer gauge and the gauge laser bonded around a hole are shown in Figure 8(a) and (b).

3.6 Features of the technology

(i) Self-contained strain/fatigue monitoring system; (ii) minimal skilled labor required; (iii) nonlinear, multicomponent strain sensor: two orthogonal extensional strains and shear strain from a single gauge; (iv) resting system error $<5\,\mu\varepsilon$; (v) measurement repeatability: $<30\,\mu\varepsilon$; (vi) accuracy increases as strain increases; (vii) temperature compensation enabled; (viii) no fragile electrical components, connection free; (ix) line-of-sight target acquisition; (x) material independent (works on plastics, metals, composites, etc.); (xi) gauge application process does not denigrate material/mechanical properties of host material; and (xii) gauge contains binary encoded information (for gauge location, serial number, or asset management) up to 4 billion unique numbers for each gauge.

3.7 Cold-working validation using optical strain gauge

A 7075-T6 specimen with five fastener holes was used to demonstrate the application of the technology to validate cold working [11]. Wire-free Direct Measurements, Inc. (DMI) gauges were applied around each hole as shown in Figure 9 (front) and baseline readings made and recorded. The specimen

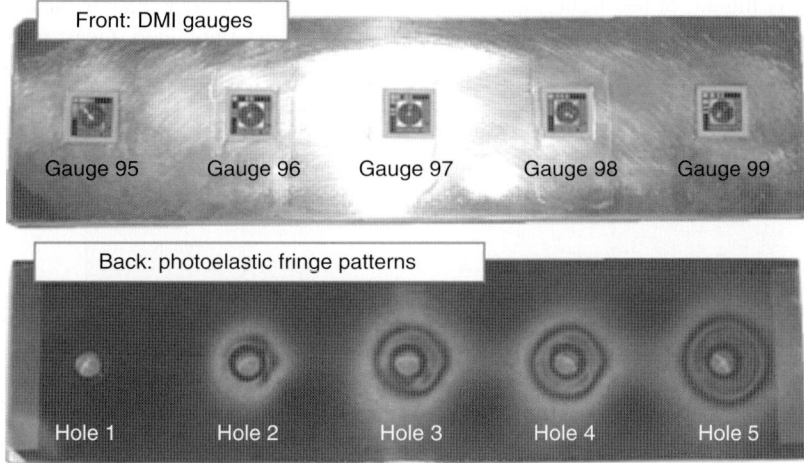

Figure 9. 7075-T6 test specimen front and back.

Table 1. Cold work by hole and gauge

Hole #	CX type	DMI gauge #
1	None	95
2	2%CX (under expanded)	96
3	CB minimum (minimum standard)	97
4	CB maximum (maximum standard)	98
5	CA nominal (over expanded)	99

CX, Cold Work Expansion

was cold worked by Fatigue Technology Inc. (FTI). Prescribed expansions were achieved according to Table 1. After FTI completed cold-work expansion, each gauge was read and the residual strain was measured and recoded. This summary presents the results of the strain measurements and shows the applicability of DMI technology to cold-work expansion.

Residual strain measurements were made on both the inner and outer boundary of each gauge and measured strain components E1, E2, E3, and E4 were recorded according to Figure 9. The holes and corresponding gauge codes are shown in Table 1 and Figure 10.

The DMI gauge outer and inner boundary measurements are shown in Figures 11 and 12. This demonstrates that the residual strain magnitudes follow the trend in photoelastic patterns in Figure 9. Note that the right-hand side fringe order is slightly higher in holes 3, 4, and 5. Accordingly, the strain component E2 measured on the right-hand side of gauges 97, 98, and 99 is also slightly higher. In this figure, gauge 98 fails to indicate higher E2 components due to inner

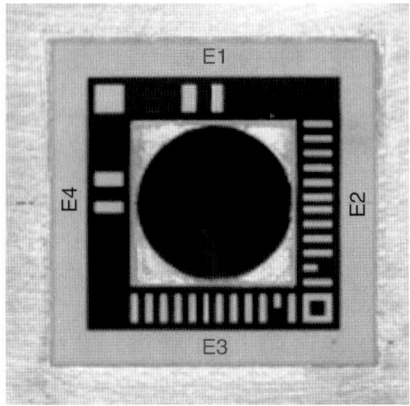

Figure 10. Gauge 96 surrounding hole 2.

boundary gauge damage during cold-work expansion. Nonetheless, the remaining components give sufficient indication of the degree of cold working. In addition, DMI gauges can be produced with dimensions that better match holes sizes, thereby avoiding damage.

4 CONCLUDING REMARKS

Operational load sensors that measure force, torque, pressure, and strain have been successively employed in operating mechanical systems for many years. Specific applications of accelerometers that are used in HUMS systems for rotary wing aircraft measure data associated with exceedance monitoring, rotor track and balance, operational usage, regime recognition, and drivetrain diagnostics. Strain gauges are ideal for dynamic load measurements such as the

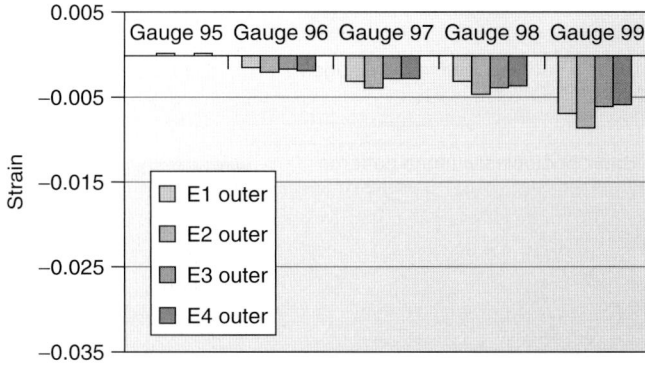

Figure 11. Outer boundary gauge residual strains.

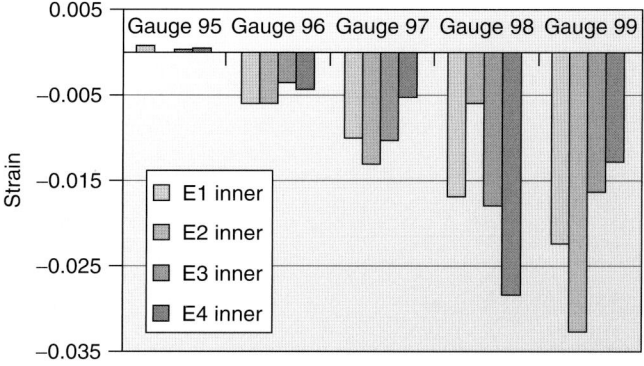

Figure 12. Inner boundary gauge residual strains.

axial load application for the pitch link illustrated in this article. The electrical resistance strain gauge has been an important sensor for measurements in full-scale fatigue tests in both fixed and rotary wing aircraft. The need to develop small scale gauges of this type for operational aircraft has led to the development of energy harvesting wireless strain gauges such as the one by Microstrain. Strain gauge applications of this type measure strains that can be calibrated to axial, bending, pressure, and torque loading. The application of the optical strain gauge while used to measure strains due to cold working are not a measure of loads directly. Cold working of fastener holes increases aircraft fatigue life; however, due to overloads, these compressive residual strains can be relaxed, thus reducing the expected fatigue life. The optical strain gauge measures the compressive strain relaxation and is thus an indirect measure of the effects of overload conditions.

REFERENCES

[1] Curtis D. *Johnson Process Control Instrumentation Technology*. Prentice-Hall: Upper Saddle River, NJ, 1993.

[2] Brandt G, Moon S, Miller SM. Manuever regime recognition development and verification for H-60 structural monitoring. *Presented at the American Helicopter Society 63rd Annual Forum*. Virginia Beach, VA, 1–3 May 2006.

[3] Brandt G. Moon S. Development of a fatigue tracking program for navy rotary wing aircraft. *Presented at the American Helicopter Society 50th Annual Forum*. Washington, DC, 11–13 May 1994.

[4] Dora R, Baker T, Hess R. Application of the IMD HUMS to the UH-60A Blackhawk. *Presented at the American Helicopter Society 58th Annual Forum*. Montreal, QC, 11–13 June 2002.

[5] Hiatt DS, Hayden R. Concepts for certifying a data-driven HUM system. *Presented at the American Helicopter Society 58th Annual Forum*. Montreal, QC, 11–13 June 2002.

[6] Kobayashi AS (ed). *Handbook on Experimental Mechanics*, Society for experimental Mechanics, Prentice-Hall: Englewood Cliffs, NJ, 1987.

[7] Sohn H, et al. A Review of Structural Health Monitoring Literature from 1996–2001. Los Alamos National Laboratory, report LA-13976-MS, 2003. http://www.findarticles.com/p/articles/mi_qa5348/is_200403/ai_n21346075/pg_4.

[8] Dally JW, Riley WF. *Experimental Stress Analysis*, Fourth Edition. College House Enterprises, LLC: Knoxville, TN, 2005.

[9] Maley S, Plets J, Phan ND. US Navy roadmap to structural health and usage monitoring—the present and future. *Presented at the American Helicopter Society 63rd Annual Forum*. Virginia Beach, VA, 1–3 May 2007.

[10] Novozhilov V. *Foundations of the Non-Linear Theory of Elasticity*. Graylock Press: New York, 1953.

[11] Ranson WF, Vachon RI. Crack detection and growth monitoring in holes using DMI SR-2 technology. *Paper # 17 2007 SEM Annual Conference*. Springfield, MA, 3–6 June 2007.

Chapter 54

Nondestructive Evaluation/Nondestructive Testing/Nondestructive Inspection (NDE/NDT/NDI) Sensors—Eddy Current, Ultrasonic, and Acoustic Emission Sensors

Mark Blodgett, Eric Lindgren, Shamachary Sathish and Kumar V. Jata

Air Force Research Laboratory, Wright Patterson Air Force Base, Dayton, OH, USA

1 Introduction	1003
2 Eddy-current Sensing	1004
3 Ultrasonic Sensors	1008
4 Summary	1010
References	1011

1 INTRODUCTION

Nondestructive evaluation (NDE) exploits the measurement of changes in physical properties of materials using methods that leave the material in undisturbed state after measurements. The basic NDE methods most often exploit electromagnetic, elastic, and thermal properties of materials. Elastic properties are often measured using acoustic/ultrasonic wave propagation methods, while the electrical and magnetic properties can be measured using eddy current and microwave propagation. The thermal property measurements are often performed using infrared detection.

The eddy-current method utilizes a coil energized by a sinusoidal electromagnetic signal. When the coil is brought near an electrically conductive material, the changing magnetic field induces eddy currents. The eddy currents oppose the changing magnetic field causing changes in the impedance of the coil. The magnitude of the impedance change is directly related to the electrical conductivity of the material. Presence of defects in a material will significantly affect eddy-current generation that is detected as impedance changes in the coil. The frequency of the electromagnetic signal and the electrical conductivity of the material determine the penetration depth of the eddy currents. The resolution and size of the defect that can be detected depends on the diameter of the eddy-current probe. Since eddy current can be generated only in electrically conductive materials, the technique is excellent for crack and defect

Encyclopedia of Structural Health Monitoring. Edited by Christian Boller, Fu-Kuo Chang and Yozo Fujino © 2009 John Wiley & Sons, Ltd. ISBN: 978-0-470-05822-0.

detection in metallic materials and structures. This article describes in detail the basic principles, coil design, application to defect detection, and advanced developments in material characterization using eddy current.

In general, ultrasonic waves propagated into a material or structure measure the elastic properties (velocity/modulus, attenuation/damping). Whenever the propagating ultrasonic waves are obstructed by defects, the energy is reflected, transmitted, and scattered. Examination of the reflected/transmitted/scattered signals is used to locate and identify the defects in the material. The frequency and wavelength of the ultrasonic wave determine the size of the defect that can be detected and the attenuation in the material determines the penetration depth into the material. In systems that use scanning mode of operation, the physical dimensions of the transducer limits the resolution. One of the major advantages of ultrasonic NDE is the ability of ultrasonic waves to propagate through any type of material. The section on ultrasonic sensors describes in detail the generation, detection, and the interaction of ultrasonic waves with defects in varieties of materials for NDE application.

Some NDE sensors are also used to detect signals emitted by a material when a component is under external load. Strain energy is released when excessive deformation, crack initiation, and certain types of crack propagation occur, and this energy is then converted to acoustic energy. The sudden bursts of acoustic energy can be listened to by an acoustic emission sensor. The acoustic emission monitoring is a passive listening method and the sensors can be attached to the structure permanently and can be used for periodic or continuous interrogation of the material. This method is gaining popularity in structural health monitoring applications.

2 EDDY-CURRENT SENSING

2.1 Background

Eddy currents are closed loops of induced current circulating in planes perpendicular to the magnetic flux, making them useful for nondestructive inspection (NDI), thickness gauging, and position sensing. Typically, eddy currents travel parallel to the exciting coil's windings and their spread is limited by the size of the inducing magnetic field. Eddy currents concentrate in the near-surface layers adjacent to the excitation coil and their strength decreases with distance from the coil according to Maxwell's equations, which show that eddy-current density decreases exponentially with depth, known as the *skin effect*. The depth that eddy currents penetrate into a material is affected by the frequency of the excitation current, the electrical conductivity, and the magnetic permeability of the specimen (Figure 1). The depth of penetration decreases with increasing frequency and increasing conductivity and magnetic permeability. The depth at which eddy-current density has decreased to 1/e, or approximately 37% of the surface density, is called the *standard depth of penetration*, δ. "Standard" penetration denotes the use of a plane-wave approximation to the electromagnetic field interaction with the test sample and is unaffected by the size of the coil. The eddy-current inspection methodology is widely regarded as the most common electromagnetic NDE tool and is used in many applications for detection and quantification of defects in strength critical structures like pipelines and aircraft, thickness measurements, coating measurements, and various applications involving measurement of conductivity.

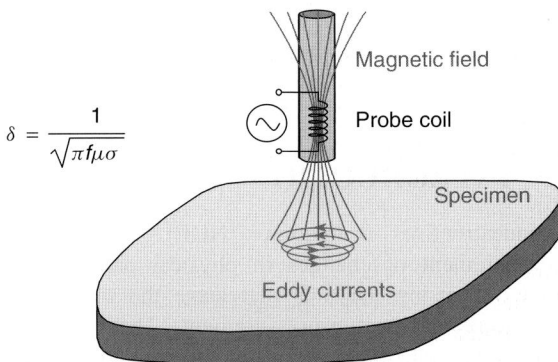

$$\delta = \frac{1}{\sqrt{\pi f \mu \sigma}}$$

Figure 1. An illustration describing the interaction of the alternating current excited eddy-current probe coil with a conducting specimen to generate eddy currents. The standard penetration is shown as, δ (mm), where $\pi = 3.14$, f = frequency (Hz), μ = magnetic permeability (H mm^{-1}), and σ = electrical conductivity (%IACS) International Annealed Copper Standard.

2.2 Early discoveries

Contemporary eddy-current sensors and instrumentation are the result of decades of development and have an even longer history in terms of the fundamental principles on which the technology is founded. Electromagnetism was discovered in 1820 by Hans Christian Oersted when he demonstrated the effect of electricity in a wire on a compass needle. He discovered that the wire conducting electric current generated invisible magnetic lines of force, which when placed above the compass needle caused it to swing perpendicular to the direction of the current flow and to swing in the opposite direction upon reversal of the current flow. Of course, magnets also have invisible lines of force and several years later, Michael Faraday and Joseph Henry made nearly simultaneous independent discoveries of electromagnetic induction, in 1831. Both men used iron bars bent into a circle and wrapped in copper wire to create an electromagnet, and then in different arrangements demonstrated the principle of electromagnetic induction, namely that the induced current is proportional to the rate of change of magnetic flux through a coil. Some time later, Lèon Foucault discovered in 1851 that a changing magnetic field produces a circulating current upon intersection with a conductor. These swirls or eddies of current produce magnetic fields that oppose the applied magnetic field, which created them in the first place and are only present when the applied magnetic field is either physically moving or changing in strength. Capping off the early discovery era is the influential work of D. E. Hughes [1], who demonstrated in 1879 that eddy currents can be used as a basis to compare and sort materials by virtue of differences in electrical parameters, such as conductivity and permeability. Hughes showed that the principle of electromagnetic induction can be used to measure properties of coinage and the relative conductivities of metals and also demonstrated techniques for enhancing sensitivity. By combining these seminal early discoveries, it is possible to begin to see the foundation of modern eddy-current testing take shape, given (i) an alternating current (ac) used to excite a probe coil that gives rise to a changing magnetic field; (ii) when placed near a conductor the energized probe produces eddy currents in the solid, which have their own magnetic fields; and (iii) the magnetic fields of the induced eddy currents oppose the applied magnetic field thereby affecting the impedance of the probe coil, which is the basis of most modern eddy-current testing.

2.3 Measurement and testing applications

Modern eddy-current testing has evolved over the course of more than a century of research and development and finds its origins in metal sorting and comparing. During this long development period tremendous advances have been made in sensors, electronics, integrated circuitry, microprocessors, and computers leading us to today's handheld, portable eddy-current NDE instrumentation geared for field applications. In the latter part of the 1930s, Friedrich Förster, the person widely regarded as the forefather of modern eddy-current testing, began to make significant advances to the field by developing theory for complex impedance analysis, experimental equipment, calibration procedures, and nondestructive techniques to quantitatively test the properties of metals [2]. These early developments led to the production of commercially available electromagnetic induction testing equipment in the United States in the 1950s. Today's eddy-current test devices and instrumentation are available for a wide range of industrial applications such as aerospace (*see* **Chapter 90**; **Chapter 96**; **Chapter 97**; **Chapter 115**), energy (*see* **Chapter 59**; **Chapter 147**; **Chapter 148**), marine (*see* **Chapter 142**; **Chapter 144**), and automobile (*see* **Chapter 146**) industries. An extensive accounting of contributions by early researchers in this field is documented in the NDT Handbook on Electromagnetic Testing [3].

Eddy-current testing techniques are primarily used to measure material properties such as conductivity and permeability, discriminating components based on the presence of flaws or discontinuities, and precise measurement of displacement, thickness, positioning, and proximity. Many critical applications require automated computer-controlled handling of eddy-current sensors and test objects due to the large scale of the inspection area, which allows the inspection area to be mapped to the component requiring inspection. For example, turbine engine components are designed to optimize aircraft performance, while accommodating the adverse effects of high temperatures and dynamic loading conditions. In critical components such as turbine disks, the hardware is closely monitored throughout its service life using eddy current and other NDIs. Some United States Air Force ((USAF)) turbine engines are entirely disassembled and components are subjected to intensive NDIs to ensure that dimensional tolerances are met and surfaces are free from life-limiting flaws. Many individual components are cleaned and prepared for eddy-current inspections in an effort to detect surface-breaking fatigue cracks, fretting damage, and foreign object damage along with other detrimental features such as dents and gouges. Moreover, with appropriate calibration standards and tracking methods based on the initial fabrication conditions, microstructure, and alloy composition, quantitative eddy-current conductivity measurements could be potentially made to complement eddy-current flaw inspections, which are based on probability-of-detection standards, for the same critical engine components.

Eddy-current NDE techniques have been practiced for several decades, primarily as a means of flaw detection in the near-surface of metallic parts. For turbine engine applications, eddy-current testing is the primary method used to inspect the surface for life-limiting surface flaws, and a great array of probes have been manufactured to allow comprehensive inspections of these complex components. The optimal frequency range of a given probe, and therefore its sensitivity and effective penetration depth, may be selected by manipulating the probe's diameter, the number of wire turns in the coil, and the impedance matching circuitry. Eddy-current inspection is based on the electromagnetic induction principle, i.e., that a changing magnetic field will induce electrical currents in a nearby conductor, which in turn will load the coil and affect the phase and magnitude of its impedance. Accurate measurement of these eddy-current loading effects on the probe's impedance is the basis of most eddy-current measurements and provides the means to evaluate materials for near-surface cracking and inclusions, heat treatment verification, and thickness gauging. It is well known that eddy-current conductivity is affected by a number of things including residual stress, chemical composition, microstructure, hardness, surface roughness, and temperature. Therefore, it is essential to use stable quantitative test equipment and reliable procedures to assure integrity of the acquired experimental data and repeatability of the measurements. Calibration and reference standards are also essential for accurate eddy-current measurements and typically consist of materials with either known conductivities (or permeabilities for magnetic materials) for materials property characterization or with known flaw sizes for defect detection and characterization.

Eddy-current inspection is often used to detect corrosion, erosion, cracking, and other changes in tubing. Heat exchangers and steam generators, which are used in power plants, have thousands of tubes that must be prevented from leaking. This is especially important in nuclear power plants where reused, contaminated water must be prevented from mixing with freshwater that will be returned to the environment. The contaminated water flows on one side of the tube and the freshwater flows on the other side. The heat is transferred from the contaminated water to the freshwater and the freshwater is then returned back to its source, which is usually a lake or river. It is very important to keep the two water sources from mixing, so power plants are periodically shut down so that the tubes and other equipment can be inspected and repaired. The eddy-current test method provides high-speed inspections for these types of applications.

2.4 Basic measurement instrumentation

Most modern eddy-current testing equipment consist of a probe coil, an ac source, a voltmeter (or ammeter), and a display for analysis of the data. Most instruments used to measure eddy-current NDI data are based on the electromagnetic principle of mutual induction, which can be illustrated as an electrical circuit between the energized eddy-current probe and the conductive object undergoing testing. Since eddy currents generate their own magnetic fields that affect the primary field of the coil, it is essential that the instrument is capable of measuring the changes in resistance and inductive reactance of the probe coil at a given frequency. The mutual inductance between the probe and the test object is affected by electrical conductivity, magnetic permeability, and liftoff, which is the distance between the

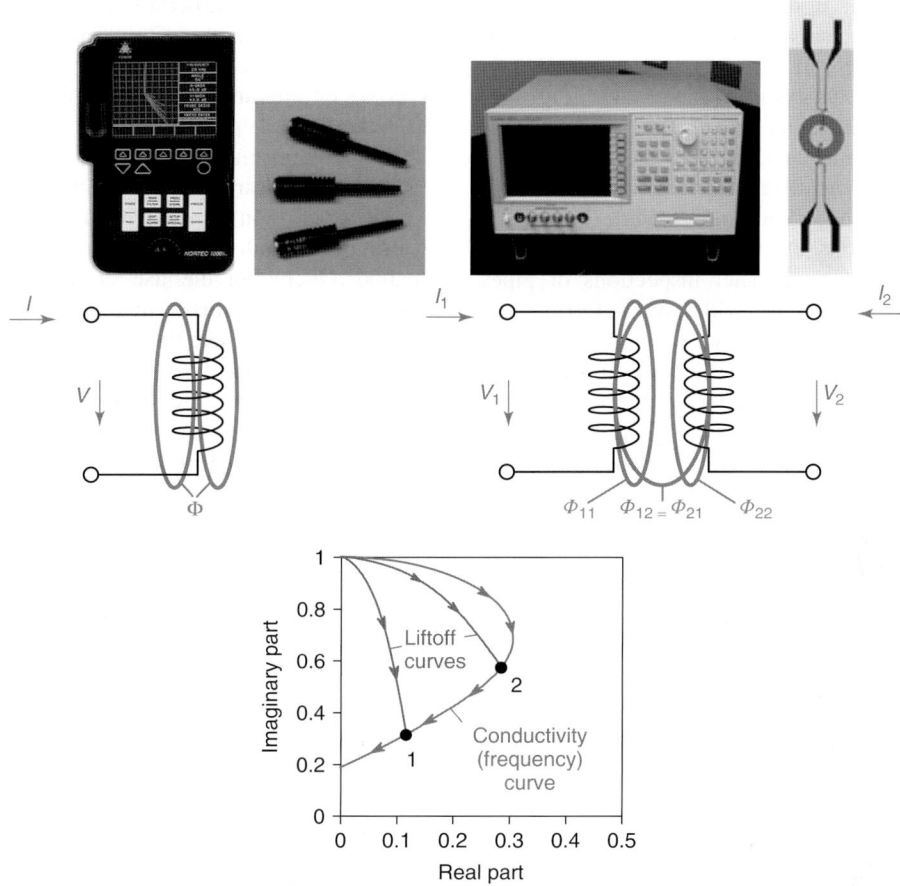

Figure 2. Some commonly available eddy-current probes, flexible coil and associated instrumentation; resonance eddy-current probe circuits and an impedance plane diagram showing the result of an eddy-current scan of a conducting material including the effects of lifting the probe from the surface of the material.

inspection object and the probe. Most commercially available eddy-current instruments have a broad range of excitation frequencies from, for example, 100 Hz to 6 MHz. The basis for the eddy-current measurement is the (Maxwell) ac bridge, which is used to measure unknown inductances in terms of calibrated resistance and capacitance. Since the phase shift between inductors and capacitors is $90°$ out of phase, capacitive impedance can be used to balance out an inductive impedance if they are in opposite legs of the ac bridge. The eddy-current instrument display is essentially an impedance plane with resistance along the abscissa and inductive reactance along the ordinate and the measured impedance is the vector addition of the two components.

Eddy-current probes typically consist of a primary coil, which is capable of being brought into close proximity with the test object, and an identical reference coil housed in the body of the probe to accommodate temperature changes and minimize thermal drift (Figure 2). Probes are essentially resonant circuits, but operated well below the resonance to avoid stray capacitance and other parasitic affects. A wide variety of eddy-current probes exist today for different types of inspection applications including absolute probes for conductivity measurement or thickness gauging, differential probes for crack detection complex geometry, and reflection probes for materials characterization. Hybrid probes exist for applications such as detection of tiny surface cracks in aerospace components, where a split-D type of transducer might be used, which consists of a drive coil encompassing two D-shaped sensing coils. Other types of eddy-current probes also exist, which might use giant magnetoresistive elements, Hall-effect sensors, or flexible thin-film sensors for specialized applications (*see* **Chapter 58**; **Chapter 63**). Similarly, a wide variety of probe configurations exist for surfaces, bolt-holes, and inner- and outer-diameter inspections of pipes and tubing. Often, probes will have a ferrite core to intensify the magnetic lines of force near the core or take advantage of shielding to limit the spread of the probes magnetic field and therefore that of the eddy currents. In most applications, sensing the eddy currents involves assessing the impedance changes of a detector coil using an ac bridge where the impedance of the probe is compared to known impedances, forming a balancing arm in the bridge circuit. High sensitivity is achieved by matching the impedance of the probe to the impedance of the measuring instrument, otherwise known as *nulling the instrument*. The fundamental science of electromagnetic nondestructive test methods and applications for engineers and students can be found in [4].

3 ULTRASONIC SENSORS

Ultrasound is the commonly used technique in NDE to detect the presence of defects or determine the elastic properties of materials. Ultrasound is a linear elastic mechanical wave that propagates at frequencies greater than 20 kHz. Typically, frequencies between 1 and 10 MHz are used in production NDE, but frequencies in the range of hundreds of megahertz have been used in specialized inspections. The most common method to generate and detect ultrasonic waves is to place a sensor in contact with the material being evaluated. These sensors are called *transducers* and use a variety of material properties to generate and detect the mechanical wave. The most common class of transducers is based on piezoelectricity. However, other types of transducers can be based on magnetostrictive, electromagnetic, or capacitive behavior of the element in the transducer, where the element refers to the component of the transducer that physically moves to generate and detect the ultrasonic mechanical waves and converts this energy into electrical signals that can be viewed and/or recorded on oscilloscopes or similar displays. This article will provide an overview of piezoelectricity and a review of the key components that are typically required in a contact piezoelectric transducer. It will also provide a brief description of other methods to generate and detect ultrasound, plus a summary of noncontact methods, including the use of lasers. It is important to note that this article is a short summary of these items. There is a substantial volume of literature, including books, monographs, and conferences, dedicated to the design, development, and evaluation of ultrasonic transducers [5–8] and the reader is encouraged to explore these for additional information.

Piezoelectricity is described as a material property that converts electrical signals to mechanical deformation and *vice versa*. This behavior was first observed by Jacques and Pierre Curie in 1880 by detecting the presence of an electric charge when certain materials were compressed. Later work indicated that inverse behavior can occur. This behavior is caused by the crystallographic structure of the piezoelectric crystal. The most commonly cited natural piezoelectric material is quartz, which has a hexagonal crystal structure [9]. To obtain the piezoelectric behavior, single crystals of this material are needed and they have to be cut along specific axes to obtain the desired deformation. Figure 3 illustrates the application of an electric field to a single hexagonal crystal to generate a longitudinal ultrasonic wave and Figure 4 shows how a transverse, or shear, wave is generated from a single crystal cut along a different plane.

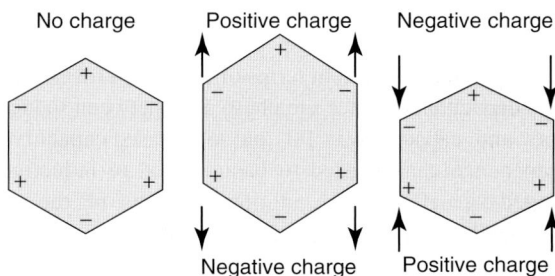

Figure 3. A schematic demonstrating the generation of longitudinal ultrasonic waves by applying an electrical pulse to a hexagonal quartz crystal.

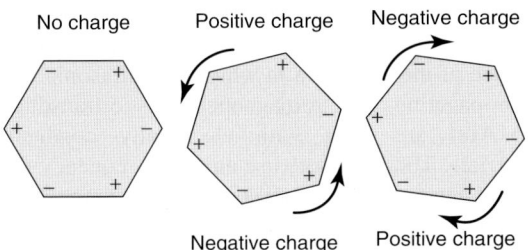

Figure 4. A schematic demonstrating the generation of shear, or transverse, ultrasonic waves by applying an electrical pulse to a hexagonal quartz crystal.

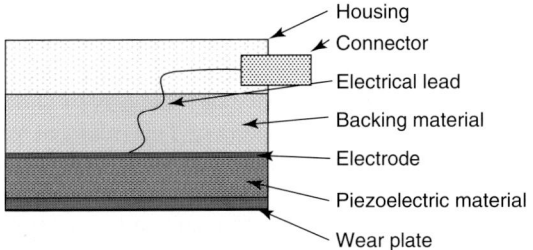

Figure 5. A cross section of a typical commercial transducer showing the typical components of such a transducer.

When an electrical spike is applied to the crystal, it will deform to either attract or repel the electrical charge in response to the asymmetric balance of charges in the signal crystal. The linear deflection of the single crystal generates a longitudinal wave, whereas the rotational deformation will generate a shear, or transverse, wave. Note that a much thicker couplant capable of supporting shear motion must be used when generating or detecting shear waves using a piezoelectric element. Alternative methods, such as those using angle-beam techniques, generate shear waves using refraction, which can occur with less viscous couplants, such as water or inert gels.

Subsequent developments in piezoelectric materials resulted in the discovery and development of ceramic materials, such as lead zirconate titanate (PZT) and lead metaniobate. An advantage of these materials is that they do not need to be single crystals to demonstrate piezoelectric behavior. These materials are poled at elevated temperatures to align their piezoelectric properties, which yield a deformation in these materials when they are exposed to an electrical field. Using this approach, the physical motion of these materials enables the preparation of transducers that can generate and detect both longitudinal and shear waves. This eliminates the need for careful sectioning of a ceramic crystal, as required for quartz, but has the disadvantage that these materials can only be used at temperatures below the depolarization temperature, which can range between 130 and 575 °C. These materials are very sensitive to surface displacements, which make them ideal transmitters and receiver of ultrasonic energy that remains in the linear elastic regime [10].

Commercially available transducers typically use these materials as their active element. Common features of commercial transducers are shown in a representative cross section of a transducer in Figure 5. The transducer is typically a metallic case. The active element is protected from the inspection surfaces by a wear plate, which is made of a material that minimizes wear and matches the acoustic impedance of the active element as much as possible. Electrical leads are coupled to either surface of the piezoelectric material to apply the voltage spike or tone burst to excite the element. A significant amount of the volume of a transducer is dedicated to the dampening material that minimizes the resonance of the piezoelectric element. The dampening material decreases the number of cycles the piezoelectric material will resonate, which assists in resolving multiple reflections from the material being inspected. For example, this minimizes the interference of these signals when trying to detect damage close to the location of the transducer. Note that electrical wire failure and disbanding of the dampening material are the most common causes of failure in a transducer, which would lead to a diminished response or a perturbation in the received signal from damage.

A more recent development in transducer design is the emergence of composite elements. These transducers, which have been commercially available for approximately 10 years, combine a polymer material together with the piezoelectric material, usually to assist in the dampening process. The polymer-based material is commonly introduced into the transducer by cutting thin channels in the ceramic element and filling the channels with the polymer material. The polymer can dampen the vibration of the ceramic while minimizing the effect on the sensitivity of the ceramic when compared to a traditional dampening backing layer. Thus, composite transducers are frequently used when high sensitivity and resolution are needed.

A multitude of other materials and methods have been explored to generate and detect ultrasonic waves. Several polymer materials, most notably poly(vinylidene fluoride) (PVDF), have been shown to have piezoelectric properties. This material can be configured into relatively thin films and retains its piezoelectric behavior when flexed, making it a desired sensor when trying to conform to a complex geometry. These materials are typically very effective detectors of ultrasonic signals, but are not very efficient transmitters. In addition, their response can be very dependent on temperature, requiring these sensors to be used in a stable environment.

The transducers described above are used in most bulk material ultrasonic NDE applications. However, for acoustic emission (AE), the performance requirements of transducers change as these sensors are tailored to detect ultrasonic signals emitted from crack propagation, delaminations, and similar types of damage. To maximize the detection of these signals, AE transducers typically have a response that is very flat as a function of frequency. With conventional piezoelectric transducers, this performance characteristic is typically obtained by increasing the dampening of the piezoelectric element. Therefore, these sensors are commonly larger than conventional transducers and frequently have broadband preamplifiers to amplify the dampened signal. An alternative approach is to make the contact area of the sensing element very small, which can be exemplified by conical transducers and pinducers. With these small contact areas, extra dampening material is not required as the sensor functions as a point detector. Owing to the requirement that these sensors be very broadband, they are also very inefficient and are not commonly used as transmitters. Note that other broadband sensors, such as PVDF, can also be very effective AE sensors.

Alternative materials have been used for generation and detection of ultrasonic signals. Magnetostrictive materials convert magnetic fields into mechanical deformation by the alignment of their magnetic moments [11]. These materials are commonly used for sound navigation and ranging (SONAR) applications as they are very effective and efficient at low frequencies. Capacitive transducers use changes in capacitance between the sensor and the material in which the ultrasonic wave is propagating to sense ultrasonic signals [12]. As this behavior can occur through air, these transducers do not require couplant, but typically must be in close proximity of the surface of the specimen in which the ultrasonic wave is propagating. Electromagnetic acoustic transducers (EMATs) are used with electrically conductive materials. These sensors use an radio frequency (RF) coil to generate eddy currents in the specimen. When the coil is placed inside a permanent magnetic field, the eddy currents interact with the magnetic field to generate mechanical waves in the material. Note that these three effects occur in reverse order when the ultrasonic signal is detected. Several efforts have used piezoelectric materials for air-coupled transducers. They are most effective in a through transmission mode for composite materials, which have a lower acoustic impedance when compared to metallic structures.

Another noncontact method that has been researched extensively is the use of lasers to generate and detect ultrasound. For generation, it is common to use a short pulse laser that causes very rapid and localized heating to thermoelastically generate a mechanical wave. If too much energy is deposited on the surface, the generation method becomes ablative, which will cause material to be removed from the surface. Detection is typically performed using an optical interferometer, such as Michelson or Fabry–Perot. Extensive literature exists that explores multiple laser-based variations to generate and detect ultrasonic signals [9].

A final type of transducer to consider is array transducers. These transducers include multiple piezoelectric elements sufficiently small to be considered as point sources of ultrasonic signals. These elements are excited in a controlled manner to use the principle of superposition to direct the effective wavefront in different directions. There are limits to the amount an ultrasonic wave can be directed and controlled, which is determined by the material, the number and size of the individual elements, and the physics of ultrasound. However, the use of arrays is growing in popularity, especially in structures that do not contain an excessive number of scattering features that can interfere with the superposition of the ultrasonic waves generated from the point sources of ultrasound.

4 SUMMARY

This article described some introductory ideas of eddy-current coils used for material characterization and detecting damage in materials and structures. Similarly, ultrasonic transducers were described for the generation and detection of ultrasonic waves. There are extensive volumes of literature that describe the material properties and physics that enable these NDE transducers to work as designed and the reader is strongly encouraged to explore this literature before embarking on research and development efforts in this field.

REFERENCES

[1] Hughes DE. Induction balance and experimental researches therewith. *The London, Edinburgh and Dublin Philosophical Magazine and Journal of Science. Fifth Series*, 1879, Vol. 8, No. 46, pp. 50–57.

[2] Förster F. The first picture: A Review of the Initial Steps in the Development of Eight Branches of Nondestructive Material Testing. *Materials Evaluation* 1938 **41**(3):1477–1488.

[3] McMaster RC, McIntire P, Mester ML (eds). *Nondestructive Testing Handbook*, 2nd Edition. The American Society for Nondestructive Testing, 1986, Vol. 4.

[4] Libby HL. *Introduction to Electromagnetic Nondestructive Test Methods*. Wiley-Interscience, 1971.

[5] Silk MG. *Ultrasonic Transducers for Nondestructive Testing*. Adam Hilger, 1984.

[6] Krautkramer J, Krautkramer H. *Ultrasonic Testing of Materials*. Springer-Verlag, 1990.

[7] Bray DE, Stanley RK. *Nondestructive Evaluation*. McGraw-Hill, 1989.

[8] Van Vlack LH. *Elements of Materials Science and Engineering*. Addison-Wesley, 1980.

[9] Kossoff G. The effects of backing and matching on the performance of piezoelectric ceramic transducers. *IEEE Transactions on Sonics and Ultrasonics* **SU-13**(1):20–30.

[10] Savage HT, Clark AE, McMaster OD. *Rare Earth-Iron Magnetostrictive Materials and Devices Using These Materials*, US Patent Number 4,308,474, December 1981.

[11] Schindel DW, Hutchins DA, Zou L, Sayer M. Air-coupled capacitance transducers. *IEEE Transactions on Ultrasonics Ferroelectrics and Frequency Control* **42**(1):42–50.

[12] Scruby CB, Drain LE. *Laser Ultrasonics Techniques and Applications*. Taylor & Francis, 1990.

Chapter 55
Piezoelectric Wafer Active Sensors

Lingyu Yu and Victor Giurgiutiu
University of South Carolina, Columbia, SC, USA

1 Introduction	1013
2 PWAS Principles	1014
3 PWAS Ultrasonic Transducers	1016
4 PWAS High-frequency Modal Sensors	1021
5 Novel PWAS Under Development	1024
6 Summary and Conclusions	1026
References	1026

1 INTRODUCTION

In recent years, piezoelectric wafer active sensors (PWAS) permanently attached to the host structures have been used for guided-wave generation and detection during the structural health monitoring (SHM) process. PWAS are inexpensive transducers that operate on the piezoelectric principle. PWAS are used for SHM employing the following three methods: (i) modal analysis and transfer function; (ii) electromechanical (E/M) impedance; and (iii) wave propagation. The use of PWAS for high-frequency local modal sensing with the E/M impedance method

Encyclopedia of Structural Health Monitoring. Edited by Christian Boller, Fu-Kuo Chang and Yozo Fujino © 2009 John Wiley & Sons, Ltd. ISBN: 978-0-470-05822-0.

as well as for damage detection with Lamb wave propagation has been pursued by many researchers [1]. PWAS are no more expensive than conventional high-quality resistance strain gauges. However, PWAS performance by far exceeds that of conventional resistance strain gauges because they can be used as active interrogators. PWAS can be used in high-frequency applications at hundreds of kilohertz and beyond. In summary, PWAS can be used in several ways:

1. **Piezoelectric resonator**
A PWAS has the property of performing mechanical resonances under direct electrical excitation; thus, very precise frequency standards can be created with a simple setup consisting of the PWAS and a signal generator. The resonant frequencies depend only on the wave speed in the PWAS material and the geometric dimensions. Precise frequency values can be obtained through machining the PWAS geometry.

2. **High bandwidth strain exciter and detector**
As an exciter, a PWAS directly converts electrical energy into mechanical energy; thus, it can easily induce vibrations and waves in the substrate material. PWAS acts very well as an embedded generator of waves and vibration. High-frequency waves and vibrations are easily excited with input signals as low as 10 V. As a sensor, PWAS directly converts

mechanical energy to electrical energy. The conversion effectiveness increases with the signal frequency. In the kilohertz range, signals of the order of hundreds of millivolt are easily obtained. No conditioning amplifiers are needed; the PWAS can be directly connected to a high-impedance measurement instrument, such as a digitizing oscilloscope. These dual-sensing and excitation characteristics of PWAS justify their name of "active sensors". A particularly fruitful application is the PWAS phased-array approach, which uses steered guided waves to effectively scan large areas of thin-wall structures from a single location.

3. High-frequency modal sensor

The PWAS can directly measure the high-frequency modal spectrum of the support structure. This is achieved with the E/M impedance method, which reflects the mechanical impedance of the support structure into the real part of the electrical impedance measured at PWAS terminals. The high-frequency characteristic of this method, which has been proven to operate at hundreds of kilohertz and beyond, cannot be achieved with conventional modal instrumentation techniques. Thus, PWAS are the sensors of choice for high-frequency modal measurement and analysis.

2 PWAS PRINCIPLES

This section addresses the behavior of a bonded (constrained) PWAS when excited by an alternating electric voltage. As shown in Figure 1(a), PWAS are small and unobtrusive and utilize the coupling between in-plane strain and transverse electric field. A 5 mm × 5 mm × 0.2 mm square PWAS weighs about 50 mg and costs about $10.

2.1 1D PWAS analysis

Consider a PWAS of length l_a, width b_a, and thickness t_a that is undergoing longitudinal expansion (u_1) induced by the thickness polarization electric field (E_3), as shown in Figure 1(b). The electric field is produced by the application of a harmonic voltage, $V(t) = \hat{V}e^{j\omega t}$, between the top and bottom surfaces (electrodes). The resulting electric field, $E_3 = V/t$, is assumed to be uniform over the PWAS. Assume that h, b, l have widely separated values ($h \ll b \ll l$) such that the length, width, and thickness motions are practically uncoupled. The motion predominantly in the longitudinal direction (x_1) will be considered (1D assumption). PWAS operate on the piezoelectric principle that couples the electrical and mechanical variables in the material in the form

$$S_1 = s_{11}^E T_1 + d_{31} E_3 \quad (1)$$

$$D_3 = d_{31} T_1 + \varepsilon_{33}^T E_3 \quad (2)$$

where S_1 is the strain, T_1 is the stress, D_3 is the electrical displacement (charge per unit area), s_{11}^E is the mechanical compliance at zero field, ε_{33}^T is the dielectric constant at zero stress, and d_{31} is the induced strain coefficient (mechanical strain per unit electric field). When PWAS is bonded to the structure, the structure constrains the PWAS motion with a structural stiffness (k_{str}); that is to say, PWAS are elastically constrained as shown in Figure 1(b). In this model, the overall structural stiffness applied to the PWAS is split into two equal components, $2k_{str}$ each, applied to the ends of the PWAS, such that

$$k_{total} = \left[(2k_{str})^{-1} + (2k_{str})^{-1}\right]^{-1} = k_{str} \quad (3)$$

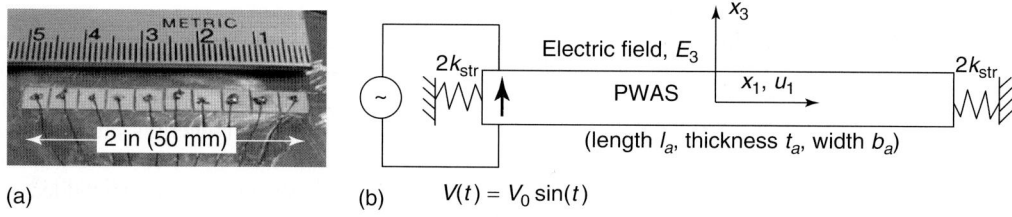

Figure 1. Embedded piezoelectric wafer active sensors: (a) a phased array composed of 10 PWAS, each 5-mm square, permanently attached to the host structure and (b) a one-dimensional bonded PWAS under electric excitation, constrained by structural stiffness k_{str}.

Note that the effective structural stiffness, k_{str}, is a frequency-dependent complex quantity reflecting the structural dynamics. The boundary conditions applied at the ends of the PWAS balance the resulting stress $(T_1 b_a t_a)$ with the spring reaction force $(2k_{\text{str}} u_1)$

$$T_1\left(\frac{l_a}{2}\right) b_a t_a = -2k_{\text{str}} u_1\left(\frac{l_a}{2}\right),$$
$$T_1\left(-\frac{l_a}{2}\right) b_a t_a = 2k_{\text{str}} u_1\left(-\frac{l_a}{2}\right) \qquad (4)$$

Using the Newton's law of motion $(T_1' = \rho \ddot{u}_1)$ and the strain-displacement relation $(S_1 = u_1')$, the axial wave equation can be obtained:

$$\ddot{u}_1 = c_a^2 u_1'' \qquad (5)$$

2.1.1 PWAS responses

Introduce the notations: induced strain $S_{\text{ISA}} = d_{31}\hat{E}_3$, displacement $u_{\text{ISA}} = S_{\text{ISA}} l = (d_{31}\hat{E}_3) l$; wave number γ, $\gamma = \omega/c$; wavelength λ, $\lambda = cT = c/f$, $f = \omega/2\pi$, quasi-static PWAS stiffness $k_{\text{PWAS}} = A_a/s_{11}^E l_a$, piezoelectric material wave speed $c_a^2 = 1/\rho s_{11}^E$, and $\frac{\partial}{\partial x_1}() = ()'$, $\frac{\partial}{\partial t}() = (\dot{})$, the general solution $u_1(x,t) = \hat{u}(x)e^{j\omega t}$ of equation (5) can be obtained as

$$\hat{u}(x) = \frac{1}{2} u_{\text{ISA}} \frac{\sin \gamma x}{\gamma l \cos(\gamma l/2)/2 + \gamma \sin(\gamma l/2)} \qquad (6)$$

if the system determinant is nonzero $(\Delta \neq 0)$

$$\Delta = \begin{vmatrix} \phi \cos\phi + r \sin\phi & -(\phi \sin\phi - r \cos\phi) \\ \phi \cos\phi + r \sin\phi & (\phi \sin\phi - r \cos\phi) \end{vmatrix}$$
$$= 2(\phi \cos\phi + r \sin\phi)(\phi \sin\phi - r \cos\phi) \qquad (7)$$

The electrical responses of PWAS under harmonic electric excitation can be represented as the admittance (Y), which is defined as the ratio between the current and the voltage and the admittance (Z), which is the inverse of admittance.

$$Y = \frac{\hat{I}}{\hat{V}} = j\omega C\left[1 - k_{31}^2\left(1 - \frac{1}{r + \phi \cot\phi}\right)\right] \qquad (8)$$

$$Z = \frac{\hat{V}}{\hat{I}} = \frac{1}{j\omega C}\left[1 - k_{31}^2\left(1 - \frac{1}{r + \phi \cot\phi}\right)\right]^{-1} \qquad (9)$$

with $\phi = \gamma l/2$, $k_{31}^2 = d_{31}^2/(s_{11}^E \varepsilon_{33}^T)$ (E/M coupling coefficient), $C = \varepsilon_{33}^T b_a l_a/t_a$ (stress-free PWAS capacitance), and r as the structural stiffness ratio. For free PWAS, $r = 0$ ($k_{\text{str}} = 0$) and full constrained PWAS, $r \to \infty$ ($k_{\text{str}} \to \infty$). When the oscillation frequency is so low that the dynamic effects inside the PWAS are negligible, quasi-static conditions are encountered with $\phi = 0$, i.e., $\gamma l = 0$.

2.1.2 PWAS resonances

The determinant in equation (7) is zero when the first or second parenthesis is zero, corresponding to two types of PWAS resonances, mechanical resonances ($\phi \sin\phi - r \cos\phi = 0$) or E/M resonances ($\phi \cos\phi + r \sin\phi = 0$). The E/M resonances are specific to piezoelectric materials. They reflect the coupling between the mechanical and electrical variables. During E/M resonances, there are two possibilities that arise:

- *Resonance*, when $Y \to \infty$; that is, $Z = 0$. This resonance is associated with the situation in which a device draws very large current when excited harmonically with a constant voltage at a given frequency.
- *Antiresonance*, when $Y = 0$; that is, $Z \to \infty$. This antiresonance is associated with the situation in which a device under constant voltage excitation draws almost no current.

2.2 2D circular PWAS

The 2D constrained PWAS behavior can be analyzed using constitutive equations in cylindrical coordinates (r, θ, z),

$$S_{rr} = s_{11}^E T_{rr} + s_{12}^E T_{\theta\theta} + d_{31} E_z \qquad (10)$$
$$S_{\theta\theta} = s_{12}^E T_{rr} + s_{11}^E T_{\theta\theta} + d_{31} E_z \qquad (11)$$
$$D_z = d_{31}(T_{rr} + T_{\theta\theta}) + \varepsilon_{33}^T E_z \qquad (12)$$

The wave equation in polar coordinates is

$$\frac{\partial^2 u_r}{\partial r^2} + \frac{1}{r}\frac{\partial u_r}{\partial r} - \frac{u_r}{r^2} = \frac{1}{c_p^2}\frac{\partial^2 u_r}{\partial t^2} \qquad (13)$$

where $c_p = \sqrt{1/[\rho s_{11}^E(1-v_a^2)]}$, $v_a^2 = s_{12}^E/s_{11}^E$. Equation (13) provides a general solution in terms of Bessel functions of the first kind (J_1) in the form

$$u_r(r,t) = A J_1\left(\frac{wr}{c}\right) e^{j\omega t} \qquad (14)$$

where the coefficient A is determined from the boundary conditions. When PWAS is bonded to a structure, at the boundary $r = r_a$, the boundary condition is

$$T_{rr}(r_a) t_a = k_{str}(\omega) u_r(r_a) \qquad (15)$$

The electric impedance is calculated as the ratio between voltage and current amplitudes

$$Z(\omega) = \left\{ j\omega C \left(1 - k_p^2\right) \left[1 + \frac{k_p^2}{1 - k_p^2} \frac{(1 + v_a) J_1(\phi_a)}{\phi_a J_0(\phi_a) - (1 - v_a) J_1(\phi_a) - \chi(\omega)(1 + v_a) J_1(\phi_a)} \right] \right\}^{-1} \qquad (16)$$

with $\phi_a = \omega r_a/c$, $\chi(\omega) = k_{str}(\omega)/k_{PWAS}$ (dynamic stiffness factor), and $k_p^2 = 2d_{31}^2/[s_{11}^E(1-v_a)\varepsilon_{33}^T]$ (planar coupling factor). Equation (16) predicts that the E/M impedance spectrum can be measured by the impedance analyzer at the embedded PWAS terminals during an SHM process and it allows for direct comparison between calculated predictions and experimental results. The structural dynamics is reflected in equation (16) through the dynamic stiffness factor $\chi(\omega)$, which is a measure of the dynamic stiffness of the structure.

3 PWAS ULTRASONIC TRANSDUCERS

For embedded nondestructive evaluation (NDE) applications, PWAS can be used as embedded ultrasonic transducers. PWAS act as both Lamb wave exciters and detectors (Figure 2). PWAS couple their in-plane motion with the particle motion of Lamb waves on the material surface. The in-plane PWAS motion is excited by the applied oscillatory voltage thought the d_{31} piezoelectric coupling.

The PWAS ultrasonic transducer operation is fundamentally different from that of conventional ultrasonic probes

1. PWAS achieve Lamb wave excitation and sensing through surface "pinching" (in-plane strains), while conventional ultrasonic probes excite through surface "tapping" (normal stress).
2. PWAS are strongly coupled with the structure and follow the structural dynamics, while conventional ultrasonic probes are relatively free from the structure and follow their own dynamics.
3. PWAS are nonresonant wideband devices, while conventional ultrasonic probes are narrowband resonators.

Optimum Lamb wave excitation and detection happen when the PWAS length is an odd multiple of the half wavelength of particle wave modes. Geometric tuning can be obtained through matching between

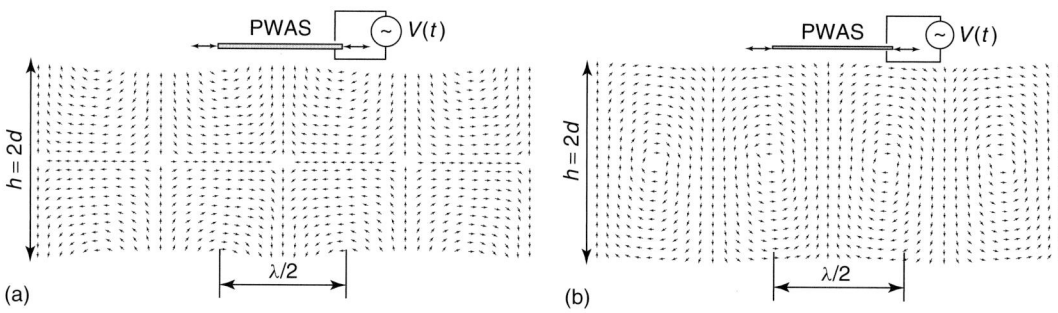

Figure 2. PWAS interaction with Lamb waves in a plate: (a) S_0 mode and (b) A_0 mode.

their characteristic direction and the half wavelength of the excited Lamb wave mode. Circular PWAS excite omnidirectional Lamb waves that propagate in circular wave fronts. Omnidirectional Lamb waves are also generated by square PWAS, though their pattern is somewhat irregular in proximity to the PWAS. At far enough distance ($r \gg a$), the wave front generated by square PWAS is practically identical with that generated by circular PWAS.

3.1 Shear-layer interaction between PWAS and the host structure

The transmission of actuation and sensing between the PWAS and the structure is achieved through the adhesive layer. The adhesive layer acts as a shear layer, in which the mechanical effects are transmitted through shear effects. As shown in Figure 3(a), a thin-wall structure of thickness t and elastic modulus E is attached with a PWAS of thickness t_a and elastic modulus E_a to the upper surface through a bonding layer of thickness t_b and shear modulus G_b. The PWAS length is l_a while the half length is $a = l_a/2$. In addition, $t = 2d$. Upon application of an electric voltage, the PWAS experiences an induced strain of $\varepsilon_{ISA} = d_{31}V/t$. The induced strain is transmitted to the structure through the bonding layer interfacial shear stress (τ). Upon analysis, we obtain the PWAS displacement,

$$u_a(x) = \frac{\alpha}{\alpha + \psi}\varepsilon_{ISA} a \left(\frac{x}{a} + \frac{\psi}{\alpha} \frac{\sinh \Gamma x}{(\Gamma a)\cosh \Gamma a} \right) \quad (17)$$

the interfacial shear stress in bonding layer,

$$\tau(x) = \frac{t_a}{a}\frac{\psi}{\alpha + \psi} E_a \varepsilon_{ISA} \left(\Gamma a \frac{\sinh \Gamma x}{\cosh \Gamma a} \right) \quad (18)$$

and the structural displacement at the surface,

$$u(x) = \frac{\alpha}{\alpha + \psi}\varepsilon_{ISA} a \left(\frac{x}{a} - \frac{\sinh \Gamma x}{(\Gamma a)\cosh \Gamma a} \right) \quad (19)$$

where $\psi = (Et)/(E_a t_a)$, and α is a parameter that depends on the stress and strain distribution across the thickness. Under static and low-frequency dynamic conditions (i.e., uniform stress and strain for the symmetric deformation and linear stress and strain for antisymmetric deformation), elementary analysis yields $\alpha = 4$. For Lamb wave modes, which have complex stress and strain distributions across the thickness, the parameter α varies from mode to mode. These equations apply for $|x| < \alpha$. The shear-lag parameter $[\Gamma^2 = (G_b/E_a)(1/t_a t_b)((\alpha + \psi)/\psi)]$ controls the x distribution. The effect of the PWAS is transmitted to the structure through the interfacial shear stress of the bonding layer. A small shear-stress value in the bonding layer produces a gradual transfer of strain from the PWAS to the structure,

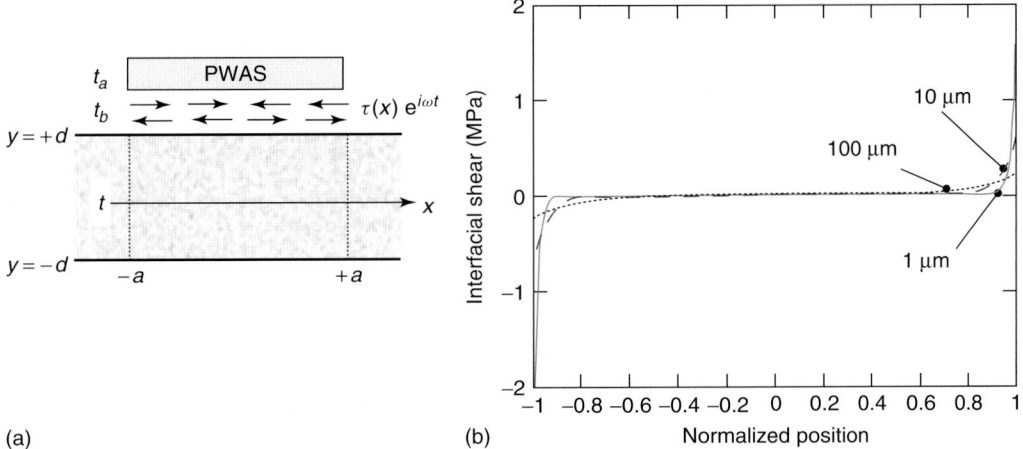

Figure 3. Shear-lag interaction between PWAS and structure: (a) interaction between the PWAS and the structure and (b) variation of shear-lag transfer mechanism with bond thickness.

whereas a large shear stress produces a rapid transfer. Because the PWAS ends are stress free, the build up of strain takes place at the ends, and it is more rapid when the shear stress is more intense. For large value of Γa, the shear transfer process becomes concentrated toward the PWAS ends. Figure 3(b) presents the results of the simulations on an APC-850 PWAS ($E_a = 63$ GPa, $t_a = 0.2$ mm, $l_a = 7$ mm, and $d_{31} = -175$ mm kV^{-1}) bonded to a thin-wall aluminum structure ($E = 70$ GPa and $t = 1$ mm) using cynoacrylate adhesive ($G_b = 2$ GPa) of various thicknesses, $t_b = 1, 10, 100$ μm. It reveals that a relatively thick bonding layer produces a low Γa value—a slow transfer over the entire span of the PWAS (the 100-μm curves)—whereas, a very thin bonding layer produces a very rapid transfer (the 1-μm curves) that is confined to the ends, i.e.,

$$\tau(x) = a\tau_0[\delta(x-a) - \delta(x+a)] \quad (20)$$

The shear-lag analysis indicates that in the limit, as $\Gamma a \to \infty$, all the load transfer can be assumed to take place at the PWAS actuator ends. This leads to the concept of ideal bonding, also known as the *pin-force model*, in which all the load transfer takes place over an infinitesimal region at the PWAS ends, and the induced-strain action is assumed to consist of a pair of concentrated forces applied at the ends. The ideal bonding stress is then given by

$$\tau(x) = a\tau_a[\delta(x-a) - \delta(x+a)] \quad (21)$$

Using the shear-lag model, the energy transferred from PWAS to the structure can be found by analyzing either the elastic energy in the structure or work done by the shear stresses at the structural surface.

3.2 PWAS Lamb waves excitation and reception

Lamb waves, also known as *guided plate waves*, are a type of ultrasonic waves and are guided between two parallel free surfaces, such as the upper and lower surfaces of a plate (*see* **Chapter 3**). Lamb waves can exist in two basic types: symmetric (designated as $S_0, S_1, \ldots$) and antisymmetric (designated as $A_0, A_1, \ldots$). For each propagation type, a number of modes exist, corresponding to the solutions of the Rayleigh–Lamb equation. The symmetric Lamb waves resemble the axial waves while the antisymmetric Lamb waves resemble the flexural waves. Lamb waves are highly dispersive and their speed depends on the product of frequency and the plate thickness. Details of Lamb wave theory can be found in [1, 2]. Traditional generation of Lamb waves has been through a wedge probe, and modification of the wedge angle and excitation frequency allows the selective tuning of various Lamb wave modes. Another traditional way is to cause excitation through a comb probe, in which the comb pitch is matched with the half wavelength of the targeted Lamb mode. However, both the wedge and comb probes are relatively large and expensive and not appropriate for installation in large numbers in an aerospace structure as part of an SHM system. Being smaller, lighter, more affordable, PWAS could be deployed into the structure being permanently wired and could interrogate at will.

3.2.1 Lamb wave excitation by PWAS

The excitation of Lamb waves with PWAS is studied by considering the excitation applied by the PWAS through a surface stress $\tau = \tau_0(x)e^{j\omega t}$ applied to the upper surface of a plate in the form of shear-lag adhesion stresses over the interval $(-a, +a)$. Applying a space domain Fourier transform analysis of the basic Lamb wave equations to yield the strain wave and displacement wave solutions [3], we have

$$\varepsilon_x(x,t)|_{y=d} = -i\frac{a\tau_0}{\mu}\left[\sum_{\xi^S}\sin(\xi^S a)\frac{N_S(\xi^S)}{D'_S(\xi^S)}\right.$$
$$\times e^{i(\xi^S x - \omega t)} + \sum_{\xi^A}\sin(\xi^A a)$$
$$\left.\times \frac{N_A(\xi^A)}{D'_A(\xi^A)}e^{i(\xi^A x - \omega t)}\right]$$

$$N_S = \xi\beta(\xi^2 + \beta^2)\cos(\alpha d)\cos(\beta d),$$
$$D_S = (\xi^2 - \beta^2)^2\cos(\alpha d)\sin(\beta d)$$
$$+ 4\xi^2\alpha\beta\sin(\alpha d)\cos(\beta d)$$
$$N_A = \xi\beta(\xi^2 + \beta^2)\sin(\alpha d)\sin(\beta d),$$

$$D_A = (\xi^2 - \beta^2)^2 \sin(\alpha d) \cos(\beta d)$$
$$+ 4\xi^2 \alpha \beta \cos(\alpha d) \sin(\beta d) \quad (22)$$

where ξ^S and ξ^A are the zeros of D_S and D_A, respectively. We can note that these are the solutions of the Rayleigh–Lamb equation. Raghavan and Cesnik [4] extended these results to the case of a circular transducer coupled with circular-crested Lamb waves and proposed corresponding tuning prediction formulae based on Bessel functions:

$$\varepsilon_r(r,t)|_{z=d} = \pi \frac{\tau_0 a}{\mu} e^{i\omega t} \left[\sum_{\xi^S} J_1(\xi^S a) \xi^S \frac{N_S(\xi^S)}{D_S'(\xi^S)} \right.$$
$$\times H_1^{(2)}(\xi^S r) + \sum_{\xi^A} J_1(\xi^A a) \xi^A$$
$$\left. \times \frac{N_A(\xi^A)}{D_A'(\xi^A)} H_1^{(2)}(\xi^A r) \right] \quad (23)$$

3.2.2 PWAS Lamb wave tuning

An important characteristic of PWAS, which distinguishes them from conventional ultrasonic transducers, is their capability of tuning into various guided-wave modes. A comprehensive study of these prediction formulae in comparison with experimental results has recently been performed by Bottai and Giurgiutiu [5]. Simulation plot of the equation (22) is presented in Figure 4(a) using a 7-mm square PWAS installed on 1.07-mm-thick 2024-T3 aluminum alloy plate. Note that the efficient PWAS length for a 7-mm PWAS has been verified as 6.4 mm [5]. Equation (22) contains the $\sin(\xi a)$ behavior that displays maxima when the PWAS length $l_a = 2a$ equals an odd multiple of the half wavelength, and minima when it equals an even multiple of the half wavelength. A complex pattern of such maxima and minima emerges, since several Lamb modes, each with its own different wavelength, coexist at the same time. The plot in Figure 4(a) shows that at 210 kHz, the amplitude of the A_0 mode goes through zero, while that of the S_0 is close to its peak. This represents an excitation "sweet spot" for S_0 Lamb waves. Experimental results confirming this prediction are presented in Figure 4(b).

3.3 PWAS Lamb wave structural health monitoring

As active sensors, PWAS interact directly with the structure and find its state of health and reliability through the use of ultrasonic Lamb waves [6]. Similar to the conventional ultrasonic transducers, PWAS can operate in pitch–catch, pulse–echo, or be wired into phased array to implement structural scanning.

3.3.1 Pitch–catch method

The pitch–catch method detects damage from the changes in the Lamb waves traveling through a

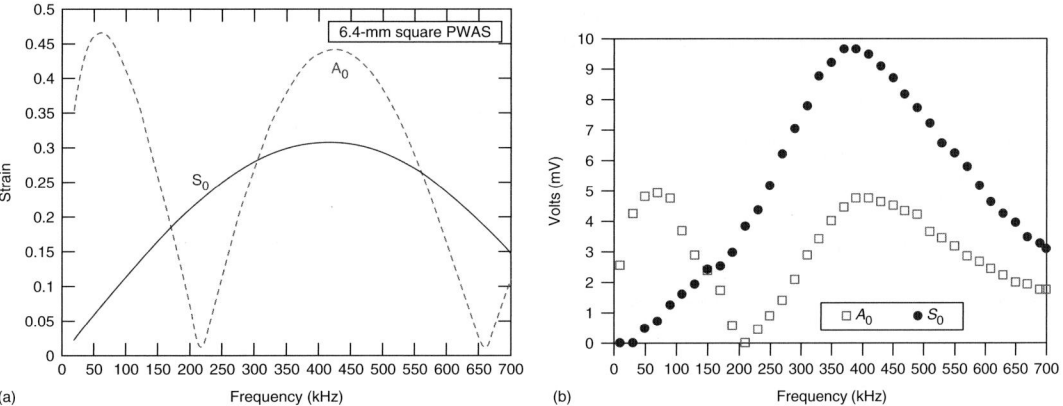

Figure 4. PWAS Lamb wave tuning using a 7-mm square PWAS placed on 1.07-mm 2024-T3 aluminum alloy plate: (a) prediction with equation (22) for 6.4-mm effective sensor length and (b) experimental results.

damaged region. The method uses the transducers in pairs: one as transmitter, and the other as receiver. The pitch–catch SHM method has been used extensively [7–10]. Changes occurring owing to damage in the properties of the transmitted medium induce changes in the signal transmitted through the medium. Of particular interest is the time reversal of the pitch–catch SHM method. In the PWAS time reversal procedure [11], an input signal is reconstructed at an exciter point if an output signal recorded at a receiver point is reversed in time domain and emitted back to the original source point. The reversion is conducted in the frequency domain and depends on the structure transfer function $G(\omega)$. When the reversed signal is reconstructed, it is compared with the original transmission signal to detect the damage occurrence along the transmission–reception line. Figure 5(a) shows the block diagram of PWAS Lamb wave time reversal procedure. With the time reversal method, the original status of the structure is no longer needed for damage detection, so it is known as a *baseline-free* methodology. In PWAS time reversal procedure, it is important to use a single Lamb mode such that the transfer function $G(\omega)$ can remain constant [12]. A simulation with 16-count Hanning window smoothed transmission signal has been conducted in a 1-mm aluminum plate using 7-mm round PWAS. The result of S_0 mode excitation at 290 kHz is shown in Figure 5(b) with noticeable residual waves, while the result of A_0 mode at 30 kHz in Figure 5(c) demonstrating no residual wave packets beside the reconstructed wave.

3.3.2 Pulse–echo crack detection

The embedded PWAS pulse–echo method follows the general principles of conventional Lamb wave NDE. A PWAS transducer attached to the structure acts as both transmitter and detector of guided Lamb waves traveling in the structure. The wave sent by the PWAS is partially reflected at the crack. The echo is captured at the same PWAS, which acts as receiver. To ensure the crack detection, an appropriate Lamb wave mode must be selected. It has been verified that the S_0 Lamb waves can give much better reflections from the through-the-thickness cracks than the A_0 Lamb waves [3]. The selection of such a wave is achieved through the Lamb wave tuning. A wave propagation experiment was conducted on an aircraft panel to illustrate crack detection through the pulse–echo method (Figure 6). A baseline signal was first collected, and then after the simulated crack was added, another reading was recorded. By comparing

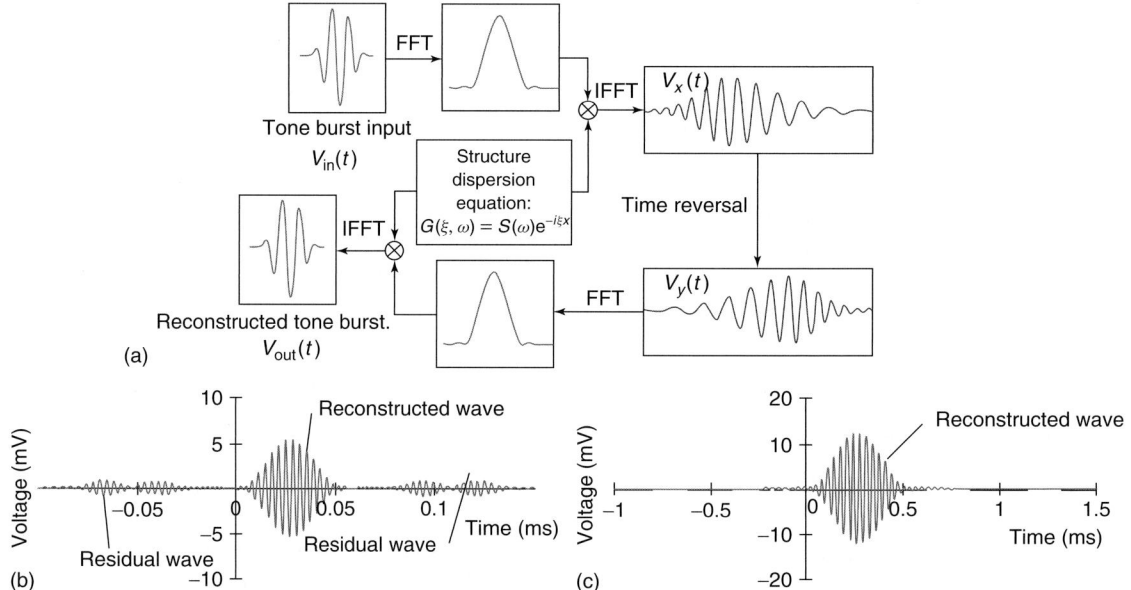

Figure 5. PWAS Lamb wave time reversal procedure. (a) Flowchart block diagram; (b) S_0 mode time reversal at 290 kHz showing residuals; and (c) A_0 mode time reversal at 30 kHz with no residual wave packets present.

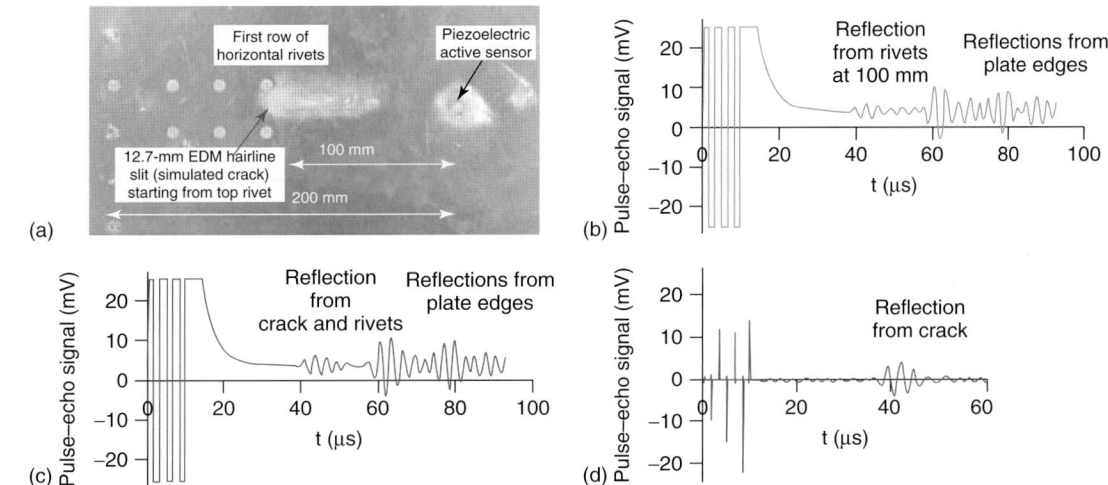

Figure 6. PWAS pulse–echo crack-detection experiment: (a) plate with built-in rivets and a simulated 12.7-mm crack; (b) baseline signal containing reflections from the plate edges and the rivets at 100 mm; (c) reading containing reflections from the crack, plate edges, and rivets; and (d) subtracted signal containing reflection from the crack alone.

the difference between the reading and baseline, the reflection caused by the crack could be extracted and detected.

3.3.3 Phased arrays for large area imaging

PWAS can also be wired as phased arrays to detect damage in thin-wall structures (*see* **Chapter 16**). Using guided-wave PWAS phased-array methods, wide coverage could be achieved from a single location. PWAS phased arrays have been developed for thin-wall structures (e.g., aircraft shells, storage tanks, large pipes, etc.) that use Lamb waves to cover a large surface area through beam steering from a central location [13]. Its principle of operation is derived from two general concepts:

1. the generation of tuned guided Lamb wave with PWAS;
2. the principles of conventional phased-array radar.

The embedded ultrasonic structural radar (EUSR) algorithm [13] using PWAS is different from the conventional phased-array approach in two aspects:

1. it uses structurally embedded PWAS transducers;
2. it works in virtual time, not in real time.

The latter observation is very important for SHM, because it allows the phased-array benefits without the drawback of real-time multichannel phased excitation equipment. Whereas real-time phased-array transducers require heavy and complex multichannel phasing equipment, the virtual-time approach adopted by the EUSR method can be carried out with only one channel and very simple equipment. Figure 7(a) shows the typical setup of using a nine 7-mm square PWAS array to detect a broadside crack in a 1-mm-thick aluminum plate. On the investigated structure, the EUSR method captures and stores signals of an M-PWAS array in a round-robin fashion, i.e., one PWAS being activated as transmitter, while all the other PWAS act as receivers. Thus a matrix of $M \times M$ elemental signals is generated. In the post-processing phase, the elemental signals are assembled into the synthetic beamforming response using the synthetic beamforming algorithm and the scanning result is given as a 2D planar image (Figure 7b).

4 PWAS HIGH-FREQUENCY MODAL SENSORS

Modal analysis and dynamic structural identification are an intrinsic part of current engineering practice. Structural frequencies, damping, and modes shapes identified through this process are subsequently used to predict dynamic response, avoid resonance, and even monitor structural change that are indicative of

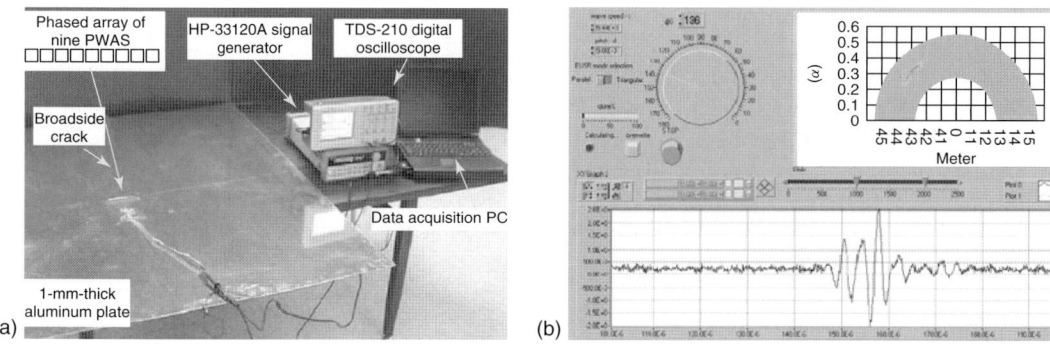

Figure 7. EUSR PWAS phased-array damage detection: (a) equipment setup and (b) EUSR scanning image.

incipient failure. The advantage of using PWAS for damage detection resides in their very high frequency capability, which exceeds by orders of magnitudes the frequency capability of conventional modal analysis sensors. They can be permanently attached to the structural surface and could form sensor and actuator arrays that permit effective modal identification in a wide frequency band. Thus PWAS are able to detect subtle changes in the high-frequency structural dynamics at local scales. Such local changes in the high-frequency structural dynamics are associated with the presence of incipient damage, which would not be detected by conventional modal analysis sensors that operate at low frequencies.

4.1 Analytical model

Consider a 1D structure with a PWAS attached to its surface (Figure 8a). Upon activation, the PWAS expands by $\varepsilon_{\text{PWAS}}$, generating a reaction force F_{PWAS} from the beam onto the PWAS and an equal and opposite force from PWAS onto the beam (Figure 8b).

The force excites the beam. As the PWAS is electrically excited with a high-frequency harmonic signal, it induces elastic waves into the beam structure. The elastic waves travel sideways into the beam structure and set it into oscillation. In a steady-state regime, the structure oscillates at the PWAS excitation frequency. The dynamic stiffness presented by the structure to the PWAS depends on the internal state of the structure, excitation frequency, and the boundary conditions as

$$k_{\text{str}}(\omega) = \frac{\hat{F}_{\text{PWAS}}(\omega)}{\hat{u}_{\text{PWAS}}(\omega)} \quad (24)$$

where $\hat{F}_{\text{PWAS}}(\omega)$ is the reaction force and $\hat{u}_{\text{PWAS}}(\omega)$ is the displacement amplitude at frequency ω. The symbol $\hat{}$ signifies the complex amplitude of a time-varying function. Because the size of the PWAS is very small with respect to the size of the structure, equation (24) represents a pointwise structure

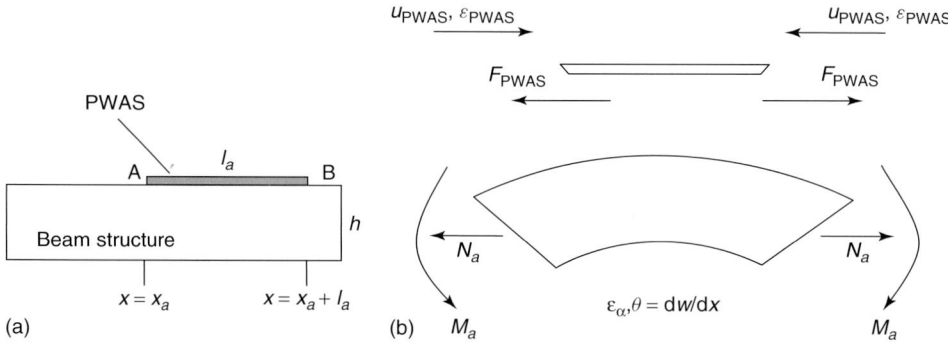

Figure 8. Interaction between a PWAS and a beamlike structural substrate: (a) geometry and (b) forces and moments.

stiffness. The response of the structural substrate to the PWAS excitation can be deduced from the general theory of beam vibrations. However, note that the PWAS excitation departs from the typical textbook formulation since it activates a pair of self-equilibrating axial forces and bending moments that are separated by a small finite distance l_{PWAS}. Kinematic analysis finally gives the dynamic structural stiffness as

$$k_{\text{str}}(\omega) = \rho A \left\{ \sum_{n_u} \frac{[U_{n_u}(x_a + l_a) - U_{n_u}(x_a)]^2}{\omega_{n_u}^2 + 2j\varsigma_{n_u}\omega_{n_u}\omega - \omega^2} + \left(\frac{h}{2}\right)^2 \sum_{n_w} \frac{[W'_{n_w}(x_a + l_a) - W'_{n_w}(x_a)]^2}{\omega_{n_w}^2 + 2j\varsigma_{n_w}\omega_{n_w}\omega - \omega^2} \right\}^{-1}$$
(25)

where n_u, ω_{n_u}, $U_{n_u}(x)$ and n_w, ω_{n_w}, $U_{n_w}(x)$ represent the axial and flexural vibrations frequencies and mode shapes, respectively.

4.2 Simulation and experimental results

The analytical model was used to perform numerical simulations that directly predict the E/M impedance and admittance signature at PWAS terminals during structural identification. Numerical simulations were performed with steel beams assuming damping coefficient at 1% over a modal subspace that incorporated all modal frequencies in the frequency bandwidth of interest. Finite element method (FEM) was also conducted to predict the structural natural frequencies using (i) the purely mechanical response via conventional structural elements and (ii) directly the E/M response via coupled-field elements [14]. The stiffness k_{str} in equation (25) is used to get the structural stiffness and calculate the E/M impedance. Subsequent experiments were performed to verify these predictions using a small steel beam ($100 \times 8 \times 2.6$, in millimeter) ($E = 200\,\text{GPa}$, $\rho = 7750\,\text{kg m}^{-3}$) instrumented with 7-mm square PWAS ($t_a = 0.22\,\text{mm}$) placed at $x_a = 40\,\text{mm}$ from one end (Figure 9(a) and (b)). During the experiments, recording of the E/M impedance real part spectra was performed with the HP4194A impedance analyzer in the $1 \sim 30\,\text{kHz}$ range. Numerical and FEM simulations and experimental results are given in Figure 9(c) and (d). Note that the experimental results are consistent with the predictions regarding the natural frequencies. Comparing the two FEM methods, it is noted that the impedance data obtained from the coupled-field analysis was more close to the experimental data, validating the possibility of using

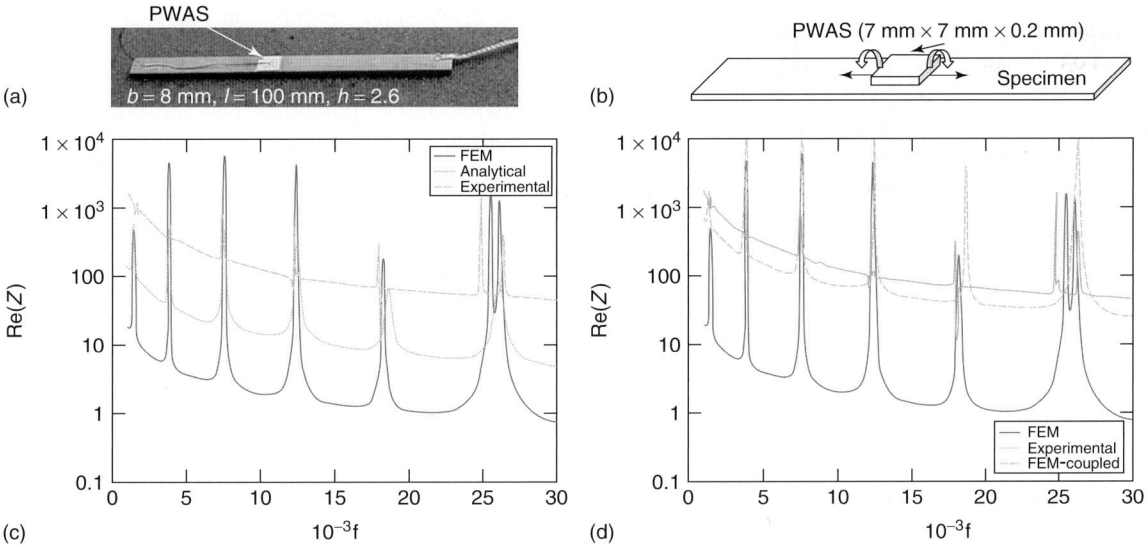

Figure 9. 1D structure simulation and experiments: (a) narrow-beam specimen; (b) beam structure layout for simulation analysis; (c) real part impedance spectra using structure FEM method; and (d) real part impedance spectra using coupled-field FEM method.

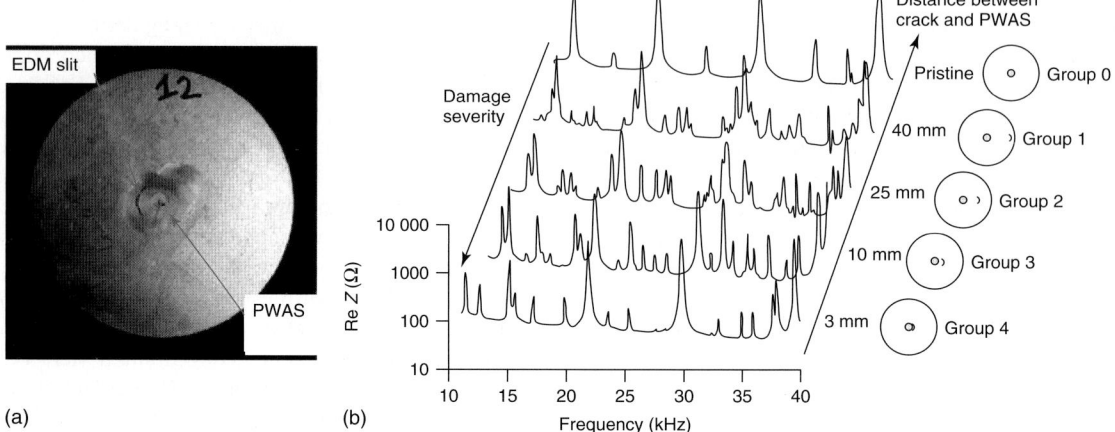

Figure 10. Experiments on dependence of the E/M impedance spectra on the location of damage on metallic plate specimen, E/M impedance in 0.5 ∼ 40 kHz frequency range: (a) the aluminum plate specimen with PWAS installed in the center and (b) E/M impedance spectra at various crack situations.

this method to analyze and simulate the structure with PWAS [14].

4.3 Damage detection with PWAS modal sensors

PWAS and the associated structural dynamics identification methodology based on the E/M impedance response are ideally suited for small machinery parts that have natural frequencies in the kilohertz range [15–17]. PWAS are able to detect subtle changes in the high-frequency structural dynamics at local scales. Such local changes in the high-frequency structural dynamics are associated with the presence of incipient damage, which would not be detected by conventional modal analysis sensors that operate at low frequencies. The E/M impedance of a 7-mm round PWAS has been measured with an HP4194A impedance analyzer at various crack situations on an aluminum metallic plate to assess the crack-detection capabilities, as shown in Figure 10.

5 NOVEL PWAS UNDER DEVELOPMENT

PWAS transducers have been proven to be valuable enablers of SHM systems. As a small embedded sensing device, PWAS have also shown great potential for many other applications such as biomedical applications (bio-PWAS) for *in vivo* monitoring of the bodily reaction to implants [18]. However, the bonded interface between the PWAS and the structure is often the durability weak link for SHM applications. The bonding layer is susceptible to environmental ingression that may lead to loss of contact with the structural substrate. The bonding layer may also induce acoustic impedance mismatch with detrimental effects on damage detection. Additionally, prefabricated piezoelectric sensors such as PWAS do not fit well on surfaces with complex geometry. Better durability may be obtained from a built-in sensor that is incorporated into the structure.

5.1 *In situ* fabricated PWAS

Polymer-based piezoelectric paints have been investigated as a potential substitute for PWAS in certain sensing applications. A new type of smart material, piezoelectric paint that cures at ambient temperature, has been developed by Zhang and Li [19] as sensing material for ultrasonic-based NDE (*see* **Chapter 57**). Piezoelectric paint is a composite piezoelectric material that is composed of tiny piezoelectric particles randomly dispersed within a polymer matrix phase with adjustable material properties that are not easily attainable in a single-phase material. Through judicious selection of the polymer matrix, the composite properties of the piezoelectric paint can be tailored to meet the specific requirements of an application condition. Because of its ease of application, piezoelectric paint can be easily deposited onto the complex surface of host structures; the paint uniformly cures at ambient temperature. On the basis of the preliminary results from the experimental study and finite element simulation, piezoelectric paint sensor appears to have a great potential for use as distributed sensors for fatigue crack monitoring in metallic structures.

5.2 Thin-film nano-PWAS

An alternative way to incorporate PWAS into the structure is the ferroelectric thin-film PWAS that uses nanofabrication technology to produce the sensors directly on the structural substrate [20]. Ferroelectric thin films have been shown to have piezoelectric properties that are those of single-crystal ferroelectrics, which are an order of magnitude better than common piezoceramics. In addition, the thin films require much smaller poling voltage/power. Thin film can get the same strain as that obtained by the PWAS with a much lower voltage, and through layering, the power of the device can be amplified many times. Figure 11 illustrates the thin-film multilayer PWAS development process [20]. C-axial preferred oriented ferroelectric $BaTiO_3$ thin films have been successfully fabricated on Ni metal tapes with a thin NiO buffered layer by pulsed laser ablation. Ferroelectric polarization measurements have shown the hysteresis loop at room temperature in the film with a large remnant polarization, indicating that the ferroelectric domains have been created in the as-deposited BTO films. These excellent properties in this piezoelectric thin film indicate that the as-fabricated BTO films show promise in the development of SHM systems. Hudak *et al.* [21] successfully pursued the alternative path of magnetostrictive thin-film PWAS and achieved enhanced thin-film performance by optimizing architecture through multilayered films, which demonstrated a 4 times stronger receiving signal in pulse–echo experiments. The guided waves have been used in pulse–echo mode to clearly detect

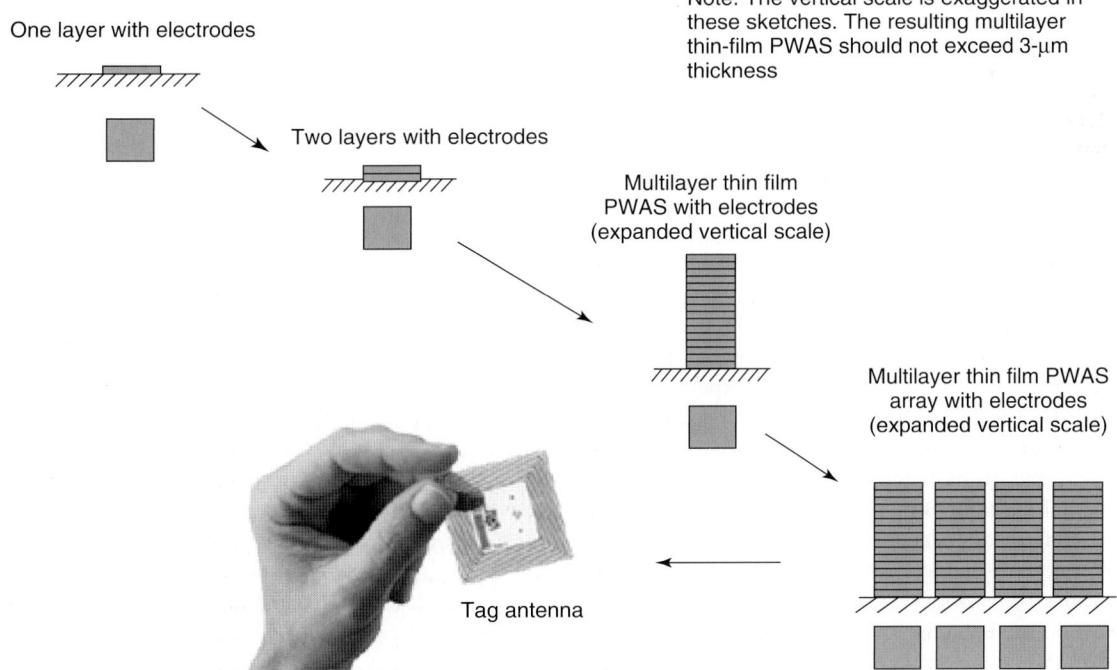

Figure 11. Blueprint of multilayer thin-film PWAS array.

1.52 mm × 7.62 mm defects in laboratory specimen at temperatures up to 554 °F.

6 SUMMARY AND CONCLUSIONS

This article has presented the use of PWAS as resonators, ultrasonic transducers, and high-frequency modal sensors for ultrasonic SHM. The PWAS transducer can be permanently attached to structures for *in situ* health monitoring to determine the structural state of health and predict the remaining life. After reviewing the PWAS operational principles, the PWAS piezoelectric properties (actuation and sensing) were presented in connection with Lamb wave excitation and detection. It was shown analytically and verified experimentally that Lamb wave mode tuning can be achieved by the judicious combination of PWAS dimensions, frequency values, and Lamb wave mode characteristics. For example, the A_0 and S_0 Lamb modes could be separately tuned by using the same PWAS installation but different frequency bands. Next, the use of embedded PWAS transducers as modal sensors was introduced for dynamic structural identification. The PWAS are able to detect subtle changes in the high-frequency structural dynamics at local scales, which are associated with the presence of the incipient damage, which would not be detected by conventional modal analysis sensors that operate at low frequencies. Finally, with all the success of using PWAS transducer for *in situ* SHM, several on-going novel PWAS developments were introduced, including the *in situ* fabricated PWAS and thin-film nano-PWAS.

Although remarkable progress has been made in using PWAS for NDE, considerable work remains to be done. To increase the acceptance of this emerging technology, the refining of the theoretical analysis and calibration against well-planned experiments is needed. However, to deploy the PWAS transducers to *in situ* SHM applications, several hurdles have still to be overcome. In particular, the operational and environmental variations of the monitored structure need to be addressed. In addition, a better understanding of the micromechanical coupling between PWAS and structure for various Lamb wave modes must be achieved. The behavior of the bonding layer between the PWAS and the structure must be clarified as well, such that the predictable and repeatable results are achieved. The durability of this bond under extended environmental exposure must be determined. Last, but not least, the signal analysis methods must be developed to achieve probability of detection values at least comparable with that of conventional NDE methods.

REFERENCES

[1] Giurgiutiu V, Lyshevski SE. *Micro Mechatronics: Modeling, Analysis, and Design with MATLAB*, ISBN 084931593X. CRC Press, 2004.

[2] Rose JL. *Ultrasonic Waves in Solid Media*. Cambridge University Press: Cambridge, 1999.

[3] Giurgiutiu V. Embedded ultrasonics NDE with piezoelectric wafer active sensors. *Journal of Instrumentation, Mesure, Metrologie* 2003 **3**(3–4):149–180.

[4] Raghavan A, Cesnik CES. Modeling of piezoelectric-based Lamb-wave generation and sensing for structural health monitoring. In *Proceedings of SPIE—Smart Structures and Materials 2004: Sensors and Smart Structures Technologies for Civil, Mechanical, and Aerospace Systems*, Liu S-C (ed). SPIE, July 2004; Vol. 5391, pp. 419–430.

[5] Bottai G, Giurgiutiu V. Simulation of the Lamb wave interaction between piezoelectric wafer active sensors and host structure. In *Proceedings of SPIE—Smart Structures and Materials 2005: Sensors and Smart Structures Technologies for Civil, Mechanical, and Aerospace Systems*, Tomizuka M (ed). SPIE, May 2005; Vol. 5765, pp. 259–270.

[6] Staszewski WJ. *Health Monitoring of Aerospace Structures: Smart Sensor Technologies and Signal Processing*. Bobs Books (JRM): Wayne, NJ, 2004.

[7] Chang FK. Built-in damage diagnostics for composite structures. *Proceedings of the 10th International Conference on Composite Structures (ICCM-10)*. Whistler, 14–18 August 1995; Vol. 5, pp. 283–289.

[8] Ihn JB, Chang FK. Multicrack growth monitoring at riveted lap joints using piezoelectric patches. In *Proceedings of SPIE—Smart Nondestructive Evaluation for Health Monitoring of Structural and Biological Systems*, Kundu T (ed). SPIE, June 2002; Vol. 4702, pp. 29–40.

[9] Lin X, Yuan FG. Diagnostic Lamb waves in an integrated piezoelectric sensor/actuator plate: analytical and experimental studies. *Journal of Smart Materials and Structures* 2001 **10**:907–913.

[10] Diamanti K, Soutis C, Hodgkinson JM. Piezoelectric transducer arrangement for the inspection of large

composite structures. *Journal of Composites, Part A* 2007 **38**:1121–1130.

[11] Kim S, Sohn H, Greve D, Oppenheim I. Application of a time reversal process for baseline-free monitoring of a bridge steel girder. *International Workshop on Structural Health Monitoring*. Stanford, CA, 15–17 September 2005.

[12] Xu B, Giurgiutiu V. Lamb waves time-reversal method using frequency tuning technique for structural health monitoring. In *Proceedings of SPIE—Sensors and Smart Structures Technologies for Civil, Mechanical, and Aerospace Systems*, Tomizuka M (ed). SPIE, 2007; Vol. 6529, pp. 65290R1–65290R12.

[13] Giurgiutiu V, Bao J, Zagrai AN. *Structural Health Monitoring System Utilizing Guided Lamb Waves Embedded Ultrasonic Structural Radar*, US Patent, Patent #US6996480B2, February 2006.

[14] Liu W, Giurgiutiu V. Finite element simulation of piezoelectric wafer active sensors for structural health monitoring with couple-field elements. In *Proceedings of SPIE—Sensors and Smart Structures Technologies for Civil, Mechanical, and Aerospace Systems*, Tomizuka M (ed). SPIE, 2007; Vol. 6529, pp. 65293R1–65293R13.

[15] Liang C, Sun FP, Rogers CA. Coupled electro-mechanical analysis of adaptive material system-determination of the actuator power consumption and system energy transfer. *Journal of Intelligent Material Systems and Structures* 1994 **5**:12–20.

[16] Sun FP, Liang C, Rogers CA. Experimental modal testing using piezoceramic patches as collocated sensors-actuators. *Proceedings of the 1994 SEM Spring Conference and Exhibits*. Baltimore, MD, 6–8 June 1994.

[17] Chaudhry Z, Joseph T, Sun FP, Rogers CA. Local area health monitoring of aircraft via piezoelectric actuator/sensor patches. *Proceedings of the SPIE North American Conference on Smart Structures and Materials*. Orlando, FL, February 26–March 3 1995.

[18] Bender JW, Friedman HI, Giurgiutiu V, Watson C, Fitzmaurice M. The use of biomedical sensors to monitor capsule formation around soft tissue implants. *Annals of Plastic Surgery* 2006 **56**(1):72–77.

[19] Zhang Y, Li X. Piezoelectric paint sensor for ultrasonic NDE. In *Proceedings of SPIE—Sensors and Smart Structures Technologies for Civil, Mechanical, and Aerospace Systems*, Tomizuka M (ed). SPIE, 2007; Vol. 6529, pp. 652904-1–652904-12.

[20] Lin B, Giurgiutiu V, Yuan Z, Liu J, Chen C, Bhalla AS, Guo R. Ferroelectric thin-film active sensors for structural health monitoring. In *Proceedings of SPIE—Sensors and Smart Structures Technologies for Civil, Mechanical, and Aerospace Systems*, Tomizuka M (ed). SPIE, 2007; Vol. 6529, pp. 65290I1–65290I8.

[21] Hudak SJ, Lanning BR, Light GM. A thin-film sensor for monitoring materials damage at elevated temperature. *Proceedings of AeroMat 2005*. Orlando, FL, June 2005.

Chapter 56
Damage Presence/Growth Monitoring Sensors

Hua Gu[1] and Ming L. Wang[2]

[1] Advanced Structures Group, Caterpillar Production System Division, Caterpillar Inc., Peoria, IL, USA
[2] Department of Civil and Materials Engineering, University of Illinois, Chicago, IL, USA

1 Introduction	1029
2 Design and Fabrication of a PVDF IDT	1030
3 Experimental Setup	1031
4 Results and Discussion	1033
5 Conclusions	1033
Acknowledgments	1033
References	1034

1 INTRODUCTION

Most damage detection methods are based on the fact that damage will change the stiffness, mass, or energy dissipation properties of a system. As a result, the measured dynamic response will change as well. Most global damage detection methodologies suffer from lack of sensitivity. Since damage is a local phenomenon, it may not tremendously influence the global response of a structure, which usually deals with lower frequencies measured during vibration tests. Other environmental and operational variations, such as varying temperature, moisture, loading conditions, and the complexity of the structure itself, make damage detection more challenging [1].

The employment of Lamb waves in nondestructive testing has attracted more and more attention in structural health monitoring research community. Lamb waves are able to propagate over a long distance with very little amplitude loss. If a receiving transducer is positioned at a remote location on the structure, the received signal contains information about the integrity of the line between the transmitting and receiving transducers. The test thereby monitors a line rather than a point and so considerable testing time may potentially be saved [2, 3].

Perhaps the most important work in Lamb wave inspection is conducted by three groups of researchers—Peter Cawley's group from the Imperial College, London, Joseph Rose's group from the Pennsylvania State University, and Victor Giurgiutiu's group from the University of South Carolina. Work from these researchers can be found in numerous literatures [2–34].

The excitation and reception of Lamb waves in structures can be carried out by using interdigitated transducers (IDTs). These transducers have a comb structure composed of sets of periodically distributed fingers with a spatial period equal to the excited

Encyclopedia of Structural Health Monitoring. Edited by Christian Boller, Fu-Kuo Chang and Yozo Fujino © 2009 John Wiley & Sons, Ltd. ISBN: 978-0-470-05822-0.

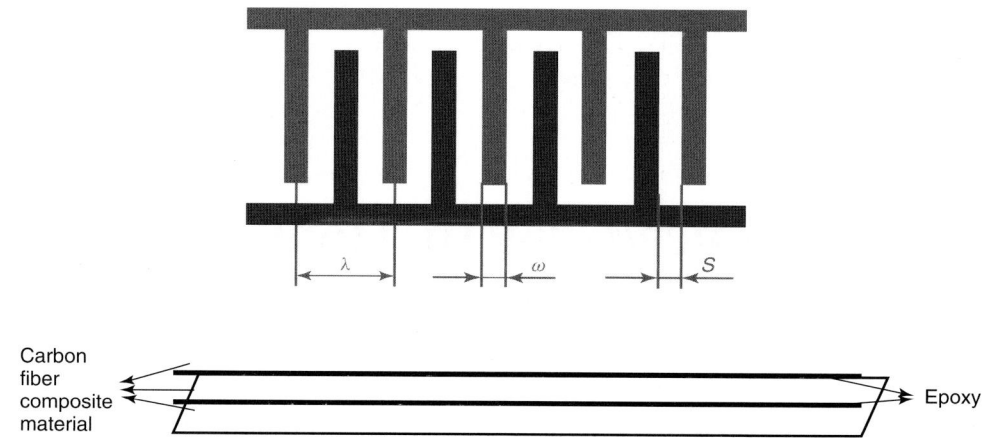

Figure 1. Schematic of a PVDF IDT.

Lamb wavelength [12, 35]. It is a common application to have an interdigital electrode on the surface of a piezoelectric material form an IDT. Owing to its broad bandwidth and low unit price, polyvinyldine fluoride (PVDF) film has envisioned itself to be a suitable substrate for surface and guided-wave transducers that are required to couple electrical energy with the piezoelectric material to detect the wave. The advantages of using PVDF IDTs are as follows. Firstly, PVDF IDTs are low profile and unobtrusive compared with other types of Lamb wave devices. Secondly, PVDF-based IDTs are flexible, so that they can be used on convex or concave surfaces such as pipes and pressure vessels [7].

Lamb wave signals are usually complex and nonstationary. At least two Lamb wave modes can propagate at a given frequency. Conventional fast Fourier transform (FFT) is unable to reveal the nature of Lamb wave signals containing frequency information that varies with time. To solve this problem, time–frequency representations (TFRs) (*see* **Chapter 26**) are commonly employed for signal analysis in Lamb wave applications. The continuous wavelet transform (CWT) is used as a TFR in this work.

2 DESIGN AND FABRICATION OF A PVDF IDT

A successfully designed PVDF IDT will have a finger pattern sitting on top of a PVDF film as indicated in Figure 1. λ, w, and s represent wavelength, finger width, and finger spacing, respectively. These parameters along with the length of the transducer are the criteria that need to be considered during a design process.

It has been proved that as long as $50\% \leq \frac{w}{w+s} \leq 90\%$ holds true, there is little difference among different finger widths [19]. As for the length of the transducer, a single pair of electrodes/fingers cannot excite surface acoustic waves very efficiently. Therefore, several pairs of electrodes/fingers are commonly seen in an interdigital transducer. A long transducer with many pairs of electrodes/fingers tends to be efficient for exciting and receiving signals only over a narrow frequency range. Consequently, it is also easier to generate a certain wave mode. However, it is practically not possible to build a sensor with infinite length. Besides, the more finger pairs the sensor has, the narrower the bandwidth of the sensor becomes. It is undesirable to design a sensor with very narrow bandwidth that limits its applications. Therefore, it is common to have IDTs with five or six pairs of electrodes/fingers.

The wave length λ is determined depending on the dispersion curves of the materials to be tested. Figure 2 displays the phase velocity dispersion curves of the tested structure used in this study. It is a laminated structure composed of 3 layers of carbon fiber reinforced polymer (CFRP) material. As the dispersion curves have explained themselves, the wave propagation in this structure, composed of anisotropic material, is much less dispersive than that

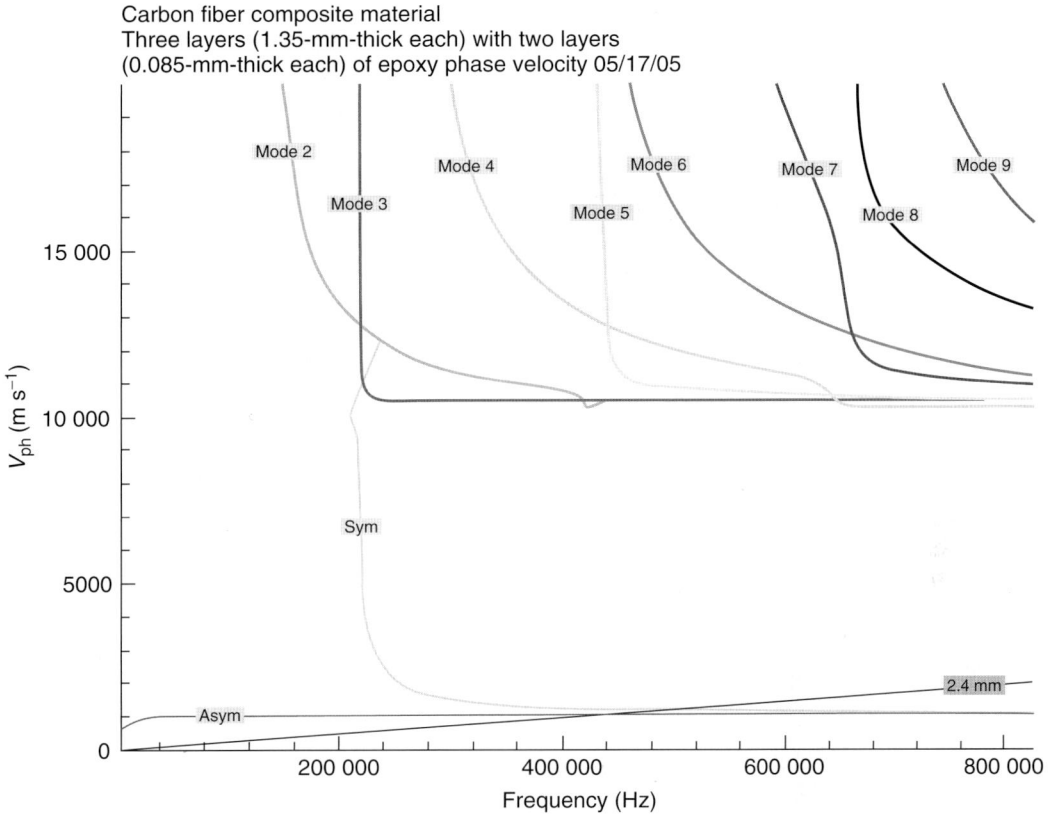

Figure 2. Phase velocity of laminated CFRP plate.

in an isotropic material, such as steel. Since the wave length of IDT is determined by $\lambda = \frac{c_{ph}}{f}$, a straight design line starting from the origin of phase velocity diagram is drawn to help the design process. The slope of the line represents a certain wavelength that if intersects with a dispersion mode will excite that mode at the corresponding frequency. To simplify the wave signal that is to be received, a sensor with a wavelength of 2.4 mm is designed, which ideally only generates symmetric and antisymmetric wave modes in a large frequency band.

The fabrication of PVDF IDTs is realized by using the etch-back photolithography technology adopted from semiconductor industry, which is carried out in a photolithography bay inside a clean room. PVDF films with gold coating on one of the surfaces are acquired from manufacturer. The reason for choosing gold is twofold. Firstly, backing with a high-density material such as gold enhances the performance of PVDF film at low frequencies. Secondly, gold has superb conductivity. After the surface is cleaned, a thin layer of photoresist is evenly spun on top of the gold coating. The finger pattern design is carefully printed out on transparencies and transferred to the photoresist layer through exposing. After being developed, the pattern then stays on the film. The finger pattern is revealed by etching out gold not covered by the resist and the transducer is finalized by removing the photoresist.

3 EXPERIMENTAL SETUP

The test sample structure is 915 mm long, 51 mm wide, and 4.22 mm thick. Through voids of different sizes have been introduced to the middle layer of the laminated structures. They are 0, 1, and 5 mm wide. Figure 3 shows the schematic of the experimental setup. A tone burst sinusoidal signal with a Hanning

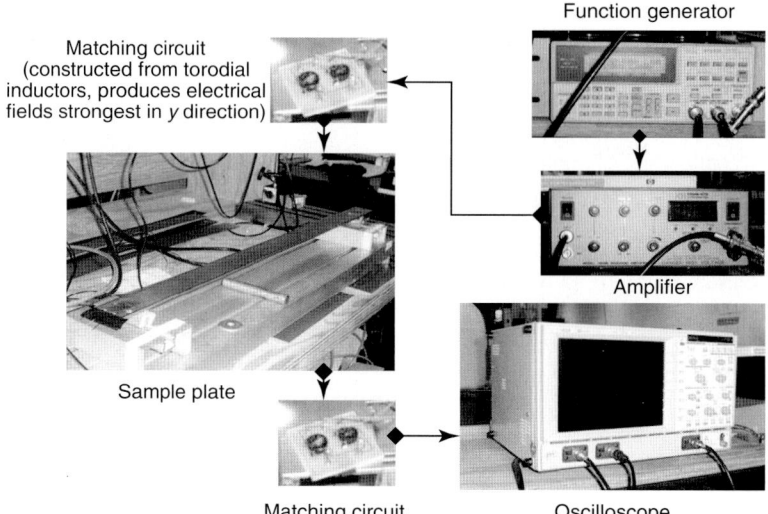

Figure 3. Schematic of the experimental setup.

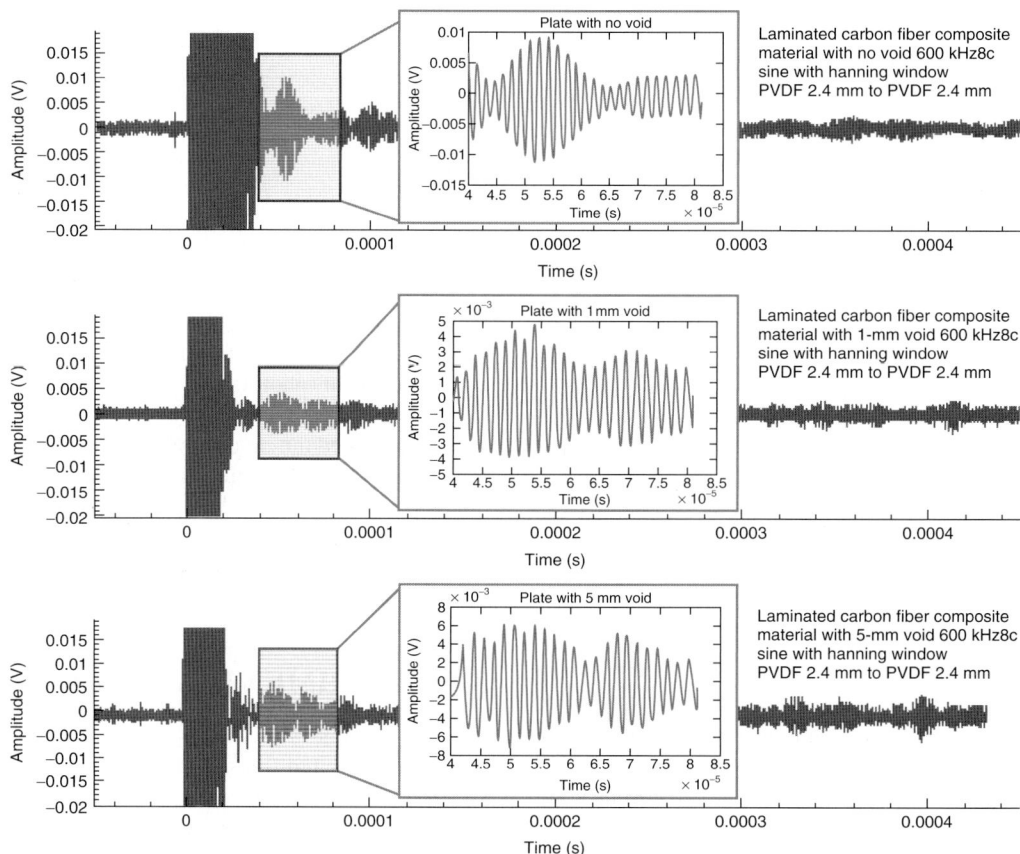

Figure 4. Response from PVDF IDTs on CFRP plates with different void sizes.

window is sent to the function generator as the excitation signal. The signal is then amplified by a high-voltage amplifier before passing through a custom-designed circuit. This circuit applies antiphase voltage on these two groups of fingers of PVDF IDT. As a result, these fingers move in opposite directions, and Lamb waves are generated. After propagating on the plate, wave signals are acquired by another piece of PVDF IDT functioning as a receiver. Another customized matching circuit similar as before is used and converts antiphase voltage into a single-phase voltage before the wave signals are eventually sent to the oscilloscope.

4 RESULTS AND DISCUSSION

Signals received from PVDF IDTs on CFRP plates with no void, 1-mm void, and 5-mm void are plotted in Figure 4. Part of each signal has been extracted and downsampled to reduce the amount of data the computer has to deal with as well as de-noise the signals. Conventional FFT is unable to reveal the nature of signals containing frequency information that varies with time, such as Lamb wave signals. TFRs are usually used for Lamb wave signal processing, among which the CWT is one of the most widely used TFRs for Lamb wave signals. CWT uses a window function called *mother wavelet* to chop the original signal into small sections (*see* **Chapter 27**). Wavelet transform is then applied on each section. The CWT has been performed on the downsampled signals as illustrated in Figure 5.

The results in Figure 5 reveal the fact that if no voids exist in a plate, all the wave modes are integrated in a single pack; once voids appear, wave modes tend to separate from each other; and they get more separated as the size of the void increases.

5 CONCLUSIONS

A PVDF IDT has been built and used for damage detection in structural health monitoring by generating and receiving Lamb waves. The fabrication of the sensor is achieved on the basis of the etch-back photolithography technology, which gives the sensor a monolithic structure. Results have shown the ability of this PVDF IDT to detect the existence and the severity of damage with the assistance from advanced digital signal-processing methods such as CWT.

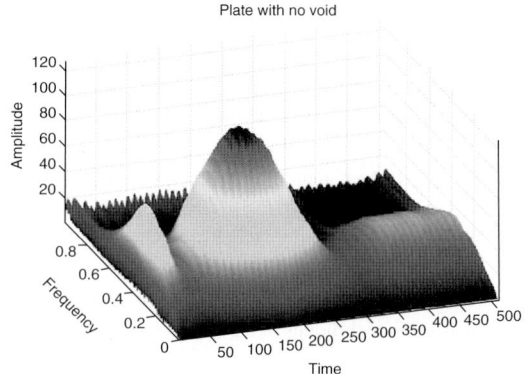
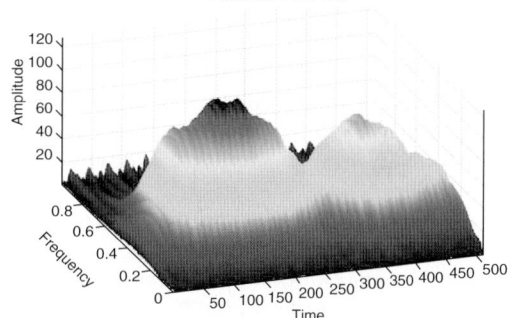
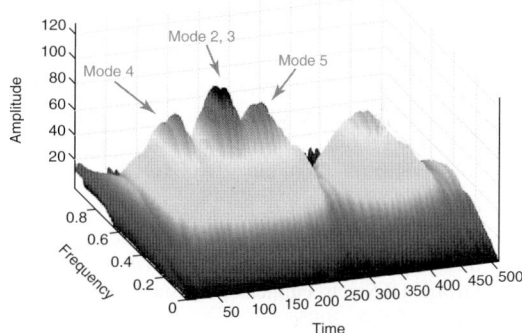

Figure 5. CWT of signals from plates with different void sizes.

ACKNOWLEDGMENTS

This research is supported by the National Science Foundation under grants CMS-0220027. This support is gratefully acknowledged.

REFERENCES

[1] Sohn H, Farrar CR, Hemez FM, Shunk DD, Stinemates DW, Nadler BR. *A Review of Structural Health Monitoring Literature: 1996–2001.* Los Alamos National Laboratory, 2003.

[2] Alleyne DN, Cawley P. The interaction of Lamb waves with defects. *IEEE Transactions on Ultrasonics, Ferroelectrics, and Frequency Control* 1992 **39**(3):381–397.

[3] Alleyne DN, Cawley P. Optimization of Lamb wave inspection techniques. *NDT and E International* 1992 **25**(1):11–22.

[4] Dalton RP, Cawley P, Lowe MJS. The potential of guided waves for monitoring large areas of metallic aircraft fuselage structure. *Journal of Nondestructive Evaluation* 2001 **20**(1):29–46.

[5] Monkhouse RSC, Wilcox PD, Cawley P. Flexible interdigital PVDF transducers for the generation of Lamb waves in structures. *Ultrasonics* 1997 **35**(7):489–498.

[6] Monkhouse RSC, Wilcox PD, Lowe MJS, Dalton RP, Cawley P. The rapid monitoring of structures using interdigital Lamb wave transducers. *Smart Materials and Structures* 2000 **9**(3):304–309.

[7] Wilcox PD, Cawley P, Lowe MJS. Acoustic fields from PVDF interdigital transducers. *IEE Proceedings—Science, Measurement and Technology* 1998 **145**(5):250–259.

[8] Wilcox PD, Monkhouse RSC, Cawley P, Lowe MJS, Auld BA. Development of a computer model for an ultrasonic polymer film transducer system. *NDT and E International* 1998 **31**(1):51–64.

[9] Cawley P, Simonetti F. Structural health monitoring using guided waves-potential and challenges. *The 5th International Workshop on Structural Health Monitoring.* DEStech Publications, Stanford University: Stanford, CA, 2005.

[10] Pilarski A, Rose JL. A transverse-wave ultrasonic oblique-incidence technique for interfacial weakness detection in adhesive bonds. *Journal of Applied Physics* 1988 **63**(2):300–307.

[11] Rose JL, Nestleroth JB, Balasubramaniam K. Utility of feature mapping in ultrasonic non-destructive evaluation. *Ultrasonics* 1988 **26**(3):124–131.

[12] Giurgiutiu V, Zagrai A. Damage detection in thin plate and aerospace structures with the electromechanical impedance method. *Structural Health Monitoring—An International Journal* 2005 **4**(2):99–118.

[13] Ditri JJ, Rose JL. Excitation of guided elastic wave modes in hollow cylinders by applied surface tractions. *Journal of Applied Physics* 1992 **72**(7):2589–2597.

[14] Younho C, Hongerholt DD, Rose JL. Lamb wave scattering analysis for reflector characterization. *IEEE Transactions on Ultrasonics, Ferroelectrics, and Frequency Control* 1997 **44**(1):44–52.

[15] Shin HJ, Rose JL. Guided wave tuning principles for defect detection in tubing. *Journal of Nondestructive Evaluation* 1998 **17**(1):27–36.

[16] Rose JL, Avioli MJ, Mudge P, Sanderson E. Guided wave inspection potential of defects in rail. *NDT and E International* 2004 **37**(2):153–161.

[17] Hay TR, Rose JL. Fouling detection in the food industry using ultrasonic guided waves. *Food Control* 2003 **14**(7):481–488.

[18] Rose JL. Guided wave nuances for ultrasonic nondestructive evaluation. *IEEE Transactions on Ultrasonics, Ferroelectrics, and Frequency Control* 2000 **47**(3):575–583.

[19] Rose JL, Pelts SP, Quarry MJ. A comb transducer model for guided wave NDE. *Ultrasonics* 1998 **36**(1–5):163–169.

[20] Hay TR, Rose JL. Flexible PVDF comb transducers for excitation of axisymmetric guided waves in pipe. *Sensors and Actuators, A* 2002 **100**:18–23.

[21] Li J, Rose JL. Implementing guided wave mode control by use of a phased transducer array. *IEEE Transactions on Ultrasonics, Ferroelectrics, and Frequency Control* 2001 **48**(3):761–768.

[22] Rose JL. A baseline and vision of ultrasonic guided wave inspection potential. *Journal of Pressure Vessel Technology* 2002 **124**(3):273–282.

[23] Giurgiutiu V, Zagrai AN. Characterization of piezoelectric wafer active sensors. *Journal of Intelligent Material Systems and Structures* 2000 **11**:959–975.

[24] Giurgiutiu V, Zagrai AN. Embedded self-sensing piezoelectric active sensors for online structural identification. *ASME Journal of Vibration and Acoustics* 2002 **124**(1):116–125.

[25] Giurgiutiu V. Review of smart-materials actuation solutions for aeroelastic and vibration control. *Journal of Intelligent Material Systems and Structures* 2000 **11**:525–544.

[26] Giurgiutiu V, Reynolds A, Rogers CA. Experimental investigation of e/m impedance health monitoring for spot-welded structural joints. *Journal of Intelligent Material Systems and Structures* 1999 **10**:802–812.

[27] Zagrai AN, Giurgiutiu V. Electro-mechanical impedance method for crack detection in thin plates. *Journal of Intelligent Material Systems and Structures* 2001 **12**:709–718.

[28] Giurgiutiu V, Zagrai A, Bao JJ. Piezoelectric wafer embedded active sensors for aging aircraft structural health monitoring. *An International Journal of Structural Health Monitoring* 2002 **1**(1):41–61.

[29] Giurgiutiu V, Zagrai A, Bao JJ. Embedded active sensors for in-situ structural health monitoring of thin-wall structures. *ASME Journal of Pressure Vessel Technology* 2002 **124**(3):293–302.

[30] Giurgiutiu V. Embedded NDE with piezoelectric wafer active sensors in aerospace applications. *Journal of Materials* 2003; Special issue on NDE, http://www.tms.org/pubs/journals/JOM/0301/Giurgiutiu/Giurgiutiu-0301.html.

[31] Giurgiutiu V. Embedded ultrasonics NDE with piezoelectric wafer active sensors. *Journal Instrumentation* 2003 **3**(3–4):149–180.

[32] Giurgiutiu V. Tuned Lamb wave excitation and detection with piezoelectric wafer active sensors for structural health monitoring. *Journal of Intelligent Material Systems and Structures* 2005 **16**(4):291–306.

[33] Yu L, Giurgiutiu V. Advanced signal processing for enhanced damage detection with embedded ultrasonics structural radar using piezoelectric wafer active sensors. *Smart Structures and Systems—An International Journal of Mechatronics, Sensors, Monitoring, Control, Diagnosis, and Maintenance* 2005 **1**(2):185–215.

[34] Giurgiutiu V, Cuc A. Embedded NDE for structural health monitoring, damage detection, and failure prevention. *Shock and Vibration Reviews* 2005 **37**(2):83–105.

[35] Viktorov IA. *Rayleigh and Lamb Waves—Physical Theory and Applications*. Plenum Press: New York, 1967.

Chapter 57
Piezoelectric Paint Sensors for Ultrasonics-based Damage Detection

Yunfeng Zhang
Department of Civil and Environmental Engineering, University of Maryland, College Park, MD, USA

1 Introduction	1037
2 Composition and Fabrication of Piezoelectric Paint	1038
3 Review of Piezoelectric Paint Sensor Development	1041
4 Ultrasonic Sensing Characteristics of Piezoelectric Paint	1043
5 Conclusions	1046
References	1047

1 INTRODUCTION

Structural defect inspection and corresponding retrofit actions lead to a prolonged life and enhanced reliability of structural systems. However, in practice, manual inspection is still the most commonly used approach; this can be quite costly and time consuming, and therefore unsuitable for rapid assessment of structural conditions. It is thus desirable to replace manual inspection with on-line structural defect detection techniques. For example,

Encyclopedia of Structural Health Monitoring. Edited by Christian Boller, Fu-Kuo Chang and Yozo Fujino © 2009 John Wiley & Sons, Ltd. ISBN: 978-0-470-05822-0.

existing nondestructive evaluation (NDE) techniques for fatigue crack detection include dye penetrants, eddy current, radiography, acoustic emission (AE), magnetic particle inspection, ultrasonic testing, etc. All these NDE techniques have limited use for on-line fatigue crack monitoring due to either one or a combination of the following problems: accessibility, automation, bulky size, power supply, environmental noise, and long-term durability. In particular, most NDE instruments give results that need to be interpreted by a skilled operator, with a potential for subjective mistakes.

In the past decade or so, sensors made of piezoelectric materials have gained increasing popularity in the field of structural health monitoring (e.g. [1, 2]; see **Chapter 156**). For example, recent research works by Ayres *et al.* [3], Park *et al.* [4], Giurgiutiu [5], and Yu [6] (*see* **Chapter 55**) on piezoelectric active wafer sensor have made it quite a promising technique for NDE applications. Piezoelectric materials can be broadly classified into three major categories [7]: ferroelectric ceramics, piezoelectric polymers (e.g., polyvinylidene fluoride, PVDF), and piezoelectric composites. Because of their electromechanical coupling property, piezoelectric materials have been widely used for sensing and actuation applications [8–11]. Piezoelectric ceramics are perhaps the most popular piezoelectric materials for sensing purpose. As of today, most piezoelectric sensors are made of piezoelectric ceramics like lead zirconate titanate (PZT). However, limitations of the mechanical properties of PZT such as brittleness and cracking at large deformations could impede its use in certain applications. Additionally, prefabricated PZT wafers do not fit well on surfaces with complex geometry such as welded joints.

To overcome these problems associated with conventional piezoelectric materials such as PZT, polymer-based piezoelectric paints have been investigated by a few researchers as a potential substitute for piezoelectric ceramics in certain sensing applications [12–22]. Piezoelectric paint typically consists of tiny piezoelectric particles randomly dispersed within a polymer matrix and therefore belongs to the 0–3 piezoelectric composite family. Table 1 summarizes the composition and piezoelectric activity of the piezoelectric paint formulations developed by different researchers. Like piezoelectric ceramics, piezoelectric paint can be used for ultrasonic NDE. Piezoelectric paint that can be directly deposited onto the surface of host structures has several advantages over conventional piezoelectric ceramics. This article reviews the past and current research work in the area of piezoelectric paint as well as its potential use as sensing material for ultrasonic NDE, particularly distributed AE sensor or ultrasonic guided wave–based embedded sensor in metal or fiber reinforced polymer (FRP) structures. Although a comprehensive experimental program still needs to be carried out to fully investigate the potential of piezoelectric paint sensor technology, it can be concluded on the basis of preliminary experimental results that piezoelectric paint sensor appears to provide a promising inexpensive NDE technique for ultrasonic-based monitoring of structural defects such as fatigue cracks in structures.

2 COMPOSITION AND FABRICATION OF PIEZOELECTRIC PAINT

Piezoelectric composite materials consisting of ferroelectric ceramics and polymer [26–31] have received considerable interest as sensing elements because of their favorable material properties that often cannot be obtained in single-phase materials. Through judicious selection of the polymer matrix, the composite properties of the piezoelectric paint can be tailored to meet the specific requirements of application conditions. For example, piezoelectric paint might be more suitable for use in fiber reinforced polymer composite structures because of its improved acoustic impedance matching property compared to piezoelectric ceramics.

The arrangement of the component phases within a composite is critical for the electromechanical properties of composites [22]. Newnham et al. [32] have developed the concept of connectivity to describe the manner in which the individual phases are self-connected. In a diphasic system, there are 10 types of connectivities in which each phase is continuous in zero, one, two, or three dimensions. The 10 connectivities are denoted as the following: 0–0, 0–1, 0–2, 0–3, 1–1, 1–2, 2–2, 1–3, 2–3, and 3–3. It is conventional for the first digit to refer to the piezoelectrically-active phase. Piezoelectric paint typically consists of tiny piezoelectric particles mixed within polymer matrix and therefore belongs to the "0–3" piezoelectric composite. The "0–3" means that the piezoelectrically active ceramic particles are randomly dispersed in a three dimensionally connected polymer matrix. Conceivably, the advantages of "0–3" piezoelectric composites over other connectivity types are their ease of fabrication into complex shapes including large flexible thin sheets, extruded bars and fibers, and molded shapes; they may conform to any curved surface. In general, the 0–3 composite family has been found to exhibit high hydrostatic piezoelectric voltage coefficients and "figure of merit" when compared to the properties of conventional single-phase materials [12].

Piezoelectric paint is comprised of three major components: piezoelectric ceramics powder with average particle size in the range of $3-20\,\mu m$, a polymer binder to carry the powder in suspension during application and bind it together on curing, and additives including defoamer, dispersants, and surfactants to enhance the paint mixing, deposition, and curing properties. The properties of 0–3 composites are strongly dependent on both the piezoelectric and polymer phases utilized, as well as the fabrication method employed [26]. Among all piezoelectric materials, PZT, $BaTiO_3$, $PbTiO_3$, $LiTaO_3$ ceramics,

Table 1. Summary of piezoelectric paint properties in prior research (number in bracket indicates volume fraction of the respective constituent phases)

Researchers	Ingredients		Fabrication method		Piezoelectric activity
	Ceramics	Polymer	Application/cuing	Poling	d_{33} (pCN^{-1})
Hanner et al. [12]	PZT/PbTiO$_3$ (60–70%)	Acrylic copolymer in suspension form or polyurethane	Spreading/24-h drying in air +24-h drying in vacuum oven at 110 °C	Oil bath poling at 100 °C with 10–15 MV m^{-1} or corona poling	26–39
Egusa and Iwasawa [13]	PZT (53%)	Epoxy	Screen printing/3-day drying in air	Conventional poling at 20 MV m^{-1} for 30 min	Unknown
Wenger et al. [14]	PTCa (65 or 60%)	P(VDF-TrFE) or epoxy	Hot-rolling/unknown curing method	Conventional poling at 10–25 MV m^{-1} for 30 min at 100 °C	33 or 30
Badcock and Birt [17]	PZT-5H (60%)	Epoxy	Spreading/unknown curing method	Oil bath poling at 19 MV m^{-1} for 5 min at 60 °C	16.5
Sakamoto et al. [18]	PZT (59%)	Polyurethane	Pressed at 20 MPa into thick film	Conventional poling at 3 MV m^{-1} for 1 h at 100 °C	12
Li et al. [19]	PZT (35–70%)	Cement	Spread into 3-mm-thick film/cured in concrete curing room (23 °C, 100% humidity) for 26 days	Oil bath poling at 20 MV m^{-1} for 1 h at 150 °C	7–33
Hale et al. [23]	PZT (40%)	Acrylic paint	Spray painting/unknown curing method	Conventional poling at 60 °C up to 1 h	20
Zhang and Li [24]	PZT (30%)	Epoxy	Spreading/cured at elevated temperature 50 °C for 3 days	Conventional poling (5 MV m^{-1}) at 70 °C for 30 min	5
Kobayashi et al. [25]	PZT	Unknown sol–gel solution	Spray/drying and firing by heat gun	Corona poling at 120 °C for 10 min	30
Lahtinen et al. [22]	PZT (60%)	Unknown sol–gel solution	Spreading with paint applicator/unknown curing method	Conventional poling (12.5 MV m^{-1}) at room temperature for 15 min	Unknown

and the mentioned ceramics with different dopings have been used for 0–3 composites, while PZT is by far the most popular active ingredient for piezoelectric paints. Marin-Franch et al. [33] used calcium modified lead titanate/poly(ether ketone ketone) (PTCa/PEKK) composites to detect AE when mounted on carbon fiber reinforced composite panels. PTCa ceramic is selected because it has the appropriate piezoelectric properties together with a low Curie temperature 260 °C, so it can be readily poled at temperatures appropriate to the polymer.

Han et al. [34] developed a colloid processing method to improve the microstructural homogeneity and decrease the chance of void formation in 0–3 composites. With this technique, piezoceramic powder was dispersed in a dilute polymer solution, allowing a polymer coating to be absorbed onto the powder surface. Colloidally processed composites composed of coprecipitated PT-BF powder and Eccogel polymer were measured to have the largest d_{33} (65 pC N^{-1}) and highest figure of merit ($dg = 6000 \times 10^{-15}$ m^2 N^{-1}) of all "true" 0–3 composites [26]. Han et al. also studied the effect of the particle size on dielectric and piezoelectric properties of 0–3 composites. Lau et al. [35] studied incorporation of nanosized PT powder into a poly(vinylidene fluoride - trifluoroethylene) [P(VDF-TrFE) 70/30 mol%] matrix to form a 0–3 nanocomposite with a 0.2 volume fraction of PT. A thin film of 5-μm thickness was prepared by spin coating the composite on a glass substrate. The transducer was 1.2 mm in diameter with a center frequency of 40 MHz.

The polymer matrix material is critical to the ease of fabrication and performance characteristics of piezoelectric paint sensor. For example, to investigate the effect of the polymer on resistivity and dielectric properties of 0–3 composites, Han et al. [34] prepared PT composites with Eccogel polymer, PVDF copolymer, and ethylene-propylenediene monomer (EPDM) polymer. It was found that although higher poling conditions could be applied to the PVDF copolymer and EPDM composites, the highest d_{33} value was obtained from the epoxy composites. The higher electrical conductivity of the polymer matrix may have created more electric flux paths between the ceramic particles. This in turn increased the electric field acting on the ceramic filler and made poling of the ceramic easier [26]. The epoxy gave the highest figure of merit ($d_h g_h = 5600 \times 10^{-15}$ m^2 N^{-1}) and the EPDM gave the lowest figure of merit (600×10^{-15} m^2 N^{-1}).

The poling of ferroelectric materials to induce piezoelectricity is an important stage in the fabrication of piezoelectric paint. Poling can be carried out by two rather different techniques—the conventional dc poling method and the Corona discharge method. Poling of composites having a polymer matrix with 0–3 connectivity is especially difficult because the electric field within the high-dielectric-constant grains is far smaller than in the low-dielectric-constant polymer matrix [36]. Since most polymers have a lower dielectric constant compared with piezoelectric ceramic materials, most of the applied electric field will pass through the lower dielectric constant phase. Moreover, local breakdown at weak spots short-circuits the electrodes and prevents further poling since very large electric fields are required to pole these types of composites. One way to resolve this difficulty with poling is to introduce a third conductive phase between the piezoelectric particles. Sa-Gong et al. [37] prepared such composites by adding carbon, germanium, or silicon to PZT. Sakamoto et al. [18] doped 0–3 piezoelectric composites of PZT powder and vegetable-based polyurethane (PU) with small amounts of carbon powder (vol% in the range of 0.5–2.0%) to ease the poling by creating a continuous electric flux path between PZT grains.

The corona poling technique has been proposed by Waller and Safari [36] to pole flexible piezoelectric composites such as piezoelectric paints. This technique has been successful in poling PVDF films. In the Corona poling method, electric charge from a corona point is sprayed onto the unelectroded surface of the sample, creating an electric field between the sample faces. Because of the absence of electrodes, there is no short-circuiting of the sample at weak spots. This is particularly beneficial for poling of piezoelectric paint sensors because short-circuited piezoelectric paint sensors have to be manually removed from the host structure after poling. Generally speaking, heating lowers the coercive field and makes poling easier. The piezoelectric properties of ceramics and composites poled by the Corona method are comparable or better than those poled by the conventional poling technique.

3 REVIEW OF PIEZOELECTRIC PAINT SENSOR DEVELOPMENT

In their early studies of thin-film 0–3 polymer/piezoelectric ceramic composites, Hanner *et al.* [12] prepared two formulations of piezoelectric paints—one based on an acrylic copolymer and the other based on a PU and the electrical properties of these piezoelectric paints were characterized. Both formulations were loaded with 60–70% volume fraction of PZT and a coprecipitated $PbTiO_3$. The paint samples, with a film thickness ranging between 200 and 500 μm, were placed in a vacuum oven at 110 °C for 24 h to remove residual water or solvent. Poling of the paints was accomplished by both the conventional oil bath and the corona discharge techniques. The values of the piezoelectric charge constant d_{33} for the PZT-filled composites were found to be 25–28 pC N^{-1}, while those for the coprecipitated $PbTiO_3$/acrylic copolymer composites were 35–38 pC N^{-1}.

Egusa and Iwasawa [13] prepared an epoxy resin-based piezoelectric paint with 53 vol% PZT powder as filler. The paints were spread onto one side of a 30-mm-wide aluminum beam. The 152-μm-thick paint film was then cured in air at room temperature for at least three days or at 150 °C for 45 min. The effects of poling field, film thickness, and cure temperature on the paint's piezoelectric activity were investigated [38]. The piezoelectric sensitivity of the paint film as an AE sensor was also evaluated in their study [39]. A nearly flat frequency response of the paint film to AE waves was observed in the frequency range above 0.3 MHz.

Hale and Tuck [15], White *et al.* [20], and Hale *et al.* [23] fabricated and tested a piezoelectric paint-based strain sensor for use in structural vibration monitoring. The paint formulation comprised of milled PZT ceramic powder mixed in a water-based acrylic paint, based with PZT powder concentrations of up to 80% by weight (approximately 40% by volume). The cured paint was poled at 600 V for 1 h. For trouble-free spraying and high-quality paint films, it was found necessary to perform high-shear mixing for half an hour prior to coating application. It was found from the tests by White *et al.* [20] that the most successful spray system was a miniature DeVilbiss 0.8-mm air-atomizing spray gun operating at approximately 1.7 bar using a gravity feed paint cup. The piezoelectric charge coefficient d_{33} was found to vary between 10 and 30 pC N^{-1} with a predominance around 20 pC N^{-1} [23].

Paints belonging to the class of 0–3 piezoelectric composites, consisting of a ferroelectric ceramic powder of calcium modified PTCa dispersed in a polymer matrix, have been fabricated and their ferroelectric properties have been investigated by Wenger *et al.* [14]. Two formulations of 0–3 composites, one with a copolymer of P(VDF-TrFE) and the other with a thermosetting epoxy resin, were studied. The composite of the ceramic with the epoxy, 60/40 vol% PTCa/epoxy, was prepared by gradually adding the ceramic powder to the resin while stirring continuously to ensure an even mixture [40]. The ceramic/copolymer composite PTCa-P(VDF-TrFE) with a 65/35 volume fraction was prepared by a hot-rolling technique. The composite films were subsequently poled in a dc field of $1–2.5 \times 10^7$ V m^{-1} for 30 min at 100 °C in an insulating silicone oil bath. The d_{33} coefficients for the composites of PTCa/P(VDF-TrFE) and PTCa/epoxy are 33 and 26 pC N^{-1} respectively. Surface-mounted AE sensors were fabricated using the composite films thus obtained and their frequency response was evaluated using a face-to-face technique over the frequency range of 300 kHz to 50 MHz. Embedded transducers, constructed from the piezoelectric composite films were used to detect plate waves within a laminate glass/epoxy plate measuring 304.8 mm × 304.8 mm × 1.9 mm. It was found in their study [14] that, in the case of the embedded transducers, the PTCa/epoxy films seem to be the better choice of transducer material, producing signals comparable in amplitude to those of an embedded PTCa/P(VDF-TrFE) film but more clearly defined.

Sakamoto *et al.* [18] studied a flexible piezoelectric composite with 0–3 connectivity, made from PZT power and castor oil-based PU, which was doped with a small amount of fine-grained carbon powder to facilitate poling at relatively low electric field in a short time span. The carbon particles located between the PZT particles help create a continuous electric flux path. The PZT powder was mixed with carbon by a vibrator for 30 min prior to introducing this mixture in the PU matrix. The composite was placed between two paraffin papers and pressed at room temperature. The applied pressure was about 20 MPa and it was possible to obtain samples in the thickness range of 250–350 μm. All the samples were poled with a 3 MV m^{-1} electric field at 100 °C for 1 h. After poling, the d_{33} coefficient was measured and it was observed that the highest value of the d_{33} coefficient (around 12 pC N^{-1}) was achieved with PZT/C/PU samples with 59/1/40 vol% composition. On adding 1.0 vol% of carbon, the d_{33} coefficient increased by 25% in comparison with the composite without the semiconductor phase. The disadvantage associated with carbon doping appears to be the relatively low electrical breakdown voltage of about 6 MV m^{-1}. The PZT/C/PU composite has shown the ability to detect both extensional and flexural modes of simulated AE at a distance of up to 8.0 m from the source [41].

Badcock and Birt [17] have examined the use of piezoelectric 0–3 composite as embedded Lamb wave sensors for damage detection in carbon fiber reinforced composite plates (E-glass/913 and T800/924 aerospace composites). The 0–3 composite consists of PZT-5H powder dispersed in an epoxy matrix. Cured PZT/epoxy materials with a PZT volume fraction of 60% was immersed in an oil bath for poling at a predefined temperature in the range of 50–120 °C. It was found that the greatest poling efficiency is achieved at 60 °C (most samples were poled at a field of 19 kV mm^{-1} for 5 min), with sensitivities of around 17 pC N^{-1} for the piezoelectric charge constant d_{33}. Comparative study of three different transducers—a conventional PZT ultrasonic transducer, a 500-μm-thick PVDF element, and a 600-μm-thick piezoelectric 0–3 composite film—were carried out to receive the Lamb wave signal at a fixed distance from the transmitting transducer. It was found the piezoelectric 0–3 composites were twice as sensitive when compared with PVDF films.

Li et al. [19] developed cement-based 0–3 piezoelectric composites containing up to 70 vol% of PZT particles by normal mixing and spreading method. Cement and PZT particles were first thoroughly mixed together without adding water, and then water and superplasticizer were added to the mixture. The mixture was then spread on a glass plate to form a disk with a diameter of ~10 mm and thickness of ~3 mm. The samples were put in the curing room at a temperature of 23 °C and relative humidity of 100% for 28 days before measurement. Poling was conducted in a silicon oil bath at a temperature of ~150–160 °C for 1 h. The d values were found to vary from 7.2 pC N^{-1} for the 35 vol% of PZT formulation to 33.4 pC N^{-1} for the 70 vol% of PZT formulations. The cement-based 0–3 piezoelectric composites were shown to have a slightly higher piezoelectric factor and electromechanical coefficient than those of 0–3 PZT/polymer composites with a similar content of PZT particles.

A sol–gel spray technique was used by Kobayashi et al. [25, 42] to produce thick-film broadband ultrasonic transducers that are based on piezoelectric 0–3 composites. The piezoelectric particles of PZT and LiTaO$_3$ were dispersed in a sol–gel solution with a viscosity suitable for spray painting [42]. The fabrication process involved spray coating, firing at 420 °C using a heat gun or gas torch, and annealing at 650 °C to optimize the PZT sol–gel. Multiple layers were made to reach the desired transducer thickness. After sputtering the platinum top electrodes, the piezoelectric films of a 50–60-μm thickness were poled at 380 °C with a 7–9 MV m^{-1} electric field. The films were observed to be able to withstand more than 10 thermal cycles between the room temperature and 250 °C without being detached from the steel rod substrate. The d_{33} values of the films were measured to be 0.1 pC N^{-1}. Ultrasonic pulse–echo signals with a signal-to-noise ratio of more than 25 dB have also been received by the PZT/LiTaO$_3$ composite films at elevated temperatures up to 250 °C. More recently, Kobayashi et al. [25] used the sol–gel spray technique to deposit PZT/PZT sol–gel composite directly onto selected substrates including graphite/epoxy, aluminum, and steel. The Corona poling technique was used to pole the cured film for 10 min at 120 °C. The piezoelectric constants d_{33} and d_{31} of the films were 30 × 10^{-12} and −26 × 10^{-12} m V^{-1}, measured by an optical interferometer and optical coherence tomography, respectively.

Zhang and Li [24] developed an epoxy-based flexible piezoelectric paint that cures at room temperature. The paint film from this recent formulation is very compliant after curing, as can be seen from the bent shape of the paint sample in Figure 1. PZT-5A powder was chosen as the active piezoelectric material in this paint formulation. The epoxy resin used in this study is a liquid resin comprised mainly of diglycidyl ether of bisphenol A (DGEBPA). Additives such as dispersing agents and defoamers were also added to the paint mixture during the paint grinding and mixing process. Curing of the paint was

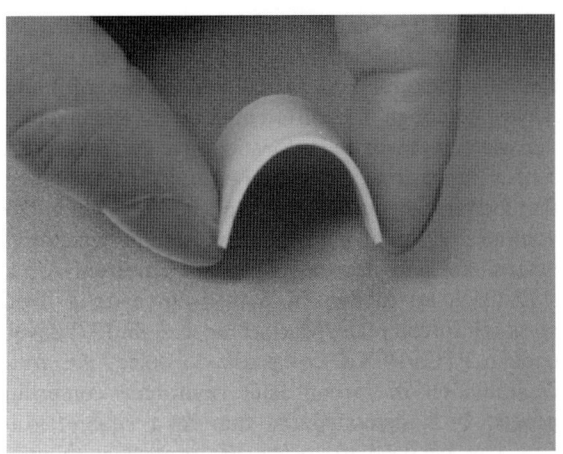

Figure 1. Sample of flexible piezoelectric paint.

done at elevated temperature (50 °C) for three days before applying conductive paint-based electrode to the top side of the paint. The volume fraction of PZT power in cured paint was about 30%. After the curing and electroding process had been completed, poling of the piezoelectric paint was carried out to activate its piezoelectric effect. In this study, poling was performed using a conventional electrode poling device at a temperature of 70 °C for 30 min at an electrical field of 5 MV m^{-1}. The average value of d_{33} was measured to be 5 pC N^{-1}. The ultrasonic sensing characteristic of the piezoelectric paints from this study is described in the next section.

Lahtinen et al. [22] developed a piezopaint-based vibration sensor by mixing piezoceramic powder with epoxy resin, which has been shown to be a very sensitive vibration monitoring device. The paint was mixed with 60 vol% PZT pigment concentration. The PZT powder was PZ27 produced by Ferroperm Electroceramics A/S, Denmark. The size range of the powder particles was 0–25 μm, with average particle size of 5 μm. The paint was applied with a laboratory applicator and the dry film thickness was 80 μm. Poling of the piezopaint layer was carried out by applying a 1-kV potential over the piezo layer for a period of 15 min at room temperature. Dynamic loading test of the piezopaint vibration sensors was conducted both in the laboratory and in field conditions. The thermal stability of the piezopaint sensor is excellent compared with piezoelectric polymers because of the fact that the piezoelectricity is a property of a ceramic material in the composite. The thermal stability is limited to the properties of the polymer matrix and the electrode materials. The thermal stability could possibly be further enhanced by applying more stable polymer matrix materials such as Polyetheretherketone (PEEK).

Hale and Lahtinen [43] conducted a program of testing to evaluate the susceptibility of piezoelectric paint to environmental degradation and so to evaluate its suitability for use in shock and vibrations sensors on large outdoor structures. Controlled weathering trials show that the sensors drop in sensitivity over the first few months from manufacture, but thereafter maintain a constant sensitivity irrespective of exposure to sunlight, rain, or frost, etc. Field trials on river crossing bridges in United Kingdom and Finland have shown that the sensors can survive harsh outdoor conditions and remain functional for six years, with no sign of an end to their lives at that time.

4 ULTRASONIC SENSING CHARACTERISTICS OF PIEZOELECTRIC PAINT

Like all other piezoelectric materials, piezoelectric paint can produce electric voltage signals proportional to the applied mechanical deformation in its film plane. Without loss of generality, we assume that piezoelectric paint sensor is only subjected to unidirectional strain in its film plane. The dielectric displacement D_3 is related to the generated electric charge by the following relationship,

$$q = \iint D_3 \, dA_3 \qquad (1)$$

where dA_3 is the differential electrode area in the 1–2 plane of the piezoelectric paint. The electric charge generated by the sensor is

$$q = d_{31} Y_C b_C \int_{l_c} \varepsilon_1 \, dl \qquad (2)$$

where Y_C is the Young's modulus of the piezoelectric paint, and l_c and b_c are the length and width of the piezoelectric paint sensor, respectively.

Sirohi and Chopra [44] studied the behavior of PZT and PVDF strain sensor over a frequency range of 5–500 Hz. In their work, a typical piezoelectric patch sensor is considered as a parallel plate capacitor, which stores the electric charge generated by the piezoelectric sensor when mechanical deformation is applied. It is shown that if only one-dimensional in-plane deformation is considered, the voltage generated across the electrodes of the capacitor can be related to the average in-plane strain by a sensor sensitivity parameter and the sensor capacitance. If a charge amplifier is connected to the sensor electrode, then the output voltage is

$$V_{\text{out}}(t) = -\frac{q}{C_f} = -\frac{d_{31} Y_C b_C}{C_f} \int_{lc} \varepsilon_1 \, dl \quad (3)$$

Therefore, the sensor output voltage is proportional to the integral of the in-plane strain in the 1-direction.

AE consists of a propagating elastic wave generated by a sudden release of energy within a material (see, e.g. [45–47]; see **Chapter 4**; **Chapter 13**). In recent years, AE has been shown to provide real-time information on the progression of damage within metallic structures. The elastic wave generated by structural damage propagates through the solid to the surface where it can be detected by surface-mounted sensors. The sensor captured AE waveforms contain information about the damage source, like location, size and type, etc.; it is possible to extract such information by analysis of those waveforms. In a manner different from ultrasonic active sensing, which intentionally excite elastic waves in a solid, AE sensors passively listen for acoustic signals generated by crack initiation and progression. AE has been proved to be a useful NDE method for the investigation of local damage in structural members.

The AE sensing capability of piezoelectric paints has been verified by a number of researchers including Egusa and Iwasawa [39], Wenger *et al.* [14, 40], Sakamoto *et al.* [41], Kobayashi *et al.* [25, 42], Marin-Franch *et al.* [33], and Li and Zhang [48]. Egusa and Iwasawa [39] evaluated the sensitivity of piezoelectric paint film with a PZT volume fraction of 53% to film thickness and poling field as an AE sensor in the frequency range of 0–1.2 MHz. Wenger *et al.* [14] examined the suitability of piezoelectric 0–3 composites embedded into glass-reinforced laminate plates as AE sensors. Piezoelectric bimorph sensors for embedded AE sensing in glass–epoxy laminate platelike structures have been fabricated and investigated by Wenger *et al.* [40]. The bimorphs can differentiate between two types of plate wave propagation—flexural and extensional modes of Lamb waves, although the differentiation is attributed to the wavelength of the acoustic wave relative to the sensor dimensions. Sakamoto *et al.* [41] have shown that the piezoelectric 0–3 composite film with a PZT volume fraction of 49% can successfully detect both extensional and flexural modes of simulated AE at a distance up to 8 m from the source on a fiberglass reinforced plate. Marin-Franch *et al.* [33] developed a PTCa/PEKK composite to detect AE from delamination in carbon fiber reinforced composite panels. It is demonstrated that in a square array they can be used to locate AE sources with good accuracy.

We recently conducted a series of ultrasonic tests including pitch–catch ultrasonic test and pencil break test to examine the sensing capability of piezoelectric paint sensor over ultrasonic frequency range. The piezopaint sensor loaded with 30% by volume of PZT powder was deposited to a 609.6 mm × 609.6 mm × 1.59 mm aluminum plate. A layer of conductive silver paint was applied on top of the piezopaint having a size of 7 mm × 7 mm. The paint film thickness was 1 mm. After curing and electroding, poling of the piezoelectric paint was carried out to activate its piezoelectric effect. A 7 mm × 7 mm × 0.2 mm piezoelectric ceramic patch was also bonded along the centerline of the piezoelectric paint sensors as the actuator to excite Lamb waves in the aluminum plate. A five-cycle windowed toneburst excitation signal was generated by a function generator at various frequencies. Generally, guided waves propagating through metal structure surface are highly dispersive and tend to decay very fast with the increase of the propagation distance. Piezoelectric paint sensors were used to measure the guided-wave propagation at corresponding sensor locations. An AD745 high-speed operational amplifier circuit with a voltage gain of 100 was connected to the piezoelectric paint sensor as a preamplifier and then the amplified signal was recorded with a Tektronix TDS 2024 oscilloscope connected to a computer via a general purpose interface bus (IEEE488) (GPIB) adapter.

In our test, the PZT wafer was excited by 10 V peak-to-peak voltage signal, and ultrasonic lamb waves were received by a piezoelectric paint sensor

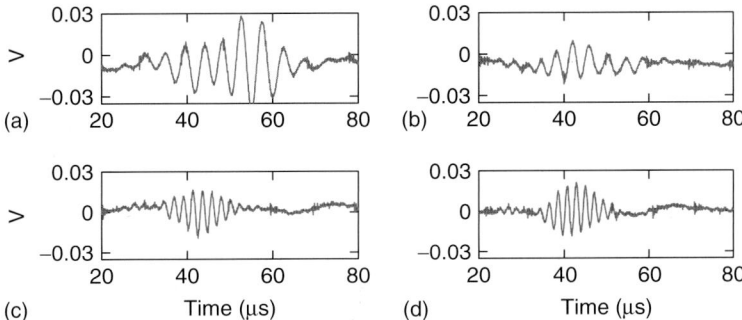

Figure 2. Raw data received by piezoelectric paint sensor: (a) 40-kHz excitation, (b) 60-kHz excitation, (c) 80-kHz excitation, and (d) 100-kHz excitation.

which was 25 mm away from excitation. The received raw time traces at excitation frequencies of 40, 60, 80, and 100 kHz are shown in Figure 2. The amplitude of the received signals at piezoelectric paint sensors was in the range of 20 mV peak-to-peak. Before raw sensor data can be interpreted to determine structural health condition, signal processing is necessary to ease the interpretation process. This is especially true for piezoelectric paint sensor because of its low sensitivity and hence low signal-to-noise ratio compared with a PZT sensor. In the present research, discrete wavelet transform (DWT) technique was used to filter the signal captured by piezoelectric paint sensor. Figure 3 shows the DWT-filtered signals (using a db (Daubechies) 6 level 5 wavelet) at piezoelectric paint sensors using wavelet transform. The filtered signals clearly show the propagation of Lamb waves over time. This experimental study conforms that piezoelectric paint sensor can be used to detect ultrasonic wave motion in platelike structures.

To verify the capability of piezoelectric paint sensor as an AE sensor, we performed pencil break tests on the same aluminum plate. Pencil break source (also called *Hsu-Neilsen source*) is a common approach to simulate AE due to its simplicity and reproducibility. In our study, a pencil lead with a 0.5-mm diameter was broken by pressing against the aluminum plate to simulate AE signal from a microcrack. The same data-acquisition equipment as in the ultrasonic test was used here. For comparison purpose, a 7 mm × 7 mm × 0.2 mm PZT sensor has been placed right next to the piezoelectric paint sensor on the aluminum plate to capture the AE signal. The same amplifier used in the ultrasonic test was applied to the piezoelectric paint sensor. The pencil lead was fractured at the same distance of 127 mm (∼5 in.) away from both the PZT and piezoelectric paint sensor. Figure 4(a) plots the captured AE signals collected at a sampling rate of 2.5 MHz. After filtering with the db 6 level 5 wavelet, the filtered

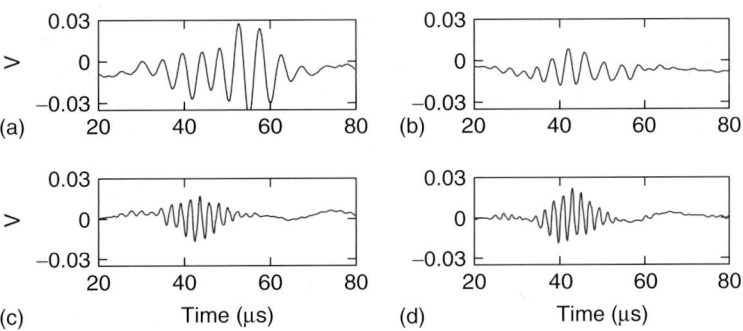

Figure 3. Filtered signal received by piezoelectric paint sensor: (a) 40-kHz excitation, (b) 60-kHz excitation, (c) 80-kHz excitation, and (d) 100-kHz excitation.

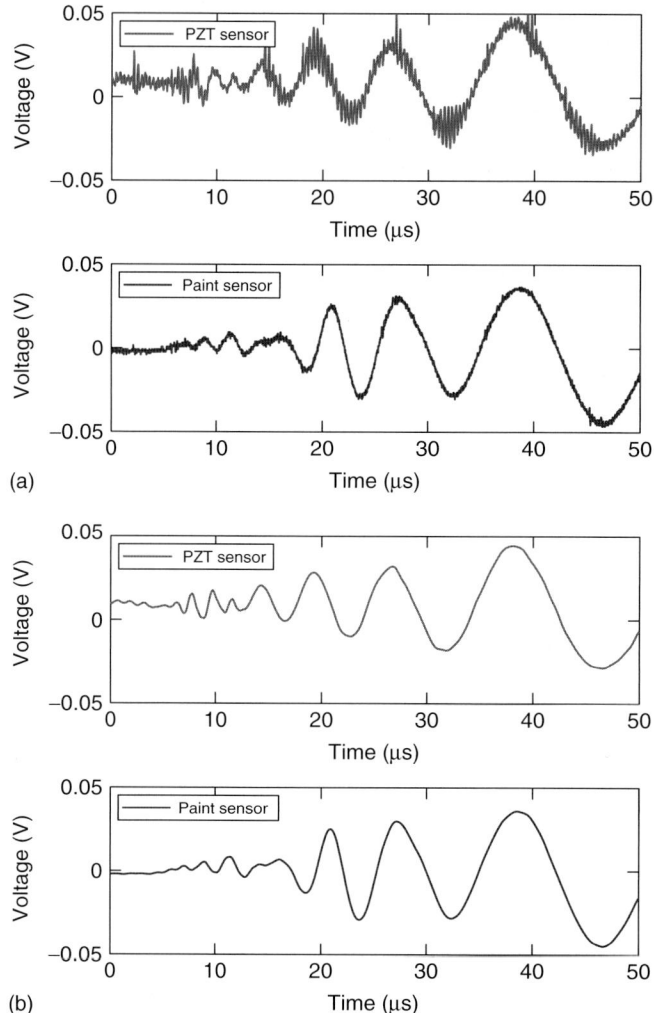

Figure 4. Acoustic emission signal received by piezoelectric sensors in pencil break test: (a) raw data and (b) DWT-filtered data.

AE signals are shown in Figure 4(b). The piezoelectric paint sensor (with amplifier) gave a peak-to-peak response close to that of the PZT sensor (without amplifier) both at the amplitude of 70 mV peak-to-peak. The two signals agree with each other very well.

5 CONCLUSIONS

The state of the art in piezoelectric paint sensor technology and its potential application to ultrasonic-based damage detection are reviewed in this paper. Piezoelectric paint is a composite piezoelectric material with adjustable material properties that are not easily attainable in a single-phase material. Through judicious selection of the polymer matrix, the composite properties of the piezoelectric paint can be tailored to meet the specific requirements of an application condition. Various formulations for piezoelectric paints have been developed over the last 20 years, and its ability to sense ultrasonic signals has been experimentally verified. Compared with conventional piezoelectric ceramics and piezoelectric polymers, piezoelectric paint has several advantages for ultrasonic sensing application: (i) it enables true distributed sensing since piezoelectric paint can be applied to the surface of the structure in potential hot spots; (ii) painting represents an inexpensive way to apply this novel sensor technology over large surface areas of a host structure, especially for those with curved surfaces or complex geometries; and (iii) piezoelectric paint is more thermally stable than piezoelectric polymer such as PVDF. Additionally, a thickness range varying between 60 μm and 2 mm for piezoelectric paint film can be obtained, while the thickness of PVDF film is limited because of its high poling voltage.

Preliminary results from ultrasonic tests including pitch–catch tests and simulated AE tests that were carried out to verify the ultrasonic sensing capability of piezoelectric paint are reported in this paper and based on these preliminary results, piezoelectric paint sensor appears to offer a promising embedded sensor technology for ultrasonic-based NDE of structural defects or damages in either metallic or composite structure. However, extensive work still needs to be done to fully investigate the performance of piezoelectric paint sensor technology before this technique can be satisfactorily used in practical applications.

REFERENCES

[1] Giurgiutiu V, Cuc A. Embedded non-destructive evaluation for structural health monitoring, damage detection, and failure prevention. *The Shock and Vibration Digest* 2005 **37**(2):83–105.

[2] Jata KV. Piezoelectric transducers. *Encyclopedia of Structural Health Monitoring*. John Wiley & Sons: Chichester, 2008.

[3] Ayres JW, Lalande F, Chaudhry Z, Rogers CA. Qualitative impedance based health monitoring of civil infrastructures. *Smart Materials and Structures* 1998 **7**(5):599–605.

[4] Park G, Sohn H, Farrar CR, Inman DJ. Overview of piezoelectric impedance-based health monitoring and path forward. *The Shock and Vibration Digest* 2003 **35**(6):451–463.

[5] Giurgiutiu V. Embedded ultrasonics NDE with piezoelectric wafer active sensors. *Journal of Instrumentation, Mesure, Metrologie* 2003 **3**(3–4):149–180.

[6] Yu L. Piezoelectric wafer active sensors. *Encyclopedia of Structural Health Monitoring*. John Wiley & Sons: Chichester, 2008.

[7] Ikeda T. *Fundamentals of Piezoelectricity*. Oxford University Press: Oxford, 1990.

[8] Polla DL, Francis LF. Processing and characterization of piezoelectric materials and integration into microelectromechanical systems. *Annual Review of Materials Science* 1998 **28**:563–597.

[9] Damjanovic D, Muralt P, Setter N. Ferroelectric sensors. *IEEE Sensors Journal* 2001 **1**(3):191–206.

[10] Niezrecki C, Brei D, Balakrishnan S, Moskalik A. Piezoelectric actuation: state of the art. *The Shock and Vibration Digest* 2001 **33**(4):269–280.

[11] Gautschi G. *Piezoelectric Sensorics*. Springer-Verlag: Berlin, 2002.

[12] Hanner KA, Safari A, Newnham RE, Runt J. Thin film 0–3 polymer/piezoelectric ceramic composites: piezoelectric paints. *Ferroelectrics* 1989 **100**:255–260.

[13] Egusa S, Iwasawa N. Piezoelectric paints: preparation and application as built-in vibration sensors of structural materials. *Journal of Materials Science* 1993 **28**:1667–1672.

[14] Wenger MP, Blanas P, Shuford RJ, Das-Gupta DK. Acoustic emission signal detection by ceramic/polymer composite piezoelectrets embedded in glass-epoxy laminates. *Polymer Engineering and Science* 1996 **36**(24):2945–2954.

[15] Hale JM, Tuck J. A novel thick-film strain transducer using piezoelectric paint. *Proceedings of the Institution of Mechanical Engineers, Part C* 1999 **213**:613–622.

[16] Papakostas T, White N. Screen printable polymer piezoelectrics. *Sensor Review* 2000 **20**(2):135–138.

[17] Badcock RA, Birt EA. The use of 0–3 piezocomposite embedded Lamb wave sensors for detection of damage in advanced fiber composites. *Smart Materials and Structures* 2000 **9**:291–297.

[18] Sakamoto WK, de Souzab E, Das-Gupta DK. Electroactive properties of flexible piezoelectric composites. *Materials Research* 2001 **4**(3):201–204.

[19] Li Z, Zhang D, Wu K. Cement-based 0–3 piezoelectric composites. *Journal of the American Ceramic Society* 2002 **85**(2):305–313.

[20] White JR, de Poumeyrol B, Hale JM, Stephenson R. Piezoelectric paint: ceramic-polymer composites for vibration sensors. *Journal of Materials Science* 2004 **39**(9):3105–3114.

[21] Zhang Y. In situ fatigue crack detection using piezoelectric paint sensor. *Journal of Intelligent Material Systems and Structures* 2006 **17**(10):843–852.

[22] Lahtinen R, Muukkonen T, Koskinen J, Hannula S-P, Heczko O. A piezopaint-based sensor for monitoring structure dynamics. *Smart Materials and Structures* 2007 **16**:2571–2576.

[23] Hale JM, White JR, Stephenson R, Liu F. Development of piezoelectric paint thick-film vibration sensors. *Proceedings of the Institution of Mechanical Engineers, Part C* 2005 **219**:1–9.

[24] Zhang Y, Li X. Test-bed implementation of piezopaint-based acoustic emission sensor for crack initiation monitoring. *Proceedings of the 4th International Conference on Bridge Maintenance, Safety, and Management (IABMAS'08)*. Seoul, 13–17 July 2008; accepted for publication.

[25] Kobayashi M, Jen C-K, Moisan JF, Mrad N, Nguyen SB. Integrated ultrasonic transducers made by the sol-gel spray technique for structural health monitoring. *Smart Materials and Structures* 2007 **16**:317–322.

[26] Safari A. Development of piezoelectric composites for transducer. *Journal de Physique III* 1994 **4**:1129–1149.

[27] Chilton JA. Electroactive composites. In *Special Polymers for Electronics and Optoelectronics*, Chilton JA, Goosey MT (eds). Chapman & Hall: London, 1995.

[28] Gomez TE, Espinosa FM, Levassort F, Lethiecq M, James A, Ringgard E, Millar CE, Hawkins P. Ceramic powder—polymer piezocomposites for electroacoustic transduction: modeling and design. *Ultrasonics* 1998 **36**:907–923.

[29] Tresseler JF, Alkoy S, Newnham RE. Piezoelectric sensors and sensor materials. *Journal of Electroceramics* 1998 **2**(4):257–272.

[30] Dias CJ, Igreja R, Marat-Mendes R, Inacio P, Marat-Mendes JN, Das-Gupta DK. Recent advances in ceramic-polymer composite electrets. *IEEE Transactions on Dielectrics and Electrical Insulation* 2004 **11**(5):35–40.

[31] Akdogan EK, Allahverdi M, Safari A. Piezoelectric composites for sensor and actuator applications. *IEEE Transactions on Ultrasonics, Ferroelectrics, and Frequency Control* 2005 **52**(5):746–775.

[32] Newnham RE, Skinner DP, Cross LE. Connectivity and piezoelectric-pyroelectric composites. *Materials Research Bulletin* 1978 **13**:525–536.

[33] Marin-Franch P, Martin T, Fernandez-Perez O, Tunnicliffe DL, Das-Gupta DK. Evaluation of PTCa/PEKK composites for acoustic emission detection. *IEEE Transactions on Dielectrics and Electrical Insulation* 2004 **11**(1):50–55.

[34] Han KH, Safari A, Riman RE. Colloid processing for improved piezoelectric properties of flexible 0–3 ceramic/polymer composites. *Journal of the American Ceramic Society* 1991 **74**(7):1699–1702.

[35] Lau ST, Li K, Chan HLW. PT/P(VDF-TrFE) nanocomposites for ultrasonic transducer applications. *Ferroelectrics* 2004 **304**:19–22.

[36] Waller D, Safari A. Corona poling of PZT ceramics and flexible piezoelectric composites. *Ferroelectrics* 1988 **87**:189–195.

[37] Sa-Gong G, Safari A, Jang SJ, Newnham RE. Poling flexible piezoelectric composite. *Ferroelectrics Letters* 1986 **5**:131–142.

[38] Egusa S, Iwasawa N. Piezoelectric paints as one approach to smart structural materials with health-monitoring capabilities. *Smart Materials and Structures* 1998 **7**:438–445.

[39] Egusa S, Iwasawa N. Application of piezoelectric paints to damage detection in structural materials. *Journal of Reinforced Plastics and Composites* 1996 **15**:806–817.

[40] Wenger MP, Blanas P, Shuford RJ, Das-Gupta DK. Characterization and evaluation of piezoelectric composite bimorphs for *in situ* acoustic emission sensors. *Polymer Engineering and Science* 1999 **39**(3):508–518.

[41] Sakamoto WK, Marin-Franch P, Tunnicliffe D, Das-Gupta DK. Lead zirconate titanate/polyurethane (PZT/PU) composite for acoustic emission sensors. *IEEE Annual Conference on Electrical Insulation and Dielectric Phenomena*. Kitchener, 2001; pp. 20–23.

[42] Kobayashi M, Olding TR, Sayer M, Jen C-K. Piezoelectric thick film ultrasonic transducers fabricated by a sol-gel spray technique. *Ultrasonics* 2002 **39**:675–680.

[43] Hale JM, Lahtinen R. Piezoelectric paint: effects of harsh weathering on aging. *Plastics Rubber and Composites* 2007 **36**(9):419–422.

[44] Sirohi J, Chopra I. Fundamental understanding of piezoelectric strain sensors. *Journal of Intelligent Material Systems and Structures* 2000 **11**(4):246–257.

[45] Ono K. Current understanding of mechanisms of acoustic emission. *Journal of Strain Analysis* 2005 **40**(1):1–14; special issue.

[46] Gorman MR. Acoustic emission in structural health monitoring. *Encyclopedia of Structural Health Monitoring*. John Wiley & Sons: Chichester, 2008.

[47] Wevers M. Applications of acoustic emission for structural health monitoring: a review. *Encyclopedia of Structural Health Monitoring*. John Wiley & Sons: Chichester, 2008.

[48] Li X, Zhang Y. Piezoelectric paint sensor for ultrasonic NDE. *Proceedings of the Smart Structures Technologies for Civil, Mechanical, and Aerospace Systems*. SPIE: San Diego, CA, 19–21 March 2007; SPIE Vol. 6529.

Chapter 58

Eddy-current *in situ* Sensors for SHM

Neil Goldfine, Vladimir Zilberstein, Darrell Schlicker and Dave Grundy
JENTEK Sensors, Inc., Waltham, MA, USA

1 Introduction	1051
2 Examples of Advanced Eddy-current SHM Implementations	1054
3 Fatigue Monitoring	1055
4 Corrosion Monitoring	1059
5 Stress Monitoring	1060
6 Temperature Monitoring	1061
7 Composite Disbond/Delamination Monitoring	1061
8 Conclusions	1063
End Notes	1063
References	1063

1 INTRODUCTION

This article complements **Chapter 20**. The focus is on surface-mounted and embedded eddy-current sensors, arrays, and sensor networks for structural

Encyclopedia of Structural Health Monitoring. Edited by Christian Boller, Fu-Kuo Chang and Yozo Fujino © 2009 John Wiley & Sons, Ltd. ISBN: 978-0-470-05822-0.

health monitoring (SHM). The reader is expected to have some knowledge of eddy-current methods.

Generally, eddy-current sensors include both scanning sensor arrays and sensors mounted at selected critical locations. This article describes primarily capabilities of surface-mounted and embedded eddy-current sensors for detection and monitoring of damage states as well as for monitoring of usage state variables such as stresses and temperatures.

Eddy-current SHM sensors described here are either permanently or temporarily placed in close proximity of, directly on, or within a structure at selected locations to provide enhanced observability of damage and usage state variables—*specifically to support life management decisions*. Arrays, as defined here, are local sensor constructs with a common drive (primary winding) and multiple one-dimensional or two-dimensional sensing elements that form an array in a plane positioned parallel to the surface of a component. One such sensor is the JENTEK MWM®-Array (MWM stands for meandering winding magnetometer) [1–10]. Networks of sensors and sensor arrays can be distributed throughout a structure to provide coverage of critical locations. Either (i) portable data-acquisition and analysis units, (ii) central onboard electronics, or (iii) distributed electronics modules can be used to acquire data from eddy-current SHM sensor networks.

When used for damage detection and monitoring, embedded or surface-mounted eddy-current sensors perform a function similar to the function of eddy-current sensors for nondestructive testing (NDT) applications described in **Chapter 20**. The advantage of permanently mounting these eddy-current sensors and arrays is that they can be used to inspect difficult-to-access locations without requiring disassembly or surface preparation. The principal disadvantages of such permanently mounted sensors for damage detection and monitoring are that (i) once in place, they cannot be scanned to sample areas that have no damage for relative comparisons with defect indications; (ii) they are limited in surface imaging resolution by the size of the individual sensing elements in an array (note that imaging resolution should not be confused with the crack detection capability since detectable cracks may be either bigger or smaller than the size of the sensing element depending on the signal-to-noise ratio); (iii) they cannot be removed for calibration in air or on standards at the time of the inspection; and (iv) in general, they can only detect damage directly under the sensor (or sensing elements of an array)[a].

Historically, SHM successes have been limited to usage and diagnostic state monitoring. There has been wide use of temperature, pressure, strain, and vibration sensing onboard aircraft and on other high value assets. This includes SHM implementations in aviation, automotive, energy, and many other market sectors. For example, vibration sensing, using inductive sensors, is common. These more traditional SHM uses are not addressed in this article. However, some of these more traditional uses can gain from the sensor technology described here. One example is weight and balance monitoring for aircraft that is limited by the ability of strain gauges or other load monitoring methods to provide reliable measurement of loads over long service periods. Attempts to implement reliable systems have generally not been successful. This article describes the approaches that can address the need for improved stress/load monitoring of both static and dynamic stresses.

To enable *individual component life management* beyond the push over the last few decades for individual aircraft tracking, two new, or at least recently reinvigorated, thrusts in SHM development have emerged. These include the need (i) to replace, or at least supplement NDT, which is typically performed with off-board sensors/probes, by direct onboard damage and condition monitoring, both continuous and intermittent, in difficult-to-access critical locations and (ii) to provide direct monitoring of local stress/strain, temperature, and environmental conditions at the critical locations that are most likely to accumulate damage or experience changes associated with relevant events, e.g., overloads and foreign object damage (FOD) for metals and composites.

To enable life extension and predictable performance for subsystems and components of high-value assets, such as aircraft, rotorcraft, ships, pipelines, bridges, etc., the often-stated goal of replacing or at least reducing the use of conventional NDT by introducing onboard SHM for *direct damage and material condition monitoring* comes from (i) the need to mitigate life cycle cost escalation while improving readiness for aging aircraft and (ii) the need, on new platforms and in platform upgrades, to improve mission readiness/capability rates, reduce field and depot logistics footprints, to enable more aggressive component/platform designs using less conservative safety margins and using advanced materials (composites, functional coatings, multi-material systems), to reduce overall aircraft weight by enabling application of damage tolerance methods by making difficult-to-access areas inspectable, to avoid collateral damage associated with disassembly for inspection, better manage mission loads, to alert to events such as excessive FOD and consequences of inadequate maintenance actions, and to remove the human from the field maintenance loop.

Thus, *the expanded use of onboard sensors for SHM is inevitable*, particularly when advanced surface-mounted and embedded sensors have proven, for some applications, to be far less invasive, with lower total implementation costs, than conventional off-board NDT methods in both logistics support requirements and practical impact on asset readiness/availability.

This article addresses the need for enhanced monitoring of (i) damage states, (ii) usage states, (iii) diagnostic states, including (iv) detection of upset events.

Damage states are scalar, vector, or multidimensional quantities that (i) can be used in models to predict damage behavior progression, e.g., degree of relaxation of compressive residual stresses intentionally introduced to mitigate damage such as fatigue or stress corrosion cracking or (ii) provide a measure of one or more damage features (e.g., crack size, material loss from corrosion, or composite disbond/delamination area).

Usage state variables of interest include stresses and temperatures that directly affect cumulative exposure to mechanical and/or thermal conditions at critical locations on individual components, for which damage evolution needs to be monitored. Enhanced stress, temperature, and environmental monitoring are continually identified as necessary to achieve the promise of SHM for aircraft and rotorcraft.

Diagnostic states of interest include geometric misalignment, vibration levels, thermal effects, and overload events. The focus of new developments in this area of SHM development is to enable onboard or temporarily installed monitoring capability for such diagnosis at reduced cost and with limited logistics support requirements.

Upset *event detection* through monitoring of *event-dependent states* of a material (e.g., impact or overheating in metallic or composite components) is often identified as a separate requirement that must be addressed. Event detection is particularly important to components managed on a safe-life basis (as practiced by the US Navy).

Of course, there is an inevitable overlap and interdependence in the above definitions, for example, between some diagnostic states and upset events. Upset event identification is certainly a part of diagnostics. Also, detected upset events should be accounted for in the assessment of usage.

This article specifically discusses eddy-current sensor capabilities demonstrated over the last decade to provide material damage monitoring, including fatigue, corrosion, and overloads, as well as stress and temperature monitoring. The focus of this article is on the MWM and MWM-Array eddy-current sensors [11–15, 18–22], since these sensors address the above categories of interest (damage, diagnostics, and usage).

For detection and monitoring of damage or parameters of interest with the available mountable eddy-current sensors, such as MWM sensors or MWM-Arrays, portable units plug into onboard sensors for inspection or short duration monitoring at difficult-to-access locations. These portable units can support onboard NDT, diagnostic, or usage survey function.

Onboard instrumentation, when available, would provide continuous monitoring capabilities for damage, diagnostics, and usage recording. Such onboard resources could be queried in parallel, through prescribed multiplexing. Alternatively, they can be allocated using a concept named *neural plasticity* by analogy to a phenomenon related to human neurological development. Neural plasticity is a concept that enables the use of limited onboard resources that learn and grow in capability with experience by reallocating limited resources.

Onboard SHM applications of eddy-current sensors are beginning to transition to operational fleet use, having been proven extensively in the laboratory and more recently in full-scale component and full-scale aircraft tests. Extensive flight testing of eddy-current sensors for fatigue monitoring and stress monitoring with portable data-acquisition systems is expected to occur within the next few years, followed by transition to fully integrated onboard systems using eddy-current sensors. One example of a potential application would be fatigue-critical locations on landing gear components (*see* **Chapter 114**).

In **Chapter 20**, the concept of mapping and tracking of fatigue damage using time-sequenced scanning with advanced eddy current methods is introduced. For example, in some critical areas in aircraft components, such as engine disk slots, where early damage detection using onboard SHM sensors is not feasible, the only means for early detection is mapping and tracking with improved scanning NDT. On the other hand, integration of onboard SHM sensor data with such mapping and tracking, when practical, would significantly enhance health monitoring capabilities.

In general, (i) onboard SHM sensors for damage detection and monitoring, (ii) traditional scanning NDT methods, (iii) advanced NDT using new mapping and tracking methods for damage evolution, and (iv) enhanced direct load monitoring capability, using onboard SHM sensors, are all needed to implement next-generation adaptive life management programs.

One advantage of using MWM-Array eddy-current sensors in support of adaptive life management is that eddy-current measurements in scanning mode and with onboard SHM sensors are complementary (providing self-consistent data formats). For fatigue damage, for example, each of these methods provides a measure of the effective electrical conductivity or magnetic permeability variation, with the damage level of interest, for the material under test.

One proposed adaptive life management approach, described by Goldfine *et al.* [13] and modified here to include mounted "onboard" sensors, is outlined in the flow chart in Figure 1.

This approach relies on the availability, reliability, and reproducibility of the NDT scanning/imaging capability and surface-mounted or embedded capability for both coupons and in-service components. Until recently, there were no depot/field inspection tools that could meet this requirement. Now, not only the MWM-Array sensors can meet this need and enable the adaptive life management approach, but other advanced methods, including Ultrasonic Testing (UT), laser UT, thermography, and digital radiography, offer the potential to deliver the previously unattainable level of reproducibility needed to support mapping and tracking of relatively early damage behavior. Also, onboard MWM-Array, UT, and alternative SHM sensors are becoming available. We anticipate that many such technologies will be proven and reach commercial maturity during the next decade.

In the near term, the approach described above is envisioned as an integration of advanced scanning NDT (mapping and tracking) with periodic or scheduled data from onboard SHM sensors recorded using portable (not onboard) data-acquisition units. In a more broadly integrated approach, onboard SHM sensors might provide continuous monitoring, with onboard electronics, along with periodic NDT scanning of selected locations. The goal is to provide the required observability of damage, diagnostics, usage state variables, and significant events.

For aluminum and titanium structural components in aircraft and rotorcraft, there are practical onboard SHM sensing alternatives to scanning NDT inspection methods. This article focuses on these applications for onboard SHM damage monitoring. Also, this article describes the value of surface-mounted and embedded sensors for laboratory fatigue and corrosion testing. Furthermore, new magnetic (eddy current) stress gauge capabilities are described for direct load monitoring.

2 EXAMPLES OF ADVANCED EDDY-CURRENT SHM IMPLEMENTATIONS

In the following five main sections, examples of detection/monitoring of damage and measurements related to diagnostics as well as to usage state variables are described within the following categories: (i) fatigue, (ii) corrosion, (iii) stress, (iv) temperature, (v) mechanical damage, and (vi) composite disbonds/delaminations. These sections rely on MWM-Array implementations to frame the discussion of *in situ* eddy current sensor applications, since the authors are most informed on this advanced sensor implementation. Some of these sections include a discussion of performance evaluation needs and methods for surface-mounted and embedded SHM sensors.

The focus here is on the use of permanently surface-mounted or embedded sensors. However,

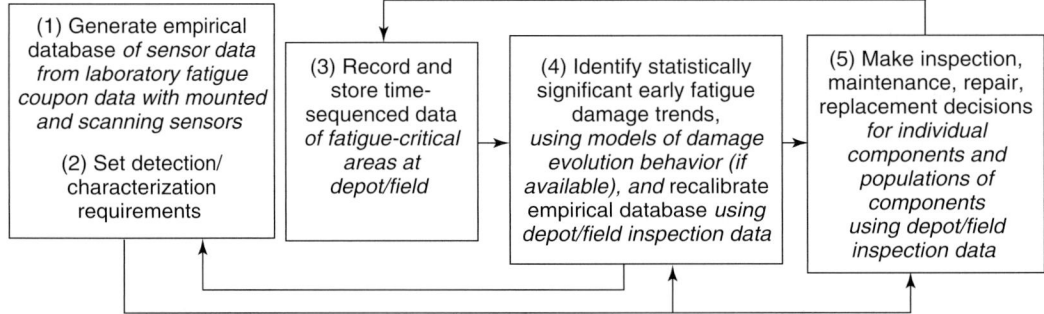

Figure 1. Flow chart of proposed adaptive life management framework.

some discussion is included on integrated methods that combine such SHM data with more conventional NDT data, including mapping and tracking using repeated NDT type imaging.

3 FATIGUE MONITORING

3.1 Fatigue case study 1: linear MWM-Arrays inside bolt holes

In **Chapter 20**, an example of a mapping and tracking application for crack detection and monitoring using scanning MWM-Array bolt-hole inspection, with C-scan imaging, is provided. In Figure 8 of **Chapter 20**, permanently mounted MWM-Array data is also provided for the same coupon test.

As shown in Figures 7 and 8 of the same article (*see* **Chapter 20**), the combination of MWM-Array scanning and permanently mounted sensor data provides valuable insight into the crack growth behavior. This format of coupon testing, using both sensing modes, is useful for generating and calibrating damage behavior databases, as well as for generating real-crack specimens that can be used for sensor performance verification and for building probability of detection curves for both scanning and permanently mounted sensors.

3.2 Fatigue case study 2: detection of small fatigue cracks at a dimple

Also, in **Chapter 20** and in [15], a method of detecting cracks at mechanical damage sites was described (*see* Figure 9, **Chapter 20**). Figure 2

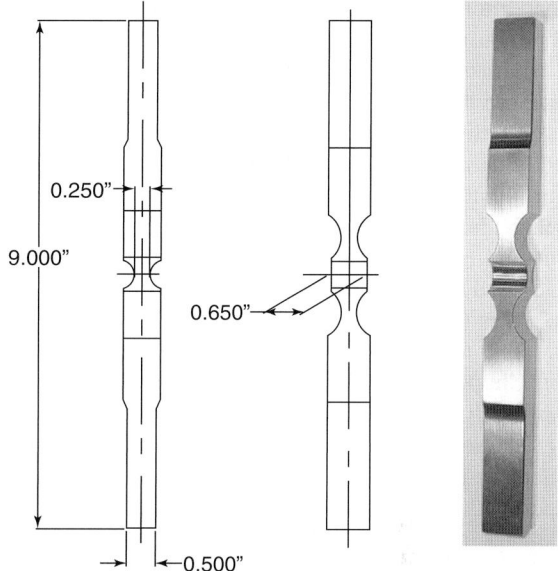

Figure 2. Schematic and photograph of a customized Ti-6Al-4V coupon with exposed fatigue-critical surfaces [8].

shows the customized titanium coupon designed for that test. In Figure 3(a) and (b), a special scanner is shown for periodically scanning the fatigue-critical area on the coupon to simulate in-service time-sequenced NDT imaging. Figure 3(c) shows a 7-channel MWM-Array designed for this purpose. Figure 2 of **Chapter 20** illustrates the measurement grid approach developed by Goldfine *et al.* [5] for this purpose. This method enables real-time conversion of thousands of digital eddy-current measurements into two images: (i) conductivity image—used to detect cracks and (ii) liftoff (sensor proximity to the material surface) image—used for self-diagnostics to ensure

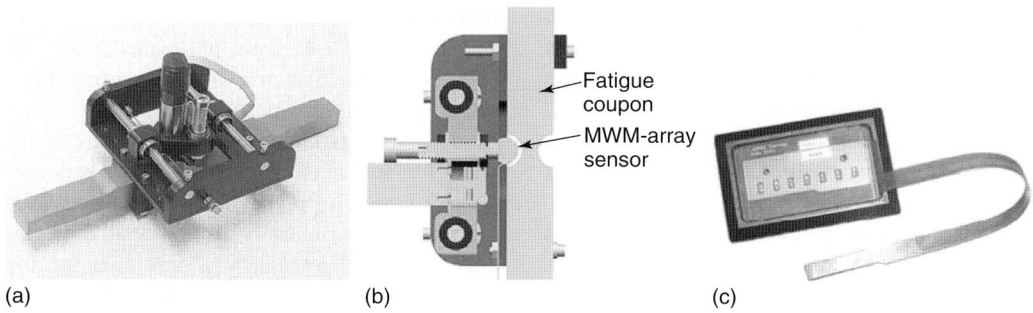

Figure 3. (a) Photograph of the coupon scanner used in support of the fatigue testing, (b) solid model representation of the coupon scanner showing the sensor position, and (c) MWM-Array FA43 sensor.

that liftoff is within an acceptable range at each of the thousands of points within an image.

Furthermore, this approach uses "air calibration," as described in ASTM Standard Practice E2338-04 [16], to improve robustness compared to conventional eddy-current testing (ET) methods that use crack (or simulated crack, e.g., EDM (electrical discharge machining) notch) standards to calibrate. This proven method is now a standard practice at US Navy and US Air Force bases, and is in use by foreign military services (FMS) as well.

This test and the data provided in Figure 9 of **Chapter 20** illustrate the value of a combined approach that integrates NDT and permanently mounted sensor data. This combination could be implemented in fatigue tests and on actual structural members. For example, when localized mechanical damage or a small crack is detected using a scanning array, the damage can be either repaired or left in place, and then a permanently mounted sensor could be applied at this location to monitor initiation and/or growth of cracks at the location. One goal is to estimate the probability of failure before the next inspection for the purpose of scheduling inspection intervals and enabling life extension. The coupon fatigue test data and in-service data obtained from such combined monitoring can also be used to calibrate databases of damage evolution behavior.

Eventually, this should also enable reliable life extension through restoration of parts with localized damage using laser shock processing, low plasticity burnishing, laser additive manufacturing, or other repair methods. It is anticipated that substantial (possibly over additional 100% of design life) life extension for many components is possible if such "health control" actions are implemented.

One key attribute is the portability of experience from one component to the next. If the entire process must be repeated for each new component, then this is not a practical approach. Thus, coupon tests and NDT databases must be sufficiently generic to adapt (i.e., recalibrate) with limited effort for the next application.

Reliable mapping and tracking with MWM-Arrays has been demonstrated for titanium alloy components, in coupon fatigue tests and on engine components. With the use of titanium alloys on both commercial and military aircraft and rotorcraft, the need to detect early fatigue damage and deliver on-condition maintenance is critical. For example, titanium alloys are typically notch sensitive. Thus, small nicks, dings, and scratches introduced during handling, assembly or in service can initiate cracks at fatigue-critical surfaces earlier than predicted by design life estimates. Blending of such surface defects is sometimes allowed followed by, for example, re-shot peening to enhance remaining life. However, many parts are not re-shot peened because of quality control issues; thus, their fatigue resistance is not fully restored. In other cases, components are simply replaced, if surface defect depths exceed a prescribed limit (often just a few thousandths of an inch). This is extremely costly. Thus, methods are needed to provide early detection of damage for critical titanium alloy components, as well as to image and size surface defects to support on-condition maintenance decisions. The combination of surface-mounted sensor monitoring and mapping and tracking has proven to be very effective for such applications. Moreover, surface-mounted sensor monitoring followed by a one-time MWM-Array scanning of the suspect areas can be sufficiently effective, as well.

3.3 Fatigue case study 3: embedded MWM-Arrays for crack initiation and growth monitoring

In a series of fatigue tests, MWM-Arrays embedded between two aluminum alloy layers (Figure 4) monitored crack growth in both 4-hole and 10-hole lap joint tests. These MWM-Arrays retained their integrity as they were removed and reused for a number of successive tests to failure of the lap joints. The four-hole tests were performed under NAVAIR (The Naval Air Systems Command) funding, and the ten-hole tests were performed under US Air Force funding for Lockheed Martin to address a specific need for the F-16 aircraft as described previously [17]. A variety of precrack configurations were tested with larger ("primary") precracks at the center hole and smaller ("secondary") precracks at the other holes. The crack tip progression data are shown in Figure 5. The lines indicate the progression of crack tips from one hole towards another as a function

Eddy-current in situ Sensors for SHM 1057

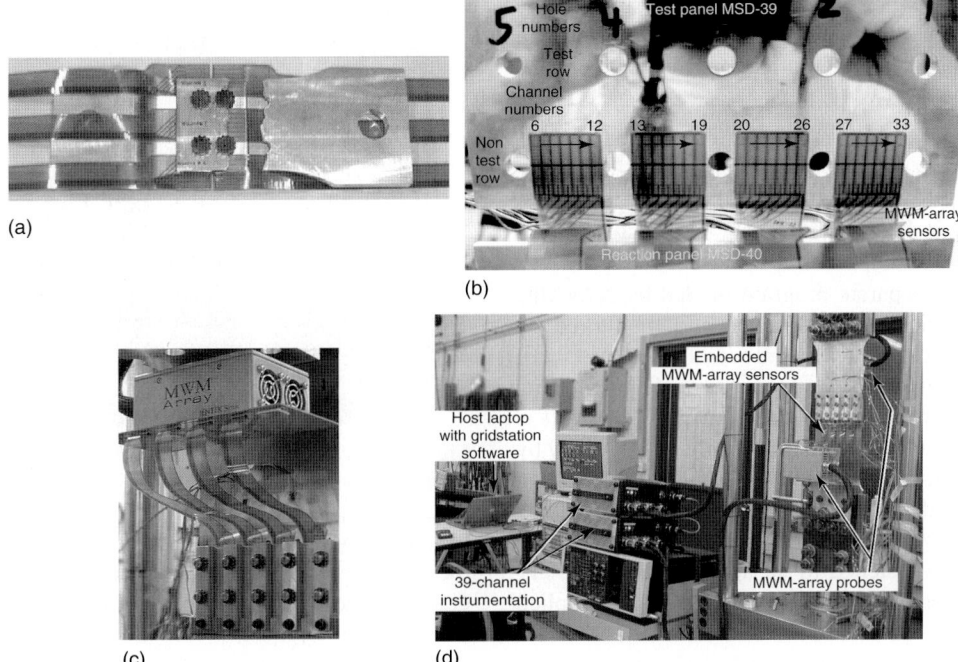

Figure 4. (a) Four-hole lap joint test specimen with embedded MWM-Arrays, (b) the 10-hole specimen with embedded MWM-Arrays shown prior to bolting up, (c) the 10-hole specimen mounted in the load frame, and (d) laboratory fatigue test setup for the 10-hole test.

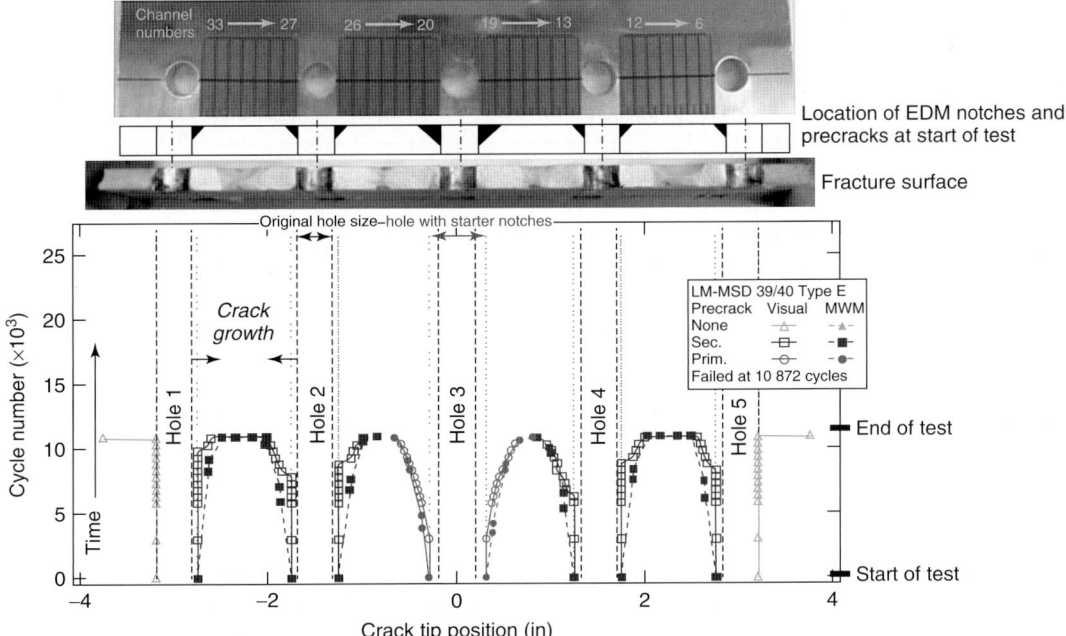

Figure 5. Representative multiple crack growth monitoring data. The specimen had primary (prim.) precracks at hole 3 and secondary (sec.) precracks at the other holes.

of number of cycles. The MWM-Arrays successfully monitored crack growth throughout the test with the sensors embedded at the buried interface between the metal plates.

3.4 Fatigue case study 4: buried cracks at bolt holes with MWM-Rosettes

Also, under a separate program funded by NAVAIR, subsurface cracks were detected and monitored by MWM-Rosettes, which were configured as smart washers and operated in a continuously monitoring mode at low frequencies (see Figure 6). This test is part of a NAVAIR funded program that is to follow by installation on two high-time aircraft for flight testing of this method.

For the test described in Figure 7, two MWM-Rosettes were fastened to an approximately 0.2-in.-thick aluminum coupon using bolts and nuts supplied

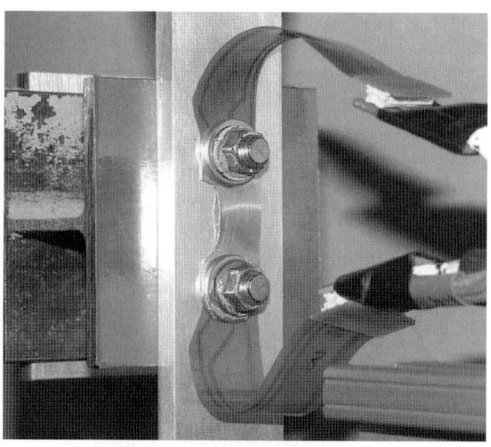

Figure 6. Photograph of the MWM-Rosette fatigue test setup.

by the Navy. The coupon was manually notched on the far side to promote the initiation of a fatigue crack on the side of the coupon opposite to that of

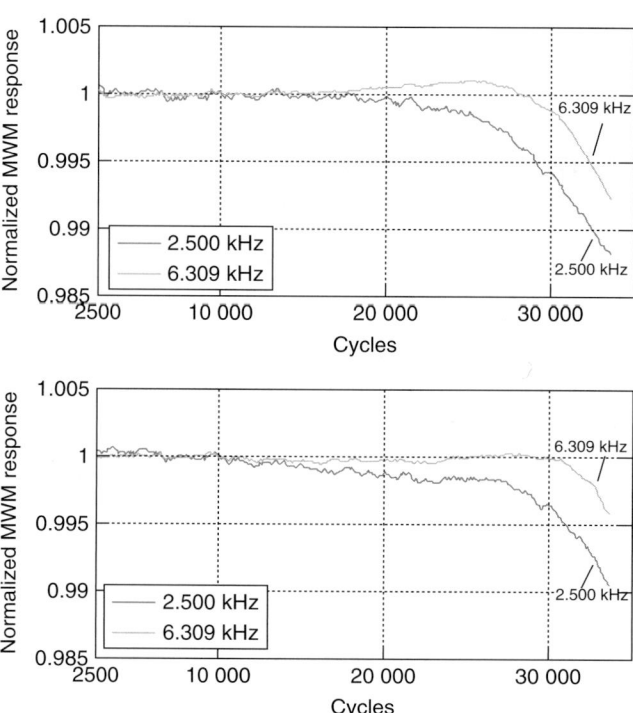

Figure 7. Normalized response of MWM-Rosette in the top hole (top plot) and bottom hole (bottom plot) acquired during a coupon fatigue test. The test data displayed is from approximately cycle 2500 to cycle 33 600. Both sensors were still functional after two fatigue tests and the accumulation of 90 000 total fatigue cycles. The "radial by axial" size of the cracks was 0.10 in. × 0.15 in. (top) and 0.11 in. × 0.14 in. (bottom).

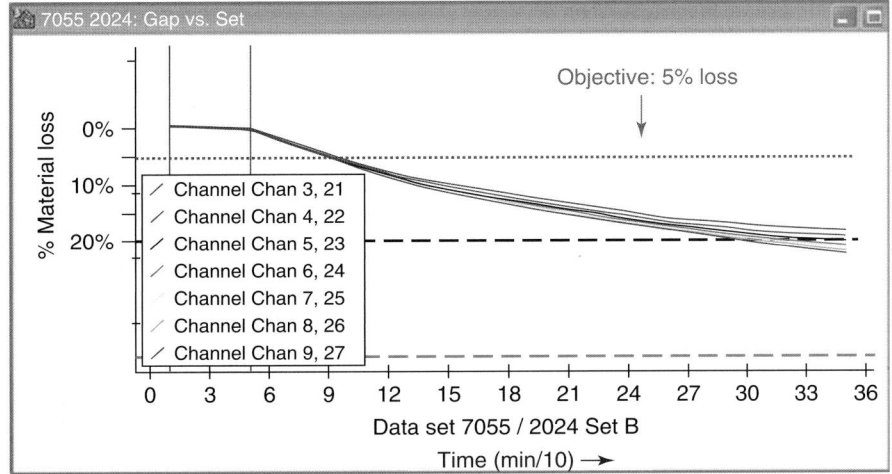

Figure 8. MWM-Array detection and monitoring of hidden corrosion during an accelerated corrosion test.

the MWM-Rosettes. Results of the fatigue test are shown in Figure 7. Cracks initiated at the notches at both holes. The crack extension was then monitored by the MWM-Rosettes. The test was stopped at 33 600 cycles, and acetate replicas of the cracked surfaces were taken. Measurements on the acetate replicas taken after the test showed that at the top hole, the crack extended radially 0.10 in. along the surface and axially 0.15 in. into the hole. At the bottom hole, the crack extended radially 0.11 in. along the surface and axially 0.14 in. into the hole. The cracks produced a reduction of 0.95 to 1.1% in the MWM signal. The MWM-Rosette was able to both detect the buried cracks and monitor their growth.

To illustrate the durability of the MWM-Rosettes operating in the smart washer configuration, a second fatigue test was performed using the same sensors. The sensors accumulated over 90 000 fatigue cycles and were still operable at the end of the test. The results of these tests provide an excellent example of the fatigue monitoring capability of MWM-Array networks.

4 CORROSION MONITORING

Hidden corrosion detection and monitoring has also been demonstrated in various tests using both mounted and scanning MWM-Array sensors [18–20]. The goal is to provide onboard detection and monitoring of hidden corrosion in difficult-to-access locations with mounted, large footprint MWM-Array sensors as well as hidden corrosion detection and monitoring in large accessible areas using wide-area scanning sensor arrays. Figure 8 shows the results of an accelerated corrosion testing program in which hidden corrosion was detected rather early and monitored with a one-dimensional MWM-Array mounted on the opposite side of the wall [19].

Figure 9 shows a surface-mountable two-dimensional MWM-Array for hidden corrosion detection and monitoring. This array is designed to monitor remaining wall thickness under its 6 in. × 6 in. footprint as metal is lost on the far side due to corrosion. Figure 10(a) shows data on a measurement grid for

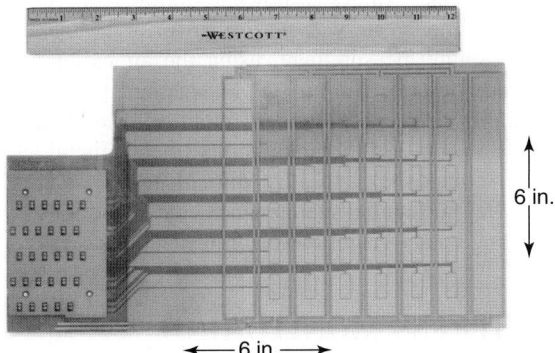

Figure 9. A two-dimensional MWM-Array for onboard monitoring of hidden corrosion.

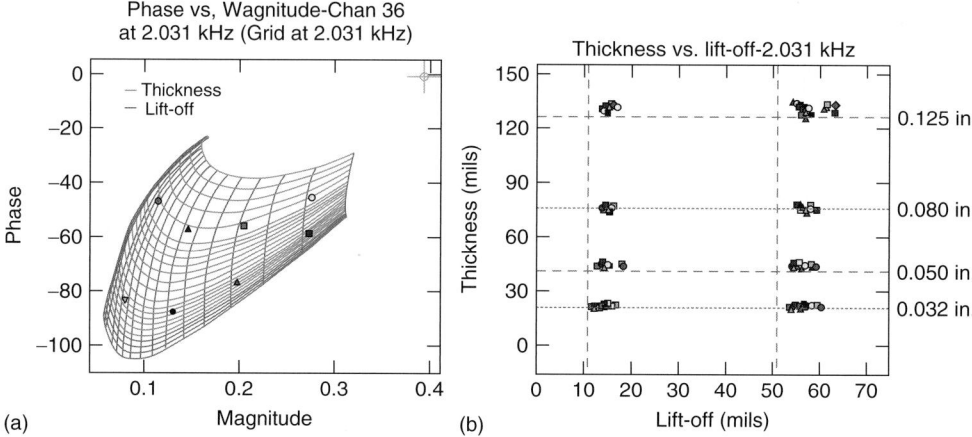

Figure 10. (a) Hidden corrosion monitoring results displayed on a metal (aluminum alloy) layer thickness—lift-off measurement grid and (b) comparison of metal layer thickness measured by MWM-Array with nominal values.

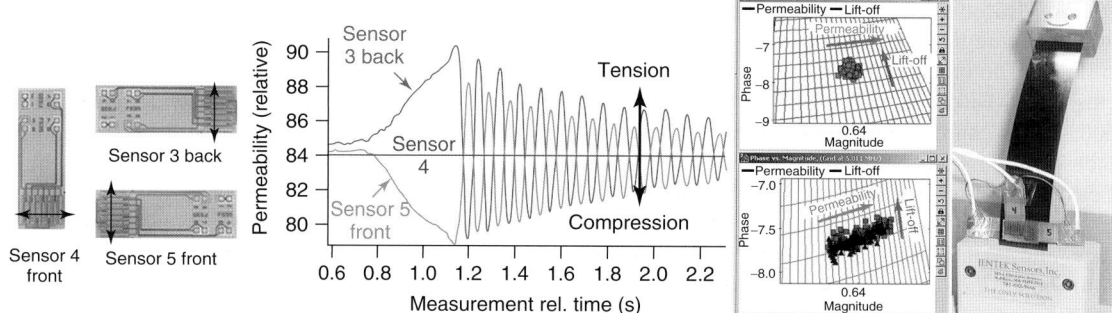

Figure 11. Demonstration of dynamic stress monitoring in a steel strip using networked MWM sensors.

remaining metal layer thickness and liftoff, demonstrating independent measurement of these two variables. The comparison of estimated layer thicknesses with "nominal" values is shown in Figure 10(b). Only a calibration in air was performed prior to making this absolute measurement, using the air calibration methods described in an ASTM standard [16]. This capability to provide onboard monitoring of hidden corrosion loss in fuel tanks, joints (metal–metal and metal–composite), and other difficult-to-access areas can offer substantial maintenance cost reductions by avoiding costly visual corrosion inspections.

5 STRESS MONITORING

Demonstrated capability of MWM sensors and MWM-Arrays to monitor stresses in steel components [8, 21, 22] permits monitoring of dynamic stresses in rotorcraft components using a network of these sensors called *magnetic stress gauges*. This capability is illustrated here for a vibrating steel strip, as shown in Figure 11. The magnetic stress gauges mounted on opposite sides of the strip detect the alternating tensile and compressive stresses and correctly indicate their phase relationship as well as gradual reduction of stresses due to damping. The third magnetic stress gauge was mounted orthogonal to the other two gauges and, consequently, was insensitive to the motion, since it responds to transverse stress that is zero in the case of plane stress, i.e., sufficiently close to the edges of the strip.

In separate overload tests on an actual steel component loaded well outside the elastic range, we have found, however, that a magnetic stress gauge mounted orthogonal to the principal stress *can* provide an indication of an overload event, which affects permeability in both directions as a result of the changes in residual stresses caused by stress redistribution due to plastic deformation that resulted from such an overload event. Stresses can also be monitored in components fabricated from nonferrous alloys as long as selected regions are coated with a ferromagnetic coating. Dynamic stress monitoring can be performed in a noncontact mode, as has been demonstrated in laboratory tests.

6 TEMPERATURE MONITORING

As shown in Figure 12, a deep penetration MWM-Array has been used to demonstrate the capability to monitor temperature of an aluminum plate separated from the sensor by air gap and an intermediate aluminum plate [14]. This capability of either single- or multiple-frequency MWM-Arrays eddy-current sensors to monitor temperature noncontact and/or through metal layers offers a unique value for the SHM of propulsion systems, energy systems, and other systems with high internal temperatures. Because these sensors require only a conducting path with remote instrumentation, they can also be used to operate in harsh environments or within processing facilities. Figure 13 illustrates that, if the temperature of the intermediate plate is not properly accounted for, the measurement of the buried plate temperature is incorrect. Thus, as illustrated in Figure 13, the use of a three-dimensional lattice (database) for the estimation of three unknowns (the conductivity of the two aluminum plates, and the gap between them) is essential. The relationship between temperature and conductivity is stable up to several hundred degrees. Thus, this is a practical tool for many applications.

7 COMPOSITE DISBOND/ DELAMINATION MONITORING

MWM-Arrays have also demonstrated the capability to provide monitoring of buried disbonds/delaminations in graphite fiber composites. As shown in Figure 14, the MWM-Array was used to monitor the initiation and growth of damage in a composite–composite joint test performed at Lockheed Martin. This test performed under NAVAIR funding demonstrated the capability of the MWM-Array to monitor disbond growth through over 0.2-in.-thick composite material, during the test, using a surface-mounted MWM-Array.

Unfortunately, eddy-current sensors only detected changes to the fibers (positions, directions, fiber cracking, and bulk conductivity changes associated with density and contacts). Thus, matrix damage is not detectable with this method. Other methods such as ultrasonics and thermography offer the

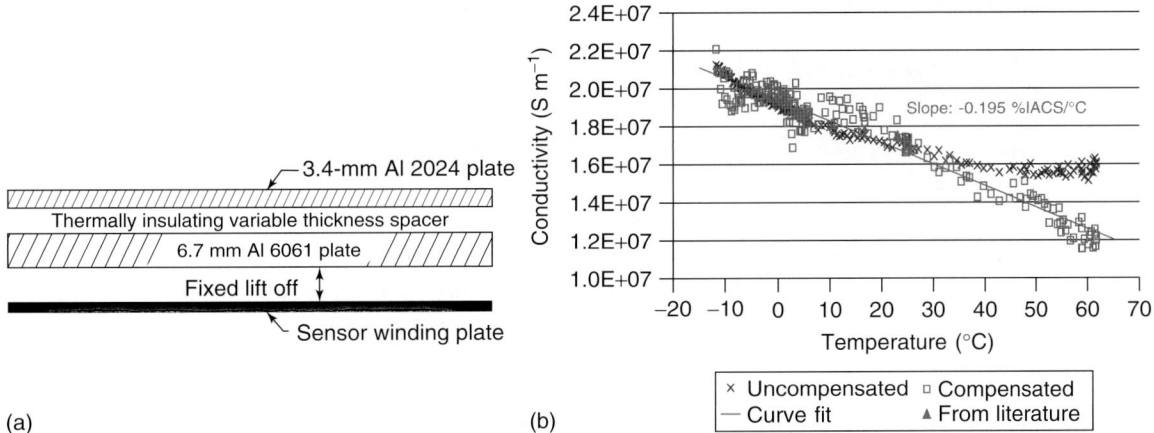

Figure 12. (a) Schematic of the demonstration setup and (b) response of MWM-Array with and without correction for the temperature variation of the intermediate aluminum plate. Note that the temperature of the two plates is measured independently using the MWM-Array at one frequency with two different-shaped magnetic field excitations [14].

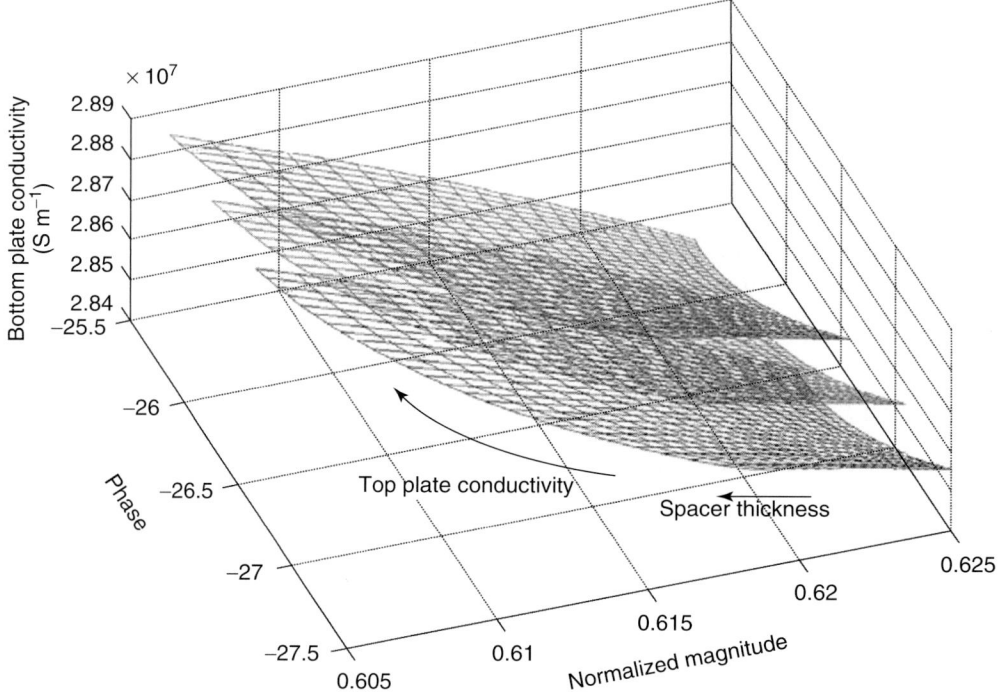

Figure 13. Lattice (precomputed database) for independent estimation of two plate temperatures (via conductivities) and the gap between them.

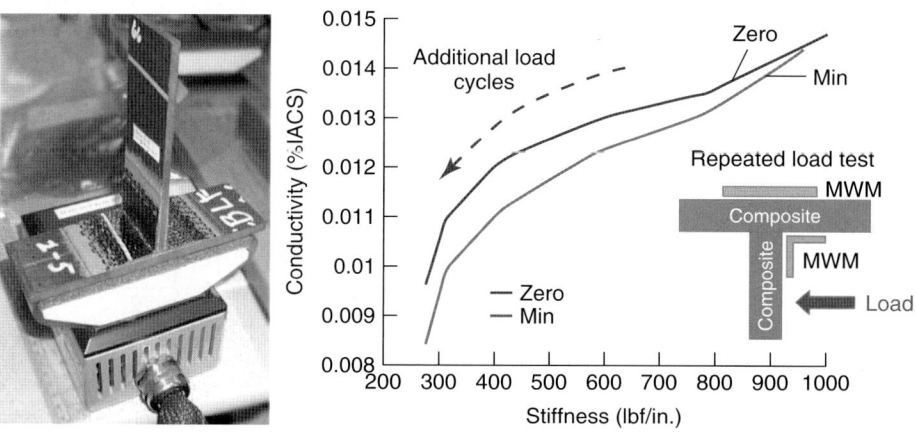

Figure 14. Composite–composite joint test performed at Lockheed Martin. MWM-Array measurements of conductivity are strongly correlated with the decreasing stiffness of the specimen as it undergoes load cycling.

capability to monitor matrix and fiber damage. One interesting method being demonstrated under ongoing NASA funding is Magneto-Thermography™. This promising method, developed by JENTEK Sensors, Inc., takes advantage of the demonstrated capability of MWM-Arrays to monitor buried fiber temperatures, to produce thermography images. This magneto-thermographic method, still in the research stage, can be implemented in a scanning or permanently mounted mode.

8 CONCLUSIONS

Eddy-current *in situ* health monitoring methods are now advancing and transitioning into use. These methods are now used in laboratory fatigue and corrosion test monitoring. In the next couple of years, the broader use of these methods are expected to accelerate owing to planned on-aircraft testing of fatigue sensor networks, using smart washer configurations of the MWM-Rosettes for buried cracks and linear MWM-Arrays for surface-breaking cracks.

Magnetic stress gauges also offer promise for near-term implementation. Full-scale testing and laboratory testing have proven the capability of this method to deliver reliable stress monitoring. This method is also expected to transition into use within the next few years.

Finally, adaptive life management, combining both installed fatigue monitoring and scanning NDT data, is expected to deliver the promise of cost savings and reduced logistics footprints for a variety of future and legacy platforms.

This is an exciting time for both offboard NDT and onboard SHM.

END NOTES

[a.] Note that for materials that exhibit significant variation of effective electrical properties, e.g., electrical conductivity or magnetic permeability with applied or residual stress measured by MWM, the potential to monitor remote damage is possible—by relating such damage occurrence to measured stress changes.

REFERENCES

[1] Goldfine N. Magnetometers for improved materials characterization in aerospace applications. *Materials Evaluation* 1993 **51**(3):396–405.

[2] Sheiretov Y. *Deep Penetration Magnetoquasistatic Sensors*, Ph.D. thesis. Department of Electrical Engineering and Computer Science, Massachusetts Institute of Technology: Cambridge, MA, June 2001.

[3] Goldfine N. *Hidden Feature Characterization Using a Database of Sensor Responses*, US Patent Number 7,161,351 B2, January 2007.

[4] Goldfine N, Zilberstein V, Grundy D, Weiss V, Washabaugh A. *Fabrication of Samples Having Predetermined Material Conditions*, US Patent Number 7,106,055 B2, September 2006.

[5] Goldfine N, Schlicker D, Washabaugh A, Grundy D, Zilberstein V. *Process Control and Damage Monitoring*, US Patent Number 7,095,224, B2, August 2006.

[6] Schlicker D, Goldfine N, Washabaugh A, Walrath K, Shay I, Grundy DC, Windoloski M. *Test Circuit Having Parallel Drive Segments and a Plurality of Sense Elements*, US Patent Number 7,049,811 B2, May 2006.

[7] Shay I, Goldfine N, Washabaugh A, Schlicker D. *Magnetic Field Sensor Having a Switchable Drive Current Spatial Distribution*, US Patent Number 6,992,482 B2, January 2006.

[8] Goldfine N, Schlicker D, Washabaugh A, Zilberstein V, Tsukernik V. *Surface Mounted and Scanning Spatially Periodic Eddy-Current Sensor Arrays*, US Patent Number 6 952 095 B1, October 2005.

[9] Schlicker D, Goldfine N, Washabaugh A, Walrath K, Shay I, Grundy D, Windoloski M. *Eddy Current Sensor Arrays Having Drive Windings with Extended Portions*, US Patent Number 6,784,662 B2, August 2004.

[10] Goldfine N, Melcher J. *Apparatus and Methods for Obtaining Increased Sensitivity, Selectivity, and Dynamic Range in Property Measurements Using Magnetometers*, US Patent Number 5,629,621, May 1997.

[11] Goldfine N, Melcher J. *Magnetometer Having Periodic Winding Structure and Material Property Estimator*, US Patent Number 5,453,689, September 1995.

[12] Goldfine N. Meandering winding magnetometers: the basics. *1992 ASNT Fall Conference*. Chicago, IL, November 1992.

[13] Goldfine N, Windoloski M, Zilberstein V, Contag G, Phan N, Davis R. Mapping and tracking of damage in titanium components for adaptive life management. *10th Joint NASA/DoD/FAA Conference on Aging Aircraft*. Atlanta, GA, 16–20 April 2007.

[14] Shay I, Zilberstein V, Washabaugh A, Goldfine N. Remote temperature and stress monitoring using low frequency inductive sensing. *SPIE Conference, NDE/Health Monitoring of Aerospace Materials and Composites*. San Diego, CA, 2003.

[15] Goldfine N, *et al.* Damage and usage monitoring for vertical flight vehicles. *AHS 63rd Annual Forum and*

Technology Display. Virginia Beach, VA, 1–3 May 2007.

[16] ASTM Standard Practice E2338-04, *Characterization of Coatings Using Conformable Eddy-Current Sensors without Coating Reference Standards*. ASTM International, Book of Standards, 2004; Vol. 03.

[17] Ball D, Sigl K, McKeighan P, Veit A, Grundy D, Washabaugh A, Goldfine N. An experimental and analytical investigation of multi-site damage in mechanically fastened joints. *USAF Aircraft Structural Integrity Program (ASIP) Conference*. Memphis, TN, December 2004.

[18] Goldfine N, Grundy D, Washabaugh A, Schlicker D, Sheiretov Y, Hugeunin C, Lovett T, Roach D. Corrosion and fatigue monitoring sensor networks. *Structural Health Monitoring Workshop*. Palo Alto, CA, September 2005.

[19] Weiland H, Moran J, Bovard F, Grundy D, Zilberstein V, Lorilla I, Schlicker D, Goldfine N. Corrosion monitoring of lap joints using MWM-array sensors. *ASM AeroMat*. Seattle, WA, May 2006.

[20] Goldfine N, *et al.* Corrosion detection and prioritization using scanning and permanently mounted MWM eddy-current arrays. *Tri-Service Corrosion Conference*. San Antonio, TX, January 2002.

[21] Goldfine N, Grundy D, Washabaugh A, Craven C, Weiss V, Zilberstein V. Fatigue and stress monitoring with magnetic sensor arrays. *2006 Annual Society for Experimental Mechanics (SEM) Conference*. St. Louis, MO, June 2006.

[22] Zilberstein V, Fisher M, Grundy D, Schlicker D, Tsukernik V, Vengrinovich V, Goldfine N, Yentzer T. Residual and applied stress estimation from directional magnetic permeability measurements with MWM sensors. *ASME Journal of Pressure Vessel Technology* 2002 **124**:375–381.

Chapter 59
Fiber-optic Sensor Principles

Kara Peters
Department of Mechanical and Aerospace Engineering, North Carolina State University, Raleigh, NC, USA

1 Introduction	1065
2 Lightwave Propagation in Optical Fibers	1065
3 Lightwave Sources	1069
4 Sensing Mechanisms	1071
5 Mechanical Properties of Optical Fibers	1073
6 Integration of Optical Fiber Sensors in Structural Components	1075
7 Conclusions	1078
End Notes	1078
Related Articles	1078
References	1079

1 INTRODUCTION

Optical fiber sensors present numerous advantages for the measurement of strain, temperature, humidity, pressure, and other parameters. For a review of their application to structural health monitoring, *see* Measures [1] and **Chapter 109**. Optical fiber sensors are immune to electromagnetic interference, do not present an ignition hazard, are lightweight, relatively unobtrusive, and are durable and resistant to corrosion (with appropriate coatings). Additionally, a large number of sensors can be multiplexed into a single optical fiber, drastically reducing the amount of cabling required to access the sensors. This can be an important weight and fire hazard reduction. This article presents the fundamental principles common to all optical fiber sensors, as well as general issues concerning their application to structural health monitoring systems. More detailed descriptions of specific optical fiber sensors can be found in **Chapter 60** and **Chapter 61**.

2 LIGHTWAVE PROPAGATION IN OPTICAL FIBERS

Optical fibers generally consist of a fused silica (SiO_2) core and cladding. During drawing of the optical fiber, dopants are added to the silica to provide an index of refraction distribution throughout the cross section of the optical fiber [2]. Typical core diameters are $5-10\,\mu m$ for single-mode optical fibers and greater than $50-200\,\mu m$ for multimode optical fibers. The cladding diameter for almost all optical fibers is standardized at $125\,\mu m$ to permit standardized couplers. Small diameter fibers ($80\,\mu m$) have also been developed for sensing applications [3]. The small diameter makes the fiber sensor less invasive

Encyclopedia of Structural Health Monitoring. Edited by Christian Boller, Fu-Kuo Chang and Yozo Fujino © 2009 John Wiley & Sons, Ltd. ISBN: 978-0-470-05822-0.

when the fiber is embedded in a host material system and increases the sensitivity of the sensor to applied loads.

Standard optical fibers are coated with a UV-cured acrylate to prevent moisture from entering the fiber and make the fiber more durable for handling. Humidity can induce microcracking in the fiber and cause premature failure of the fiber. A standard diameter for acrylate coatings is 250 μm. For optical fiber sensors, different coatings may be applied such as polyimide, which is stiffer, to provide better protection and strain transfer and is able to withstand higher working temperatures. Removal of the polyimide coating for connectors is considerably more difficult, however, than removal of acrylate coatings. Sensing applications also often require bare optical fibers or coatings that are specially fabricated to increase sensitivity of the optical fiber to outside parameters such as humidity (*see* **Chapter 61**) [4–7].

2.1 Step-index fibers

The cross section of a step-index optical fiber is shown in Figure 1. The index of refraction distribution is only a function of the radius with

$$n(r,\theta) = n_1 \quad r \leq a$$
$$n(r,\theta) = n_2 \quad r > a \quad (1)$$

In order for propagation of guided lightwaves to occur $n_1 > n_2$.

The index of refraction difference between the core and cladding is typically less than 1%, therefore we can use the weakly guiding assumption to describe lightwave propagation through the fiber [8]. This assumption states that lightwave can be divided into transverse electric (TE) and transverse magnetic (TM) fields that are orthogonal to one another as they propagate. The two fields also have the same propagation constant, i.e., phase shift per unit distance. Therefore, we can study either the TE or TM mode individually and apply the same results to the other component. Applying the weakly guiding assumption, each component of the lightwave propagating along the fiber in the z direction satisfies the scalar wave equation,

$$\psi(r,\theta,z,t) = \bar{\psi}(r,\theta)\,e^{i(\omega t - \beta z)} \quad (2)$$

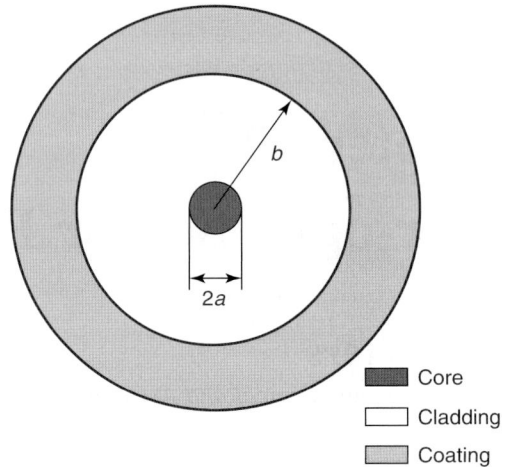

Figure 1. Cross section of step-index optical fiber. Figure not to scale.

where $\bar{\psi}$ is the energy distribution in the plane perpendicular to the propagation direction, ω is the angular frequency ($\omega = 2\pi c/\lambda$ where c is the speed of light in a vacuum and λ is the free space wavelength of the lightwave), and β is the propagation constant of the mode (phase shift per unit length). We assume an infinite diameter cladding and solve over equation (2) in each of the index of refraction regions defined by equation (1). Next, we apply continuity conditions at $r = a$ and find a number of distinct guided modes,[a] which are commonly referred to as the *linearly polarized* (LP) modes.

The energy distribution of the $LP_{\ell m}$ mode in the cross section of the fiber can be written as

$$\bar{\psi}(r,\theta) = \begin{cases} A_0[J_\ell(pr)/J_\ell(pa)]\cos(\ell\theta) & r < a \\ A_0[K_\ell(\gamma r)/K_\ell(\gamma a)]\cos(\ell\theta) & r > a \end{cases} \quad (3)$$

where $J_\ell(r)$ is the Bessel function of the first kind of order ℓ, $K_\ell(r)$ is the modified Bessel function of the first kind of order ℓ, $p = \sqrt{4\pi^2 n_1^2/\lambda^2 - \beta^2}$ and $\gamma = \sqrt{\beta^2 - 4\pi^2 n_2^2/\lambda^2}$ [8]. For each choice of ℓ, there exists zero, one, or multiple solutions to β within the guided range $n_2 < (\beta\lambda)/(2\pi) < n_1$. We order these solutions beginning with the highest value, and thus the $LP_{\ell m}$ corresponds to the mth solution. The energy distribution of equation (3) is different for each mode, due to the different values of β and therefore p and γ. The intensity distribution, $I(r,\theta) = |\bar{\psi}(r,\theta)|^2$, for several of the lowest

Fiber-optic Sensor Principles **1067**

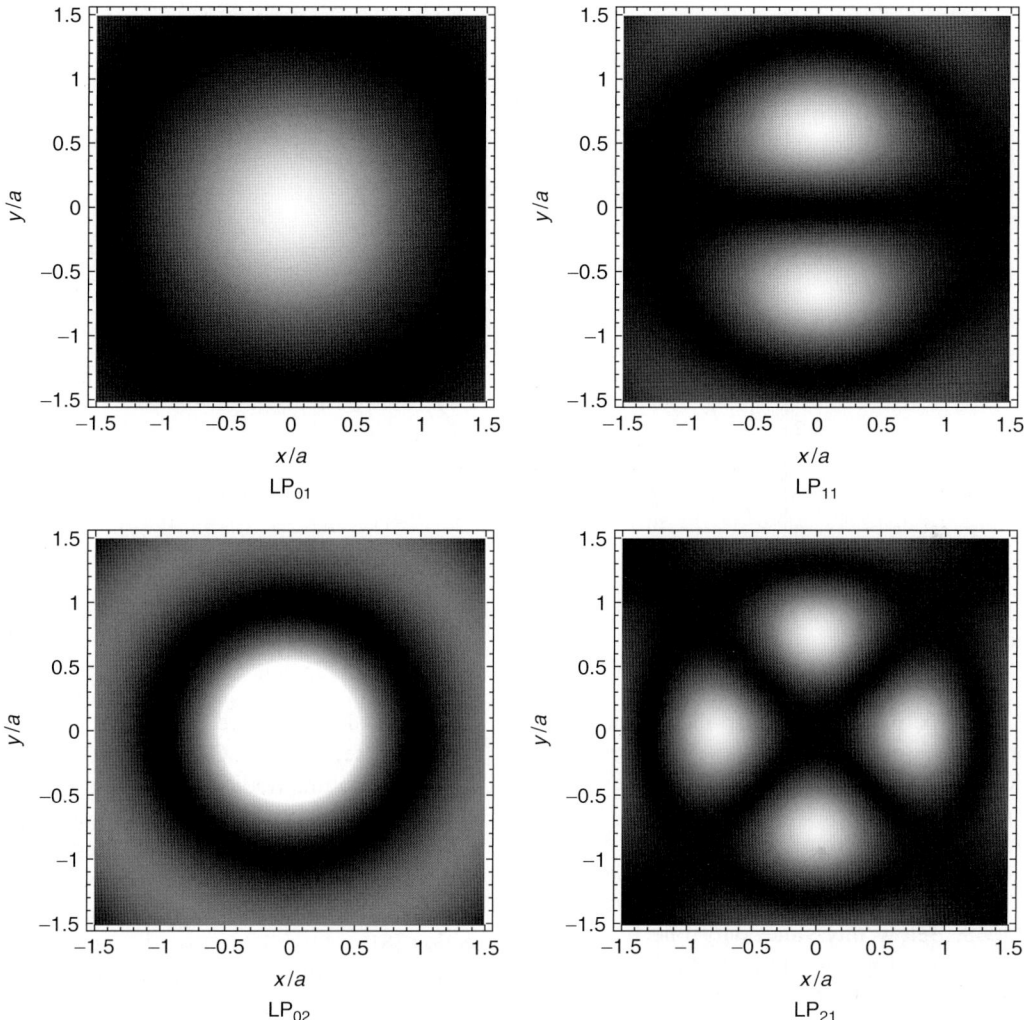

Figure 2. Intensity distribution plots for first four modes (LP_{01}, LP_{11}, LP_{02}, and LP_{21}) for step-index fiber. Radius of core is a. White area represents maximum intensity. [Reproduced from Ref. 2. © Cambridge University Press, 1998.]

propagating modes are plotted in Figure 2. For each propagating mode, we can also define an effective index of refraction, $n_{\text{eff}} = \beta\lambda/(2\pi)$, where n_{eff} corresponds to the index of refraction of an equivalent homogeneous material for which the planar wave would propagate with the same propagation constant β as through the step-index fiber above. For guided modes, $n_2 < n_{\text{eff}} < n_1$. The effective index of refraction, or mode propagation constant β, will play a critical role in the response of many optical fiber sensors (*see* **Chapter 60**; **Chapter 61**).

The number of modes that can propagate at a particular wavelength depends upon the normalized frequency at that wavelength, V,

$$V = \frac{2\pi a}{\lambda}\sqrt{n_1^2 - n_2^2} \qquad (4)$$

Figure 3 plots a normalized propagation constant b,

$$b = \frac{n_{\text{eff}}^2 - n_2^2}{n_1^2 - n_2^2} \qquad (5)$$

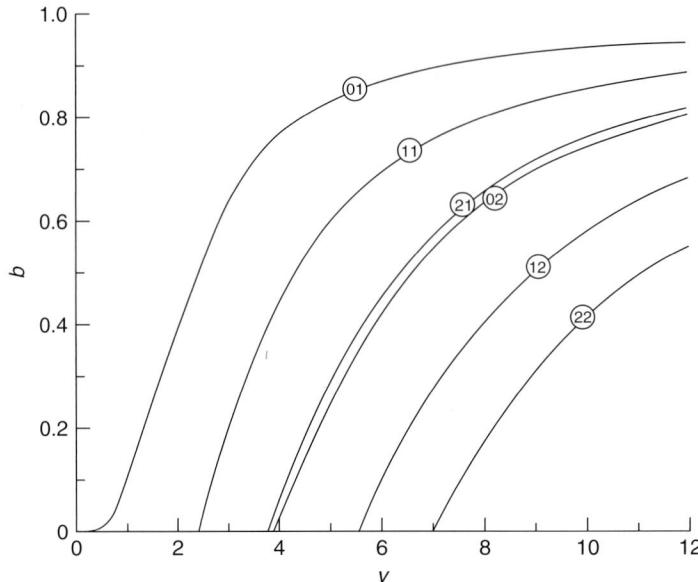

Figure 3. Propagation constant versus normalized frequency for modes LP_{01}, LP_{11}, LP_{21}, LP_{02}, LP_{12}, and LP_{22} of a step-index fiber [2].

for the first six modes of a step-index fiber as a function of the normalized frequency. The normalization allows one to apply Figure 3 to any step-index fiber parameters. As can be seen in Figure 3, each mode has a cutoff value of V such that it will not propagate at any value of V lower than the cutoff value. The cutoff value for the LP_{11} mode is at $V = 2.4048$. Below this value only one mode, the LP_{01} mode, can propagate through the fiber. Under these conditions, the fiber is referred to as a *single-mode fiber*. The LP_{01} is referred to as the *fundamental mode* as it has the highest power density in the core of the optical fiber. For many sensing applications, it is important to utilize fibers that are single mode at the wavelength to be interrogated so that the mode of propagation and coupling is well defined. Additionally, the fundamental mode is the least affected by bending losses, as the energy is concentrated in the center of the fiber. The input light is then coupled to individual modes.

For optical fibers operating at $V > 2.4048$, input lightwaves are coupled into multiple modes. The specific power distribution between the multiple modes is a function of the coupling. Multimode sensors do exist and are generally easier to practically couple to instrumentation as the fiber core is larger and the modes are more spread out throughout the fiber cross section (*see* **Chapter 60**) [9–14]. As the normalized natural frequency V increases, the number of modes rapidly increases, with a good approximation of $N \simeq V^2/2$ for $V \gg 1$. Therefore, when a large number of modes are present, one can assume a uniform distribution of energy throughout the cross section [8].

Finally, as the cladding diameter is not actually infinite, additional mode solutions are possible due to the interface condition between the outer diameter of the fiber and the surrounding medium. The form of these modes depends upon the index of refraction of the surrounding medium, n_3. First, consider the case of $n_3 < n_2$ such as when the fiber has been stripped of any coating and is exposed to air. Solving equation (2), taking into account the cladding diameter yields additional discrete guided modes, called *cladding modes*, with β values below those of the guided modes. The power distributions of the cladding modes are spread over a much larger area of the cross section of the optical fiber, therefore they are far more sensitive to imperfections and microbends in the drawn silica. The cladding modes can be excited locally in long period grating sensors and exploited for the measurement of cure monitoring, chemical or environmental sensing (*see* **Chapter 61**) [15].

2.2 Other fiber types

Although the step-index fiber is the easiest for which to analyze mode propagation, most commercially available optical fibers have different index distributions as shown in Figure 4 [2]. Such index distributions provide better dispersion properties, important for long-haul telecommunication networks. While most optical fiber manufacturers do not provide detailed information on the index distributions, they do provide data such as the core radius, design operating wavelength, and n_{eff} at common wavelengths (see, for example, [16]). From this information, the user can calculate the response of the optical fiber sensor using the same principles as for the step-index fiber (*see* **Chapter 60**).

A second important class of optical fibers applied for sensing is high-birefringence fibers, also referred to as *polarization maintaining* (PM) fibers (see examples in Figure 5) [17]. PM fibers propagate two separate, LP fundamental modes (LP_{01}) at two separate propagation constants, β_1 and β_2. The modes are polarized about orthogonal axes, referred to as the *fast and slow axes*. As the polarization axes are orthogonal, the two modes do not interfere with one another as they propagate along the optical fiber. These orthogonal modes can be used to independently measure multiple parameters (*see* **Chapter 61**). The fiber types shown in Figure 5 can be divided into two categories. The first group shown in Figure 5(a) and (b), the elliptical core fiber and D-fiber, are PM fibers owing to the fact that the fibers are not geometrically axisymmetric. These fibers are considered mechanically homogeneous. The second group of fibers shown in Figure 5(c–e) achieve birefringence due to the lack of geometric symmetry as well as residual stresses induced by the stress-applying portion (SAP) surrounding the core. This SAP has different mechanical and thermal expansion properties than the fused silica, inducing large residual stresses on the fiber core during cooling of the fiber after it is drawn. The residual stresses induce birefringence at the core through the photoelastic effect. The magnitude of birefringence due to the residual stresses is much higher than the geometric effect, and therefore the fiber types of Figure 5(c–e) are more suitable for sensor applications as the measurands will be easier to distinguish. Owing to the presence of the SAP, these fibers are considered to be mechanically inhomogeneous.

3 LIGHTWAVE SOURCES

Lightwave sources for optical fiber sensors generally fall into three categories: light emitting diodes (LEDs), laser diodes, and superluminescent diodes

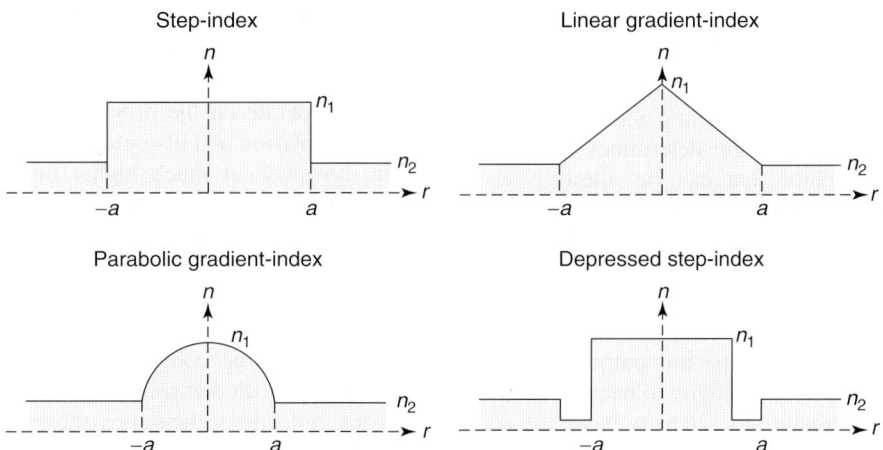

Figure 4. Common fiber index of refraction profiles.

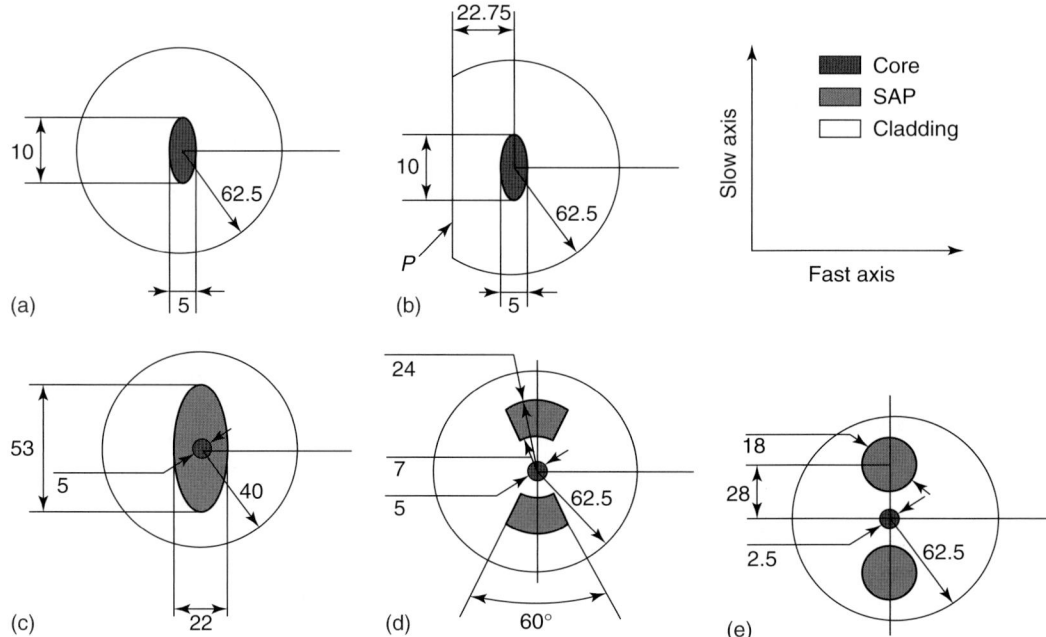

Figure 5. Geometry of common PM fiber types: (a) elliptical core fiber; (b) D-fiber; (c) elliptical core SAP fiber; (d) bow-tie fiber; and (e) panda fiber. The slow and fast axes are also indicated. All dimensions are shown in micrometers.

(SLDs). A summary of each of these is provided in this section; for further information, see [1, 2, 18]. Each of these light sources has advantages and disadvantages based on the output bandwidth, power, coherence length, and cost. The coherence length is defined as the distance over which the amplitude modulation falls off to a certain percentage of the original lightwave amplitude and can be estimated by

$$L_c = \frac{c}{d\upsilon} \quad (6)$$

where $d\upsilon$ is the spectral frequency width of the light source. The coherence length determines the optical path length difference that can be allowed when interfering with multiple lightwaves from the same original source (*see* **Chapter 60**). Additional light sources exist such as gas lasers; however, it is not practical to couple them into optical fibers for field applications.

LEDs provide a broad spectral output at a relatively low cost and low sensitivity to back reflections. LEDs are most suitable for intensity-based sensors. However, their low-coherence length makes them not suitable for interferometric applications other than low-coherence interferometry. LEDs exist in two forms: surface emitting diodes and edge emitting diodes. The surface emitting diodes provide an output spectrum that is difficult to focus in a single direction. Therefore, they are more suitable for coupling to large core size multimode optical fibers than to single-mode fibers. Edge emitting diodes provide a more easily focused output for coupling into optical fibers; however, the total amount of optical power input to the optical fiber tends to be low.

SLDs operate on the principle of amplified spontaneous emission and also provide a broadband spectrum, however, at much higher output power than LEDs. The angular narrowing of the output of the SLD also allows improved coupling into single-mode optical fibers, with output powers on the order of 8 mW.

Laser diodes emit a very narrow bandwidth spectrum with a large coherence length at high output powers. Laser diodes sources are highly susceptible to back reflections; however, these can be reduced or eliminated by applying antireflection coatings to or angling the end of the coupled optical fiber or through optical isolators. Laser diodes also require accurate temperature control to control the spectral output. One of the advantages of laser diodes is that they can be coupled with single-mode optical fibers through microlenses to produce fiber laser diodes that operate with high output powers, on the order of 10 mW. By combining the fiber laser diode with a Fabry–Perot cavity, one can produce an extremely strong, narrow bandwidth output. Additionally, the cavity length can be scanned to provide a laser with a narrow bandwidth, wavelength tunable output.

4 SENSING MECHANISMS

External sensing parameters are generally encoded into information within lightwaves propagating through an optical fiber in one of four manners, summarized in Figure 6.

4.1 Intensity modulation

The intensity of the lightwave propagating through an optical fiber can be modified through microbending of the optical fiber, a change in coupling from the fundamental mode to other nonguided modes, fracture of the optical fiber or a change in power coupled into the fiber (*see* **Chapter 60**). In general, the input power entering the optical fiber is held constant, while the intensity of the light transmitted through the fiber or reflected back to the input of the fiber is measured with a photodetector. Measuring the intensity of the propagating lightwave is relatively simple; however, light sources themselves often fluctuate in intensity. Feedback control loops can be applied to reduce these fluctuations, or the illuminating lightwave divided to create a reference intensity, as shown in Figure 7(a). Intensity-based sensors provide absolute measurements, meaning that they are insensitive to power interruptions between data collections. One drawback to intensity-based sensors, however, is that they cannot be multiplexed for sensor networks.

4.2 Phase modulation

The application of strain or temperature to an optical fiber changes the optical path length traversed by the lightwave propagating through the fiber, $n_{\text{eff}}L$

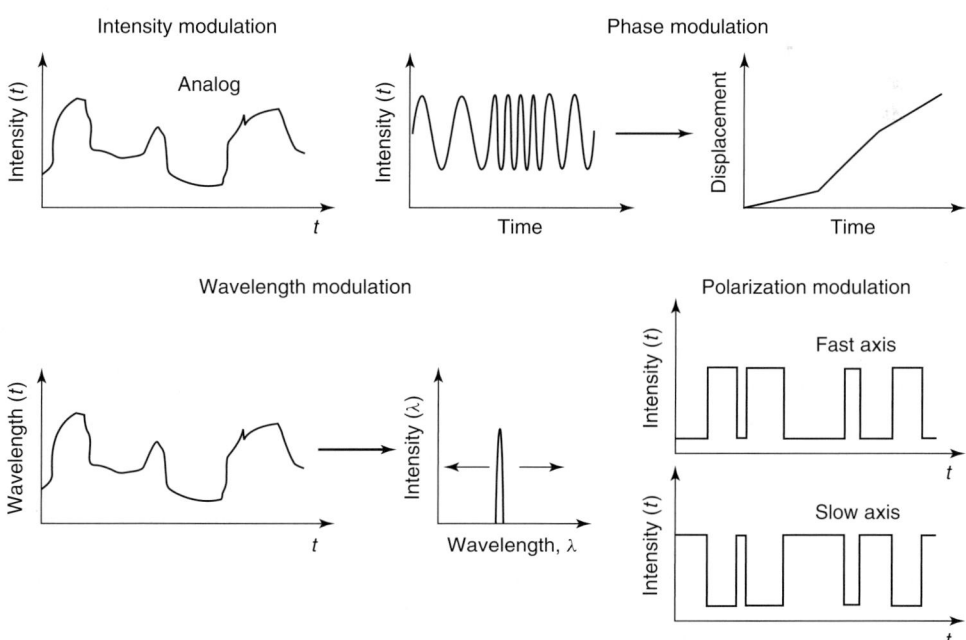

Figure 6. Output signals of optical fiber sensors using four primary sensing mechanisms.

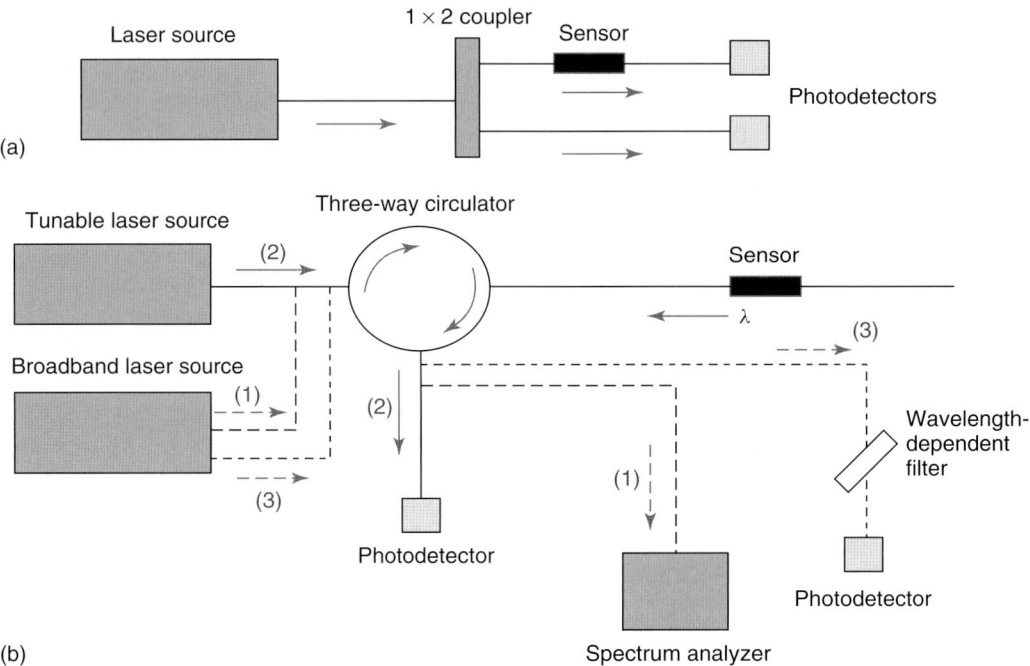

Figure 7. Sensor interrogation schemes for (a) intensity-based sensors and (b) wavelength (frequency) based sensors. For (b), three different possible schemes are shown. See text for details.

(*see* **Chapter 60**). As the phase of the lightwave cannot be measured directly, the lightwave is generally recombined with a reference lightwave from the same laser source (so that the two are coherent). When the reference lightwave is not exposed to the external parameters, their relative phase shift between the signals can be related to the applied strain or temperature. The phase modulation is extremely sensitive to strain and therefore can provide very accurate measurements [1]; however, the measurement of phase shift is not absolute owing to the signal periodicity and is therefore affected by power interruptions between data collections. Additionally, fluctuations in the laser source intensity can be misinterpreted as phase shifts. Several systems are now commercially available to alleviate these problems, each with relative advantages and disadvantages (*see* **Chapter 60**). Interferometric measurements also cannot generally be multiplexed. The reference signal can be completely sheltered from external parameters or used for compensation of unwanted measurands such as temperature.

4.3 Spectral modulation

External sensing parameters can also be converted into spectral information of the lightwave, for example, using fiber Bragg gratings (*see* **Chapter 61**) or Fabry–Perot interferometers (*see* **Chapter 60**). Such sensors typically act as filters, transmitting certain wavelengths and radiating or reflecting others. Changes in external parameters are therefore converted into a wavelength shift of the transmitted or reflected spectrum of the sensor. A typical signal is shown in the example of Figure 6, for which the reflected wavelength is changing with time. The wavelength encoded signal must then be interpreted, which is commonly performed in using one of three methods: (i) by launching a broadband of light into the fiber and applying a spectrum analyzer to select the reflected wavelength; (ii) by applying a tunable laser for which the output wavelength can be scanned while the reflected intensity is measured by a photodetector; or (iii) by launching a broadband of light and using a wavelength-dependent filter to identify the reflected wavelength (Figure 7b). Other, more complex methods based on three-way couplers or fast Fourier transforms are also available for data acquisition at higher scan rates or multiplexing sensors (*see* **Chapter 61**).

4.4 Polarization modulation

A final method to encode sensing information for transmission through an optical fiber is through the polarization state of the propagating lightwave [1, 19, 20]. For the example shown in Figure 6, the power of a lightwave propagating through a PM fiber is transferred between the mode polarized about the fast axis and the mode polarized about the slow axis. The power in each mode can be determined by applying a polarizing filter to the output signal. As mentioned above, the two modes propagating through a PM fiber do not normally transfer power as they propagate since they are orthogonally polarized. However, external stimuli such as pressure or twisting of the optical fiber will induce transfer between the two modes.

5 MECHANICAL PROPERTIES OF OPTICAL FIBERS

It is important to know the strength and attenuation characteristics of optical fibers to apply them for sensing applications where severe loading conditions may occur. This is particularly true for many structural health monitoring applications. Additionally, attenuation properties often determine the maximum number of sensors that can be applied in a single network or the physical distance over which the network may span.

5.1 Stiffness and strength

Standard fused silica fibers are linear elastic until failure with the following material properties: the elastic modulus, $E = 72$ GPa, Poisson's ratio, $\nu = 0.20$, and thermal expansion coefficient, $\alpha = 5.5 \times 10^{-7}/°C$ [21]. Fused silica optical fibers are extremely brittle and their tensile strength follows a Weibull distribution. The cumulative survival probability, P, of a fiber of length L at an applied axial stress level S is therefore

$$P(S, L) = \exp\left[-\left(\frac{S}{S_0}\right)^m \left(\frac{L}{L_0}\right)\right] \quad (7)$$

where m is the material Weibull parameter and S_0 is the inert strength of the fiber (under pristine conditions), measured for a length of fiber L_0 and a survival probability of 36.8% [22]. The medial strength, σ_{med} corresponds to a failure probability $P = 0.5$. Another important material property is the corrosion susceptibility, n, which determines the growth rate of a microcrack in the fiber,

$$\frac{dw}{dt} = AK_I^n \quad (8)$$

where w is the crack length and A is a constant. Typical measured values for standard silica optical fibers are $m = 112, n = 14, \sigma_{med} = 5.13$ GPa [23]. Kapron and Yuce [22] tested a large set of standard optical fibers in static and dynamic fatigue. The bending strength of optical fibers can be described similarly by a Weibull distribution; however, a general rule is that standard silica sensor fibers can be safely bent to a radius of 5 mm [24].

A second common material class used for optical fibers is polymers. Polymer optical fibers (POFs) have considerably lower stiffness than silica fibers, with $E = 2.4–3.0$ GPa, $\nu = 0.34$ [25]. Their stress–strain response is also extremely sensitive to strain rate, temperature, humidity, and hysteresis effects. POFs are more flexible in bending those silica fibers; therefore, they can be handled and embedded without a protective coating.

5.2 Attenuation and dispersion

The two major factors that determine the maximum length of signal transmission in an optical fiber are the attenuation and dispersion properties of the fiber. For a fixed length of optical fiber, the attenuation, $\bar{\alpha}$, is defined as the ratio of power input, P_i, to the power output, P_o, of a given length of optical fiber and is measured in decibels [2],

$$\bar{\alpha} = 10 \log_{10} \frac{P_i}{P_o} \quad (9)$$

On the other hand, dispersion (or broadening of propagating pulses) is generally due to the wavelength-dependent properties of silica and, therefore, the wavelength-dependent propagation constants. For telecommunication applications these factors limit the length of an optical fiber that can be used; however, for sensing applications these limits are rarely an issue, except for large structural applications.

For fused silica optical fibers, three main phenomena contribute to attenuation [2, 8]:

- Rayleigh scattering occurs owing to inhomogeneities in the amorphous glasslike silica. The resulting attenuation is proportional to λ^{-4}.
- The infrared absorption properties of silica essentially prevent transmission at wavelengths above 1600 nm.
- OH impurities in the fused silica create "water" absorption peaks at wavelengths of 950, 1240, and 1390 nm.

The resulting attenuation spectrum of a silica optical fiber can be seen in Figure 8. Three primary transmission windows are therefore used in telecommunication applications: (i) around 850 nm with an approximate attenuation of 2 dB km^{-1}; (ii) around 1300 nm with an approximate attenuation of 0.4 dB km^{-1} (here the material dispersion is practically zero); and (iii) around 1550 nm with an approximate attenuation of 0.25 dB km^{-1}. The transmission windows at 1550 and 1300 nm are most commonly used since the attenuation is the lowest of the three windows at 1550 nm, and at 1300 nm the material dispersion is practically zero. The same three transmissions windows are typically used for optical fiber sensors owing to the relative low cost, wide selection and high quality of components including optical fibers, couplers, and laser sources available for telecommunications applications.

POFs considerably demonstrate different attenuation characteristics than fused silica. Most noticeably, the magnitude of attenuation is orders of magnitude larger than that of silica [25]. Therefore, POFs are most often used for short connector applications that require a flexible fiber. Secondly, whereas the attenuation of silica decreases with wavelength in the infrared region, the attenuation of the polymer materials rapidly increases with wavelength. Therefore,

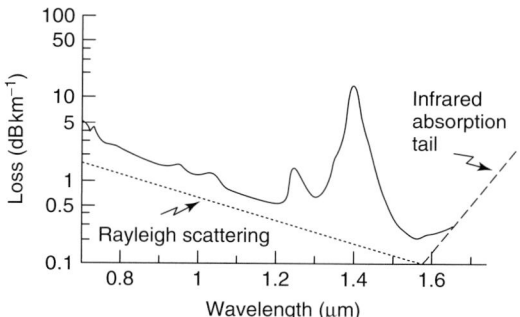

Figure 8. Material attenuation losses as a function of wavelength for fused silica (SiO_2) [2]. [Reproduced with permission from Ref. 2. © Cambridge University Press, 1998.]

POF sensors typically operate in the visible wavelength ranges rather than in the infrared ranges. Most commercially available POFs are multimode fibers in the visible wavelengths; however, recent advances in fabrication techniques have produced single-mode POFs for visible wavelength transmission [26–29] (*see* **Chapter 62**).

Other important factors contributing to signal losses in optical fiber sensors are splice losses, component losses, and bending losses. Splice losses depend upon the type of splice applied; fusion splices yield much lower losses than mechanical splices for example. Component losses appear in network components such as couplers or circulators. Finally, one important difference between optical fiber sensor networks and electrical sensor networks is that the optical fiber cannot be bent to an arbitrary radius of curvature. The bend loss in a step-index fiber per unit

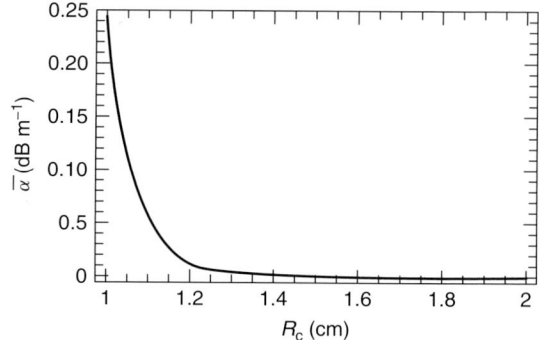

Figure 9. Bending loss as a function of bend radius of curvature ($a = 4\,\mu m$, $\lambda = 1550\,nm$, $n_1 = 1.460$, and $n_2 = 1.453$).

fiber length can be approximated by

$$\alpha = 4.343 \left(\frac{\pi}{4aR_c}\right)^{1/2} \left[\frac{pa}{K_1(\gamma a)}\right]^2 \frac{1}{(\gamma a)^{3/2}}$$
$$\times \exp\left[-\frac{2(\gamma a)^3}{3k_0^2 a^3 n_1^2} R_c\right] \quad (10)$$

where R_c is the radius of curvature of the bent fiber and p and γ were defined for equation (3) [2]. Equation (10) is plotted for typical single-mode fiber properties in Figure 9, from which one can see that the bending loss increases exponentially for small radii of curvature.

6 INTEGRATION OF OPTICAL FIBER SENSORS IN STRUCTURAL COMPONENTS

In this section, we summarize the general issues to consider when integrating sensors into structural components, particularly embedded sensors. These include coating options, material compatibilities, strain transfer, and structural integrity. Specific issues related to composite laminates and concrete structures are included.

Various coating materials have been applied for embedded applications including the standard acrylate, which is relatively pliable and allows the fiber to slide within the coating, to reduce crimping of the fiber. However, the acrylate coating does not provide good strain transfer between the sensor and the host material owing to the low stiffness of the coating and the fact that the fiber debonds from the coating well before failure occurs in the host material [30]. As an alternative, polyimide coatings are stiffer and well bonded to the optical fiber. Polyimide coatings provide excellent strain transfer and excellent durability of the sensor during high temperature (up to 600 °C) and high-pressure fabrication conditions [21]. The main disadvantage to polyimide coatings is that removal of the coating for coupling, splicing, etc. must be done chemically. Bare optical fibers have also been embedded for good stress–strain transfer when durability concerns are not as important. Several researchers have calculated optimal coating properties to minimize the obtrusivity of embedded sensors

[31, 32] or maximize the measurement capabilities of embedded sensors considering induced strains and their criticality for damage initiation [33]. For composites with other than polymer matrices, surface bonding of the optical sensor is often possible. For example, Figure 10 shows a gold coated silica optical fiber bonded to a titanium matrix composite using a Ni-based plasma spray.

One of the challenges to apply embedded optical fibers for structural health monitoring applications is to calculate the stress–strain state in the host material as a function of the stress–strain state in the optical fiber sensor. A variety of models have been developed for the stress–strain transfer to optical fiber sensors with and without coatings embedded in isotropic materials [35–38]. These models are based on shear-lag analysis, originally derived for stress transfer between reinforcing fibers in fiber reinforced composites. Li *et al.* [39] and LeBlanc [40, 41] extended the shear-lag analysis to include plastic yielding of the coating and debonding at the coating–fiber interface. For the application of optical fiber sensors embedded in orthotropic materials, Van Steenkiste and Springer [42] treated the optical fiber as an inclusion in the composite cross section and applied elasticity to calculate the relationship between the strain in the host material and strain in the optical fiber. Each of these approaches assumes that the strain field is slowly varying; therefore, the average strain at the location of the sensor is sufficient to predict

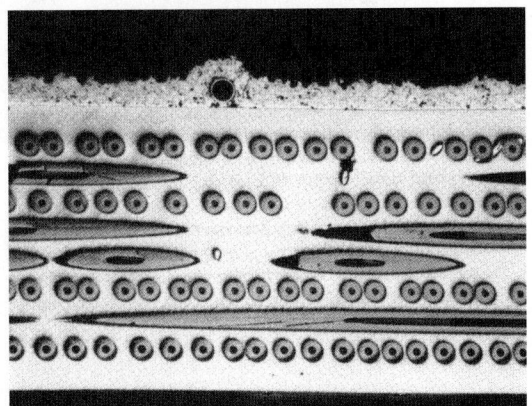

Figure 10. Gold coated silica optical fiber surface bonded to titanium matrix composite with silicon carbide reinforcing fibers using Ni-based plasma spray (image scale 32×) [34]. [Reproduced with permission from Ref. 34. © Cambridge University Press, 1998.]

its response. For more rapidly varying strain fields such as near the location of damage, Prabhugoud and Peters [43] incorporated the shear-lag theory to predict the response of a sensor located near matrix cracking in a unidirectional composite ply.

Finally, one of the major advantages of optical fiber sensors for structural health monitoring of composite structures is that large networks of sensors can be relatively unobtrusively embedded into the laminate for measurement throughout the lifetime of the structure. Owing to their high multiplexing capability (*see* **Chapter 61**), a single optical fiber strand can contain a large number of sensors, requiring only a single ingress and egress location. It is necessary to protect the optical fiber well and eliminate strong stress concentrations that would otherwise appear at the ingress or egress points since the optical fibers are extremely fragile in bending. Commonly, this can be accomplished using polymer cabling; however, for some structural applications more advanced strategies are required [44].

6.1 Laminated composites

Optical fiber sensors are well suited for strain monitoring in polymer matrix–fiber reinforced composites because the failure strain of the optical fiber is significantly larger than that of typical reinforcing fibers such as carbon [45]. A typical carbon–fiber reinforced epoxy composite has carbon reinforcing fibers of $5-10\,\mu\text{m}$ diameter and ply thicknesses

Figure 11. Micrograph of embedded optical fibers at different orientation to reinforcing fibers. Air voids next to optical fibers can be seen for orientations at $60°$ and $90°$ [47]. [Reproduced from Ref. 47. © Sage Publications, 2006.]

of 120–140 μm. Therefore, the optical fiber size (without coating) is on the order of one-ply thickness. When embedded between partially cured (prepreg) plies during the lamination fabrication, the presence of the optical fiber therefore introduces a perturbation to the local material system, typically resulting in a resin-rich zone referred to as a *resin eye* which forms around the fiber as seen in Figure 11. This resin eye enhances the local stresses surrounding the optical fiber and potentially induces local matrix cracking in adjacent plies or delamination [31]. For optical fiber sensors embedded within a single ply, the interphase region between the host epoxy and the embedded sensor can also play a significant role in the durability of the composite. Sirkis and Lu [46] used electron backscatter and acoustic emission to measure interphase thicknesses. They also determined that the residual stresses present in the optical fiber sensor after laminate fabrication can be significant.

Sirkis and Lu [46] and Jensen *et al.* [48, 49] demonstrated that optical fibers embedded parallel to the reinforcing fibers do not degrade the axial tensile strength of the laminate; however, the laminate transverse tensile strength and axial compressive strength are reduced by 20–70%. Embedding optical fibers at orientations other than parallel to the reinforcing plies does decrease the axial tensile strength of the composite [50, 51]. Kim *et al.* [51] demonstrated that there was not a significant decrease in axial tensile strength until the relative angle reached 45°, after which point the strength decreases rapidly with relative angle. They attributed this decrease in strength to the aspect ratio of the resin eye, which increased with relative angle. This was later confirmed by Shivakumar [47] (Figure 11) who also observed the presence of air voids next to the optical fiber at high relative angles. Similar observations on the laminate strengths have been made for fatigue loading conditions. The presence of the optical fiber does not noticeably reduce the fatigue life for laminates in tension–tension fatigue [52, 53], however, does create a significant decrease in fatigue life for tension–compression fatigue [53].

Optical fiber sensors embedded between multiple plies of two-dimensional woven composites demonstrate additional challenges. In addition to the resin eye formation, the presence of the optical fiber also introduces additional stresses owing to the undulations in the woven plies surrounding the optical

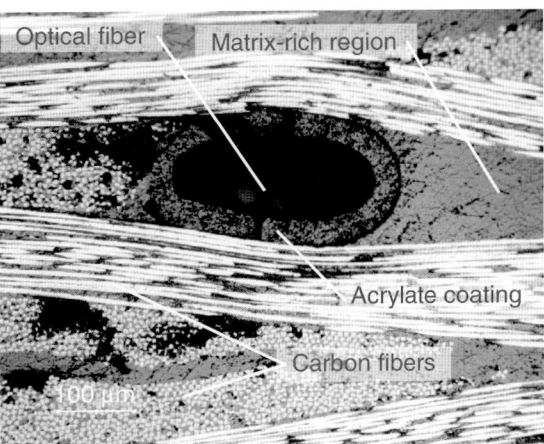

Figure 12. Optical micrograph of embedded optical fiber with acrylate coating in two-dimensional woven graphite–epoxy composite after impact loading. Optical fiber has fractured during polishing, therefore all of surface is not visible.

fiber as can be seen in Figure 12. Lebid *et al.* [54] measured varying residual stresses along the optical fiber due to the undulations and observed that this must be accounted for when interpreting the sensor response. Optical fiber sensors do survive when embedded in woven composites when high temperature and pressure are applied during the laminate fabrication, although the transmission losses due to the undulations are more significant than those for nonwoven laminates [55].

Finally, researchers have also demonstrated the advantages of embedding optical fiber sensors in composite laminates for impact detection and identification of the resulting internal damage. Similar to the previous loading conditions, optical fibers embedded parallel to the reinforcing fibers showed little or marginal effects on the extent and form of damage induced in the laminates [55–57]. Fibers embedded perpendicular to the reinforcing fibers increased the extent of damage, once again due to the presence of large matrix-rich regions [57].

6.2 Concrete infrastructures

Another promising application for optical fiber sensors is for long-term monitoring of civil infrastructures, specifically concrete structures. Researchers have demonstrated that optical fiber sensors can

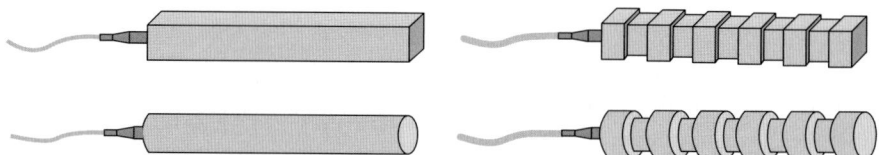

Figure 13. Regular and corrugated preembedded concrete bar sensors [67]. [Reproduced from Ref. 67. © Elsevier, 2000.]

be embedded during casting of concrete structures and provide accurate strain measurements under a variety of conditions [58, 59]. In addition to strain measurements, optical fiber sensors have been used for the measurement of crack opening displacements [60] and chloride penetration [61]. The properties of concrete are significantly different than those of polymer matrix composites; therefore several different issues become important.

For example, the gauge length of the optical fiber sensor must be chosen to be larger than the aggregate size so as to provide useful measurements [62]. Bending of the optical fiber around the individual aggregates can also induce significant signal loss over long sections of the optical fiber [63]. Another challenge to embedding optical fiber sensors in concrete can be the placement and protection of the sensor during the casting process, particularly for large-scale structures. Researchers have bonded optical fibers to rebars prior to casting to prevent movement of the optical fiber during cure of the concrete [64–66]. An alternative idea is to precast the sensor into smaller concrete blocks and then place these precast blocks into the larger structure [67]. Such blocks can be cast with a variety of surface features to improve bonding between the precast block and the concrete structure as shown in Figure 13.

Another significant challenge to embedding optical fiber sensors in concrete structures is the high alkali and moisture content of the concrete, which can degrade the coating and fiber-coating bonding over time [63, 68]. Further, once the coating has been removed, the hydroxide ions diffuse into the optical fiber surface and induce microcracking, leading to eventual failure of the optical fiber. Habel *et al.* [69] and Leung *et al.* [68] studied the long-term performance of potential optical fiber coatings including acrylate, which was rapidly degraded, and polyimide and Tefzel–silicone, which provided excellent long-term protection for the sensor. Alternatively, Kuang *et al.* [70] embedded POFs, which do not degrade as rapidly with moisture and demonstrated that they are compatible with concrete.

7 CONCLUSIONS

Fiber-optic sensors are versatile sensors for structural health monitoring applications owing to the fact that they are durable, immune to electromagnetic interference and corrosion, can be multiplexed into large sensor networks, and can be embedded in a variety of materials including composite laminates and concrete. External parameters for sensing are generally encoded in the lightwave signal through intensity, polarization states, wavelength or phase information. A variety of optical fiber types and coatings have been applied for structural health monitoring applications. The prediction of the response of optical sensors to external parameters requires an understanding of the basic principles of lightwave propagation and its sensitivity to physical changes applied to the optical fiber, which have been introduced in this chapter. For the analysis of specific optical fiber sensors such as fiber-optic interferometers, fiber Bragg gratings, and Brillouin scattering sensors *see* **Chapter 60** and **Chapter 61**. For more recent technologies for fiber-optic sensors, *see* **Chapter 62**.

END NOTES

[a.] By guided we mean that the energy remains confined within the optical fiber as it propagates.

RELATED ARTICLES

Chapter 17: Lamb Wave-based SHM for Laminated Composite Structures

Chapter 155: Reliable Use of Fiber-optic Sensors

REFERENCES

[1] Measures RM. *Structural Monitoring with Fiber Optic Technology*. Academic Press: San Diego, CA, 2001.

[2] Ghatak A, Thyagarajan K. *Introduction to Fiber Optics*. Cambridge University Press: Cambridge, UK, 1998.

[3] Takeda N, Okabe Y. Durability analysis and structural health management of smart composite structures using small-diameter fiber optic sensors. *Science and Engineering of Composite Materials* 2005 **15**:1–12.

[4] Lee ST, Gin J, Nampoori VPN. A sensitive fibre optic pH sensor using multiple sol-gel coatings. *Journal of Optics, A* 2001 **3**:355–359.

[5] Kronenberg P, Rastogi PK, Giaccari P, Limberger HG. Relative humidity sensor with optical fiber Bragg gratings. *Optics Letters* 2002 **27**:1385–1387.

[6] Corres JM, Arregui FJ, Matias IR. Design of humidity sensors based on tapered optical fibers. *Journal of Lightwave Technology* 2006 **24**:4329–4336.

[7] Zilbermann I, Meron E, Maimon E, Soifer L, Elbaz L, Korin E, Bettelheim A. Tautomerism in N-confused porphyrins as the basis of a novel fiber-optic humidity sensor. *Journal of Porphyrins and Phthalocyanines* 2006 **10**:63–66.

[8] Buck JA. *Fundamentals of Optical Fibers*. John Wiley & Sons: Hoboken, NJ, 2004.

[9] Kosaka T, Takeda N, Ichiyama T. Detection of cracks in FRP by using embedded plastic optical fiber. *Materials Science Research International* 1999 **5**:206–209.

[10] Xie GP, Keey SL, Asundi A. Optical time-domain reflectometry for distributed sensing of the structural strain and deformation. *Optics and Lasers in Engineering* 1999 **32**:437–447.

[11] Doyle C, Martin A, Liu T, Wu M, Hayes S, Crosby PA, Powell GR, Brooks D, Fernando G. In-situ process and conditioning monitoring of advanced fibre-reinforced composite materials using optical fibre sensors. *Smart Materials and Structures* 1998 **7**:145–158.

[12] Jiang MZ, Gerhard E. A simple strain sensor using a thin film as a low-finesse fiber-optic Fabry-Perot interferometer. *Sensors and Actuators A* 2001 **88**:41–46.

[13] Kuang KSC, Cantwell WJ, Scully PJ. An evaluation of a novel plastic optical fibre sensor for axial strain and bend measurements. *Measurement Science and Technology* 2002 **13**:1523–1534.

[14] Han W, Wang AB. Mode power distribution effect in white-light multimode fiber extrinsic Fabry-Perot interferometric sensor systems. *Optics Letters* 2006 **31**:1202–1204.

[15] James SW, Tatam RP. Optical fibre long-period grating sensors: characteristics and application. *Measurement Science and Technology* 2003 **14**:R49–R61.

[16] Corning SMF-28: Optical Fiber Product Information, Corning Incorporated: New York, 2003.

[17] Emslie C. Polarization maintaining fibers. In *Specialty optical fibers handbook*, Méndez A, Morse TF (eds). Academic Press Amsterdam, NL, 2007.

[18] Udd E. Light sources. In *Fiber Optic Sensors*, Udd E (ed). John Wiley & Sons: Hoboken, NJ, 2006.

[19] Murukeshan VM, Chan PY, Seng OL, Asundi A. On-line health monitoring of smart composite structures using fiber polarimetric sensor. *Smart Materials and Structures* 1999 **8**:544–548.

[20] Spillman WB. Multimode polarization sensors. In *Fiber Optic Sensors*, Udd E (ed). John Wiley & Sons: Hoboken, NJ, 2006.

[21] Carmen GP, Sendeckyj GP. Review of the mechanics of embedded optical sensors. *Journal of Composites Technology & Research* 1995 **17**:183–193.

[22] Kapron FP, Yuce HH. Theory and measurement for predicting stressed fiber lifetime. *Optical Engineering* 1991 **30**:700–708.

[23] Varelas D, Costantini DM, Limberger HG, Salathé RP. Fabrication of high-mechanical-resistance Bragg gratings in single-mode optical fibers with continuous-wave ultraviolet laser side exposure. *Optics Letters* 1998 **23**:397–399.

[24] Annovazziolodi V, Donati S, Merlo S, Zapelloni G. Statistical analysis of fiber failures under bending-stress fatigue. *Journal of Lightwave Technology* 1997 **15**:288–293.

[25] Zubia J, Arrue J. Plastic optical fibers: an introduction to their technological processes and applications. *Optical Fiber Technology* 2001 **7**:101–140.

[26] Bosc D, Toinen C. Full polymer single-mode optical fiber. *IEEE Photonics Technology Letters* 1992 **4**:749–750.

[27] Kuzyk MG, Garvey DW, Canfield BK, Vigil SR, Welker DJ, Tostenrude J, Breckon C. Characterization of single-mode polymer optical fiber and

electrooptic fiber devices. *Chemical Physics* 1999 **245**:327–340.

[28] Jiang C, Kuzyk MG, Ding JL, Jons WE, Welker D. Fabrication and mechanical behavior of dye-doped polymer optical fiber. *Journal of Applied Physics* 2002 **92**:4–12.

[29] Silva-Lopez M, Fender A, MacPherson WN, Barton JS, Jones JDC. Strain and temperature sensitivity of a single-mode polymer optical fiber. *Optics Letters* 2005 **30**:3129–3131.

[30] Uskokovi PS, Bala I, Rakin M, Puti S, Srekovi M, Aleksi R. Stress field analysis in composites laminates with embedded optical fiber. *Materials Science Forum* 2000 **352**:177–182.

[31] Barton EN, Ogin SL, Thorne AM, Reed GT. Optimisation of the coating of a fibre optical sensor embedded in a cross-ply laminate. *Composites Part A* 2002 **33**:27–34.

[32] Hadjiprocopiou M, Reed GT, Hollaway L, Thorne AM. Optimization of fibre coating properties for fiber optic smart structures. *Smart Materials and Structures* 1996 **5**:441–448.

[33] Jarlås R, Levin K. Location of embedded fiber optic sensors for minimized impact vulnerability. *Journal of Intelligent Material Systems and Structures* 1999 **10**:187–194.

[34] Henkel DP. Microstructure of high temperature smart materials. *Proceedings of Smart Structures and Materials.* SPIE, 1993; Vol. 1916, pp. 97–108.

[35] Pak YE. Longitudinal shear transfer in fiber optic sensors. *Smart Materials and Structures* 1992 **1**:57–62.

[36] Yang HT, Wang ML. Optical fiber sensor system embedded in a member subjected to relatively arbitrary loads. *Smart Material and Structures* 1995 **4**:50–58.

[37] Duck G, LeBlanc M. Arbitrary strain transfer from a host to an embedded fiber-optic sensor. *Smart Materials and Structures* 2000 **9**:492–497.

[38] Ansari F, Libo Y. Mechanics of bond and interface shear transfer in optical fiber sensors. *Journal of Engineering Mechanics* 1998 **124**:385–394.

[39] Li Q, Li G, Wang G, Ansari F, Asce M, Liu Q. Elasto-plastic bonding of embedded optical fiber sensors in concrete. *Journal of Engineering Mechanics* 2002 **128**:471–478.

[40] LeBlanc MJ. Study of interfacial interaction of an optical fibre embedded in a host material by in situ measurement of fibre end displacement—Part 1: theory. *Smart Materials and Structures* 2005 **14**:637–646.

[41] LeBlanc MJ. Study of interfacial interaction of an optical fibre embedded in a host material by in situ measurement of fibre end displacement—Part 2: experiments. *Smart Materials and Structures* 2005 **14**:647–657.

[42] Van Steenkiste RJ, Springer GS. *Strain and Temperature Measurement with Fiber Optic Sensors*. Technomic Publishing: Lancaster, PA, 1997.

[43] Prabhugoud M, Peters K. Efficient simulation of Bragg grating sensors for implementation to damage identification in composites. *Smart Materials and Structures* 2003 **12**:914–924.

[44] Kang HK, Park JW, Ryu CT, Hong CS, Kim CG. Development of fibre optic ingress/egress methods for smart composite structures. *Smart Materials and Structures* 2000 **9**:149–156.

[45] Levin K, Jarlås R. Vulnerability of embedded EFPI-sensors to low-energy impacts. *Smart Materials and Structures* 1997 **6**:369–382.

[46] Sirkis JS, Lu IP. On interphase modeling for optical-fiber sensors embedded in unidirectional composite systems. *Journal of Intelligent Material Systems and Structures* 1995 **6**:199–209.

[47] Shivakumar K, Emmanwori L. Mechanics of failure of composite laminates with an embedded fiber optic sensor. *Journal of Composite Materials* 2004 **38**:669–680.

[48] Jensen DW, Pascual J, August JA. Performance of graphite/bismaleimide laminates with embedded optical fibers. Part II: uniaxial compression. *Smart Materials and Structures* 1992 **1**:31–35.

[49] Jensen DW, Pascual J, August JA. Performance of graphite/bismaleimide laminates with embedded optical fibers. Part I: uniaxial tension. *Smart Materials and Structures* 1992 **1**:24–30.

[50] Kim MS, Lee CS, Hwang W. Effect of the angle between optical fiber and adjacent layer on the mechanical behavior of carbon/epoxy laminates with embedded fiber-optic sensor. *Journal of Materials Science Letters* 2000 **19**:1673–1675.

[51] Lau KT, Chan CC, Zhou LM, Jin W. Strain monitoring in composite-strengthened concrete structures using optical fibre sensors. *Composites, Part B* 2001 **32**:33–45.

[52] Lee DG, Mitrovic M, Friedman A, Carman GP, Richards L. Characterization of fiber optic sensors for structural health monitoring. *Journal of Composite Materials* 2002 **36**:1349–1366.

[53] Badcock RA, Fernando GF. An intensity-based optical fibre sensor for fatigue damage detection in advanced fibre-reinforced composites. *Smart Materials and Structures* 1995 **4**:223–230.

[54] Lebid S, Habel W, Daum W. How to reliably measure composite-embedded fibre Bragg grating sensors influenced by transverse and point-wise deformations? *Measurement Science and Technology* 2004 **15**:1441–1447.

[55] Pearson JD, Zikry MA, Prabhugoud M, Peters K. Global-local assessment of low-velocity impact damage in woven composites. *Journal of Composite Materials* 2007 **41**:2759–2783.

[56] Sirkis JS, Chang CC, Smith BT. Low-velocity impact of optical-fiber embedded laminated graphite-epoxy panels. Part 1: macroscale. *Journal of Composite Materials* 1994 **28**:1347–1370.

[57] Jeon BS, Lee JJ, Kim JK, Huh JS. Low velocity impact and delamination buckling behavior of composite laminates with embedded optical fibers. *Smart Materials and Structures* 1999 **8**:41–48.

[58] Masri SF, Agbabian MS, Abdelghaffar AM, Higazy M, Claus RO, Devries MJ. Experimental study of embedded fiberoptic strain-gauges in concrete structures. *Journal of Engineering Mechanics* 1994 **120**:1696–1717.

[59] Bin LinY, Chern JC, Chang KC, Chan YW, Wang LA. The utilization of fiber Bragg grating sensors to monitor high performance concrete at elevated temperature. *Smart Materials and Structures* 2004 **13**:784–790.

[60] Yuan LB, Ansari F. Embedded white light interferometer fibre optic strain sensor for monitoring crack-tip opening in concrete beams. *Measurement Science and Technology* 1998 **9**:261–266.

[61] Fuhr PL, Huston DR, MacCraith B. Embedded fiber optic sensors for bridge deck chloride penetration measurement. *Optical Engineering* 1998 **37**:1221–1228.

[62] Bonfiglioli B, Pascale G. Internal strain measurements in concrete elements by fiber optic sensors. *Journal of Materials in Civil Engineering* 2003 **15**:125–133.

[63] Zeng XD, Bao XY, Chhoa CY. Strain measurement in a concrete beam by use of the Brillouin-scattering-based distributed fiber sensor with single-mode fibers embedded in glass fiber reinforced polymer rods and bonded to steel reinforcing bars. *Applied Optics* 2002 **41**:5105–5114.

[64] Spammer SJ, Fuhr PL, Nelson M, Huston D. Rebar-epoxied optical fiber Bragg gratings for civil structures. *Microwave and Optical Technology Letters* 1998 **18**:214–219.

[65] Davis MA, Bellemore DG, Kersey AD. Distributed fiber Bragg grating strain sensing in reinforced concrete structural components. *Cement and Concrete Composites* 1997 **19**:45–57.

[66] Maaskant R, Alavie T, Measures RM, Tadros G, Rizkalla SH, GuhaThakurta A. Fiber-optic Bragg grating sensors for bridge monitoring. *Cement and Concrete Composites* 1997 **19**:21–33.

[67] Yuan LB, Jin W, Zhou LM, Lau KT. The temperature characteristic of fiber-optic pre-embedded concrete bar sensor. *Sensors and Actuators A* 2001 **93**:206–213.

[68] Leung CKY, Darmawangsa D. Interfacial changes of optical fibers in the cementitious environment. *Journal of Materials Science* 2000 **35**:6197–6208.

[69] Habel WR, Hofmann D, Hillemeier B. Deformation measurements of mortars at early ages and of large concrete components on site by means of embedded fiber-optic microstrain sensors. *Cement and Concrete Composites* 1997 **19**:81–102.

[70] Kuang KSC, Akmaluddin CWJ, Thomas C. Crack detection and vertical deflection monitoring in concrete beams, using plastic optical fibre sensors. *Measurement Science and Technology* 2003 **14**:205–216.

Chapter 60

Intensity-, Interferometric-, and Scattering-based Optical-fiber Sensors

Kara Peters

Department of Mechanical and Aerospace Engineering, North Carolina State University, Raleigh, NC, USA

1 Introduction	1083
2 Intensity-based Optical-fiber Sensors	1083
3 Interferometric Optical Fiber Sensors	1084
4 Scattering-based Optical-fiber Sensors	1093
5 Conclusions	1094
Related Articles	1094
References	1095

1 INTRODUCTION

Optical-fiber sensors present numerous advantages for the measurement of strain, temperature, humidity, pressure, and other parameters. This article focuses on sensors that use the intrinsic properties of optical fibers themselves as a transducer. By evaluating the behavior of a lightwave propagating through an optical fiber, strain and temperature fields can thus be inferred from microbending, microfracture, or scattering losses that occur along the fiber. Alternatively,

Encyclopedia of Structural Health Monitoring. Edited by Christian Boller, Fu-Kuo Chang and Yozo Fujino © 2009 John Wiley & Sons, Ltd. ISBN: 978-0-470-05822-0.

the phase shift of the propagating lightwave can be used to determine the strain or temperature profile along the fiber through the photoelastic effect. In comparison, fiber Bragg gratings and other optical-fiber sensors are reviewed in **Chapter 61** and **Chapter 62**.

A summary of the fundamental principles for optical fibers, useful to understand the sensing phenomena described in this article, can be found in **Chapter 59**. General issues concerning the application of all the sensors described in this article to structural health-monitoring systems are also presented in **Chapter 59** and **Chapter 109**.

2 INTENSITY-BASED OPTICAL-FIBER SENSORS

Intensity-based optical-fiber sensors are potentially the simplest in-fiber sensors to apply. Typically, these are based on the measurement of the loss of intensity transmitted through an optical fiber. Such sensors can be interrogated with low-cost lightwave sources and detection systems such as light emitting diodes (LEDs) and photodetectors. Measures *et al.* [1, 2] presented the first example of an intensity-based optical-fiber sensor for the

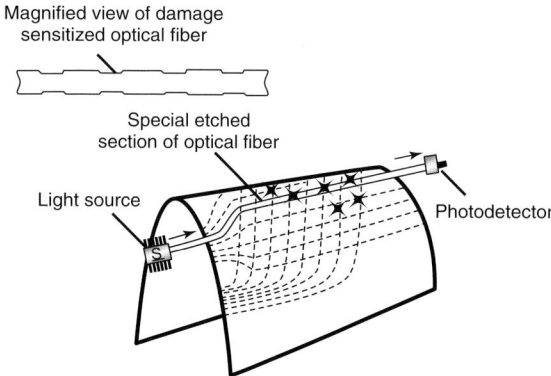

Figure 1. Intensity-based damage-sensitized optical fiber embedded as a two-dimensional sensor network in a composite aircraft wing. [Reproduced with permission from Ref. 3. © Elsevier, 2001.]

measurement of structural damage. As shown in Figure 1, the authors sensitized discrete regions of the optical fiber by etching the cladding to reduce the diameter of the optical fiber. The optical fibers were then embedded to form a two-dimensional grid network in an aircraft wing. The optical fiber in these regions was thus sensitive to damage at approximately the same level as the surrounding aramid fiber reinforced composite. During impact testing of the wing, microfractures in the optical fiber produced the bleeding of light from the fibers. By measuring the total intensity of the lightwave transmitted through each fiber, the researchers could therefore determine the extent of damage within the composite.

Intensity-based sensors that load the optical fiber through microbending have been developed for the measurement of strain, temperature, pressure, or other parameters [4–7]. Significant microbending along the optical fiber creates transmission losses that increase rapidly with the bend radius of curvature (*see* **Chapter 59**). Figure 2 presents a schematic of a typical microbend sensor design. As tensile strain is applied (for the example of Figure 2), the microbending applied to the optical fiber is reduced and therefore the transmitted intensity increases. Conversely, if compressive strain is applied, the transmitted intensity is reduced. To prevent errors due to fluctuations in the intensity of the light source, a reference optical fiber and a dual photodetector are also included. While relatively inexpensive, the accuracy of intensity-based sensors can be limited.

3 INTERFEROMETRIC OPTICAL FIBER SENSORS

Butter and Hocker [8] designed and demonstrated the first in-fiber strain sensor. This "fiber-optics strain gauge" was based on the measurement of the change in phase shift of a lightwave propagating through an optical fiber due to axial strain applied to the

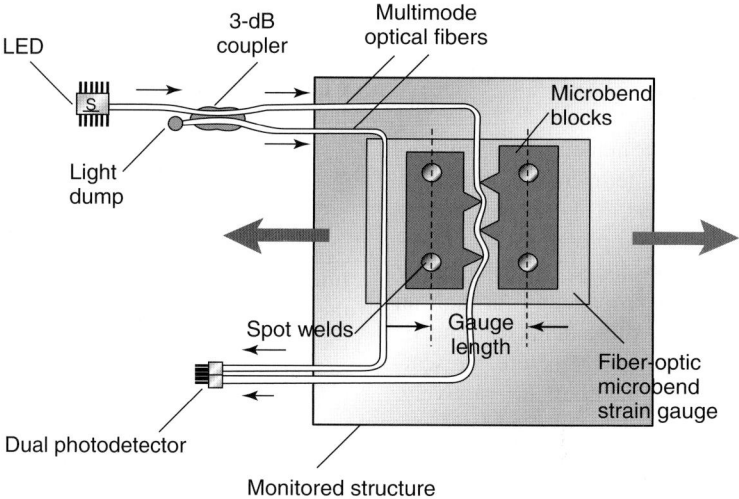

Figure 2. Schematic of a typical intensity-based microbend. [Reproduced with permission from Ref. 3. © Elsevier, 2001.]

optical fiber. This section derives the change in phase shift of the lightwave due to mechanical and thermal loads and presents common interferometer configurations for the measurement of this change in phase shift. Finally, more recent distributed interferometer sensor devices based on low-coherence interferometery and stimulated Brillouin scattering are also described.

3.1 Strain and temperature sensitivity

We assume that the optical-fiber material is an optically isotropic single crystal with an index of refraction in the unstressed state of n_0 (this would be the effective index of refraction of the fundamental mode, see **Chapter 59**). Since the fiber cross section is a two-dimensional surface, we define the orthogonal coordinate system with the axes p and q shown in Figure 3. These axes are the principal optical axes, which correspond to the propagation axes for the fiber. We write the field vector for an electromagnetic plane wave of wavelength λ, propagating in the axial, 1, direction (Figure 3) as

$$\mathbf{E} = A^p \mathbf{S}^p \sin\left[\omega t - \frac{2\pi n^p}{\lambda} x_1\right] + A^q \mathbf{S}^q \sin\left[\omega t - \frac{2\pi n^q}{\lambda} x_1\right] \quad (1)$$

where $\mathbf{S}^p$ and $\mathbf{S}^q$ are orthogonal unit vectors in the 2–3 plane in the direction of the principle optical axes, ω is the angular frequency of the lightwave, and A^p and A^q are the amplitudes of the orthogonal components [9]. The propagating lightwave is thus split into two orthogonal lightwaves, both with the same wavelength and frequency as the original wave, but each experiencing a separate index of refraction, n^p and n^q.

The vector field, $\mathbf{E}$, must satisfy the planar wave equation

$$\mathbf{R} \times (\mathbf{R} \times \mathbf{B}\mathbf{E}) + \frac{1}{n^2}\mathbf{E} = 0 \quad (2)$$

where n is the index of refraction experienced by the field and $\mathbf{R}$ is the unit vector in the direction of propagation; in this case, $\mathbf{R} = (1, 0, 0)$. The tensor $\mathbf{B}$ is the material dielectric impermeability tensor,

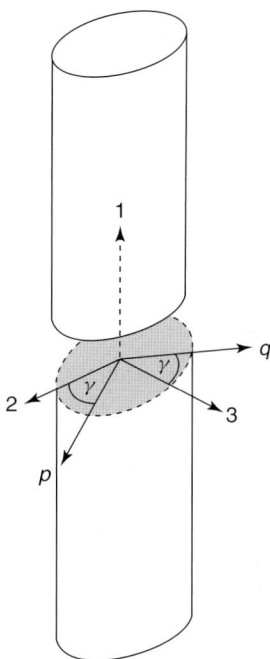

Figure 3. Sketch of optical fiber with coordinate systems highlighted on an arbitrary cross section ($p - q$ are the principal optical axes; 1–2–3 are the applied strain directions).

which is directly related to the index of refraction of the material.

$$\mathbf{B} = \begin{bmatrix} B_1 & B_6 & B_5 \\ B_6 & B_2 & B_4 \\ B_5 & B_4 & B_3 \end{bmatrix} \quad (3)$$

For an isotropic material, $B_1 = B_2 = B_3 = n_0$, $B_4 = B_5 = B_6 = 0$ [10]. Since the vector field is propagating in the 1 direction, $\mathbf{E}$ has spatial components in the 2 and 3 directions, $\mathbf{E} = (0, E_2, E_3)$. Substituting $\mathbf{E}$ and $\mathbf{R}$ into equation (2) and expanding yields the two equations

$$\begin{bmatrix} B_2 - 1/n^2 & B_4 \\ B_4 & B_3 - 1/n^2 \end{bmatrix} \begin{pmatrix} E_2 \\ E_3 \end{pmatrix} = 0 \quad (4)$$

Nontrivial solutions for E_2 and E_3 occur only when the determinate of the matrix on the left-hand side is zero. Applying this condition and solving for n yields

$$\frac{1}{n^2} = \frac{(B_2 + B_3) \pm \sqrt{(B_2 - B_3)^2 + 4B_4^2}}{2} \quad (5)$$

In general, there are therefore two solutions to n, which we previously labeled as n^p and n^q. For the isotropic fiber, we simply have $n^p = n^q = n_0$; therefore, there is only one propagation solution.

The next step is to calculate the change in phase shift of the lightwave propagating through a length L of the optical fiber. As the lightwave, **E**, propagates through the optical fiber, it experiences a total phase shift, φ,

$$\varphi = \beta L = \frac{2\pi}{\lambda} nL \qquad (6)$$

β is thus the propagation constant or phase shift per unit length, $\beta = 2\pi n/\lambda$. Once strain or thermal loading is applied to the optical fiber, both the length of the fiber and the effective index of refraction change, altering the phase shift of the propagating lightwave [8],

$$\Delta\varphi = \frac{2\pi}{\lambda}(\Delta n L + n \Delta L) \qquad (7)$$

The change in length of the optical fiber is directly related to the axial strain, ε_1, and the coefficient of thermal expansion, α, as $\Delta L = \varepsilon_1 L + \alpha \Delta T$.

We now calculate the term Δn of equation (7). The photoelastic effect describes the change in optical properties of the material due to strain and temperature. Since the components of the tensor **B** are dependent on the material indices of refraction, we can write $B_i = B_i^0 + \Delta B_i$ and apply this to equation (5). Simplifying this expression and using the isotropic values for B_i^0, we find

$$\frac{1}{n^2} = \frac{1}{n_0^2} + \frac{(\Delta B_2 + \Delta B_3) \pm \sqrt{(\Delta B_2 - \Delta B_3)^2 + 4\Delta B_4^2}}{2} \qquad (8)$$

Assuming the applied strains are small, we can apply the linear thermoelastic strain-optic effect, which can be written as

$$\Delta B_i = p_{ij}(\varepsilon_j - \alpha_j \Delta T) + W_i \Delta T \qquad (9)$$

where the summation convention is applied and the strain components are written in compact notation [9]. Here we consider that all six components of strain can be applied to the optical fiber; more specific loading cases will be considered later. The terms W_i are the changes in B_i due to temperature measured at a constant stress state, $W_i = \partial B_i/\partial \Delta T|_{\sigma=\text{const.}}$. The components of the **p** matrix are commonly referred to as the *Pockel's constants* for the material. For an isotropic material, **p** has the form [10]

$$\mathbf{p} = \begin{bmatrix} p_{11} & p_{12} & p_{12} & 0 & 0 & 0 \\ p_{12} & p_{11} & p_{12} & 0 & 0 & 0 \\ p_{12} & p_{12} & p_{11} & 0 & 0 & 0 \\ 0 & 0 & 0 & \frac{(p_{11}-p_{12})}{2} & 0 & 0 \\ 0 & 0 & 0 & 0 & \frac{(p_{11}-p_{12})}{2} & 0 \\ 0 & 0 & 0 & 0 & 0 & \frac{(p_{11}-p_{12})}{2} \end{bmatrix} \qquad (10)$$

Similarly, for an isotropic material,

$$\mathbf{W} = \begin{bmatrix} -\frac{2}{n_0^3}\frac{dn_0}{dT} \\ -\frac{2}{n_0^3}\frac{dn_0}{dT} \\ -\frac{2}{n_0^3}\frac{dn_0}{dT} \\ 0 \\ 0 \\ 0 \end{bmatrix} \qquad (11)$$

Typical values for fused silica fibers are $p_{11} = 0.17$, $p_{12} = 0.36$, $dn_0/dT = 1.2 \times 10^{-5}/°C$ [9]. Combining equations (8–11), we find the general expression for the index change due to strain and temperature loading

$$\frac{1}{n^2} = \frac{1}{n_0^2} + p_{12}\varepsilon_1 + \frac{(p_{11}+p_{12})}{2}(\varepsilon_1 + \varepsilon_2)$$
$$\pm \frac{(p_{11}-p_{12})}{2}\sqrt{(\varepsilon_2 - \varepsilon_3)^2 + 4\varepsilon_4^2}$$
$$- \frac{2}{n_0^3}\frac{dn_0}{dT}\Delta T - \alpha \Delta T(p_{11} + 2p_{12}) \qquad (12)$$

where the strain components are the total strain due to both the thermal and mechanical loading.

Considering the specific case of pure axial tension without thermal loading, for which $\varepsilon_1 = \varepsilon$, $\varepsilon_2 = \varepsilon_3 = -\nu\varepsilon$, we find

$$\frac{1}{n^2} = \frac{1}{n_0^2} + [p_{12} - \nu(p_{11} + p_{12})]\varepsilon \qquad (13)$$

Calculating $n - n_0$ and linearizing with respect to strain, we find

$$\Delta n = -\frac{1}{2}n_0^3 \left[p_{12} - \nu(p_{11} + p_{12}) \right] \varepsilon \quad (14)$$

and substituting into equation (7),

$$\Delta\varphi = \frac{2\pi}{\lambda} n_0 L \varepsilon \left\{ 1 - \frac{1}{2}n_0^2 \left[p_{12} - \nu(p_{11} + p_{12}) \right] \right\}$$

$$= \frac{2\pi}{\lambda} n_0 L \varepsilon (1 - p_e) \quad (15)$$

where p_e is referred to as the *effective photoelastic constant* for applied axial strain [8]. The phase shift is therefore linearly proportional to the change in length of the optical fiber, or the applied strain.

Adding the effect of a temperature change ΔT to equation (15), where the axial elongation ε is now the total axial elongation, we find

$$\Delta\varphi = \frac{2\pi}{\lambda} n_0 L \left\{ \varepsilon - \frac{1}{2}\varepsilon n_0^2 \left[p_{12} - \nu(p_{11} + p_{12}) \right] \right.$$
$$\left. + \frac{1}{n_0}\frac{dn_0}{dT}\Delta T + \frac{n_0^2}{2}(p_{11} + 2p_{12})\alpha\Delta T \right\} \quad (16)$$

[9]. Typical material property values for fused silica are $\nu = 0.16$ and $\alpha = 0.5 \times 10^{-6}/\,°C$.

The above equations consider a constant strain field applied along the gauge length of the optical fiber. In general, the fiber can be mounted in any orientation on a surface; therefore, these strain components can also vary along the length of the gauge. Sirkis and Haslach [11] first calculated the total phase shift, φ, in the optical fiber due to a known strain field. We define the variable s to be the variable along the path length of the fiber. We find the total phase shift by integrating the local phase shift along the optical fiber

$$\varphi = \int_0^L \left(\frac{d\varphi}{ds}\right) ds$$

$$= \left(\frac{2\pi}{\lambda}\right) \int_0^L n_0 (1 - p_e)(1 + \varepsilon_n) \, ds \quad (17)$$

where ε_n is the strain component tangent to the fiber path at each location. Sirkis and Haslach also showed that transfer of shear stress and transverse stresses was negligible for surface-mounted sensors; however, these components could be significant for sensors embedded in materials. Details on the inclusion of these strain components for embedded sensors can be found in [9, 12].

3.2 Conventional interferometers

Interferometric sensors present an extremely high sensitivity to external parameters. Additionally, as the optical fiber itself is used as the measurement device, such sensors are beneficial for simplicity and cost as compared to other optical fiber–based sensors. When using an optical fiber as strain sensor, it is not possible to measure the phase shift directly, and therefore we typically measure the interference between the sensor fiber and a second, reference fiber, which is not exposed to the environmental changes and therefore has a constant phase shift. Additionally, this second fiber can be exposed to only some of the loading, for example, temperature, so as to provide compensation during the measurements. The interferometric measurement of phase shifts for in-fiber sensors parallels that for classical free-space interferometers. The two most commonly applied interferometric arrangements are the Mach–Zehnder and Michelson interferometers [13]. Figures 4 and 5 show each of these classical interferometers and their equivalent for in-fiber sensors.

Assuming that the light source has a high coherence length and polarization effects are compensated between the two fibers, the average intensity of the interference pattern varies sinusoidally, as shown in Figure 6. The cyclic form of the intensity measurement presents two challenges when the intensity is near one of the quadrature points, i.e., where $dI/d\varphi = 0$. The first challenge is that the direction of fringe movement cannot be determined at the quadrature point (directional ambiguity) and the second is that the measurement sensitivity goes to zero at the same point (signal fading). A variety of signal-processing techniques have been applied to remove these difficulties, generally categorized into passive and active demodulation. Each of these categories can be further divided on the basis of whether the output signal is divided into multiple branches at the same frequency (homodyne) or different frequencies (heterodyne) [14]. One example of a passive homodyne demodulation is shown in Figure 6. For

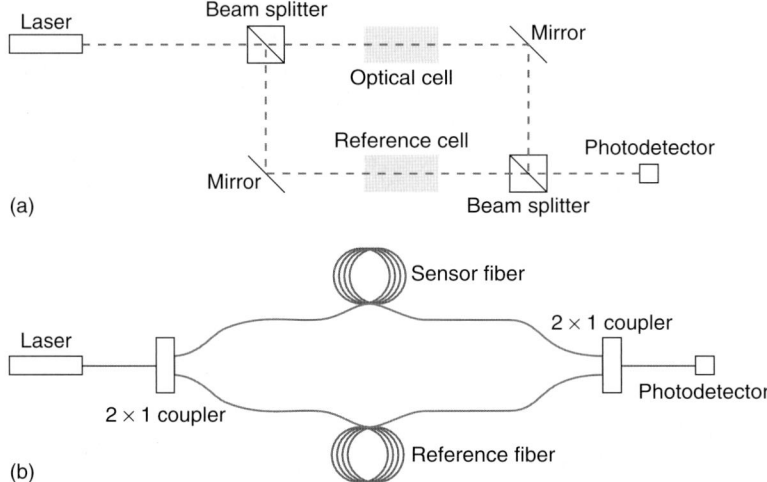

Figure 4. Schematic of Mach–Zehnder interferometer: (a) free-space optics version measures the difference in optical path length through the optical cell and reference cell and (b) in-fiber version measures the difference in optical path length through sensor fiber and reference fiber.

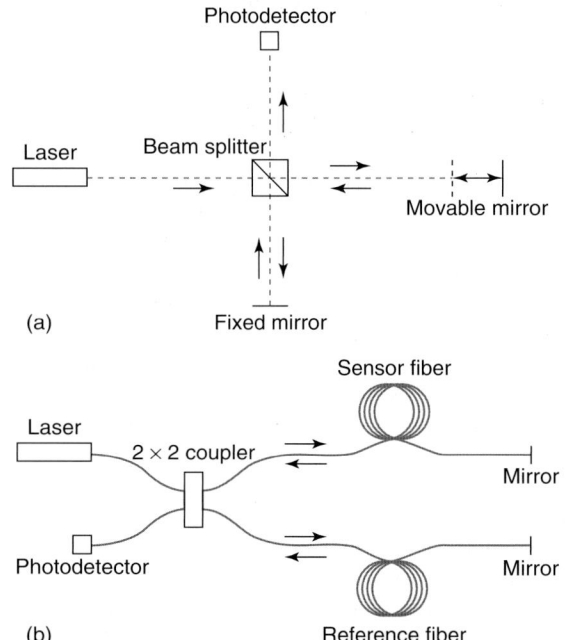

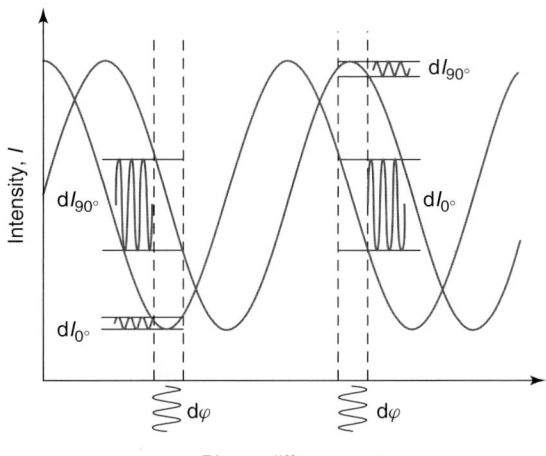

Figure 6. Intensity signals as a function of interferometer phase difference for two sensor signals, phase shifted by $0°$ and $90°$. When one signal is as a quadrature point ($dI/d\varphi = 0$), the other is at the location of maximum sensitivity. [Reproduced from Ref. 14. © John Wiley & Sons Inc., 2006.]

Figure 5. Schematic of Michelson interferometer: (a) free-space optics version and (b) in-fiber version.

this example, the sensor lightwave is divided into two channels, with one phase shifted by $90°$, and interfered with the reference lightwave. As seen in Figure 6, when one signal is near a quadrature point, the other is at the point of maximum sensitivity, removing both the signal fading and directional ambiguity issues. A second popular example of passive demodulation is the use of a 3×3 coupler, exploiting the phase shifts between the various outputs [15]. Active homodyne demodulation involves actively loading the reference fiber to keep the interferometer signal 90° from the quadrature point (or at the quadrature point for some examples), for example, by wrapping the reference fiber around a piezoelectric cylinder [16]. Kersey [17] reviews multiplexing strategies for interferometric optical-fiber sensors, including time-division multiplexing and frequency-division multiplexing examples.

The measurement of phase shift is not an absolute measurement. Specifically, in order to know the current strain level, the system must be continually operated throughout the lifetime of the structure. This presents significant challenges for monitoring of structures since power interruptions could invalidate all later measurements. One solution to this problem is through low-coherence interferometry, which is described in the following section.

3.3 Low-coherence interferometers

One technique to provide an absolute measurement is through the use of low-coherence interferometry, also referred to as *white-light interferometry*. This measurement technique is particularly useful for large structural applications as the spatial resolution is significantly less than the previous interferometers. However, the system would not be affected by power interruptions that may occur over the lifetime of the structure. Until recently, low-coherence interferometry had also been limited to quasi-static applications; however, it has been demonstrated successfully in many field applications. Bock *et al.* [18] applied low coherence in high-birefringence fibers (*see* **Chapter 59**) for the measurement of pressure. Inaudi *et al.* [19] embedded low-coherence interferometric sensors in a concrete structure to measure shrinkage of the concrete during cure and later strain during loading of the structure. Yuan and Ansari [20] applied a similar embedded low-coherence interferometric sensor to measure crack-tip openings in a concrete structure.

The fundamental measurement concept applied to low-coherence interferometric sensors is to interrogate a primary interferometer consisting of a sensor and reference arm (referred to as the *sensing interferometer*) with a second similar interferometer for which the optical path difference between the two arms can be scanned (referred to as the *scanning interferometer*). A typical low-coherence interferometer configuration for optical-fiber sensors is shown in Figure 7. Interference fringes are observed only when the optical path difference in the scanning interferometer is within the coherence length of the laser or when the optical path difference between the two interferometers is within the same coherence length. By selecting a laser source with an extremely low-coherence length such as a superluminescent diode or multimode laser diode, the displacement of the sensing interferometer can be determined within a reasonable precision [21].

The interference between two beams from a low-coherence source can be described by

$$I = I_0 \left[1 + V(nx) \cos\left(\frac{2\pi}{\lambda} nx\right) \right] \quad (18)$$

where nx is the optical path difference in the interferometer, I_0 is the mean intensity, and V is the variation in fringe visibility as a function of optical path difference for the laser source (typically a Gaussian profile) [22]. This equation yields a wave packet whose width is approximately equal to the coherence length and centered around the condition $nx = 0$, as seen in Figure 8. For the interferometer of Figure 7, the optical path difference in the sensing interferometer to be measured is ΔL_1. The movable mirror in the scanning interferometer is displaced while the interference pattern between the two interferometers is measured. When $\Delta L_2 \cong 0$, the strongest interference occurs (as seen in Figure 8), and this condition is used as a reference point. For $\Delta L_2 \cong +\Delta L_1$, the phase imbalance in the sensing interferometer is offset by the phase imbalance of the scanning interferometer for a portion of the lightwave. Thus, interference also occurs at a lower maximum intensity. From these two measurements, ΔL_1 can be determined, which is the optical path difference in the original sensing interferometer. As in the previous section, this optical path difference can then be converted into the physical change in length of the sensing fiber, ΔL, through [19]

$$\Delta L = \frac{\Delta L_1}{n_0 (1 - p_e)} \quad (19)$$

Inaudi *et al.* [19] demonstrated a low-coherence optical-fiber sensor using a LED with a coherence length of 30 μm. By calculating the center of

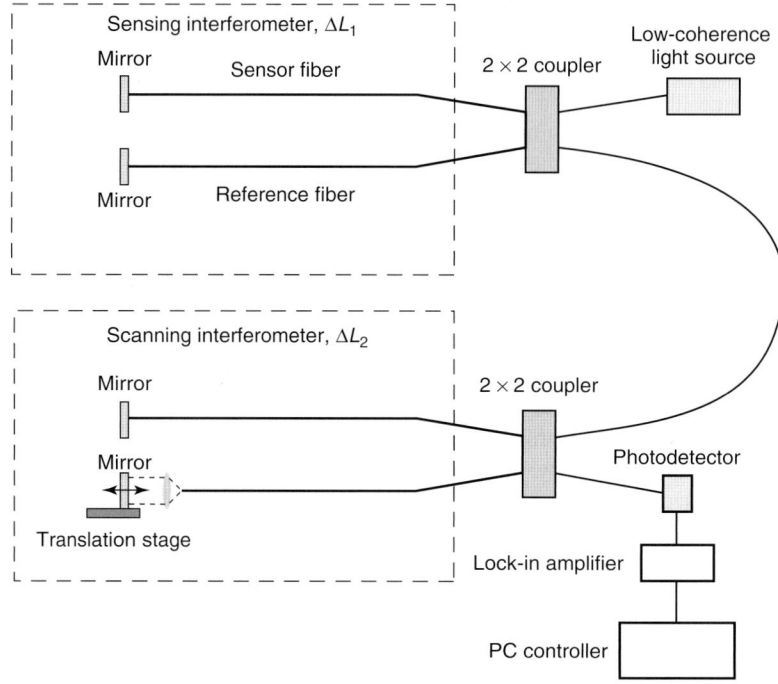

Figure 7. Typical low-coherence in-fiber interferometer configuration. [Reproduced with permission from Ref. 19. © Elsevier, 1994.]

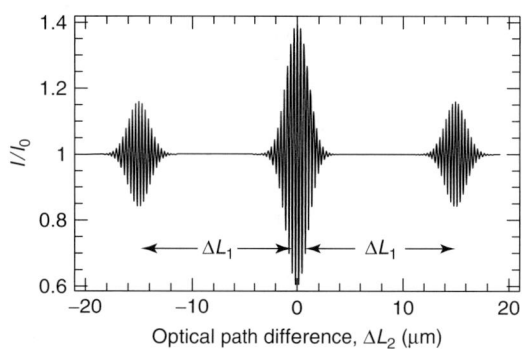

Figure 8. Interference signal from double Michelson interferometer with low-coherence source shown in Figure 7. [Reproduced with permission from Ref. 19. © Elsevier, 1994.]

gravity of the interference pattern, they were able to increase the displacement resolution to 10 μm. Rao and Jackson [23] provide an excellent review of signal processing for low-coherence interferometric sensors, as well as methods to increase the accuracy of the maximum point on the interference envelope. Meggitt et al. [22] replaced the translational stage for the scanning mirror by wrapping an optical fiber around a piezoelectric cylinder. Therefore, the optical path difference could be scanned by applying an electrical current to the cylinder, without the need for externally moving components.

To increase the spatial resolution of the basic low-coherence interferometer for structural measurements, Inaudi [24] implemented partial reflectors along the sensing arm to provide multiple interference lengths. By placing the reflectors sufficiently apart and scanning all possible interference distances, it is possible to use the same interrogation system for multiple sensors. A typical output from this system, demonstrating the interference lengths for each sensor, is shown in Figure 9. Later, Yuan and Ansari [25] multiplexed several sensing lengths using partial reflectors and a low-coherence source; however, they applied N balance arms for N sensors and switched between the balance arms to measure each sensor. Most recently, Lloret et al. [26] applied amplitude modulation of the low-coherence light source at high frequencies to obtain dynamic strain measurements up to 100 Hz.

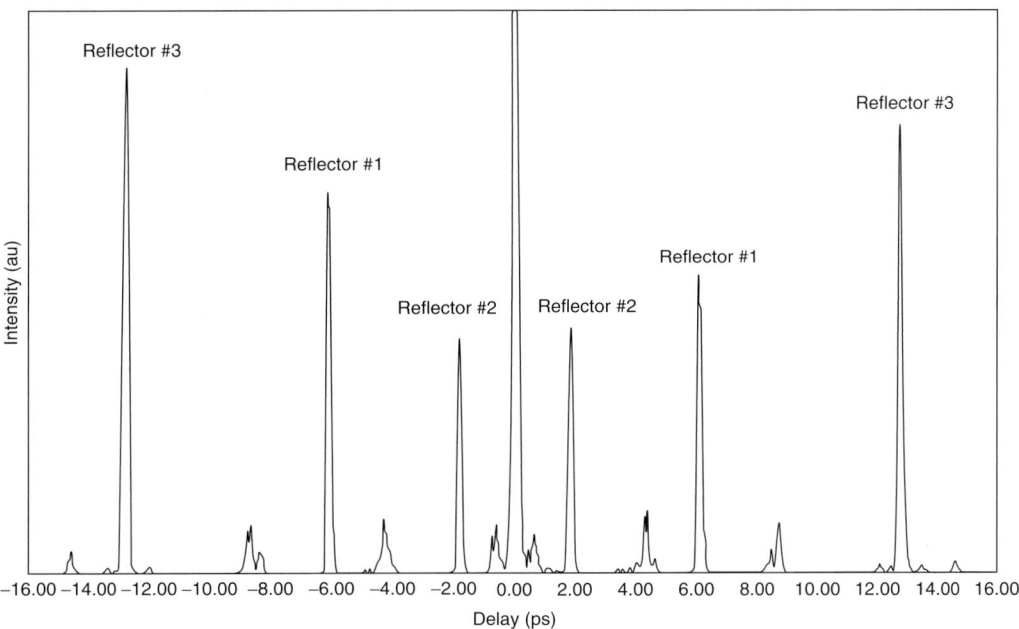

Figure 9. Interference signal from multiplexed low-coherence interferometric sensors (only envelope of demodulation signal is plotted). [Reproduced with permission from Ref. 24. © SPIE, 1995.]

3.4 Fabry–Perot interferometers

A second absolute interferometer configuration is the Fabry–Perot interferometer (FPI), first demonstrated by Murphy *et al.* [27]. FPI sensors provide local strain and temperature measurements. FPI sensors have been surface mounted on composites for the measurement of acoustic emission signals [28] and impact detection [29]. They have also been embedded in composite structures for the detection and measurement of fatigue crack propagation in bonded composite patch repairs [30], delamination and buckling [31], process-induced residual stresses [32], and the development of impact damage [33]. They have also been combined with fiber Bragg grating sensors for the independent measurement of strain and temperature [34] (*see* **Chapter 61**). The miniaturization of FPI sensors is also discussed in **Chapter 62**.

A schematic of a Fabry–Perot in an optical fiber is shown in Figure 10(a). The sensor cavity of length L is formed between two partial reflectors with reflectivities r_1 and r_2. These partial reflectors are typically partial mirrors, but fiber Bragg gratings have also been used [35]. For high reflectivity values, the FPI acts as a "multipass" interferometer whose transmission characteristics are wavelength dependent. We can assume that no losses are incurred at the reflectors so that the transmissivities $t_1 = 1 - r_1$ and $t_2 = 1 - r_2$. The cavity itself can be a section of the optical fiber (as shown in Figure 10a) for which the sensor is termed *intrinsic*. Extrinsic FPIs can be formed using external air gaps or other mediums to be analyzed. An example of an extrinsic Fabry–Perot interferometer (EFPI) is shown in Figure 10(b). Extrinsic FPI sensors are larger than intrinsic sensors and therefore more intrusive; however, they are easier to fabricate and have less inherent noise. They also demonstrate low transverse strain sensitivities and low apparent thermal strains [36].

To calculate the response of the FPI, we define the index of refraction of the cavity medium as n. Considering a plane wave (arriving from the left in Figure 10a) that has passed through the first reflector, we can write its time-averaged (or steady-state) amplitude as $E = E_0 \exp(i\beta z)$ (*see* **Chapter 59**). When this wave reaches the second partial reflector $E = E_0 \exp(i\beta L)$, the transmitted and reflected portions of the waves are

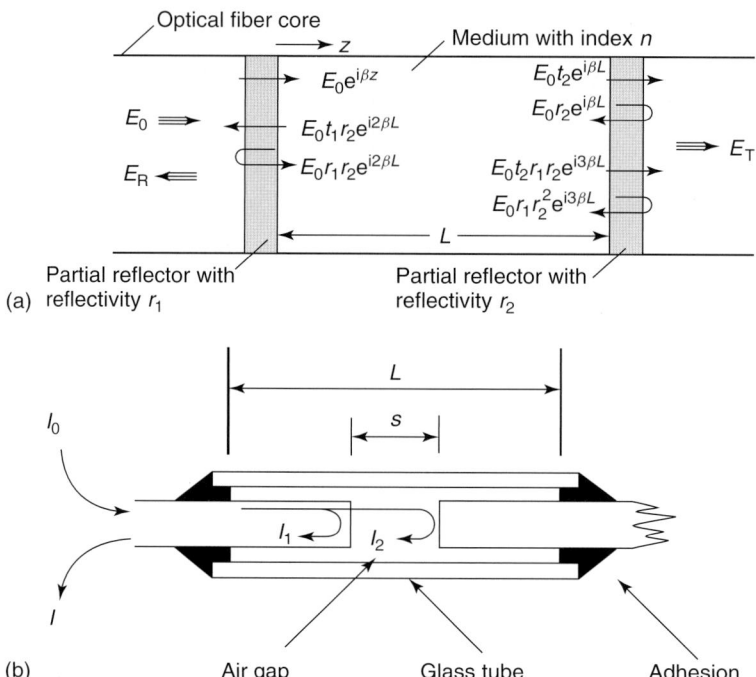

Figure 10. (a) Representation of multipass interference in a Fabry–Perot interferometer and (b) extrinsic Fabry–Perot interferometer. [Reproduced from Ref. 31. © Sage Publications, 2000.]

$E_t = t_2 E_0 \exp(i\beta L)$ and $E_r = r_2 E_0 \exp(i\beta L)$ respectively. The reflected portion then reaches the first partial reflector, at which point $E = r_2 E_0 \exp(i2\beta L)$. The reflected portion of this wave reaches the second reflector, at which point $E_t = t_2 r_1 r_2 E_0 \exp(i2\beta L)$ and $E_r = r_1 r_2^2 E_0 \exp(i2\beta L)$.

Continuing these calculations for an infinite number of passes, we find the total lightwave exiting the FPI to the right,

$$E_T = t_2 E_0 \exp(i\beta L) \left[1 + r_1 r_2 \exp(2i\beta L) \right.$$
$$\left. + r_1^2 r_2^2 \exp(4i\beta L) + \cdots \right]$$
$$= \frac{E_0 t_2 \exp(i\beta L)}{1 - r_1 r_2 \exp(2i\beta L)} \quad (20)$$

We could also calculate the total lightwave reflected from the cavity, $E_R = E_0 - E_T$. The intensity of the transmitted lightwave is then

$$I_T = \frac{1}{2} \varepsilon_0 c |E_T|^2 = \frac{I_0 t_2^2}{(1-R)^2 [1 + \mathcal{F} \sin^2(\beta L)]} \quad (21)$$

where ε_0 is the permittivity of the vacuum ($\varepsilon_0 = 8.854 \times 10^{-12}\,\mathrm{F\,m^{-1}}$), c is the speed of light in a vacuum, $R^2 = r_1 r_2$, and $I_0 = \varepsilon_0 c E_0^2 / 2$ is the intensity of the input lightwave. The coefficient $\mathcal{F}$ is referred to as the *coefficient of finesse* of the FPI and determines the response characteristics of the FPI. $\mathcal{F}$ is given by the expression

$$\mathcal{F} = \frac{4R}{(1-R)^2} \quad (22)$$

Figure 11 plots the normalized transmitted intensity of the FPI as a function of wavelength. The peaks in the transmission spectrum correspond to the condition $L = m\lambda/2n$, where m is an integer. By measuring the wavelength location of one of these peaks, one can determine the length of the cavity and therefore the applied strain, displacement, pressure, or temperature. For this measurement strategy, it is best to have high finesse values to narrow the peaks as much as possible [37]. Spectral measurements provide absolute strain or temperature measurements without the problems of directional ambiguity and signal fading previously described.

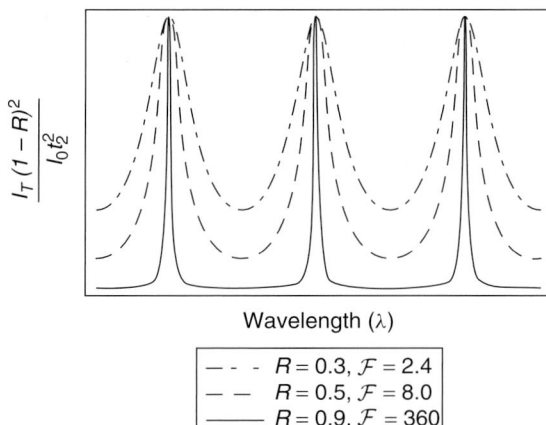

Figure 11. Typical normalized transmitted intensity spectrum for a Fabry–Perot interferometer. Three different finesse values are plotted.

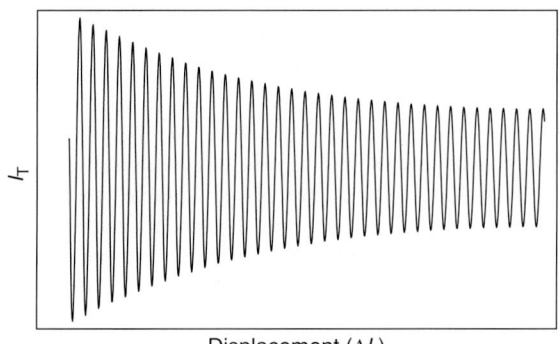

Figure 12. Transmitted intensity for a low-finesse Fabry–Perot interferometer as a function of displacement of one of the reflectors ($R = 0.1$, $\mathcal{F} = 0.49$).

High-speed interrogation of spectral sensors is discussed in **Chapter 61**. Rao [38] presents a review of multiplexing and signal interrogation of FPI sensors.

Alternatively, low-finesse FPI sensors can also be applied for high-resolution, dynamic strain sensing [27]. For very low values of reflectivity (and finesse), the FPI acts as essentially a two-beam interferometer. A plot of the response of a low-finesse FPI as a function of the displacement of one of the mirrors is shown in Figure 12. For small displacements, this sensor can be interrogated similar to the Mach–Zehnder or Michelson configurations previously described. Various signal interrogation methods have been applied to remove the direction ambiguity including quadrature phase shifted signals [27], path-matched differential interferometry [39], and low-coherence interferometry [40, 41].

4 SCATTERING-BASED OPTICAL-FIBER SENSORS

A final property of optical fibers that has been exploited for the measurement of strain and temperature is the intrinsic scattering property of the fused silica material. As a lightwave propagates through the optical fiber, an extremely small portion is backscattered owing to local inhomogeneities in the material and therefore the index of refraction. Several forms of scattered waves can be detected, including Rayleigh, Brillouin, and Raman components. Figure 13 shows a typical scattering spectrum for an optical fiber. As can be seen, the Rayleigh component is the largest component of the scattering spectrum. Brillouin scattering occurs due to interactions with acoustic waves known as *phonons* and is therefore frequency shifted (through the Doppler effect) as compared to the Rayleigh scattering component. This frequency shift is dependent upon the local density of the fused silica and is therefore linearly related to applied strain or temperature [42]. The Brillouin backscattered light is an extremely weak signal and difficult to measure outside of the laboratory; however, this scattering can be amplified by stimulating acoustic waves through the use of a second pulsed laser connected to the opposite end of the optical fiber. By scanning the frequency of the pulsed laser, the relative frequency shifts at which the stimulated

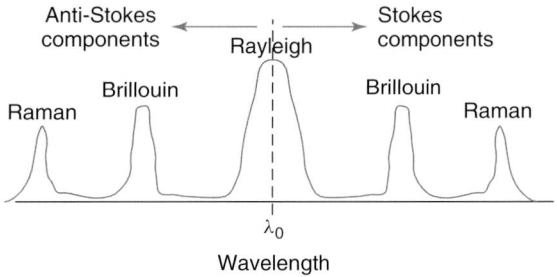

Figure 13. Spectrum of intrinsic material scattering for fused silica. Stokes and anti-Stokes components are indicated.

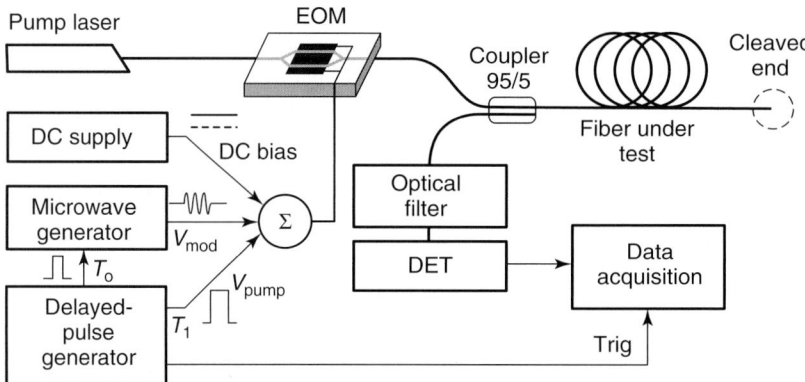

Figure 14. Experimental setup for distributed Brillouin gain spectrum analysis. [Reproduced with permission from Ref. 43. © 1996, OSA Publishing.]

Brillouin scattering occurs can be identified and converted into the applied strain or temperature. To determine the location of each Brillouin scattering event, the time of arrival of the scattered light is interrogated. In this manner, the strain or temperature distribution along an optical fiber can be measured. A schematic of a typical stimulated Brillouin scattering data-acquisition system is shown in Figure 14. Bao *et al.* [42] first demonstrated the use of stimulated Brillouin scattering as a temperature and strain sensor, achieving a strain and temperature resolution of $20\,\mu\varepsilon$ and $2\,°C$ with a spatial resolution of 5 m, over a sensing length of 22 km. Later, Niklès *et al.* [43] improved the strain and temperature resolution to $5\,\mu\varepsilon$ and $0.25\,°C$ using a single laser source, however, with a spatial resolution of 35 m.

Brillouin scattering–based optical-fiber sensors provide a unique distributed sensing capability using only the intrinsic properties of the optical fiber itself. Generally, such sensors have been applied for the monitoring of large-scale structures due to the limits on the spatial resolution. Example applications include composite cure sensing [44], monitoring of steel beam and composite fuselage structures [45, 46], and damage identification in sandwich composites [47]. At the same time, researchers have recently applied signal-processing techniques and hardware configurations to reduce the spatial resolution substantially. For example, Song *et al.* [48] recently demonstrated a spatial resolution of 1.6 mm, applying Brillouin optical correlation domain analysis.

5 CONCLUSIONS

As reviewed in this chapter, optical fibers can be applied as strain and temperature sensors through their intrinsic properties. Intensity-based optical-fiber sensors offer low-cost, simple solutions to complete optical measurements. For more precise measurements, a variety of interferometer configurations and interrogation methods are also available, including variations for high-speed and absolute measurements. Finally, low-coherence interferometry and stimulated Brillouin scattering–based sensors permit distributed measurements for large structural applications. As compared to their electrical counterparts, optical-fiber sensors can require more complicated and expensive instrumentation. On the other hand, they can be readily multiplexed into large sensor networks with a single data-acquisition system and can therefore be less obtrusive to the structure. Additionally, the distributed examples discussed in this article emphasize that simple optical fibers can be used to instrument large structures without the need to monitor a large number of sensors.

RELATED ARTICLES

Chapter 17: Lamb Wave-based SHM for Laminated Composite Structures

Chapter 155: Reliable Use of Fiber-optic Sensors

REFERENCES

[1] Measures RM, Glossop NDW, Lymer J, LeBlanc M, West J, Dubois S, Tsaw W, Tennyson RC. Structurally integrated fiber optic damage assessment system for composite-materials. *Applied Optics* 1989 **28**:2626–2633.

[2] LeBlanc M, Measures RM. Impact damage assessment in composite-materials with embedded fiber-optic sensors. *Composites Engineering* 1992 **2**:573–596.

[3] Measures RM. *Structural Monitoring with Fiber Optic Technology*. Academic Press: San Diego, CA, 2001.

[4] Lagakos N, Bucaro JA. Fiber optic microbend sensor. *ISA Transactions* 1998 **27**:19–24.

[5] Donlagic D, Culshaw B. Microbend sensor structure for use in distributed and quasi-distributed sensor systems based on selective launching and filtering of the modes in graded index multimode fiber. *Journal of Lightwave Technology* 1999 **17**:1856–1868.

[6] Xie GP, Keey SL, Asundi A. Optical time-domain reflectometry for distributed sensing of the structural strain and deformation. *Optics and Lasers in Engineering* 1999 **32**:437–447.

[7] Pandey NK, Yadav BC. Embedded fibre optic microbend sensor for measurement of high pressure and crack detection. *Sensors and Actuators, A* 2006 **128**:33–36.

[8] Butter CD, Hocker GB. Fiber optics strain-gauge. *Applied Optics* 1978 **17**:2867–2869.

[9] Van Steenkiste RJ, Springer GS. *Strain and Temperature Measurement with Fiber Optic Sensors*. Technomic Publishing: Lancaster, PA, 1997.

[10] Nye JF. *Physical Properties of Crystals*. Oxford Science Publications: Oxford, 1985.

[11] Haslach HW, Sirkis JS. Surface-mounted optical fiber strain sensor design. *Applied Optics* 1991 **30**:4069–4080.

[12] Peters KJ, Washabaugh PD. Balance technique for monitoring *in situ* structural integrity of prismatic structures. *American Institute of Aeronautics and Astronautics Student Journal* 1997 **35**:869–874.

[13] Rogers AJ. Essential optics. In *Optical Fiber Sensors: Principles and Components*, Dakin J, Culshaw B (eds). Artech House: Boston, MA, 1988.

[14] Dandridge A. Fiber optic sensors based on the Mach-Zehnder and Michelson interferometers. In *Fiber Optic Sensors*, Udd E (ed). John Wiley & Sons: Hoboken, NJ, 2006.

[15] Koo KP, Tveten AB, Dandridge A. Passive stabilization scheme for fiber interferometers using (3x3) fiber directional couplers. *Applied Physics Letters* 1982 **41**:616–618.

[16] Jackson DA, Priest R, Dandridge A, Tveten AB. Elimination of drift in a single-mode optical fiber interferometer using a piezoelectrically stretched coiled fiber. *Applied Optics* 1980 **19**:2926–2929.

[17] Kersey AD. Distributed and multiplexed fiber optic sensors. In *Fiber Optic Sensors*, Udd E (ed). John Wiley & Sons: Hoboken, NJ, 2006.

[18] Bock WJ, Urbanczyk W, Wojcik J, Beaulieu M. White-light interferometric fiber-optic pressure sensor. *IEEE Transactions on Instrumentation and Measurement* 1995 **44**:694–697.

[19] Inaudi D, Elamari A, Pflug L, Gisin N, Breguet J, Vurpillot S. Low-coherence deformation sensors for the monitoring of civil-engineering structures. *Sensors and Actuators, A* 1994 **44**:125–130.

[20] Yuan L, Ansari F. Embedded white light interferometer fibre optic strain sensor for monitoring crack-tip opening in concrete beams. *Measurement Science and Technology* 1998 **9**:261–266.

[21] Liu T, Brooks D, Martin A, Badcock R, Ralph B, Fernando GF. A multi-mode extrinsic Fabry-Perot interferometric strain sensor. *Smart Materials and Structures* 1998 **7**:550–556.

[22] Meggitt BT, Hall CJ, Weir K. An all fibre white light interferometric strain measurement system. *Sensors and Actuators, A* 2000 **79**:1–7.

[23] Rao YJ, Jackson DA. Recent progress in fibre optic low-coherence interferometry. *Measurement Science and Technology* 1996 **7**:981–999.

[24] Inaudi D. Coherence multiplexing of in-line displacement and temperature sensors. *Optical Engineering* 1995 **34**:1912–1915.

[25] Yuan L, Ansari F. White-light interferometric fiber-optic distributed strain-sensing system. *Sensors and Actuators, A* 1997 **63**:177–181.

[26] Lloret S, Rastogi P, Thévenaz L, Inaudi D. Measurement of dynamic deformations using a path-unbalance Michelson-interferometer-based optical fiber sensing device. *Optical Engineering* 2003 **42**:662–669.

[27] Murphy KA, Gunther MF, Vengsarkar AM, Claus RO. Quadrature phase-shifted, extrinsic Fabry-Perot optical fiber sensors. *Optics Letters* 1991 **16**:273–275.

[28] Duke JC, Cassino CD, Childers BA, Prosser WH. Characterization of an extrinsic Fabry-Perot interferometric acoustic emission sensor. *Materials Evaluation* 2003 **61**:935–940.

[29] Akhavan F, Watkins SE, Chandrashekhara K. Prediction of impact contact forces of composite plates using fiber optic sensors and neural networks. *Mechanics of Composite Materials and Structures* 2000 **7**:195–205.

[30] Seo DC, Kwon IB, Lee JJ. Fatigue crack growth monitoring by optical fiber sensors in smart composite patch repairs. *Key Engineering Materials* 2006 **321–323**:286–289.

[31] Park JW, Ryu CY, Kang HK, Hong CS. Detection of buckling and crack growth in the delaminated composites using fiber optic sensor. *Journal of Composite Materials* 2000 **34**:1602–1623.

[32] Lawrence CM, Nelson DV, Bennett TE, Spingarn JR. An embedded fiber optic sensor method for determining residual stresses in fiber-reinforced composite materials. *Journal of Intelligent Material Systems and Structures* 1998 **9**:788–799.

[33] Liu TY, Cory J, Jackson DA. Partially multiplexing sensor network exploiting low coherence interferometry. *Applied Optics* 1993 **32**:1100–1103.

[34] Ferreira LA, Ribeiro ABL, Santos JL, Farahi F. Simultaneous measurement of displacement and temperature using a low finesse cavity and a fiber Bragg grating. *Journal of Lightwave Technology* 1996 **11**:1519–1521.

[35] Legoubin S, Douay M, Bernage P, Niay P, Boj S, Delevaque E. Free spectral range variations of grating-based Fabry-Perot filters photowritten in optical fibers. *Journal of the Optical Society of America A* 1995 **12**:1687–1694.

[36] Sirkis JS, Brennan DD, Putman MA, Berkoff TA, Kersey AD, Friebele EJ. In-line fiber étalon for strain measurement. *Optics Letters* 1993 **18**:1973–1975.

[37] Bhatia V, Murphy KA, Claus RO, Tran TA, Greene JA. Recent developments in optical-fiber-based extrinsic Fabry-Perot interferometric strain sensing technology. *Smart Materials and Structures* 1995 **4**:246–251.

[38] Rao YJ. Recent progress in fiber-optic extrinsic Fabry-Perot interferometric sensors. *Optical Fiber Technology* 2006 **12**:227–237.

[39] Sirkis J, Berkoff TA, Jones RT, Singh H, Kersey AD, Friebele EJ, Putnam MA. In-line fiber etalon (ILFE) fiber-optic strain sensors. *Journal of Lightwave Technology* 1995 **13**:1256–1263.

[40] Bhatia V, Schmid CA, Murphy KA, Claus RO, Tran TA, Greene JA, Miller MS. Optical fiber sensing technique for edge-induced and internal delamination detection in composites. *Smart Materials and Structures* 1995 **4**:164–169.

[41] Chang CC, Sirkis J. Absolute phase measurement in extrinsic Fabry-Perot optical fiber sensors using multiple path match conditions. *Experimental Mechanics* 1997 **37**:26–32.

[42] Bao X, Webb DJ, Jackson DA. Combined distributed temperature and strain sensor based on Brillouin loss in an optical fiber. *Optics Letters* 1994 **19**:141–143.

[43] Niklès M, Thévenaz L, Robert PA. Simple distributed fiber sensor based on Brillouin gain spectrum analysis. *Optics Letters* 1996 **21**:758–760.

[44] Bao X, Huang C, Zeng X, Arcand A, Sullivan P. Simultaneous strain and temperature monitoring of the composite cure with a Brillouin-scattering-based distributed sensor. *Optical Engineering* 2002 **41**:1496–1501.

[45] Bao X, DeMerchant M, Brown A, Bremner T. Tensile and compressive strain measurement in the lab and field with the distributed Brillouin scattering sensor. *Journal of Lightwave Technology* 2001 **19**:1698–1704.

[46] Yari T, Nagai K, Takeda N. Aircraft structural-health monitoring using optical fiber distributed BOTDR sensors. *Advanced Composite Materials* 2004 **13**:17–26.

[47] Murayama H, Kageyama K, Naruse H, Shimada A. Distributed strain sensing from damaged composite materials based on shape variation of the Brillouin spectrum. *Journal of Intelligent Material Systems and Structures* 2004 **15**:17–25.

[48] Song KY, He Z, Hotate K. Distributed strain measurement with millimeter-order spatial resolution based on Brillouin optical correlation domain analysis. *Optics Letters* 2006 **31**:2526–2528.

Chapter 61
Fiber Bragg Grating Sensors

Kara Peters
Department of Mechanical and Aerospace Engineering, North Carolina State University, Raleigh, NC, USA

1 Introduction	1097
2 Fiber Bragg Grating Parameters	1098
3 Fabrication and Strength of Fiber Bragg Grating Sensors	1100
4 Multiplexing and Interrogation of FBG Sensor Networks	1101
5 Multiple Axis Sensing	1102
6 Nonuniform Sensing	1104
7 Long Period Gratings	1106
8 Conclusions	1106
Related Articles	1107
References	1107

1 INTRODUCTION

Fiber Bragg grating (FBG) sensors have many advantages for strain sensing in structural health monitoring applications including their small size, the potential to multiplex hundreds of sensors with a single ingress/egress fiber, their immunity to electromagnetic interference, and their corrosion resistance [1]. FBG sensors have been applied for a variety of structural health monitoring applications including spacecraft [1, 2], bonded aircraft repairs [3–5], cryogenic composite tanks [6, 7], highway and railway bridges [8, 9], fiber reinforced polymer (FRP) composite bridges [10], FRP strengthened concrete [11], offshore platforms [12], and nuclear reactors [13]. Researchers have also applied FBG sensors for modal-based damage identification [14, 15], the identification of multiple failure modes in composite laminates [16–18], and monitoring of welded and composite joints [19, 20].

The FBG sensor, shown in Figure 1, is a permanent periodical perturbation in the index of refraction of the optical fiber core. This index modulation can be written mathematically as

$$n_{\text{eff}}(z) = n_{\text{eff}} + \overline{\delta n_{\text{eff}}} \left\{ 1 + v \cos \left[\frac{2\pi}{\Lambda} + \phi(z) \right] \right\} \quad (1)$$

where v is the fringe visibility, Λ the grating period, $\phi(z)$ the grating chirp function (which describes any variation in the grating period), n_{eff} the effective index of refraction of the fiber for the fundamental mode, and $\overline{\delta n_{\text{eff}}}$ the "dc" average index change [21]. Hill et al. [22] first fabricated permanent Bragg gratings in an optical fiber as a filter for telecommunication applications.

The FBG sensor acts as a "wavelength-dependent filter" as shown in Figure 1. When a broad spectrum of wavelengths is passed through the FBG, a narrow bandwidth of wavelengths is reflected, while all

Encyclopedia of Structural Health Monitoring. Edited by Christian Boller, Fu-Kuo Chang and Yozo Fujino © 2009 John Wiley & Sons, Ltd. ISBN: 978-0-470-05822-0.

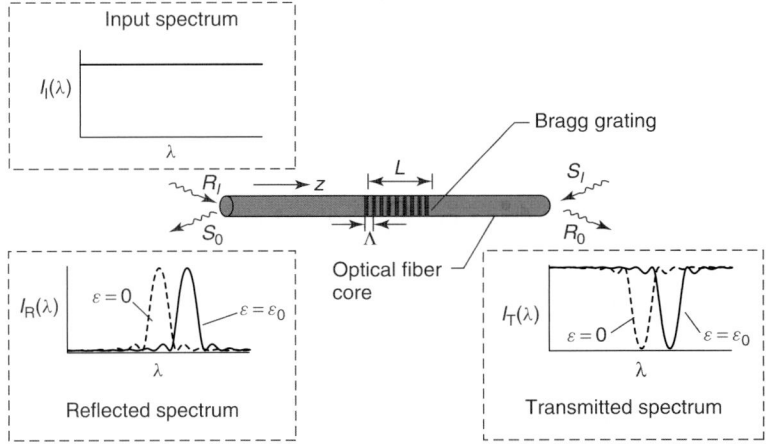

Figure 1. Optical fiber Bragg grating sensor. Reflected and transmitted spectra are shown for a broadband input spectrum. Dashed line is for unstrained FBG and solid line is for strained FBG. [Reproduced with permission. © Springer, 2002.]

others are transmitted. The wavelength at maximum reflectivity is referred to as the Bragg wavelength, λ_B, and is determined by the condition

$$\lambda_B = 2n_{\text{eff}}\Lambda \qquad (2)$$

As axial strain, ε, is applied to the FBG, the Bragg wavelength shifts to lower wavelengths (compression) or higher wavelengths (tension). The applied strain is thus encoded in the FBG Bragg wavelength shift. If the applied axial strain is not constant along the length of the grating, the spectrum will distort as well as shift. Section 6 describes how this distortion can be interpreted to measure the applied strain distribution along the FBG.

2 FIBER BRAGG GRATING PARAMETERS

The phenomenon of wavelength selective reflection is due to coupling between various modes propagating through the FBG. For short period FBGs ($\Lambda < 1\,\mu$m), coupling occurs between a forward propagating fundamental (LP_{01}) mode and a backward propagating LP_{01} mode (*see* **Chapter 59**). The coupling condition can be described through the coupled mode equations

$$\frac{dR}{dz} = i\hat{\sigma} R(z) + i\kappa S(z)$$

$$\frac{dS}{dz} = -i\hat{\sigma} S(z) - i\kappa R(z) \qquad (3)$$

where $R(z)$ and $S(z)$ are the amplitudes of the forward and backward propagating lightwaves, respectively, and z is the coordinate along the axis of the fiber [21]. The coefficients $\hat{\sigma}$ and κ are defined as

$$\hat{\sigma} = \frac{2\pi}{\lambda}\left(n_{\text{eff}} + \overline{\delta n_{\text{eff}}}\right) - \frac{\pi}{\Lambda} - \frac{1}{2}\frac{d\phi}{dz}$$

$$\kappa = \frac{\pi}{\lambda}\nu\overline{\delta n_{\text{eff}}} \qquad (4)$$

where λ is the wavelength of the propagating lightwaves.

For a grating with a constant period (i.e., nonchirped grating), we have a constant grating chirp function, $\phi(z) = \phi_0$. Therefore, the parameters $\hat{\sigma}$ and κ are constants and equation (3) can be solved analytically. We find the solution to equation (3) for the reflectivity coefficient,

$$r = \left|\frac{S(-L/2)}{R(-L/2)}\right|^2 = \frac{\sinh^2\left(L\sqrt{\kappa^2 - \hat{\sigma}^2}\right)}{\cosh^2\left(L\sqrt{\kappa^2 - \hat{\sigma}^2}\right) - \frac{\hat{\sigma}^2}{\kappa^2}} \qquad (5)$$

This solution is plotted for different values of κL in Figure 2. One can see in Figure 2 that for the case of strong coupling ($\kappa L = 6$) the grating is supersaturated. For sensing applications, FBGs with a

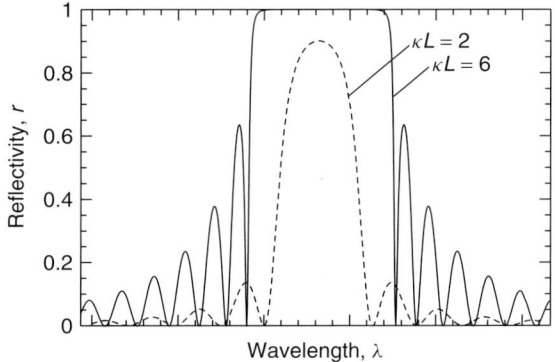

Figure 2. Reflectivity as a function of wavelength for FBG with coupling coefficients $\kappa L = 2$ and $\kappa L = 6$ ($n_{\text{eff}} = 1.46$, $\overline{\delta n_{\text{eff}}} = 3 \times 10^{-6}$, $\nu = 1$, $\lambda_B = 1550$ nm).

narrow bandwidth and high reflectivity (i.e., $\kappa L = 2$ in Figure 2) produce the largest signal-to-noise ratio. The maximum reflectivity of the grating, r_{max}, and bandwidth between the first zero crossings, $\Delta \lambda_0$, are found from equation (5),

$$r_{\text{max}} = \tanh^2(\kappa L) \tag{6}$$

$$\frac{\Delta \lambda_0}{\lambda_B} = \sqrt{\left(\frac{\nu \delta n_{\text{eff}}}{n_{\text{eff}}}\right)^2 + \left(\frac{2\Lambda}{L}\right)^2} \tag{7}$$

Procedures such apodization are often applied to the FBG sensor during fabrication to reduce the secondary peaks and narrow the bandwidth [23].

As strain is applied to the FBG sensor, the Bragg wavelength shifts according to

$$\Delta \lambda_B = 2 \left(\Delta n_{\text{eff}} \Lambda_0 + n_{\text{eff}} \Delta \Lambda_0\right) \tag{8}$$

where Δn_{eff} is due to the strain–optic effect and $\Delta \Lambda_0$ is due to the change in period of the grating. For the specific case of pure axial loading we find (*see* **Chapter 60**),

$$\Delta \lambda_B = \lambda_B \left[1 - \frac{n_{\text{eff}}^2}{2}(p_{12} - \nu p_{12} - \nu p_{11})\right] \varepsilon \tag{9}$$

in terms of the Pockel's constants of silica, p_{11} and p_{12}. Defining the effective photoelastic constant for axial strain, $p_e = n_{\text{eff}}^2(p_{12} - \nu p_{11} - \nu p_{12})/2$ we can write,

$$\frac{\Delta \lambda_B}{\lambda_B} = (1 - p_e)\varepsilon \tag{10}$$

Thus, the shift in Bragg wavelength is linearly related to the applied axial strain. Typical values for silica optical fibers are $p_{11} = 0.12$, $p_{12} = 0.27$, and $p_e = 0.22$ [24].

Similar to electrical resistance strain gauges, FBG sensors are also sensitive to temperature through

$$\frac{\Delta \lambda_B}{\lambda_B} = 2 \left(\frac{\partial n_{\text{eff}}}{\partial T} \Lambda_0 + \frac{\partial \Lambda}{\partial T} n_{\text{eff}}\right) \Delta T = (\alpha + \zeta) \Delta T \tag{11}$$

where α is the thermal expansion coefficient and ζ is the thermo-optic coefficient of the optical fiber material. A typical value for fused silica is $(\alpha + \zeta) = 6.67 \times 10^{-6}/°C$ [24]. It is important to note that the thermal sensitivity of an FBG sensor is considerably higher than that of its electrical strain gauge counterpart, increasing the need for thermal compensation for strain measurement applications.

One method of temperature compensation is to write two FBGs at different Bragg wavelengths, λ_{B1} and λ_{B2}, and to measure the Bragg wavelength shifts in each grating due to the applied strain and temperature. The strain and temperature can thus be independently determined from the equation

$$\begin{pmatrix} \Delta \lambda_1 \\ \Delta \lambda_2 \end{pmatrix} = \begin{pmatrix} K_{\varepsilon 1} & K_{T1} \\ K_{\varepsilon 2} & K_{T2} \end{pmatrix} \begin{pmatrix} \varepsilon \\ \Delta T \end{pmatrix} \tag{12}$$

where the K terms are the sensitivities of each grating to strain and temperature. Xu et al. [25] fabricated a two FBG sensor with $\lambda_{B1} = 850$ nm and $\lambda_{B2} = 1300$ nm. Experimental measurements demonstrated a 6.5% difference in $K\varepsilon$, and a 9.8% difference in K_T for the two FBGs. One difficulty with this approach is that λ_{B1} and λ_{B2} must be sufficiently far apart to obtain a significant difference in the FBG sensitivities and therefore an accurate calculation of the strain and temperature when inverting equation (12). This often requires two separate laser sources for the sensor interrogation.

3 FABRICATION AND STRENGTH OF FIBER BRAGG GRATING SENSORS

Several methods are commonly used for the fabrication of FBGs, each based on the exposure of the optical fiber core to ultraviolet (UV) laser light [26]. The UV exposure increases the index of refraction of the silica locally through the process of photosensitivity. Prior to exposure, the optical fiber is typically doped to increase the photosensitivity of the silica, e.g., through high-pressure hydrogen loading of the fiber. Molecular hydrogen diffuses in the central core region due to its small size. This increase in photosensitivity is not a permanent effect since the hydrogen diffuses out over time. However, if the fiber is exposed to UV radiation during this period, the hydrogen molecules react in the silica Si–O–Ge sites forming OH species and UV bleachable germanium oxygen deficiency centers, which are responsible for the enhanced photosensitivity [23].

Several fabrication techniques are commonly applied for FBG sensors. For excellent reviews see [23] and [27]. The interferometric technique is based on a bulk interferometer that splits the incoming UV light into two beams and then recombines them to form an interference pattern that is focused onto the optical fiber core (Figure 3). The interference pattern exposure induces a refractive index modulation in the fiber core. The periodicity of the refractive index modulation is related to the UV source wavelength, λ_{UV}, by

$$\Lambda = \frac{\lambda_{UV}}{2\sin(\theta/2)} \quad (13)$$

By changing the intersecting angle θ between the two writing beams, it is possible to write Bragg gratings for almost any wavelength. However, the stability of the interference pattern is extremely difficult to maintain due to mechanical vibrations that may be present in the bulk interferometer.

Because of its simplicity, the phase mask technique is the most widely used method of FBG fabrication (see Figure 4). The phase masks themselves may be formed holographically or by electron-beam lithography from fused silica. A near-field fringe pattern is produced by the interference of the plus and minus first-order diffracted beams. The period of the fringes produced is one-half of the phase mask period, Λ_{pm}.

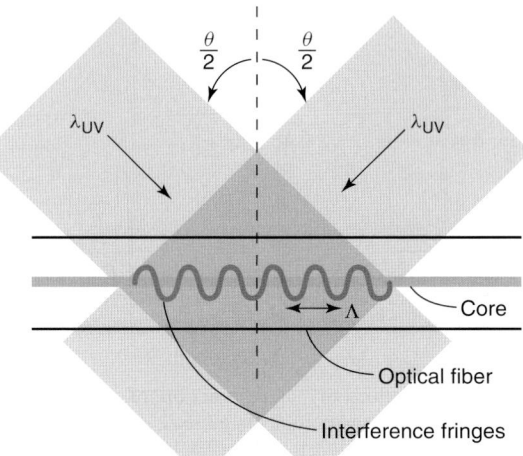

Figure 3. Bulk interferometer arrangement for the fabrication of fiber Bragg gratings.

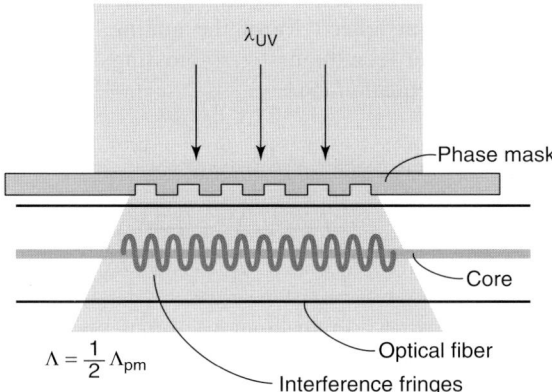

Figure 4. Phase mask arrangement for the fabrication of fiber Bragg gratings.

The simplicity of the phase mask writing technique provides a robust and inherently stable method for producing FBGs. Since the fiber is usually placed close to the mask, the sensitivity to mechanical vibrations is minimized. The main disadvantage of the phase mask technique, however, is the need to produce a phase mask for each desired wavelength. This limitation can be partially overcome by applying tension to the phase mask to vary the period and provide a range of possible Bragg wavelengths (typically ±5% of the original Bragg wavelength).

Stripping the coating from an optical fiber prior to writing the FBG is necessary as the coating will reduce the amount of UV light arriving at

Table 1. Summary of FBG fabrication methods and resulting strengths

Reference	Fiber Bragg grating fabrication method	σ_{med} (Gpa)	m
Wei et al. [28]	Pristine fiber	4.95	66.6
	Fiber after chemical stripping	3.38	3.1
	Pulsed laser exposure + chemical stripping	2.52	3.2
Varelas et al. [29]	Pristine fiber + chemical stripping	5.13	112
	Continuous wave UV exposure + chemical stripping	>5	57
Askins et al. [30]	Pristine reference fiber (Corning SMF-28)	5.6–6.0	90
	Single pulse exposure, writing during draw tower process prior to coating	5.4	45
Gu et al. [31]	One-step process of writing through polysiloxane buffer, no stripping required	3.76	13.2
Han et al. [32]	LPG fabrication via residual stress relaxation with CO_2 laser (no UV irradiation required)	4.9	10.5

the optical fiber core. However, mechanical stripping of the coating introduces microcracks at the surface of the optical fiber and can significantly reduce the strength of the fiber at the location of the FBG. Researchers have developed several techniques to increase the strength of FBGs, including chemically stripping the coating from the optical fiber and writing the grating in the optical fiber during the drawing of the fiber prior to coating. A summary of the resulting FBG strength properties taken from the literature is listed in Table 1. For definitions of the Weibull modulus and median strength, *see* **Chapter 59**. Additionally, description of common coating types for FBG sensors for durability in structural health monitoring applications are provided in **Chapter 59**.

4 MULTIPLEXING AND INTERROGATION OF FBG SENSOR NETWORKS

One of the strongest advantages of FBG sensors over other available strain or temperature sensors for structural health monitoring applications (*see* **Chapter 54**; **Chapter 57**; **Chapter 64**) is the fact that they can be multiplexed into a large sensor network. For structural health monitoring applications, having only a single or limited ingress/egress points and lead cables is a notable advantage because it can significantly reduce the weight of the lead cables and disruptions to the structure itself [1].

Several approaches to multiplexing large FBG sensor networks have been applied in which the sensors are interrogated using wavelength division multiplexing (WDM), time division multiplexing (TDM), and combinations thereof (*see* **Chapter 59**). For single FBG sensors, the most common method to measure the peak shifts is to pass the reflected signal from the FBG through a wavelength-dependent filter that outputs a different intensity based on the input wavelength. Examples of wavelength-dependent filters include Fabry–Perot cavities [33], acousto-optic filters [34], chirped Bragg gratings [35, 36], long period gratings (LPGs) [37], and WDM couplers [38, 39]. These techniques have been expanded to WDM FBG sensor networks by scanning the filters to cover the peak wavelengths of multiple sensors. For example, Davis et al. [40] demonstrated interrogation of 60 FBGs in 2.5 s with 50 averages per sensor.

The interrogation of a large number of FBGs along a single optical fiber is limited by the power reflected from each grating, since each FBG reflects a portion of the wavelengths in the vicinity of the Bragg wavelength. For WDM applications, reducing the spacing between each FBG reduces the time required to scan the network; however, it also increases the cross talk losses between FBGs. For TDM applications, low reflectivity gratings ($r_{max} \cong 1\%$) should be used to allow for a large number of multiplexed sensors. By combining WDM with TDM, as shown in Figure 5, one can reduce the power loss per FBG,

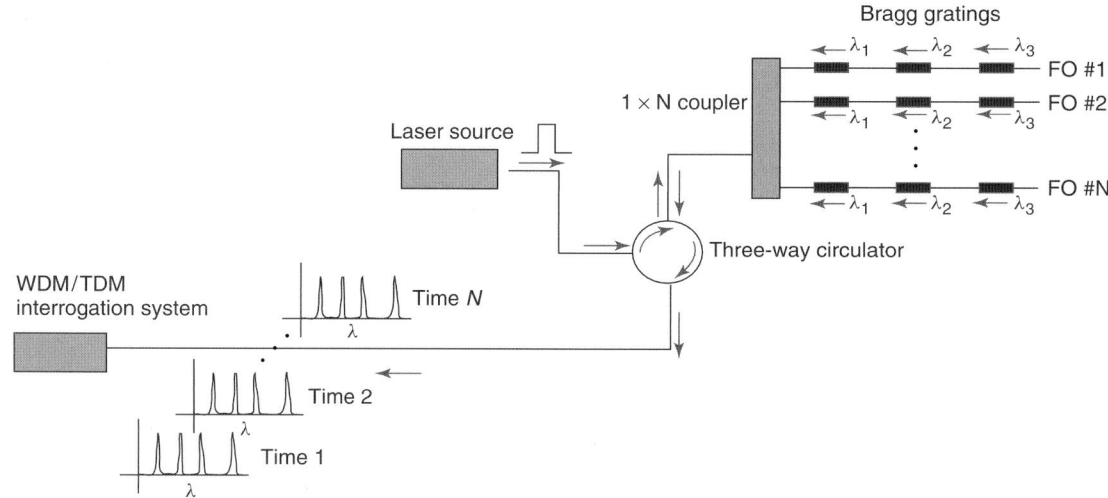

Figure 5. Combined wavelength division multiplexing and time domain multiplexing for FBG sensor network.

as well reduce the total wavelength range that must be scanned to interrogate the entire network [24]. In this case, each parallel network of FBGs has the same wavelength range, while the laser output is pulsed and the particular network being interrogated determined by the time of arrival of the signal.

Alternative approaches to interrogating FBG sensor networks include expanding the output-reflected spectrum via a plane grating or prism to a linear charge-coupled device (CCD) array that would then operate as a CCD spectrometer. In this manner, Askins *et al.* [41] achieved interrogation of 20 sensors at 3.5 kHz. A network of FBG sensors that have the same initial Bragg wavelength can also be interrogated quasi-statically through interpretation of the inverse Fourier transformation of the reflected spectrum [42]. Finally, interferometric methods have also been applied for high-speed interrogation of FBG sensor networks [43–45].

5 MULTIPLE AXIS SENSING

One unique characteristic of the FBG as a strain sensor is its ability to measure and distinguish between multiple strain components. Researchers have fabricated FBG rosettes, following the same strategy as for electrical resistance strain gauges [46, 47]. However, the FBG itself can also be used to monitor axial and transverse strain components through the birefringence that occurs in the optical fiber due to applied transverse loads. Figure 6 shows an example of three independent loads applied to the optical fiber, along with the resulting three principal strain components at the center of the fiber core. The effect of the axial load (P_1) is to shift the reflected peak to higher or lower wavelengths; however, the effect of the transverse loads (P_2 and P_3) is to create peak splitting due to the fast and slow axes that develop in the optical fiber (for a discussion of birefringence, *see* **Chapter 59**). A typical example of peak-splitting behavior is also shown in Figure 6.

In general, FBGs written in polarization maintaining (PM) optical fibers exhibit enhanced discrimination between multiple loading components due to geometrical and stress-induced birefringence in the fibers (*see* **Chapter 59**). The stress-induced portion of the birefringence can be due to residual stresses and/or the applied transverse loading components. This birefringence effect has been exploited for the independent measurement of multiple strain components and/or temperature changes using single or multiple Bragg gratings [48]. This capability makes the FBG sensor ideal for the monitoring of residual stresses during curing of FRP laminated composites as well as further monitoring of internal stresses during loading of the laminates [17, 49–52]. Additionally, by writing two FBGs with different Bragg wavelengths, λ_{B1} and λ_{B2}, at the same

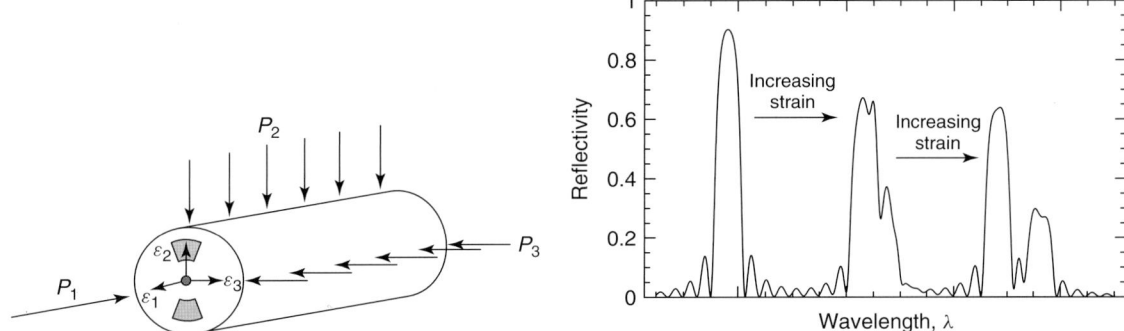

Figure 6. Multiaxis loading applied to FBG sensor written into bow-tie PM fiber. The graph plots the effects of the multiaxis loading on a single FBG (the peak on the left-hand side is the initial peak before load is applied).

location and monitoring the peak wavelength shifts for the fast and slow axes of each grating, one can independently calculate the three principle strains as well as a uniform temperature change applied to the sensor [48],

$$\begin{pmatrix} \Delta\lambda_{1f} \\ \Delta\lambda_{1s} \\ \Delta\lambda_{2f} \\ \Delta\lambda_{2s} \end{pmatrix} = \begin{pmatrix} K_{11f} & K_{12f} & K_{13f} & K_{1Tf} \\ K_{11s} & K_{12s} & K_{13s} & K_{1Ts} \\ K_{21f} & K_{22f} & K_{23f} & K_{2Tf} \\ K_{21s} & K_{22f} & K_{23f} & K_{2Ts} \end{pmatrix} \begin{pmatrix} \varepsilon_1 \\ \varepsilon_2 \\ \varepsilon_3 \\ \Delta T \end{pmatrix}$$
(14)

Generally, the relative sensitivities of the FBG written in a PM fiber are calibrated because of the complexity of predicting these sensitivities and the lack of detailed material information for the particular PM fibers used. Urbanczyk et al. [53], Chehura et al. [54], Lawrence et al. [48], and Bosia et al. [55] performed experimental measurements of the wavelength shifts in FBGs due to multi-axis loading in various PM fiber types. Chehura et al. [54] also performed a comparative experimental study of various PM fiber types to determine which fiber type was optimal for multiaxis strain sensing and temperature monitoring. The authors determined that amongst the specific, commercially available PM fibers studied, the elliptical core stress-applying portion (SAP) fiber provided the maximum sensitivity to transverse loading, while the Panda fiber provided the maximum sensitivity to temperature loading. However, the particular elliptical core SAP fiber used had a significantly smaller cladding diameter than the other fibers, which the authors note was the cause of the increased transverse load sensitivity. The transverse load and temperature sensitivities of the FBG can also be increased through the simultaneous measurement of the response of both the fundamental LP_{01} and LP_{11} modes; however, propagating the LP_{11} mode in the FBG presents substantial challenges [56].

To predict the response of an FBG written in a PM fiber to transverse loading, Kim et al. [57] considered the PM fiber to be optically and mechanically homogeneous with orthotropic material properties. As in later models, the assumption is made that most of the energy of the fundamental mode propagating in the fiber is contained in the core; therefore, the principal strains at the center of the fiber are sufficient to estimate the wavelength shift. Sirkis [58] later related the change in Bragg wavelengths, $\Delta\lambda_{B1}$ and $\Delta\lambda_{B2}$, to the principal strains at the center of the core: ε_1, ε_2, and ε_3 (see Figure 6),

$$\frac{\Delta\lambda_{B1}}{\lambda_{B1}} = \varepsilon_1 - \frac{1}{2}n_{\text{eff}}^2\left(p_{11}\varepsilon_2 + p_{12}\varepsilon_3 + p_{12}\varepsilon_1\right)$$

$$\frac{\Delta\lambda_{B2}}{\lambda_{B2}} = \varepsilon_1 - \frac{1}{2}n_{\text{eff}}^2\left(p_{12}\varepsilon_2 + p_{11}\varepsilon_3 + p_{12}\varepsilon_1\right) \quad (15)$$

Lawrence et al. [48] and later Bosia et al. [55] modeled the mechanical inhomogeneities in the optical fiber due to the SAP and applied the finite element method to calculate the strains at the center of the fiber due to transverse loading. Both the experimental and numerical studies of Lawrence et al. [48] and Bosia et al. [55] demonstrate that for a PM fiber, the shift in Bragg wavelength is nonlinear with transverse load for certain loading angles. Prabhugoud and Peters [59, 60] performed finite element analyses of FBGs written in PM fibers analysis incorporating the photoelastic-induced index distribution throughout the cross section of the optical fiber and calculating its contribution to the wavelength shift of the FBG. Comparison of this method with the previous assumption of principal strains at the center of the core in equation (15) for a variety of PM fiber types and loading cases reveals that the approximation of equation (15) is sufficient for most loading cases; however, its accuracy varies between the different fiber types.

6 NONUNIFORM SENSING

Distortion of the grating spectrum due to strain gradients has been observed in several applications of embedded FBG sensors [61–74]. An example of experimentally measured spectral distortion due to the highly nonuniform strain field near a notch tip is shown in Figure 7 [61]. This sensitivity of the sensor response to the form of the strain profile is unique to the optical FBG since other strain gauges (e.g., the classical electrical strain gauge) average the applied strain over the gauge length. This section presents both the forward prediction of the FBG spectral response due to the applied nonuniform strain field and the reverse calculation of the applied strain field from the spectral response.

6.1 Prediction of FBG response

The T-matrix approximation, first introduced by Yamada and Sakuda [75], is widely used to model Bragg gratings with nonconstant properties as explained in Figure 8. The primary advantage of the T-matrix method is that it is computationally efficient as compared to direct numerical integration of equation (3) [21]. This approach divides the grating into M smaller sections each with uniform coupling properties. Defining R_i and S_i to be the amplitudes $R(z)$ and $S(z)$ after each lightwave traverses the ith section, the propagation through this uniform section can be described in the form of a transfer matrix

$$\begin{bmatrix} R_i \\ S_i \end{bmatrix} = F_i \begin{bmatrix} R_{i-1} \\ S_{i-1} \end{bmatrix} \qquad (16)$$

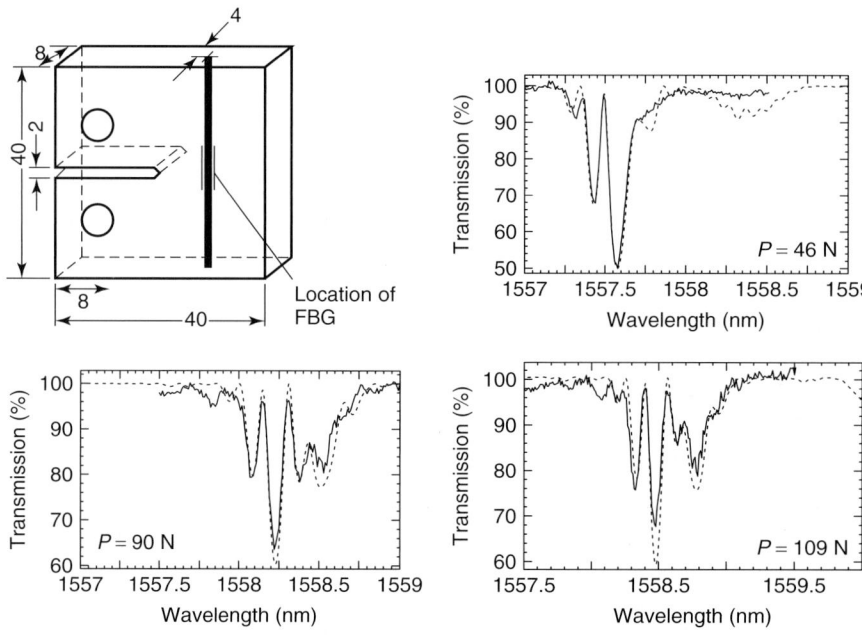

Figure 7. Response of FBG embedded in a compact tension specimen. Location of 10-mm grating is shown in the sketch of the specimen. Experimental grating spectral data in transmission (solid lines) are shown for three load levels (P). The simulated spectra calculated using the transfer matrix method are also plotted (dashed lines). [Reproduced with permission. © Springer, 2001.]

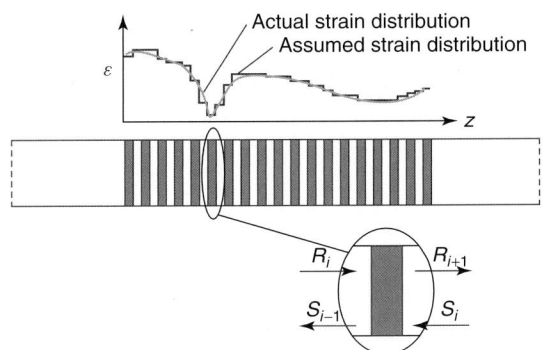

Figure 8. Schematic of transfer matrix approximation for FBG subjected to nonuniform strain field. [Reproduced with permission. © Springer, 2001.].

The optical transfer matrix can be derived as

$$F_i = \begin{bmatrix} \cosh(\gamma_B \Delta z) - i\dfrac{\hat{\sigma}}{\gamma_B}\sinh(\gamma_B \Delta z) & -i\dfrac{\kappa}{\gamma_B}\sinh(\gamma_B \Delta z) \\ i\dfrac{\kappa}{\gamma_B}\sinh(\gamma_B \Delta z) & \cosh(\gamma_B \Delta z) + i\dfrac{\hat{\sigma}}{\gamma_B}\sinh(\gamma_B \Delta z) \end{bmatrix} \quad (17)$$

where $\gamma_B = \sqrt{\kappa^2 - \hat{\sigma}^2}$ and Δz is the length of the section. For the entire grating, the combined optical transfer matrix can then be written as

$$\begin{bmatrix} R(-L/2) \\ S(-L/2) \end{bmatrix} = F \begin{bmatrix} R(L/2) \\ S(L/2) \end{bmatrix} \quad (18)$$

where $F = F_M \cdot F_{M-1} \cdots F_1$. The reflectivity as a function of wavelength can be calculated from equation (18). A limitation of the T-matrix approximation, however, is that the number of sections M cannot be arbitrarily large since several grating periods are required for complete coupling. However, $M \cong 100$ is typically more than sufficient to accurately model chirped gratings.

For axial strain-sensing applications, the T-matrix method was first applied to a Bragg grating subjected to a nonuniform strain distribution by Huang *et al.* [76]. Huang *et al.* [76] approximated the applied strain as a piecewise continuous function, calculating the average period in each grating segment due to the applied strain and substituting this local period into the coupling coefficient for the T-matrix method. The validity of this approximation was later demonstrated by Peters *et al.* [77] for various nonuniform strain fields. Prabhugoud and Peters [78] demonstrated that this prediction works well for small magnitudes of strain gradients. For an arbitrary nonuniform strain field one can enter the strain field into the coupled mode equations or transfer matrix through an "equivalent" period that accounts for both geometrical and photoelastic effects,

$$\Lambda(z) = \Lambda_0 \left[1 + (1 - p_e)\varepsilon(z) + (1 - p_e)z\varepsilon'(z)\right] \quad (19)$$

where ( )′ refers the derivative with respect to z [78]. A similar expression for the equivalent period for cases where the grating is originally chirped can be found in Prabhugoud and Peters [78].

6.2 Inversion of measured spectrum

When the axial strain applied to the sensor is nonconstant along its gauge length, one can no longer simply measure the shift in wavelength at maximum reflectivity to obtain useful strain data. Rather, the calculation of the applied strain profile from the spectral response becomes an inverse problem. The most commonly used inversion techniques for optical FBG strain sensor spectra are reviewed by Huang *et al.* [76]. The simplest technique is the intensity-spectrum-based approach, which requires only the amplitude information from the complex reflected spectrum. Although suitable for some applications, this technique is only valid for monotonically varying strain fields and averages out rapidly changing strain fields. A second approach based on the Fourier transform is more suitable to invert nonmonotonic strain profiles. However, this technique requires both intensity and phase information from the reflected spectrum, which significantly increases the cost and complexity of sensor data collection. Several other techniques to invert the spectral information based on heuristic approaches have been developed, including iterative solutions to the coupled mode equations (3) [79], inverse scattering algorithms (*see* **Chapter 26**) [80–82], time–frequency transform methods [83], neural networks (*see* **Chapter 32**) [84], and genetic algorithms [85, 86]. Such techniques do not identify the direction of the strain profile; however,

Kitcher et al. [87] have demonstrated the presence of directional dependent losses in FBGs, which could be used to identify the orientation of the strain field.

7 LONG PERIOD GRATINGS

While short period FBGs reflect or transmit lightwaves at certain wavelengths based on coupling between counterpropagating core modes, LPG sensors are based on coupling between the forward propagating core mode and forward propagating cladding modes (see **Chapter 59** for a description of cladding modes). While both of these and other coupling phenomena occur for both types of grating sensors, the large period of the LPG (100 μm < Λ < 1 mm) results in cladding mode coupling at wavelengths in the near-IR range. Because of the fact that cladding modes attenuate rapidly in optical fibers, the LPG coupling is not measurable from the reflected spectrum, but rather appears as multiple loss peaks in the transmission spectrum (see Figure 9). The phase matching condition for each of these loss peaks is given by

$$\lambda_i = \left[n_{\text{eff}} - n_{\text{clad}}^{(i)} \right] \Lambda \qquad (20)$$

where $n_{\text{clad}}^{(i)}$ is the effective index of refraction of the ith cladding mode [88]. The maximum transmission loss for the ith cladding mode is given by

$$T_i = 1 - \sin^2(\kappa_i L) \qquad (21)$$

where κ_i is the coupling coefficient for the mode (similar to κ for the short period FBG).

Because of the larger scale of the LPG features as compared to that of short period FBGs, a variety of fabrication procedures have been demonstrated to write LPGs in silica fibers including photosensitivity [89], electric arc discharge [90], etching of the fiber using hydrofluoric acid [91], and CO_2 lasers [92]. Additionally, nonpermanent LPGs have been fabricated in optical fibers by applying mechanical pressure to the fiber with an undulating surface to create a photoelastic effect–induced change in index within the fiber core [93, 94].

While LPG sensors themselves have primarily been applied to the measurement of environmental or

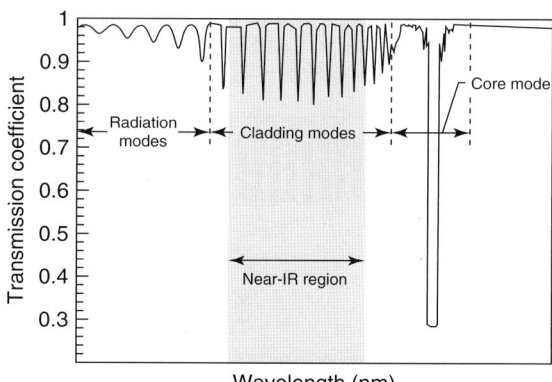

Figure 9. Theoretical transmission spectrum for long period FBG. Portion of spectrum in near-IR region is indicated.

chemical parameters because of their strong sensitivity to the index of refraction of the surrounding material system, researchers have also combined them with other optical fiber sensors for multiple parameter sensing. LPGs react to applied strain and temperature fields in much the same manner as short period FBGs, including peak splitting due to transverse loading. A significant difference between the two sensors is that the temperature sensitivity of LPGs typically ranges from 3 nm/100 °C to 10 nm/100 °C, an order of magnitude greater than that of FBGs [88]. Additionally, the strain and temperature sensitivity can be tuned by writing LPGs with different properties. Bhatia et al. [95] designed a "strain insensitive" LPG for which the photoelastic effects, Poisson contraction, and period extension as a function of strain canceled each other out, leading to an independent temperature sensor. Patrick et al. [96] fabricated a hybrid FBG/LPG sensor for independent strain and temperature measurements. Kim et al. [97] fabricated a hybrid LPG/Fabry–Perot interferometer (see **Chapter 60**) for the simultaneous measurement of refractive index and temperature.

8 CONCLUSIONS

FBGs are versatile and lightweight sensors that can be multiplexed into sensor networks and embedded into various material systems for the measurement of strain and temperature. For structural health monitoring applications, FBG sensors offer the advantages of immunity to electromagnetic interference and the ability to measure multiaxis strain fields and local strain distributions. As compared to electrical resistance or piezoelectric strain gauges, these advantages come at the cost of expense and data acquisition rate limits for large networks. At the same time, advances in telecommunication networking and devices have led to recent advances in these areas (for example, high-speed tunable microelectromechanical systems (MEMS) filters). Recent developments including the topics of FBG sensors for high-temperature applications and FBGs in polymer and photonic crystal optical fibers can be found in **Chapter 62**.

RELATED ARTICLES

Chapter 17: Lamb Wave-based SHM for Laminated Composite Structures
Chapter 109: Fiber-optic Sensors
Chapter 155: Reliable Use of Fiber-optic Sensors

REFERENCES

[1] Friebele EJ, et al. Optical fiber sensors for spacecraft applications. *Smart Materials and Structures* 1999 **8**:813–838.

[2] Ecke W, Latka I, Willsch R, Reutlinger A, Graue R. Fibre optic sensor network for spacecraft health monitoring. *Measurement Science and Technology* 2001 **12**:974–980.

[3] McKenzie I, Jones R, Marshall IH, Galea S. Optical fibre sensors for health monitoring of bonded repair systems. *Composite Structures* 2000 **50**:405–416.

[4] Sekine H, Fujimoto SE, Okabe T, Takeda N, Yokobori T. Structural health monitoring of cracked aircraft panels repaired with bonded patches using fiber Bragg grating sensors. *Applied Composite Materials* 2006 **13**:87–98.

[5] Li HCH, Beck F, Dupouy O, Herszberg I, Stoddart PR, Davis CE, Mouritz AP. Strain-based health assessment of bonded composite repairs. *Composite Structures* 2006 **76**:234–242.

[6] Mizutani T, Takeda N, Takeya H. On-board strain measurement of a cryogenic composite tank mounted on a reusable rocket using FBG sensors. *Structural Health Monitoring—An International Journal* 2006 **5**:205–214.

[7] Kang DH, Kim CU, Kim CG. The embedment of fiber Bragg grating sensors into filament wound pressure tanks considering multiplexing. *NDT&E International* 2006 **39**:109–116.

[8] Chan THT, Yu L, Tam HY, Ni YQ, Chung WH, Cheng LK. Fiber Bragg grating sensors for structural health monitoring of Tsing Ma bridge: background and experimental observation. *Engineering Structures* 2006 **28**:648–659.

[9] Zhang W, Gao JQ, Shi B, Cui HL, Zhu H. Health monitoring of rehabilitated concrete bridges using distributed optical fiber sensing. *Computer-Aided Civil and Infrastructure Engineering* 2006 **21**:411–424.

[10] Kister G, Winter D, Badcock RA, Gebremichael YM, Boyle WJO, Meggitt BT, Grattan KTV, Fernando GF. Structural health monitoring of a composite bridge using Bragg grating sensors. part 1: evaluation of adhesives and protection systems for the optical sensors. *Engineering Structures* 2007 **29**:440–448.

[11] Lau KT, Yuan LB, Zhou LM, Wu JS, Woo CH. Strain monitoring in FRP laminates and concrete beams using FBG sensors. *Composite Structures* 2001 **51**:9–20.

[12] Ren L, Li HN, Zhou J, Li DS, Sun L. Health monitoring system for offshore platform with fiber Bragg grating sensors. *Optical Engineering* 2006 **48**:084401-1–084401-9, Art. No. 084401.

[13] Bin Lin Y, Lin TK, Chen CC, Chiu JC, Chang KC. Online health monitoring and safety evaluation of the relocation of a research reactor using fiber Bragg grating sensors. *Smart Materials and Structures* 2006 **15**:1421–1428.

[14] Todd MD, Johnson GA, Vohra ST. Deployment of a fiber Bragg grating-based measurement system in a structural health monitoring application. *Smart Materials and Structures* 2001 **10**:534–539.

[15] Cusano A, Capoluongo P, Campopiano S, Cutolo A, Giordano M, Felli F, Paolozzi A, Caponero M. Experimental modal analysis of an aircraft model wing by embedded fiber Bragg grating sensors. *IEEE Sensors Journal* 2006 **6**:67–77.

[16] Ling HY, Lau KT, Cheng L. Determination of dynamic strain profile and delamination detection of composite structures using embedded multiplexed fibre-optic sensors. *Composite Structures* 2005 **67**:317–326.

[17] Takeda N, Okabe Y. Durability analysis and structural health management of smart composite structures using small-diameter fiber optic sensors. *Science and Engineering of Composite Materials* 2005 **15**:1–12.

[18] Pearson JD, Zikry MA, Prabhugoud M, Peters K. Global-local assessment of low-velocity impact damage in woven composites. *Journal of Composite Materials* 2007 **41**:2759–2783.

[19] Herszberg I, Li HCH, Dharmawan F, Mouritz AP, Nguyen M, Bayandor J. Damage assessment and monitoring of composite ship joints. *Composite Structures* 2005 **67**:205–216.

[20] Kim MH. A smart health monitoring system with application to welded structures using piezoceramic and fiber optic transducers. *Journal of Intelligent Material Systems and Structures* 2006 **17**:35–44.

[21] Erodgan T. Fibre grating spectra. *Journal of Lightwave Technology* 1997 **15**:1277–1294.

[22] Hill KO, Fujii Y, Johnson DC, Kawasaki BS. Photosensitivty in optical fiber waveguides—application to reflection filter fabrication. *Applied Physics Letters* 1978 **32**:647–649.

[23] Kashyap R. *Fiber Bragg Gratings*. Academic Press San Diego, CA, 1999.

[24] Kersey AD, Davis MA, Patrick HJ, LeBlanc M, Koo KP, Askins CG, Putnam MA, Friebele EJ. Fiber grating sensors. *Journal of Lightwave Technology* 1997 **15**:1442–1463.

[25] Xu MG, Archambault JL, Reekie L, Dakin P. Discrimination between strain and temperature effects using dual-wavelength fibre grating sensors. *Electronics Letters* 1994 **30**:1085–1087.

[26] Meltz G, Morey WW, Glenn WH. Formation of Bragg gratings in optical fibers by a transverse holographic method. *Optics Letters* 1989 **14**:823–825.

[27] Othonos A, Kalli K. *Fiber Bragg Gratings: Fundamentals and Applications in Telecommunications and Sensing*. Artech House Norwood, MA, 1999.

[28] Wei CY, Ye CC, James SW, Tatam RP, Irving PE. The influence of hydrogen loading and the fabrication process on the mechanical strength of optical fiber Bragg gratings. *Optical Materials* 2002 **20**:241–251.

[29] Varelas D, Costantini DM, Limberger HG, Salathé RP. Fabrication of high-mechanical-resistance Bragg gratings in single-mode optical fibers with continuous-wave ultraviolet laser side exposure. *Optics Letters* 1998 **23**:397–399.

[30] Askins CG, Putnam MA, Patrick HJ, Friebele EJ. Fibre strength unaffected by on-line writing of single-pulse Bragg gratings. *Electronics Letters* 1997 **33**:1333–1334.

[31] Gu XJ, Guan L, He YF, Zhang HBB, Herman PR. High-strength fiber Bragg gratings for a temperature-sensing array. *IEEE Sensors Journal* 2006 **6**: 668–671.

[32] Han YG, Han WT, Lee BG, Paek UC, Chung YJ. Temperature sensitivity control and mechanical stress effect of boron-doped long-period fiber gratings. *Fiber and Integrated Optics* 2001 **20**:591–600.

[33] Kersey AD, Berkoff TA, Morey WW. Multiplexed fiber Bragg grating strain-sensor system with a fiber Fabry–Perot wavelength filter. *Optics Letters* 1993 **18**:1370–1372.

[34] Xu MG, Geiger H, Archambault JL, Reekie L, Dakin JP. Novel interrogating system for fiber Bragg grating sensors using an acoustooptic tunable filter. *Electronics Letters* 1993 **29**:1510–1511.

[35] Davis MA, Kersey AD. Matched-filter interrogation technique for fiber Bragg grating arrays. *Electronics Letters* 1995 **31**:822–823.

[36] Betz DC, Thursby G, Culshaw B, Staszewski WJ. Acousto-ultrasonic sensing using fiber Bragg gratings. *Smart Materials and Structures* 2003 **12**: 122–128.

[37] Fallon RW, Zhang L, Everall LA, Williams JAR, Bennion I. All-fibre optical sensing system: Bragg grating sensor interrogated by a long-period grating. *Measurement Science and Technology* 1998 **9**:1969–1973.

[38] Davis MA, Kersey AD. All fiber Bragg grating strain sensor demodulation technique using a wavelength division coupler. *Electronics Letters* 1994 **30**:75–77.

[39] Zhang Q, Brown DA, Kung H, Townsend JE, Chen M, Reinhart LJ, Morse TF. Use of highly overcoupled couplers to detect shifts in Bragg wavelength. *Electronics Letters* 1995 **31**:480–482.

[40] Davis MA, Bellemore DG, Putnam MA, Kersey AD. Interrogation of 60 fiber Bragg grating sensor with μstrain resolution capability. *Electronics Letters* 1996 **32**:1393–1394.

[41] Askins CG, Putnam MA, Williams GM, Friebele EJ. Stepped-wavelength optical fiber Bragg grating arrays fabricated in line on a draw tower. *Optics Letters* 1994 **19**:147–149.

[42] Froggatt M, Moore J. Distributed measurement of static strain in an optical fiber with multiple Bragg gratings at nominally equal wavelengths. *Applied Optics* 1998 **37**:1741–1746.

[43] Koo KP, Kersey AD. Bragg grating-based laser sensors systems with interferometric interrogation and wavelength-division multiplexing. *Journal of Lightwave Technology* 1995 **13**:1243–1249.

[44] Todd MD, Johnson GA, Chang CC. Passive, light intensity-independent interferometric method for fibre Bragg grating interrogation. *Electronics Letters* 1999 **35**:1970–1971.

[45] Todd MD, Johnson GA, Althouse BL. A novel Bragg grating sensor interrogation system utilizing a scanning filter, a Mach-Zehnder interferometer and a 3 × 3 coupler. *Measurement Science and Technology* 2001 **12**:771–777.

[46] Magne S, Rougeault S, Vilela M, Ferdinand P. State-of-strain evaluation with fiber Bragg grating rosettes: application to discrimination between strain and temperature effects in fiber sensors. *Applied Optics* 1997 **36**:9437–9447.

[47] Betz DC, Thursby G, Culshaw B, Staszewski WJ. Advanced layout of a fiber Bragg grating strain gauge rosette. *Journal of Lightwave Technology* 2006 **24**:1019–1026.

[48] Lawrence CM, Nelson DV, Udd E, Bennett T. A fiber optic sensor for transverse strain measurements. *Experimental Mechanics* 1999 **39**:202–209.

[49] Kang HK, Kang DH, Hong CS, Kim CG. Simultaneous monitoring of strain and temperature during and after cure of unsymmetric composite laminate using fibre-optic sensors. *Smart Materials and Structures* 2003 **12**:29–35.

[50] Leng JS, Asundi A. Real-time cure monitoring of smart composite materials using extrinsic Fabry–Perot interferometer and fiber Bragg grating sensors. *Smart Materials and Structures* 2002 **11**:249–255.

[51] Murukeshan VM, Chan PY, Ong LS, Seah LK. Cure monitoring of smart composites using fiber Bragg grating based embedded sensors. *Sensors and Actuators, A* 2000 **79**:153–161.

[52] O'Dwyer MJ, Maistros SM, James SW, Tatam RP, Partridge IK. Relating the state of cure to the real-time internal strain development in a curing composite using in-fibre Bragg gratings and dielectric sensors. *Measurement Science and Technology* 1998 **9**:1153–1158.

[53] Uranczyk W, Chmielewska E, Bock WJ. Measurements of temperature and strain sensitivities of two-mode Bragg gratings imprinted in a bow-tie fibre. *Measurement Science and Technology* 2001 **12**:800–804.

[54] Chehura E, Ye CC, Staines SE, James SW, Tatam RP. Characterization of the response of fibre Bragg gratings fabricated in stress and geometrically induced high birefringence fibres to temperature and transverse load. *Smart Materials and Structures* 2004 **13**:888–895.

[55] Bosia F, Giaccari P, Botsis J, Facchini M, Limberger HG, Salathé R. Characterization of the response of fibre Bragg grating sensors subjected to a two-dimensional strain field. *Smart Materials and Structures* 2003 **12**:925–934.

[56] Doyle C, Martin A, Liu T, Wu M, Hayes S, Crosby PA, Powell GR, Brooks D, Fernando GF. In-situ process and conditioning monitoring of advanced fibre-reinforced composite materials using optical fibre sensors. *Smart Materials and Structures* 1998 **7**:145–158.

[57] Kim KS, Kollàr L, Springer GS. A model of embedded fiber optic Fabry–Perot temperature and strain sensors. *Journal of Composite Materials* 1993 **27**:1618–1662.

[58] Sirkis JS. Unified approach to phase-strain-temperature models for smart structure interferometric optical fiber sensors: part 1, development. *Optical Engineering* 1993 **32**:752–761.

[59] Prabhugoud M, Peters K. Finite element model for embedded fiber Bragg grating sensor. *Smart Materials and Structures* 2006 **15**:550–562.

[60] Prabhugoud M, Peters K. Finite element analysis of multi-axis strain sensitivities of Bragg gratings in PM fibers. *Journal of Intelligent Material Systems and Structures* 2007 **18**:861–873.

[61] Peters K, Studer M, Botsis J, Iocco A, Limberger H, Salathé R. Embedded optical fiber Bragg grating sensor in a nonuniform strain field: measurements and simulations. *Experimental Mechanics* 2001 **41**:19–28.

[62] Studer M, Peters K, Botsis J. Method for determination of crack bridging parameters using long optical fiber Bragg grating sensors. *Composites Part B* 2003 **34**:347–359.

[63] Ling HY, Lau KT, Chen L, Chow KW. Embedded fibre Bragg grating sensors for non-uniform strain sensing in composite structures. *Measurement Science and Technology* 2005 **16**:2415–2424.

[64] Ling HY, Lau KT, Su Z, Wong ETT. Monitoring mode II fracture behavior of composite laminates using embedded fiber-optic sensors. *Composites, Part B* 2007 **38**:488–497.

[65] Kang DH, Park SO, Hong CS, Kim CG. The signal characteristics of reflected spectra of fiber Bragg

[65] grating sensors with strain gradients and grating lengths. *NDT&E International* 2005 **38**:712–718.

[66] Kuang KSC, Kenny R, Whelan MP, Cantwell WJ, Chalker PR. Embedded fibre Bragg grating sensors in advanced composite materials. *Composites Science and Technology* 2001 **61**:1379–1387.

[67] Guemes JA, Menéndez JM. Response of Bragg grating fiber-optic sensors when embedded in composite laminates. *Composites Science and Technology* 2002 **62**:959–966.

[68] Prabhugoud M, Peters K. Efficient simulation of Bragg grating sensors for implementation to damage identification in composites. *Smart Materials and Structures* 2003 **12**:914–924.

[69] Rao YJ, et al. Simultaneous strain and temperature measurement of advanced 3-D braided composite materials using an improved EFPI/FBG system. *Optics and Lasers in Engineering* 2002 **38**:557–566.

[70] Yashiro S, Takeda N, Okabe T, Sekine H. A New approach to predicting multiple damage states in composite laminates with embedded FBG sensors. *Composite Science and Technology* 2005 **65**:659–667.

[71] Yashiro S, Okabe T, Toyama N, Takeda N. Monitoring damage in holed CFRP laminates using embedded chirped FBG sensors. *International Journal of Solids and Structures* 2007 **44**:603–613.

[72] Takeda N, Yashiro S, Okabe T. Estimation of the damage patterns in notched laminates with embedded FBG sensors. *Composites Science and Technology* 2006 **66**:684–693.

[73] Yashiro S, Okabe T, Takeda N. Damage identification in a holed CFRP laminate using a chirped fiber Bragg grating sensor. *Composites Science and Technology* 2007 **67**:286–295.

[74] Minakuchi S. Real-time detection of debonding between honeycomb core and facesheet using a small-diameter FBG sensor embedded in adhesive layer. *Journal of Sandwich Structures and Materials* 2007 **9**:9–33.

[75] Yamada M, Sakuda K. Analysis of almost-periodic distributed feedback slab waveguides via a fundamental matrix approach. *Applied Optics* 1987 **26**:3474–3478.

[76] Huang S, Ohn M, LeBlanc M, Measures RM. Continuous arbitrary strain profile measurements with fiber Bragg gratings. *Smart Materials and Structures* 1998 **7**:248–256.

[77] Peters K, Pattis P, Botsis J, Giaccari P. Experimental verification of response of embedded optical fiber Bragg grating sensors in non-homogeneous strain fields. *Optics and Lasers in Engineering* 2000 **33**:107–119.

[78] Prabhugoud M, Peters K. Modified transfer matrix formulation for Bragg grating strain sensors. *Journal of Lightwave Technology* 2004 **22**:2302–2309.

[79] Peral E, Capmany J, Marti J. Iterative solution to the Gel'fand-Levitan-Marchenko coupled equations and application to synthesis of fiber gratings. *IEEE Journal of Quantum Electronics* 1996 **32**:2078–2084.

[80] Feced R, Zervas M, Muriel M. An efficient inverse scattering algorithm for the design of nonuniform fiber Bragg gratings. *IEEE Journal of Quantum Electronics* 1999 **35**:1105–1115.

[81] Giaccari P, Limberger HG, Salathé RP. Local coupling-coefficient characterization in fiber Bragg gratings. *Optics Letters* 2003 **28**:598–600.

[82] Chapeleau X, Leduc D, Lupi C, Le NyR. Experimental synthesis of fiber Bragg gratings using optical low coherence reflectometry. *Applied Physics Letters* 2003 **82**:4227–4229.

[83] Azana J, Muriel M. Fiber Bragg grating period reconstruction using time-frequency signal analysis and application to distributed sensing. *Journal of Lightwave Technology* 2001 **19**:646–654.

[84] Paterno AS, Silva JCC, Milczewski MS, Arruda LVR, Kalinowski HJ. Radial-basis function network for the approximation of FBG sensor spectra with distorted peaks. *Measurement Science and Technology* 2006 **17**:1039–1045.

[85] Casagrande F, Crespi P, Grassi AM, Luilli A, Kenny R, Whelan MP. From the reflected spectrum to the properties of a fiber Bragg grating: a genetic algorithm approach with application to distributed strain sensing. *Applied Optics* 2002 **41**:5238–5244.

[86] Gill A, Peters K, Studer M. Genetic algorithm for the reconstruction of Bragg grating sensor strain profiles. *Measurement Science and Technology* 2004 **15**:1877–1884.

[87] Kitcher DJ, Nand A, Wade SA, Jones R, Baxter GW, Collins SF. Directional dependence of spectra of fiber Bragg gratings due to excess loss. *Journal of Optical Society of America A* 2006 **23**:2906–2911.

[88] James SW, Tatam RP. Optical fibre long-period grating sensors: characteristics and application. *Measurement Science and Technology* 2003 **14**: R49–R61.

[89] DeVries M, Bhatia V, D'Alberto T, Arya V, Claus RO. Photoinduced grating-based optical fiber sensors

for structural analysis and control. *Engineering Structures* 1998 **20**:205–210.

[90] Frazao O, Romero R, Rego G, Marques RVS, Salgado HM, Santos JL. Sampled fibre Bragg grating sensors for simultaneous strain and temperature measurement. *Electronics Letters* 2002 **38**:693–695.

[91] Lin CY, Wang LA, Chern GW. Corrugated long-period fiber gratings as strain, torsion, and bending sensors. *Journal of Lightwave Technology* 2001 **19**:1159–1168.

[92] Wang YP, Wang DN, Jin W. CO2 laser-grooved long period fiber grating temperature sensor system based on intensity modulation. *Applied Optics* 2006 **45**:7966–7970.

[93] Mishra V, Singh N, Jain SC, Kaur P, Luthra R, Singla H, Jindal VK, Bajpai RP. Refractive index and concentration sensing of solutions using mechanically induced long period grating pair. *Optical Engineering* 2005 **44**:094402-1–094402-4, Art. No. 094402.

[94] Cho JY, Lim JH, Lee KS. Optical fiber twist sensor with two orthogonally oriented mechanically induced long-period grating sections. *IEEE Photonics Technology Letters* 2005 **17**:453–455.

[95] Bhatia V, Campbell DK, Sherr D, Ten Eyck GA, Murphy KA, Claus RO. Temperature-insensitive and strain-insensitive long-period grating sensors for smart structures. *Optical Engineering* 1997 **36**: 1872–1876.

[96] Patrick HJ, Williams GM, Kersey AD, Pedrazzani JR, Vengsarkar AM. Hybrid fiber Bragg grating/long period fiber grating sensor for strain/temperature discrimination. *IEEE Photonics Technology Letters* 1996 **8**:1223–1225.

[97] Kim DW, Shen F, Chen XP, Wang AB. Simultaneous measurement of refractive index and temperature based on a reflection-mode long-period grating and an intrinsic Fabry–Perot interferometer sensor. *Optics Letters* 2005 **30**:3000–3002.

Chapter 62
Novel Fiber-optic Sensors

Kara Peters
Department of Mechanical and Aerospace Engineering, North Carolina State University, Raleigh, NC, USA

1 Introduction	1113
2 Extreme Temperature Sensors	1113
3 Polymer Large-deformation Sensors	1115
4 Microstructured Fiber Sensors	1118
5 Multicore Optical Fiber Sensors	1120
6 MEMS Optical Fiber Sensors	1121
7 Conclusions	1122
Related Articles	1122
References	1122

Encyclopedia of Structural Health Monitoring. Edited by Christian Boller, Fu-Kuo Chang and Yozo Fujino © 2009 John Wiley & Sons, Ltd. ISBN: 978-0-470-05822-0.

1 INTRODUCTION

Fiber-optic sensors have been applied for a variety of structural health monitoring applications because of their immunity to electromagnetic interference, low weight, small size, and multiplexing capabilities (*see* **Chapter 109**). Some of the articles in this encyclopedia review optical fiber concepts useful for understanding fiber-optic sensors (*see* **Chapter 59**), as well as the basic operating principles for intensity-based, interferometric, and fiber Bragg grating (FBG) sensors (*see* **Chapter 60**; **Chapter 61**). The goal of this article is to present recent, novel concepts for fiber-optic sensors. Most of these sensors are based on recent advances in optical fiber technology, including single-crystal sapphire, single-mode polymer, and microstructured and multicore optical fibers. Such advances allow the use of optical fiber sensors for high-temperature and large-strain applications.

2 EXTREME TEMPERATURE SENSORS

A variety of factors limit the usable temperature range for fiber-optic sensors including the material limits of the optical fibers themselves and temperature limits for any adhesives used in packaging. Another important limit for FBG sensors (*see* **Chapter 61**) is that the index modulation in the FBG written through photosensitivity can be erased at approximately 200 °C. This limitation could be important for many composite applications, as processing temperatures of polymer matrix–based composites may exceed this limit. Other matrix materials have even higher processing temperatures. Coating the optical fiber with polyimide does increase the temperature survival range for the FBG to 500 °C [1]; however, this may not be suitable for many high-temperature applications. The temperature limit for

FBG erasing is well below the material limit for fused silica, typically 1000–1100 °C. As one solution, Luo et al. [1] developed a "Bragg stack" by depositing quarter-wavelength-thickness layers on the end of a polyimide-coated silica optical fiber. The alternating layers of silicon nitrite and silicon-rich silicon nitrite acted as a Bragg diffraction grating, which could be interrogated in reflection through the optical fiber. The Bragg stack was applied as a temperature sensor and was demonstrated to survive to at least 800 °C with a 2 °C temperature resolution.

An alternative fabrication method was applied by Lowder et al. [2], who fabricated FBGs through first etching a significant portion of the cladding on the flat surface of a D-fiber. Then a surface relief FBG close to the core of the optical fiber was etched. Similar methods have been applied to write long period gratings (see **Chapter 61**). As the FBG is created mechanically, rather than chemically as in the case of photosensitive FBGs, the FBG can withstand significantly higher temperatures. Lowder et al. [2] demonstrated the linearity of the sensor up to 1100 °C.

To overcome the temperature limit of the silica itself, researchers have also developed optical fiber sensors based on single-crystal sapphire fibers, which have a melting point of 2050 °C [3–5]. Tong et al. [4] demonstrated a single-crystal sapphire fiber temperature sensor based on blackbody radiation for high-temperature applications; however, they discovered that surface contamination can significantly degrade the measurements at high temperatures. Wang et al. [3] and Xiao et al. [5] designed extrinsic Fabry–Perot white-light interferometer sensors, coupling the lightwave from a single mode to a multimode sapphire fiber and achieving a strain resolution of 0.2 microstrain up to 1004 °C. Zhu and Wang [6] later extended the sapphire fiber sensor by bonding the single-crystal sapphire fiber to the surface of a sapphire wafer whose optical path length through the thickness changes with temperature. By further splicing the sapphire fiber to the standard silica fiber lead, removing the need for optical adhesives, the authors demonstrated a temperature resolution of 0.4 °C up to 1170 °C.

At the other end of the usable temperature range for fiber-optic sensors, Zeisberger et al. [7] demonstrated that FBGs can be applied for accurate strain measurements at extremely low temperatures, specifically in the temperature range of 30–300 K. The measured wavelength shift of an FBG due to only the thermo-optic effect reported by the authors is plotted in Figure 1. For these measurements, the FBG was constrained so as to not permit thermal expansion.

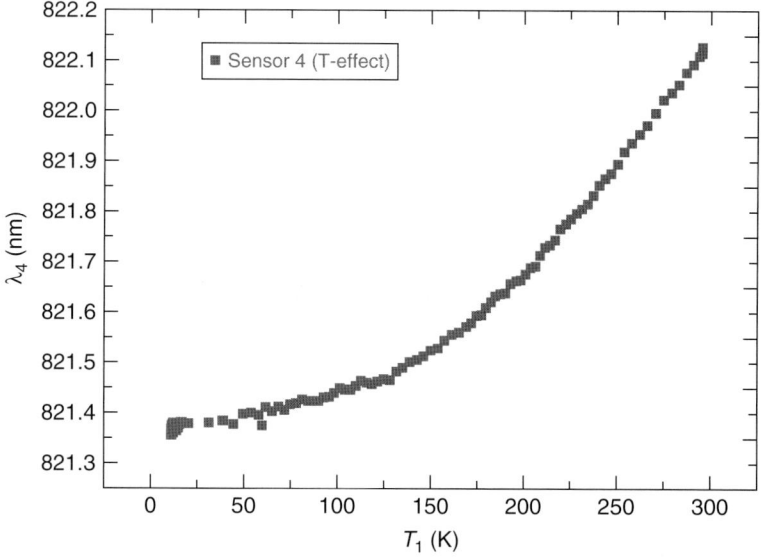

Figure 1. Measured Bragg wavelength of FBG as a function of temperature (change due to thermo-optic effect only). [Reproduced with permission from Ref. 7. © 2005, IEEE.]

The response of the FBG to temperature is linear to approximately 250 K, and then strongly nonlinear at lower temperatures.

3 POLYMER LARGE-DEFORMATION SENSORS

As demonstrated by the large number of optical fiber sensors and their applications (*see* **Chapter 60**; **Chapter 61**; **Chapter 109**), optical fiber sensors are extremely versatile. However, the brittle nature of silica limits their maximum strain range to approximately 1–2%, particularly for FBGs [8]. This measurement range is comparable with electrical resistance strain gauges. Polymer optical fibers (POFs) on the other hand offer a much larger potential strain measurement range. In general, the fabrication of POFs is not as well controlled as for silica fibers; therefore, their performance has not been comparable. At the same time, intrinsic material losses in polymers are orders of magnitude higher than for silica. Therefore, POFs have primarily been used for short, flexible optical connections and as sensors operating through bending or environmental intensity losses. However, recent advances in the fabrication of single-mode POFs with minimal attenuation levels have changed the potential for POF sensors [9–13].

Figure 2 plots the inherent material losses for several optical polymers and silica. The material losses in polymethylmethacrylate (PMMA) generally increases with wavelength, and therefore POFs are often designed to operate at wavelengths below the near-IR wavelengths used with silica fibers. Fabricating single-mode fibers at these lower wavelengths requires smaller core diameters in order to keep the normalized frequency, V, in the single-mode range (*see* **Chapter 59**). Kuzyk *et al.* [9] first fabricated PMMA optical fibers with a dye-doped core that were single mode at an operating wavelength, $\lambda =$

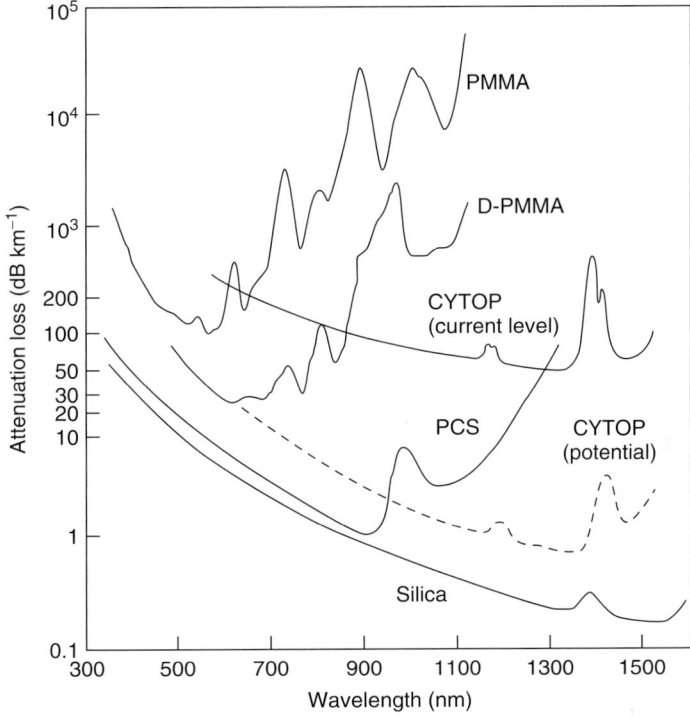

Figure 2. Transmission loss spectra for various polymer optical fiber materials: PMMA, D-PMMA (deuterated), CYTOP, and PCS. Loss spectra for silica is also plotted for comparison. [Reproduced with permission from Ref. 14. © 2001, Elsevier.]

1300 nm. These fibers demonstrated attenuation levels of approximately $0.3\,\mathrm{dB\,cm^{-1}}$, close to the intrinsic material attenuation for PMMA at that wavelength. Bosc and Toinen [10] later fabricated single-mode PMMA optical fibers with attenuation levels of $0.25\,\mathrm{dB\,cm^{-1}}$ at $\lambda = 1550\,\mathrm{nm}$ and $0.05\,\mathrm{dB\,cm^{-1}}$ at $\lambda = 850\,\mathrm{nm}$. Finally, Garvey et al. [11] improved the process to create single-mode POFs at $\lambda = 1060\,\mathrm{nm}$ with an attenuation level of $0.18\,\mathrm{dB\,cm^{-1}}$.

3.1 Sensing properties of POFs

Silva-López et al. [15] first measured the sensitivity of dye-doped single-mode POFs to strain and temperature. The POFs were designed to be single mode at 850 nm with an acrylic cladding ($n = 1.4905$) and doped PMMA core ($n = 1.4923$). The authors operated the fibers with a visible light source at $\lambda = 632\,\mathrm{nm}$, outside of the single-mode region; however, they only observed the fundamental, LP_{01}, mode propagating through the fiber. Using a Mach–Zender interferometer arrangement and loading the optical fiber on a translation stage, they measured a phase sensitivity to displacement of $131 \times 10^5\,\mathrm{rad\,m^{-1}}$ and temperature sensitivity of $-212\,\mathrm{rad\,m^{-1}\,K^{-1}}$. The phase sensitivity to displacement measured is in good agreement with the properties of bulk PMMA. When compared with silica optical fibers, the sensitivity to displacement of the PMMA fiber is 14% higher than that of the silica, which is an advantage for strain sensing. The difference is primarily due to the large difference in Poisson's ratio and the photo-elastic constants (see **Chapter 60**). On the other hand, the temperature sensitivity of the PMMA fiber is negative, whereas the temperature sensitivity of silica optical fibers is positive. The temperature sensitivity was also about 25% more than predicted. The maximum operating temperature for the fiber was predicted to be in the 80–120 °C range.

The measurements of Silva-López et al. [15] provide useful information on the behavior of POF sensors and can be used to predict the response of a variety of sensors based on them. However, the measurements were made in the strain range of 0–0.04% strain, which is a limited portion of the strain range over which they have potential to be applied. For this reason, Kiesel et al. [16] derived a formulation to predict the phase sensitivity to strain

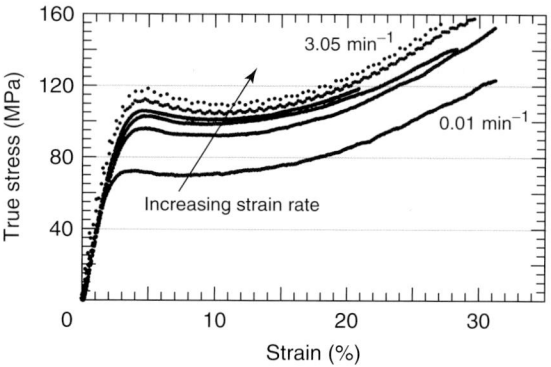

Figure 3. Measured true stress–strain curves for PMMA POF at various strain rates. Strain rates plotted are 0.01, 0.30, 0.60, 0.90, 1.22, and $3.05\,\mathrm{min^{-1}}$. [Reproduced with permission from Ref. 16. © IOP Publishing, Ltd., 2007.]

of the dye-doped PMMA POF for the full usable range of the sensor ($\cong 0–6\%$) and measured the mechanical properties of the fiber in this range. The formulation includes both the finite deformation of the optical fiber and nonlinear photoelastic effects, both potentially significantly important over the usable strain range. The measured true stress–strain response curve of the PMMA POF is plotted in Figure 3 for multiple strain rates in the range of $0.01–3.05\,\mathrm{min^{-1}}$. One can see from Figure 3 that the response of the sensor is very sensitive to strain rate and that the yield strain increases with strain, typical of polymer fibers [13]. The failure strain of the POF was around 30%, while transmission of the coherent lightwave for sensing was demonstrated to be 15.8%, well beyond the yield strain of the POF [17]. Using the mechanical properties measured from the single-mode POF samples, Kiesel et al. [16] predicted the phase sensitivity of the POF over a strain range of 0–6%. This sensitivity is plotted in Figure 4 with the response of a silica fiber for comparison. The important differences between POF sensor and silica sensor from Figure 4 are as follows: (i) the PMMA has a high strain sensitivity throughout the strain range, (ii) the nonlinearity in the PMMA phase response is of the opposite sign and an order of magnitude greater than for silica, and (iii) the importance of including the nonlinearity in the phase response is evident at a much lower strain value ($\cong 1\%$) for PMMA than for silica ($\cong 3\%$). Additional effects are expected to be important when applying

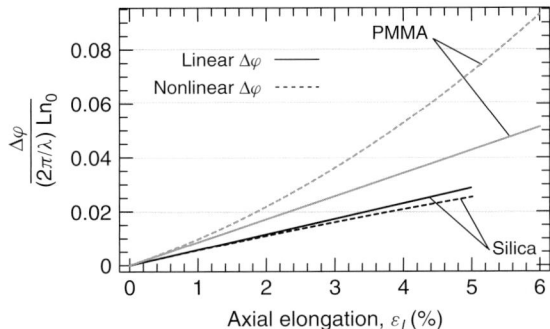

Figure 4. Predicted normalized phase shift of silica optical fiber and PMMA POF, applying linear and nonlinear deformation [16]. For specific material parameters used in calculation, see. [Reproduced with permission from Ref. 16. © IOP Publishing, Ltd., 2007.]

POFs as large-strain sensors, including material hysteresis, creep effects (especially at elevated temperatures), and the anisotropy of the polymer.

3.2 Fiber Bragg gratings in POFs

Peng and Chu [12] first demonstrated the writing of FBGs in POFs through the same photosensitivity process used for writing FBGs in silica optical fibers. The FBGs were fabricated by UV exposure of PMMA single-mode optical fibers (fabricated by the authors) with dye-doped cores to increase the photosensitivity of the polymer. Peng and Chu [12] wrote FBGs operating in near-IR wavelengths ($\lambda_B \cong 1570$ nm) with reflectivities >80% and a bandwidth of 1 nm. By applying strain to the FBG, the authors shifted the Bragg wavelength by a total of 73 nm (Figure 5), which is an order of magnitude larger than that previously achieved with FBGs in silica optical fibers. The large wavelength shift was partially due to the increased sensitivity of the POF described in the previous section and partially due to the large yield strain of the POF (Peng and Chu report 6.1% yield strain for the fiber used [12]). Applying thermal loading, the authors tuned the FBG over 20 nm; however, erasing of the FBG occurred when the grating was exposed to thermal loads for extended periods of time [18].

Liu et al. [19] later wrote FBGs in CYTOP optical fibers, which have a lower intrinsic material loss than PMMA (Figure 2). The sensitivity of the FBGs in CYTOP fibers was not as high as for those in PMMA fibers, for example, the gratings could only be tuned thermally over 10 nm. However, thermal erasing of the gratings did not occur when they were exposed to thermal loads for long durations.

Liu et al. [18] combined FBG sensors in silica and PMMA optical fibers in series to independently measure strain and temperature, through the equation

$$\begin{pmatrix} \Delta\lambda_p \\ \Delta\lambda_s \end{pmatrix} = \begin{pmatrix} K_{\varepsilon p} & K_{Tp} \\ K_{\varepsilon s} & K_{Ts} \end{pmatrix} \begin{pmatrix} \varepsilon \\ \Delta T \end{pmatrix} \quad (1)$$

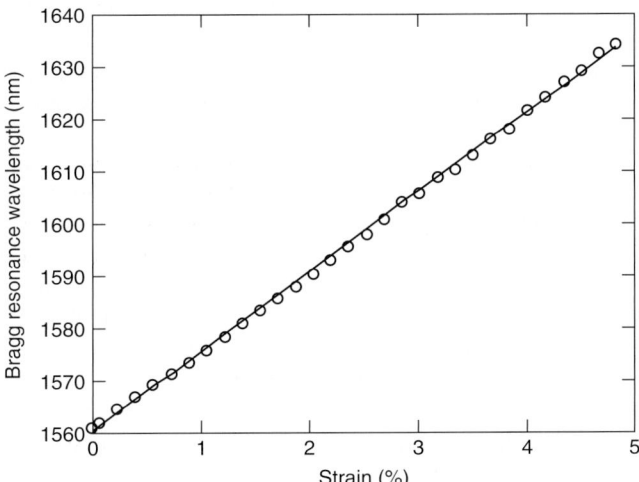

Figure 5. Measured Bragg wavelength shift due to axial strain for FBG written in polymer optical fiber. [Reproduced with permission from Ref. 12. © Taylor & Francis, Ltd., 2000.]

where the K matrix is constructed of the sensitivities of the Bragg wavelength of the FBGs in the polymer ($\Delta\lambda_p$) and silica ($\Delta\lambda_s$) fibers to strain and temperature (*see* **Chapter 61**). These measurements are significantly better than those previously obtained using two FBGs in silica fibers at different wavelengths (*see* **Chapter 61**) due to the fact that the sensitivity to strain and temperature is very different for the polymer and silica. The measured strain sensitivities were $K_{\varepsilon 1} = 1.48\,\mathrm{pm}\,\mu\varepsilon^{-1}$ for the PMMA FBG and $K_{\varepsilon 2} = 1.15\,\mathrm{pm}\,\mu\varepsilon^{-1}$ for the silica FBG. Additionally, the temperature sensitivity of the silica FBG is positive, while the temperature sensitivity of the PMMA FBG is negative.

Challenges to applying polymer FBGs for large-strain sensing include high attenuation in the fibers, difficulties in preparing the fibers for coupling, and thermal erasing of the gratings. This last challenge could be important for many structural health monitoring applications, but is still not fully understood. Similar to the writing process for FBGs in silica fibers, the grating depth increases with exposure time until a threshold is reached, at which point damage to the polymer fiber occurs and transmission losses are introduced in the fiber [20]. On the other hand, Liu *et al.* [21] observed that, when FBGs were written with lower power UV exposures, the FBG peak depth increased, reached a maximum, remained constant, then began to erase during the exposure time. Once the FBG was completely erased, the UV exposure was stopped and the FBG reappeared over a period of 8 h, then was permanent and stable. The authors speculate that the heating of the fiber during the UV exposure temporarily changed the index of refraction, counteracting the change due to photosensitivity.

4 MICROSTRUCTURED FIBER SENSORS

Microstructured optical fibers, also referred to as *photonic crystal fibers*, have been fabricated in a variety of forms including solid core, "holey" fibers, as shown in Figure 6. For a first approximation, the holey fiber acts as an index-guiding waveguide in the same manner as conventional solid optical fibers (*see* **Chapter 59**). The air holes in the cladding region reduce the effective index of the cladding and

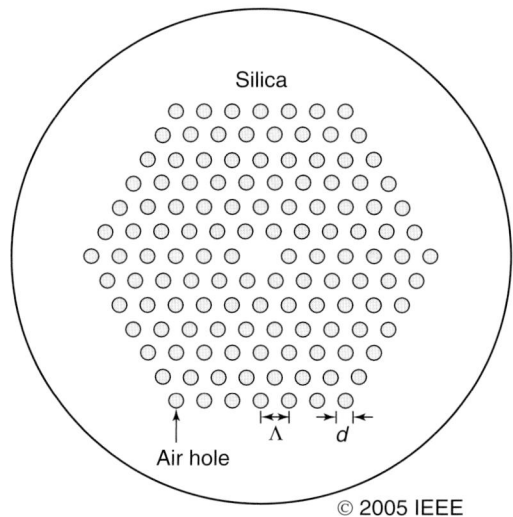

Figure 6. Schematic representation of a typical microstructured holey optical fiber. Λ and d are the hole pitch and diameter respectively. [Reproduced with permission from Ref. 24. © IOP Publishing Ltd., 2005.]

therefore provide much stronger confinement of the lightwave as it propagates through the fiber [22]. Such microstructured optical fibers are typically fabricated by drawing a preform consisting of capillary tubes with a solid silica rod replacing the center tube to form the solid core. The complex geometry of these holey fibers provides both new sensing possibilities and enhanced response characteristics for conventional fiber-optic sensors applied to these fibers. For example, Ju *et al.* [23] fabricated an interferometric sensor based on a holey fiber and observed a nonmonotonic temperature dependence on wavelength.

Nasilowski *et al.* [25] and Bock *et al.* [26] applied microstructured holey fibers for the measurement of temperature and pressure. Through measurement of the group refractive index change, Bock *et al.* [26] estimated the temperature, axial strain, and pressure sensitivity of such fibers. The primary differences observed as compared to conventional fibers were that the reduced Poisson contraction of the cross section (i) reduced the strain sensitivity slightly and (ii) changed the pressure sensitivity to be negative rather than positive as for solid fibers. Nasilowski *et al.* [25] applied a high-birefringence holey fiber, created by introducing several defects to form a

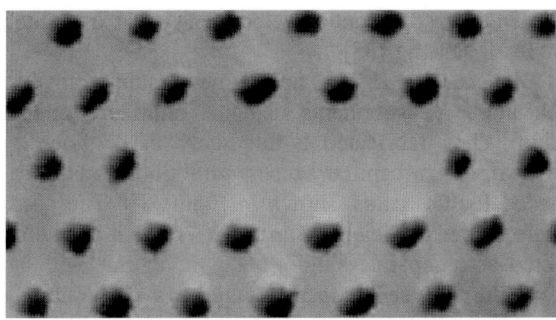

Figure 7. Scanning-electron-microscope image of the core of high-birefringence holey fiber showing the triple defect core. [Reproduced with permission from Ref.23. © Springer Verlag, 2005.]

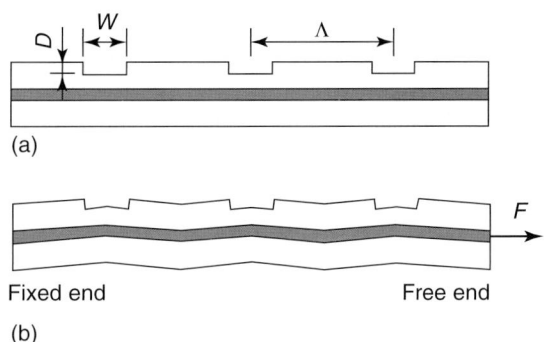

Figure 8. Schematic of CO_2 laser carved LPG (a) before and (b) after axial force is applied to the fiber. Λ, D, and W are the grating pitch, depth, and, width of the grooves, respectively. [Reproduced with permission from Ref. 27. © OSA Publishing, 2006.]

noncircular core, as a pressure sensor. An image of the fiber core and surrounding region is shown in Figure 7. An important feature of high-birefringence holey fibers is that strong birefringence can be obtained without inducing thermal residual stresses through a different material region in the fiber (*see* **Chapter 59**). From a sensing perspective, this birefringence is therefore stable with temperature and does not introduce temperature errors into high-birefringence sensors [25].

Several researchers have written long period fiber Bragg gratings (LPGs) in microstructured fibers through CO_2 laser or electric arc etching [27–29]. These LPGs demonstrate an excellent strain sensitivity, with little or no temperature sensitivity, meaning that temperature compensation is not required [27, 29]. When bending the optical fiber, He *et al.* [28] observed mode splitting and bending directional dependence in the output of the LPG, phenomena not observed with LPGs written in conventional solid silica optical fibers. The directional dependence is due to the fact that the in-plane stiffness of the cross section has been significantly weakened due to the presence of the holes. The response of the LPG does appear to strongly depend on the fabrication method, as Petrovic *et al.* [29] did not observe such phenomena in similar tests. Wang *et al.* [27] also found that, during axial stretching of the optical fiber, the periodic grooves in the LPGs created local microbending around the grooves (Figure 8). This microbending increased the sensitivity of the LPG to axial strain. Demonstrating the fabrication of grating structures through photosensitivity, Dobb *et al.* [30] successfully wrote FBGs in PMMA holey fibers operating at $\lambda_B = 1569$ nm with a 0.5 nm bandwidth. The FBGs were fabricated in both few-mode and single-mode holey fibers.

Zou *et al.* [31] demonstrated independent strain and temperature measurements through stimulated Brillouin scattering in microstructured optical fibers (*see* **Chapter 60**). In contrast to scattering in conventional solid fibers, several peaks were observed in the scattered frequency spectrum due to independent interactions between the lightwave and stimulated acoustic waves occurring in the graded-Ge-doped core region, the intermediate silica region, and the microstructured cladding region. The response of the peak in the core region was the same as for a doped solid optical fiber; however, the other peaks had differing sensitivities to temperature and strain.

One final example of a microstructured fiber sensor is the tapered holey fiber sensor demonstrated by Villatoro *et al.* [32]. The authors tapered the middle section of a holey fiber to a diameter of 28 μm (the original diameter was 125 μm), which caused the holes to collapse in the cross section. The output lightwave intensity from the optical fiber was then an interference pattern between the multiple modes of the tapered waist [33]. These interference peaks shifted linearly with strain and were insensitive to temperature up to 180 °C.

5 MULTICORE OPTICAL FIBER SENSORS

Multicore optical fibers have great potential for the measurement of strains and curvatures over large, flexible structures such as aircraft wings or towed hydrophone arrays. When multiplexed, these local strain and curvature measurements can be used to determine the shape of the flexible structure. The concept for curvature measurements based on differential strain measurements between core pairs is shown in Figure 9 for a four-core optical fiber. The presence of multiple cores within a single fiber can provide more stable curvature measurements than those from separate single-mode fibers because more accurate spacing is maintained between the cores [34]. The choice of spacing between the cores can be an important design consideration for multicore sensors. Close spacing of the cores makes the sensor more compact as well as less sensitive to temperature gradients across the fiber cross section. However, reducing the core spacing also increases the coupling between modes propagating through the cores and reduces the accuracy of curvature measurements [35]. Methods to couple the multicore fiber to multiple single-mode fibers for data acquisition include graded index (GRIN) lenses and fan-outs [35]. Multicore optical fibers have also been applied for Doppler differential velocimetry [36] and pitch and roll sensing [35].

Blanchard *et al.* [34] first demonstrated strain and curvature measurements using a multicore optical fiber. They fabricated a three-core holey fiber by drawing silica capillary tubes with silica rods introduced for the cores, shown in Figure 10. The cores were placed sufficiently far apart such that each core acted as a separate single-mode waveguide with no coupling between propagating lightwaves (assuming that no twisting is applied). Light was launched into the optical fiber, creating a far-field interface pattern at the end of the 23-cm sensor length, which was then projected onto a Charge-Coupled Device (CCD) camera. The optical phase difference of each core was extracted from the Fourier transform of the interference pattern. On the basis of the differential phase differences, the sensor demonstrated a strain resolution of 7.4 nε and bend angle resolution of 100 μrad. MacPherson *et al.* [38] later demonstrated that the phase sensitivity to displacement ($\Delta\phi/\Delta L$) of the multicore holey fiber is approximately the same as a solid fiber. Using only a two-core holey fiber, they obtained a bend angle resolution of 170 μrad.

MacPherson *et al.* [37] wrote FBGs into each of the cores of the four-core solid optical fiber of Figure 9

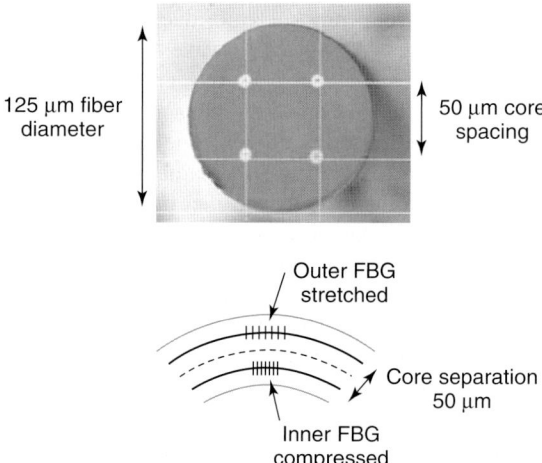

Figure 9. Cross section of multicore fiber illuminated with white light and principle of curvature measurement from FBG pairs in different cores. [Reproduced with permission from Ref. [37]. © IOP Publishing Ltd., 2006.]

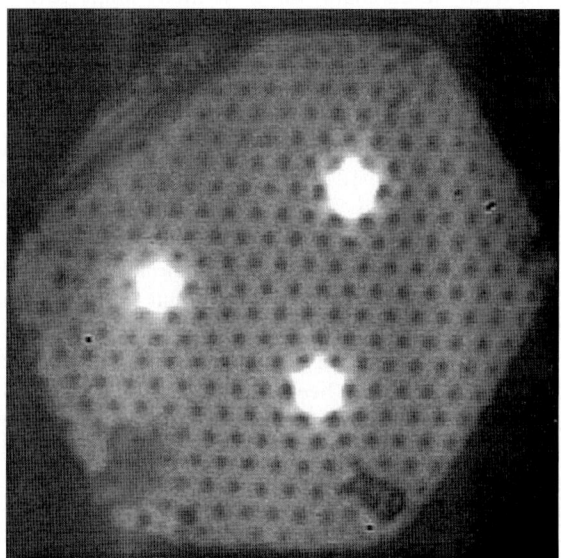

Figure 10. SEM image of the end of a photonic crystal fiber near-field pattern at the output of the three-core fiber with all cores illuminated at 633 nm. [Reproduced with permission from Ref. 34. © IOP Publishing Ltd., 2000.]

for shape monitoring of a tunnel structure. Multiplexing a large number of FBGs along the length of the optical fiber allows for a distributed measurement of the curvature, which can be integrated to obtain the shape of the structure [39]. MacPherson *et al.* obtained a shape resolution of ±0.1 mm between two sets of FBGs spaced over a length of 5 mm along the fiber. Flockhart *et al.* [40] and Cranch *et al.* [41] used the same fiber configuration, but wrote sets of FBG pairs in each core to act as a low-finesse Fabry–Perot interferometer to enhance the static and dynamic resolution of the shape measurement. Applying differential interferometric demodulation between the grating pairs, the authors obtained a strain resolution of $0.6\,\text{n}\varepsilon\,\text{Hz}^{-1/2}$ and curvature resolution of $0.012\,\text{km}^{-1}\,\text{Hz}^{-1/2}$ for measurement frequencies above 1 Hz, a 30 times improvement over the previous curvature measurements.

6 MEMS OPTICAL FIBER SENSORS

The ability to multiplex a large number of sensors into a single optical fiber is one of the key advantages to optical fiber sensors for structural health monitoring. Additionally, their relatively small size makes them reasonable to embed or surface mount to a variety of structures (*see* **Chapter 59**). However, more and more sensors are being developed for applications where size is critical, thus, necessitating microelectromechanical systems (MEMSs) fabrication techniques. To date, MEMS technology has primarily been applied to optical fiber sensors for Fabry–Perot temperature and pressure sensors (*see* **Chapter 60**). Watson *et al.* [42] reviews the state of the art for low-profile external Fabry–Perot diaphragm sensors. However, due to their extrinsic nature, the outside diameters of these sensors are generally larger than the outside diameter of the optical fiber itself. To reduce the outside sensor diameter of end-face Fabry–Perot pressure sensors, Watson *et al.* [42] fabricated a MEMS Fabry–Perot sensor. The authors ablated the end of an optical fiber with an excimer laser, and then bonded an aluminized polycarbonate foil diaphragm to the end of the fiber, which can be seen in Figure 11. The laser ablation produced a good quality cavity; however, the authors

(a)

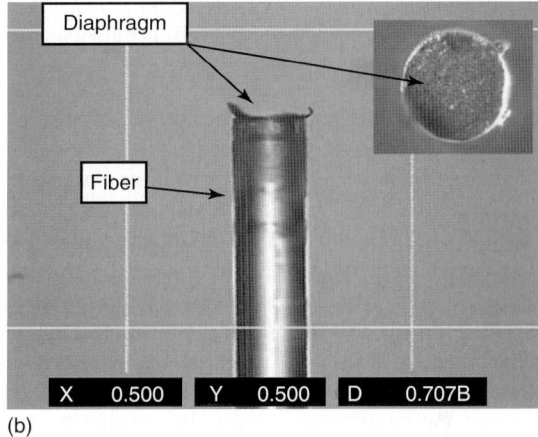

(b)

Figure 11. (a) A 70-μm-diameter cavity ablated into the end of a single-core fiber and (b) photograph of the side view of a single-core fiber sensor (the inset shows the top view). [Reproduced with permission from Ref. 42. © OSA Publishing, 2006.]

encountered some difficulties due to tearing of the diaphragm during bonding and poor adhesion of the diaphragm.

The requirement to adhere the diaphragm with an adhesive glue also limits the temperature range over which such MEMS sensors can be applied [43]. Abeysinghe *et al.* [43], therefore, developed a Fabry–Perot temperature and pressure sensor without adhesives, also applying a MEMS fabrication technology. They micromachined the end of the optical fiber through wet etching to create a cavity similar to the last example, but then bonded a crystalline silicone membrane to the end face of the optical

Figure 12. Scanning-electron-microscopy image of the fiber-top cantilever (before evaporation of the silver layer) [44]. Dimensions: length $\cong 112\,\mu\text{m}$, width $\cong 14\,\mu\text{m}$, and thickness $\cong 3.7\,\mu\text{m}$. [Reproduced with permission from Ref.44. © American Institute of Physics, 2006.]

fiber through anodic bonding. The authors experimentally demonstrated the excellent performance of the MEMS sensor in a 25–300 °F thermal environment.

At a smaller scale, Iannuzzi et al. [44] and Deladi et al. [45] machined a thin rectangular beam out of the cleaved edge of an optical fiber, as shown in Figure 12. The vertical displacement of the cantilever was determined by measuring the amplitude of interference of laser light reflected by the fiber–air interface and reflected by the cantilever. The authors prepared the cantilevered sensor by coating a standard single-mode optical fiber with a thin metallic layer (5-nm Cr, 20-nm Pd) and then machining the cantilever using focused ion beam machining. A 100-nm layer of silver was then deposited on the sensor to increase the reflectivities. For a further description of focused ion beam machining applied to optical fiber sensors, see Nellen and Brönnimann [46]. The authors demonstrated that the subnanometer displacement resolution of the cantilever beam sensor was comparable with results from atomic force microscopy.

7 CONCLUSIONS

This article reviews recent advances in fiber-optic sensor technology based on surface relief gratings and sapphire optical fibers for high-temperature applications; single-mode POFs for large-deformation sensors; and microstructured and multicore optical fibers. Applications of MEMs fabrication technique to fiber-optic sensors are also presented. These advances have extended the range for optical fiber sensors such that they can now be uniquely applied for structural health monitoring applications in extreme environments.

RELATED ARTICLES

Chapter 17: Lamb Wave-based SHM for Laminated Composite Structures

Chapter 155: Reliable Use of Fiber-optic Sensors

REFERENCES

[1] Luo F, Hernández-Cordero J, Morse TF. Multiplexed fiber-optic Bragg stack sensors (FOBSS) for elevated temperatures. *IEEE Photonics Technology Letters* 2001 **13**:514–516.

[2] Lowder TL, Smith KH, Ipson BL, Hawkins AR, Selfridge RH, Schultz SM. High-temperature sensing using surface relief fiber Bragg gratings. *IEEE Photonics Technology Letters* 2005 **17**:1926–1928.

[3] Wang A, Gollapudi S, May RG, Murphy KA, Claus RO. Sapphire optical fiber-based interferometer for high-temperature environmental applications. *Smart Materials and Structures* 1995 **4**:147–151.

[4] Tong LM, Shen YH, Ye LH. Performance improvement of radiation-based high-temperature fiber-optic sensor by means of curved sapphire fiber. *Sensors and Actuators, A* 1999 **75**:35–40.

[5] Xiao H, Deng J, Pickrell G, May RG, Wang A. Single-crystal sapphire fiber-based strain sensor for high-temperature applications. *Journal of Lightwave Technology* 2003 **21**:2276–2283.

[6] Zhu Y, Wang A. Surface-mount sapphire interferometric temperature sensor. *Applied Optics* 2006 **45**:6071–6076.

[7] Zeisberger M, Latka I, Ecke W, Habisreuther H, Litzkendorf D, Gawalek W. Measurement of the thermal expansion of melt-textured YBCO using optical fibre grating sensors. *Superconductor Science and Technology* 2005 **18**:S202–S205.

[8] Nellen PhM, Mauron P, Frank A, Sennhauser U, Bohnert K, Pequignot P, Bodor P, Brändle H. Reliability of fiber Bragg grating based sensors for downhole applications. *Sensors and Actuators, A* 2003 **103**:364–376.

[9] Kuzyk MG, Paek UC, Dirk CW. Guest-host polymer fibers for nonlinear optics. *Applied Physics Letters* 1991 **59**:902–904.

[10] Bosc D, Toinen C. Full polymer single-mode optical fiber. *IEEE Photonics Technology Letters* 1992 **4**:749–750.

[11] Garvey DW, et al. Single-mode nonlinear-optical polymer fibers. *Journal of the Optical Society of America A* 1996 **18**:2017–2023.

[12] Peng GD, Chu PL. Polymer optical fiber photosensitivities and highly tunable fiber gratings. *Fiber and Integrated Optics* 2000 **19**:277–293.

[13] Jiang CH, Kuzyk MG, Ding JL, Johns WE, Welker DJ. Fabrication and mechanical behavior of dye-doped polymer optical fiber. *Journal of Applied Physics* 2002 **92**:4–12.

[14] Zubia J, Arrue J. Plastic optical fibers: an introduction to their technological processes and applications. *Optical Fiber Technology* 2001 **7**:101–140.

[15] Silva-López M, et al. Strain and temperature sensitivity of a single-mode polymer optical fiber. *Optics Letters* 2005 **30**:3129–3131.

[16] Kiesel S, Peters K, Hassan T, Kowalsky M. Behaviour of intrinsic polymer optical fibre sensor for large-strain applications. *Measurement Science and Technology* 2007 **18**:3144–3154.

[17] Kiesel S, Peters K, Hassan T, Kowalsky M. Large deformation in-fiber polymer optical fiber sensor. *IEEE Photonics Technology Letters* 2008 **20**:416–418.

[18] Liu HB, Liu HY, Peng GD, Chu PL. Strain and temperature sensor using a combination of polymer and silica fibre Bragg gratings. *Optics Communications* 2003 **219**:139–142.

[19] Liu HY, Peng GD, Chu PL. Thermal stability of gratings in PMMA and CYTOP polymer fibers. *Optics Communications* 2002 **204**:151–156.

[20] Liu HY, Liu HB, Peng GD, Chu PL. Observation of type I and type II gratings behavior in polymer optical fiber. *Optics Communications* 2003 **220**:337–343.

[21] Liu HB, Liu HY, Peng GD, Chu PL. Novel growth behaviors of fiber Bragg gratings in polymer optical fiber under UV irradiation with low power. *IEEE Photonics Technology Letters* 2004 **16**:159–161.

[22] Zolla F, Renversez G, Nicolet A, Kuhlmey B, Guenneau S, Felbacq D. *Foundations of Photonic Crystal Fibres*. Imperial College Press: London, 2005.

[23] Ju J, Wang Z, Jin W, Demokan MS. Temperature sensitivity of a two-mode photonic crystal fiber interferometer sensor. *IEEE Photonics Technology Letters* 2006 **18**:2168–2170.

[24] Saitoh K, Koshiba M. Numerical modeling of photonic crystal fibers. *Journal of Lightwave Technology* 2005 **23**:3580–3590.

[25] Nasilowski T, et al. Temperature and pressure sensitivities of the highly birefringent photonic crystal fiber with core asymmetry. *Applied Physics B* 2005 **81**:325–331.

[26] Bock WJ, Urbanczyk W, Wojcik J. Measurements of sensitivity of the single-mode photonic crystal holey fibre to temperature, elongation and hydrostatic pressure. *Measurement Science and Technology* 2004 **15**:1496–1500.

[27] Wang YP, Xiao L, Wang DN, Jin W. Highly sensitive long-period fiber-grating strain sensor with low temperature sensitivity. *Optics Letters* 2006 **31**:3414–3416.

[28] He Z, Zhu Y, Du H. Effect of macro-bending on resonant wavelength and intensity of long-period gratings in photonic crystal fiber. *Optics Express* 2007 **15**:1804–1810.

[29] Petrovic JS, Dobb H, Mezentsev VK, Kalli K, Webb DJ, Bennion I. Sensitivity of LPGs in PCFs fabricated by an electric arc to temperature, strain and external refractive index. *Journal of Lightwave Technology* 2007 **25**:1306–1312.

[30] Dobb H, Webb DJ, Kalli K, Argyros A, Large MCJ, Van Eijkelenborg MA. Continuous wave ultraviolet light-induced fiber Bragg gratings in few- and single-mode microstructured polymer optical fibers. *Optics Letters* 2005 **30**:3296–3298.

[31] Zou L, Bao X, Chen L. Distributed Brillouin temperature sensing in photonic crystal fiber. *Smart Materials and Structures* 2005 **14**:S8–S11.

[32] Villatoro J, Minkovich VP, Monzon-Hernández D. Temperature-independent strain sensor made from tapered holey fiber. *Optics Letters* 2006 **31**(3):305–307.

[33] Minkovich VP, Monzon-Hernandez D, Villatoro J, Sotsky AB, Sotskaya LI. Modeling of holey fiber tapers with selective transmission for sensor applications. *Journal of Lightwave Technology* 2006 **24**:4319–4328.

[34] Blanchard PM, et al. Two-dimensional bend sensing with a single, multi-core optical fibre. *Smart Materials and Structures* 2000 **9**:132–140.

[35] MacPherson WN, Flockhart GMH, Maier RRJ, Barton JS, Jones JDC, Zhao D, Zhang L, Bennion I. Pitch and roll sensing using fibre Bragg gratings in multicore fibre. *Measurement Science and Technology* 2004 **15**:1642–1646.

[36] MacPherson WN, Jones JDC, Mangan BJ, Knight JC, Russell PStJ. Two-core photonic crystal fibre for Doppler difference velocimetry. *Optics Communications* 2003 **223**:375–380.

[37] MacPherson WN, *et al.* Tunnel monitoring using multicore fibre displacement sensor. *Measurement Science and Technology* 2006 **17**:1180–1185.

[38] MacPherson WN, *et al.* Remotely addressed optical fibre curvature sensor using multicore photonic crystal fibre. *Optics Communications* 2001 **193**:97–104.

[39] Duncan RG, Froggatt ME, Kreger ST, Seeley RJ, Gifford DK, Sang AK, Wolfe MS. High accuracy fiber-optic shape sensing. In *Proceedings of the SPIE Smart Sensor Systems and Networks: Phenomena, Technology, and Applications for NDE and Health Monitoring (SPIE Vol. 6530)*, Peters K (ed). SPIE: Bellingham, WA, 2007, pp. 65301S-1–65301S-11.

[40] Flockhart GMH, Cranch GA, Kirkendall CK. Differential phase tracking applied to Bragg gratings in multi-core fibre for high accuracy curvature measurement. *Electronics Letters* 2006 **42**:390–391.

[41] Cranch GA, Flockhart GMH, MacPherson WN, Barton JS, Kirkendall CK. Ultra-high sensitivity two-dimensional bend sensor. *Electronics Letters* 2006 **42**:520–522.

[42] Watson S, Gander MJ, MacPherson WN, Barton JS, Jones JDC, Klotzbuecher T, Braune T, Ott J, Schmitz F. Laser-machined fibers as Fabry-Perot pressure sensors. *Applied Optics* 2006 **45**:5590–5596.

[43] Abeysinghe DC, Dasgupta S, Jackson HE, Boyd JT. Novel MEMS pressure and temperature sensors fabricated on optical fibers. *Journal of Micromechanics and Microengineering* 2002 **12**:229–235.

[44] Iannuzzi D, Deladi S, Gadgil VJ, Sanders RGP, Schreuders H, Elwenspoek MC. Monolithic fiber-top sensor for critical environments and standard applications. *Applied Physics Letters* 2006 **88**:053501-1–053501-3.

[45] Deladi S, Iannuzzi D, Gadgil VJ, Schreuders H, Elwenspoek MC. Carving fiber-top optomechanical transducers from an optical fiber. *Journal of Micromechanics and Microengineering* 2006 **16**:886–889.

[46] Nellen PhM, Brönnimann R. Milling micro-structures using focused ion beams and its application to photonic components. *Measurement Science and Technology* 2006 **17**:943–948.

Chapter 63
Electric and Electromagnetic Properties Sensing

Michel B. Lemistre
Laboratoire SATIE/CNRS, Ecole Normale Supérieure de Cachan, Cachan, France

1 Introduction	1125
2 Theoretical Considerations	1126
3 Electrical Capacitance Sensors	1130
4 Electromagnetic Sensors for Composite Materials	1131
5 Conclusion	1135
Related Articles	1135
References	1135

1 INTRODUCTION

In the domain of aeronautics, particularly in the area concerning composite structures, electromagnetic techniques are little used for structural health monitoring (SHM). This could be explained by the fact that the main researchers working in the field of SHM have mechanical engineering training. One possible exception concerns eddy currents techniques, but these are mainly used for metallic structures [1]. In fact, in the case of composite structures, eddy currents techniques are impossible or very difficult to use and generally give poor results, these kinds of structures being either purely dielectric or poor conductors (e.g., carbon epoxy). However, another method also based on eddy currents, but using a low frequency holographic technique [2–4], gives good results on carbon–epoxy structures; nevertheless, this method is not easily transposable into the SHM domain, the external equipment required being too complicated. However, there is a possible application for SHM by measurement of electrical resistance or electrical potential [5, 6]. One other possibility is a dynamic capacitive method [7, 8] uniquely used in the domain of civil engineering.

A new family of electromagnetic techniques, which makes it possible to obtain good information on the health of structures made of composite (i.e., carbon fiber reinforced plastic (CFRP) and glass fiber reinforced plastic (GFRP)) has been developed. These techniques consist of determining the state of the health of a structure by measurement of its two electrical parameters, the electric conductivity σ and/or the dielectric permittivity ε, since damages induces locally significant variations of these parameters. The goal of this article is to explain these techniques and to give their field of application.

First, we shall provide a recall of the electromagnetic theory necessary for gaining a full understanding of the electromagnetic techniques developed.

Encyclopedia of Structural Health Monitoring. Edited by Christian Boller, Fu-Kuo Chang and Yozo Fujino © 2009 John Wiley & Sons, Ltd. ISBN: 978-0-470-05822-0.

1126 Sensors

Next, the various techniques will be explained with examples of application. The transposition of these techniques in the domain of SHM are fully explained in **Chapter 80**.

2 THEORETICAL CONSIDERATIONS

2.1 Surface impedance

Let us consider a plane structure made of a material having the following electrical properties:

- magnetic permeability $\mu = \mu_0$;
- dielectric permittivity $\varepsilon = \varepsilon_0 \varepsilon_r$;
- electric conductivity σ;
- thickness d.

The structure is illuminated by a plane wave with an inclined incidence (see Figure 1).

Electric and magnetic components of the incident wave $\vec{p}_i$ are $\vec{E}_i$ and $\vec{H}_i$ respectively, $\vec{E}_r$ and $\vec{H}_r$ being the components of the refracted wave $\vec{p}_r$; θ_1 and θ_2 are the angles of incidence and refraction, respectively; N_1 is the refractive index of the external medium and N_2 is the refractive index of the structure. θ_1 and θ_2 are linked by the following relation:

$$N_1 \sin \theta_1 = N_2 \sin \theta_2 \qquad (1)$$

If one considers that the external medium is in free space, $N_1 = 1$. Taking into account the complex relative permittivity of the material $\varepsilon_r^* = \varepsilon'_r - j\varepsilon''_r$, N_2 the index of the material is given by

$$N_2 = \sqrt{\varepsilon_r^*} = \sqrt{\varepsilon'_r - j\frac{\sigma}{\omega \varepsilon_0}} \qquad (2)$$

If one considers the material as a good conductive material (i.e., $\sigma \gg j\omega\varepsilon$), N_2 can be reduced to the following:

$$N_2 = \sqrt{-j\frac{\sigma}{\omega \varepsilon_0}} \qquad (3)$$

then the relation (1) becomes

$$\sin \theta_2 = \frac{1}{N_2} \sin \theta_1 \qquad (4)$$

or

$$\cos \theta_2 = \left(1 - \left(\frac{1}{N_2}\right)^2 \sin^2 \theta_1\right)^{\frac{1}{2}}$$

$$= \left(1 + j\frac{\omega \varepsilon_0}{\sigma} \sin^2 \theta_1\right)^{\frac{1}{2}} \qquad (5)$$

For good conductive materials, neglecting the second term of the parenthesis, the relation (5) yields

$$\cos \theta_2 \cong 1 \qquad (6)$$

Therefore, the wave penetrates through the material perpendicular to the surface of the structure ($\theta_2 = 0$). Figure 2 shows the field's configuration inside the structure.

The two fields $\vec{E}(z)$ and $\vec{H}(z)$ can be written as follows:

$$\vec{E}(z) = \vec{E}_a \exp(-\gamma z) + \vec{E}_b \exp(+\gamma z) \qquad (7a)$$

$$\vec{H}(z) = \left(\vec{n} \times \vec{E}_a\right)\frac{\exp(-\gamma z)}{Z}$$
$$+ \left(\vec{n} \times \vec{E}_b\right)\frac{\exp(+\gamma z)}{Z} \qquad (7b)$$

with Z being the wave impedance inside the material defined by the relation

$$Z = \sqrt{\frac{\mu_0}{\varepsilon^*}} = (j+1)\sqrt{\frac{\mu f \pi}{\sigma}} = \frac{j+1}{\sigma \delta} \qquad (8)$$

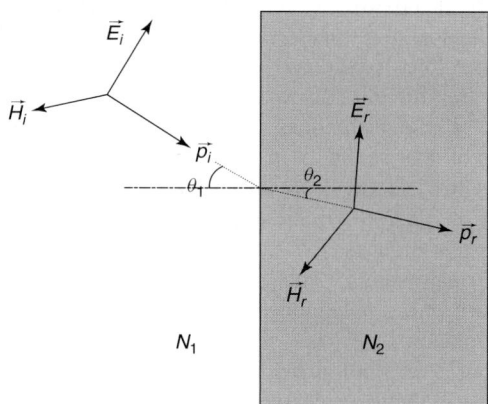

Figure 1. Illumination of a conductive material.

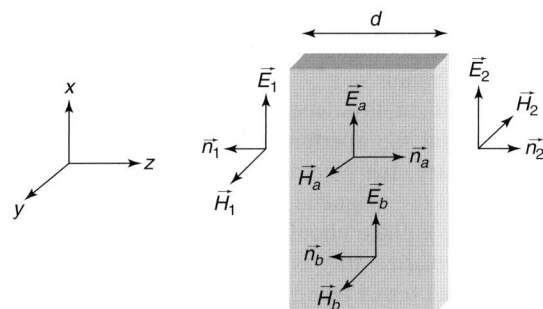

Figure 2. Field configuration inside a structure of thickness d.

and γ the propagation constant given by

$$\gamma = j\omega\sqrt{\varepsilon^*\mu_0} = (j+1)\sqrt{\mu f \pi \sigma} = \frac{j+1}{\delta} \quad (9)$$

where $\vec{n}$ is the unit vector in the z direction (i.e., the thickness of the material), f the frequency of the incident wave, and δ the skin depth defined by the following relation:

$$\delta = \frac{1}{\sqrt{\mu f \pi \sigma}} \quad (10)$$

Electric and magnetic fields tangential to the structure, on each one of its faces, $\vec{E}_1$, $\vec{H}_1$ and $\vec{E}_2$, $\vec{H}_2$ (Figure 2), are given by the following equations:

$$\vec{E}_1 = \frac{Z}{th(\gamma d)}\vec{n}_1 \times \vec{H}_1 + \frac{Z}{sh(\gamma d)}\vec{n}_2 \times \vec{H}_2 \quad (11a)$$

$$\vec{E}_2 = \frac{Z}{sh(\gamma d)}\vec{n}_1 \times \vec{H}_1 + \frac{Z}{th(\gamma d)}\vec{n}_2 \times \vec{H}_2 \quad (11b)$$

One can define a surface current density $\vec{J}_s$, which is the integral of the volume current density $\vec{J}$ in the thickness of the structure. One can write the boundary equations:

$$\vec{J}_s = \vec{n}_1 \times (\vec{H}_1 - \vec{H}_2) = \vec{n}_2 \times (\vec{H}_2 - \vec{H}_1) \quad (12)$$

For low frequencies such as $d \ll \delta$, that is, $|\gamma d| \ll 1$, one can write the following approximations:

$$sh(\gamma d) \cong \gamma d \quad (13a)$$

and

$$th(\gamma d) \cong \gamma d \quad (13b)$$

Relations (11a) and (11b) yield

$$\vec{E}_1 = \vec{E}_2 = \vec{E}_{tg} = \frac{1}{\sigma d}\vec{J}_s \quad (14)$$

E_{tg} being the electric field tangential to the surface of the structure, it is now possible to define the surface impedance Z_s as

$$\vec{E}_{tg} = Z_s \vec{J}_s \quad (15)$$

with

$$Z_s = \frac{1}{\sigma d} \quad (16)$$

One can see that at low frequencies, the current density is distributed uniformly in the material. One can then neglect the skin effect and resonances inside the material. The condition $|\gamma d| \ll 1$ (or $\delta > d$) defines the domain of application of the concept of surface impedance; the material is then called *electrically thin*. In the case of $\delta < d$, the material is called *electrically thick*; losses by attenuation inside the material and by reflection are then the main phenomena.

The surface impedance is given in "square ohm" (symbol Ω_c); it is the impedance of an electrically thin square sample, having a conductivity σ. One can represent the equivalent electric diagram to the surface impedance by Figure 3.

2.2 Diffraction by a circular aperture

For an incident plane wave, Bethe [9] has given an approximate analytical representation of diffracted

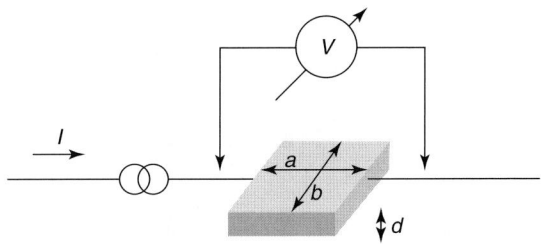

Figure 3. Equivalent electric diagram of the surface impedance.

fields by a small circular aperture in a structure that is considered as infinitely conductive (a good conductive metal such as aluminum can be considered as infinitely conductive), having an infinite surface (i.e., large compared with the diameter of the aperture), with the following assumptions:

- The size of the aperture is small compared with the wavelength of the incident field.
- The fields are calculated at a large distance compared with the size of the aperture.

Bethe has given the concept of "short-circuit fields" $\vec{E}_{cc}$ and $\vec{H}_{cc}$, which represent the fields on the aperture loaded by a perfectly conductive material. These fields are defined in the following manner:

$$\vec{E}_{cc} = 2\vec{E}_0 \quad (17a)$$

and

$$\vec{H}_{cc} = 2\vec{H}_0 \quad (17b)$$

where $\vec{E}_0$ and $\vec{H}_0$ are the orthogonal electric component and the tangential magnetic component of the incident field, respectively. The diffracted fields by the aperture are the sum of the radiated fields by an electric dipole having a moment $\vec{P}_e$ and a magnetic dipole having a moment $\vec{P}_m$. These two dipoles model the orthogonal and tangential fields, respectively as shown in Figure 4.

Let us introduce the concept of "polarizability"; the dipolar moments are related to short-circuit fields by the following relations:

$$\vec{P}_e = \varepsilon \alpha_e \vec{E}_{cc} \quad (18a)$$

and

$$\vec{P}_m = -\overline{\overline{\alpha}}_m \vec{H}_{cc} \quad (18b)$$

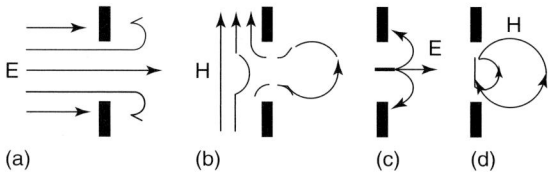

(a) (b) (c) (d)

Figure 4. Incident fields and equivalent dipoles: (a) orthogonal electric field, (b) tangential magnetic field, (c) equivalent electric dipole, and (d) equivalent magnetic dipole.

where α_e and $\overline{\overline{\alpha}}_m$ are the electric polarizability and the magnetic polarizability, respectively, given by an aperture of radius a.

The polarizability depends on the geometry of the structure and the aperture. For a plane aperture of any geometry, the electric polarizability is a scalar number and the magnetic polarizability is a tensor of rank 1. For a plane circular aperture of radius a, the magnetic polarizability is a diagonal tensor with the two diagonal terms being equal.

$$\alpha_e = \frac{2}{3}a^3 \quad (19a)$$

$$\alpha_{mxy} = \alpha_{myx} = 0 \quad (19b)$$

$$\alpha_{mxx} = \alpha_{myy} = \alpha_m = \frac{4}{3}a^3 \quad (19c)$$

With Bethe's assumptions, Casey [10] has calculated the moment of the magnetic dipole for a plane circular aperture having a radius a, loaded by a nonperfect conductive material such as carbon epoxy. Casey defines a resistance of "electrical gasket" R_g as the contact resistance between the conductive structure and the material loading the aperture. R_g is given by the product ohm × meter (length of contact between the material and the structure). Casey's method involves the solution of an integral equation. Two solutions are proposed—the first one being exact leads to a semianalytical expression of the magnetic dipole, including a coefficient that must be calculated numerically; the second one, obtained by an approximate method, leads to the dipolar moment $\vec{P}_m$ under the form of the transfer function of a first order low pass filter.

$$\frac{\vec{P}_m}{\vec{P}_{m0}} = \frac{1}{1 + j\dfrac{f}{f_c}} \quad (20)$$

In this expression, $\vec{P}_{m0}$ is the magnetic dipolar moment given by a free aperture, f represents the frequency of the incident wave, and f_c is the cutoff frequency given by the following relation:

$$f_c = \frac{3}{8\mu_0} \frac{Z_s}{a} \left(1 + \frac{R_g}{aZ_s}\right) \quad (21)$$

It should be noted here that the cut-off frequency is a function of the surface impedance Z_s and thus the conductivity σ of the material (relation 16). This cutoff frequency is called *Casey's frequency*.

2.3 Polarization of dielectrics

All materials that have a conductivity $\sigma \leq 10^{-20}$ S·m^{-1} at room temperature (≈ 300 K) are called *dielectrics*. On a large scale, dielectrics seem to be electrically neutral. However, on a microscopic scale, dielectrics show an "assembly" of elementary electric dipoles having a random space orientation. For a large number of dipoles, one can consider dielectrics as statistically neutral. If a dielectric material is subjected to an electric field, elementary dipoles tend to orient in the direction of the incident electric field; one can define a polarization vector per unit volume $\vec{P}$ as

$$\vec{P} = Nq\vec{\delta} \qquad (22)$$

with q elementary charges (per atom or molecule), separated by a distance $\vec{\delta}$ and N the number of atoms (or molecules) per unit volume. The product $q\vec{\delta}$ represents the elementary dipolar moment $\vec{p}$ for each atom (or molecule); this dipolar moment is related to the local electric field $\vec{E}_0$ by $\vec{p} = \alpha \vec{E}_0$. In this relation, the proportionality factor α is called *polarizability*, and it is a function of the electrical characteristics of the medium (i.e., the dielectric) and more precisely of its dielectric relative permittivity ε_r, defined by the Clausius–Mossoti's relation [11]:

$$\alpha = \frac{3}{4\pi N}\left(\frac{\varepsilon_r - 1}{\varepsilon_r + 2}\right) \qquad (23)$$

The polarization phenomenon in dielectric medium results from three different sources: electronic polarization, ionic polarization, and orientation polarization. These three sources admit polarizability coefficients α_e, α_i, and α_o, respectively; the real part of the coefficient α is the sum of the three real parts of elementary polarizability.

Electronic polarization arises because the center of local electronic charge cloud around the nucleus is displaced under the action of the electric field $\vec{P}_e = N\alpha_e \vec{E}_0$.

Ionic polarization occurs in ionic materials because the electric field displaces positives and negatives ions in opposite directions $\vec{P}_i = N\alpha_i \vec{E}_0$.

Orientation polarization can occur in materials composed of molecules that have permanent electric dipoles. The alignment of these dipoles depends on temperature and leads to an "orientational polarizability" per molecule $\alpha_o = \frac{p^2}{3KT}$, where p is the permanent dipolar moment per molecule, K is the Boltzmann constant, and T is the temperature.

Because of the different nature of these three polarization processes, the response of a dielectric solid to an applied electric field will strongly depend on the frequency of the field. The resonance of the electronic excitation takes place in the ultraviolet part of the electromagnetic spectrum; the characteristic frequency of the ions' vibration is located in the infrared, while the orientation of dipoles requires fields of much lower frequencies (below 10^9 Hz). This response to electric field of different frequencies is shown in Figure 5.

For a low-frequency excitation (i.e., 1 kHz to 10 MHz), one can consider the response of the dielectric medium to be quasi-static. So, it is possible to describe the electric field inside the dielectric by the following equation:

$$\nabla \cdot \vec{E} = \frac{\rho_l + \rho_p}{\varepsilon_0} \qquad (24)$$

where ρ_l is the free charge density and ρ_p an apparent density of charges due to the polarization phenomenon. The field $\vec{E}$ can be considered as the resultant of two components: the field resulting from the free charges $\vec{E}_f$ and the field resulting from the polarization phenomenon $\vec{E}_p$, that is, $\vec{E} = \vec{E}_f + \vec{E}_p$.

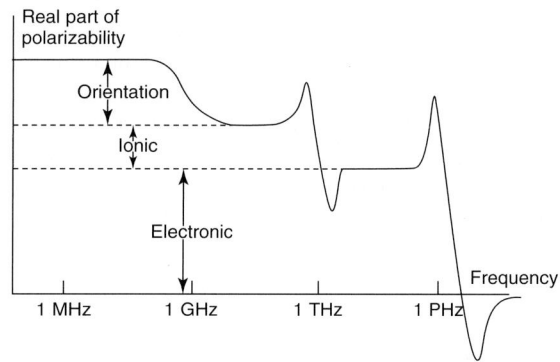

Figure 5. Frequency dependence of the different contribution to polarizability. From Handbook of Chemistry and Physics.

Then

$$\nabla \cdot \vec{E} = \nabla \cdot \vec{E}_f + \frac{\rho_p}{\varepsilon_0} \qquad (25)$$

One can define a polarization vector $\vec{P}$ by $\rho_p = -\nabla \cdot \vec{P}$ (see [12]). Equation (25) can be rewritten as

$$\vec{E} = \vec{E}_f + \frac{\vec{P}}{\varepsilon_0} \qquad (26)$$

The vector $\vec{P}$ is linked with the incident field $\vec{E}_i$ by the following relation:

$$\vec{P} = \chi_e \varepsilon_0 \vec{E}_i \qquad (27)$$

where χ_e is the electric susceptibility and is linked with the relative electric permittivity ε_r by the expression

$$\varepsilon_r = 1 + \chi_e \qquad (28)$$

Taking into account relations (26–28), one has

$$\vec{E} = \vec{E}_f + (\varepsilon_r - 1) \vec{E}_i \qquad (29)$$

However, in most of dielectrics, the term due to the free charges can be neglected and the value of the electric field $\vec{E}$ is reduced to the polarization term $(\varepsilon_r - 1) \vec{E}_i$.

3 ELECTRICAL CAPACITANCE SENSORS

Let us consider a dielectric material having an electric relative permittivity ε_r, in the presence of an electric field $\vec{E}_i$, the material being considered in a macroscopic manner (i.e., quasi-isotropic). The electric field $\vec{E}_i$ induces a phenomenon of polarization inside the material (see Section 2.3), characterized by the vector $\vec{P}$. The total electric field $\vec{E}_T$ measured at point A (Figure 6) can be written as

$$\vec{E}_T = \vec{E}_i + \frac{\vec{P}}{\varepsilon_0} = \vec{E}_i + (\varepsilon_r - 1) \vec{E}_i \qquad (30)$$

After subtraction of the incident electric field $\vec{E}_i$, one can obtain directly the value of ε_r. Note that, the frequency of excitation must be lower than the cutoff

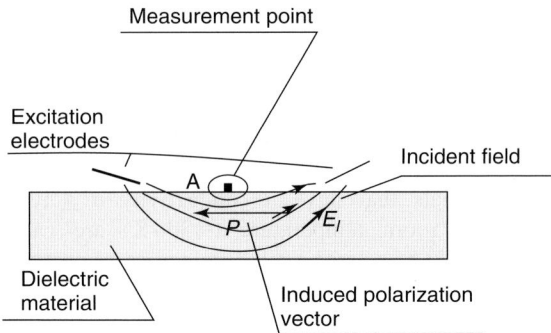

Figure 6. Configuration of measurement applied to dielectric materials.

frequency of the orientation polarization phenomenon (i.e., 10 MHz).

This method can be applied to all dielectric composite materials such as GFRP, sandwich, etc. An electric sensor has been designed [13] allowing to detect some defects in dielectric composites. This sensor performs a differential measurement between two adjacent zones (Figure 7).

Figure 8 shows an example of detection performed on a sample of glass epoxy sandwich with a lack of foam in the middle of the lower part. Figure 8(b) shows the electric image obtained by scanning the material.

This sensor allows designing a system of SHM dedicated to glass epoxy or sandwich structures and more generally all dielectric composite structures (*see* **Chapter 80**).

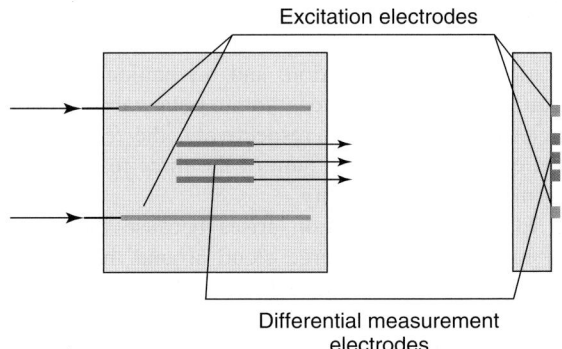

Figure 7. Differential electric capacitive sensor for detection of damages.

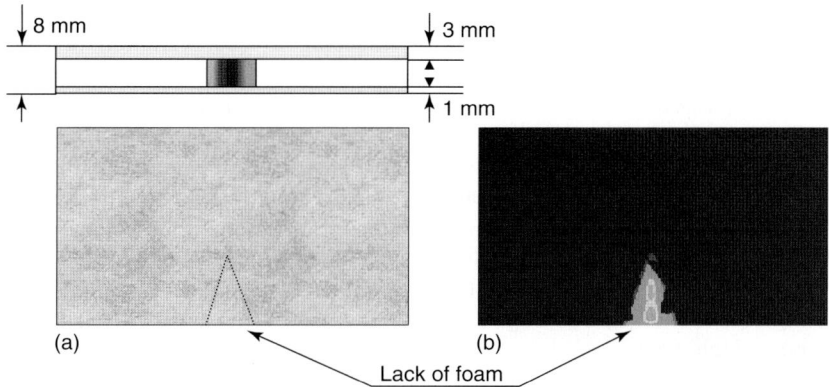

(a)   (b)

Lack of foam

Figure 8. Example of damage detection in a sandwich structure.

4 ELECTROMAGNETIC SENSORS FOR COMPOSITE MATERIALS

4.1 Magnetic sensor

The conductive materials such as metallic structures having a conductivity σ between 10^7 and 10^8 S·m^{-1} will not be considered here. For such materials, the effectiveness of classical methods using eddy currents is well established and is not necessary to demonstrate here. Here we consider composite materials having a mean conductivity σ about 10^4 S·m^{-1} such as CFRP, so-called nonperfectly conductive materials.

One has seen, in Section 2.2 that the magnetic dipolar moment $\vec{P}_\mathrm{m}$ of an aperture loaded by a nonperfectly conductive material, can be represented in the form of a transfer function of a first-order low pass filter. The dipolar moment being directly proportional to the magnetic field, the terms $\vec{P}_\mathrm{m}$ and $\vec{P}_\mathrm{m0}$ can be replaced by $\vec{H}$ and $\vec{H}_0$, respectively in relation (20):

$$\frac{\vec{H}}{\vec{H}_0} = \frac{1}{1 + j\dfrac{f}{f_\mathrm{c}}} \qquad (31)$$

where $\vec{H}_0$ is the magnetic field measured through a free aperture and $\vec{H}$ is the magnetic field measured through a loaded aperture by a conductive material. The cut-off frequency f_c being a function of the surface impedance Z_s, one has direct access to the value of the conductivity σ of the considered material by using the relation (21).

Let us now consider a local excitation by a near magnetic field (e.g., with a Hertz loop), and a local measurement of the resulting magnetic field, as shown in Figure 9; one can omit the infinite conductive plane and the contact resistance R_g.

An analytical calculus gives a new transfer function between $\vec{H}$ and $\vec{H}_0$:

$$\frac{\vec{H}(f,r)}{\vec{H}_0(f,r)} = \left(1 + \left(\frac{r}{a}\right)^2\right)^{\frac{3}{2}} \int_0^\infty \frac{u^2}{u + j\dfrac{f}{f_\mathrm{c}}} \\ \times J_1(u)\,\mathrm{e}^{-u\frac{r}{a}}\,\mathrm{d}u \qquad (32)$$

with r being the distance between the two loops, a the radius of the emission loop, and J_1 the Bessel function of first order, and the cutoff frequency f_c

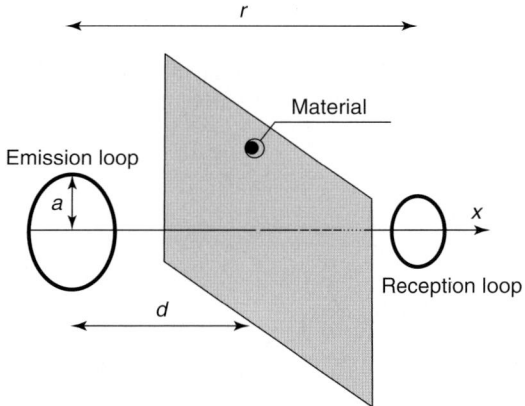

Figure 9. Excitation with near magnetic field.

having the following expression:

$$f_c = \frac{1.4}{\pi \mu_0 \sigma a e} \quad (33)$$

where a represents the radius of the emission loop and e the thickness of the material. The distance between the emission loop and the material d has no influence on the transfer function; it is possible to put the material immediately "after" the reception loop as shown in Figure 10—that is, allowing to test the material only on a single face.

However, in this case, the transfer function $\vec{H}/\vec{H}_0$ is given by the sum of the two following terms T_1 and T_2:

$$T_1 = 1 - \frac{\left(1 + \left(\frac{r}{a}\right)^2\right)^{\frac{3}{2}}}{\left(1 + \left(\frac{2d-r}{a}\right)^2\right)^{\frac{3}{2}}} \quad (34a)$$

$$T_2 = \left(1 + \left(\frac{r}{a}\right)^2\right)^{\frac{3}{2}} \int_0^\infty \frac{u^2}{u + j\frac{f}{f_c}}$$

$$\times J_1(u) e^{-u\frac{2d-r}{a}} du \quad (34b)$$

One can state that when f tends toward infinity, the function $T_1 + T_2$ tends toward the constant T_1. So, if the distance d increases, the value of the constant T_1 becomes close to the value of the transfer function before attenuation (i.e., $f < f_c$); this is the problem of the "lift-off" well known in the eddy current techniques.

The goal of this kind of analysis is not necessarily to measure the exact value of the conductivity of a material under test, but to detect damages inducing a local variation of this conductivity. A differential sensor [13, 14] measuring the "contrast" of the conductivity between two adjacent zones has been designed (Figure 11). This sensor compares the magnitude of the magnetic field between the two zones, 1 and 2 (Figure 11); when the two magnetic fields are different, the voltage V_{out} is nonzero.

The excitation frequency f is set between two characteristic frequencies: the skin frequency f_s and the Casey's frequency f_c (relation 33). The excitation frequency must be smaller than the skin frequency f_s but greater than the Casey's frequency f_c to have a significant variation of the measured magnetic field due to the variation of the local conductivity σ (relation 34). Figure 12 shows the evolution of these two frequencies as a function of the thickness of a quasi-isotropic carbon–epoxy multilayer structure.

By scanning a structure, one can build an image in which damaged areas appear clearly; an example is given in Figure 13. This figure shows the magnetic

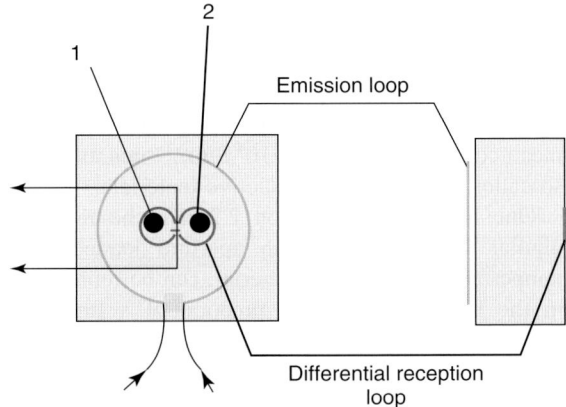

Figure 11. Differential magnetic sensor.

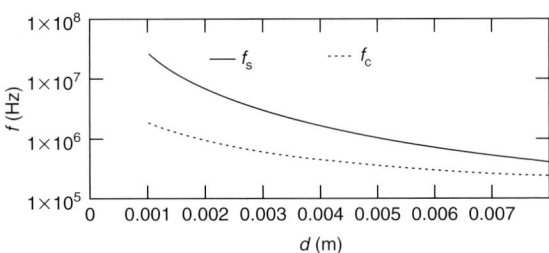

Figure 12. Evolution of f_s and f_c as a function of the thickness d.

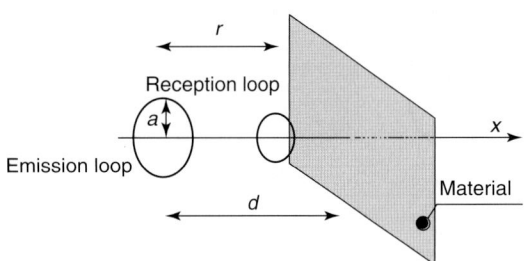

Figure 10. Excitation and measurement on a single face.

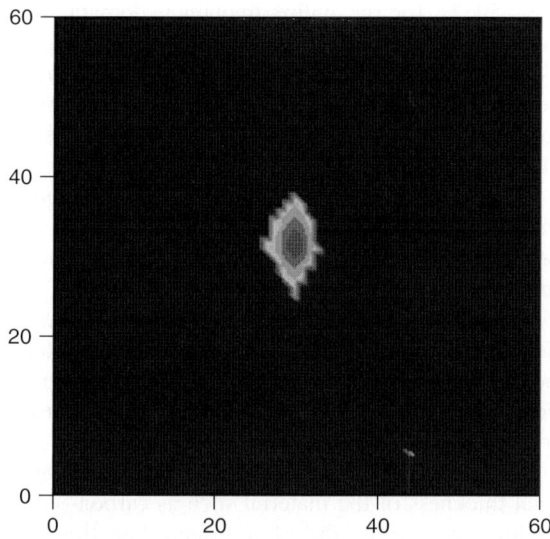

Figure 13. Magnetic image performed on a carbon–epoxy sample including a delamination.

image (i.e., σ contrast) performed on a quasi-isotropic carbon–epoxy multilayer sample of 60 mm × 60 mm × 2 mm dimension, including a delamination and fiber break. The delamination has been performed by a calibrated impact with energy 3 J.

4.2 Hybrid electromagnetic sensor

One can consider that a carbon–epoxy structure is made up of two different media: a conductive medium, the carbon fibers (conductivity $\sigma \approx 10^4$ S·m^{-1}), and a dielectric medium, the resin (relative dielectric permittivity $\varepsilon_r \approx 4$).

The local electric field $\vec{E}_1$ induced inside a conductive structure by a magnetic induction $\vec{B}$, can be represented by the following Maxwell's equation:

$$\nabla \times \vec{E}_1 = -\frac{\partial \vec{B}}{\partial t} \quad (35)$$

This electric field itself induces a current density $\vec{J} = \sigma \vec{E}_1$. However, in a carbon–epoxy media, the current density $\vec{J}$ is the sum of two terms: $\vec{J} = \vec{J}_c + \vec{J}_d$. The first term $\vec{J}_c$ (conductive current density) is due to the conductivity of the carbon fibers; the second one $\vec{J}_d$ (displacement current density) is a transient term due to the polarization phenomenon in the resin epoxy. Let us consider the quasi-static hypothesis, frequency below 10^7 Hz (i.e., below the cutoff frequency of the orientational polarization phenomenon); taking into account equation (26), the measured local electric field $\vec{E}_m$ can be written as follows:

$$\vec{E}_m = \frac{\vec{J}_c}{\sigma} + \frac{\vec{P}}{\varepsilon_0} \quad (36)$$

where $\vec{P}$ is the polarization vector due to the resin epoxy. With the relation (29) one can write

$$\vec{E}_m = \frac{\vec{J}_c}{\sigma} + \vec{E}_1 (\varepsilon_r - 1) \quad (37)$$

So the measured electric field $\vec{E}_m$ is a function of the conductivity σ of the medium and also of its relative dielectric permittivity ε_r. This technique based on the magnetic induction (i.e., eddy currents) and on the analysis of the resulting electric field is called *hybrid electromagnetic method*.

The possibility to have simultaneous access to the main electric properties of the medium presents a great potential for carbon–epoxy structures used in the aeronautical domain, for detecting main damages that are found in these structures. These damages can be classified into three categories:

- The damages having a mechanical origin—they are provoked by impacts, inducing delaminations and generally, fiber breakage. These critical damages are common for carbon (fibers); so in term of electrical properties, they induce uniquely a local variation of the conductivity.
- The thermal damages arise either from proximity to a hot body or from an electrical impact (i.e., spark, lightning). In the first case, one can detect the damage uniquely by conductivity variation. In the second case, if the burn is light, there is not much σ variation, but only significant ε_r variation due to the pyrolysis phenomenon of the resin. Nevertheless, this kind of damage generally affects the two electrical parameters.
- The damages resulting from liquid ingress (water, oil, fuel) can be critical because of the fact that the liquid can start a chemical reaction and weaken the structure. This kind of damage induces uniquely a variation in ε_r due to the

presence of a new medium having a different dielectric permittivity (i.e., the liquid).

From this hybrid concept, a new sensor has been designed [15]. This sensor can be considered as a combination of the electric capacitance sensor and the magnetic sensor, including some improvements. Figure 14(a) presents a photograph of a probe based on this technique, designed with a dielectric parallelepiped (3 cm × 3 cm × 1 cm) including on one face an inductive coil and a differential dipole for the measurement of the electric field on the opposite face. Figure 14(b) shows a schematic view of this probe.

The inductive coil appears as a Moebius loop made with a hard coaxial cable (Figure 14c).

This geometry allows the formation of a double induction loop with preservation of real impedance equal to the characteristic impedance of the coaxial (i.e., 50 Ω), for the entire frequency domain, from 100 kHz to 10 MHz. The measurement unit appears as double crossed dipole (Figure 14d) that allows to perform differential measurements with a sensitivity to the two orthogonal components of the tangential electric field $\vec{E}_x$ and $\vec{E}_y$.

The operating frequency f is not submitted to the same imperative as magnetic measurement. It is only necessary to operate with a frequency below the orientational polarization phenomenon cutoff (i.e., $f < 10^7$ Hz). However, the magnitude of the electric field induced inside the material is directly proportional to the frequency of the inductive magnetic field (equation 35), and it is preferable to use a frequency as high as possible. Nevertheless, too high a frequency does not allow penetration through the total thickness of the material such as carbon–epoxy multilayer, due to the skin effect. For this kind

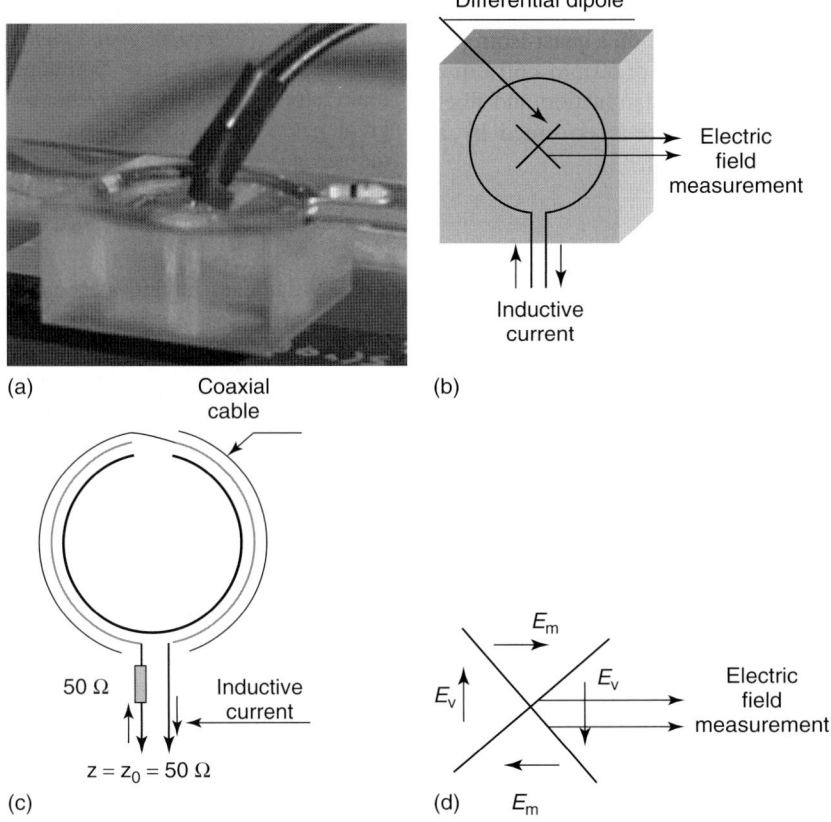

Figure 14. Hybrid electromagnetic probe: (a) photograph of the probe, (b) global view, (c) inductive coil, and (d) electric field sensor.

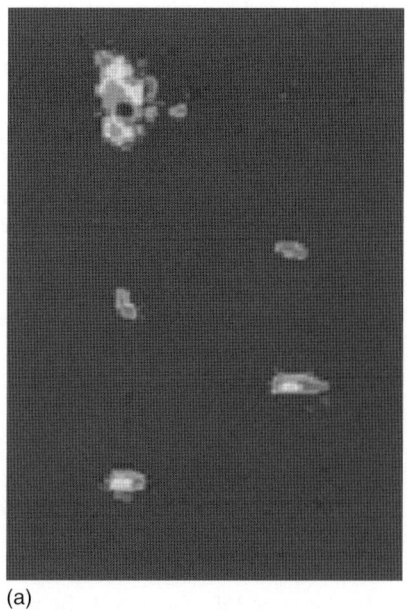 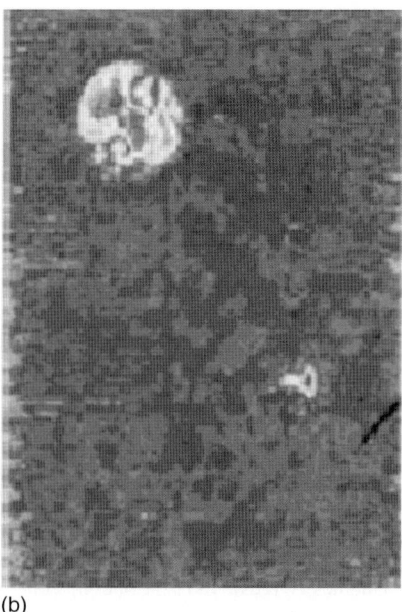

Figure 15. Investigation of a composite sample including various burns: (a) electromagnetic image and (b) ultrasonic C-scan image.

of material, the good frequency domain is between 100 kHz and 10 MHz.

Figure 15(a) shows the electromagnetic image of a quasi-isotropic sample of carbon epoxy (200 mm × 150 mm × 4 mm) with various burns. These results are compared with a C-scan ultrasonic image shown in Figure 15(b). Only the electromagnetic hybrid investigation is able to detect light burns.

5 CONCLUSION

The examples presented are results of various methods of signal processing used to extract relevant information, particularly the methods using wavelet transform. It is a fact that the resulting signal is very weak and the signal/noise ratio is close to 1. So, the main process for data reduction consists of performing a noise reduction by discrete wavelet transform (DWT) by using Donoho's method [16]. With regard to electromagnetic techniques, one can say that it is possible to detect practically all kinds of damage and their severity. However, the sensitivity of these methods is lower than the classical ultrasonic C-scan method as regards damages having a mechanical origin, such as light delaminations without fiber breaking, because of the fact that there is no variation of electric conductivity and the very weak variation of dielectric permittivity induced by the air layer of the delamination is not significant. Conversely, these techniques show greatest sensitivity as regards to other kinds of damages (i.e., thermal damages, liquid ingress, etc.). One of the interests of these methods lies in the fact that it is very easy to transpose them in the domain of SHM and it is possible to design a fully integrated SHM system for composite structures.

RELATED ARTICLES

Chapter 57: Piezoelectric Paint Sensors for Ultrasonics-based Damage Detection

Chapter 59: Fiber-optic Sensor Principles

REFERENCES

[1] Goldfine NJ, Zilberstein VA, Schlicker DE, Sheiretov Y, Walrath K, Washabaugh AP, Van Otterloo S. Surface mounted periodic field eddy currents sensors

for Structural Health Monitoring. *Proceedings of SPIE* 2001 **4335**:20–34.

[2] Madaoui N, Savin A, Premel D, Venard O, Grimberg R. An approach for quantitative nondestructive evaluation of discontinuities in flat conductive materials using eddy currents. 5th International Workshop on Electromagnetic Nondestructive Evaluation. IA, August 1999.

[3] Grimberg R, Savin A, Premel D, Mihalache O. Nondestructive evaluation of the severity of discontinuities in flat conductive materials using eddy currents transducer with orthogonal coils. *IEEE Transaction on Magnetics* 2000 **35**(1):299–331.

[4] Grimberg R, Premel D, Lemistre MB, Balageas DL, Placko D. Compared NDE of damages in graphite epoxy composites by electromagnetic methods. *Proceedings of SPIE* 2001 **4336**:65–72.

[5] Salvia M, Abry J.C. SHM using electrical resistance. In *Structural Health Monitoring*, Balageas D, Fritzen CP, Guemes A (eds). ISTE: London, 2006; Chapter 5, pp. 379–405.

[6] Abry J.C, Choi YK, Chateauminois A, Dalloz B, Giraud G, Salvia M. In situ monitoring of damage in CFRP laminates by means of AC and DC measurements. *Composite Science and Technology* 2001 **61**(6):855–864.

[7] Derobert X, Iaquinta J. Capacitive Methods for Structural Health Monitoring in Civil Engineering. In *Structural Health Monitoring*, Balageas D, Fritzen CP, Guemes A (eds). ISTE: London, 2006; Chapter 7, pp. 463–489.

[8] Iaquinta J. Contribution of capacitance probes for the inspection of external prestressing ducts. *Proceedings of the 16th World Conference on Nondestructive Testing*. Montréal, Canada, 2004.

[9] Bethe HA. Theory of diffraction by small holes. *Physical Review* 1944 **7–8**:163–175.

[10] Casey KF. Low frequency electromagnetic penetration of loaded apertures. *IEEE Transaction on Electromagnetic Compatibility* 1981 **23**(4):367–377.

[11] Coelho R, Aladenize B. *Les Diélectriques*. Hermès: Paris, 1993.

[12] Feynman RP. *Electromagnétisme*. InterEditions: Paris, 1984; Vol. 2, 217–227.

[13] Lemistre MB. Electromagnetic structural health monitoring for composite materials. In *Structural Health Monitoring, The Demands and Challenges*, Chang FK (ed). CRC Press, 2001, 1281–1290.

[14] Lemistre MB, Gouyon R, Balageas D. Electromagnetic localization of defects in carbon epoxy materials. *Proceedings of SPIE* 1998 **3399**:89–96.

[15] Lemistre MB, Deom A. Détection de brûlures dans les composites à base de carbone. In *Nouvelles Méthodes D'instrumentation*, Lavoisier H (ed). 2004; Vol. 2, 305–312.

[16] Donoho D, Johnstone I. *Ideal Denoising in an Orthonormal Basis Chosen from a library of Bases*. C.R. French Academy of Science: Paris, 1994; Serie I.

Chapter 64
Directed Energy Sensors/Actuators

James L. Blackshire
Air Force Research Laboratory, Wright Patterson Air Force Base, OH, USA

1 Introduction	1137
2 Physical Principles	1139
3 Directed Energy Sensing Methods	1144
4 Conclusions	1150
References	1150

1 INTRODUCTION

The interaction of electromagnetic energy with matter represents one of the most common and useful methods for inspecting and assessing a material or structure. Visual inspections, in fact, continue to be one of the most widely used and effective methods for characterizing the surface properties of a material or system [1]. By studying the reflective properties of a surface, for example, structural damage in the form of surface-breaking cracks, corrosion, and disbonds can be distinguished from undamaged areas based on surface roughness or topographic variations, which tend to reflect visible light differently, providing a simple means for detecting damage easily with the naked eye.

This article is a US government work and is in the public domain in the United States of America. Copyright © 2009 John Wiley & Sons, Ltd in the rest of the world. ISBN: 978-0-470-05822-0.

Of course, the characterization of surface properties alone represents a somewhat limited capability for understanding the structural health state of a material system. With the fundamental scientific discovery in the mid-seventeenth century that the electromagnetic spectrum is actually a continuum of frequencies/wavelengths, and the subsequent discovery of X rays [2], microwaves [3], and radar [4], the material penetration capabilities of electromagnetic radiation were revealed with profound implications. In particular, the ability to "see" into or through a visibly opaque material became possible using radiography at high frequencies ($\sim 10^{18}$ Hz) and microwave imaging at low frequencies ($\sim 10^{11}$ Hz).

The present article is concerned with the use of *directed energy* as an inspection tool for use in structural health monitoring applications. Directed energy sensing can be loosely defined as "the controlled insertion of electromagnetic energy into a material or system, where the observation of electromagnetic–material interactions is used to characterize the structural health of the system". At the core of this definition is the electromagnetic–material interaction, which uses the changes induced in an electromagnetic wave/field by a material system as its measurand. These changes can involve phase, amplitude, frequency, directionality, polarization, time of flight, and many other phenomena that are usually associated with wave phenomenon. It is in this basic theme that the article is organized, where fundamental principles related to electromagnetic fields/waves, directed energy beam characteristics, radiation–material interactions, and damage sensing are covered. An initial historical overview is first given followed by a brief description of the state-of-the-art directed energy sensing methods. A general overview of each of the major directed energy inspection methods is then provided, where physical principles of each method are covered along with specific measurement examples.

1.1 Historical background

Electromagnetic radiation has been at the forefront of scientific discovery for several millenniums. What began as a need for understanding the stars and heavens has grown to a scientific and technological discipline, which touches every part of human life and existence. Much of what we know about visible light and its interaction with materials was first studied in the Renaissance period (fourteenth to seventeenth centuries) by Descartes [5], Newton [6], Fermat [7], Snell [8], Huygens [9], Fresnel [10], and others [11–13]. The principles of light reflection, refraction, scattering, and diffraction were identified and systematically studied in much of this early work. The concept of "light waves" versus "light particles" was also debated during this period, which is currently known as the *wave-particle duality* of electromagnetic waves and photons of energy [14]. Both of these concepts play an important role in directed energy sensing methods.

For the next 200 years, theoretical and experimental work defined the inner workings of electromagnetism, culminating in the work of Maxwell [15] who demonstrated mathematically that electric and magnetic fields travel through space, in the form of waves, and at the constant speed of light. In 1861, Maxwell wrote his four-part publication in the *Philosophical Magazine* called "On Physical Lines of Force", where he first proposed that light was a form of energy composed of both electric and magnetic phenomena [15]. This fundamental discovery made the basic connection between electromagnetic waves and electronic material states possible, where electromagnetic fields exert a force (the Lorentz force) on the charged particles in a material system. The set of four equations known as *Maxwell's equations* describes this interrelationship between electric fields, magnetic fields, electric charges, and electric currents, forming the foundation of classical electromagnetism and electrodynamics. The material property concepts of permittivity, permeability, and complex dielectric were also defined, providing a means for characterizing electromagnetic–material interactions, where Maxwell's equations can be used in conjunction with the dielectric properties of a material to understand how light interacts with a material in a classical sense.

Around this same time (1860), the term *blackbody* was introduced by Kirchhoff [16], which relates the electromagnetic radiation emission properties of an object at increasing temperatures. In effect, the amount of electromagnetic radiation (and its corresponding wavelength) emitted by a black body source is directly related to its temperature. Black bodies above ~700 K (430 °C) produce radiation at visible wavelengths starting at red, going through orange, yellow, and white before ending up at blue as the temperature increases. This discovery represented a key connection between electromagnetic and thermodynamic physics (*see* **Chapter 19**).

The early work of Hertz [17] and Roentgen [2] in the 1880s–1890s extended the useful range of the electromagnetic spectrum to radio frequencies and to X-ray frequencies, respectively. Earlier in 1886, Hertz developed the dipole antenna receiver, which helped to establish the photoelectric effect, and represented the first practical instrument for producing and receiving electromagnetic wave energy (at ultra high frequency (UHF) radio frequencies). At the other end of the electromagnetic spectrum, Roentgen discovered X rays on November 8, 1895 when he observed a fluorescent glow from crystals near a cathode-ray tube. His systematic characterization of the penetrating nature of the radiation emitted by cathode-ray tubes ushered in the scientific field of radiography. The first practical use of electromagnetic energy as an inspection tool soon followed when X-ray fluoroscopes were used in 1896 to inspect postal packages, porcelain materials, precious stones, and simple metal welds [2] (*see* **Chapter 54**).

In 1900, Planck presented a paper at the German Physical Society in which he proposed the revolutionary scientific thought of quantum mechanics, which describes submicron particles as being composed of discrete states of energy [18]. In that work, Planck used the blackbody radiation concepts of Kirchhoff [16] and Wien [19] to mathematically describe the distribution of electromagnetic energy emitted by the different modes of charged oscillators in matter—what is known today as *Planck's law* for blackbody radiation. Building on these original ideas, in 1905 Einstein proposed that light can act as individual, discrete energy states—what we now know to be *quantum* of radiant energy or photons [20]. Einstein further described a new relationship between the energy and frequency of a photon as $E = h\nu$, where h is known as *Planck's constant* [20]. The foundation of electromagnetic–material interactions at a quantum mechanical level was based on these early ideas, which were verified by the end of the 1920s by Bohr, Born, Heisenberg, Schrödinger, De Broglie, Pauli, Dirac, and others. The work of Bohr, in particular, set forth in 1913 the fundamental quantum mechanical relationship between atomic states and electromagnetic energy emission and absorption, where he predicted the wavelengths of emission for a hydrogen atom [21]. The light emitted (and absorbed) by an atom, molecule, or material is now understood to arise from the transition of electrons between discrete energy states in a material system, where electromagnetic energy is either given to or taken from the system. The fundamental nature of electromagnetic–material interactions was, therefore, found to be both discrete in nature—quantum mechanics and photon energies—and continuous in nature—electromagnetic wave interactions with dielectric material properties (*see* **Chapter 63**).

1.2 State of the art

The use of directed energy methods for material inspections and structural health monitoring has seen significant progress and advancement in the past decade. The state of the art now permits detailed characterization of numerous materials with unprecedented spatial resolutions and damage sensitivity levels [22]. Perhaps, the most impressive are the three-dimensional imaging approaches (X-ray computed tomography [23], optical coherence tomography [24], microwave holography [25], and terahertz time-domain spectroscopy [26]), which have recently shown capabilities for probing materials with thicknesses of more than 25.4 cm (10 in.) (e.g., ceramic foams) and at spatial resolutions approaching 1 μm [27] (*see* **Chapter 54**).

The detailed characterization of surface and near-surface damage (e.g., surface-breaking cracks) using laser ultrasound [28], thermosonics [29], and optical coherence tomography [24] methods have also produced damage sensitivity levels, which now permit damage initiation, damage growth, and damage precursor indications to be measured. The whole-field imaging of structural microcracks using thermosonics, in particular, has been shown to be a powerful tool for imaging cracks, disbonds, and delaminations in complex geometry aerospace structures with damage features approaching 10 μm in size [29] (*see* **Chapter 65**).

The detection of hidden damage has also been reported recently using penetrating, nonionizing radiation in the terahertz [26], microwave [30], and millimeter-wave frequency bands, where damage hidden beneath aerospace coatings, insulation, composite, and ceramic materials has been reported [26, 30]. Significant advances in active and passive thermography have also been made in a large part due to improvements in focal plane array camera technologies, which have dramatically improved spatial resolution, thermal sensitivity, and noise rejection levels [31–33] (*see* **Chapter 19**).

2 PHYSICAL PRINCIPLES

Directed energy sensing fundamentally involves the interaction of electromagnetic energy with a material system. The sensing process can involve a number of different physical processes depending on the type of electromagnetic radiation used and the specific characteristics of the materials involved (*see* **Chapter 63**). A decrease in X-ray beam intensity, for example, can be attributed to an increase in material density or sample thickness based on material absorption principles, while a phase shift in a reflected laser beam at visible wavelengths may be due to surface motions of a vibrating material surface. In this section, a description of the physical principles involved in directed energy sensing is given, where similarities and differences between the electromagnetic frequencies are highlighted.

2.1 Electromagnetic radiation

Electromagnetic radiation propagates through space as an oscillating electric and magnetic field. The two field components are in phase, and oscillate at right angles to each other and to the direction of propagation. In 1861, Maxwell derived what is referred to as the electromagnetic *wave equation*, which governs the propagation of electromagnetic energy in free space and within a given material [34]. In general, electromagnetic radiation can be classified according to the frequency of the wave: radio waves, millimeter waves, microwaves, terahertz radiation, infrared radiation, visible light, ultraviolet radiation, and X rays. Figure 1 shows a diagram of the electromagnetic spectrum, where key frequency ranges are highlighted along with technologies associated with each frequency range.

An electromagnetic wave, like other wave phenomena, can be characterized by its wavelength (the distance from a point on one cycle to the corresponding point on the next cycle) or its frequency (the number of oscillations per second). In a vacuum, all electromagnetic waves travel at the same speed, the speed of light, $c = 299\,792\,458\,\mathrm{m\,s^{-1}}$. The wavelength, λ, and frequency, ν, of an electromagnetic wave are related by the equation

$$\nu = \frac{c}{\lambda} \quad (1)$$

which holds true for all forms of electromagnetic radiation. The energy of an electromagnetic wave is related to its frequency and wavelength by the relationship

$$E = h\nu = h\left(\frac{c}{\lambda}\right) \quad (2)$$

where h is a constant known as *Planck's constant*.

2.2 Radiation and matter

Depending on the frequency used, electromagnetic energy can interact with matter in very different ways. In a classical sense, electromagnetic energy interacts with materials as a localized field or wave phenomenon through material properties like the complex dielectric, the index of refraction, and the absorption coefficient. In a quantum mechanical sense, photons of electromagnetic energy interact with specific energy states in a material through

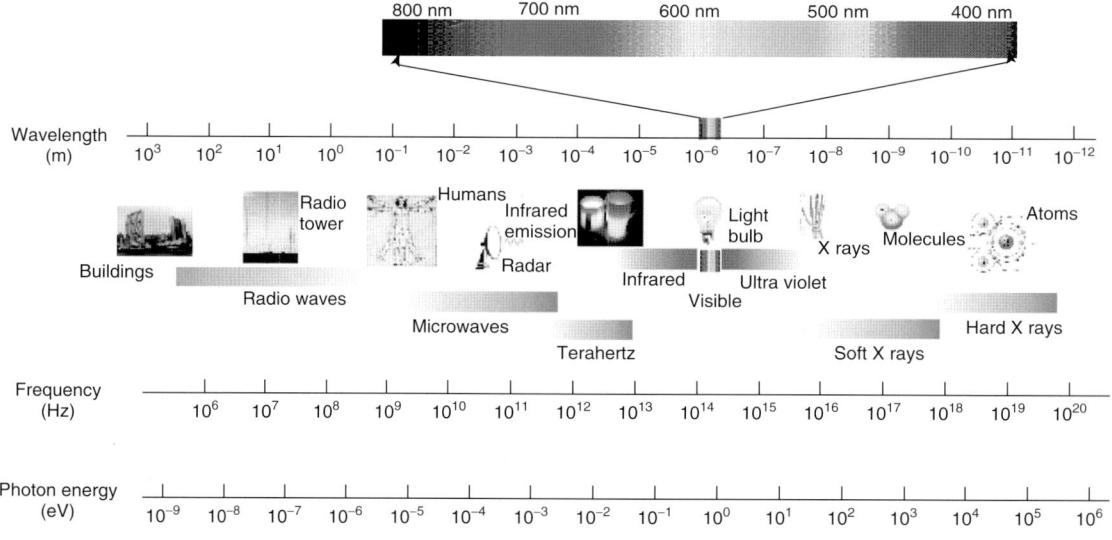

Figure 1. Electromagnetic spectrum with associated length scales and technologies.

probability functions and allowable transitions. Both of these concepts play a role in directed energy sensing (*see* **Chapter 63**).

In the classical approach, matter consists of a positive charge centers surrounded by clouds of negative charge [35]. In the presence of an electromagnetic field, the charge cloud distorts, which constitutes a simple dipole through the dipole moment. Mathematically, this process can be described using the dielectric properties of the material through a complex quantity denoted by

$$\varepsilon = \varepsilon' - j\varepsilon'', \varepsilon_r = \varepsilon'_r - j\varepsilon''_r, \varepsilon = \frac{\varepsilon}{\varepsilon_0} \quad (3)$$

where ε is called the absolute complex dielectric constant of the medium, ε_r is the relative complex dielectric constant, ε' is the permittivity (the ability of the material to store electromagnetic energy), ε'' is the dielectric loss factor (the ability of the material to absorb electromagnetic energy), and $\varepsilon_0 = 8.85419 \times 10^{-12}$ (F m^{-1}) is the permittivity of free space [30].

For a plane electromagnetic wave propagating in the z direction and polarized along the x axis inside a medium with a dielectric constant of $\varepsilon_r = \varepsilon'_r - j\varepsilon''_r$, the electric field intensity at any point can be described by the relationship [30]

$$\vec{E}(z) = E_0 e^{-(\alpha+j\beta)z} \hat{a}_x = E_0 e^{-\gamma z} \hat{a}_x,$$
$$\alpha = k_0 |Im\{\sqrt{\varepsilon_r}\}| \quad \beta = k_0 Re\{\sqrt{\varepsilon_r}\} \quad (4)$$

where E_0 is the electric field intensity at $z = 0$, α is the absorption constant, β is the phase constant, $\gamma = \alpha + j\beta$ is the propagation constant, and $k_0 = 2\pi/\lambda_0$ is the wave number in free space, and λ_0 is the wavelength in free space. For a plane electromagnetic wave traveling through a thickness, d, of material, the product $\gamma \cdot d$ determines the amount of attenuation and phase shift that the wave will experience [30].

When an electromagnetic wave traverses a boundary between two media, the concept of index of refraction becomes useful. Similar to the complex dielectric concept, the complex refractive index of a material determines how much the wave speed is reduced inside a medium and how much absorption loss occurs through the expression

$$\tilde{n} = n - ik \quad (5)$$

where n is the index of refraction, k is the extinction coefficient, and n is related to the material's relative permittivity, ε_r, and relative permeability, μ_r, through the expression $n = \sqrt{\varepsilon_r}\sqrt{\mu_r}$. The mathematical expressions describing the electromagnetic wave reflection and transmission at a boundary are referred to as the *law of reflection* and *Snell's law*, which are depicted schematically in Figure 2, and are given by the expressions: $\theta_i = \theta_r$ and $n_i \sin\theta_i = n_t \sin\theta_t$, respectively, and where $\theta_{i,r,t}$ are the angles of incidence, reflectance, and transmission.

In addition to the reflection, refraction, and transmission of light, the absorption and scattering of electromagnetic energy by a material represents a key aspect of most directed energy measurement approaches (*see* **Chapter 60**). In particular, the scattering of wave energy becomes important when the wavelength of an electromagnetic wave approaches (Mie scattering), or becomes larger than (Rayleigh scattering), the size of a scattering object.

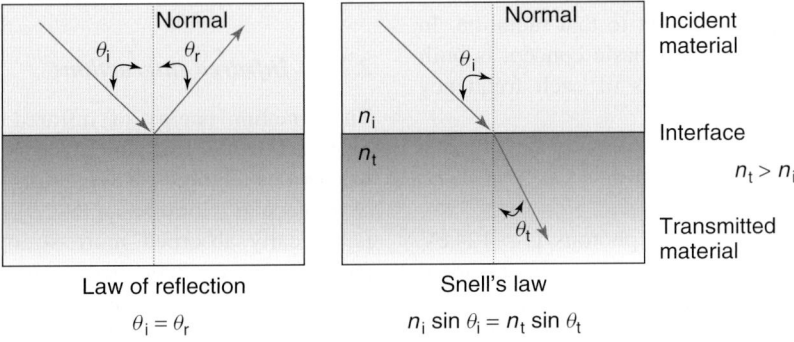

Figure 2. Reflection and refraction of electromagnetic radiation at a material interface.

For an electromagnetic wave incident on a single small scattering object, the intensity, I, of the scattered light can be written as follows:

$$I \cong I_0 \frac{1+\cos^2\theta}{2R^2}\left(\frac{2\pi}{\lambda}\right)^4\left(\frac{n^2-1}{n^2+2}\right)^2\left(\frac{d}{2}\right)^6 \quad (6)$$

where R is the distance from the particle, θ is the scattering angle, n is the refractive index of the scattering object, and d is the nominal size of the object. An important feature of equation (6) is the intensity of the scattered energy scaling inversely with the fourth power of the wavelength.

The absorption of electromagnetic energy by a material is traditionally described by Beer's law, which states that a logarithmic dependence exists between the transmission of light through a substance, the concentration/density of the substance, and the length of material that the light travels through. If the linear absorption (or attenuation) coefficient is defined as μ, then Beer's law can be written as follows:

$$I_z = I_0 \exp-[\mu z] \quad (7)$$

where I_z is the intensity of the beam at a distance z and I_0 is the intensity at the specimen surface. The primary factor that determines how radiation will be absorbed by a material is based on the atoms and molecules that make up the material. At a very fundamental level, the energy of an atomic and molecular system is quantized (i.e., made up of discrete levels). If a particular electromagnetic radiation energy matches one of these quantized energy states, then a strong interaction will result (e.g., absorption, reflection, refraction, and scattering), and if there are no available energy levels that match the energy of the incident radiation, then the material will typically be transparent to that radiation. In the following subsections, this basic concept is built upon, where the major features of each frequency range are covered.

2.2.1 X-ray interactions

The quantum energy of X-ray photons is $\sim$124 eV and greater, which is much too high to be absorbed in electron transitions between states for most atoms and molecules [36–39]. Because of this fact, most X rays penetrate through materials, with only the occasional X ray knocking an electron completely out of an atom/molecular system. During this process, the X ray can give up all of its energy to the electron (photoionization), or it can give up part of its energy (Compton scattering). A third possibility also exists if the X ray has sufficient energy, resulting in the creation of an electron–positron pair. A few electron volts of photon energy are typically required to eject an electron and ionize an atom, which places the threshold for ionization somewhere in the ultraviolet region of the electromagnetic spectrum. The X-ray wavelength range is $\sim$10 nm and shorter, while the frequency range is $\sim 3 \times 10^{16}$ Hz or greater. X-ray energies below 10 keV are typically referred to as *soft* X rays, which are used for porous or thin material measurements, while more energetic X rays above 10 keV are referred to as *hard* X rays for dense, thick material measurements.

2.2.2 Visible and ultraviolet light interactions

The quantum energy of visible and ultraviolet photons is in the range $\sim$1.65–124 eV, which is the dominant range of energies for elevating bound electrons to higher energy levels within an atomic or molecular system [36–39]. There are typically many available states in most material systems, so visible and ultraviolet light are strongly absorbed by most materials. The shorter ultraviolet wavelengths can reach the ionization energy for some molecules, which permits them to act in a similar fashion to "soft" X rays, while the net result of an absorption event by nonionizing visible radiation is generally just to heat the material sample. The wavelength range for visible–ultraviolet radiation is from $\sim$750 to 10 nm, while the frequency range is from $\sim 4 \times 10^{14}$ to 3×10^{16} Hz.

2.2.3 Infrared interactions

The quantum energy of infrared photons is in the range 0.001–1.7 eV, which corresponds to the range of energies required for separating the quantum states of molecular vibrations [36–39]. Infrared radiation is typically absorbed more strongly than terahertz and microwave frequencies, but less strongly than visible light. The result of an infrared absorption event is the heating of the material due to increased molecular vibrational activity. Infrared radiation can penetrate further into most materials relative to visible light due in part to its longer wavelength and in part to the reduced number of available quantum states for infrared energy coupling with the material. The wavelength range for infrared radiation is from $\sim 30\,\mu m$ to 750 nm, while the frequency range is $\sim 0.003\text{--}4 \times 10^{14}$ Hz.

In addition to the absorption of electromagnetic energy, infrared energy can flow or move within a material (thermal diffusion), and it can be radiated or emitted by a material according to Planck's law for blackbody radiation. The thermal transport or diffusion of heat energy within a material is important for infrared directed energy measurements, where active heating is used (e.g., pulsed thermography). For passive measurements, the emission of thermal radiation from a material is important. This "emissivity" depends on factors such as temperature, emission angle, and wavelength [31–33] (*see* **Chapter 19**).

2.2.4 Terahertz, microwave, and millimeter-wave interactions

The quantum energy of terahertz, microwave, and millimeter-wave photons is in the range $\sim 0.00001\text{--}0.001$ eV, which is in the range of quantum state energies for molecular rotation and torsion [26, 30, 36–39]. Terahertz energy is absorbed more strongly than microwaves, but less strongly than infrared light. The interaction of terahertz/microwave radiation with matter results in the rotation of molecules and the production of heat as a result of that molecular motion. Conductors strongly absorb microwaves (and lower frequencies) because they cause electric currents to form, which quickly heats the material. Most dielectric materials are largely transparent to both terahertz and microwaves energies/frequencies. Because the quantum energies of terahertz and microwave radiation are approximately million times lower than those of X rays, they do not produce ionization and are, therefore, a safer method for characterizing materials when depth penetration is needed. Most microwave applications fall in the range 3000–30 000 MHz (3–30 GHz). Current microwave ovens operate at a nominal frequency of 2450 MHz, a band assigned by the Federal Communications Commission (FCC). There are also some amateur and radio navigation uses of the 3–30 GHz range. The wavelength range for terahertz radiation is from $\sim 30\,\mu m$ to 1 mm, while the frequency range is from $\sim 3 \times 10^{11}$ to 1×10^{12} Hz. The wavelength range for microwave (and millimeter-waves) radiation is ~ 1 mm and greater, while the frequency range is $\sim 3 \times 10^{11}$ Hz and smaller.

2.3 Directed energy measurements

As stated earlier, directed energy sensing methods use electromagnetic energy to probe and characterize a material or structure. In most cases, active illumination is used to irradiate the material at some standoff distance, and a light-sensitive detector is used to collect the transmitted, reflected, or scattered electromagnetic energy from the material system (*see* **Chapter 60**). Material characterization or damage identification is then accomplished by analyzing the collected signals and applying the appropriate physics to extract material information of interest.

Figure 3 depicts the three primary approaches used in directed energy sensing measurements. If the electromagnetic radiation is not transmissive (or partially transmissive), then single-sided measurements are typically used (Figure 3a and b), where electromagnetic energy reflected from the material surface or scattered from within the near-surface region of the material is collected and analyzed. Most ultraviolet, visible, and infrared wavelength methods fall into this category. Some terahertz time-domain spectroscopy and near-field microwave methods also use the single-sided detection method depicted in Figure 3(a). If the electromagnetic radiation is transmissive, then volumetric measurements are possible using the through-transmission measurement approach depicted in Figure 3(c).

A wide variety of sources are available from broadband radiating sources (e.g., thermal radiation blackbody sources as previously described), to X-ray tubes, to microwave horns, to laser sources. An equally broad range of detectors are also available from semiconductor diodes, photomultiplier tubes, to photosensitive films and electronic imaging arrays. In most cases, the goal of a directed energy measurement is to collect the electromagnetic radiation energy and convert it to an electrical

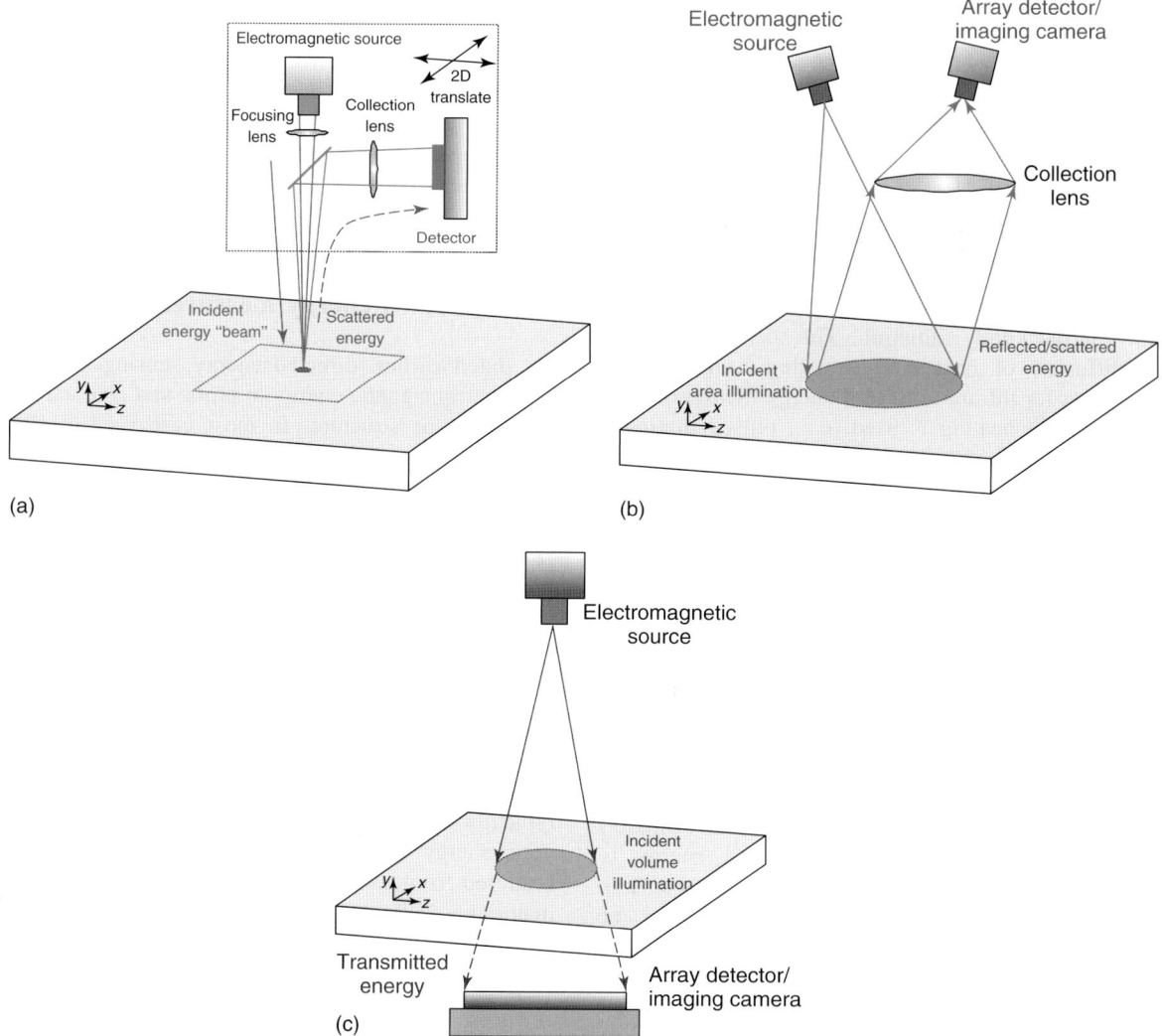

Figure 3. Directed energy measurement approaches: (a) single-sided, single-point measurement; (b) single-sided, imaging measurement; and (c) through-transmission, volumetric measurement.

signal, which can then be further analyzed. Simple measurements involve detecting raw intensity, while more sophisticated measurements keep track of phase, amplitude, and frequency. Two-dimensional imaging systems and three-dimensional computed tomographic systems also keep track of spatial position. Raster scanning of the single-point measurement system depicted in Figure 3(a) permits two-dimensional information to be obtained, while lenses and other beam manipulation systems permit whole-field images to be captured.

3 DIRECTED ENERGY SENSING METHODS

In this section, a variety of directed energy measurement examples are provided for electromagnetic energy in the X-ray, visible, infrared, terahertz, and microwave frequency bands. The examples show only a very small sampling of the available directed energy methods, which currently number in the hundreds to thousands. The interested reader is encouraged to refer to the numerous cited articles, and also to the books by Cartz [40], Demtroder [41], Scruby and Drain [28], Mittleman [26], and Zoughi [42] for additional methods and more detailed information.

3.1 Radiographic methods

Radiography is one of the five traditional nondestructive evaluation methods [40] (*see* **Chapter 54**). In recent years, technological advances in sources, detectors, and analysis methods have resulted in significant improvements in spatial resolution and sensitivity levels [43]. In addition, computed tomographic approaches have provided unmatched three-dimensional measurement capabilities for complex geometry structures and dense material samples [44]. Although different types and levels of radiographic energy can be used, X rays are most often used for most materials, while film radiography is quickly being replaced by electronic detector systems.

In most X-ray measurements, the basic setup depicted in Figure 3(c) is used, where the output signal is related to the material density according to an expression similar to equation (7). Subtle variations in material composition can also be characterized when dual-energy approaches are used [45]. Figure 4 depicts a series of measurements taken with a Phillips MGC-03 film radiography system, which represents a medium-resolution X-ray capability with a nominal 0.4-mm focal spot size [46]. The current, exposure time, and distance for the measurements depicted in Figure 4 correspond to 5 mA, 1.5 min, and 1.016 m (40 in.), respectively. The material sample (a laser-machined reference standard) was placed on the film and exposed to the X rays at a certain kilovolt energy level. The X-ray film was subsequently digitized. The image on the far left was taken at 55 kV energy level, and shows a series of laser-machined triangle features with increasing sizes and depths. The five images on the right correspond to a magnified view of the smallest (1 mm) and deepest (260 μm) triangle taken at increasing kilovolt energy levels (55–80 kV). As the kilovolt energy level increased, the signal-to-noise ratio decreased from 27.11 at 55 kV to 2.7 at 80 kV, while the image contrast remained almost constant at a ratio of ~1.2 : 1.0.

The ability to create fully three-dimensional characterizations of solid structures represents one of the most important capabilities for radiography, where X-ray computed tomography can be used. An example of an X-ray computed tomography measurement is provided in Figure 5(a), where a cross-sectional cut is depicted for the reference standard depicted in Figure 4. In this case, an ARACOR Tomoscope system was used, which uses a 225-kV, 50-μm microfocus X-ray source, and a fiber-optic scintillator/charge-coupled device (CCD)

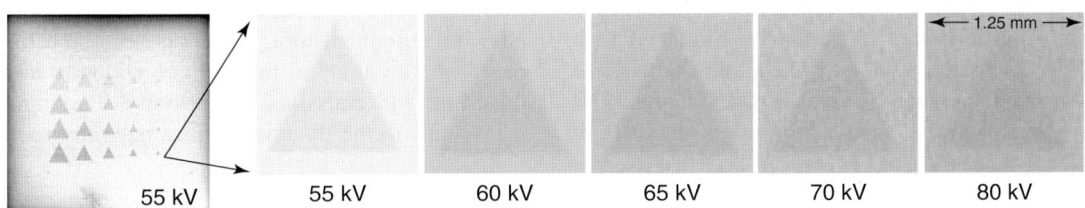

Figure 4. Radiographic through-transmission images of laser-machined triangle reference sample.

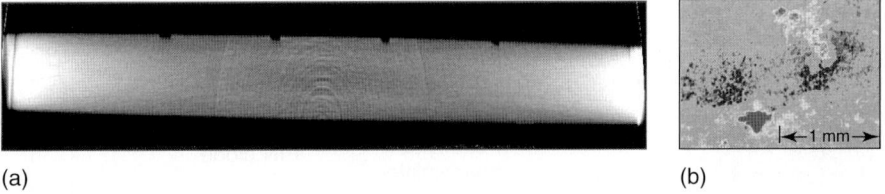

Figure 5. (a) Radiographic computer tomography measurement result showing a cross-section image through the laser-machined triangle sample and (b) X-ray measurement of corrosion.

image sensor system for detection. Thickness changes as small as 2% of the equivalent thickness were measured with an accuracy of ±1%. Figure 5(b) shows a similar measurement capability, where 1–5% corrosion material loss has been characterized [46].

3.2 Visible radiation and laser methods

Visible radiation represents one of the most prolific directed energy methods in use today. Spectroscopic measurements, in particular, provide a particularly useful characterization tool for identifying atomic and molecular states for a wide range of applications. As a material and system health monitoring capability, visible wavelength radiation has found use in a number of methods. Optical interferometry and holography, for example, provide a means for measuring surface topography and dynamic vibrations in a noncontact manner (*see* **Chapter 60**). Figure 6 provides an example of an interferometric measurement system capable of measuring vibration features on a material surface with micron spatial resolution levels in the 25 kHz to 20 MHz frequency bandwidth range.

When combined with an ultrasonic wave or vibration source, the system depicted in Figure 6 can be used to characterize surface damage with microscopic precision. An example of such a measurement is provided in Figure 7, where the motion field of a surface-acoustic wave (SAW) has been imaged with the laser vibrometry system to detect and characterize a microscopic surface-breaking crack feature. The crack feature is measured as a "brightness" increase in the lower right image field due to increased vibration levels near the crack [47].

The crack image in Figure 7 was made by raster scanning the laser probe position in a two-dimensional manner, similar to the depiction in Figure 3(a). At each scan position, the out-of-plane motions were captured by the laser interferometry system, which was collected and analyzed to provide a displacement versus time signal level at each position on the material surface. Similar measurements have been made with a moderate intensity pulsed laser to excite elastic waves in the material. This type of directed energy measurement system is termed *laser ultrasound*, which has found use recently as an effective means for assessing the structural health status of composite materials. Similar to other directed energy methods, the laser ultrasound approach provides a noncontact measurement capability that can be done at a standoff distance.

3.3 Infrared methods

Thermographic imaging is a relatively new nondestructive evaluation (NDE) technology, which uses thermal differences between a material defect and its local surroundings as a noncontact and full-field NDE measurement tool (*see* **Chapter 19**). Infrared cameras with higher sensitivities and resolutions are currently helping to transition the technology from a novelty to a comprehensive and quantitative NDE measurement tool. Materials are characterized based on variations in thermal absorption, transport, diffusion, and emission. In a passive thermography measurement, the material surface naturally radiates,

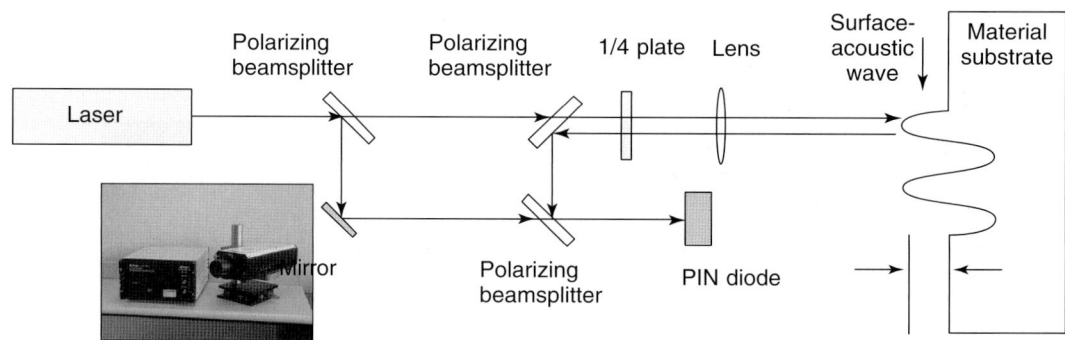

Figure 6. Laser interferometry system for measuring surface vibrations and displacements.

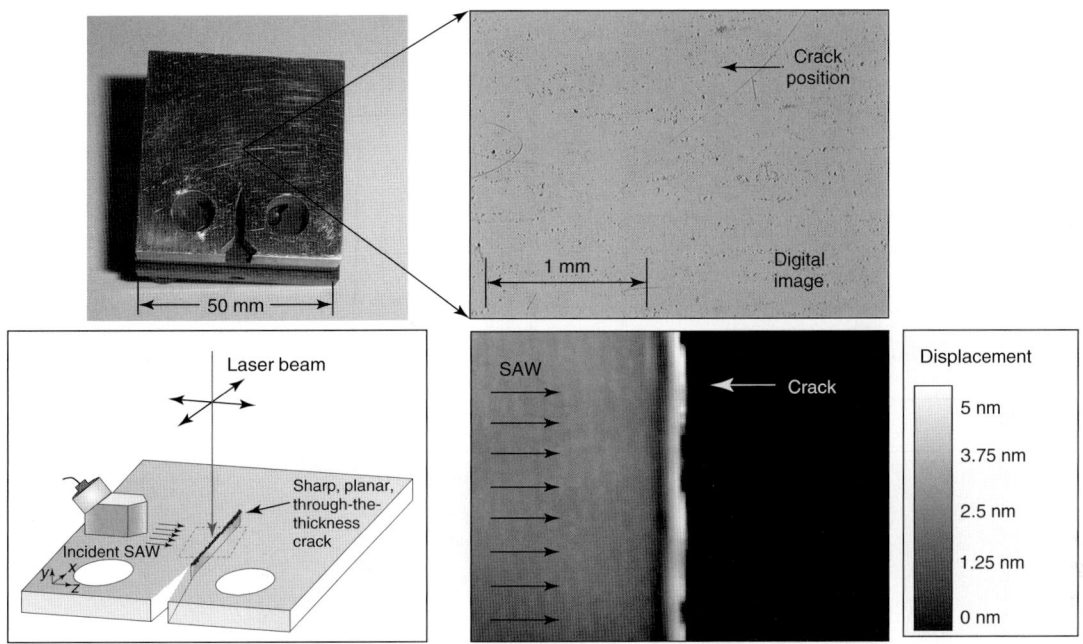

Figure 7. Laser interferometry measurement of surface-acoustic wave interaction with crack.

scatters, and reflects infrared energy that can be imaged by the thermal camera. Differences in thermal emissivity are imaged as differences in brightness by the camera, allowing corrosion, voids, and other damage to be detected and characterized.

In certain instances, thermographic measurements can be made in a spectral band-pass window that allows the infrared energy to propagate efficiently through a coating layer to probe the material substrate underneath [48]. This in fact is the case for many aerospace coating materials. Figure 8 depicts a spectral transmission window for a 3-mil-thick paint layer (~75 microns) between the 2 and 12 μm wavelength range, which occurred between 3.5 and 5.8 μm, peaking at 5.2 μm. By using a midwave infrared camera sensitive to 3–5 μm thermal energies, and/or using band-pass filters in that wavelength range, damage (e.g., hidden corrosion) can be imaged directly through the paint with a simple mid-IR camera system (Figure 8a).

The combination of ultrasonics and thermal imaging has also recently shown promise as a new directed energy sensing method for detecting microscopic cracks, composite disbonds, and delaminations [29]. The method termed *thermosonics* or *sonic-IR* uses an infrared camera to monitor the local heating that can occur in a material when vibrational energy causes frictional rubbing of surfaces at crack boundaries or disbonded/delamination locations. Figure 9(a) depicts a typical thermosonics system, which was used to detect a microscopic fatigue crack (Figure 9c) in a turbine engine blade (Figure 9b).

3.4 Terahertz methods and microwave methods

Terahertz and microwave measurement systems, like thermal imaging systems, are relatively new technologies. For directed energy measurements, terahertz and microwave methods provide a unique ability for penetrating through most dielectric materials. When this is the case, both methods can provide a volumetric measurement capability similar to X-ray measurements without the safety hazard problems. In addition, simple and effective imaging capabilities may be possible soon, where direct imaging through various dielectric material layers may be possible with the appropriate choice of directed energy frequency/wavelength.

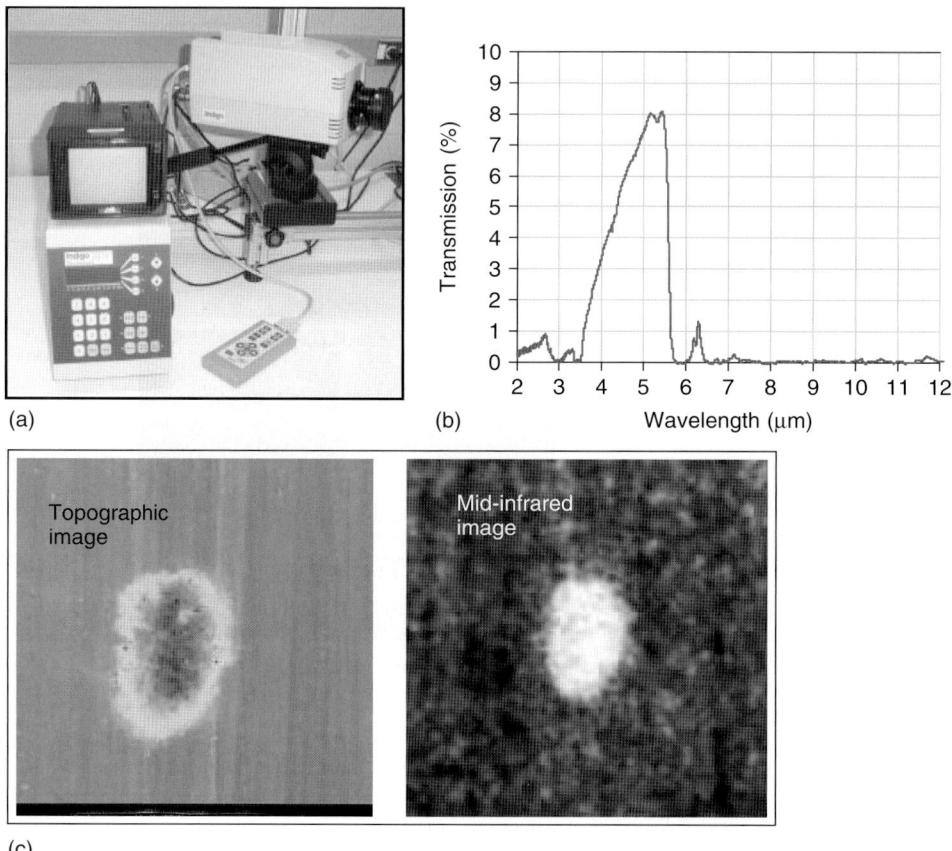

Figure 8. (a) Midinfrared camera system, (b) midinfrared transmission window through aerospace coating, and (c) passive infrared image of corrosion feature hidden under the coating.

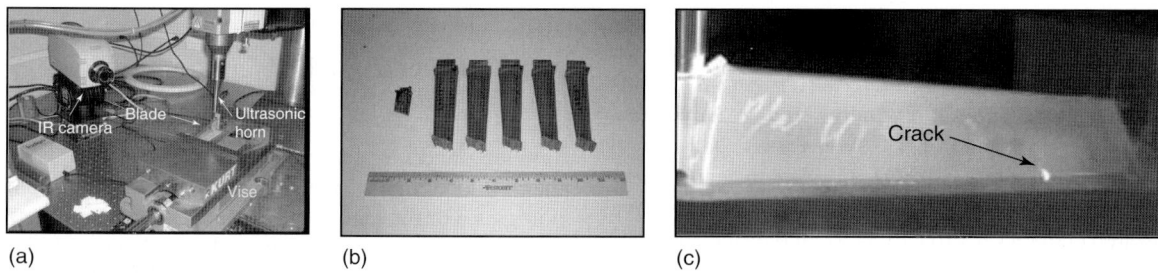

Figure 9. (a) Thermosonics system, (b) turbine engine blades, and (c) crack measurements.

An example of an evanescent microwave probe measurement taken through an aerospace coating is provided in Figure 10. An evanescent microwave probe is a coaxial resonator system that operates nominally at 1–4-GHz frequencies. The probe depicted in Figure 10(a) was resonant at ∼2.485 GHz, and had a sharp coupling tip that permitted near-surface, evanescent field measurements, which enhanced spatial resolution levels and measurement sensitivity [48]. By raster scanning the probe over the sample surface, an image can be produced, which represents changes in dielectric properties of

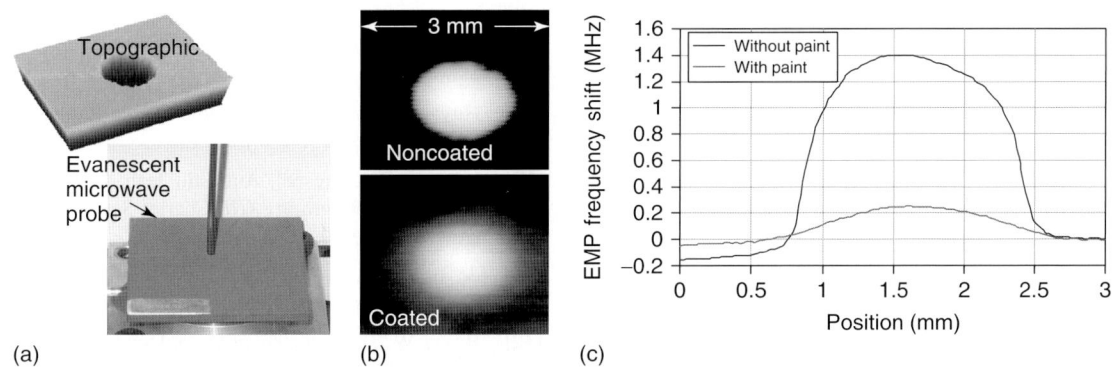

Figure 10. (a) Evanescent microwave probe system, (b) microwave measurements taken with and without coating layer present, and (c) comparison of response signals across corrosion pit.

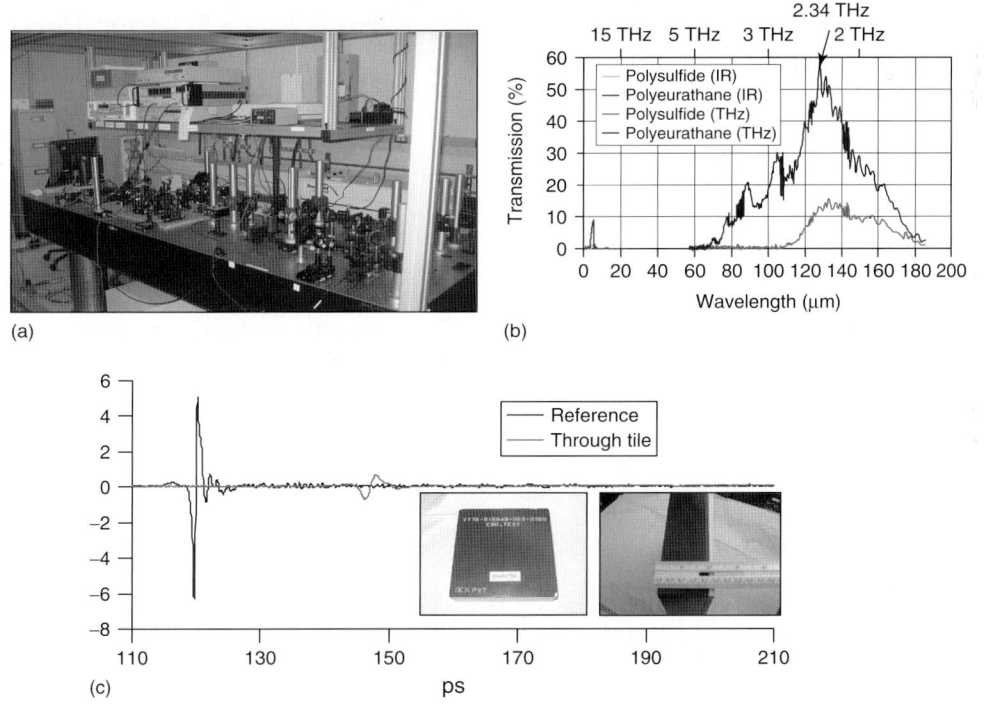

Figure 11. (a) Terahertz time-domain spectroscopy system, (b) terahertz spectral transmission through aerospace coatings near 2.3 THz, and (c) terahertz transmission through foam material.

the material according to equations (3) and (4). The presence of a hidden corrosion pit feature through an aerospace coating is depicted in Figure 10(b) and (c).

Terahertz measurement systems have also become available recently, with the promise of microwave-type penetration capabilities in reduced sizes and with improved spatial resolutions. Terahertz time-domain spectroscopy systems, in particular, have recently shown promise for measuring damage and material state through coatings, foam materials, ceramics, and composites [48]. Figure 11 depicts a typical time-domain spectroscopy system along with two measurement examples showing terahertz transmission properties through aerospace coating materials

(Figure 11b), and thermal protection system foam insulation materials (Figure 11c).

4 CONCLUSIONS

Directed energy sensing methods are becoming more capable and are finding more uses for nondestructive evaluation and structural health monitoring applications. Recent technological breakthroughs in numerous electromagnetic radiation sources and detectors are providing sensing capabilities across the entire electromagnetic spectrum from X rays to millimeter waves and beyond. In this brief review article, the key underlying principles of directed energy sensing are presented with a special emphasis on material inspections, nondestructive evaluation, and structural health monitoring applications. Specific examples are provided for directed energy sensing with X rays, laser ultrasonics, passive thermal imaging, thermosonics, evanescent microwaves, and terahertz time-domain spectroscopy. It is anticipated that directed energy sensing will continue to expand in capabilities and uses in the future, providing improved noncontact inspection capabilities for many years to come.

REFERENCES

[1] Matzkanin G, Easter J. *NDE of Hidden Corrosion—A Report Update; NTIAC Report*, NTIAC-SR-04-03. Nondestructive Testing Information Analysis Center: San Antonio, TX, 2004.

[2] Roentgen WC. On a new kind of ray. *Nature* 1896 **53**:274.

[3] Pozar DM. *Microwave Engineering*. Addison-Wesley: Boston, MA, 1993.

[4] Buderi R. *The Invention that Changed the World: The Story of Radar from War to Peace*. Simon & Schuster: New York, 1996.

[5] Descartes R. *The Geometry*. Dover Publications: New York, 1954, (1637).

[6] Newton I. *Opticks*. Dover Publications: New York, 1979, (1704).

[7] Mahoney MS. *The Mathematical Career of Pierre de Fermat*. Princeton University Press: Princeton, NJ, 1994, pp. 1601–1665.

[8] Struik DJ. Snel, Willebror. In *Dictionary of Scientific Biography XII*. Charles Scribner's Sons: New York, 1980.

[9] Huygens C. *Treatise of Light*. Dover Publications: New York, 1962, (1690).

[10] Silliman RH. Fresnel, Augustin Jean. In *Dictionary of Scientific Biography XIII*. Charles Scribner's Sons: New York, 1990.

[11] de Broglie L. *Matter and Light: The New Physics*. Dover Publications: New York, 1959.

[12] Bragg W. *The Universe of Light*. Dover Publications: New York, 1959.

[13] Sabra AI. *Theories of Light from Descartes to Newton*. Cambridge University Press: Cambridge, MA, 1981.

[14] Eisberg R, Resnick R. *Quantum Physics of Atoms, Molecules, Solids, Nuclei, and Particles, Second Edition*. John Wiley & Sons: New York, 1985, pp. 59–60.

[15] Maxwell JC. On physical lines of force. *Philosophical Magazine* 1861 **21**:161–175.

[16] Boltzmann L, Kirchhoff GR. *Populäre Schriften*. Verlag von J.A. Barth: Leipzig, 1905.

[17] Buchwald JZ. *The Creation of Scientific Effects: Heinrich Hertz and Electric Waves*. University of Chicago Press: Chicago, IL, 1994.

[18] Planck M. *Treatise on Thermodynamics*. Dover Publications: New York, 1922.

[19] Rüchardt E. Zur Erinnerung an Wilhelm Wien bei der 25. Wiederkehr seines Todestages. *Naturwissenschaften* 1955 **42**(3):57–62.

[20] Einstein A. On a heuristic viewpoint concerning the production and transformation of light. *Annalen der Physik* 1905 **17**:132–148.

[21] Bohr N. On the constitution of atoms and molecules. *Philosophical Magazine* 1913 **6**(26):1–25.

[22] Morre P, McIntire P (eds). *Nondestructive Testing Handbook, Special Nondestructive Testing Methods*. ASNT Press: New York, 1995; Vol. 9.

[23] Morgan CL. *Basic Principles of Computed Tomography*. University Park Press: Baltimore, MD, 1983.

[24] Schmitt J. Optical coherence tomography (OCT): a review. *IEEE Selected Topics in Quantum Electronics* 1999 **5**(4):1205–1215.

[25] Case J, Randazzo A, Pastorino M, Zoughi R. Evaluation of reconstruction error in microwave holographic imaging with reduced data sets. *Proceedings*

[26] Mittleman D (ed). *Sensing with Terahertz Radiation.* Springer: New York, 2003.

[27] Chen H, Kersting R, Cho G. Terahertz apertureless scanning near-field optical microscopy with nanometer resolution. *Applied Physics Letters* 2003 **83**:15.

[28] Scruby C, Drain L. *Laser Ultrasonics—Techniques and Applications.* Adam Hilger: Bristol, 1990.

[29] Favro L, Thomas R, Han X, Ouyang Z, Newaz G, Gentile D. Sonic infrared imaging of fatigue cracks. *International Journal of Fatigue* 2001 **23**:471–476.

[30] Zoughi R, Ganchev S. *Microwave Nondestructive Evaluation; NTIAC Report*, NTIAC-95-01. Nondestructive Testing Information Analysis Center: San Antonio, TX, 1995.

[31] Shull J. *Nondestructive Evaluation: Theory, Techniques, and Applications.* Marcel Dekker: New York, 2002.

[32] Maldague X. *Infrared Methodology and Technology.* CRC Press: Boca Raton, FL, 1994.

[33] Gaussorgues G. *Infrared Thermography.* Springer: Berlin, 1994.

[34] Maxwell JC. *Electricity and Magnetism.* Dover Publications: New York, 1959, (1861).

[35] Jackson JD. *Classical Electrodynamics.* John Wiley & Sons: New York, 1962.

[36] Chen S, Kotlarchyk M. *Interaction of Photons and Neutrons with Matter.* World Scientific Publishing: Hackensack, NJ, 1997.

[37] Moseley P, Crocker A. *Sensor Materials.* CRC Press: Boca Raton, FL, 1996.

[38] MacDonald N. *Nuclear Structure and Electromagnetic Interactions.* Plenum Press: New York, 1965.

[39] Weider R, Sells R. *Elementary Modern Physics.* Allyn & Bacon: Boston, MA, 1973.

[40] Cartz L. *Nondestructive Testing.* ASM International: Materials Park, OH, 1996.

[41] Demtroder W. *Laser Spectroscopy—Basic Concepts and Instrumentation.* Springer-Verlag: Berlin, 1998.

[42] Zoughi R. *Microwave Nondestructive Testing and Evaluation.* Kluwer Academic Publishers: Boston, MA, 2000.

[43] Zoofan B, Rokhlin S. Microradiographic detection of corrosion pitting. *Materials Evaluation* 1998 **56**:191–194.

[44] Hagemaier D. Aerospace radiography—the last three decades. *Materials Evaluation* 1985 **43**:1262–1283.

[45] Guillemaud R, Robert-Coutant C, Darboux M, Gagelin J, Dinten J. Evaluation of dual-energy radiography with a digital X-ray detector. *Proceedings of SPIE* 2001 **4320**:469–478.

[46] Blackshire J, Hoffmann J, Kropas-Hughes C, Tansel I. Microscopic NDE of hidden corrosion. *Proceedings of SPIE* 2003 **5045**:93–103.

[47] Blackshire J, Sathish S. Near-field ultrasonic scattering from surface-breaking cracks. *Applied Physics Letters* 2002 **80**:3442–3444.

[48] Blackshire J, Buynak C, Steffes G, Marshall R. Nondestructive evaluation through aircraft coatings: a state-of-the-art assessment. *9th Joint FAA/DoD/NASA Aging Aircraft Conference.* Atlanta, GA, 6–9 March 2006.

Chapter 65

Full-field Sensing: Three-dimensional Computer Vision and Digital Image Correlation for Noncontacting Shape and Deformation Measurements

Michael A. Sutton
Department of Mechanical Engineering, University of South Carolina, Columbia, SC, USA

1 Introduction	1153
2 Key Technologies Used in Three-dimensional Computer Vision (3D-DIC) for Structural Measurements	1156
3 Basic Concepts in Three-dimensional Computer Vision (3D-DIC) for Structural Measurements of Deformation and Shape	1157
4 Application of the Technique for Large Component Measurements	1162
5 Conclusions and Remarks	1164
Acknowledgments	1165
End Notes	1165
References	1165

Encyclopedia of Structural Health Monitoring. Edited by Christian Boller, Fu-Kuo Chang and Yozo Fujino © 2009 John Wiley & Sons, Ltd. ISBN: 978-0-470-05822-0.

1 INTRODUCTION

1.1 Area of application

The area of application is known as *three-dimensional digital image correlation* (3D-DIC), a noncontacting measurement method capable of accurately quantifying small or large three-dimensional surface displacements and surface strains on specimens ranging from a 10^{-3} to 10 m or larger.

1.2 Motivation

With the increase in computer processing speed, the ability to computationally predict a wide range of complex phenomena has increased dramatically (*see* **Chapter 2**; **Chapter 3**; **Chapter 4**; **Chapter 5**; **Chapter 6**; **Chapter 7**; **Chapter 8**; **Chapter 9**; **Chapter 10**; **Chapter 42**; **Chapter 43**; **Chapter 44**; **Chapter 45**; **Chapter 46**). Phenomena such as stable crack growth, dynamic crack growth,

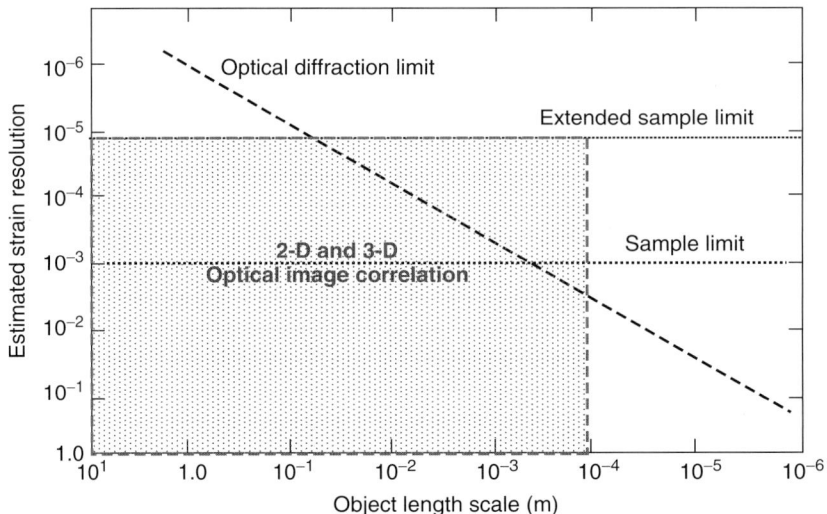

Figure 1. Schematic relating strain resolution to object dimensions. The graph focuses primarily on small-scale 2-D measurements. IC denotes *image correlation*, which appears to be a viable method for measurements over several orders of magnitude in out length scale. Optical IC includes both 2-D and 3-D methods. (The graph shown is a modified version of one developed by Prof. K.S. Kim, Brown University.)

mechanical response of complex structures, impact and blast loading, as well as a wide range of other phenomenon can be simulated with modern computational algorithms and associated software.

Until recent years, the ability to obtain accurate, full-field measurements under general conditions including (i) three-dimensional motions, (ii) large deformations, and (iii) high rate loading has been limited by the methods available. To validate the simulations, one approach is to compare specific predictions to experimental measurements. For example, if a fracture criterion is assumed as part of a simulation (*see* **Chapter 8**; **Chapter 9**; **Chapter 10**), then measurements such as the following can be directly compared with theoretical predictions to provide quantitative measures of the quality of the predictions:

- load-crack extension
- crack-tip strain field during crack extension
- crack opening displacement.

Figure 1 presents one view[a] of the relationship between strain sensitivity and specimen length scale for digital image correlation (IC) methods. As shown in this figure, for structural measurements on components on the order of $100\,\mu m$ or larger, optical IC methods offer investigators the ability to quantify strains on the order of 10^{-4} or larger. Though originally developed for 2-D measurement methods (e.g., 2-D IC), stereovision concepts have been used to formulate a general, noncontacting method designated *three-dimensional digital image correlation* (3D-DIC) that has an accuracy similar to that of two-dimensional digital image correlation (2D-DIC), while extending the capability to 3-D motion measurements on 3-D objects. Applications in the past two decades have shown conclusively that 3D-DIC is capable of making full-field deformation measurements on curved or planar specimens in a wide range of applications, with strain accuracy on the order of 10^{-4}.

1.3 Historical background

A cursory review of the literature indicates that the earliest developments in image-based shape measurements resulted in the formation of the field of photogrammetry. Originally focused on extracting height/shape information through comparative photography, it appears that some of the first work in the area of "image correlation" was performed by Gilbert Hobrough in the 1950s. Hobrough compared analog representations for photographs to register features from various views [1], and later designed and built an instrument to "correlate high-resolution reconnaissance photography with high precision survey photography in order to enable more precise measurement of changeable ground conditions" [2], thereby being one of the first investigators to attempt a form of digital IC to extract height information from the IC/matching process.

As digitized images became available in the 1960s and 1970s, researchers in artificial intelligence and robotics began to develop vision-based algorithms and stereovision methodologies in parallel with photogrammetry applications in aerial photography (see Rosenfeld [3] for an extensive bibliography). As noted by Rosenfeld, engineering applications for shape and deformation measurements using digital images were either nonexistent or rare up to 1980.

While digital image analysis methods were undergoing explosive growth in many areas, much of the field of experimental solid mechanics was focused on applying recently developed laser technology. Holography [4–6], laser speckle [7], laser speckle photography [8, 9], laser speckle interferometry [10], speckle shearing interferometry [11], moire' methods [12, 13], holographic interferometry [14] and fiber optic sensors (*see* **Chapter 2**; **Chapter 3**; **Chapter 4**; **Chapter 5**; **Chapter 6**; **Chapter 7**; **Chapter 8**; **Chapter 9**; **Chapter 10**; **Chapter 42**; **Chapter 43**; **Chapter 44**; **Chapter 45**; **Chapter 46**) are typical examples of the type of measurement techniques developed for use with coherent light sources. In all cases except fiber optic sensing, the measurement data (surface slopes and displacements) was most often embedded in the photographic medium, typically in the form of a fringe pattern. Since the photographic recording process is generally nonlinear, resulting in difficulties in extracting partial fringe positions with high accuracy, the most common process employed by experimental mechanicians was a laborious determination of estimates for fringe center locations at a few points.

Given the difficulties encountered by experimental mechanicians during the postprocessing of photographically recorded measurement data, it was natural for researchers to employ recent progress in digital imaging technology and develop (i) methods for digitally recording images containing measurement data, (ii) algorithms to analyze the digital images and extract the measurement data, and (iii) approaches for automating the entire process.

One of the earliest papers that proposed the use of computer-based image acquisition and deformation measurements in material systems was written by Peters and Ranson in 1982 [15]. Originally envisioned as a method for use with ultrasonic waves, the authors suggested comparing small regions (known as *subsets*) from each of the digitally recorded ultrasonic images before and after deformation. Using fundamental continuum mechanics concepts governing the deformation of small areas, subset matching throughout each image was proposed to obtain a dense set of full-field, two-dimensional displacement measurements. Over the next decade, the basic concepts were extended and applied to optical images of an undeformed and deformed object. The proposed methodology was modified and refined, resulting in a set of numerical algorithms that were validated through a combination of experimental and computational studies [16–23]. The 2D-DIC method has been used to measure crack-tip strain fields [24–27], creep deformations at elevated temperature [28], and tensile deformations of thin paper sheets [29, 30]; a high contrast random pattern was applied to the paper sheets using Xerox toner power. For these measurements, a random pattern and incoherent illumination were used to obtain high-contrast, white-light speckle images. By selecting subsets from the pattern and comparing deformed and undeformed patterns, the matching process is used to obtain full-field displacements.

More recently, investigators have begun to probe the fundamentals of the image-matching process and quantify the potential accuracy of the method. For example, Schreier *et al.* [31] showed that intensity interpolation must be performed to improve the accuracy of the displacement measurements. In follow-on on studies, Schreier and Sutton [32] have shown that quadratic shape functions provide some advantages when performing the matching process, especially for nonuniform strain fields. Relative to the importance of distortions in pattern matching, Schreier *et al.* [33] developed and applied nonparametric distortion measurement and removal methodologies.

Since 2D-DIC requires predominantly in-plane displacements and strains, relatively small out-of-plane motion of the object will change the magnification and introduces errors in the measured in-plane displacement. To overcome this limitation, stereovision principles developed for robotics, photogrammetry, and other shape and motion measurement applications were modified and used by Chao *et al.* to successfully develop and apply a two-camera stereovision system for the measurement of three-dimensional crack-tip deformations [34–36]. To overcome some of the key limitations of the method (square subsets remained square in both cameras, mismatch in the triangulation of corresponding points, and a calibration process that was laborious and time consuming), the stereovision method was modified to include (i) the effects of perspective on subset shape and (ii) constraints on the analysis to include the presence of epipolar lines [37]. The method has been used successfully in several small- and large-scale applications [38–40].

2 KEY TECHNOLOGIES USED IN THREE-DIMENSIONAL COMPUTER VISION (3D-DIC) FOR STRUCTURAL MEASUREMENTS

Figure 2 shows several actual stereovision systems that have been used successfully to measure 3-D shape and 3-D displacement fields. Key technologies used to construct 3D-DIC measurement systems include the following:

- High-resolution, scientific-grade, charge coupled device (CCD) or complementary metal oxide semiconductor (CMOS) cameras

  - Typical cameras record in monochrome with 8 bits.
  - Spatial resolution is on the order of 1024 × 1024 pixels.
  - Pixels are square in physical dimension.

- Quality optical lenses for imaging during experiments

  - Nikon, Canon, Sigma, and Schneider lenses are typical brands used.
  - Lenses with focal lengths in the range from f19 to f200 are commonly used.

- Stable, durable camera support structures

  - Heavy-duty tripods, or equally strong structures, to support the cameras.
  - Rigid cross members to maintain the relative positions of cameras during an experiment.

- Translation stages to adjust camera position(s)

  - Digitally controlled stages for automation of adjustment process have been used successfully.
  - Rigid cross members with cameras attached can be moved as a unit without affecting the calibration.

- Temporal synchronization unit for simultaneous multicamera image acquisition during calibration and experiment

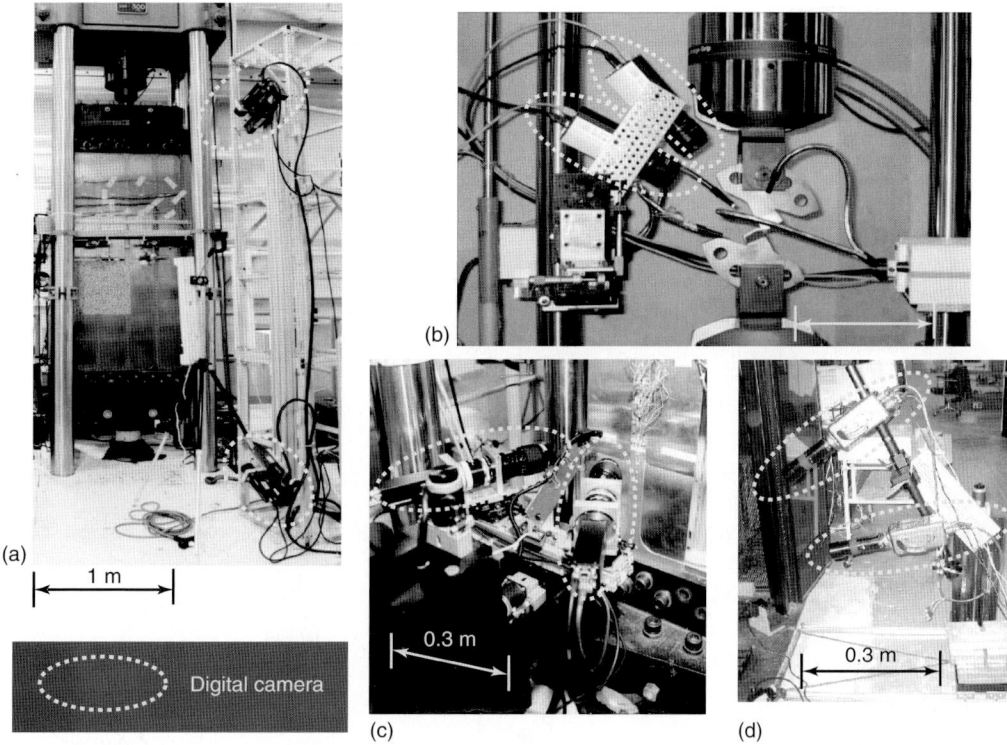

Figure 2. Typical stereovision systems used for laboratory measurements; (a) Pulnix cameras attached to 3-m gantry structure to view 1-m-wide specimen in 300-kips (1334 kN) loading frame; (b) Q-imaging cameras mounted to loading frames via translation stages to view specimen—the small angle between the cameras is due to space constraints for this application; (c) Pulnix cameras viewing rivet holes in 0.3-m-wide specimen undergoing tensile loading; and (d) high-speed Phantom V7.1 cameras viewing specimen mounted inside drop tower.

- Computer
  - Image acquisition and storage.
  - Postprocessing of images.
- Software[b]
  - Image acquisition.
  - Postprocessing of images.
  - Profile, displacement, and strain measurements.
  - Graphical presentation of measurements.
- Lighting to maintain adequate pattern contrast during calibration and experimentation.

Figure 3 presents a flow chart for a typical experiment using a two-camera stereovision system to acquire images of an object subjected to known mechanical loading and environmental conditions. In practice, the computer control system may also serve as (a) the image storage device and (b) the platform for postprocessing and data analysis.

3 BASIC CONCEPTS IN THREE-DIMENSIONAL COMPUTER VISION (3D-DIC) FOR STRUCTURAL MEASUREMENTS OF DEFORMATION AND SHAPE

A stereovision system consists of at least two views of the object from different orientations and positions. Figure 4 presents a schematic containing the key parameters related to the imaging process.

1158 *Sensors*

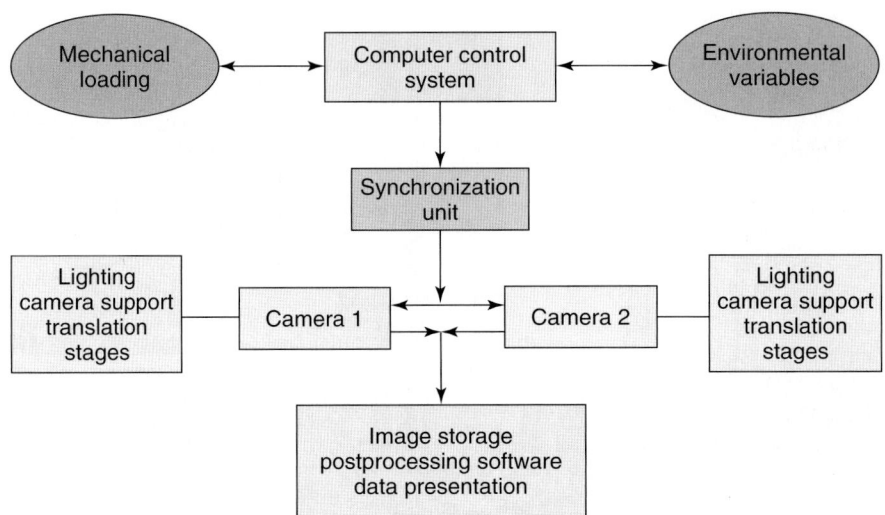

Figure 3. Flow chart for acquisition of synchronized images using a typical stereovision system.

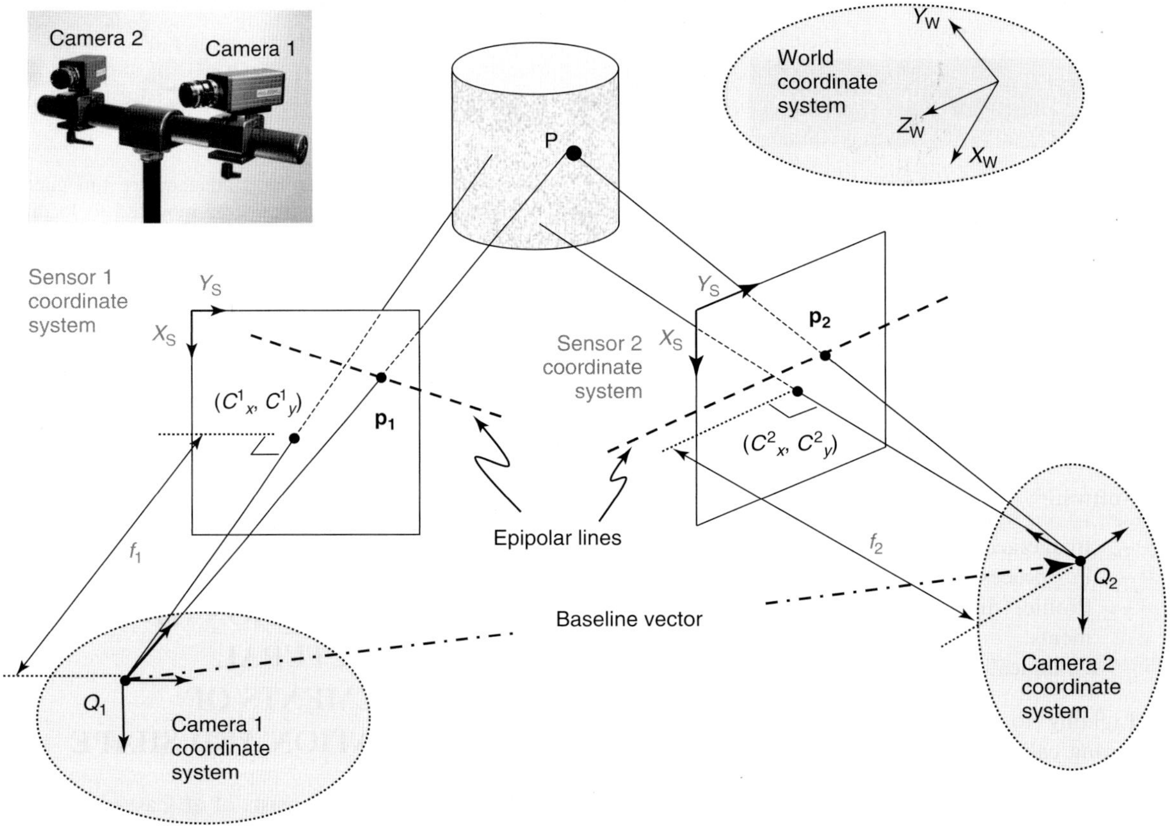

Figure 4. Actual stereovision system and schematic showing pinhole camera models for each camera.; the parameters in the model must be determined to convert the image positions ($\mathbf{p}_1$ and $\mathbf{p}_2$) into an accurate 3-D location of the point $\mathbf{P}$.

In this study, each image system is modeled as a pinhole camera. Intrinsic parameters for each camera include (i) the focal length, f, (ii) the image plane center, (C_x, C_y), (iii) lens distortion coefficient, κ, and (iv) scale factors, λ_x and λ_y relating metric distance on the object to pixel position in the image plane. Extrinsic parameters for each camera include (i) three independent components of the rotation matrix [**R**] and (ii) three components of the translation vector to orient the world coordinate system (WCS), with axes (X_W, Y_W, Z_W), relative to a camera coordinate system (CCS) located at a camera pinhole, either Q_1 and/or Q_2.

In practice, the WCS axes oftentimes are defined to be at a specific object position during the calibration process. For example, if a two-dimensional planar grid is used for calibration then (i) the plane of the grid lines can be assumed to be the plane $(X_W, Y_W, 0)$, (ii) the grid intersection at the lower left corner can be defined to be the origin $(0, 0, 0)$, and (iii) the Z_W axis is perpendicular to the planar grid.

In a similar manner, each CCS is generally located at the pinhole, Q. In many cases, the Z_C axis is aligned with the optical axis through the points Q–C, while X_C and Y_C are oriented to align with the camera sensor axes.

The final coordinate system defines the sensor coordinate system (SCS) with (X_S, Y_S) aligned with the row and column directions of the sensor plane, and Z_S perpendicular to the sensor plane. Located in the retinal or image plane of the pinhole camera, the SCS has units of pixels for (X_S, Y_S) to define sensor locations.

Assuming the rows and columns in the sensor plane are orthogonal, transformations among the WCS, CCS, and SCS are performed to develop the relationship between sensor plane coordinates (X_S, Y_S) of point **p** and a 3-D position of a point **P** at (X_W, Y_W, Z_W). The resulting scalar form can be written as

$$\begin{cases} X_S = C_x + f\lambda_x \dfrac{R_{11} X_W + R_{12} Y_W + R_{13} Z_W + t_x}{R_{31} X_W + R_{32} Y_W + R_{33} Z_W + t_z} \\ Y_S = C_y + f\lambda_y \dfrac{R_{21} X_W + R_{22} Y_W + R_{23} Z_W + t_y}{R_{31} X_W + R_{32} Y_W + R_{33} Z_W + t_z} \end{cases} \quad (1)$$

In equation (1), [**R**] is the rotation matrix and has three independent angles to define all components in [**R**].

Since equation (1) assumes an ideal, undistorted imaging system, improved accuracy in the measured positions can be obtained by including the effect of radial lens distortion at any point in the sensor plane. Such distortions are typically defined as the difference between undistorted and distorted sensor plane position using a cubic function of the radial distance from the image center. Defining the distortion vector, **d**, and the distorted image position by (X^d_S, Y^d_S), then the corrected position, (X_S, Y_S), can be written as follows:

$$\begin{aligned} (X_S, Y_S) &= (X^d_S - d_x, Y^d_S - d_y) \\ d_x &= \kappa[(X^d_S - C_x)^2 + (Y^d_S - C_y)^2]^{3/2} \bullet \cos(\zeta) \\ d_y &= \kappa[(X^d_S - C_x)^2 + (Y^d_S - C_y)^2]^{3/2} \bullet \sin(\zeta) \\ \mathbf{r}(\mathbf{p}) &= [|r| \bullet \cos(\zeta)]\mathbf{e}_x + [|r| \bullet \sin(\zeta)]\mathbf{e}_y \\ &= \text{vector location of point } \mathbf{p} \text{ relative} \\ &\quad \text{to image center } (C_x, C_y) \\ \zeta &= \text{counterclockwise polar angle from} \\ &\quad X_S \text{ axis, with origin at } (C_x, C_y) \\ &\quad \text{with } (-\pi < \zeta \leq \pi) \\ \kappa &= \text{ratial distortion coefficient} \\ (\mathbf{e}_x, \mathbf{e}_y) &= \text{unit vectors in } X_S \text{ and } Y_S \\ &\quad \text{directions, respectively} \end{aligned} \quad (2)$$

It is worth noting that all digitized intensity patterns are recorded at integer pixel locations. Since the distortion-corrected positions typically form a nonuniform grid at noninteger pixel positions, interpolation of the pixel values is required to make measurements with optimal, subpixel accuracy.

3.1 Camera calibration

As shown in equations (1 and 2), there are 11 independent parameters for each camera to be determined during the calibration process; six extrinsic parameters ($t_x, t_y, t_z, \theta_x, \theta_y$, and θ_z) and five intrinsic parameters ($C_x, C_y, f\lambda_x, f\lambda_y$, and k). A typical camera-calibration process uses a calibration target with known grid spacing. During the calibration process, the target is translated and/or rotated in three dimensions, with images acquired by both cameras at each position during the motion sequence. By locating at least three noncollinear points in each deformed image, the extrinsic parameters for each view (orientation and translation) are estimated and used to give initial locations for corresponding grid points throughout the view.

Defining an image-based objective function in the form

$$E = \sum_{i=1}^{M} \sum_{j=1}^{N} \{(X_{S\ \text{measured}}^{ij} - X_{S\ \text{model}}^{ij})^2 + (Y_{S\ \text{measured}}^{ij} - Y_{S\ \text{model}}^{ij})^2\} \quad (3)$$

where $X_{S\ \text{model}}$ and $Y_{S\ \text{model}}$ are given by equations (1) and (2), and the measured locations of features in the calibration standard ($X^d{}_S{}^{ij}$, $Y^d{}_S{}^{ij}$) are extracted from $j = 1, 2, \ldots, N$ pixel locations by analyzing all $i = 1, 2, \ldots, M$ images of the calibration standard. Though several approaches have been used to perform the nonlinear optimization of equation (3) for each camera, approaches such as steepest descent, Newton–Raphson, or a combined method such as Levenberg–Marquardt, have been used effectively.

3.2 Three-dimensional measurements

If one considers a specific point, **P**, with position (X_W, Y_W, Z_W), then equation (1) relates this position to the corresponding image location (X_S, Y_S) in the sensor plane for each camera; the image points are denoted by $\mathbf{p_1}$ and $\mathbf{p_2}$ for cameras 1 and 2, respectively, in Figure 4. When equation (1) is applied to both calibrated cameras, there are four equations with three unknowns, specifically the 3-D position of the point **P**. An optimal solution can be obtained through a least-square process to locate the best estimate for the position (X_W, Y_W, Z_W). By repeating this process for each position of interest, a full field of 3-D points can be determined at each load level. The difference between the initial position, $\mathbf{P_0}$, and the deformed position at time t, $\mathbf{P_0(t)}$, is the displacement vector with components $\{(U(\mathbf{P};t), V(\mathbf{P};t), W(\mathbf{P};t))\}$.

3.3 Image-based correlation for subset matching

To determine the corresponding image location (X_S, Y_S) in the sensor plane of each camera with utmost accuracy (this is essential for accurate 3-D measurement), the procedure used in this study is to perform optimal matching of image plane subsets; this is shown schematically in Figure 5. Defining an image of the object in one of the cameras (e.g., camera 1 in Figure 1) as the "reference" image, the matching process is typically completed by selecting a dense set of subregions (subsets) in the reference image and performing digital IC to identify the corresponding intensity pattern for each subset in (i) the other view of the undeformed object and (ii) both views of the deformed object (*see* **Chapter 30** for pattern recognition concepts).

Though a variety of matching metrics can be used to optimally locate the point, a normalized cross-correlation is oftentimes used. Since a nonlinear search process is required to locate the best match, most applications use an initial estimate for the parameters to initiate the search process. Typically, initial estimates for the rotation and translations of a subset are determined visually by locating the pixel positions for at least three, noncollinear, matching points and using these approximate matches to initiate the search process. As with the camera-calibration problem, a wide range of optimization methods have been used successfully.

3.4 Overall procedure for 3D vision-based measurements

The following procedure is typical for most practical situations:

1. Set up the stereovision system in preparation for the actual experiment. Cameras and lenses are selected to obtain high spatial resolution over the required visual field of view. For a stereovision system, the cameras and lenses typically are matched to simplify the setup and calibration procedures.
2. Arrange the cameras to meet the physical constraints in the application. In many cases, the

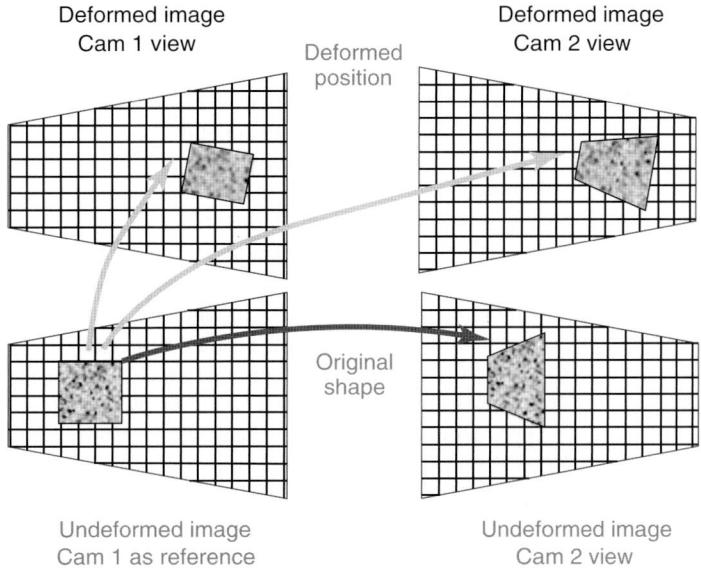

Figure 5. Schematic of image-matching process required to obtain the 3-D position of the object given corresponding images in different views. Camera 1 and camera 2 undeformed views are used to obtain initial 3-D shape of the object. Camera 1 and camera 2 deformed views are combined with camera 1 reference view to obtain the deformed 3-D position of the object.

position of the cameras is dictated by the layout (e.g., laboratory configuration).

3. Carry out a complete calibration of the stereovision system, using several motions of a calibration target. During this process, it is recommended that image acquisition be synchronized between both cameras.
4. Remove the calibration grid, install the specimen in the loading system, and verify that the loading is synchronized with the image acquisition process so that both cameras acquire images simultaneously during the experiment.
5. Postprocess all images to determine shape and deformations. After acquiring N pairs of images during the loading process, the analysis process requires that the images from one camera be designated as the *reference* images (e.g., camera 1).
6. To determine the initial shape (see Figure 5), subsets are selected in camera 1 and IC is performed to locate the matching position in the initial image for camera 2. Using the dense set of center-point locations obtained by the correlation process, the procedure described in Section 3.2 is used to obtain a dense set of 3-D points to define the initial object shape. For each loading state, cross-camera image matching and 3-D point generation also obtains the object shape in any configuration.
7. To determine the true 3-D position of a deformed body (see Figure 5, all four images are used as shown), subsets are selected in camera 1 and IC is performed (i) to locate the matching position in the initial image for camera 2, (ii) to locate the matching position in the deformed state as viewed by camera 1, and (iii) to locate the matching position in the same deformed image as viewed by camera 2. Using the dense set of center-point locations obtained by the correlation process, the procedure described in Section 3.2 is used to obtain a dense set of 3-D points for points in both the initial and deformed configurations. These are used to define the 3-D displacement vector for each point **P** after undergoing deformation. For each loading state, the process is repeated to obtain the displacement field for each deformed state, generating $\{(U(\mathbf{P};t), V(\mathbf{P};t), W(\mathbf{P};t))\}$.

4 APPLICATION OF THE TECHNIQUE FOR LARGE COMPONENT MEASUREMENTS

Figure 6 shows (i) a 305-mm-wide, 2.3-mm-thick, center-notched Al2024-T3 sheet with notch length/specimen width, $a/w = 1/3$, in the L–T orientation, as well as the tensile grip components, (ii) speckle pattern on the bottom half of the specimen as viewed by the left camera, and (iii) the stereocamera arrangement for acquiring 3-D shape and deformation data on the specimen as it is subjected to far-field tensile loading in a 446-kN tensile loading frame. References [38–40] provide details for similar experiments on a 0.61-m-wide panel; the configuration of cameras and specimen in these previous experiments is shown in Figure 2(a).

Canon lenses with a focal length of 28 mm were attached to Pulnix 9701 cameras with 8-bit intensity resolution and 768×484 spatial resolution, and used to image the specimen. As shown in Figure 6, the cameras were separated horizontally with a viewing angle difference of $\approx 53°$. To obtain similar spatial resolution in the horizontal and vertical directions, the camera bodies were rotated $90°$ so that 768 pixels in the CCD sensor are aligned with the longer vertical direction of the sheet. Magnification is ≈ 1.5 pixels/mm. Owing to specimen size, the global speckle patterns were applied using a self-adhesive vinyl sheet with a random pattern of appropriate size for the magnification printed on the visible surface.

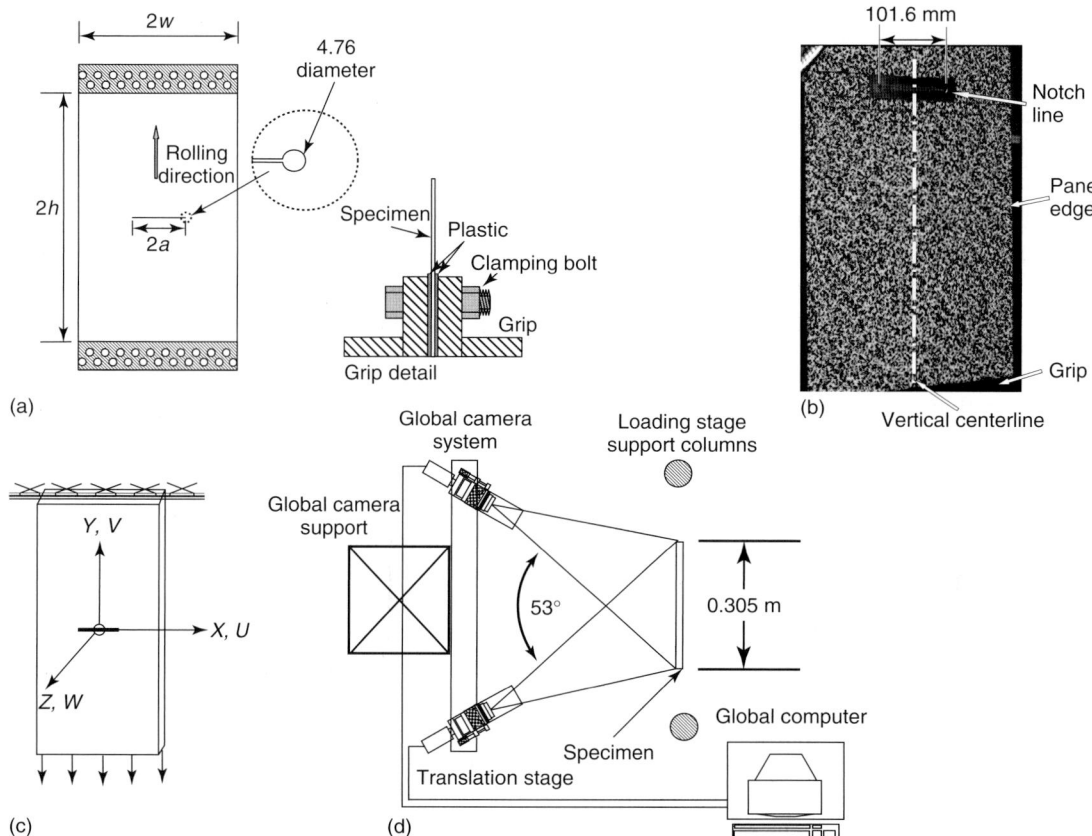

Figure 6. (a) Uniaxial tensile specimen with grips; in this work, $w = 0.1525$ m, $h = 0.330$ m, and $a/w = 1/3$; (b) speckle pattern on the bottom half of the specimen as viewed by the left camera; (c) specimen-based coordinate system to display output results; and (d) schematic of camera arrangement for the experiment. The cameras are mounted horizontal to the gantry system, at a distance of ≈ 1.2 m from specimen.

Camera calibration is performed using images of a calibration grid with known grid spacing. Owing to the physical size of the specimen field of view, a specially designed glass grid with grid spacing of 101.6 mm is used. To simplify the calibration process, the large glass grid is held stationary and the camera is moved approximately perpendicular to its sensor plane to acquire additional image(s). Using (i) the known spacing of the grid, (ii) the known movement(s) of the camera, and (iii) the location of the grid intersections, as extracted from the calibration images, nonlinear optimization is used to find the camera parameters that best describe the position and operating characteristics of the camera. The process is then repeated, without moving the grid, for the second camera. With this reduced motion process, measurement error had a standard deviation of ±0.01 mm (±0.015 pixels) with peak-to-peak values of ±0.05 mm (±0.075 pixels)[c]. After completing the calibration process, both CCS were transferred to the specimen shown in Figure 6(d).

The experiment was conducted using grip displacement control with a ramp rate of 2.83 µm s^{-1}. The experiment was paused each time the axial load increased by an increment of 2.22 kN, to acquire image data. This process was continued until the panel failed, at a load just beyond 116 kN. Load and ram displacement data were recorded every 5 s throughout the experiment.

4.1 Results

Figure 7 presents the applied load–vertical displacement data for the specimen, where the axial displacement is obtained by averaging the vision-based vertical displacement along the horizontal grip line located near the bottom of speckled image in Figure 6.

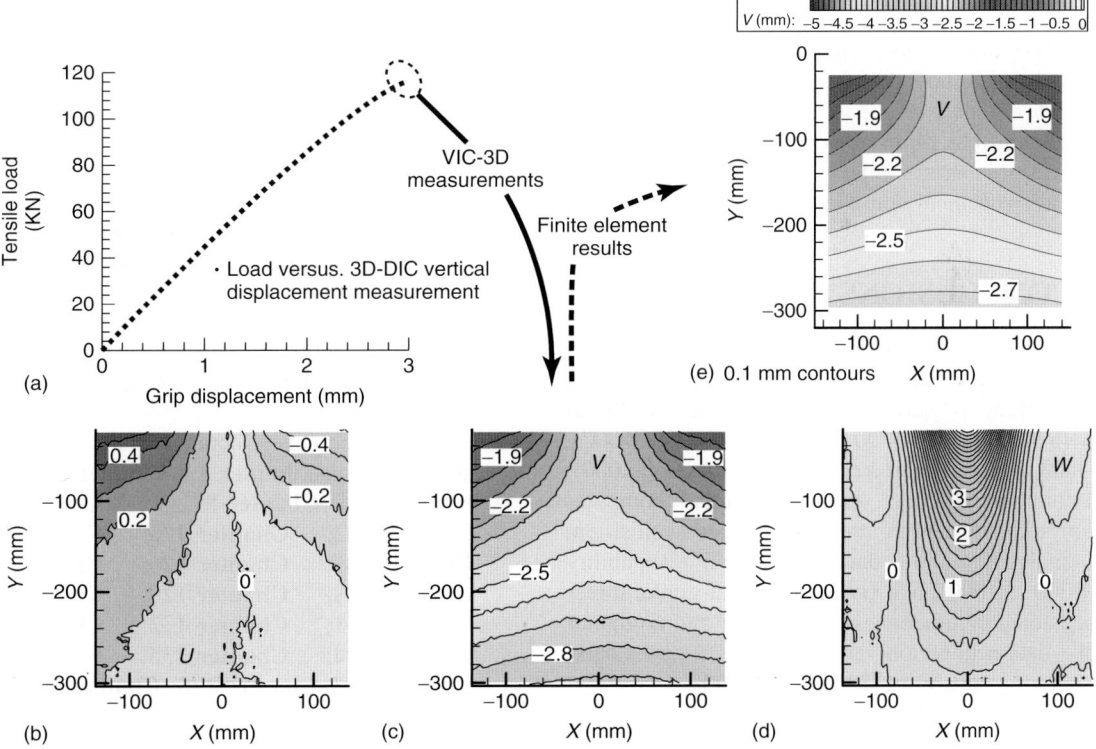

Figure 7. (a) Measured load–displacement data for specimen; (b–d) measured full-field horizontal (U), vertical (V), and out-of-plane (W) displacement fields at maximum tensile load (116 kN); (e) finite element prediction for vertical displacement field ($V(x, y)$) at maximum tensile load, assuming uniform displacement along lower boundary. All displacements are in millimeters. Measurement region is the lower half of specimen shown in Figure 6.

Inspection shows that the load–displacement data is slightly nonlinear for $P > 50$ kN, coinciding with the onset of increased out-of-plane displacements along the notch line.

Also shown in Figure 7 are the full-field plots of the displacement components in the lower half of the sheet, with (i) U, parallel to the notch line; (ii) V, perpendicular to the notch line; and (iii) W, perpendicular to the specimen surface.

As expected, the U-displacement field is antisymmetric relative to the Y axis, with both sides of the sheet moving approximately 0.40 mm toward the centerline. The V-displacement field is symmetric relative to the Y axis and nominally negative, since the bottom edge of the specimen undergoes applied downward displacement (see Figure 6(c)). As expected, the bottom edge of the specimen undergoes nearly constant displacement, confirming that the "rigid" grip was performing as expected. The W-displacement field is also symmetric relative to the Y axis, with maximum out-of-plane motion of 6 mm occurring along the centerline.

To show that the measured displacement fields are consistent with model predictions, Figure 7(e) shows a direct comparison between the V-displacement fields obtained by finite element analysis and stereovision measurements; similar comparisons are obtained for all displacement components.

5 CONCLUSIONS AND REMARKS

At the time of this article, both software and hardware are available commercially; Hardware costs include (i) scientific grade, 8–12 bit digital camera with 1000 × 1000 pixel array ($4000–6000); (ii) heavy-duty support structures for camera ($1000); (iii) commercial-grade, two-dimensional computer vision software and computer with data-acquisition, data-analysis, and data-presentation capability ($10 000–20 000); and (iv) commercial-grade, three-dimensional computer vision software and computer with data-acquisition, data-analysis, and data-presentation capability ($40 000–65 000).

The combination of modern digital image processing with stereovision imaging offers a unique opportunity to obtain quantitative deformation data on specimens that are undergoing a combination of in-plane and out-of-plane deformations. As long as the specimen remains within the depth of field for both cameras, the method is capable of measuring full-field 3-D shape, 3-D displacements, and surface strains on planar or curved specimen surfaces. The method has been used to make measurements in a wide range of applications including the following:

1. surface strains in excess of 100% on a highly ductile elastic–plastic metallic materials;
2. thermal strains on metallic specimens heated to 700 °C (see **Chapter 19** for thermal imaging methods, **Chapter 149**);
3. thermal and mechanical strains on specimens being imaged in a scanning electron microscope;
4. deformations on curved fuselage structure during both internal pressurization and external dynamic impact (see **Chapter 7**; **Chapter 8**; **Chapter 9**; **Chapter 10**; **Chapter 44**; **Chapter 46**; **Chapter 54**; **Chapter 56**; **Chapter 82**; **Chapter 83**; **Chapter 84**; **Chapter 86**; **Chapter 87**; **Chapter 88**; **Chapter 89**; **Chapter 91**; **Chapter 92**; **Chapter 93**; **Chapter 94**; **Chapter 95**; **Chapter 96**; **Chapter 97**; **Chapter 98**; **Chapter 101**; **Chapter 102**; **Chapter 105**; **Chapter 111**; **Chapter 116**.)
5. concrete, asphalt, polymers, and fiber-reinforced composites during mechanical loading (see **Chapter 17**; **Chapter 42**; **Chapter 111**; **Chapter 153**).
6. civil engineering structures including bridges and joints (see **Chapter 13**; **Chapter 84**; **Chapter 87**; **Chapter 89**; **Chapter 121**; **Chapter 122**; **Chapter 123**; **Chapter 124**; **Chapter 125**; **Chapter 126**; **Chapter 127**; **Chapter 128**; **Chapter 129**; **Chapter 130**; **Chapter 131**; **Chapter 132**; **Chapter 133**; **Chapter 134**; **Chapter 135**; **Chapter 136**; **Chapter 137**; **Chapter 138**; **Chapter 139**).

ACKNOWLEDGMENTS

The author wishes to thank Dr Jeffrey D. Helm for his tireless efforts in completing the bulk of the wide panel experimental work at NASA Langley Research Center. In addition, the technical and editorial support of Dr Hubert Schreier, Dr Stephen R. McNeill, and Dr Junhui Yan in completing this manuscript is deeply appreciated. The financial support of (i) Dr Charles E. Harris, Dr Robert S. Piascik, and Dr James C. Newman, Jr. at NASA Langley Research Center, (ii) Dr Oscar Dillon, Dr Clifford Astill, and Dr Albert S. Kobayashi, former NSF Solid Mechanics and Materials Program Directors, (iii) Dr Bruce LaMattina at the Army Research Office through several grants including ARO 50408-EG-DPS, (iv) Dr Kenneth Chong through NSF CMS-0201345, and (v) the University of South Carolina, Office of Research, are gratefully acknowledged. Finally, the support provided by Correlated Solutions, Inc., through the granting of access to their commercial software for our internal use is deeply appreciated. Through the unwavering technical and financial assistance of all these individuals and organizations, the true potential of image correlation methods is now being realized.

END NOTES

[a.] The original version was developed by Prof. K.S. Kim, Brown University, with emphasis on methods available for measurements on reduced-length scale specimens.
[b.] All image analysis in this work was performed using VIC-2D and VIC-3D software developed by Correlated Solutions Inc.; 120 Kaminer Way, Columbia, SC 29205, www.correlatedsolutions.com.
[c.] Modifications to this procedure have been developed and converted into commercial code. In the version developed by Correlated Solutions, Inc., the grid can be moved and rotated freely, acquiring images simultaneously by all stereo cameras and using equation (3) to efficiently convert grid images into calibration parameter sets for both cameras.

REFERENCES

[1] Doyle FJ. The historical development of analytical photogrammetry. *Photogrammetric Engineering* 1964 **XXX**:259–265.

[2] *The Photogrammetric Record* Gilbert Louis Hobrough. 2003 **18**(104):337–340.

[3] Rosenfeld A. From image analysis to computer vision: an annotated bibliography, 1955–1979. *Computer Vision and Image Understanding* 2001 **84**:298–324.

[4] Gabor D. Microscopy by reconstructed wavefronts. *Proceedings of the Royal Society* 1949 **A197**:454–487.

[5] Haines K, Hildebrand BP. Contour generation by wavefront construction. *Physics Letters* 1965 **21**: 422–423.

[6] Leith EN, Upatnieks J. Reconstructed wavefronts and communication theory. *Journal of the Optical Society of America* 1962 **25**:1123–1130.

[7] Dainty JC (ed). *Laser Speckle and Related Phenomena*, Springer-Verlag: Berlin, 1975.

[8] Archbold E, Burch JM, Ennos AE. Recording of in-plane surface displacements by double exposure speckle photography. *Optica Acta* 1970 **17**:883–898.

[9] Luxmoore AR, Amin FAA, Evans WT. In-plane strain measurement by speckle photography: a practical assessment of the use of Young's fringes. *Journal of Strain Analysis* 1974 **9**:26–34.

[10] Mallik S, Roblin ML. Speckle pattern interferometry applied to the study of phase objects. *Optics Communications* 1972 **6**:45–49.

[11] Leendertz JA, Butters JN. An image shearing speckle pattern interferometer for measuring bending moments. *Journal of Physics E: Scientific Instruments* 1973 **7**:1107–1110.

[12] Post D. White light Moiré interferometry. *Applied Optics* 1979 **24**:4163–4167.

[13] Post D, Ifju P, Han BT. *High Sensitivity Moiré*. Springer-Verlag, 1994.

[14] Vest CM. *Holographic Interferometry*. John Wiley & Sons, 1979.

[15] Peters WH, Ranson WF. Digital imaging techniques in experimental stress analysis. *Optical Engineering* 1982 **21**(3):427–431.

[16] Sutton MA, Wolters WJ, Peters WH, Ranson WF, McNeill SR. Determination of displacements using an improved digital correlation method. *Image and Vision Computing* 1983 **1**(3):133–139.

[17] Chu TC, Ranson WF, Sutton MA, Peters WH. Applications of digital-image-correlation techniques to experimental mechanics. *Experimental Mechanics* 1985 **25**(3):232–244.

[18] Peters WH, Zheng-Hui HE, Sutton MA, Ranson WF. Two-dimensional fluid velocity measurements by use of digital speckle correlation techniques. *Experimental Mechanics* 1984 **24**(2):117–121.

[19] Sutton MA, Chae TL, Turner JL, Bruck HA. Development of a computer vision methodology for the analysis of surface deformations in magnified images, ASTM STP 1094. In *MiCon 90: Advances in Video Technology for Microstructural Control*, Vander Voort GF (ed). ASTM: Conshohocken, PA, 1991, pp. 109–132.

[20] Sutton MA, Cheng MQ, Peters WH, Chao YJ, McNeill SR. Application of an optimized digital correlation method to planar deformation analysis. *Image and Vision Computing* 1986 **4**(3):143–150.

[21] Sutton MA, McNeill SR, Helm JD, Chao YJ. Advances in two-dimensional and three-dimensional computer vision. In *Photomechanics, Topics in Applied Physics*, Rastogi PK (ed). Springer Verlag: Berlin, 2000.

[22] Bruck HA, McNeill SR, Sutton MA, Peters WH. Digital image correlation using Newton-Raphson method of partial differential correction. *Experimental Mechanics* 1989 **29**(3):261–267.

[23] Sutton MA, Turner JL, Chae TL, Bruck HA. Full field representation of discretely sampled surface deformation for displacement and strain analysis. *Experimental Mechanics* 1991 **31**(2):168–177.

[24] Amstutz BE, Sutton MA, Dawicke DS. Experimental study of mixed mode I/II stable crack growth in thin 2024-T3 aluminum. *ASTM STP 1256 on Fatigue and Fracture*, ASTM: Conshohocken, PA, 1995; Vol. 26, pp. 256–273.

[25] Han G, Sutton MA, Chao YJ. A study of stationary crack tip deformation fields in thin sheets by computer vision. *Experimental Mechanics* 1994 **34**(2):751–761.

[26] Han G, Sutton MA, Chao YJ. A study of stable crack growth in thin SEC specimens of 304 stainless steel. *Engineering Fracture Mechanics* 1995 **52**(3):525–555.

[27] Liu J, Sutton MA, Lyons JS. Experimental characterization of crack tip deformations in Alloy 718 at High Temperatures. *ASME Journal of Engineering Materials and Technology* 1998 **20**(1):71–78.

[28] Lyons JS, Liu J, Sutton MA. High-temperature deformation measurements using digital-image correlation. *Experimental Mechanics* 1996 **36**(1):64–70.

[29] Chao YJ, Sutton MA. Measurement of strains in a paper tensile specimen using computer vision and digital image correlation – Part 1: data acquisition and image analysis system. *Tappi Journal* 1988 **70**(3):173–175.

[30] Chao YJ, Sutton MA. Measurement of strains in a paper tensile specimen using computer vision and digital image correlation – Part 2: tensile specimen test. *Tappi Journal* 1988 **70**(4):153–156.

[31] Schreier HW, Braasch J, Sutton MA. Systematic errors in digital image correlation caused by intensity interpolation. *Optical Engineering* 2000 **39**(11):2915–2921.

[32] Schreier HW, Sutton MA. Systematic errors in digital image correlation due to undermatched subset shape functions. *Experimental Mechanics* 2002 **42**(3):303–310.

[33] Schreier HW, Garcia D, Sutton MA. Advances in light microscope stereo vision. *Experimental Mechanics* 2004 **44**(3):278–288.

[34] Luo PF, Chao YJ, Sutton MA. Computer vision methods for surface deformation measurements in fracture mechanics. *ASME-AMD Novel Experimental Methods in Fracture* 1993 **176**:123–133.

[35] Luo PF, Chao YJ, Sutton MA, Peters WH. Accurate measurement of three-dimensional deformations in deformable and rigid bodies using computer vision. *Experimental Mechanics* 1993 **33**(2):123–132.

[36] Luo PF, Chao YJ, Sutton MA. Application of stereo vision to three-dimensional deformation analyses in fracture experiments. *Optical Engineering* 1994 **33**(3):981–990.

[37] Helm JD, McNeill SR, Sutton MA. Improved 3-D image correlation for surface displacement measurement. *Optical Engineering* 1996 **35**(7):1911–1920.

[38] Helm JD, Sutton MA, Dawicke DS, Hanna G. Three-dimensional computer vision applications for

aircraft fuselage materials and structures, *Proceedings of 1st Joint DoD/FAA/NASA Conference on Aging Aircraft*, Ogden, Utah, 1997; Vol. 1, pp. 1327–1341.

[39] Helm JD, Sutton MA, McNeill SR. Deformations in wide, center-notched, thin panels, part I: three-dimensional shape and deformation measurements by computer vision. *Optical Engineering* 2003 **42**(5):1293–1305.

[40] Helm JD, Sutton MA, McNeill SR. Deformations in wide, center-notched, thin panels, part II: Finite element analysis and comparison to experimental measurements. *Optical Engineering* 2003 **42**(5):1306–1320.

Chapter 66
Global Navigation Satellite Systems (GNSSs) for Monitoring Long Suspension Bridges

Xiaolin Meng[1] and Wei Huang[2]

[1] Institute of Engineering Surveying and Space Geodesy, University of Nottingham, Nottingham, UK
[2] Intelligent Transportation Systems Research Centre, Southeast University, Nanjing, China

1 A Brief Introduction to the Global Positioning System	1169
2 GPS for Structural Health Monitoring (SHM)	1172
3 Implementation of GNSS Centered Sensor Systems for SHM	1176
4 Future Vision	1183
References	1184

1 A BRIEF INTRODUCTION TO THE GLOBAL POSITIONING SYSTEM

1.1 GPS constellation

The full term of the well-known acronym GPS is NAVSTAR global positioning system, where NAVSTAR stands for NAvigation System with Time And Ranging [1]. GPS is a satellite-based navigation and positioning system that was designed in the early 1970s by the US military to allow soldiers to autonomously and continuously determine their position within 10–20 m of accuracy, at any point on the earth's surface and under any weather conditions. Providing precise timing information is another important function of GPS. GPS was originally utilized even for the military operations of US military forces; the last two decades have seen a rapid expansion of civilian GPS user groups. The current GPS configuration consists of three segments: the space segment—a constellation of 24 (nominal) satellites distributed in six orbital planes of 55° inclination to the equator with an altitude of 20 200 km above the earth's surface as shown in Figure 1; the control segment—comprises a master control station, worldwide distributed monitor stations, and ground control stations; and the user segment—anyone who receives and uses GPS signal with any type of GPS-enabled devices. More detailed information about GPS segments can be found in [1, 2].

1.2 GPS measurements

Each GPS satellite transmits at least two carrier signals in an L-band, such as L1 and L2 carriers, and more signals are available for the modernized

Encyclopedia of Structural Health Monitoring. Edited by Christian Boller, Fu-Kuo Chang and Yozo Fujino © 2009 John Wiley & Sons, Ltd. ISBN: 978-0-470-05822-0.

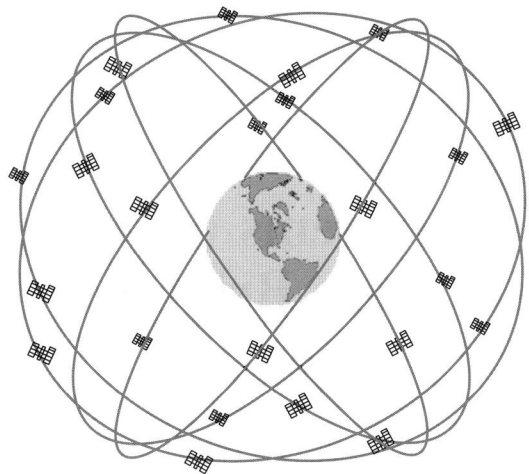

Figure 1. Current GPS constellation (http://www.faa.gov/about/office_org/headquarters_offices/ato/service_units/techops/navservices/gnss/).

GPS satellites. The L1 and L2 carriers are modulated by ranging codes and navigation information such as orbital parameters. The carriers together with ranging codes are used mainly to determine the distance from the antenna of a user's receiver to the GPS satellites in view [3]. These uncorrected distances are called *code* or *carrier pseudoranges* since they include the true geometric distance between the receiver antenna and the tracked satellites plus small range corrections to account for the orbital and clock errors, signal delays caused by the atmosphere, and multipath due to the signal reflection by the surroundings in the vicinity of the GPS antenna [1]. For determining a three-dimensional (3D) location of a GPS antenna, simultaneous tracking to three GPS satellites is adequate. However, since a GPS receiver is normally equipped with a much cheaper clock, compared with the atomic ones used by the GPS satellites, tracking to at least one more satellite is required to solve the clock bias of the GPS receiver for achieving a higher positioning accuracy.

1.3 GPS relative positioning

A prerequisite for achieving GPS positioning accuracy of a few meters (code pseudorange) to centimeters or even subcentimeters (carrier phase) is the effective cancelation of various ranging errors. The number of satellites tracked by a GPS receiver and their spatial distribution are also major affecting factors to the achievable positioning accuracy. Generally, the GPS relative positioning technique involves a pair of GPS receivers simultaneously tracking at least four well-distributed identical satellites for solving the position of an unknown stationary GPS antenna and at least five satellites for a moving antenna. For canceling out spatially correlated GPS ranging errors, one GPS receiver must be installed on a precisely measured benchmark called a *reference station* to help solve the unknown coordinates of another receiver called a *rover*, either in a stationary mode or a continuous motion mode. When relative positioning is utilized in surveying, differencing the GPS ranging measurements made to two satellites simultaneously by a reference receiver and a rover receiver could eliminate both the satellite clock and the receiver clock biases. Carrier and/or code pseudorange measurements can be used in GPS relative positioning and the positioning solutions in the form of 3D coordinate time series can be obtained either in real time or through postprocessing of collected GPS measurements.

1.4 Achievable positioning accuracy and major inhibitors to GPS positioning accuracy

Both code and carrier phase pseudorange measurements are "polluted" by errors that originate from GPS receivers, GPS satellites, signal propagation medium, and observation environment [1, 3]. In the data processing stage, incorrect processing models adopted and/or bugs in the software packages also cause positioning errors. There are many studies in this area on how to effectively reduce, mitigate, and eventually eliminate GPS error sources through mathematical modeling, software development, and digital signal-processing approaches, and hardware design and configuration. Some error sources, which are observation environment oriented and change from place to place, prove very difficult to be modeled or mitigated. For instance, pervasive multipath effect, which is a major error source for both the carrier phase and the code-pseudorange-based GPS positioning, makes the effort for achieving highly accurate GPS positioning elusive. Most GPS antenna will not be able to detect which signals are directly transmitted by the satellites and which are reflected from the objects surrounding a GPS antenna. As a result of the complexity of objects surrounding an antenna, there is virtually no versatile model or physical antenna design to effectively reduce multipath, which introduces a maximum error of up to about 5 cm for carrier-phase-based positioning, a quarter of its L1 wavelength, and several meters for code-pseudorange-based GPS positioning.

1.5 Integer ambiguity resolution for high accuracy positioning

Centimeter or even subcentimeter positioning accuracy can be achieved through using carrier-phase-based GPS relative positioning under the condition of successful resolution of the initial integer ambiguities. When carrier phase measurements are employed in positioning, only a fraction of the wavelength of the carrier phase is recorded by a GPS receiver, but the whole number of carrier wavelengths or cycles between a GPS antenna and the tracked satellites remains unknown. Reliable and quick resolution of the unknown number of cycles from a tracked satellite to a GPS antenna is vital in using carrier phase measurements for subcentimeter positioning. Until the early 1990s, GPS was only used to monitor static deformation of structures with low dynamics, for example, dams and low buildings. The least squares adjustment, Kalman filtering, statistical process, and other search techniques can resolve this ambiguity and determine the most probable solution [4, 5]. Furthermore, the invention of integer ambiguity on-the-fly (OTF) algorithms by different researchers around the world in the beginning of 1990s made the real-time kinematic global positioning system (RTK GPS) a viable tool for monitoring more flexible structures such as long suspension bridges, high-rise buildings, and slender TV or communications towers [6, 7], or for tracking high-speed moving objects such as low earth orbit (LEO) satellites, because of its high accuracy, productivity, flexibility, and simplicity in data processing (conducted by the embedded receiver firmware). The positioning solutions produced by the internal processor of a GPS receiver can be streamed to a control center via cable connection or through wireless communications for further analysis and visualization of resulting solutions. However, the collected raw carrier phase measurements, which are recorded with memory cards inside the GPS receivers, can be decoded as American Standard Code for Information Interchange (ASCII) data files in a receiver independent exchange (RINEX) format. These data files can be postprocessed in a kinematic mode by using the OTF approach with any commercial or household GPS postprocessing software to output each epoch positioning solution. More accurate positioning results can be obtained because of the possibility to edit and clean GPS measurements.

1.6 Communication links

Radio modems provide wireless communications between GPS reference receivers and rover receivers for carrying out real-time relative positioning. When a reference GPS receiver broadcasts the corrections via its radio modem transmitter, an unlimited number of rover GPS receivers can pick up these data via their own radio modems. The transmission range depends on the power of the radio modem transmitter and also on the terrain and the radio antenna setup. There are many different radio modems in the market and ultrahigh frequency (UHF), very high frequency (VHF), and spread spectrum are the most commonly used in RTK GPS positioning. Government authorization may be required for using certain types of radio modems. In some countries, there are bands that are allocated for public use without the need for any special authorization. For instance, the 900-MHz band in the United States and 2.4 GHz in most European countries are allowed for spread spectrum communications without any special authorization (but there are limitations on the amount of power that one can use to transmit signals, for instance, in the United Kingdom, only 0.5-W carrier power is allowed). More recently, some GPS manufacturers adopted cellular communication technology or the third-generation (3G) wideband digital networks, such as global system for mobile communications (GSM) and general packet radio service (GPRS), as an alternative GPS communication link. Since subscribers only pay for the data amount that they have actually transmitted or received, the use of GPRS technology is more flexible and much cheaper compared with a GSM communication link. Dedicated local area network (LAN) can also be utilized to transmit GPS corrections from reference stations to the rovers over the Internet. Satellite-based wireless communication approach is becoming popular and has already been employed to transmit corrections to the rovers in the area where there is no GSM/GPRS coverage and the use of radio modems is constrained by the local terrain and transmission power limitation.

2 GPS FOR STRUCTURAL HEALTH MONITORING (SHM)

2.1 Brief history of GPS for SHM

According to [8], the process of implementing a damage detection strategy for aerospace, civil, and mechanical engineering infrastructures is referred to as *structural health monitoring* (SHM) and a damage is defined as changes to the material and/or geometric properties of these systems, including changes to the boundary conditions and system connectivity, which adversely affect the performance of the systems. The SHM process involves the observation of a system over time, using periodically sampled dynamic response measurements from an array of sensors, the extraction of damage-sensitive features from these measurements, and the statistical analysis of these features to determine the current state of system health. The output of this process is periodically updated information regarding the ability of the structure to perform its intended function in light of the inevitable aging and degradation resulting from operational environments. An SHM system should have the capacity to provide, in near real time, reliable information regarding the integrity of the structure after earthquakes or blast loading.

Recent advances in GPS positioning, computer science, telecommunications technologies, and advanced digital signal processing have made GPS a much more robust, reliable, convenient, accurate, and cost-effective tool for the deformation monitoring of natural and artificial structures. To date, GPS is widely used to monitor volcano eruptions [9, 10], crustal movements [11, 12], vertical land movements [13], landslides [14], earth structures [15], dams [11], buildings [16–21], and bridges [22–29]. In general, these applications can be roughly categorized into three levels, i.e. large, medium, and local scales, according to the separations of the GPS stations. A variety of regional continuous global positioning system (CGPS) networks, such as EUREF Permanent Network (as of March 04, 2007, it has 199 permanent GNSS tracking stations, which cover the European continent); the OS Net of the Ordnance Survey in the United Kingdom, which consists of more than 100 permanent stations; and the Geographical Survey Institute's GPS Network, which comprises 1224 GPS real-time stations in Japan (www.euref.eu; www.ordnancesurvey.co.uk; www.gsi.go.jp), are typical examples of the first category applications. These networks play a vital role in monitoring crustal movement, which is crucial for the evaluation process of seismic and volcanic activity as well as the extraction of geodynamic information. On a medium scale, GPS-based volcano eruption monitoring presents a good example. The last level of GPS-based monitoring applications takes place on a variety of natural or artificial structural deformation and deflection monitoring in the local scale, which is the main scope discussed in this article (*see* **Chapter 119**).

2.2 Case studies of GPS for SHM of long suspension bridges

Under different loading conditions, a suspension bridge generally experiences two distinct types of deformations, i.e., the long-term movement caused by the foundation settlement, bridge-deck creep, and stress relaxation; and the short-term motion of the bridge, or bridge deflection, such as those activated by wind, tidal current, earthquake, or traffic [27, 30]. The latter deformation is recoverable in most cases and the bridge will resume to its original status from the deformation with the release of external forces.

When GPS is utilized to monitor bridges, it has the capacity to detect two different kinds of deformations simultaneously [25, 31]. Analyzing the response of a long bridge to the short-term irregular loading is much more important in terms of risk level of major damage and its significance for the improvement of design code. However, it is more difficult to be measured compared with the identification of long-term foundation settlement, concrete creep, loss of prestress, and thermal expansion or contraction. In monitoring foundation displacement, averaging GPS time series of a few hours or more can be used, producing positioning accuracy of a few millimeters, since error sources such as multipath (an unwanted indirect signal reflected by the surroundings before reaching a GPS antenna), which characterizes as a low period of movement, could be effectively removed. To measure the deformations occurring over a time interval of a few seconds, there is a much shorter or even no averaging time available for mitigating errors; the time interval may not even be long enough for resolving the critical initial integer ambiguities. This significantly restricts the usage of GPS positioning in monitoring short-term movements of more flexible structures. Also, to measure short-term effects, sampling rates of GPS sensors must be significantly increased, which will incur heavier onboard data processing and communication overloads if RTK GPS positioning is required.

In the last decade or so, with rapid advances in GPS technology, digital signal processing (DSP), telecommunications, and computing science, as well as with the continuous efforts conducted by the researchers around the world, GPS positioning is gradually being accepted by the civil engineering community as a viable monitoring tool and has started to play an important role in SHM. This is mainly due to the advent of OTF integer ambiguity resolution technology. As discussed in the previous section of this article, solving the unknown number of cycles from a tracked satellite to the GPS antenna is vital in using carrier phase measurements for subcentimeter positioning, which is a prerequisite for GNSS-based SHM. The significant increase in the GPS sampling rate from 1 Hz to rates higher than 50 Hz currently is another reason for the acceptance of GPS-based SHM of long suspension bridges.

Structural aging and degradation resulting from changes in the materials and operational environments such as significantly increased traffic volume and weight, as well as the boundary conditions and system connectivity, and flaws in design are the main reasons for many recent bridge closures and failures around the world. Transport agencies and researchers have strong interests in finding appropriate sensor systems and computational models to implement on-line monitoring and diagnosing systems to detect and locate the changes to the materials and/or geometric properties of bridges. Of various monitoring tasks, measuring short-term dynamic behavior of such structures is of particular interest.

The Tsing Ma Bridge in Hong Kong, China, is the sixth longest suspension bridge in the world and it is the longest single-span suspension bridge that carries both road and rail traffic [32]. The Tsing Ma Bridge and other two cable-stayed linking bridges are perhaps the only bridges around the world that are equipped with the most comprehensive and sophisticated sensor systems, forming a part of six whole SHM components. The six integrated modules include the sensory system, the data acquisition and transmission system, the data processing and control system, the structural health evaluation system, the structural health data management system, and the inspection and maintenance system.

The main objective of the Tsing Ma SHM system is to monitor the loading and structural parameters so that the performance of the bridge under current and future loading conditions can be evaluated and predicted. Such evaluated results will facilitate the planning of bridge inspection activities, and help the determination of not only the causes of damages, but also the extent of remedial works, once the damage is identified [33, 34].

There are nine different types of sensors to form the sensory systems, which include anemometers, thermometers, servo-type accelerometers, dynamic weigh-in-motion sensors, GPS receivers, leveling sensing stations, displacement transducers, weldable strain gauges and monitoring CCTV cameras, making a total of 848 sensors on the Tsing Ma and two other adjacent cable-stayed bridges [33]. GPS technology was introduced because of its improved measurement efficiency and accuracy [29]. Of a total of 29 high-grade geodetic-type dual-frequency GPS receivers (using both L-band carries), two GPS receivers were set up as the reference stations. On the Tsing Ma Bridge, 14 GPS receivers were permanently installed as the monitoring stations at the critical locations of the bridge deck where maximum displacements are expected: four pairs on the bridge deck itself, one pair on the each side of the supporting towers, and one pair on the cable at midspan. The remaining 13 GPS receivers were installed onto two cable-stayed bridges near the Tsing Ma Bridge to form a large monitoring system in the region. Both chokering and lightweight antennas were used for the reference stations and monitoring stations, respectively. Most GPS stations were installed close to the leveling and sensing stations or to the accelerometers for validating and comparing the data sets from different monitoring sensor systems. GPS data were collected at a sampling rate of 10 Hz continuously, and optical fibers were used to transmit the positioning results from each monitoring station to a processing center for further analysis and visualization purposes.

The Akashi Kaikyo Bridge in Japan is the world's longest road suspension bridge. It was constructed using a newly developed wind- and seismic-resistant design and it is necessary to verify the design assumptions and constants during strong winds and severe earthquake loadings [25]. Navigation is extremely difficult and there have been many shipping accidents in the Akashi Strait because of the hostile environmental conditions. To keep the world's longest suspension bridge in a safe operating condition and also to study the dynamic response to various loadings, an experimental trial was carried out by using dual-frequency GPS receivers in 1999 to monitor bridge deformation in a real-time mode. The first results from the monitoring of the Akashi Kaikyo Bridge were presented in [25]. GPS receivers were set up at three locations, constituting a reference station at the anchorage on the Kobe side of the bridge and two monitoring stations on the top of a supporting tower and at midspan, to monitor three-dimensional displacements. Wind loading and temperature data were also collected simultaneously. The displacement time series of six months were analyzed together with recorded time series of wind and temperature loadings. The instantaneous gradients of displacement and temperature were calculated. The regression functions established could be used to identify future abnormal deflections after attacks of strong wind loading or earthquakes. Data from other sensors are also available for this bridge, but for a bridge of this size, three GPS receivers seem inadequate for even picking up global displacements.

China has many of the world's longest suspension bridges, which were constructed in recent years as a result of the high demand for land transport infrastructure. Structural safety has already caused great concern and many SHM systems have been installed in many new bridges. The most commonly used sensors are anemometers, accelerometers, temperature sensors, and strain gauges. Only a very few long suspension bridges in the mainland China such as the Jiangying Bridge (main span of 1385 m) and the Runyang Bridge (main span of 1490 m), both over the Yangtze River were installed with permanent GPS monitoring systems. Like the Akashi Kaikyo Bridge, only a limited number of monitoring GPS stations have been installed at the critical locations on the bridge deck and supporting towers. For instance, there are nine permanently installed dual-frequency GPS receivers forming an important part of the monitoring system for the Jiangyin Bridge. One GPS receiver is used as the reference station and other eight receivers are placed atop the two bridge towers, and at the 1/4, 1/2, and 3/4 points of the bridge span [35]. The sampling rate of the GPS receivers is set to 20 Hz.

In other countries, GPS has been used in a number of bridge monitoring studies, mostly carried out by GPS receivers temporarily installed on the bridge decks or supporting towers to collect sample data sets. For example, GPS receivers were employed to monitor the displacement of France's Normandy Suspension Bridge under controlled traffic loading to verify whether the performance of the bridge was consistent with the design specifications [36]; the Danish Road Directorate used GPS to determine the as-built geometry and assess temporal deformations of Denmark's Storebaelt Bridge and other bridges [37]; and the Applied Research Laboratory (ARL) of the University of Texas at Austin in the United States has conducted a series of research on GPS-based structural deformation monitoring since the beginning of 1990s and published many valuable articles on relevant topics [26, 38–43]. The research conducted by the ARL is perhaps the earliest practice in GPS-based bridge deformation monitoring.

The results of a three-day experiment, in which GPS technology was used to measure the motion of a cable-stayed suspension bridge over the Mississippi River near New Orleans, were presented in [26]. The goals of this experiment were to verify the feasibility of using GPS to achieve centimeter-level accuracy in bridge deflection monitoring, and to evaluate the practicability and limitations of the whole monitoring system. Twelve Trimble 4000 SST GPS receivers were used in the test. Two receivers were located on the shore at known reference sites about 10 m apart, approximately 1.6 km away from the bridge. This approach made it possible to distinguish real bridge motion from GPS-induced biases or errors through a thorough data processing. The remaining 10 GPS receivers were situated at the critical bridge sites where minimum and maximum bridge displacements were expected. A 10-s sampling rate was chosen because the expected bridge motion during a 10-s period was considered to be negligible, and this allowed all the data to be stored on receivers for the whole monitoring period. The result showed a maximum relative vertical displacement of about 7 cm caused by a change in ambient temperature during the experiment. The results revealed a strong relationship between the vertical position of the bridge roadbed and the change in ambient temperature. Multipath also made significant error contribution to the positioning solution and caused problems in the displacement analysis. It was clear that the cables and metal surroundings, the bridge sites, coupled by the passing vehicles, had created a severe multipath environment and additional multipath analysis and mitigation techniques are apparently essential for achieving high accuracy positioning. Other GPS-based bridge deformation trials were also carried out by the ARL team with the attempt to monitor more dynamic bridge performance induced by traffic and wind loadings with 10-Hz GPS receivers, identify the multipath signature and quantify the measurement quality, compare GPS-measured displacements with those of servo accelerometers, and assess the reliability and performance of system components [42].

Preliminary results from ARL and other research teams around the world proved that GPS is a viable tool for detecting both transient deformations of flexible bridge structures and long-period movements. It was also recognized that the big challenge with GPS-based bridge deformation monitoring was how to effectively eliminate or reduce the impact of multipath. Other factors restraining large-scale GPS applications to structural deformation monitoring are the high price of dual-frequency GPS receivers, the large amount of data generated during data collection when a high sampling rate is used, and the existing gaps between the demanded high precision with what can be offered by current GPS positioning technology of centimeter accuracy.

Researchers at the IESSG in the University of Nottingham have conducted a series of trials using state-of-the-art dual-frequency GPS receivers to monitor the movements of a number of long suspension bridges and other structures in the United Kingdom [22, 28, 44]. In particular, several trials were conducted on the Humber Bridge under both a controlled loading environment and using ambient vibration to monitor the bridge deflection [22, 45]. The viability of GPS technology to monitor the displacements of the Humber Bridge subjected to known loading conditions was verified. The initial research of the IESSG was focused on the viability of GPS technology to monitor both long-term bridge settlements and dynamic response to external loading, and the latest work is on the development of a systematic approach for the sensor integration, field test arrangement, data collection, algorithm, and software development, quality control, advanced signal processing, and multipath mitigation techniques, data analysis and dynamic response identification, and result visualization [46–53] (*see* **Chapter 123**; **Chapter 124**; **Chapter 125**; **Chapter 126** and **Chapter 133**).

2.3 Advantages and limitations of GPS for bridge deformation monitoring

Very few conventional bridge monitoring sensor systems have versatile capacities like GPS. In summary, the advantages in using GPS for SHM of long bridges are as follows [27, 38–41, 54, 55]:

- monitoring both static and dynamic movements;
- fully automatic, real-time, all-weather, and long-life data acquisition;
- 3D absolute displacements linked to a global datum WGS84 (World Geodetic System of 1984 —very important for foundation monitoring);
- accurate velocity and acceleration extractions from 3D displacements, which can be validated by the measurements of other sensors such as an accelerometer;
- precisely synchronized positioning solutions for all GPS monitoring sites;

- a precise time source for time stamping/synchronizing other measurements/sensors (accelerometers, weather station, strain gauge, etc.);
- continuously increasing sampling rate (50 Hz for dual-frequency GPS receivers and up to 100 Hz or higher for single frequency GPS receivers);
- no long-term positioning accuracy degradation like accelerometers;
- a data fusion platform for integrating GPS with other sensors for a more robust monitoring system;
- continuous improvement in space segments, regional augmentation systems, ground tracking facilities/algorithms, and end-user hardware (futuristic GGG—GPS, GLONASS, GALILEO—receivers will be able to track more than 80 satellites);
- in the near future, a single GGG receiver based real-time solution of millimeter positioning accuracy at a sampling rate for at least 100 Hz will be possible.

GPS also has some inherent disadvantages, which are the potential inhibitors to a wide acceptance of GPS for SHM. The main disadvantages are as follows [27, 38–41, 54, 55]:

- The high price of GPS hardware and software is the major inhibitor for a wide application of GPS for SHM; a much cheaper single frequency GPS receiver cannot fix its initial integer ambiguity, which prohibits centimeter positioning accuracy.
- Line of sight to at least five well-distributed satellites poses a current challenge for GPS owing to signal obstruction by surroundings.
- The positioning accuracy of the vertical component is worse than the horizontal one and the ratio of the positioning accuracies of two horizontal components changes with geographical locations.
- Unmodeled residual tropospheric delays, such as those induced by the height differences, might cause a few centimeter errors.
- At the moment, high level of multipath caused by surrounding reflection cannot be effectively mitigated.
- The relatively slow sampling rate is not adequate for monitoring short span bridges or the higher frequency band of a long span bridge.
- Data processing and interrogation can be complicated.
- There has not been a thorough research into the integration of GPS measurements with a computational model, nor has there been a successful demonstrator to justify the high investment in the establishment of the sensor system.

3 IMPLEMENTATION OF GNSS CENTERED SENSOR SYSTEMS FOR SHM

3.1 Reference and monitoring GPS stations

For the establishment of a permanent GPS-based bridge monitoring system, monuments need to be installed for mounting both the reference and monitoring antennas as shown in Figure 2(b, c). Different GPS antennas have different weights, physical dimensions and multipath mitigation capacity, and can

(a) (b) (c)

Figure 2. (a) A chokering antenna; (b) a reference station set up on the Jingyin Bridge in China; (c) a monitoring station atop the Akaishi Kaikyo Bridge in Japan [photos courtesy of Leica Geosystems].

be used for different GPS sites. For instance, a chokering antenna covered with a dome could be used for the reference station since its location is usually lower than the bridge deck. High multipath signature caused by the supporting towers, dense cables, passing vehicles, and buildings is evident and can be mitigated by a chokering antenna. A chokering antenna consists of a series of cylinders that stop reflected signals from the ground and from nearby reflective surfaces as shown in Figure 2(a). To significantly reduce multipath induced by the reference stations and for designing a more secure and robust monitoring system, a reference configuration comprising multiple GPS receivers is highly recommended [27]. An inexpensive and lightweight patch antenna as shown in Figure 2(c) can be used for the monitoring sites on the bridge sites. This antenna has a metallic ground plate, which can effectively stop the reflected signals coming underneath the antenna but will not be able to reject multipath from surrounding objects such as cables, towers, and passing vehicles. This is why the high monuments for monitoring sites are preferred as shown in Figure 2(c).

For temporary bridge deformation monitoring practices, antennas of the reference stations can be mounted on tripods, which are centered to the precisely measured ground marks as shown Figure 3(a) (the Forth Road Bridge). In this monitoring, two adjacent reference stations about 6 m apart were used to reduce multipath caused by the reference stations through a data processing procedure. The other reason for this reference station setup is to reduce the risk of potential hardware failure caused by either of the reference stations. The antennas of the bridge monitoring sites can be locked to the bridge handrail using specially designed clamps. A direct access to power was made available by the bridge authority. Figure 3(b) shows a typical example of a monitoring station setup on the Forth Road Bridge [51].

At the moment, the sampling rate of the operational or experimental bridge monitoring systems is set to 10 Hz or a maximum of 20 Hz. At such a sampling rate, it is estimated that the average data volume in a binary format is $7\,\text{Mb}\,\text{h}^{-1}$. For a permanent monitoring system, the GPS receivers at the monitoring sites will conduct real-time coordinate computation and then stream these continuous time series to the control center for a further analysis. These receivers might also be able to log the raw measurements into onboard flash memory cards and send the data sets to the center on a regular basis in case of some important event analysis, for instance, an earthquake. For the experimental tests, raw data is recorded into the flash memory cards and up to 4-GB cards are currently available, which means that raw data sets of more than 574 h (equivalent to an observation of 23 days) can be continuously recorded. However, it might cause computation problem with a postprocessing GPS processing software package due to the huge file size after uncompressing the binary data. It is recommended that a data set of 12 h should be used since this data size can be handled by any commercial GPS software and a standard desktop PC. The use of two flash cards at one site could assure nearly continuous measurements at each site.

(a)   (b)

Figure 3. Reference stations setup and a monitoring station on the Forth Road Bridge [51].

3.2 GPS sensor locations on the bridge

As previously discussed in this article, a widely adopted sensor configuration for most bridge monitoring systems is to install the GPS sensors onto both sides of the supporting towers, at the 1/4, 1/2, and 3/4 points of the bridge span, with the attempt to extract structural dynamics [29, 51, 56]. With a sensor configuration as described in Figure 4, special events or structural response can be easily identified. Figure 5 shows the vertical deflection of a 100-t lorry that passed each of the monitoring sites from south to north on the Forth Road Bridge. Whether this configuration is optimal and adequate for detecting global bridge deformation and deflection is still an unanswered question. Some preliminary research was conducted using a short span suspension bridge as a case study for the optimal determination of the number and locations of GPS sensor [56].

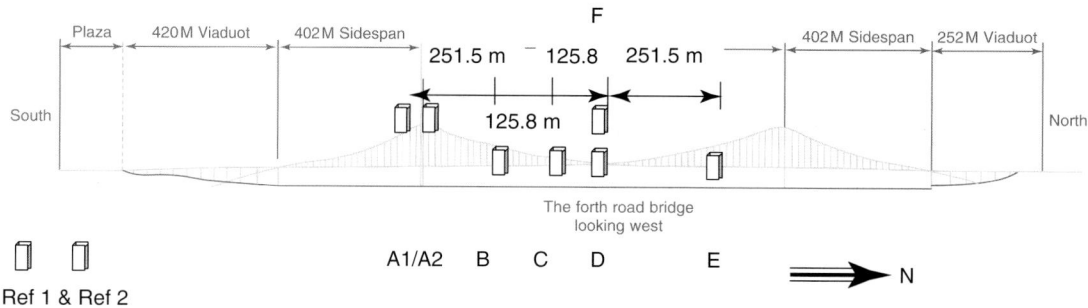

Figure 4. GPS antenna layout for monitoring the Forth Road Bridge [51].

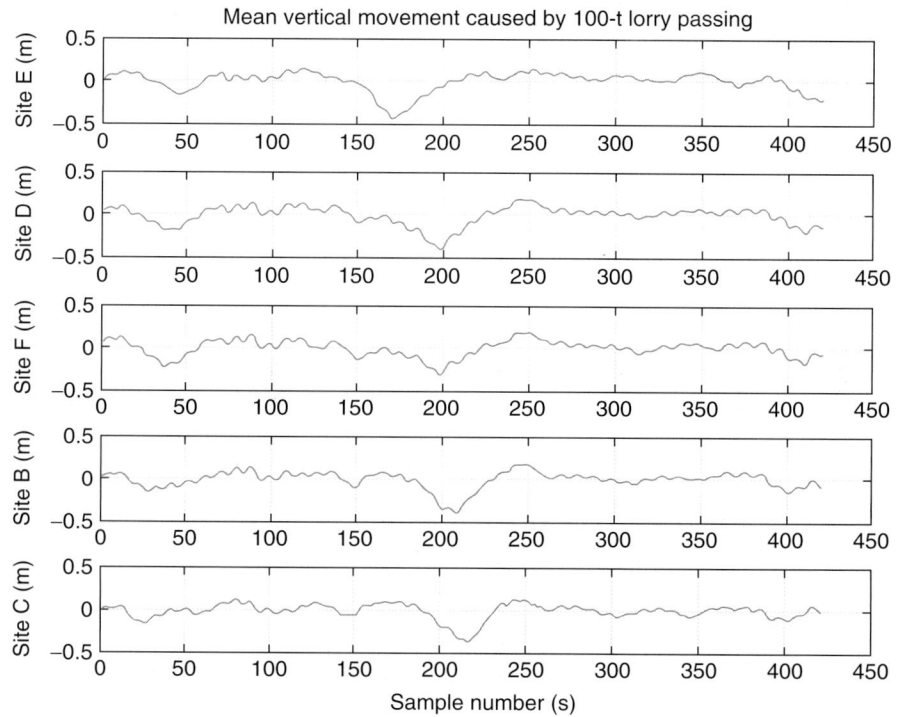

Figure 5. Vertical deflection of five GPS sites caused by a 100-t lorry passing.

3.3 Coordinates and coordinate transformation

The reference datum adopted by GPS positioning is called *the World Geodetic System of 1984*, or *WGS84*, which is a 3D, earth-centered earth-fixed (ECEF) system. Under this reference system, the coordinates determined by a GPS receiver are expressed as either geodetic coordinates (latitude, longitude, and ellipsoidal height) or Cartesian coordinates (x, y, and z). These coordinates do not make sense for many engineering applications such as analyzing 3D bridge deformation, since they are not aligned with the bridge axes. So, these geodetic coordinates within WGS84 must be transformed into rectangular grid coordinates (easting and northing). With the information about the local mean sea level, which defines a local height datum, geoid, the difference between WGS84 (ellipsoid) and the geoid can be estimated, which is called as *geoidal height*. By using this geoidal height the measured ellipsoid height of a point on the earth surface can transformed as the orthometric height, a height referred to as the local geoid. For GPS-based SHM application, using either orthometric height or ellipsoidal height does not make any difference if only one height datum is chosen for whole data analysis. However, the engineering coordinate system of a bridge, or a bridge coordinate system (BCS), is normally defined as a right-handed Cartesian coordinate system using the longitudinal axis of the bridge as x axis, the lateral direction as the y axis, and the vertical direction as z axis [27]. Hence, a second coordinate transformation for transforming the coordinates in a local coordinate system to those in a BCS involving translation and rotation is further conducted through the determination of a rotation angle of the bridge main axis from the north direction and the coordinates of BCS origin in the local grid coordinate system.

3.4 Identification of wind and thermal induced bridge deformation through GPS positioning

GPS is a very useful tool in detecting deformations induced by wind and thermal loadings. The bridge response to wind loading characterizes a short-term vibration but the deformation induced by the diurnal change of temperature is a very slow process. Figure 6(a) shows wind speed measured in a period of 48 h and Figure 6(b) shows the lateral responding deformation at a midspan point of the Forth Road Bridge. The cross-correlation coefficient between wind speed and lateral deformation is 77%, demonstrating a very high level force and response effect [49]. Figure 7 describes the relationship between the thermal loading and the vertical bridge deformation. Delay in the deformation is evident. The calculated correction coefficient is −39%. In considering the deformations caused by other affecting factors, such as traffic and wind loading, this correlation is still very high. From these two figures, the capacity of GPS for monitoring both dynamic and semistatic deformations is further confirmed.

3.5 Deformation analysis and interface with computational models

A simple bridge deformation analysis is to draw 3D deformation time series and compare these deformation curves with safety thresholds. Fast Fourier transform (FFT) algorithm can be used to identify dominant vibration frequencies and maybe the changes of these frequencies could provide some information about the change of the bridge's properties. A more complicated time-frequency analysis approach has attracted the interest of many researchers around the world and can also be employed to analyze the change of these frequencies against time [18, 27, 57]. When all the deformation time series collected from each monitoring site are gathered, it is expected that GPS can be used for the global monitoring of a bridge dynamics with the possibility to connect to localized sensors for detecting potential damage, as proposed in [34]. Hence, correct interpretation of GPS data for structural dynamics forms an important and unique part of an SHM system. From field data collection point of view, any procedures that could assure the measurement quality in a cost-effective manner should be taken. However, more study should be initiated in the near future for the interpretation and analysis of field measurements (*see* **Chapter 26**).

Some preliminary research has been carried out to extract dynamics from GPS-measured coordinate time series [58]. How the GPS measurements could

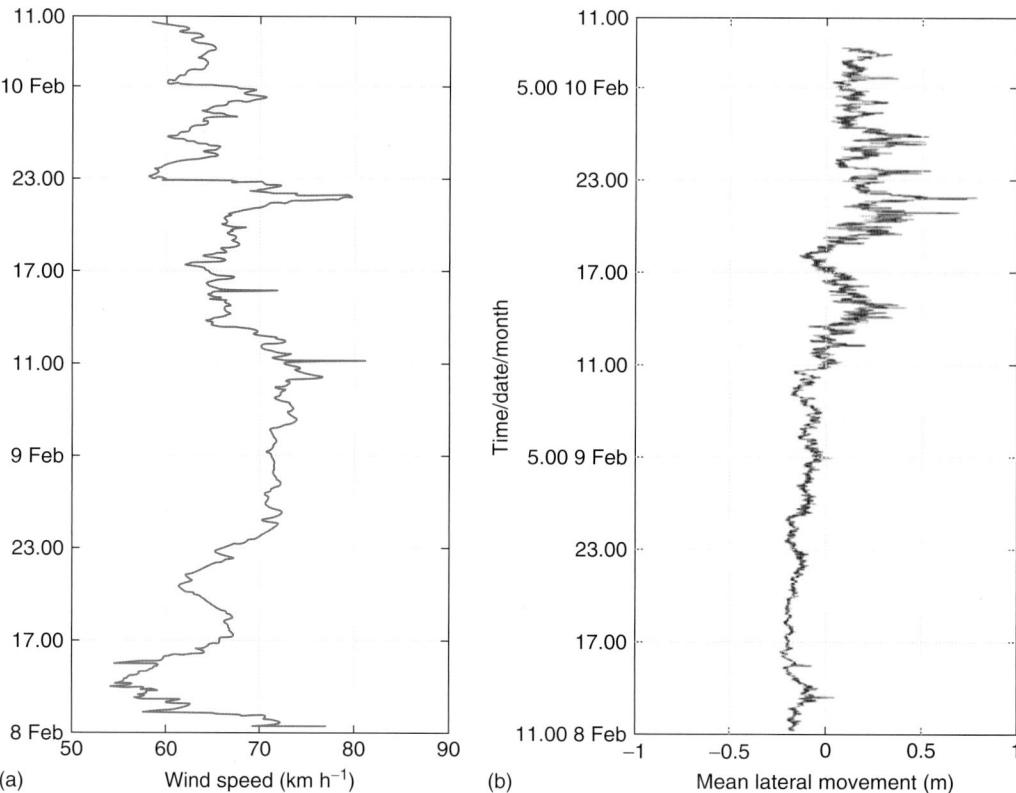

Figure 6. Wind loading and the lateral deflection of the Forth Road Bridge (a 48-h period).

be synergized with a computational model such as a finite element (FE) model is recommended in [58] and the whole loop of a practical SHM comprises the following steps:

Step 1—initial FE model creation;
Step 2—optimal sensor placement;
Step 3—bridge monitoring data collection;
Step 4—model updating (correlation of the FE model to the test data);
Step 5—further bridge monitoring data collection;
Step 6—FE model/real data comparison (comparison between the predictions of an FE model with real bridge deformations during its operational life).

3.6 Integrated sensor systems

Owing to the severe environmental conditions encountered in the bridge deflection monitoring, the instruments used must be lightweight, portable, reliable, and easy to install and the results must be easy to interpret. These are of great importance under extreme loading scenarios such as strong wind, volcanic eruption, and earthquake. At the same time, for correctly interpreting the dynamics of monitored bridges, the measurements should meet accuracy specification requirements. This means the deflections of the bridge should be measurable with available surveying instruments. For instance, to measure centimeter-level deflections the internal accuracy of a GPS receiver should be better than a few millimeters level, and multipath and other error sources should be appropriately mitigated or modeled.

Conventional surveying methods such as leveling have been used in the past to monitor static displacements of engineered structures with millimeter level or higher accuracy, and will certainly continue to be used in the future.

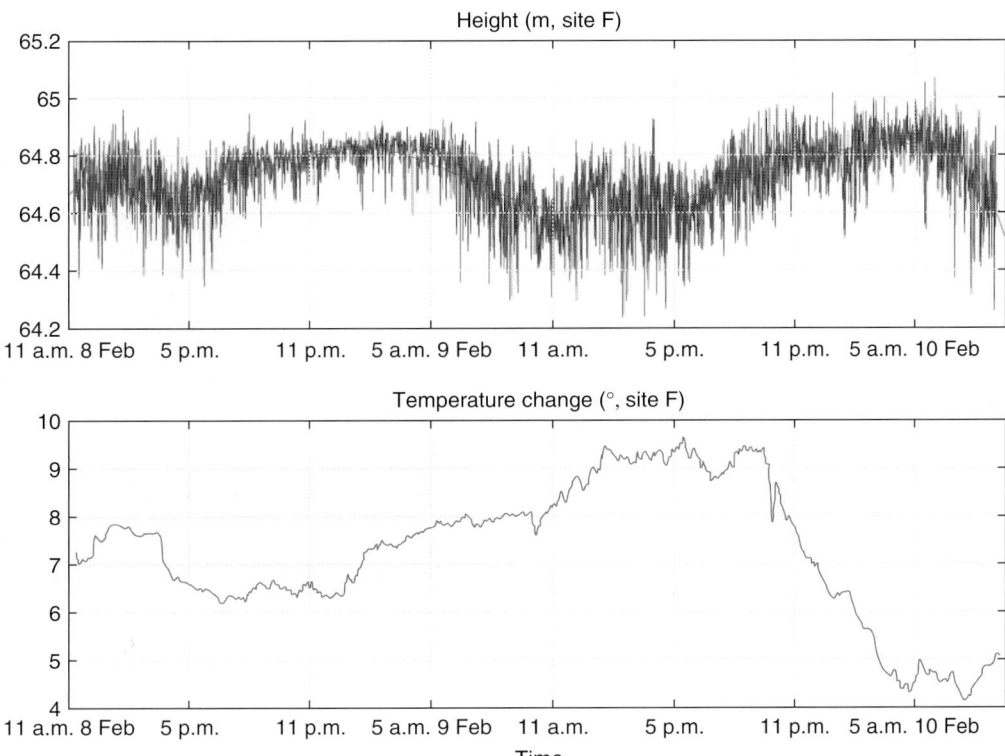

Figure 7. Diurnal temperature change versus the height variation at the midspan of the Forth Road Bridge (a 48-h period).

Modern leveling sensing stations could provide one-dimensional accuracy of about 2 mm at a sampling rate of 2.56 Hz [29]. Surveying robots, electronic distance meter (EDM), theodolites, surveying total stations, photogrammetry, and other surveying instruments could be employed to monitor structural deformation. However, the inherent disadvantages of these terrestrial surveying systems have greatly limited their applications. Previous research reveals that the main disadvantages of these surveying approaches can be summarized as follows:

- long-term intervals between measurements (days or even months);
- averaging of data over a relatively long time span (often some hours are smoothed, which leads to smoothing effects that could hide real movements of the stations);
- relatively low data sampling rate and a poor level of automation;
- batch mode analysis (data is collected, transmitted to a computer, and evaluated a few hours later).

Because of the above limitations of these terrestrial surveying methods, they cannot be employed to monitor structures with dynamic structural deflection and semistatic movements at the same time.

Accelerometer is used as an indispensable sensor in SHM of bridges for the identification of its dynamic characteristics. A triaxial accelerometer could measure three orthogonal accelerations simultaneously. Compared with other surveying systems, a triaxial accelerometer has some special advantages when it is used for bridge monitoring. For instance, the sampling rate can reach several hundreds of hertz depending upon application requirements. Triaxial accelerometers are superior to other sensors since they are not dependent on propagation of electromagnetic waves, and therefore avoid the problems of signal reflection or refraction and line-of-sight connections to the terrestrial or space objects. An accelerometer could form a completely self-contained monitoring system, utilizing only measurements of accelerations to infer the positions of the system, through integration based on the laws of motion. However, the positional drift of an accelerometer grows extremely rapid with time and can reach hundreds of meters after an interval of several hours [59]. The main error sources come from the instrumental biases and scale factor offsets and the unknown gravity of the earth. Continuous updating such as zero velocity update (ZUPT) or coordinates update (CUPT) is used to avoid error accumulation. It is the need to update that has severely restricted the wide applications of accelerometer technology as a standalone positioning method in surveying. In bridge deflection monitoring, it is impossible to conduct ZUPT; CUPT aided with GPS fixes could be a realistic option to overcome drift problem of accelerometer.

An integrated monitoring system consisting of GPS receivers with triaxial accelerometers could overcome the shortcomings of each individual sensor system and provide a much improved overall monitoring system in terms of productivity and reliability. To eliminate any potential sensor misalignment errors, the researchers in the University of Nottingham designed a cage that hosts both GPS antenna and a triaxial accelerometer as shown in Figure 8 [27]. Detailed GPS and accelerometer data fusion technique was also introduced by the same author and other researchers [30, 60].

As discussed by [47, 50, 61], current GPS constellation causes an uneven distribution of satellites in view: in the equatorial area a more scattered satellite distribution is possible, which means a nearly identical easting and northing positioning accuracies; but in the polar areas, no satellites above $45°$ elevation angle are visible, which causes degraded 3D positioning accuracies. The worst factor prohibiting wide GPS application for SHM of bridge is that the vertical positioning accuracy is always the worst component in the three coordinates and this might lead to wrong interpretation of actual bridge deformation [50]. In addition to the introduction of high-level multipath, dense cables, supporting towers, and other surroundings also cause satellite visibility problem, making the visible satellites less than adequate for starting a positioning fix. If several ground-based pseudolites could be installed with very low elevation angles,

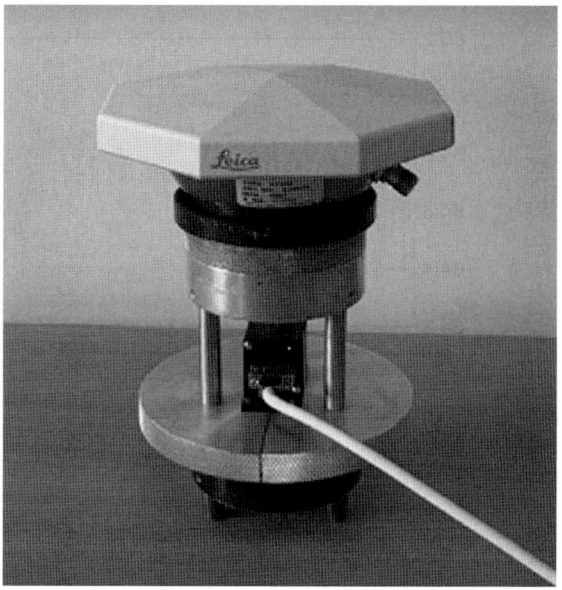

Figure 8. Device for integrated GPS antenna and a triaxial accelerometer.

it will significantly enhance the satellite geometry and positioning accuracy, especially in the vertical direction [47]. Pseudolites were used for the proof of GPS concept in the 1970s, before the launch of real GPS satellites. In the 1990s, due to the high demand in highly precise positioning solutions and the deficiency of current GPS constellation, the pseudolite concept was reinvented for augmenting GPS satellites.

A joint research has been conducted by the University of New South Wales and the University of Nottingham to verify the effect of introducing extra ground-based pseudolite transmitters for bridge deformation monitoring [48, 62–64]. Data processing of both GPS and pseudolite measurements reveals that if they are used correctly pseudolites can augment GNSS satellite geometries and improve 3D positioning accuracies to several millimeters. For the Wilford Bridge monitoring trials in Nottingham, three pseudolites were set up on the northern side of the bridge to fill the hole as shown in Figure 9 for obtaining more complete transmitter geometry. Compared with GPS-only solutions, the actual accuracy improvements in the north, east, and vertical directions are 14, 36, and 46%, respectively. The simulated accuracy improvement in the north, east,

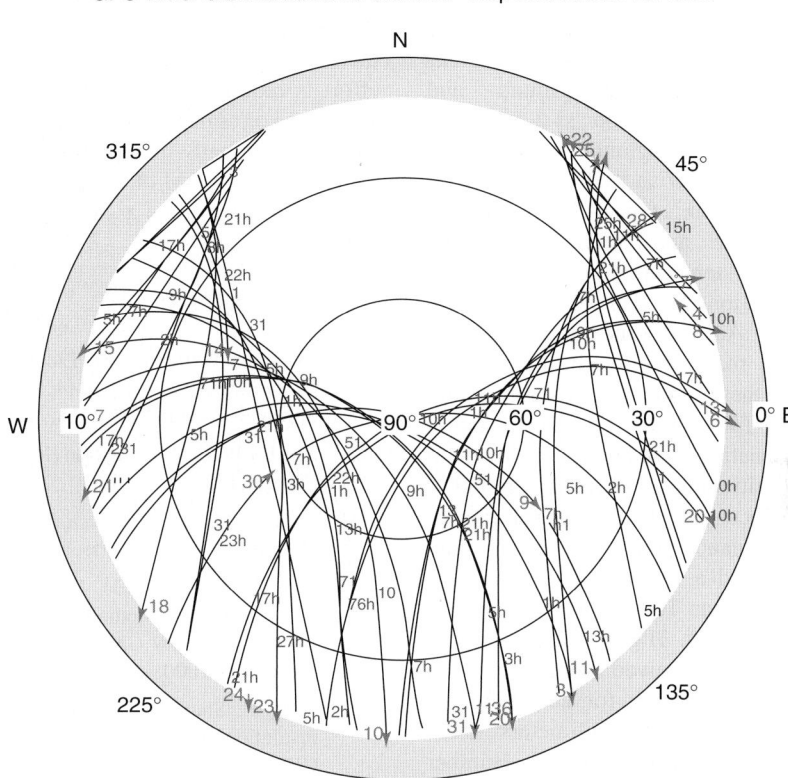

Figure 9. Skyplot of GPS satellites in Nottingham, showing the deficiency of GPS constellation.

and vertical directions are 19, 34, and 59%, which confirms the feasibility of an integrated GPS and pseudolite monitoring system for achieving more robust, accurate, and reliable results, especially in the vertical coordinate. However, since the early pseudolites transmitted on L1 frequency band, which could potentially interfere with GPS signals, the use of pseudolites was banned in many countries and its use was restrained in the United Kingdom. Collaborative research has lead to further cooperation in using a new generation pseudolite, Locatalite, which was pioneered by the Locata; it transmits at the 2.4-GHz frequency band to avoid the interference with the GPS signals [62].

Other integration approaches can also be employed in the SHM of long bridges, such as an integrated monitoring system consisting of an inertial navigation system (INS) and GPS. But the cost of this system could dramatically increase because of the high cost of an INS device.

4 FUTURE VISION

For improving positioning accuracy and availability, system integrity, and reliability, GPS is currently under its modernization through updating its space segment, improving signal quality and computation algorithms. Of course, the revival of Russia's GLONASS, and the development of European Galileo system will significantly increase signal availability and improve space geometry. China is expanding its regional navigation system, Beidou or COMPASS system, to a global coverage. Within a five-year period, the total number of navigation satellites could reach more than 110 and this will make high and pervasive real-time positioning less of a challenge. For solving the problem of signal obstruction by surroundings and making positioning in severely obstructed areas possible, R&D of regional space-, ground-, and aircraft-based SatNav augmentation systems such as Australia's Ground-Based Regional Augmentation System (GRAS), India's GPS and GEO Augmented Navigation (GAGAN), Japan's Quasi-Zenith Satellite System (QZSS) system and MTSAT Satellite-based Augmentation System (MSAS), the European Geostationary Navigation Service (EGNOS), the United State's Wide Area Augmentation Systems (WAAS), etc., is the main activity of the national space programs of these countries. Establishment of network-based RTK GNSS reference station infrastructure at a national or regional level to support improved positioning accuracy and mobility is also a recent focus in GNSS development. For instance, the Ordnance Survey of the United Kingdom has established a network GNSS facility, which consists of more than 100 permanent stations, covering the whole country. High-quality real-time RTK corrections can be received wirelessly with a GPRS-embedded GPS receiver for achieving 3D positioning accuracy of a few centimeters OTF. For monitoring the deformation of the Humber Bridge, the bridge authority only need to subscribe to the service providers for the precise (The Radio Technical Commission for Maritime Services) RTCM corrections for carrying out continuous monitoring in a more cost-effective and reliable manner. The progress in GNSS receiver and antenna technology will also make future GNSS receivers and antenna more sensible to weak signals and more robust to reject external inference and multipath. However, the following questions need to be further addressed. For instance, is 3D positioning to an accuracy in millimeters high enough for extracting the parameters of bridge deformation at sampling rate of up to 100 Hz or higher? Will network RTK GNSS with more than 110 satellites and other regional augmentation systems be able to provide more reliable, continuous, and robust positioning for monitoring bridges from several meters to several thousand meters of bridge spans? Will future wireless communications fully replace current optic-fiber or cable-based communications for both streaming RTCM corrections from the reference stations to the rover receivers and also sending the positioning solutions to the control center for further analysis? Will GNSS positioning also be able to help predict the existing life span of a bridge effectively? More effort is required to address these questions through individual or collaborative research around world and this will eventually make GNSS positioning a feasible technology for SHM of not only bridges but any civil engineering structures as well.

REFERENCES

[1] Hofmann-Wellenhof B, Lichtenegger H, Collins J. *Global Positioning System: Theory and Practice.* Springer-Verlag: New York, Wien, 1997.

[2] The Institute of Navigation (ION), *Global Navigation System*, 1980.

[3] El-Rabbany A. *Introduction to GPS: The Global Positioning System.* Artech House: London, 2002.

[4] Kim D, Langley RB. GPS ambiguity resolution and validation: methodologies, trends and issues. *Proceedings ION GNSS 2000.* Salt Lake City, UT, September 2000.

[5] Yang Y, Sharpe RT, Hatch RR. A fast ambiguity resolution technique for RTK embedded within a GPS receiver. *Proceedings ION GNSS 2006.* Fort Worth, TX, September 2006.

[6] Frei E, Beutler G. Rapid static positioning based on the fast ambiguity resolution approach FARA: theory and first results. *Manuscripts Geodaetia* 1990 **15**:325–356.

[7] Landau H, Euler HJ. On-the-fly ambiguity resolution for precise differential positioning. *Proceedings ION GPS-92.* Albuquerque, NM, September 1992.

[8] Farrar CR, Worden K. Preface. *Philosophical Transactions of the Royal Society of London, Series A* 2007 **365**:299–301.

[9] Rizos C, Han S, Ge L, Chen H-Y, Hatanaka Y, Abe K. Low-cost densification of permanent GPS networks for natural hazard mitigation: first tests on GSI's GEONET network. *Earth Planets Space* 2000 **52**:867–871.

[10] Rizos C, Han S, Roberts C, Han X. Continuously operating GPS-based volcano deformation monitoring in Indonesia: challenges and preliminary results. Geodesy beyond 2000: the challenges of the first decade. *Proceedings IAG General Assembly,*

ISBN 3-540-67002-5. Springer-Verlag: Birmingham, AL, July 1999, pp. 361–366.

[11] Hudnut KW, Behr JA. Continuous GPS monitoring of structural deformation at pacoima dam, California. *Seismological Research Letter* 1998 **69**(4):299–308.

[12] Hudnut KW, Bock Y, Galetza JE, Webb FH, Young WH. The Southern California integrated GPS network (SCIGN). *Proceedings Deformation Measurements and Analysis, 10th International Symposium on Deformation Measurements*. Orange, CA, March 2001.

[13] Teferle FN, Bingley RM, Dodson AH, Penna NT, Baker TF. Using GPS to separate crustal movements and sea level changes at tide gauges in the UK. In *Vertical Reference Systems*, Drewes H (ed). International Association of Geodesy Symposium, Springer-Verlag, 2001.

[14] Brunner FK, Hartinger H, Richter B. Continuous monitoring of landslides using GPS: a progress report. *Proceedings of Geophysical Aspects of Mass Movements*. Austrian Academy of Sciences: Vienna, 2000, pp. 75–88.

[15] Forward T, Stewart M, Penna N, Tsakiri M. Steep wall monitoring using switched antenna arrays and permanent GPS network. *Proceedings Deformation Measurements and Analysis, 10th International Symposium on Deformation Measurements*. Orange, CA, March 2001.

[16] Breuera P, Chmielewskib T, Orskic PG, Konopkad E. Application of GPS technology to measurements of displacements of high-rise structures due to weak winds. *Journal of Wind Engineering and Industrial Aerodynamics* 2002 **90**:223–230.

[17] Guo J, Ge S. Research of displacement and frequency of tall building under wind load using GPS. *Proceedings ION GPS-97*. Kansas City, MO, September 1997.

[18] Kijewski-Correa T, Kareem A, Kochly M. Experimental verification and full-scale deployment of global positioning systems to monitor the dynamic response of tall buildings. *Journal of Structural Engineering* 2006 **132**(8):1242–1253.

[19] Lovse JL, Teskey WF, Lachepelle G, Cannon ME. Dynamic deformation monitoring of tall structure using GPS technology. *Journal of Surveying Engineering* 1995 **121**(1):35–40.

[20] Nakamura SI. GPS measurement of wind-induced suspension bridge girder displacements. *Journal of Structural Engineering* 2000 **126**(12):1413–1419.

[21] Tamura Y, Matsui M, Pagnini L-C, Ishibashi R, Yoshida A. Measurement of wind-induced response of buildings using RTK-GPS. *Journal of Wind Engineering and Industrial Aerodynamics* 2002 **90**:1783–1793.

[22] Ashkenazi V, Dodson AH, Moore T, Roberts GW. Monitoring the movements of bridges by GPS. *Proceedings ION GPS-97*. Kansas City, MO, September 1997.

[23] Ashkenazi V, Dodson AH, Moore T, Roberts GW. Real time OTF GPS monitoring of the Humber Bridge. *Surveying World* 1996 **4**(4):26–28.

[24] Barnes JB, Rizos C, Wang J, Meng X, Dodson AH, Roberts GW. The monitoring of bridge movements using GPS and pseudolites. *Proceedings 11th International Symposium on Deformation Measurements*. Santorini, May 2003.

[25] Fujino Y, Murata M, Okano S, Takeguchi M. Monitoring system of the Akashi Kaikyo Bridge and displacement measurement using GPS. *Proceedings of SPIE; Proceedings Nondestructive Evaluation of Highways, Utilities, and Pipelines IV*. 2000.

[26] Leach M. Results from a bridge motion monitoring experiment. *Proceedings Sixth International Geodetic Symposium on Satellite Positioning*. The Ohio State University, 1992; pp. 801–810.

[27] Meng X. *Real-time Deformation Monitoring of Bridges Using GPS/Accelerometers*, Ph. D., The University of Nottingham: Nottingham, 2002, http://etheses.nottingham.ac.uk/archive/00000279/.

[28] Roberts GW, Dodson AH, Ashkenazi V. Twist and deflection: monitoring motion of Humber Bridge. *GPS World* 1999 **10**(10):24–34.

[29] Wong KY, Man KL, Chan WY. Monitoring Hong Kong's Bridges: real-time kinematic spans the gap. *GPS World* 2001 **12**(7):10–18.

[30] Li X, Ge L, Ambikairajah E, Rizos C, Tamura Y, Yoshida A. Full-scale structural monitoring using an integrated GPS and accelerometer system. *GPS Solutions* 2006 **10**(4):233–247.

[31] Kashima S, Yanaka Y, Mori K. Monitoring the Akashi Kaikyo Bridge: first experiences. *Structural Engineering International* 2001 **11**(2):120–123.

[32] Wikipedia, *List of World Longest Suspension Bridges*, http://en.wikipedia.org/wiki/List_of_longest_suspension_bridges (accessed Dec 2007).

[33] Wong K-Y. Instrumentation and health monitoring of cable-supported bridges. *Structural Control and Health Monitoring* 2004 **11**:91–124.

[34] Wong K-Y. Design of a structural health monitoring system for long-span bridges. *Structure and Infrastructure Engineering* 2007 **3**(2):169–185.

[35] Leica Geosystems, *Monitoring with GPS RTK Technology*, Jiangyin Bridge, http://www.leica-geo-systems.com/corporate/en/ndef/lgs_61984.htm (accessed Dec 2007).

[36] Fairweather V. Measuring bridge movement. *Civil Engineering, ASCE* 1996 **66**(6):48.

[37] Norgard P. Deformation survey of the storebaelt bridge: GPS shows its merits. *Geomatics Info* 1996 **10**(4):37–39.

[38] Duff K. Deformation monitoring with GPS, Part 1: system design and performance. *Proceedings Symposium on Surveying of Large Bridge and Tunnel Projects (FIG)*. Copenhagen, 1997.

[39] Duff K, Hyzak M. Structural monitoring with GPS. *Public Roads* 1997 **60**(4):39.

[40] Hyzak M, Leach M, Duff K. Practical application of GPS to bridge deformation monitoring. *Proceedings of the 64th Permanent Committee Meeting and Symposium of International Federation of Surveyors (FIG)*. Washington, DC, May 1997.

[41] Duff K, Nelson S. Deformation monitoring with GPS, Part 2: performance, affordability, and technology development. *Proceedings Symposium on Surveying of Large Bridge and Tunnel Projects (FIG)*. Copenhagen, 1997.

[42] Hyzak M, Leach M. Bridge monitoring by GPS. *Surveying World* 1995 **3**(3):8–11.

[43] Tolman BW, Craig BK. An integrated GPS/accelerometer system for low dynamics. *Proceedings International Symposium on Kinematic Systems in Geodesy, Geomatics and Navigation*. Banff, June 1997.

[44] Meng X, Dodson AH, Roberts GW, Cosser E. Hybrid sensor system for bridge deformation monitoring: interfacing with structural engineers. A window on the future of geodesy. *Proceedings of the International Association of Geodesy. IAG General Assembly*. Springer-Verlag: Sapporo, June 30–July 11, 2003.

[45] Cosser E, Roberts GW, Meng X, Dodson AH. Single frequency GPS for bridge deflection monitoring: progress and results. *Proceedings 1st FIG International Symposium on Engineering Surveys for Construction Works and Structural Engineering*. Nottingham, 28 June–1 July 2004.

[46] Dodson AH, Meng X, Roberts GW. Adaptive FIR filtering for multipath mitigation and its application for large structural deflection monitoring. *Proceedings of International Symposium on Kinematic Systems in Geodesy, Geomatics and Navigation (KIS 2001)*. Banff, 5–8 June 2001.

[47] Meng X, Dodson A, Roberts GW, Cosser E, Barnes J, Rizos C. Impact of GPS satellite geometry on structural deformation monitoring: analytical and empirical studies. *Journal of Geodesy* 2004 **77**:809–822.

[48] Meng X, Roberts GW, Dodson AH, Cosser E. The use of pseudolites to augment GPS data for bridge deflection measurements. *Proceedings ION GPS 2002*. Portland, ME, September 2002.

[49] Meng X, Roberts GW, Dodson AH, Meo M. GNSS for structural deflection monitoring: implementation and data analysis. *Proceedings of the 5th International Workshop on Structural Health Monitoring*. Stanford University, Stanford, CA, September 2005.

[50] Meng X, Noakes C, Dodson AH, Roberts GW. Satellite geometry and its implications for structural deformation monitoring. *Proceedings ION GPS/GNSS 2003*. Portland, ME, September 2003.

[51] Roberts GW, Brown C, Meng X. Bridge deflection monitoring—tracking millimeters across the firth of forth. *GPS World* 2006 **11**(2):26–31.

[52] Roberts GW, Meng X, Dodson AH. Using adaptive filtering to detect multipath and cycle slips in GPS/accelerometer bridge deflection monitoring data. *Proceedings FIG XXII International Congress*. Washington, DC, April 2002.

[53] Roberts GW, Meng X, Dodson AH. Integrating a global positioning system and accelerometers to monitor the deflection of bridges. *Journal of Surveying Engineering ASCE* 2004 **130**(2):65–72.

[54] Genrich JF, Bock Y. Instantaneous geodetic positioning with 10–50 Hz GPS measurements: noise characteristics and implications for monitoring networks. *Journal of Geophysical Research* 2006 **111**(B03403):1–12.

[55] Nickitopoulo A, Protopsalti K, Stiros S. Monitoring dynamic and quasi-static deformations of large flexible engineering structures with GPS: accuracy, limitations and promises. *Engineering Structures* 2006 **28**:1471–1482.

[56] Meo M, Zumpano G. On the optimal sensor placement techniques for a bridge structure. *Engineering Structures* 2005 **27**(10):1488–1497.

[57] Xu L, Guo JJ, Jiang JJ. Time-frequency analysis of a suspension bridge based on GPS. *Journal of Sound and Vibration* 2002 **254**(1):105–116.

[58] Meo M, Zumpano G, Meng X, Roberts GW, Cosser E, Dodson AH. Identification of Nottingham Wilford Bridge modal parameters using wavelet transforms. In *Proceedings of SPIE: Smart Structures*

and *Materials 2004. Modelling, Signal Processing, and Control*, Smith RC (ed). SPIE, 2004.

[59] Chen W. *Integration of GPS and INS for Precise Surveying Applications*, Ph. D., The University of Newcastle Upon Tyne, 1992.

[60] Chan WS, Xu YL, Ding XL, Dai WJ. An integrated GPS-accelerometer data processing technique for structural deformation monitoring. *Journal of Geodesy* 2006 **80**:705–719.

[61] Santerre R. Impact of GPS satellite sky distribution. *Manuscripta Geodaetica* 1991 **16**:28–53.

[62] Barnes J, Rizos C, Kanli M, Small D, Voight G, Gambale N, Lamance J. Structural deformation monitoring using locata. *Proceedings 1st FIG International Symposium on Engineering Surveys for Construction Works and Structural Engineering*. Nottingham, 28 June–1 July 2004.

[63] Barnes J, Rizos C, Lee HK, Roberts GW, Meng X, Cosser E, Dodson AH. The integration of GPS and pseudolites for bridge monitoring. In *A Window on the Future of Geodesy: Proceedings of the International Association of Geodesy. IAG General Assembly*, Sanso F (ed). Springer-Verlag: Sapporo, June 30–July 11, 2003.

[64] Dodson AH, Meng X, Roberts GW, Cosser E, Barnes J, Rizos C. Integrated approach of GPS and pseudolites for bridge deformation monitoring. *Proceedings of ENC GNSS 2003*. Graz, April 2003.

Chapter 67
Nanoengineering of Sensory Materials

Inpil Kang[1], Gunjan Maheshwari[2], YeoHeung Yun[2], Vesselin Shanov[3], Sachit Chopra[3], Jandro Abot[4], Gyeongrak Choi[5] and Mark Schulz[2]

[1] *Division of Mechanical Engineering, Pukyong National University, Busan, South Korea*
[2] *Department of Mechanical Engineering, University of Cincinnati, Cincinnati, OH, USA*
[3] *Department of Chemical and Materials Engineering, University of Cincinnati, Cincinnati, OH, USA*
[4] *Aerospace Engineering, University of Cincinnati, Cincinnati, OH, USA*
[5] *Korean Institute of Industrial Technology, Chonan-Si, South Korea*

1 Introduction	1189
2 Nanoparticle Spray-on Sensors	1192
3 Piezoresponsive Polymers Using Nanoparticles	1195
4 Electrically Conductive Cement Using Carbon Nanofibers	1200
5 Electrochemical Impedance Spectroscopy for SHM	1201
6 Nanotube Thread with Built-in Multifunctionality	1202
7 A Structural Neural System Using Carbon Nanotubes	1204
8 Future Embedded Wireless Nanosensors	1206
9 Summary and Conclusions	1207
Acknowledgments	1207
References	1208

Encyclopedia of Structural Health Monitoring. Edited by Christian Boller, Fu-Kuo Chang and Yozo Fujino © 2009 John Wiley & Sons, Ltd. ISBN: 978-0-470-05822-0.

1 INTRODUCTION

Nanoengineering is important because advanced sensors with high sensitivity, small size, and low cost may be developed on the basis of nanoscale materials. This article gives an overview of the development of nanoscale material sensors for use in structural health monitoring (SHM). The field of nanoengineering of sensors is in the beginning stages and research in this area is exciting and progressing quickly. Nanotechnology has introduced new materials such as carbon nanotubes (CNTs), zinc oxide nanobelts, nanowires (NWs), and metal or semiconductor nanoparticles that are expected to revolutionize the fields of biochemical, electrical, and mechanical sensing. Having 1–100 nm nanoscale size and a high surface to volume ratio, nanomaterials can be combined in host materials to produce changes in the electronic, photonic, catalytic, or other properties of the bulk material for use in virtually all types of sensing applications [1, 2]. Among the nanomaterials available, CNTs are promising for developing unique and revolutionary sensors owing to their structural and electrical characteristics. CNTs have high strength and high thermal and electrical conductivities and therefore can provide structural and functional capabilities simultaneously, including actuation [3] and sensing [4–6]. CNTs can be synthesized from a few microns to 1.5-cm long with nanometer diameters [7]. For applications at the macroscale, CNT smart materials are usually based on composite materials [8]. The small size of CNTs allows them to have high sensitivity in mechanical environments. Thanks to their smart material properties and many fabrication possibilities, CNTs are producing various kinds of sensors from the nano to the macroscale in size, which are described next.

In nanoscale experimentation, Tombler et al. [9] investigated the change of electrical conductivity of a single-wall carbon nanotube (SWCNT) under mechanical deformation using the tip of an atomic force microscope (AFM). Watkins et al. [10] used lithography and aligned SWCNT to fabricate an SHM sensor based on strain measurement. This microelectromechanical systems (MEMS) technique can measure small strain and it can also detect very small cracks. For macroscale sensing, Kang et al. [11] reported a CNT polymer-based piezoresistive strain sensor for SHM applications. To develop a sensor and actuator embedded in a structure, Jalili et al. [12] have been investigating next-generation functional fabrics utilizing SWCNT composites. Functional fabrics and yarns are being developed with distributed actuation/sensing capabilities using CNT-based mono- and multifilament yarns. Having high energy density storage, and the ability to be activated using low power, a nanomaterial-based sensor can be light and flexible and easily embedded into a structure. These characteristics are suitable for developing new sensors required for health monitoring of lightweight structures. *In situ* sensors can monitor the behavior of structures assessing damage or deterioration. Flexible and easily embedded CNT sensors are envisioned to be useful for *in situ* SHM for ageless vehicles that are capable of remaining in "as new" operating condition indefinitely, regardless of use. This concept has been proposed by NASA. The ageless vehicle requires structural self-assessment and repair capabilities to be carried out by distributed sensors [13].

Extraordinary recent nanoscale materials work by Baughman et al. [14–18] has been to develop artificial muscles that are powered electrically or by fuel cells, nanotube yarn and sheets that are a reinforcing material that is also used for strain sensing, and biological sensors. Using longer nanotubes to improve the properties of the spun fibers is one of the objectives of the continuing research by Baughman, Zhang, et al. Lynch et al. [19] have developed an electrical impedance tomography method for damage detection in CNT composite materials. NASA has developed nanotube sensors for mechanical, chemical, and biological applications. A good reference is the book by Meyyappan [20]. Varadan et al. have developed a strain sensor using nanotubes with a semiconductor matrix [21]. Yuan and Jin have modeled the elastic properties of CNTs [22], which may be used for modeling nanotubes in composites for strain sensing applications. A strain sensor using nanotube materials is described by Ramaratnam et al. in [23]. Getty has formed a magnetic compass using a ferromagnetic needle mechanically coupled to SWCNTs [24]. The needle is deflected in a magnetic field and bends the SWCNTs, which can cause orders of magnitude change in their resistance thus forming a tiny highly sensitive magnetic field sensor.

In the area of civil engineering, significant investigation is underway to determine if nanotubes added to cement can act as bridges across cracks and voids to transfer load and create a material more impervious to bending tensile loading [25, 26]. Nanotubes have a high strain to failure, which helps in the load transfer across cracks. Multiwall carbon nanotubes (MWCNTs), after being functionalized (chemically modified) using H_2SO_4 and HNO_3 solutions, were added to cement matrix composites. The treated nanotubes improved the flexural strength, compressive strength, and failure strain of the cement matrix composite. The porosity and pore size distribution of the composites were also reduced. The phase composition was characterized with Fourier transform infrared spectroscopy. Interfacial interactions were found between the CNTs and the hydrations in the cement (such as C–S–H and calcium hydroxide), which should produce a high bonding strength between the reinforcement and cement matrix.

There is the exciting potential to develop nanoscale hybrid materials that have designed-in properties to meet specific applications. As an example of a novel hybrid material, researchers at Argonne National Laboratory have combined the world's hardest material—diamond—with the world's strongest material—CNTs [27]. The process for "growing" diamond and CNTs together is the first successful synthesis of a diamond–nanotube hybrid material and opens the way for its use in energy-related applications and SHM owing to anticipated piezoresistive behavior.

In the general context of developing new sensory materials, there are many types of starting nanomaterials available. These 1D nanostructures include nanobelts, NWs, nanorods, nanotubes, nanonails, nanoflowers, etc. There are four commonly available nanoparticle materials that are particularly important for use in SHM and the related areas of engineering asset evaluation and condition monitoring. These nanoparticles can also be used to develop sensor systems that perform data mining at the sensor level [28, 29] and that can harvest power from ionic flow [30]. These four materials [31–36] are NWs, carbon nanofibers (CNF), carbon nanosphere chains (CNSCs), and CNT arrays, as shown in Figure 1. NWs (Figure 1a) have semiconducting, conducting, magnetic, and other properties, and have high density compared to other nanoparticles. CNF (Figure 1b) are nested MWCNTs that have a 20° conical shape, larger diameter than nanotubes, good properties, and very low cost. CNSC (Figure 1c) are carbon onions linked together, catalyst-free, highly electrically conductive when post treated, and low cost. With a specific post treatment, CNSC also have weak magnetic properties. Thus, carbon is now a member of the magnetic club. CNT arrays (Figure 1d) are parallel forests of aligned MWCNT up to 1.5-cm long with good electrical and mechanical properties.

Other upcoming nanomaterials not as commonly available are shown in Figure 2. These include telescoping MWCNTs, coiled MWCNT, alloy NWs, piezoelectric NWs, and zinc oxide NWs and nanobelts that are piezoelectric.

Properties of SWCNT are discussed briefly. Other types of nanotubes have properties that are similar but not as good as SWCNT. SWCNT is the strongest and most flexible molecular material known because of the C—C covalent bonding and seamless hexagonal network architecture. SWCNT have a Young's modulus of 1 TPa, which is above the modulus of 70 GPa for aluminum, and 700 GPA for carbon fiber. The strength-to-weight ratio of an SWCNT nanocomposite could be 4 times the same ratio for graphite/epoxy as predicted in Chapter 15 of [37]. The maximum strain of SWCNT can be $\sim 10\%$, and the thermal conductivity is $\sim 3000 \, \text{W} \, \text{mK}^{-1}$ in the axial direction. SWCNT have semiconducting or conducting electrical properties depending on the chirality (angle of twist) of the nanotube. Electrically conductive nanotubes are called *metallic tubes* and armchair nanotubes have electrical conductivity like metal while semiconducting tubes have transistor properties. Nanotubes have a high current carrying capacity and have a high aspect ratio and small tip radius of curvature ideal for field emission. SWCNT have magnetoresistive and piezoresistive properties, but negligible piezoelectric property. They also have electrochemical properties and a supercapacitance property in an electrolyte and can form an electrochemical actuator. Properties of MWCNT are similar to SWCNT but MWCNT are electrical conductors. Electrochemical actuation, a nanobearing, structural reinforcement, a telescoping actuator, and co-ax cable are some of the envisioned applications of MWCNT.

Processing materials with nanoscale features require extra care in most steps of the procedure. In the area of synthesis, researchers are continuously improving control over CNT length, diameter, and array density by careful control over substrate preparation and nanotube synthesis conditions (Chapter 5 of [37]). Dispersion of nanoparticles in polymers is another critical step that is continually being

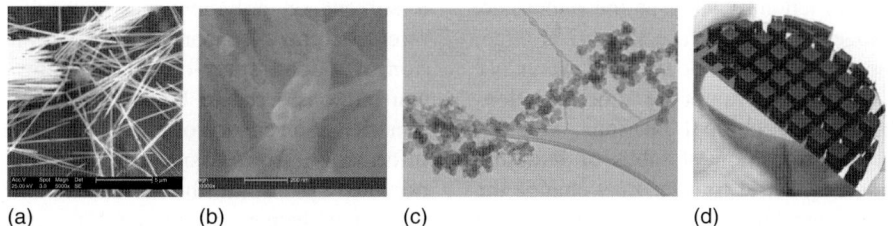

Figure 1. Nanomaterials for sensors: (a) Ni NW; (b) CNF; (c) CNSC on TEM grid; (d) MWCNT.

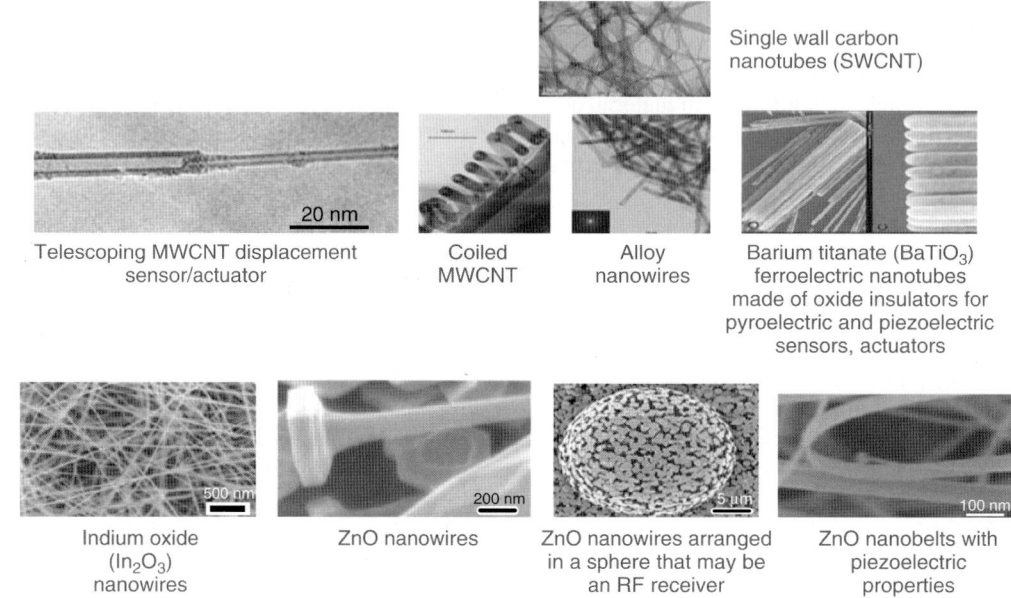

Figure 2. Less common types of nanoscale materials that may form SHM sensors.

improved [38]. High-resolution transmission electron microscopy, X-ray photoelectron spectroscopy, and environmental scanning electron microscopy are the primary tools used to characterize the nanoparticles and their smart materials. Our philosophy in developing nanoparticles and products follows from a Sherlock Holmes quote: "The world is full of obvious things which nobody by any chance ever observes." Careful observation at every step of material processing is needed in the field of nanoengineering.

Some general guidelines are discussed next to aid in the design of nanomaterials for sensors. As the size of the particle goes down, the surface area to volume ratio goes up and the properties of materials often improve. Characteristics of nanoscale materials and devices that may be helpful in sensor design are (i) viscosity increases at the nanoscale; (ii) electrostatic attraction is large at the nanoscale, particles stick together, and surface tension is large; (iii) friction depends on nanoscale surfaces, MWCNT nested nanotubes that are straight are almost friction-free translational and rotational bearings, van der Waals forces cause retraction of tubes or bundling of tubes; (iv) increased heat transfer may change the efficiency of tiny machines; (v) the melting point of nanomaterials may decrease as the thickness decreases; (vi) as size goes down, the frequency of electric circuits built using nanoparticles goes up; (vii) nanotube contact resistance is high and may be reduced by coating the nanotube ends such as with titanium and gold; and (viii) there are other electronic and optical effects described in the literature that may be useful for SHM sensing and power harvesting. In the next sections, several types of nanoparticle-based sensors are described to give an overview of the field. Since this field is rapidly changing, readers are encouraged to survey the literature for new developments.

2 NANOPARTICLE SPRAY-ON SENSORS

Many types of nanoparticles can be dispersed in a solvent or polymer and sprayed onto a structure to form a sensor. Here, we consider fabrication of a long spray-on sensor using CNTs, which can be considered to be a continuous sensor like a dendrite of a sensory or afferent neuron that conveys information from tissue and organs to the central nervous system in the human body. A structural neuron is formed using a CNT continuous sensor (a dendrite) and an analog electronic processor (the cell body) that collects inputs and generates an electrical output (fire). The output flows through a wire conductor (an axon in the biological system) to a computer (the brain) for analysis. A network of structural neurons can cover large areas and monitor a structure for damage (pain) in real time using parallel processing (as in the biological neural system) and a small number of channels of data acquisition. The CNT-based neuron can be used as an alternative to the piezoelectric ceramic-based neurons used in [39]. Piezoelectric-based neurons are described in **Chapter 147**. A structural neural system (SNS) can be built with a dispersed MWCNT solution that can be sprayed with an airbrush on a patterned surface of a structure as shown in Figure 3. The CNT neuron is a thin and narrow polymer film sensor that is sprayed, bonded, or taped onto a structure [40]. The electrochemical impedance (resistance and capacitance) of the neuron changes due to deterioration of the structure where the neuron is located.

The neuron has bulk piezoresistivity due to the CNTs in a polymer matrix, which is useful as a strain sensor for engineering applications. The polymer improves interfacial bonding between the nanotubes and enhances the strain transfer, repeatability, and linearity of the sensor. The largest contribution of piezoresistivity of the sensor may come from slippage of overlaying or bundled nanotubes in the matrix, from a macroscopic point of view. Nano interfaces of CNTs in a polymer matrix also contribute to the linear strain response compared to other microsize carbon fillers. Low weight percentages of nanotubes in a polymer not only have a percolation behavior with high sensitivity but also may have a nonlinear response. Buckypaper is formed by dispersing nanotubes and forming a film by solution evaporation. SWCNT are the best nanoparticles to form buckypaper because the van der Waals forces are large for small diameter nanoparticles. However, buckypaper is still relatively weak. Although buckypaper has a large gauge factor for strain sensing, the response is more nonlinear and the strain range is smaller as compared to using a polymer host for the nanotubes. A typical response of a nanotube-polymer sensor on a cantilever beam is shown in [11, 40] and can be modeled as $V = 0.0016 \times S$, where $V =$ voltage, $S =$ microstrain.

2.1 Strain sensing

The CNT-based neuron is modeled as a parallel R–C circuit as shown in Figure 4(a). The neuron can monitor static and dynamic strain as shown in Figure 4(a, b). This testing showed that the neuron has a bandwidth of about 20 Hz and measures the average strain over the length of the sensor. The strain sensitivity is similar to that of a strain gauge. The CNT-based neuron is a practical sensor because the monitoring signals are low voltage and form a simple circuit.

2.2 Crack detection testing

CNT-based neurons can monitor damage by two methods. The first is by the change in impedance of the sensor if a crack propagates through the sensor. In this case, a fine network of micron thin neurons could cover a structure. The second approach is to monitor the change in dynamic response of the structure. The dynamic response will change when damage occurs in the structure. A CNT neuron was tested to measure cracking on a composite beam. Deterioration due to cross-sectional damage or a crack was detected using the dynamic strain response of the neuron. The local stiffness of the structure changes due to damage, which, in turn, influences the dynamic response of the system. The CNT-based neuron can also monitor crack growth because the

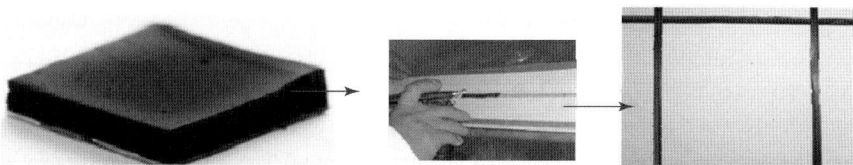

Figure 3. CNTs array dispersed in a solvent or polymer and sprayed onto a structure to form long thin continuous sensors that are analogous to a dendrite of the biological sensory neuron.

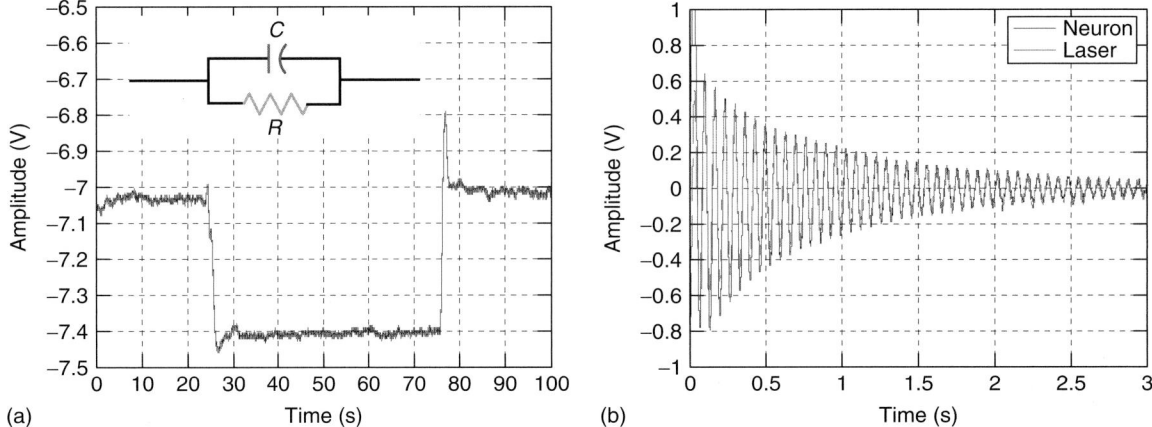

Figure 4. Modeling and strain responses of the CNT-based neuron on a glass fiber cantilever beam: (a) Electrical model and response of the CNT-based neuron under a static load; (b) dynamic strain response of the neuron during free vibration due to an initial displacement, the displacement response of the beam was measured by a laser displacement sensor (Keyence, LC-2400 Series) and almost overlays the strain response of the neuron.

nanotube film will change resistance when a crack propagates through it. Under crack propagation, the resistance of the neuron increases and the capacitance decreases, which changes the voltage response of the neuron, which can be measured using a bridge electrical circuit. The response of the neuron as a crack propagates through it is shown in Figure 5(a). The normalized crack size is defined as the ratio of crack size divided by neuron width when the crack progresses through the neuron normal to its length direction. A crack size of 100% means complete separation of the neuron by the crack. The increased resistance due to damage causes a higher amplitude voltage and the reduced capacitance induces a phase shift of the dynamic response. With this crack sensor approach, damage must change the strain at the neuron. A high density of micron size neurons can be used to detect small damage.

2.3 Corrosion detection testing

The same CNT neuron used for strain measurement can effectively measure corrosion because of the high electrochemical sensitivity of the nanotubes. Corrosion occurring on a metallic structure produces

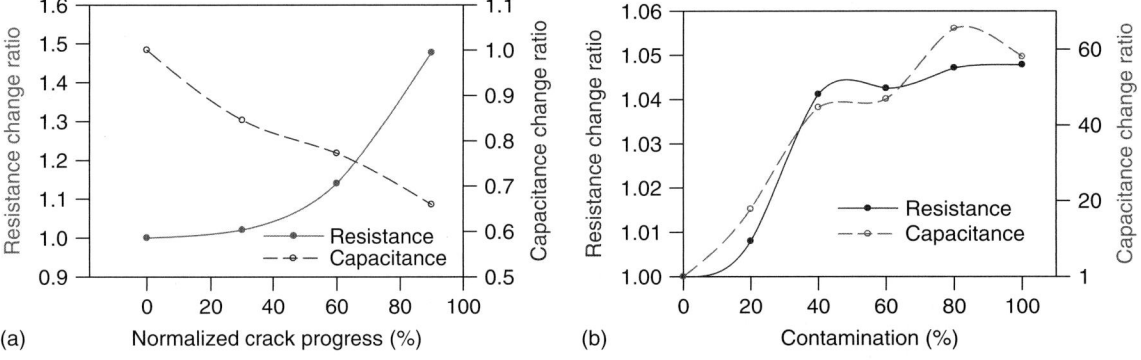

Figure 5. Response of the neuron due to cracking on a vibrating beam: (a) The variation of resistance and capacitance of the neuron as the crack propagates through the CNT neuron for the static beam; (b) The CNT neuron used as a corrosion sensor.

a diffusion layer at the interface between the structure and the CNT neuron. The corrosion ions penetrate into the CNT polymer sensor and form a double layer charge (Chapter 15 [37]) on the nanotube surface. The diffused ions also change the electrical parameters of the CNT sensor much like a doping effect. Having a large surface area, the capacitance of the CNT can be changed by orders of magnitude by the ions produced by corrosion, and the resistance will also change a small amount. A preliminary test of the corrosion sensitivity of the sensor was performed. A chemical buffer solution was used to simulate corrosion. The contamination percent is defined as the percent contaminated area on the neuron. After applying the solution, the variation in resistance and capacitance of the sensor is shown in Figure 5(b). The capacitance of the neuron sensor dramatically increased by a factor of 50 while the resistance decreased by only 5%. For composites, moisture absorption, corrosion at interfaces, and chemical degradation of the polymer may be monitored using the electrochemical response of the neuron.

As shown in the above experiments, the CNT sensor is versatile and can measure multiple physical properties of a structure. The electrical parameters of the CNT consist of a parallel resistor and capacitor and show distinct changes and trends related to physical measurements that are useful for SHM. Strain mostly causes a change of resistance. The capacitance is not affected much by strain. The sensor is moderately sensitive to changes in both resistance and capacitance when monitoring cracking. However, the sensor is dramatically sensitive to the change of capacitance when monitoring corrosion. On the basis of these sensitivities, the multifunctional capability of the neuron is a new approach for SHM. Since corrosion, infiltration of water, cracking, delamination, temperature, and change of pH can all possibly change the impedance of the nanotubes, the neuron is a sensitive barometer of the overall health of the structure. The multifunctionality is expected to effectively detect multiple symptoms of damage occurring at the same time and at the same location in complex structures. CNT neurons can be applied in the field by spraying the material onto the structure. Different types of nanoparticles (SWCNT, MWCNT, CNF, and CNSC), polymers (polymethylmethacrylate (PMMA), epoxy, polyvinyl alcohol (PVA)) and processing methods (dispersion, functionalization, vacuum, pressure, and temperature) were investigated to develop the neuron [8].

3 PIEZORESPONSIVE POLYMERS USING NANOPARTICLES

Piezoresponsive materials change properties due to strain of the material. Three types of piezoresponsive materials are generally being considered to form sensors. The materials are piezoresistive (change electrical resistance with strain), piezomagnetic (change magnetic properties with strain), and piezoelectric (produce a charge with strain). The piezoresponsive materials are usually formed by loading a polymer host material with one or more nanoparticles that have specific sensing properties. This section gives examples of different piezoresistive sensors being developed at the University of Cincinnati and by collaborators based on integrating sensor nanoparticles into polymers. Since the polymer is often used as the matrix of a composite material, the piezoresponsive polymer can be used to form self-sensing nanocomposite materials. CNSC-epoxy nanoskin and CNSC-polyurethane nanoskin are two new materials being developed to provide a means to tailor the surface properties of structures.

A general procedure to fabricate nanocomposite and sensor materials follows four steps: (i) selecting the nanoscale constituent material, synthesis, and properties characterization of nanoparticles including tubes, wires, spheres, and plates, with desired electrical, magnetic, and other properties; (ii) performing intermediate processing to prepare the material for incorporation into a host material, involves purification, thermal treatment, functionalization, spinning thread, and partial self-repair by welding or recrystallization; (iii) forming the bulk material and nanostructured smart materials including nanocomposites, sensors, films, actuators, wires, or casting into polyacrylonitrile/pitch resin (PAN/PITCH) carbon fibers; and (iv) material design for specific applications including mechanical, thermal, electrical, and environmental.

Processing of nanocomposite materials (step ii) is challenging because nanofillers have several, not all good, effects on the bulk material. Nanotubes are so small that they may affect the crystallization of the polymer and the heat transfer of the polymer affects the curing cycle. Bonding the nanoparticle to the polymer usually requires attaching interface molecules to the surface of the nanoparticle. Modification of the surface of nanoparticles is called *functionalization*, and a good overview is given in [38, 41]. Noncovalent attachment of molecules relies on van der Waals forces or polymer chain wrapping. The noncovalent bonding alters the nanoparticle surface to be compatible with the bulk polymer with the advantage that the mechanical properties of the nanoparticle are not changed. A disadvantage is that the forces between the wrapping molecule and nanoparticle are often weak, which reduces the efficiency of the load transfer between the matrix and nanoparticle.

On the other hand, covalent bonding of functional groups to the nanoparticle produces a strong connection that may improve the efficiency of load transfer between the particle and matrix. Covalent bonding is usually specific to a given system with crosslinking possibilities. On the down side, covalent bonding is done using strong acids or plasma, which will likely introduce defects in the nanoparticles, such as holes in the walls of CNTs. Defects will lower the strength of the nanoparticle and may affect the electrical and other properties of nanotubes. Ultrasonication using a tip sonicator is used to disperse nanotubes, but the sonicator may put defects into the nanotubes. A good practice is to observe the walls of the nanotubes using a high-resolution transmission electron microscope after sonication to check that the damage is small.

Using the processing methods described and different nanoparticles, a number of recipes are being developed and improved for fabricating nanocomposite multifunctional materials (Table 1). We envision that future nanocomposite materials will be custom blended for specific applications. Several nanocomposite materials in development are discussed next.

3.1 Piezoresistive epoxy

Piezoresistive epoxy is formed by adding CNF or CNSC to epoxy. CNF material is commercially available at low cost [33] and probably is the nanomaterial most commonly used to reinforce polymers. The material comes in different grades including prefunctionalized material that is chemically modified using acid treatment to improve bonding to polymers. Results of putting CNF in polymers are described in [37, 41]. The CNSC is a newer material that has

Table 1. Recipes for fabricating nanocomposite self-sensing materials

Nanoscale material	Intermediate processing	Bulk material properties
Nanotube array	Spin into thread	Strong nanocomposite self-sensing and repair
Nanotube array	Cast polymer into array (epoxy, rubber, other)	Polymer nanocomposite (piezoresistive)
Nickel NW	Disperse in polymer	Polymer nanocomposite (piezomagnetic, NDE of composites)
CNSC/CNF	Disperse in polymer	Polymer nanocomposite (piezresistive)
CNSC/CNF	Disperse in cement mix	Nanocement (piezoresistive)
ZnO nanobelt	Align in polymer	Nanocomposite (piezoelectric)
A + B + C	Disperse in D	Nano A,B,C,D (piezo)

an onion-like morphology and is highly electrically conductive and catalyst-free. The CNSC material is commercially available from Clean Technologies International [32]. Modeling nanocomposites using the simple rule of mixtures, dispersion of CNSC in the polymer, and testing CNSC/Epoxy nanocomposites are described in [42]. The basic procedure for manufacturing nanocomposites involves dispersing CNSC in epoxy resin using a shear mixer and ultrasonicator simultaneously, as shown in Figure 6(a). A compression test setup and the compression stress–strain curve are shown in Figure 6(b, c). The mechanical properties of the polymer improve a small amount by addition of 3 wt% of CNSC. The elastic modulus increased from 2.9 to 3.3 GPa, a 13% improvement, and the strength increased from 106 to 113 MPa, a 6% improvement. Dispersion of up to 5 wt% of CNSC in epoxy has been done. Plasma functionalization is also being done to improve dispersion and bonding to the polymer. Studies of the mechanical, electrical, and thermal properties of this material are under way.

3.1.1 Electrical testing of carbon nanosphere chain/epoxy nanocomposites

Nanocomposite button samples were fabricated for electrical testing. Argon plasma treated CNSC were dispersed in epoxy resin and cast in Teflon button shape molds. Epoxy Epon 862 resin with W curing agent was used. Figure 7(a) shows the resistivity of a 2-wt% CNSC-epoxy nanocomposite sample versus stress. It can be seen that the resistivity decreases initially with increasing stress and then approaches a constant value. The beginning part of the graph indicates that one can use the composite for a pressure or strain sensor. Figure 7(b) shows the resistivity of a 10-wt% CNSC-epoxy nanocomposite sample versus stress. The electrical conduction of CNSC-epoxy for

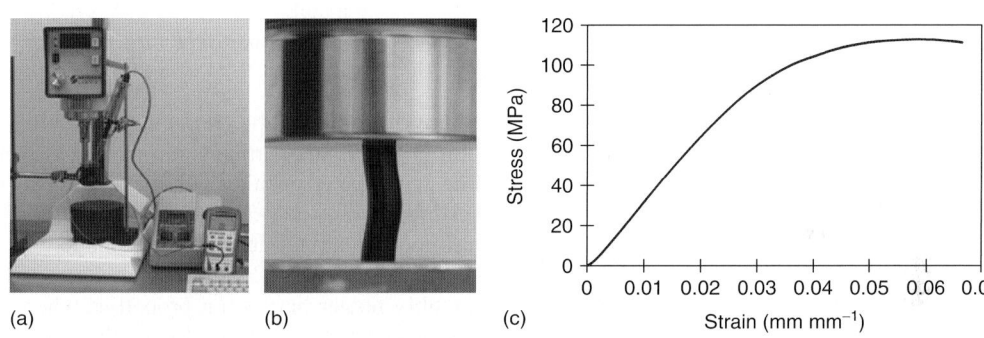

Figure 6. CNSC material processing: (a) mixer, ultrasonicator, and impedance measurement for dispersing nanoparticles; (b) nanocomposite testing; (c) stress–stain CNSC/epoxy 3 wt%.

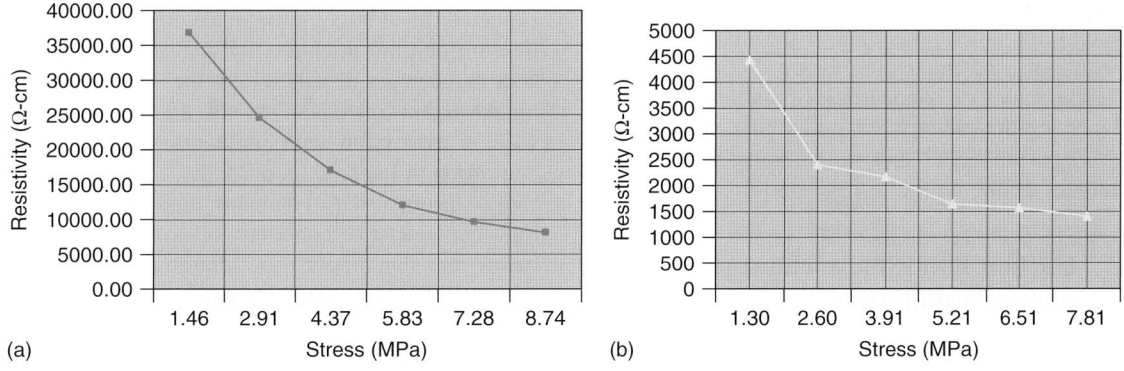

Figure 7. Resistivity versus compressive stress for CNSC/epoxy nanocomposites: (a) 2 wt% CNSC; (b) 10 wt% CNSC.

10 wt% loading is greater than for 2 wt% loading. The mechanical properties of this material have not been evaluated yet. A goal for some applications is to increase the electrical conductivity as much as possible without degrading the mechanical properties. In this case, the resistance between electrodes on the structure can be monitored to detect damage. At higher weight percentages, the viscosity of the polymer becomes very high and may prevent wetting of microfibers in a composite. To sense strain, a low weight percentage of CNSC could be used to provide a high gauge constant. High sensitivity is obtained when the loading is near the percolation level for the material, about 1 wt% or less of CNSC. Functionalization of the CNSC is under investigation to improve dispersion and to prevent reagglomeration when curing.

In general, nanoparticle-enhanced polymer design can provide electrical conductivity to the polymer and can reinforce epoxy and other matrix materials. The CNSC when compacted are highly electrically conductive, and presently about 5% by weight of CNSC can be added to the epoxy to provide electrical conductivity. A limiting factor is to not make the viscosity too high for wetting microfibers when making a laminated composite. The next step in the nanocomposite development is to investigate in detail functionalization (chemically modifying their surface) of CSNC by plasma etching to provide better dispersion and adhesion to the matrix. The plasma treatment will be controlled so that the electrical properties of the CNSC are not significantly affected. In general, a goal is to provide multifunctionality to the material, which means, for example, (i) having the greatest electrical conductivity possible, (ii) self-sensing to detect cracks, corrosion, and delamination in composites, and (iii) to do this without reducing the other properties of the matrix material. It would also be interesting to use a combination of different scales of nanoparticles (smallest to largest, SWCNT, MWCNT, CNSC, and CNF), which might stay dispersed better and might transfer load more gradually to the matrix.

3.2 Carbon nanosphere chain-polymer nanoskin materials

Carbon nanosphere chain epoxy, polyurethane (PU), and PVA nanoskin materials are discussed in this section. It has been shown that the electrical conductivity of nanoparticles depends on mechanical contact pressure. Higher pressure increases conductivity. When forming nanocomposites by dispersing nanoparticles in a polymer, it is difficult to apply contact pressure because of the shape of the mold and because the phase diagram changes (e.g., the material may not cure). Therefore, a skin material is considered because it can be held together by pressure during curing, which improves the electrical properties. CNSC are used to form the nanoskin in these examples, but other materials can be used. CNSC have high electrical conductivity after post treatment. Oxidization in air starts at about $500\,^{\circ}$C. The CNSC are thus useful for developing moderate temperature range sensors. Advantages of CNSC are low cost, large quantities of the material can be produced, and it is catalyst-free, which is useful in obtaining a large signal and to avoid toxicity. CNSC magnetic properties are under investigation.

3.2.1 Carbon nanosphere chain epoxy nanoskin

Nanoskin is formed by (i) a solution casting a CNSC film or other nanoparticles in a mold; (ii) coating the film with epoxy or other polymers; and (iii) pressure casting in a mold. The nanoskin material formed has good electrical and thermal conduction, piezoresistive and possibly piezomagnetic properties, and possibly power harvesting properties when a plasma gas or electrolyte is passed over the skin. An interesting property of nanoskin formed in this way is that it is electrically conductive on one side of the skin and electrically insulating on the other side of the skin. The film is shown in Figure 8(a). It is expected the skin could be made electrically conductive through the thickness by increasing the pressure and percentage of CNSC in the mold. Areas for improvement of CNSC-epoxy are to increase the wt% in dispersion with polymers and using higher compression casting to push the electrical conductivity closer to that of the compressed powder material. The electrical conductivity of nanoskin is quite below the conductivity of aluminum, but is being improved by postprocessing and by improved casting procedures.

The electrical resistance of a section of the epoxy nanoskin was measured by polishing the surfaces

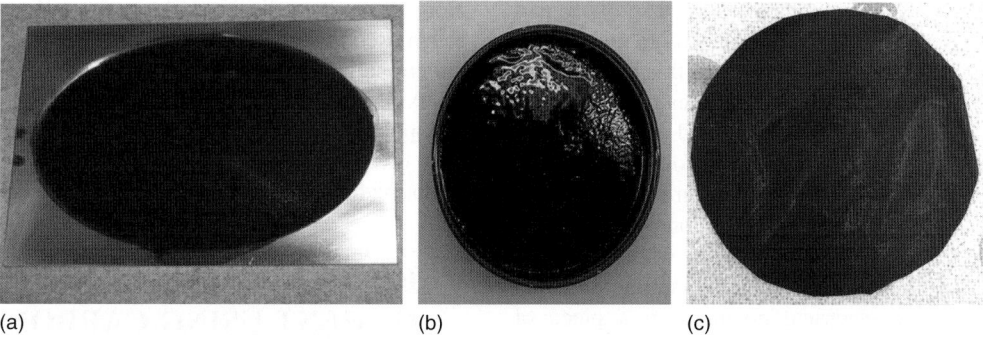

Figure 8. CNSC polymer multifunctional nanoskin: (a) bottom of mold to form CNSC-Epoxy nanoskin; (b) CNSC-Polyurethane nanoskin; (c) CNSC-Polyvinyl Alcohol nanoskin.

of the nanoskin to expose the nanoparticles and applying copper foil electrodes to the two sides of the skin. The skin was compressed and the force and resistance values were recorded and range from 64 N and 2000 Ω to 383 N and 877 Ω. This data shows that the resistance decreases with increasing load because the contact resistance of the copper and nanotube interface is reduced with pressure, and possibly because the nanoparticles are coming into closer contact in the epoxy. There is potential to greatly reduce the resistivity of the nanoskin by using a larger volume of CNSC and higher pressure during casting. The powder form of the CNSC under compaction has a resistivity of 0.5 Ω cm for the as-grown material. Post-treated material has orders of magnitude lower resistivity. The lowest resistivity possible for the nanoskin is the resistivity of the power material.

3.2.2 Carbon nanosphere chain-polyurethane nanoskin

A PU elastomer film was prepared using CNSC. The procedure was to disperse CNSC in N,N-Dimethylformamide (DMF) solvent, and evaporate the solvent in a mold leaving a thin film of CNSC on the mold. Then PU was poured over the CNSC and cured. The resulting film shown in Figure 8(b) was elastic, electrically conductive on the CSNC side, and electrically insulating on the topside. CNSC-PU material may be used for damage detection by sputtering thin film electrodes on the surface and monitoring the electrochemical impedance between the electrodes to detect cracks or corrosion. Applications of the electrically conductive PU elastomer might be an anti-icing heater film, erosion resistant coatings, and a low impedance piezoresistive sensor, and there may be medical applications such as to repair the body and build artificial organs. Electrical and mechanical characterizations are being carried out.

3.2.3 Carbon nanosphere chain-polyvinyl alcohol nanoskin

PVA nanoskin was formed by dispersing CNSC in water using a magnetic stirrer and a long mix cycle. The film was cured at room temperature in air for 48 h. The resulting film is shown in Figure 8(c). The electrical resistance of the PVA nanoskin was 150 Ω for 10% concentration of CNSC by weight. The electrical measurement was made using a two-point probe and multimeter. Resistivity will be determined using a four-point probe.

3.3 Piezoresistive and piezomagnetic nickel nanowire polymers

Nickel NWs and the general physics of NWs are discussed briefly to provide ideas for future sensor applications. NWs differ from their corresponding bulk (3D) materials because of an increased surface area to volume of the material and because of quantum confinement effects. NWs also exhibit different optical and electrical properties from the bulk material, e.g., transistor and magnetic properties, but here only the magnetic attraction and electrical conduction properties are considered for developing sensors. First, there are two basic approaches to synthesize NWs: nontemplate-assisted growth and template-assisted growth. A simple way to make NWs is to use a mold or template [43]. In this experiment, nickel NWs are grown inside the pores of an alumina filter and then the filter is removed by etching to yield magnetic NWs. The nanoporous membranes used were designed for healthcare applications including virus filtration, sample preparation, and liposome manufacture (http://www.whatman.com). These alumina membranes are manufactured by applying a large electrical potential to a piece of aluminum metal submerged in an acid. Aluminum is oxidized to alumina (Al_2O_3) and pores are created. The size of the pores depends on the applied potential. In this experiment, membranes with a pore size of 0.02 μm are used as templates.

Template-assisted growth used here requires the following steps: template preparation; filling the template with NW material; and etching away the template. Ni NWs are formed by electrochemical deposition of Ni on a cathode electrode in the electrolyte solution. One possible application for magnetic NWs filled in a polymer is for SHM of composites. The electrical or magnetic properties of the composite can be monitored and changes can be related to damage. It may be possible to adapt the eddy-current method from aircraft nondestructive evaluation (NDE) to Ni NW composites. The integration of Ni NWs in composites is in the beginning stages. Some characteristics of NWs are that they are much denser than nanotubes and initially seem more difficult to disperse. Also, the NWs affect the thermal conduction when processing the composite. Functionalization of the NWs to bond to epoxy must be explored. NWs should be much easier to align in a polymer because of their large magnetic attraction. Electrical and magnetic characterization will be performed for Ni NW nanoskin and nanocomposites.

4 ELECTRICALLY CONDUCTIVE CEMENT USING CARBON NANOFIBERS

Developing smart materials for civil infrastructure applications is an important area of research because of the huge volume of material that might be used. CNF are considered for this application because CNF are produced in large volumes in a continuous chemical vapour deposition (CVD) process using an iron catalyst. The nested carbon cones (Figure 9a) or spiral cones are electrically conductive, 70–150-nm diameter, micrometer long, have good strength, are low cost $45/kg, and are available already functionalized (PR–24–XT oxidized and debulked). Applied Sciences Inc. (Cedarville Ohio) observed very good machinability at 1–3 wt% of nanofiber in cement as evidenced by the image in Figure 9(b) of a thread on a machined cement bolt. The University of Cincinnati has pressure cast CNF into mortar mix to provide electrical conductivity and piezoresistive sensing, but brittleness of the material is a challenge.

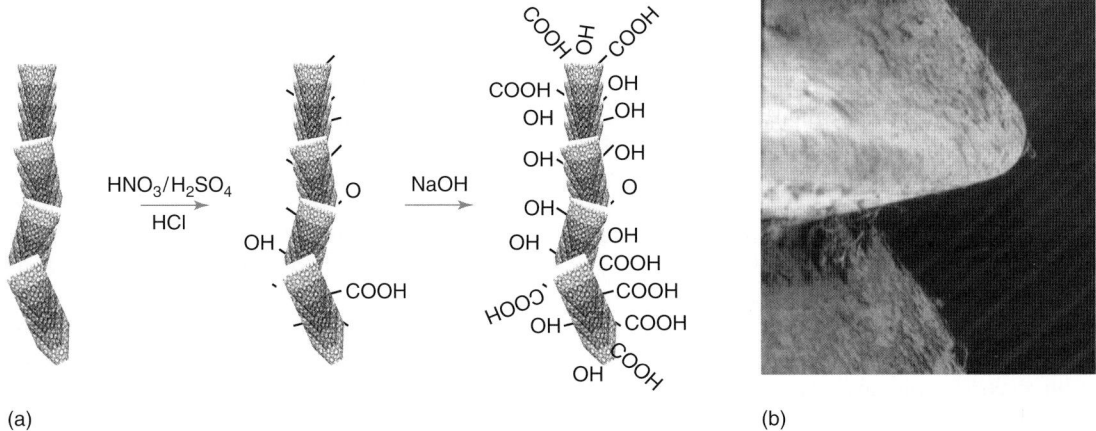

Figure 9. CNF-nanocement: (a) CNF functionalized; (b) screw threads in a nanocement bolt (image from Applied Sciences, Inc.).

The recipe to fabricate nanocement is (i) mortar mix, (ii) CNF (5% wt), (iii) ball mill 3 h, (iv) add water and mix, (v) pressure cast, (vi) heat treat, (vii) apply electrodes, and test. The CNF must be mixed with the cement as a powder because the small amount of water used would not allow a high enough percent of CNF to disperse. The cement/CNF powder is mixed by ball milling. Then water is added in slight excess and mixed to form a paste. The paste is put into a cylindrical mold and pistons are used to compress the nanocement mixture to remove air voids and excess water. Final curing is done in an oven. Electrodes are placed on the cement to compute the electrical resistance. An increase in resistance between any two electrodes indicates damage, corrosion or cracking. A specimen of the nanocement is shown in Figure 10. The resistivity decreases with increasing load until a resistivity of about $10\,\Omega\,\text{cm}$ is reached. This is an initial result and the dispersion and mechanical properties must be evaluated. Improvement in the properties of this initial sample is possible.

CNSC are the other material that seems ideal for integration into cement due to the high volume of material that can be produced at low cost and CNSC have high electrical conductivity and no catalyst. Nanotubes may be too expensive to use for reinforcing cement. Presently, high-grade cement uses chopped fiberglass with microscale fibers. It may be possible to develop hybrid cement using different types of fibers. The cement may be used to heat and melt ice, as an antenna material, and as a supercapacitor material. When wet, the cement may tend to increase corrosion of steel reinforcement bars (rebar) by providing an electrical path for corrosion to occur. This aspect has to be considered further.

5 ELECTROCHEMICAL IMPEDANCE SPECTROSCOPY FOR SHM

The analysis method to detect cracking and corrosion in nanocomposites including polymer, elastomer, and cement is electrochemical impedance spectroscopy (EIS) as shown in Figure 11.

A sine wave with a zero or nonzero dc voltage is applied to the sensor. The current is measured and the complex impedance is calculated. A redox chemical is sometimes used to reduce the impedance at a particular dc potential at which the redox reaction occurs. As an example of the use of EIS, a gold electrode was functionalized and used in an electrolyte with a redox couple. The EIS of the electrode was computed over a frequency range of 0.1 Hz–300 KHz. EIS was performed using a three-electrode cell, consisting of a gold electrode as the working electrode, an Ag|AgCl reference electrode, and a platinum wire counter electrode. EIS measurements were performed using a Gamry Potentiostat (model: PCI4/750) coupled with a Gamry EIS300

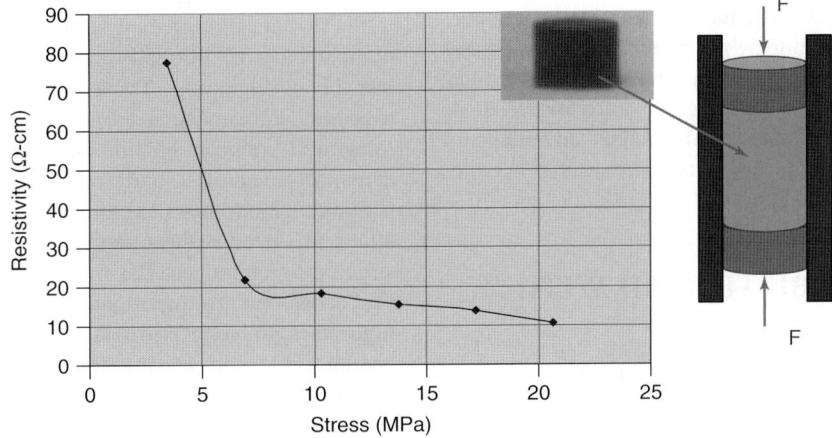

Figure 10. CNF and nanocement sample showing resistivity versus stress, the sample (inset) and schematic of pressure casting CNF in cement to improve the strength/electrical conductivity.

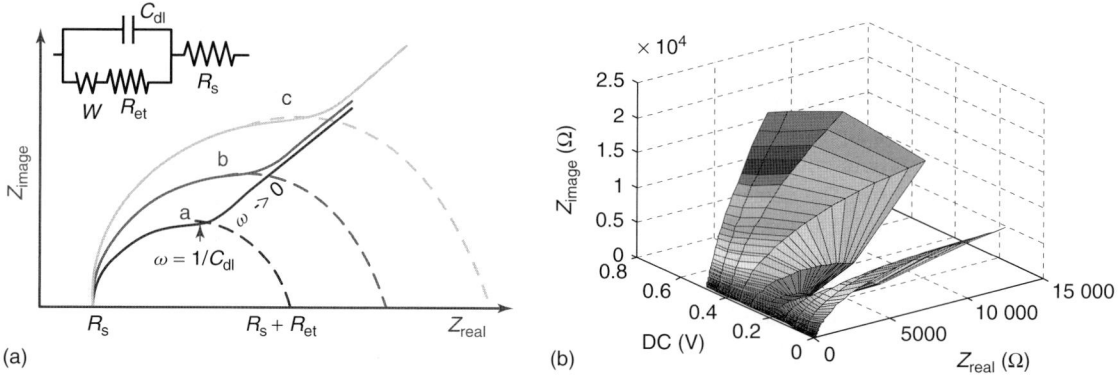

Figure 11. Electrochemical impedance spectroscopy to detect corrosion and cracking: (a) EIS model of an electrode and the response; (b) 3D EIS of a gold electrode in phosphate buffer saline with 5.0 mM K3Fe(CN)6 and 5 mM of K4Fe(CN)6. The effect of dc potential is shown.

software. Figure 11 shows the EIS response. Note that at the dc potential of 0.2 V the redox reaction greatly lowers the impedance. The EIS response can be modeled using the Randal Warburg equation $Z(\omega) = R_s + [R_{et}/(1 + \omega^2 R_{et}^2 C_{dl}^2)] - [j\omega R_{et}^2 C_{dl}/(1 + \omega^2 R_{et}^2 C_{dl}^2)]$, where $R_{et}, C_{dl}, R_s, \omega$ are the electron transfer resistance, double layer capacitance, solution resistance, and frequency, respectively. The EIS method can be adjusted to detect chemical degradation such as from corrosion, UV deterioration, hygrothermal degradation, and resistive degradation because of cracking and delamination of electrically conductive composites. A solid polymer electrolyte can also be used to further tailor the EIS signature of a composite. EIS monitoring is possibly a simpler approach than propagating waves to detect damage in large complex structures. Electrically conductive polymer, elastomer, or cement nanocomposites are multifunctional materials whose electrochemical properties change with damage or corrosion. The use of EIS is a new approach that takes advantage of the electrical conductivity properties of composites to simplify SHM.

6 NANOTUBE THREAD WITH BUILT-IN MULTIFUNCTIONALITY

Another approach for developing a multifunctional sensor material is to develop an intermediate product that can be used as a sensor material and as a fiber reinforcing material in composites. The proposed material is CNT thread. The thread is made from spinning long CNT. Long CNT are produced on wafers as shown in Figure 12. Properties of various fibers for comparison to CNT are given in [37] and show that SWCNT have higher strength, and stiffness, and are lighter than any other fiber. If a thread could be made that has similar properties to SWCNT, many new structural and electronic applications would open up. The mechanical properties of fiber nanocomposites [37] show that a polymer nanocomposite may have significantly better properties than conventional composite materials. Spinning long nanotubes into threads for reinforcement, self-sensing, and self-repair may produce a variety of versatile new smart materials.

SHM, data mining, and structural materials development have traditionally been separate enterprises, but this is changing. These disciplines are coming together to meet the needs of developing advanced integrated vehicle health monitoring systems for aerospace vehicles and other applications from nanomedicine to mars exploration. The logical way to attack this need is through the bottom-up design of a structural material that has multifunctional properties. Properties that are important for this material include high strength, high stiffness, light weight, electrical conductivity, self-sensing for damage, limited self-repair of damage, and a sensor architecture that enables data mining and other special properties depending on the application. These properties can be achieved by building a nanocomposite material based

Figure 12. Carbon nanotubes grown up to 1-cm long on a silicon wafer. The CNT are used for spinning thread. (CNT grown by First Nano Inc. using UC substrate.)

on nanotube thread. Long nanotubes and different types of nanoparticles can be blended to spin a thread. In this article, we have discussed blending short nanoparticles in a polymer. The polymer and thread may also be combined to form a nanocomposite with unique properties, such as high electrical conductivity and high strength. The blended materials approach makes all the capabilities of nanotechnology virtually available for developing enabling new materials for advanced applications. In particular, nanotube thread can bring multifunctionality to composite materials.

Designing smart materials from the bottom up can provide multifunctional material properties and practical application for SHM. However, the sensing capability should be designed into the material. In other words, all the properties of the material must be designed together. Threads are necessary to provide bulk reinforcement because CNT cannot be grown beyond approximate centimeter length, currently. Thus, centimeter-long nanotubes must be spun into thread to provide a strong bulk material. The thread can be made completely of nanotubes or it can be blended using two or more materials. For example, to provide intermediate properties and reasonable cost, long nanotubes could be blended with polyester to form the thread. In another approach, the CNT might be blended with small percentages of CNSC, or a variety of other nanoparticles, to give the desired properties of the thread. It is also possible that the CNT could be functionalized to modify their adhesion or other properties before spinning thread. The goal is to provide a strong thread with smart material properties that cannot be achieved by any other material system on earth. Multiple threads will be woven together to form a fiber. The fibers can be used to form unidirectional prepreg plies or woven into a fabric to provide two-directional properties. EIS can be used to monitor the electrical properties of the thread for damage. A smart fabric design is also possible using the thread. Smart fabric can be made by weaving the nanotube thread into cloth. Smart fabric should have several interesting applications including reinforcing fabric in composites, as tough materials for garments for soldiers, firefighters, and first-responders, as electromagnetic radiation shielding materials, and other applications. Mixing the different types of nanotube or nanoparticles and in some cases conventional thread provides a large design space to build in properties so that the material will do what we want it to.

Spinning nanoscale thread is a new area of research that is expected to become very important in the future. In general, fibers with small diameter generally have better mechanical properties. Most commercial fibers are in the micron diameter range due to economic considerations. This suggests that small diameter nanotube thread could have good mechanical properties and might replace microfibers. This idea has produced considerable interest in spinning nanotubes into thread and multistrand fibers using different approaches. Electrospinning is a method to produce continuous nanofibers from polymer solutions in high electric fields [44]. A thin polymer jet is ejected when the electric force on induced charges on the polymer liquid overcomes the surface tension. The charged jet is elongated and accelerated by the electric field, dries, and is deposited on a substrate as a random nanofiber mat. Over a hundred synthetic and natural polymers were electrospun into fibers with diameters ranging from a few nanometers to micrometers. The resulting nanofiber samples are often uniform and do not require expensive purification, and the electrospun nanofibers are continuous. Electrospinning has the potential for low-cost electromechanical control of fiber placement and integrated manufacturing of two- and three-dimensional nanofiber assemblies.

The assembly of CNT into continuous fibers has been achieved mostly through postprocessing methods. Fibers of nanotubes or nanotube-polymer blends have been drawn or spun from solutions or gels. A thread of nanotubes can be dry-drawn from an aligned assembly on a silicon substrate as a result of van der Waals interactions [17]. However, direct spinning of CNT is also being done. Nanotube thread has been formed after the pyrolysis of hexane, ferrocene, and thiophene directly in a furnace [45]. By mechanically drawing CNT directly from the gaseous reaction zone, continuous fibers were wound without an apparent limit to the length. Continuous spinning requires rapid production of high-purity nanotubes to form an aerogel in the furnace hot zone and forcible removal of the product from the reaction by continuous wind-up. Ferrocene and thiophene are used in the process. The overall question of spinning is discussed next.

6.1 To spin or not to spin

Rarely do we find noncompromising solutions to materials problems. The impasse with nanotechnology is bringing the properties of nanoscale materials to the macroscale. A key question is, how can we use CNT in the real world of airplanes and other structures for sensing and reinforcement? Spinning nanotubes into thread is one way to bring the properties of short fibers or nanotubes to the macroscale, but spinning is a compromise solution because the properties of thread are generally below the properties of the fiber or nanotube. However, thread also has advantages of energy absorption, tailorable stiffness, multicomponent material, and others as compared to nanotubes alone. When spinning, the longer the nanotube is, the smaller the thread angle becomes and possibly the greater the properties of the thread. Another compromise solution is linking nanotubes by using a binder material or by welding nanotubes to each other. There is no binder material that has mechanical properties in the order of the properties of nanotubes and unless the load path is continuous through nanotubes, the thread properties will be low. Welding nanotubes to tungsten has been done but welding nanotubes to nanotubes is difficult, based on initial investigations. There appears to be some potential to weld nanotubes to each other using a welding filler material and this technique is under investigation.

Not spinning would be the ideal solution because the full properties of nanotubes would be available at the macroscale. The problem with not spinning becomes one of synthesizing nanotubes to meter lengths. In theory, we believe that synthesis of continuous nanotubes is possible, but in practice the catalyst turns off after the nanotubes are in the centimeter-length range. Also, the quality of the nanotubes decreases with length. Considerable effort is being expended to grow the longest nanotubes possible [46, 47]. Lengths beyond the centimeter range and high quality are expected. Recently, forming nanotubes into threads or films is being done by several industries including Industrial Nano, CNT Technologies, Nanocomp, General Nano, and universities including Cambridge University, UT Dallas, and the University of Cincinnati for composites applications and possibly for use in a space elevator [48]. Spinning for simultaneous reinforcement, self-sensing, and limited self-repair may produce a versatile new smart material. Initial blend thread produced at the University of Cincinnati is shown in Figure 13. Properties of the initial thread are being evaluated. Without post processing, electrical resistivity is about $0.02\,\Omega\,cm$. This thread has promise for applications in hundreds of commercial products.

7 A STRUCTURAL NEURAL SYSTEM USING CARBON NANOTUBES

In the case of SHM of aerospace systems, many sensors are typically needed to provide sensitivity to small damage on large structures, and different

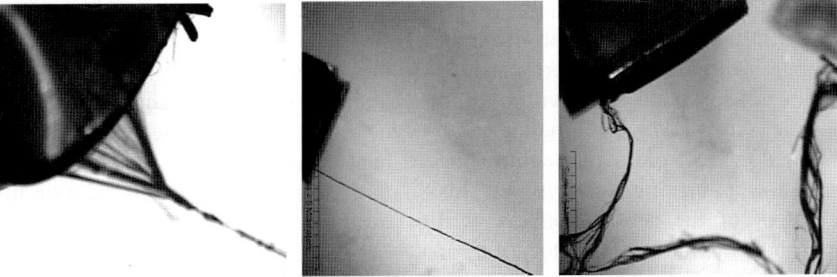

Figure 13. Long CNT called *Black Cotton* spun into thread may provide reinforcement, sensing, and partial self-repair simultaneously. Figures show spinning of thread from long CNT arrays. Two or three threads can be rolled into one yarn ∼0.2-mm diameter. A spinning machine is used to spin the thread based on the approach developed by Zhang *et al.* [17].

types of sensors are needed to monitor different types of components. Also, aerospace and other systems are becoming more and more complex. This means that onboard sensors and signal processing instrumentation are needed to simultaneously monitor the integrity of their multiple subsystems that contain structural, mechanical, and electrical components. Besides integrity monitoring, large sensor networks are also being used in qualification testing, monitoring unmanned vehicles, and for feedback control and damage mitigation of systems. In all these applications, the integration and mining of increasing amounts of data from many sensors and systems is becoming impractical based on size, weight, reliability, and cost of the monitoring and support systems. A new approach to sensing is presented next wherein only anomalous responses are obtained from sensors as opposed to longtime histories of sensor raw data.

7.1 The structural neural system

An SNS was developed for anomalous event detection and to significantly reduce the complexity and cost of SHM and data mining by selectively processing signals at the sensor level. The SNS is formed using continuous sensors and a biomimetic signal processing architecture. A continuous sensor is a one-circuit connection of multiple sensor nodes that has one output signal that is a combination of signals from the individual sensor nodes. An anomalous response from any sensor node is detected within the combined output signal from all the sensors by filtering and thresholding. Thus, a continuous sensor has only one output signal and can replace many individual sensor nodes and still detect abnormal events.

A trade-off of the continuous sensor is that the exact time response of each sensor node is not available. Only hallmark events are captured by the SNS based on the analog electronics that are used to define an anomalous event. The SNS can be customized by combining many continuous sensors into a network that is designed on the basis of the particular configuration (geometry and materials) of the component. The SNS has a biomimetic architecture that processes signals like the biological neural system—in a highly distributed massively parallel fashion. This means that many continuous sensors (which are akin to biological neurons) can operate all the time, but the neural system processes only signals that contain unusual events. This effectively reduces the burden of data mining because only anomalous events are passed to the data logging system by the SNS. In a concept future digital version of the SNS, tens of continuous sensors each with potentially 10 or more nodes (hundreds of sensors in total) can be monitored using a grid pattern and only four channels of analog to digital signal conversion. An SNS using nanotube spray-on film neurons would detect cracking and corrosion by changes in the electrochemical impedance of the neurons, thus locating and quantifying the size of the damage. More details of the SNS and testing using piezoelectric ceramic sensors are given in **Chapter 147**, and also in [49].

7.2 Multistate continuous sensors

The problem of monitoring different types of components and damage may require that several types of sensors be used in the SNS. Therefore, a recent advance in the capability of the SNS is the development of multistate continuous sensors with different types of sensor nodes. Continuous sensors can use almost any type of sensor node, but within each continuous sensor the sensor nodes must be the same type, because the analog electronics that control the output are matched to the specific sensor type. This means that the output of the sensor network will provide the same information on the location of damage and characteristics of the aberrant waveform, no matter which sensor types are used. By knowing which neuron is "firing", the waveform can be decoded into the appropriate sensor information in the computer. This allows one generic sensor network to be used for SHM, regardless of the type of sensor and system being monitored. Almost any type of sensor node such as accelerometers and strain gauges can be used to form a continuous sensor, but the interface electronics are usually specific for each sensor type. Several types of nanotube type materials can be integrated within a composite material and used as continuous sensors.

7.3 Applications of CNT thread

In many applications, the sensor material can be integrated into composite materials and provide multifunctional properties such as reinforcement, thermal conduction, and damage monitoring. In the aerospace and defense sectors, there are hundreds of likely components that could benefit from simultaneous structural improvement and monitoring using CNT thread. Use of the SNS with CNT neurons has potential for damage detection on large structures such as aircraft, building, bridges, and health monitoring of a cable for the space elevator. Leaving the Planet by Space Elevator [50] is a grand challenge problem for space exploration systems, SHM, and data mining systems. Organizations supporting the space elevator concept include NASA, Black Line Ascension, and the Spaceward Foundation. CNT thread may form a ribbon for the space elevator and EIS might be used to evaluate the condition of the ribbon. A video camera on the elevator will also monitor the ribbon. Data mining from impedance spectra and video will be needed. Long CNTs called *Black Cotton*™ are being spun into thread, which may provide reinforcement, sensing, and partial self-repair simultaneously. Potential applications of CNT are quite open. In complex composite materials, CNT thread can be put where we want it inside composite materials in sections of changing thickness and curvature, joints, around holes, in bond lines, and anywhere health monitoring is required. The thread is mostly inert and will be nanometer or micron thick and will not affect structural integrity or weight. Sensor thread can be put inside all types of structures where other sensors are impractical, such as in flexbeams and rotor blades of helicopters, in composite pressure vessels, in cement, on the surface of bridges, aircraft, space vehicles, and in elastomers.

8 FUTURE EMBEDDED WIRELESS NANOSENSORS

There are many compelling reasons for wanting to embed small sensors or devices inside composite materials. Thus far, technologically, industry has been able to develop radio frequency identification (RFID) tags that can identify a product or object, or micromotes that are battery powdered and report some physical variable. However, these devices are just the beginning. Nanoscale materials are opening up the possibility to build revolutionary devices and tiny sensors that can go inside materials (and the human body) and do what we want them to. This section presents initial analyses toward building an active microsensor that can go inside structural materials and detect damage early. Active microsensors perform electrochemical measurements and produce an electronic signal that is used to detect cracking, delamination, or electrolytes due to corrosion or water ingestion in structural materials. The active microsensor also has a feedback mechanism to increase its sensitivity or to activate a control function. The active microsensor will be built using larger electronic components first, and finally pushed down in size using nanoscale materials. Building small sensors and developing a way to communicate with them is an extraordinary difficult technological problem. To attack this problem, nanoscale electronic components (nanotubes, NWs, capacitors, inductors, solenoids, antennas, transistors, and actuators) and a four-arm nanomanipulator operating under an environmental scanning electron microscope can be used to assemble the components into prototype devices.

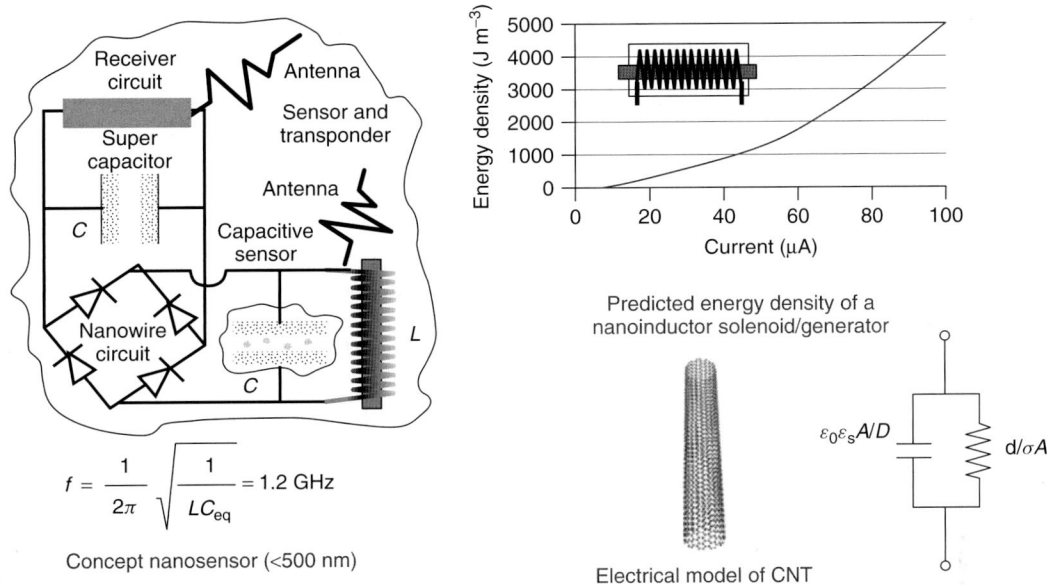

Figure 14. Initial design of a sensor transponder for asset evaluation and condition monitoring of composite structures and later for in-body detection of disease. The sensor works by receiving an RF signal, storing charge, and discharging into a transmitter circuit whose resonant frequency depends on the supercapacitance of nanotube sensors in an electrolyte.

As an example of a novel device, a transponder sensor is proposed (Figure 14) to detect damage in composites.

Developing *in situ* sensors holds great promise for making revolutionary advances in self-sensing nanocomposite materials. The circuit was analyzed using standard electronics circuit and component equations with nanoscale components. Refinement of the model is needed and details of the antenna and transistor circuit are under consideration. Once the sensor design is verified, a method to mass produce the sensors by chemical vapor deposition synthesis will be investigated.

9 SUMMARY AND CONCLUSIONS

This article introduced nanoengineering of sensor materials and discussed their potential for applications in the near future. A major advantage of sensors based on nanomaterials is that several nanomaterials can be blended to fabricate a sensor material that can do what we want it to. The blending can be in a polymer to form a polymer-based sensor material or nanoparticles can be blended in the intermediate form of a thread. The polymer and thread can be used together in composite materials to form self-sensing materials. Prospects for developing new sensors that have high sensitivity are wide open, making nanoengineering of sensors from the bottom up an exciting new field. Engineering asset evaluation and condition monitoring using nanomaterials may have hundreds of applications in components where smart nanocomposites can be used for multifunctional properties improvement, self-sensing of damage, and limited self-repair. In particular, long CNT arrays with further optimization are expected to be an enabling material for structural, electrical, sensing, and thermal applications. "Nanoizing" materials and structures is a new technological science that SHM and NDE engineers should pay attention to.

ACKNOWLEDGMENTS

This work was sponsored by Clean Technologies International Corp., North Carolina A&T SU through the ONR, the Institute for Nanoscale Science and Technology at UC, NSF grant CMS-0510823, and the Korea Institute of Industrial Technology.

Development of the SNS was supported by the National Renewable Energy Laboratory under subcontract number XCX-2-31214-01. Alan Laxson is the technical monitor of the project. Henry Westheider and Douglas Hurd built the fixtures for the smart materials casting and testing. Tom Baca, John Hurtado, and Todd Simmermacher of Sandia National Laboratories sponsored early work on health monitoring. Bradley Edwards of Industrial Nano provided suggestions about SHM of the space elevator ribbon. All this support is gratefully acknowledged.

REFERENCES

[1] Gouma P, Sberveglieri G. *Novel Materials and Applications of Electronic Noses and Tongues*, 2004, http://www.mrs.org/publications/bulletin, MRS BULLETIN/OCTOBER.

[2] Katz E, Willner I. Integrated nanoparticle–biomolecule hybrid systems: synthesis, properties, and applications. *Angewandte Chemie International Edition* 2004 **43**:6042–6108.

[3] Baughman RH, *et al.* Carbon nanotube actuators. *Science* 1999 **284**:1340–1344.

[4] Peng S, O'Keeffe J, Wei C, Cho K, Kong J, Chen R. Carbon nanotube chemical and mechanical sensors. *3rd International Workshop on Structural Health Monitoring*. Stanford, CA, 15–17 September 2003.

[5] Wood JR, Wagner HD. Single-wall carbon nanotubes as molecular pressure sensors. *Applied Physics Letters* 2000 **76**(20):2883–2885.

[6] Kong J, Frankin NR, Zhou C, Chapline MG, Peng S, Cho K, Dai H. Nanotube molecular wires as chemical sensors. *Science* 2000 **287**:622–625.

[7] Shanov VN, Schulz MJ. *Nanoworld and Smart Materials and Devices Laboratories*. University of Cincinnati, 2008, http://altmine.mie.uc.edu/mschulz/public_html/smartlab/smartlab.html.

[8] Kang I, Jung JY, Choi GR, Park H, Lee JW, Yoon KW, Yun Y, Shanov V, Schulz MJ. Developing carbon nanocomposite smart materials. *Solid State Phenomena* 2007 **119**:207–210.

[9] Tombler TW, Zhou C, Alexseyev L, Kong J, Dal H, Liu L, Jayanthl CS, Tang M, Wu SY. Reversible electromechanical characteristics of carbon nanotubes under local-probe manipulation. *Nature* 2000 **405**:769–772.

[10] Watkins AN, Ingram JL, Jordan JD, Wincheski RA, Smits JM, Williams PA. Single wall carbon nanotube-based structural health monitoring sensing materials. *NSTI Conference*. Nanotech, 2004; Vol. 3.

[11] Kang I, Schulz MJ, Kim JH, Shanov V, Shi D. A carbon nanotube strain sensor for structural health monitoring. *Smart Materials and Structures* 2006 **15**(3):737–748.

[12] Jalili N, Goswami BC, Dawson DM. Distributed sensors and actuators via electronic-textiles. *National Textile Center Research Briefs—Materials Competency*, NTC Project: M04-cL05. National Textile Center, June 2005.

[13] Abbott D, *et al. Development and Evaluation of Sensor Concepts for Ageless Aerospace Vehicles: Development of Concepts for an Intelligent Sensing System*, NASA/CR-2002-211773. NASA Langley Research Center, 2002.

[14] Ebron VH, *et al.* Fuel powered artificial muscles. *Science* 2006 **311**(5767):1580–1583.

[15] Zhang M, Shaoli F, Zakhidov AA, Lee SB, Aliev AE, Williams CD, Atkinson KR, Baughman RH. Strong, transparent, multifunctional, carbon nanotube sheets. *Science* 2005 **309**(5738):1215–1219.

[16] Baughman RH. Materials science. Playing nature's game with artificial muscles. *Science* 2005 **308**(5718):63–65.

[17] Zhang M, Atkinson KR, Baughman RH. Multifunctional carbon nanotube yarns by Downsizing an ancient technology. *Science* 2004 **306**(5700):1358–1361.

[18] Baughman RH. Materials science. Muscles made from metal. *Science* 2003 **300**(5617):268–269.

[19] Tsung-Chin H, Loh KJ, Lynch JP. Electrical impedance tomography of carbon nanotube composite materials. In *Sensors and Smart Structures Technologies for Civil, Mechanical, and Aerospace Systems, Proceedings of the SPIE*, Masayoshi T, Chung-Bang Y, Giurgiutiu V (eds). SPIE, 2007; Vol. 6529, pp. 652926.

[20] Meyyappan M. *Carbon Nanotubes: Science and Applications*. CRC Press: Boca Raton, FL, 2005.

[21] Soyoun J, Ji T, Xie J, Jining J, Varadan VK. *A Novel Strain Sensor using Carbon Nanotubes-Organic Semiconductor Matrix Composite on Polymeric Substrates; Smart Structures, Devices, and Systems III, Proceedings of the SPIE*, Al-Sarawi SF (ed). SPIE, 2007; Vol. 6414.

[22] Jin Y, Yuan FG. Simulation of elastic properties of single-walled carbon nanotubes. *Composites Science and Technology* 2003 **63**:1507–1515.

[23] Arun R, Nader J, Himanshu R. Development of a novel strain sensor using nanotube-based materials with applications to structural vibration control. In *Sixth International Conference on Vibration Measurements by Laser Techniques: Advances and Applications, Proceedings of the SPIE*, Tomasini EP (ed). SPIE, 2004; Vol. 5503, pp. 478–485.

[24] Getty S. *SWCNT Nanocompass for Next-Generation Magnetometry. Presented at SAMPE.* Materials Engineering Branch, NASA Goddard Space Flight Center, June 2007.

[25] Lia GY, Wang PM, Zhao X. Mechanical behavior and microstructure of cement composites incorporating surface-treated multi-walled carbon nanotubes. *Carbon* 2005 **43**(6):1239–1245.

[26] Makar J, Margeson J, Luh J. Carbon nanotube/cement composites—early results and potential applications. NRCC-47643. *3rd International Conference on Construction Materials: Performance, Innovations and Structural Implications*. Vancouver, 22–24 August 2005; pp. 1–10.

[27] Xiao X, Elam JW, Trasobares S, Auciello O, Carlisle JA. Synthesis of a self-assembled hybrid of ultrananocrystalline diamond and carbon nanotubes. *Advanced Materials* 2005 **17**(12):1451–1565.

[28] Schulz M. A structural neural system for data mining and anomaly detection. *Data Mining in Aeronautics, Sciences, and Exploration Systems Conference (DMASES)*. Computer History Museum. Mountain View, CA, 26–27 June 2007, http://ase.arc.nasa.gov/projects/dmases/2007/.

[29] Schulz M, Kirikera G, Yun Y, Shanov V, Mullapudi S, Maheshwari G, Allemang R. A structural neural system with multi-state sensors for integrated systems health management. *Proceedings of The 2nd World Congress on Engineering Asset Management (EAM) and The 4th International Conference on Condition Monitoring*. The Cairn Hotel, Harrogate, UK, 11–14 June 2007; pp 1739–1750.

[30] Ghosh S, Sood AK, Kumar N. Carbon nanotube flow sensors. *Science* 2003 **299**:1042–1044.

[31] The Easy Tube System, First Nano. *Carbon Nanotube Synthesis*, Ronkonkoma, NY, http://www.firstnano.com. 2007.

[32] Clean Technologies International Corporation, 2008, http://www.cleantechnano.com/CleanTechNano/.

[33] Applied Sciences, Inc. and Pyrograf Products, Inc., Cedarville, OH, 2008, www.apsci.com.

[34] Carbon Nanotechnologies, Inc., 2008, http://cnanotech.com.

[35] Nanolab, Inc., 2008, info@nano-lab.com.

[36] Shanov V, Gorton A, Yun Y, Schulz M. *Catalyst and Method for Manufacturing Carbon Nanostructured Materials*, Invention disclosure: UC 107-044 (patent pending), 17 October 2006.

[37] Schulz MJ, Kelkar A, Sundaresan M. *Nanoengineering of Structural, Functional and Smart Materials*. CRC Press, 2006.

[38] Andrews R, Weisenberger M. *Carbon Nanotube Polymer Composites: A Review of Recent Developments*. University of Kentucky Center for Applied Energy Research, 2004, www.isr.us/Spaceelevatorconference/.

[39] Kirikera GR. *An Artificial Neural System for Structural Health Monitoring*, MS thesis, University of Cincinnati, 2003.

[40] Kang I, et al. Introduction to carbon nanotube and nanofiber smart materials. *Composites Part B: Engineering* 2006 **37**(6):382–394.

[41] Koo J. *Polymer Nanocomposites: Processing, Characterization, and Applications*, First Edition, McGraw-Hill, April 18 2006.

[42] Shanov VN, Choi G, Maheshwari G, Seth G, Chopra S, Li G, Yun Y, Abot J, Schulz MJ. Structural nanoskin based on carbon nanosphere chains. *SPIE Smart Structures Conference*. San Diego CA, March 2007.

[43] Bentley AK, Farhound M, Ellis AB, Lisensky GC, Nickel A-M, Crone WC. Template synthesis and magnetic manipulation of Nickel nanowires, *Journal of Chemical Education* 2005 **82**:765–768.

[44] Spivak AF, Dzenis YA, Reneker DH. A model of steady state jet in the electrospinning process, *Mechanics Research Communications* 2000 **27**(1):37–42.

[45] Zeng LX, et al. Ultralong single-wall carbon nanotubes. *Nature Materials* 2004 **3**:673–676.

[46] Press release by NSF on long nanotube growth, 2007, http://www.nsf.gov/news/news_summ.jsp?cntn_id=108992&org=NSF&from=news.

[47] *Press Release on Long Nanotube Growth. University of Cincinnati Researchers Grow Their Longest Carbon Nanotube Ever*, 2007, http://www.uc.edu/news/NR.asp?id=4811.

[48] Black Line Ascension, Inc., 2007, www.blacklineascension.com.

[49] Kirikera GR, Shinde V, Schulz MJ, Ghoshal A, Sundaresan MJ, Allemang RJ, Lee JW. A structural neural system for real-time health monitoring of composite materials, *Structural Health Monitoring: An International Journal* 2008 **7**: 65–83.

[50] Ragan P, Edwards B. *Leaving the Planet by Space Elevator*. Lulu.com, 2006.

Chapter 68
Miniaturized Sensors Employing Micro- and Nanotechnologies

Kenneth J. Loh and Jerome P. Lynch
Department of Civil and Environmental Engineering, University of Michigan, Ann Arbor, MI, USA

1 Introduction	1211
2 Microelectromechanical System Sensors	1212
3 Nanoscale Assembly of Sensing Materials	1219
4 Conclusion	1221
References	1221

1 INTRODUCTION

As structural health monitoring (SHM) systems continue to be deployed to monitor and assess engineered structures, there has been growing interest in the miniaturization of sensing transducers used to record structural behavior. The advantages of miniaturization are multiple; for example, reduction in sensor size and weight is critically important when considering the use of SHM systems in lightweight structures such as aerospace structural systems (e.g., rockets and satellites). Smaller sensors are also easier to install, particularly if the sensor is embedded within structural elements; an example might be

Encyclopedia of Structural Health Monitoring. Edited by Christian Boller, Fu-Kuo Chang and Yozo Fujino © 2009 John Wiley & Sons, Ltd. ISBN: 978-0-470-05822-0.

thin-film sensors installed within a layered composite material (e.g., carbon fiber-reinforced polymer (CFRP) composites). In addition, miniaturization can lead to potential improvements in sensing accuracy, simultaneous to significant reductions in fabrication costs. Finally, microelectromechanical system (MEMS) sensors are more power efficient than their macroscale counterparts.

Macroscale sensor design concepts can be miniaturized by adopting emerging technologies associated with the MEMSs and nanotechnology fields. Already, MEMS has had a major impact with SHM systems widely using MEMS sensors [1]. MEMS is defined by the use of fabrication methods associated with integrated circuits (ICs) to construct mechanical structures within semiconductor substrates such as silicon (Si), germanium (Ge), and gallium arsenide (GaAs) [2]. By miniaturizing macroscale sensor transduction concepts to the micron-dimensional scale, the approach is commonly termed a *top-down* methodology. Since the initial introduction of MEMS pressure sensors in the 1960s [3], the field of MEMS has rapidly evolved over the past four decades. Today, a plethora of other miniaturized sensing transducers have been proposed by the MEMS community including accelerometers, gyroscopes, gas sensors, and ultrasonic transducers among many others [4]. These examples of MEMS sensors have been shown to offer measurement accuracies on par with macroscale counterparts. MEMS adoption of IC technologies for device fabrication allows computing and wireless communication circuits to be collocated with the MEMS sensor, thereby offering complete system-on-a-chip (SoC) solutions [2]. Furthermore, IC-based manufacturing offers fabrication of MEMS sensors by a batch process, with hundreds of devices fabricated on a single semiconducting wafer [4]. While MEMS sensors have shown tremendous promise, their market adoption has been greatest in automotive (e.g., pressure and acceleration sensing) and inertial sensing markets where high sales volume is able to amortize high fabrication costs [2, 5].

In light of the limitations of MEMS, the interdisciplinary nanotechnology field has emerged to offer chemical tools and processes that permit further miniaturization of sensors designed for SHM applications [6]. Specifically, it is now possible to design materials with specific macroscopic mechanical, electrical, and chemical properties by controlling structure and assembly at the atomistic length-scale, which is nanometers in dimension [7]. This "bottom-up" approach is in stark contrast to the "top-down" design methodology currently adopted by MEMS. Currently, molecular structures such as nanotubes [8], nanoparticles [9], and self-assembled materials [10] are finding their way into the design of sensors [11]. Clearly, future advances in miniaturized sensors will be derived from the technological developments within the nanotechnology domain.

The balance of this review is delineated into two major sections. In the first section, an overview of MEMS is provided. The general methods of MEMS fabrication are described, including bulk and surface micromachining methods. In addition, a plethora of MEMS sensors that have found use in SHM applications are presented. To illustrate the impact MEMS has had on the SHM field, applications of MEMS sensors to SHM problems investigated in the laboratory and field settings are also described (*see* **Chapter 81**). In the second section, the tools and processes of the nanotechnology field relevant to the miniaturization of sensors are presented (*see* **Chapter 67**). In essence, nanotechnology allows engineers to address many of the technological limitations currently encountered in MEMS design. A novel suite of sensors assembled at the nanoscale is highlighted to reinforce the future promise of the "bottom-up" sensor design approach for SHM.

2 MICROELECTROMECHANICAL SYSTEM SENSORS

2.1 Fabrication of MEMS structures

There are two major fabrication methods employed in MEMS design to construct three-dimensional structures, namely, bulk and surface micromachining. In the first method, bulk micromachining, the structure is built into the substrate through the removal of substrate material. In contrast, surface micromachining builds the MEMS structure on top of a substrate by depositing materials to the substrate surface. How a MEMS sensor is fabricated is very important to consider at the outset of the design process since each fabrication method uniquely constrains the final geometric shape of the MEMS device. In this section, the general steps of both fabrication approaches are delineated; methods presented are primarily for silicon, which is the preferred substrate material of most MEMS foundries. The interested reader is referred to Kovacs [2] for a more complete presentation of each fabrication approach and for information on how nonsilicon substrates such as Ge and GaAs can be integrated with the MEMS fabrication process.

2.1.1 Bulk micromachining

In bulk micromachining, three-dimensional MEMS structures are fabricated within the substrate by etching. Etching is the process by which substrate material is selectively removed to create structures within the substrate. All etching methods can be classified as part of two major etching categories: wet and dry etching.

In wet etching, liquid chemical compounds react with the silicon substrate so as to remove substrate material. In general, wet etchants are either isotropic or anisotropic depending upon their relationship with the silicon crystalline structure. For example, isotropic etchants remove silicon in all crystal directions at the same rate; the result is structures with rounded features. A common isotropic etchant is HNA, which is a solution of hydrofluoric, nitric, and acetic acids (HF, HNO_3, and CH_3COOH, respectively). In contrast, anisotropic etchants possess different reaction times depending upon the silicon crystal orientation. Using Miller index notation to denote the various planes of crystalline silicon, it is the (111) plane of silicon that is the slowest to react with anisotropic etchants. Owing to the slow etch time of the (111) plane, it is preferentially exposed during etching. In other words, the (111) plane that is exactly $54.7°$ relative to the (100) plane is exposed in the final structure after anisotropic etching. Alkali hydroxides (e.g., potassium hydroxide (KOH)) are typically used for anisotropic wet etching of silicon.

In dry etching, noble gas fluorides (e.g., xenon fluoride (XeF_2)) are introduced in a chamber containing a silicon substrate to induce isotropic substrate etching. Dry etching using noble gas fluorides can be tightly controlled through the temperature and pressure of the chamber. An additional benefit of dry etching is that traditional masking can be used to selectively etch the substrate. However, one drawback is the final etched structure is defined by high surface roughness. Another approach to dry etching is reactive ion etching (RIE). An RIE chamber consists of parallel-plate electrodes to which an alternating current (AC) is applied, resulting in an oscillatory electromagnetic (EM) field. As fluorinated gases enter the sealed chamber, the EM field strips electrons and produces highly energized ions. When a silicon wafer is placed upon one of the electrodes, the accelerating ions bombard the substrate surface with high energy, resulting in the removal of substrate material. The ion flux is generally perpendicular to the substrate, resulting in bulk micromachined structures defined by vertical side walls. For silicon, the etch gas used in RIE is generally sulfur hexafluoride (SF_6). An extension of RIE in which polymer is deposited along the sidewalls of the etched structure is termed *deep reactive ion etching* (DRIE). DRIE has gained considerable popularity in recent years owing to the high-aspect ratio structures attainable; smooth, polymer-coated vertical wall structures with aspect ratios as high as 30:1 have been achieved [2].

Both wet and dry etchings are assisted using masking and sacrificial layers. For example, silicon dioxide (SiO_2) can be thermally grown upon the surface of a silicon wafer to serve as a masking layer during wet etching. Also, photosensitive resist layers (photoresists) can be deposited upon the surface of a silicon wafer as a masking layer. Using optical lithography, portions of the resist can be exposed to ultraviolet (UV) light; after exposure, the exposed resist is dissolvable in a solvent such as acetone.

2.1.2 Surface micromachining

When employing surface micromachining for fabrication of a MEMS structure, the structure is built upon the substrate using structural thin films, including polycrystalline silicon, aluminum, and silicon nitride (Si_3N_4). Integral to surface micromachining is the incorporation of sacrificial layers of silicon dioxide and photoresists. In the fabrication of MEMS structures, both structural and sacrificial layers can be etched to selectively remove portions of each layer. Typically, wet etching of silicon dioxide is done using HF, while polycrystalline silicon is wet-etched using traditional silicon etchants. Dry etching can also be used in a surface micromachining process. In surface micromachining, MEMS structures are generally defined by large in-plane areas and small out-of-plane thicknesses; in essence, two-and-a-half dimensional (2.5D) structures are formed.

2.1.3 Fabrication of a cantilever

To illustrate the fabrication methods presented above, a simple MEMS structure widely used in the design of MEMS accelerometers is adopted. The steps included in the fabrication of a cantilever beam are detailed for both bulk and surface micromachining processes [12]. The bulk micromachining fabrication process begins with a silicon wafer upon which a layer of silicon dioxide is thermally grown. In addition, a thin layer of photoresist is deposited over the silicon dioxide. Using a photomask containing a pattern of the cantilever, lithography is used to selectively expose the resist layer (Figure 1a). Placement of the wafer in a developer solvent removes the exposed resist (Figure 1b) (in this case, a positive photoresist is employed, where areas exposed to UV light are removed by the solvent). The exposed layer of silicon dioxide can be selectively removed where there is no resist using hydrofluoric acid (Figure 1c). In the last step, the wafer is placed in a wet etchant bath (e.g., KOH) to perform anisotropic etching (Figure 1d). If a [100] wafer is used, a deep trench beneath the cantilever will result with the (111) plane revealed.

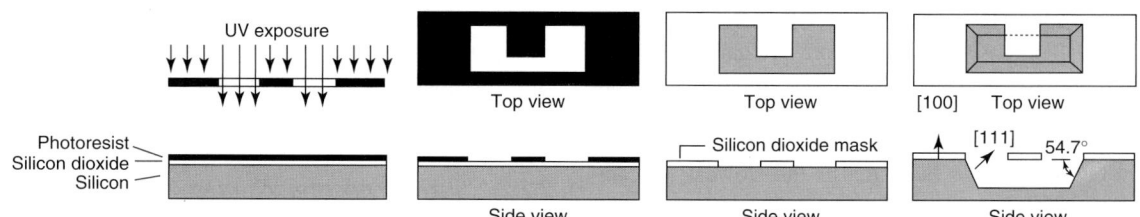

Figure 1. Fabrication of a cantilever by bulk micromachining: (a) silicon wafer with a silicon dioxide layer buried under photoresist is exposed to UV light; (b) resist is dissolved to reveal the underlying silicon dioxide layer; (c) silicon dioxide is etched to reveal the underlying silicon; (d) anisotropic wet etching of silicon results in undercutting of a cantilever element.

The surface micromachining process begins with a silicon wafer upon which silicon dioxide is thermally grown; in this process, silicon dioxide will serve as a sacrificial layer. Positive photoresist is again deposited upon the silicon dioxide and exposed to UV light so as to expose the underlying layer of silicon dioxide (Figure 2a). The silicon dioxide is etched using a wet etchant (such as HF) to reveal the underlying silicon substrate. The remaining silicon dioxide will serve as a sacrificial layer that will be removed during subsequent steps (Figure 2b). A structural layer of polycrystalline silicon is patterned and deposited upon both the silicon and silicon dioxide layers (Figure 2c). To release the cantilever, the remaining silicon dioxide beneath the cantilever is etched (Figure 2d).

2.2 Accelerometers

Accelerometers represent one of the most successful examples of a MEMS sensor in the commercial market. MEMS accelerometers have been used to measure acceleration in automotive (e.g., airbag deployment) [13], aerospace (e.g., inertial sensing) [14], and medical (e.g., pacemaker regulation) [15] applications. The design of all MEMS accelerometers consists of four major functional components within a single sensor package: proof mass, spring, damper, and readout mechanism (Figure 3a). The proof mass is a passive element that undergoes displacement due to inertial forces. The spring restrains the mass motion through a restoring force. Damping is necessary to ensure that the ratio of mass displacement to acceleration amplitude is constant up to the resonant frequency (f_R) of the accelerometer's mass–spring system. Since the displacement of the proof mass will be linearly proportional to the acceleration of the sensor package, an electromechanical transducer is needed to convert the proof mass displacement to an electrical signal. MEMS accelerometers use either capacitive, piezoresistive, or piezoelectric electromechanical transducers as readout mechanisms.

The very first MEMS accelerometer fabricated is the one proposed by Roylance and Angell [16]. The accelerometer is bulk micromachined in silicon using anisotropic wet etching. The device consists of a square proof mass connected to its silicon housing through a thin cantilever element acting as a spring. A scanning electron microscope image of the accelerometer is shown in Figure 3(b); note that the

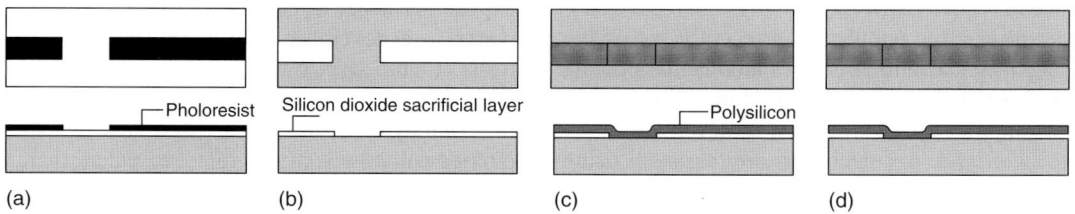

Figure 2. Fabrication of a cantilever by surface micromachining: (a) exposed resist (after lithographic exposure) is dissolved to reveal the underlying silicon dioxide layer; (b) silicon dioxide is etched to reveal the underlying silicon; (c) polycrystalline silicon is deposited over the silicon dioxide sacrificial layer; (d) selective etching of the silicon dioxide is done to release the polycrystalline silicon cantilever.

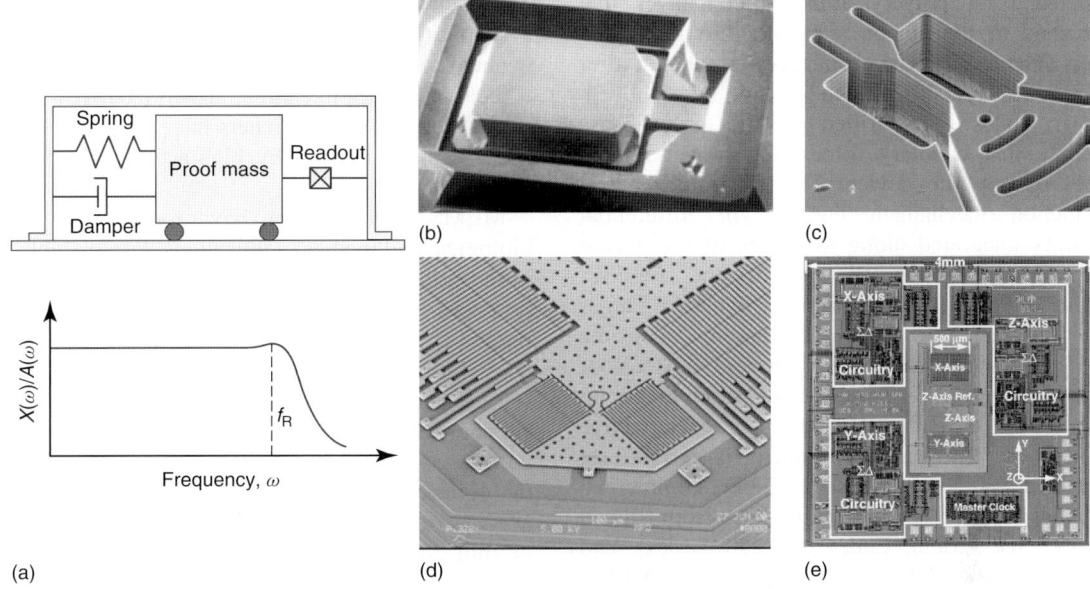

Figure 3. (a) Conceptual components of a MEMS accelerometer including a typical mass displacement–acceleration transfer function $(X(\omega)/A(\omega))$; (b) piezoresistive accelerometer by Roylance and Angell [Reproduced with permission from Ref. 16. © IEEE, 1979]; (c) high-performance planar piezoresistive accelerometer by Partridge *et al.* [Reproduced with permission from Ref. 17. © IEEE, 2000]; (d) close-up microscopic image of the Analog Devices *i*MEMS capacitive accelerometer with differential capacitors and folded springs clearly shown [Reproduced with permission from Ref. 18. © Anolog Devices, Inc., 1999]; (e) three-axis force-balanced accelerometer die with accompanying circuitry by Lemkin *et al.*. [Reproduced with permission from Ref. 19. © IEEE, 1997.]

side walls are at a 54.7° angle with the top surface of the substrate. Air within the sealed device (the silicon structure is sandwiched between two glass plates) provides viscous damping. A piezoresistor is implanted upon the top face of the cantilever to serve as the primary readout mechanism of the sensor; as the proof mass deflects, the cantilever beam experiences strain, which in turn results in a proportional change in resistance of the implanted piezoresistor. The accelerometer is reported to have a low noise floor ($0.001g$), large range ($\pm 200g$), and high bandwidth (2 kHz).

Since the seminal work by Roylance and Angell [16], a number of researchers have proposed piezoresistive MEMS accelerometers. Most recently, Partridge *et al.* [17] have prototyped a high-performance planar accelerometer for shock applications. Their accelerometer uses DRIE to define a pie-slice-shaped proof mass connected to the silicon substrate through a thin cantilever element (Figure 3c). Boron implantation in the cantilever results in a piezoresistive element for the readout of the sensor acceleration measurements. An attractive feature of the accelerometer is that the proof mass displacement is in the plane of the substrate. This restrains undesirable displacements of the proof mass during high-acceleration (shock) motions, thereby protecting the accelerometer from failure. The full dynamic range of the accelerometer is well above $10g$ with a resolution of $20\,\mu g$ at an acceleration bandwidth of 650 Hz.

While piezoresistive accelerometers are easy to fabricate and offer simple electrical interfaces, they exhibit extreme sensitivity to temperature that hampers their use in applications where environmental temperatures vary [20]. As a result, piezoresistive accelerometers have given way to MEMS accelerometers using capacitive readout mechanisms. Capacitive accelerometers have been successfully implemented in the commercial sector by Analog Devices (Norwood, MA). The Analog Devices *i*MEMS accelerometer family adopts a surface micromachining process to fabricate one- and two-axis accelerometers in polycrystalline silicon [18]. The accelerometer's square proof mass is fabricated in a polycrystalline silicon layer that is attached to the silicon substrate through folded springs (Figure 3d). The proof mass is released after a sacrificial layer of silicon dioxide is removed; "through holes" are drilled in the proof mass to ensure that capillary forces during wet release do not destroy the device. The electromechanical transduction mechanism consists of differential capacitors integrated along the sides of the square proof mass. Various version of the *i*MEMS accelerometer exist with different dynamic ranges (e.g., from ± 1.2 to $250g$). Depending upon the user's application, the bandwidth and noise floor of the sensor can be varied using discrete circuit elements installed along with the sensor. An attractive feature of the *i*MEMS accelerometer is its low cost (approximately $10 per sensor). Other commercial capacitive accelerometers include those from Silicon Designs (SD Series), Endevco (7290 Series), and Crossbow (CXL-LF Series). A force-balanced MEMS accelerometer based on a capacitive design has also been proposed by Lemkin *et al.* [19] (Figure 3e).

MEMS accelerometers have been applied to SHM applications over the past decade. In the civil engineering sector, MEMS accelerometers are low-cost alternatives to traditional seismic accelerometers (e.g., piezoelectric and force-balanced accelerometers). For example, a dense array of Crossbow CXL-LF accelerometers has been instrumented on the Geumdang Bridge (Icheon, Korea) to measure the bridge response to traffic [21]. The Analog Devices ADXL202 and Silicon Design SD2210 MEMS accelerometers have also been deployed on a pedestrian bridge on the University of California, Irvine campus [22]. The high-performance planar accelerometer proposed by Partridge *et al.* [17] has been validated to measure the acceleration of the Alamosa Canyon Bridge (southern New Mexico) [23]. It is interesting to note that in all of these cases, the MEMS accelerometers have been interfaced to battery-powered wireless sensors. This is due in part to the low-power requirements of MEMS. These applications serve as powerful examples of the success MEMS accelerometers have had in SHM.

2.3 Strain sensors

Damage (e.g., cracking) is typically localized to a specific location in a structure; as a result, measurement of global structural responses (e.g., acceleration) is generally inadequate for accurate damage detection. Damage detection can be improved if local responses like strain are utilized. Toward this end, a number of researchers have proposed MEMS strain sensors defined by high-accuracy and small-form factors.

MEMS strain sensors have found early use in biological and medical applications. For example, one of the earliest MEMS strain sensors includes a sensor proposed in 1980 for measuring strain in animal tissue [2]. The strain sensor is fabricated from silicon using bulk micromachining techniques to define the sensor geometry and to implant a piezoresistive element. Strain sensors have also been included in the design of MEMS-based neural probes to measure the forces used to penetrate tissue [24]. Recently, MEMS strain sensors proposed for measuring strain in thin films [25–27] have matured with devices now suitable for measuring strain in structural systems.

A novel MEMS strain sensor is proposed by Hautamaki *et al.* [28] for embedment in CFRP composite plates. Their sensor employs a doped polycrystalline silicon element acting as a piezoresistor placed upon a silicon nitride cantilever element (Figure 4a). Upon embedment in the epoxy matrix during construction, strain in the composite panel would be transferred to the cantilever, thereby allowing strain to be accurately measured. Their sensor exhibits a 1–2% change in resistance at $1000\,\mu\varepsilon$.

While piezoresistive MEMS strain sensors are easy to install, they exhibit sensitivity to temperature, thus requiring compensatory circuitry. In contrast, capacitive strain sensors are more stable in varying thermal environments. Aebersold *et al.* [29] propose a bulk micromachined capacitive strain sensor for measuring strain in bending structural elements. On the basis of an interdigitated finger design, the capacitive sensor is fabricated in a 150-μm-thick silicon wafer using DRIE (Figure 4b). The sensor is tested on a structural element loaded in four-point bending, resulting in a repeatable nonlinear change in capacitance as a function of strain. A high-performance MEMS strain sensor based on capacitive sensing is also proposed by Ko *et al.* [31]. Designed for measuring strain in rotating machinery, the device features a resolution of $0.09\,\mu\varepsilon$ and is able to accurately measure static and dynamic strains up to 10 kHz. While silicon is

Figure 4. MEMS strain sensors: (a) piezoresistive strain sensor fabricated on a silicon nitride cantilever for impregnation in fiber-reinforced composite panels [Reproduced with permission from Ref. 28. © IEEE, 1999]; (b) capacitive strain sensor for measurement of bending strain [Reproduced with permission from Ref. 29. © IOP Publishing Ltd., 2006]; (c) SiC MEMS resonant strain sensor for harsh environments. [Reproduced with permission from Ref. 30. © IEEE, 2007.]

a robust structural material, it can experience wear and corrosion in harsh environments. As a result, a MEMS strain sensor fabricated from polycrystalline silicon carbide (SiC) has been proposed for SHM of high-temperature components (e.g., turbine blades) [30]. This MEMS strain sensor is designed as a comb-driven tuning fork resonator whose resonant frequency varies in linear proportion to strain (Figure 4c); a resolution of 0.11-$\mu\varepsilon$ and 10-kHz bandwidth is achieved in the laboratory.

In the commercial sector, a low-cost MEMS strain sensor has been marketed by Sarcos [32]. Their uniaxial strain transducer (UAST) is a MEMS device in which strain is correlated to signals generated between electrostatic field emitters and measured by on-chip field detectors. This device has been successfully applied to monitor the health of railroad tracks as train cars traverse the track; peak strain is measured and stored for further fatigue analysis [32].

2.4 Acoustic transducers

Acoustic emission (AE) sensing is an important technology in the SHM field. AE sensing strategies seek to capture transient stress waves generated by damage initiation in a structure in order to identify and quantify the damage. Traditionally, AE requires piezoelectric ceramic transducers that are mounted to the surface of the structure to capture stress waves; however, transducers can be both costly and bulky [33]. MEMS represents an enabling technology for the reduction of both the size and cost of AE sensors. Toward that end, a variety of ultrasonic transducers fabricated as MEMS devices have been proposed in the SHM literature. The majority of MEMS sensors proposed for AE has been surface micromachined based on capacitive mechanisms. Jones et al. [34] propose the design of a silicon nitride membrane deposited on a silicon wafer to produce a resonant cavity. Deposition of electrodes on the bottom of the cavity and upon the top surface of the membrane creates a capacitor whose capacitance varies with membrane deflection. Square cavity geometries are used (1 mm^2), while the membrane thickness is adjustable from 1 to 2 μm with thinner membranes exhibiting greater sensor sensitivity. The authors have conducted various laboratory tests of the AE MEMS sensor using CFRP composite panels. The tests revealed excellent AE detection to ball drops and pencil lead-breaks on the CFRP panel.

Ozevin et al. [33] extend on the AE MEMS sensor design proposed by Jones et al. [34] by fabricating seven AE detectors on a single die; each detector is tuned to a different resonant frequency spanning from 100 kHz to 1 MHz. The readout mechanism utilized in the AE sensor is a two-plate capacitor with one plate moving in response to the AE stress wave; changes in plate location result in a measurable change in device capacitance. The AE sensor is fabricated using surface micromachining of polycrystalline silicon in the commercial multiuser MEMS processes (MUMPs) foundry. The performance of the fabricated AE MEMS sensor is compared to a commercial piezoelectric transducer during laboratory testing. Both sensors are surface-mounted to a steel beam in which a weld is included in the beam center. Cyclic three-point bending of the beam results in fatigue cracks in the weld; the performance of the MEMS sensor is shown to be comparable to the commercial piezoelectric AE sensor.

2.5 Corrosion sensors

Corrosion is a serious structural deterioration confronting many engineered structures including aircrafts, bridges, and machineries among others. Over the past few years, a number of novel corrosion sensors based on MEMS technology have been proposed in the literature. Niblock *et al.* [35] describe a MEMS fabricated corrosion sensor based on the measurement of linear polarization resistance (LPR); since LPR is correlated to the rate of corrosion, the sensor provides insight into the speed of the corrosion process. The sensor proposed is fabricated from the host metal, which is to be monitored by the MEMS corrosion sensor. The metal is machined into small 25 mm squares and ground to thicknesses as small as 50 µm. The square stock is then chemically etched to form interdigitated fingers separated by a small gap (150 µm). After formation of the fingers, a thermosetting polymer is used to fill the gaps between the electrodes. When a small potential (ΔE) is externally applied across the electrodes, a current flow at the metal surface can be measured (ΔI). The slope of the $\Delta E - \Delta I$ represents the LPR; the LPR can be used to determine the corrosion current directly. Laboratory tests of the LPR sensor reveal its ability to accurately measure the corrosion current in steel specimens.

The Southwest Research Institute (SwRI) has proposed a MEMS sensor for monitoring stress corrosion cracking [36]. Their device consists of a cantilevered beam that is fabricated from the same material for monitoring corrosion. The cantilever is notched along its center so as to allow it to crack under stress and corrosion. Measurement of the beam resistance is directly related to the splitting of the beam (which is assumed to be correlated to stress corrosion cracks developing in the host structure). For additional discussion regarding application of corrosion monitoring for aircraft applications, (*see* **Chapter 112**).

2.6 Wireless interdigital transducers

Another class of MEMS sensors more recently adopted by the SHM community is known as *surface acoustic wave (SAW) sensors*. SAW sensors are physically small and can be easily designed for wireless interrogation, thereby making them suitable for embedment within structural components. In order to generate SAWs within a piezoelectric substrate material (*see* **Chapter 52** for a discussion on piezoelectricity principles and materials), interdigitated transducers (IDTs) are patterned on the substrate surface using traditional lithography and metallization. Basically, IDTs consist of thin-film comb-shaped electrodes (typically patterned using aluminum) mounted on top of a piezoelectric element (Figure 5). When a voltage is applied to the IDT electrodes, dynamic strain is induced in the piezoelectric substrate, which in turn generates a SAW [5]. The electrode finger spacing (d) is controlled to be half of the SAW wavelength, λ. This spacing ensures that stress waves traveling in both directions from an IDT constructively interfere. The simplest IDT–SAW sensor, known as the *nondispersive delay line*, consists of a pair of IDTs. One IDT connected to an AC source acts as the SAW generator, while the other serves as the sensing element. Depending on the crystallographic cut of the piezoelectric element and the orientation of the IDT electrodes, various types of acoustic waves (e.g., Rayleigh, Love, and shear horizontal) can be generated [37].

In general, SAW sensors correlate a physical phenomenon (e.g., strain) with a change in the

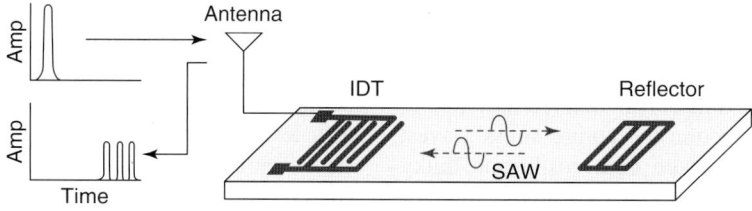

Figure 5. Surface acoustic wave (SAW) sensor with an interdigitated transducer (IDT) patterned upon the piezoelectric substrate along with a reflector element. A wireless interface receives a pulse signal. Reception of the pulse results in multiple reflections that are emitted wirelessly after a short delay.

properties of the SAW elastic wave such as velocity, v, and attenuation. Such changes can be measured by observing changes in the SAW resonant frequency, $f = v/\lambda$, as well as the phase and amplitude differences between the sensor input and output signals [37]. For example, in SAW strain sensors, strain induces a change in the separation distance between reflectors. This results in a measurable phase lag between the sensor's input–output signals, such that the measured phase lag is linearly proportional to strain [5, 38]. Many variations of SAW sensors have been proposed for gas [39], chemical [40], and acceleration [5] measurements, among others [41]. For many of the sensors proposed, antennas are integrated with the SAW sensor to accommodate input excitation and output measurements by a wireless interface [5].

3 NANOSCALE ASSEMBLY OF SENSING MATERIALS

While MEMS processing has offered a plethora of high-performance sensors for SHM, the fundamental design principle is derived from miniaturization of macroscale devices. Despite advances in lithographic patterning, etching methods, and deposition techniques, there exist technological limitations to MEMS and IC manufacturing. For example, lithographic patterning is difficult at nanometer length-scales, thereby retarding continued miniaturization of MEMS structures. Alternatively, nanotechnology offers tools and techniques that allow for the intentional assembly of novel materials at molecular length-scales (*see* **Chapter 67**). Today, the IC and MEMS industries are investing heavily in nanotechnology so that transistors and MEMS structures can continue to be miniaturized [42].

The SHM community will also be a beneficiary of the nanotechnology field. In particular, nanotechnology offers a "bottom-up" assembly approach to the design of multifunctional materials. A multifunctional material is a material system that achieves multiple functional objectives, such as the ability to take load and the capability to sense its response to loading. Integral to the development of multifunctional materials is the use of nanometer-scaled structures (e.g., carbon nanotubes (CNTs) [43], carbon black [44], and nanoparticles [45], among others) included in composite material assemblies. While many of these molecular building blocks have been employed in the design of sensors for SHM, the use of CNTs (Figure 6a) has been very popular due to their impressive physical, mechanical, and electrical properties [43, 46–53].

3.1 Multifunctional carbon nanotube materials

Since the discovery of CNTs by Iijima in 1991 [8], researchers have sought to take advantage of their impressive electrical and mechanical properties for developing novel sensors for SHM [48]. Early realization of CNTs as ideal candidates for strain sensing stems from experimentally measuring the electromechanical response of individually suspended CNTs subjected to localized atomic force microscope probe deformations [49]. Both *in situ* resistance measurements and molecular dynamics simulation results indicate that single-walled carbon nanotube (SWNT) conductance can decrease more than two orders of magnitude while remaining completely reversible over multiple deformation cycles. Although individual CNTs have demonstrated piezoresistive response to applied strain, SHM applications require macroscale sensors that can be installed and queried in engineered structures. In an effort to preserve the inherent piezoresistive properties of individual CNTs while scaling up from a nano- to a macroscale sensing material, researchers have proposed inclusion of CNTs within polymeric matrices. The result is multifunctional materials that can take considerable load [45], yet are capable of self-sensing their strain response to loading [47].

A number of researchers have embedded CNTs within polymeric matrices for strain-sensing applications. Wood *et al.* [50] embedded CNTs within a UV-curable urethane acrylate polymeric thin film. After correcting for temperature-induced strains, they showed that Raman spectra peaks decrease with applied tensile strain. Since Raman spectroscopy requires bulky equipment and is difficult to use in the field, many other researchers have proposed CNT-based "buckypaper" specimens for strain sensing [51–53]. Upon vacuum filtration of a dispersed CNT solution, buckypaper specimens can be peeled off the filter to form the final thin-film strain sensor.

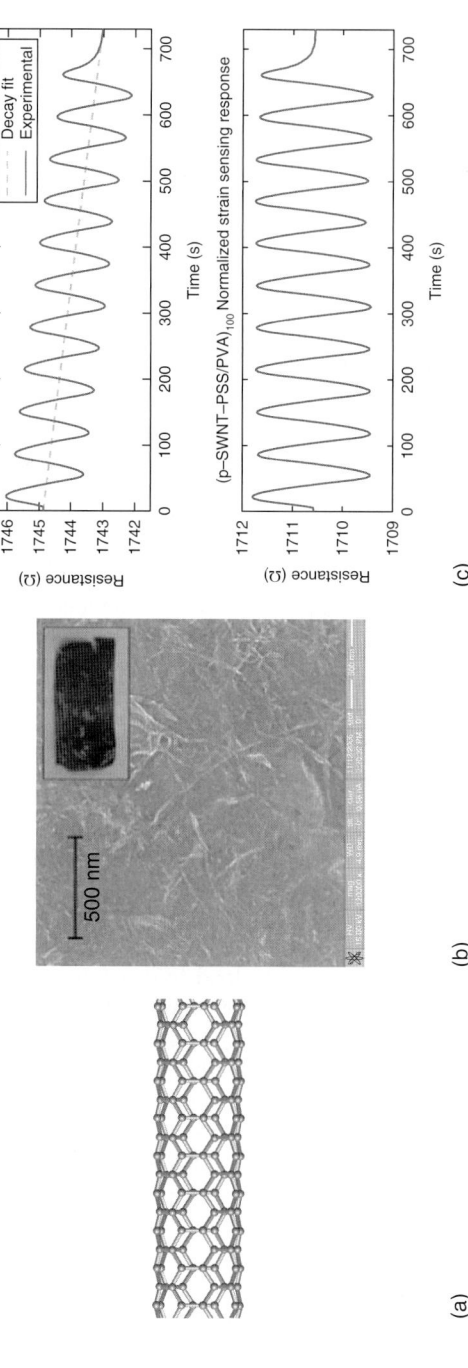

Figure 6. (a) Single-walled carbon nanotube (SWNT) [Courtesy Prof. Vincent Crespi, Penn State]; (b) microscopic view of SWNT–polymeric multifunctional material (insert: final free-standing thin film) [Reproduced with permission from Ref. 46. © IOP Publishing Ltd., 2007]; (c) variation of SWNT-based multifunctional material resistance as a function of mechanical strain. [Reproduced with permission from Ref. 47. © IOP Publishing Ltd., 2007.]

Initial studies conducted by Dharap et al. [51] have shown that, using a movable four-point probe technique, SWNT buckypapers exhibit multidirectional linear piezoresistive response up to $\pm 350\,\mu\varepsilon$ strains. Continued work by Li et al. [52] measures the Raman wave number shift of the G-band of CNT films and confirms that nanotubes within the buckypaper matrix are strained during applied tensile strains (up to $1000\,\mu\varepsilon$). Using buckypapers fabricated with SWNTs dispersed in dimethyl formamide and then mixed with poly(methyl methacrylate) (PMMA), Kang et al. [53] have derived an equivalent electrical circuit model using electrical impedance spectroscopy (EIS) to characterize the SWNT–PMMA thin films. Although SWNT–PMMA thin films exhibit a decrease in gauge factor (strain sensitivity) with increasing CNT weight concentration, pure SWNT buckypapers exhibit a gauge factor of 7 while behaving linearly within $\pm 500\,\mu\varepsilon$ strains. However, strain sensors for structural health monitoring must be robust and reliable over long periods of time. In particular, strain sensors require high sensitivity, linearity, and a wide measurable range (e.g., more than $5000\,\mu\varepsilon$).

Although a variety of carbon nanotube-based thin-film strain sensors have been presented for structural monitoring applications, each aforementioned sensor suffers from some drawbacks. Typically, buckypaper strain sensors exhibit high gauge factors but are very brittle materials; this limits their measurable strain range to within $1500\,\mu\varepsilon$. In response to these limitations, a more homogenous thin-film composite with CNTs within a polymeric matrix can be achieved by layer-by-layer (LbL) self-assembly [45–47]. LbL thin films are assembled by sequential dipping of a charged substrate in polyanionic and polycationic solutions to achieve excellent phase integration and homogenous morphology. Loh et al. [47] propose the design of multilayered thin films assembled from solutions containing SWNTs stabilized in poly(sodium 4-styrene sulfonate) (PSS) and poly(vinyl alcohol) (PVA). Free-standing thin films are deposited on a glass substrate with 50–200 layers typically deposited. Subsequent scanning electron microscopy (Figure 6b) reveals good dispersion of SWNTs within the PSS/PVA matrix. Upon application of uniaxial tensile loading, the SWNT-PSS/PVA thin film is shown to be piezoresistive with excellent linearity and high sensitivity (gauge factors span from 1 to 2) (Figure 6c).

4 CONCLUSION

MEMS is a revolutionary technology that positively impacts the field of SHM. MEMSs offer the community microscale sensors that are as accurate as macroscale counterparts at a fraction of the cost and size. Other benefits include the collocation of signal-processing circuitry on the sensor and low-power consumption. As battery-powered technologies such as wireless sensors grow in popularity, low-power MEMS sensors will serve to preserve scarce battery resources. A variety of MEMS sensors have been proposed for use in SHM applications including accelerometers, strain gauges, corrosion sensors, and acoustic transducers. Many novel design and fabrication techniques are constantly being proposed and implemented in the MEMS community to design high-performance sensing devices. On the other hand, the field of nanotechnology also promises to further improve SHM sensing technologies. Unlike the MEMS' "top-down" design methodology, nanotechnology enables molecular manipulation to achieve high-sensitivity and -selectivity, multifunctional sensors. To date, CNT-based multifunctional materials have been developed, which self-sense their response to loading and self-heal in response to damage. In future years, nanotechnology will play an increasingly greater role in the development of next-generation miniaturized sensing technologies. The combination of MEMS low-cost high-throughput fabrication techniques with the versatility of sensor designs originating from the nanotechnology domain can yield novel sensors that outperform those of the current generation.

REFERENCES

[1] Glaser SD, Li H, Wang M, Ou J, Lynch JP. Sensor technology innovation for the advancement of structural health monitoring: a strategic program of US-China research for the next decade. *Smart Structures and Systems* 2007 **3**:221–244.

[2] Kovacs GTA. *Micromachined Transducers Sourcebook*. McGraw-Hill: New York, 1998.

[3] Petersen KE. Silicon as a mechanical material. *Proceedings of the IEEE* 1982 **70**:420–457.

[4] Gardner JW. *Microsensors: Principles and Applications*. John Wiley & Sons: West Sussex, 1994.

[5] Gardner JW, Varadan VK, Awadelkarim OO. *Microsensors, MEMS and Smart Devices*. John Wiley & Sons: West Sussex, 2001.

[6] Bhushan B (ed). *Springer Handbook of Nanotechnolgy*. Springer: Berlin, 2003.

[7] Nalwa HS. *Nanostructured Materials and Nanotechnology*. Academic Press: San Diego, CA, 2002.

[8] Iijima S. Helical microtubules of graphitic carbon. *Nature* 1991 **354**:56–58.

[9] Klein DL, Roth R, Lim AKL, Alivisatos AP, McEuen PL. A single-electron transistor made from a cadmium selenide nanocrystal. *Nature* 1997 **389**:699–701.

[10] Brinker CJ, Lu Y, Sellinger A, Fan H. Evaporation-induced self-assembly: nanostructures made easy. *Advanced Materials* 1999 **11**:579–585.

[11] Mahar B, Laslau C, Yip R, Sun Y. Development of carbon nanotube-based sensors—a review. *IEEE Sensors Journal* 2007 **7**:266–284.

[12] Senturia SD. *Microsystem Design*. Kluwer Academic Press: Boston, MA, 2001.

[13] Valldorf J, Gessner W (eds). *Advanced Microsystems for Automotive Applications 2005*. Springer: Berlin, 2005.

[14] Cass S. MEMS in space. *IEEE Spectrum* 2001 **38**:56–61.

[15] Panescu D. MEMS in medicine and biology. *IEEE Engineering in Medicine and Biology Magazine* 2006 **25**:19–28.

[16] Roylance LM, Angell JB. A batch-fabricated silicon accelerometer. *IEEE Transactions on Electron Devices* 1979 **ED-26**:1911–1917.

[17] Partridge A, Reynolds JK, Chui BW, Chow EM, Fitzgerald AM, Zhang L, Maluf NI, Kenny TW. A high-performance planar piezoresistive accelerometer. *Journal of Microelectromechanical Systems* 2000 **9**:58–66.

[18] Weinberg H. Dual axis, low g, fully integrated accelerometers. *Analog Dialogue* 1999 **33**:23–24.

[19] Lemkin MA, Boser BE, Auslander D, Smith JH. A 3-axis force balanced accelerometer using a single proof-mass. *Proceedings of the International Conference on Solid-State Sensors and Actuators*, Chicago, IL. IEEE: New York, 1997; Vol. 2, pp. 1185–1188.

[20] Acar C, Shkel AM. Experimental evaluation and comparative analysis of commercial variable-capacitance MEMS accelerometers. *Journal of Micromechanics and Microengineering* 2003 **13**:634–645.

[21] Lynch JP, Wang Y, Loh KJ, Yi JH, Yun CB. Performance monitoring of the Geumdang Bridge using a dense network of high-resolution wireless sensors. *Smart Materials and Structures* 2006 **15**:1561–1575.

[22] Chung HC, Enotomo T, Loh K, Shinozuka M. Real-time visualization of bridge structural response through wireless MEMS sensors. *Proceedings of SPIE* 2004 **5392**:239–246.

[23] Lynch JP, Partridge A, Law KH, Kenny TW, Kiremidjian AS, Carryer E. Design of a piezoresistive MEMS-based accelerometer for integration with a wireless sensing unit for structural monitoring. *Journal of Aerospace Engineering* 2003 **16**:108–114.

[24] Najafi K, Hetke JF. Strength characterization of silicon microprobes in neurophysiological tissues. *IEEE Transactions on Biomedical Engineering* 1990 **37**:474–481.

[25] Guckel H, Randazzo T, Burns DW. A simple technique for the determination of mechanical strain in thin films with applications to polysilicon. *Journal of Applied Physics* 1985 **57**:1671–1675.

[26] Gianchandani YB, Najafi K. Bent-beam strain sensors. *Journal of Microelectromechanical Systems* 1996 **5**:52–58.

[27] Lin L, Pisano AP, Howe RT. A micro strain gauge with mechanical amplifier. *Journal of Microelectromechanical Systems* 1997 **6**:313–321.

[28] Hautamaki C, Zurn S, Mantell SC, Polla DL. Experimental evaluation of MEMS strain sensors embedded in composites. *Journal of Microelectromechanical Systems* 1999 **8**:272–279.

[29] Aebersold J, Walsh K, Crain M, Martin M, Voor M, Lin JT, Jackson D, Hnat W, Naber J. Design and development of a MEMS capacitive bending strain sensor. *Journal of Micromechanics and Microengineering* 2006 **16**:935–942.

[30] Azevedo RG, et al. A SiC MEMS resonant strain sensor for harsh environment applications. *IEEE Sensors Journal* 2007 **7**:568–576.

[31] Ko WH, Young DJ, Guo J, Suster M, Kuo HI, Chaimanonart N. A high-performance MEMS capacitive strain sensing system. *Sensors and Actuators, A* 2007 **133**:272–277.

[32] Lee H, Yun HB, Maclean B. Development and field testing of a prototype hybrid uniaxial strain transducer. *NDT&E International* 2002 **35**:125–134.

[33] Ozevin D, Greve DW, Oppenheim IJ, Pessiki SP. Resonant capacitive MEMS acoustic emission transducers. *Smart Materials and Structures* 2006 **15**:1863–1871.

[34] Jones ARD, Noble RA, Bozeat RJ, Hutchins DA. Micromachined ultrasonic transducers for damage detection in CFRP composites. *Proceedings of SPIE* 1999 **3673**:369–378.

[35] Niblock TGE, Surangalikar HS, Morse J, Laskowski BC, Castro-Cedeno MH, Wilson AR. Development of a commercial micro corrosion monitoring system. *Proceedings of SPIE* 2002 **4934**:179–189.

[36] Brossia CS, Hanson HS. *MEMS Sensor for Detecting Stress Corrosion Cracking*. Patent 6,925,888. U.S. Patent and Trademark Office, 2005.

[37] Polh A. A review of wireless SAW sensors. *IEEE Transactions on Ultrasonics, Ferroelectrics, and Frequency Control* 2000 **47**:317–332.

[38] Seifert F, Bulst WE, Ruppel C. Mechanical sensors based on surface acoustic waves. *Sensors and Actuators, A* 1994 **44**:231–239.

[39] Avramov ID, Voigt A, Rapp M. Rayleigh SAW resonators using gold electrode structure for gas sensor applications in chemically reactive environments. *Electronic Letters* 2005 **41**:450–452.

[40] Penza M, Antolini F, Antisari MV. Carbon nanotubes as SAW chemical sensor materials. *Sensors and Actuators, B* 2004 **100**:47–59.

[41] Reindl L, Ruppel CCW, Kirmayr A, Stockhausen N, Hilhorst MA, Balendonck J. Radio-requestable passive SAW water-content sensor. *IEEE Transactions on Microwave Theory and Techniques* 2001 **49**:803–808.

[42] Yu B, Meyyappan M. Nanotechnology: role in emerging nanoelectronics. *Solid-State Electronics* 2006 **50**:536–544.

[43] Saito R, Dresselhaus G, Dresselhaus MS. *Physical Properties of Carbon Nanotubes*. Imperial College Press: London, 1998.

[44] Donnet JB, Bansal RC, Wang MJ (eds). *Carbon Black: Science and Technology*. Marcel Dekker: New York, 1993.

[45] Kotov NA (ed). *Nanoparticle Assemblies and Superstructures*. CRC Press: Boca Raton, FL, 2006.

[46] Hou TC, Loh KJ, Lynch JP. Spatial conductivity mapping of carbon nanotube composite thin films by electrical impedance tomography for sensing applications. *Nanotechnology* 2007 **18**: 315501.

[47] Loh KJ, Kim J, Lynch JP, Kam NWS, Kotov NA. Multifunctional layer-by-layer carbon nanotube-polyelectrolyte thin films for strain and corrosion sensing. *Smart Materials and Structures* 2007 **16**:429–438.

[48] Baughman RH, Zakhidov AA, DeHeer WA. Carbon nanotubes—the route toward applications. *Science* 2002 **297**:787–792.

[49] Tombler TW, Zhou C, Alexseyev L, Kong J, Dai H, Liu L, Jayanthi CS, Tang M, Wu SY. Reversible electromechanical characteristics of carbon nanotubes under local-probe manipulation. *Nature* 2000 **405**:769–772.

[50] Wood JR, Zhao Q, Frogley MD, Meurs ER, Prins AD, Peijs T, Dunstan DJ, Wagner HD. Carbon nanotubes: from molecular to macroscopic sensors. *Physical Review B* 2000 **62**:7571–7575.

[51] Dharap P, Li Z, Nagarajaiah S, Barrera EV. Nanotube film based on single-wall carbon nanotubes for strain sensing. *Nanotechnology* 2004 **15**: 379–382.

[52] Li Z, Dharap P, Nagarajaiah S, Barrera EV, Kim JD. Carbon nanotube film sensors. *Advanced Materials* 2004 **16**:640–643.

[53] Kang I, Schulz MJ, Kim JH, Shanov V, Shi D. A carbon nanotube strain sensor for structural health monitoring. *Smart Materials and Structures* 2006 **15**:737–748.

PART 6
Systems and System Design

SECTION 1
Sensor/Actuator Network Configuration

Chapter 69
Wireless Sensor Network Platforms

Reinhard Bischoff, Jonas Meyer and Glauco Feltrin
Structural Engineering Research Laboratory, Empa, Swiss Federal Laboratories for Materials Testing and Research, Dübendorf, Switzerland

1 Introduction	1229
2 Hardware Architectures	1231
3 Software Platforms	1236
Related Articles	1237
References	1237

1 INTRODUCTION

Basically, every structural health monitoring (SHM) system is made up of various sensors measuring specific physical parameters, a data acquisition unit, and a storage device to save the acquired data. Traditional SHM systems show a starlike topology where each deployed sensor is connected via long cable runs to a central computer acting as data acquisition and storage device. The installation of such systems tends to be time consuming and therefore expensive (Figure 1). Especially in the field of civil engineering where the structures are typically large, the sensors can be located long way away from the data acquisition unit, resulting in high installation costs. These costs have proved to be a major issue, preventing a

Encyclopedia of Structural Health Monitoring. Edited by Christian Boller, Fu-Kuo Chang and Yozo Fujino © 2009 John Wiley & Sons, Ltd. ISBN: 978-0-470-05822-0.

broad application of monitoring techniques to large-scale infrastructure. Furthermore, long cable runs are prone to pick up noise, reducing the effective accuracy of the acquired data, and hence require expensive high-quality cables. Moreover, these cables are susceptible to mechanical damage involving considerable maintenance efforts. Cabled systems also tend to offer a limited flexibility in terms of rearrangement of sensors and scalability. The adoption of wireless sensor network (WSN) techniques to SHM applications promises to overcome these drawbacks.

1.1 Wireless sensor network

A WSN is essentially a computer network consisting of many small, intercommunicating computers equipped with one or several sensors. Each small computer represents a node of the network. These nodes are called sensor nodes or motes. The communication within the network is established using radio frequency transmission techniques. The sensor nodes typically form a multihop mesh network by establishing communication links to neighbor nodes. Multihop networks offer different advantages when monitoring data has to be transmitted over long distances. Mainly, the network robustness to sensor node failure and the high power efficiency [1] make multihop networks attractive for monitoring applications. Figure 2 illustrates a schematic multihop

(a)

(b)

Figure 1. Traditional, wired SHM installation.

Figure 2. Wireless sensor network deployed on a road bridge. The spots illustrate the sensor nodes, the straight lines the communication links.

network deployed on a road bridge. The network consists of several dozens of sensor nodes. Theoretically, the number of sensor nodes is unlimited. All sensor nodes are equipped with specific sensors tailored to their measurement tasks. On the one hand, these nodes act as data sources, and on the other hand, they act as relaying stations, receiving and forwarding data from adjacent nodes. One or more particular sensor nodes act as base station and represent the data sink in the network. It aggregates all the data generated within the network. In addition, the base station establishes a communication link to a data logging unit or a remote site (e.g., control center), using standard wired or wireless communication technologies like universal mobile telecommunications system (UMTS) or wireless local area network (WLAN).

The initial research into WSNs was mainly driven by military applications like battlefield reconnaissance and surveillance, nuclear, biological, and chemical attack detection, etc. These projects focused on *ad hoc*, multihop WSNs that consisted of thousands of immobile nodes randomly distributed over a large geographical area (e.g., Smart Dust). The nodes were tiny (hardly noticeable), severely resource constrained, and homogeneous (identical hard- and software). Subsequently, the emergence of civilian applications of WSNs in different fields (environmental monitoring, home automation, health applications, production, inventory, delivery control, etc.) produced a significant diversification of requirements with respect to deployment, mobility, size, cost, network topology, lifetime, etc., and therefore a flourishing of academic and commercial WSN platforms. To cope with these requirements, the platforms increased in size, computational resources, and hardware, as well as in software complexity.

The first commercial platforms appeared in the late 1990s. The most important platform was Crossbow's Rene mote, which emerged from the weC mote developed at the University of California, Berkeley, and which evolved later to the popular Mica platform. These platforms were the precursors of the recent Mica2 and MicaZ platforms (Table 1). A major reason for the popularity of Crossbow's early mote platforms was their open source policy with both hard- and software design open to the public. This policy built the base for the widespread diffusion of TinyOS as operating system for WSNs. Today, various commercial platforms with different characteristics in terms of computing resources, sensor interfaces, software architecture, etc., are available, which allow to cope with a wide spectrum of civilian applications.

2 HARDWARE ARCHITECTURES

The sensor nodes are the fundamental components of a WSN. To enable WSN-based SHM applications, the sensor nodes have to provide the following basic functionality (Figure 3):

- signal conditioning and data acquisition for different sensors;
- temporary storage of the acquired data;
- processing of the data;
- analysis of the processed data for diagnosis and, potentially, alert generation;
- self monitoring (e.g., supply voltage);
- scheduling and execution of the measurement tasks;
- management of the sensor node configuration (e.g., changing the sampling rate and reprogramming of data processing algorithms);
- reception, transmission, and forwarding of data packets;
- coordination and management of communication and networking.

2.1 General architecture

To provide the functionality described above, a sensor node is composed of one or more sensors, a signal conditioning unit, an analog-to-digital conversion module (ADC), a processing unit with memory, a radio transceiver, and a power supply (Figure 4).

If the sensor nodes are actually deployed in the field, especially in harsh environments like construction sites, they have to be protected against chemical and mechanical impacts. Therefore, an adequate packaging of the hardware is required (*see* **Chapter 81**).

2.2 Hardware platform categories

Sensor node hardware platforms can be divided into three categories [2]. Each category shows a different hardware setup matched to diverse monitoring applications.

- **Adapted general-purpose computers**
  These platforms are low-power personal computers (PCs), embedded PCs, and personal digital assistants (PDAs). These platforms mainly run on Windows CE, Linux, or other operating systems developed for mobile devices. These platforms are predominantly equipped with standard wireless communication devices like Wireless LAN (IEEE 802.11) and/or Bluetooth (IEEE 802.15.1). Because of the high processing ability and the high bandwidth communication, these platforms offer the opportunity to use higher level programming languages, which makes it easier to develop and implement software components. But in turn, they consume a considerable amount of energy and this can be prohibitive in some application scenarios. Additionally, they support networking protocols like Internet Protocol (IP). This simplifies the integration into a monitoring system.

- **Embedded sensor modules**
  These platforms are assembled from commercial off-the-shelf (COTS) Chips. Using COTS offers several benefits. These components are widely used, making them cheap because of big production quantities, and are well supported by the manufacturers and communities. The microcontroller unit (MCU) of these platforms is mostly programmed in C. This

Table 1. Selection of wireless sensor network platforms

Name		Tmote	Mica2	MicaZ	Imote2	JN5121	Sun SPOT	Agile (V-Link)	BTnode rev3
MCU	Chip manufacturer	Texas Instrument	Atmel	Atmel	Intel	OpenCores	ARM		Atmel
	Chip model	MSP430F1611	ATMega 128L	ATMega 128L	PXA271 XScale	OpenRISC1000	ARM920T		ATmega 128L
	Frequency (MHz)	8	7.383	7.383	13–416	16	180		7.383
	Type (bit)	16	8	8	32	32	32		8
	ROM, RAM (kB)	48, 10	128, 4	128, 4	32 MB, 32 MB	64, 96	4M, 512		64 + 180, 128
	Interfaces	I²C, UART, SPI	I²C, UART, SPI	I²C, UART, SPI	UART, SPI, I²C, AC97, I2S, Camera	SPI, UART			ISP, UART, SPI, I²C
Data acquisition	A/D, D/A	8, 2	8, 0	8, 0					
	A/D channels	8	8	8		4	6	8	
	Maximum sampling rate (kHz)		1	1				2	
	Resolution (bit)	12	10	10		12		12	
	D/A channels	2				2			
	Maximum sampling rate (kHz)								
	Resolution	12				11			

		1	2	3	4	5	6	7	8
Radio	Chip manufacturer	Chipcon	Chipcon	Chipcon	Chipcon		Chipcon		Zeevo, Chipcon
	Chip model	CC2420	CC1000	CC2420	CC2420		CC2420		ZV4002, CC1000
	Frequencies (kHz)	2400	310, 433 or 868/916	2400	2400	2400	2400		433 or 868/916, 2400
	Raw data rate (kbps)	250	38.4	250	250		250		
	Standard (IEEE)	802.15.4		802.15.4	802.15.4	802.15.4, ZigBee	802.15.4		Bluetooth, 802.15.1
	Range outdoor (m)[a]	125		100	30				70
External memory	Chip manufacturer	ST	Atmel	Atmel					
	Chip model Size (kB)	M25P80 1024	AT45DB41B 512	AT45DB41B 512		2048			
Power	Supply voltage min, max (V)	2.1, 3.6	2.7, 3.3	2.7, 3.3	3.2, 5	2.7, 3.6	3.7	3.2, 9	0.85, 5
	Current consumption (normal, radio off) (mA)[b]	21.8, 1.8	39, 12	29.4, 12	44–66, 31	50, 5	90, 25	25, 25	32, 12
Dimensions	(cm × cm × cm)	6.6 × 3.2 × 0.7	5.8 × 3.2 × 0.7	5.8 × 3.2 × 0.7	4.8, 3.6, 0.9	3.0, 1.8, 1.0	3.5, 2.5	7.2, 6.5, 2.4	5.8, 3.3
Manufacturer		Moteiv	Crossbow	Crossbow	Crossbow	Jennic	Sun Microsystems	Microstrain	ETH Zürich

[a] Using integrated antenna.
[b] Values declared by manufacturer or typical datasheet values (power consumption computed by individual summation of system core, flash memory, and radio component values).

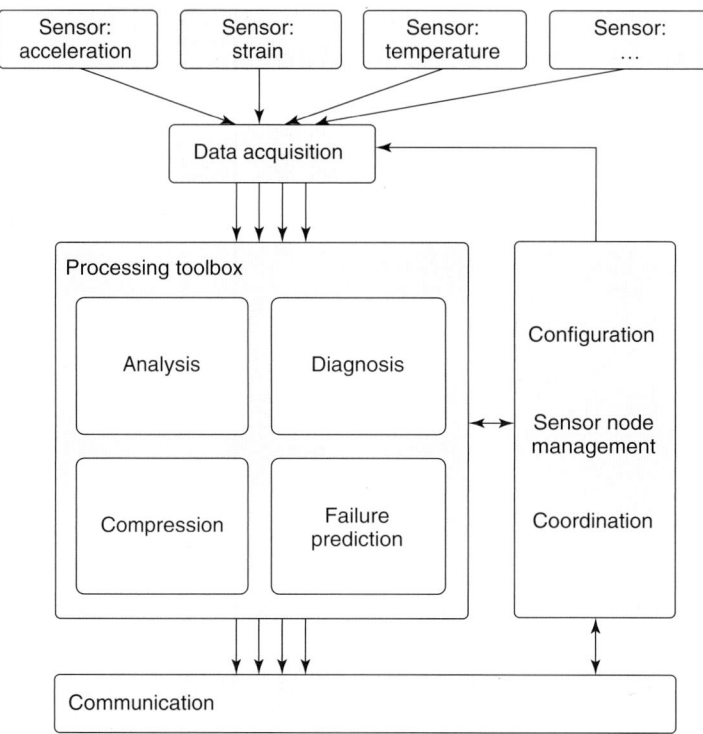

Figure 3. Basic functionality of a sensor node.

enables the development of a tight code that fits the limited memory size. Application developers have full access to hardware, but at the same time need to take care of all the resources. Examples from this category are Tmote from Moteiv, Mica2, MicaZ (Mica family), and Imote2 from Crossbow.

- **System on chip (SoC)**

These platforms integrate micro electromechanical systems (MEMS) sensors, microcontroller, and wireless transceiver technologies on one chip, an application-specific integrated circuit (ASIC). Because of this integration, platforms of this category are extreme low-power devices and have a small footprint/size. The smart dust node [3] is such an example.

2.3 Energy-related aspects

The advantages of WSNs over wired sensing systems only have an effect if an unattended operation of the

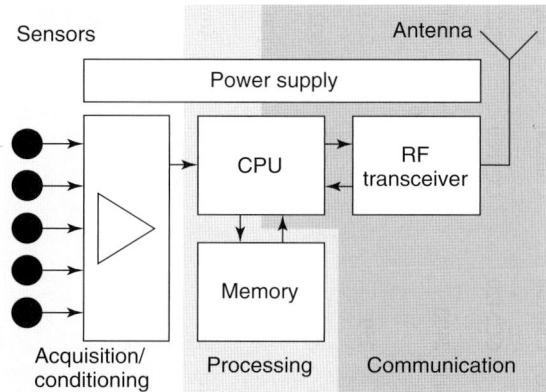

Figure 4. Sensor node (mote): hardware structure of a sensor node.

motes for a reasonably long period of time can be achieved. In terms of energy resources, this calls for a self-contained power source and has shown to be the most restrictive requirement in WSN applications. One approach to provide sufficient energy to operate the device over the desired period is to estimate the total amount of energy that will be consumed and to equip the mote with adequately dimensioned energy storage upon deployment. Although this approach is a viable solution for some lowest-duty cycle applications, the required energy storage tends to become too big for long-term SHM systems. For monitoring applications that target overall system lifetime of several years, the dissipated energy has to be regenerated from mote-extern sources. This process is referred to as energy harvesting or scavenging. These sources absorb energy from the mote's environment and convert it to electrical energy.

Numerous different types of energy storage and harvesting concepts exist, but only the most important ones for SHM systems are presented here. An overview of other, partly inconvenient concepts is given in [4].

2.3.1 Communication versus computation

In terms of power consumption, wireless data transmission is much more expensive than data processing. To extend system lifetime, it is preferable to preprocess the raw sensor readings to reduce the data items needed to be transmitted to the base station. Many recent WSN-based SHM systems transmit the raw data streams to the base station and analyze them in the traditional centralized way. Without introducing huge batteries, this is not a viable solution if a system lifetime of several months to years is targeted. For long-term monitoring applications, distributed analysis algorithms have to be introduced, which allow for decentralized data reduction or even condition assessment.

2.3.2 Energy storage devices

- **Batteries**

The most popular energy storage is batteries. Many battery types with different characteristics have been developed. Every battery has its own advantages and drawbacks and the suitable battery technology has to be selected according to the application requirements. Rechargeable batteries are utilized if energy is harvested from the mote's environment.

- **Ultracapacitors**

The features of supercapacitors lie between those of capacitors and rechargeable batteries. Supercapacitors exhibit virtually unlimited charge–discharge cycles like capacitors, but offer a much higher capacity. These components are adequate as energy storage if it is emptied and replenished in short intervals.

- **Fuel cells**

This is a more recent but promising technology. Fuel cells oxidize hydrogen or hydrocarbon fuels and convert the heat into electrical energy. Currently, the commercially available fuel cells are too big in terms of size and converted energy to be applied to motes. However, much effort is being put into the development of small fuel cells for laptops and mobile phones. These devices will suit WSN applications well.

2.3.3 Energy harvesting and scavenging devices

Because of their nature, environmental energy scavenging devices do not provide a constant energy flow. Therefore, these devices are predominantly operated in conjunction with a storage device like a supercapacitor or a rechargeable battery. It stores excess energy and provides it later, when not enough can be harvested from the environment.

- **Solar cells**

The most popular energy scavenging sources are solar cells. A reasonably small panel delivers enough energy to power a sensor node. Solar cells are predominately operated in conjunction with a supercapacitor or a rechargeable battery. This energy storage is needed to provide energy, when the panel does not. Obviously, solar cells are only an option for outdoor applications.

- **Wind mills**

More unusual energy scavenging devices are small-scale wind mills or turbines. Like solar cells, this concept is only suitable for outdoor applications.

- **Vibration**

An energy harvesting method that is considered for civil engineering applications is to convert vibration energy. Civil engineering structures contain a lot of vibration energy, but it is extremely hard to extract it. The energy levels that current prototypes provide are far to low for monitoring applications. But it could evolve to an interesting source in the future.

2.4 Overview of recent architectures

Various hardware platforms for WSNs are available today and new ones emerge regularly. This diversity offers the possibility to choose a platform that best fits the needs of a specific application. An overview of recently used platforms is given in Table 1. This table only shows a selection. Further platforms are presented in [5] and regularly updated lists in the Internet can be found at [6–9].

2.5 Tmote Sky from Moteiv Corporation

Tmote [10] from Moteiv Corporation is presented as an example of a popular WSN platform (Figure 5). Many comparable platforms with similar hardware setups exist today. All these platforms are based on the Texas Instruments microcontroller family MSP430 and the Chipcon radio CC2420.

The main components of Tmote sensor node platform are the TI MSP430F1611 microcontroller, the FTDI FT232BM USB interface, which allows for programming the microcontroller over USB, and the Chipcon CC2420 low-power radio chip for the wireless communication.

The ultra-low-power microcontroller features 10 kB of RAM and 48 kB of program memory (flash). This 16-bit processor features several power-down modes with extremely low sleep-current consumption that permits the sensor node to run for a long period of time from a limited energy resource. The MSP430 has an internal, digitally controlled oscillator (DCO) that may operate up to 8 MHz. The microcontroller may be turned on from sleep mode in 6 μs, which allows for short reaction time upon the occurrence of an event. When the DCO is off, the MSP430 is clocked from an external 32 768-Hz watch crystal.

The MSP430 has eight external 12-bit ADC ports of which six are accessible on a pin header on the Tmote. The ADC input ranges from 0 to 3.0 V. The maximum total sampling rate for all ports is 200 kHz at 12-bit resolution. The internal ADC ports may be used to monitor the internal processor temperature and the supply voltage. A variety of peripherals are available, including serial peripheral interface (SPI) and universal asynchronous receiver/transmitter (UART), enabling the communication to digital output sensors, digital I/O ports, a watchdog timer, and timers with capture and compare functionality. The I^2C port, which is also integrated into the microcontroller, is mainly used to communicate to additional sensors and signal conditioning boards. The MSP430 also includes a 2-port 12-bit digital-to-analog converter (DAC) module, a supply-voltage supervisor, and a 3-port direct memory access (DMA) controller. Detailed features of the MSP430F1611 are presented in the Texas Instruments MSP430x1xx Family User's Guide [11].

The Tmote platform is equipped with the Chipcon CC2420 radio, enabling IEEE802.15.4 standard compliant wireless communication. It offers reliable wireless communication and power management capabilities to ensure low-power consumption. The CC2420 is controlled by the TI MSP430 microcontroller through the SPI port and a series of digital I/O. The radio may be shut off by the microcontroller for reducing the power consumption. The CC2420 provides a digital receive signal strength indicator (RSSI) that may be read at any time. The programmable transmitter output power enables to optimize the power consumption. The theoretically achievable maximum data throughput rate of the system is 250 kbps, without framing and packet headers.

3 SOFTWARE PLATFORMS

Unlike general-purpose operating systems for standard PCs such as Windows or Linux, the WSN software platforms are highly tailored to the limited node hardware. These WSN software frameworks are not full-blown operating systems, since they lack a powerful scheduler, memory management, and file system support. However, these frameworks are widely referred to as *WSN operating systems*. Therefore, this term is retained in the following section.

TinyOS [12], one of the most widespread operating systems, is presented in more detail in the following section. Other operating systems developed for WSNs are Contiki [13], Mantis [14], and SOS [15].

Figure 5. Picture of a Tmote Sky form Moteiv (top view).

3.1 TinyOS

TinyOS is written in nesC [16], an extension to the C language, which supports event-driven component-based programming. The basic concept of component-based programming is to decompose the program into functionally self-contained components. These components interact by exchanging messages through interfaces. The components are event-driven. Events can originate from the environment (a certain sensor reading exceeds a threshold) or from other components, triggering a specific action. The main advantage of this component-based approach is the reusability of components.

The nesC language extension introduces several additional keywords to describe a TinyOS component and its interfaces. nesC and TinyOS are both Open Source projects supported by a fast growing community.

TinyOS has been ported to over a dozen WSN platforms (Table 1) and is also the native operating system of the presented Tmote platform. It provides a concurrency model and mechanisms for structuring, naming, and linking software components into a robust network embedded system. Today, TinyOS is a sort of *de facto* standard in WSN programming and widely used in the WSN community. As a result, a huge amount of software components for various sensors, network protocols, algorithms, and other WSN related topics is freely available on the Internet.

RELATED ARTICLES

Chapter 71: Sensor Network Paradigms

Chapter 72: Nondestructive Evaluation of Cooperative Structures (NDECS)

Chapter 76: On the Way to Autonomy: the Wireless-interrogated and Self-powered "Smart Patch" System

Chapter 77: Energy Harvesting using Thermoelectric Materials

REFERENCES

[1] Karl H, Willig A. *Protocols and Architecture for Wireless Sensor Networks*, ISBN 0-470-09510-5. John Wiley & Sons: Chichester, 2005, pp. 60–62.

[2] Zhao F, Guibas L. *Wireless Sensor Networks: An Information Processing Approach*, ISBN 1-5860-914-8. Morgan Kaufmann: San Francisco, CA, 2004, pp. 240–245.

[3] Kahn JM, Katz RH, Pister K. Mobile networking for smart dust. *ACM/IEEE International Conference on Mobile Computing and Networking (MobiCom 99)*. Seattle, WA, 17–19 August 1999.

[4] Roundy S, Steingart D, Frechette L, Wright P, Rabaey J. Power sources for wireless sensor networks. In *Proceedings of the 1st European Workshop on Wireless Sensor Networks (EWSN), Lecture Notes in Computer Science*, Karl H, Willing A, Wolisz A (eds). Springer: Berlin/Heidelberg, 2004; Vol. 2920, pp. 1–17.

[5] Lynch JP, Loh K. A summary review of wireless sensors and sensor networks for structural health monitoring. *Shock and Vibration Digest*. Sage Publications, 2005, pp. 91–128.

[6] SNM—The Sensor Network Museum.™ http://www.btnode.ethz.ch/Projects/SensorNetworkMuseum.

[7] Bokareva T. *Mini Hardware Survey*, http://www.cse.unsw.edu.au/~sensar/hardware/hardware_survey.html.

[8] Body Sensor Networks. http://ubimon.doc.ic.ac.uk/bsn/m206.html.

[9] Wireless Sensor Network (WSN) Wiki. http://wsn.oversigma.com/wiki/index.php?title=WSN_Platforms.

[10] Polastre J, Szewczyk R, Culler D. *Telos: Enabling Ultra-Low Power Research*. Information Processing in Sensor Networks/SPOTS: Berkeley, 2005.

[11] MSP430x1xx Family User's Guide. http://focus.ti.com/lit/ug/slau049f/slau049f.pdf, 2006.

[12] Levis P, *et al.* TinyOS: an operating system for wireless sensor networks. *Ambient Intelligence*. Springer-Verlag: New York, 2005.

[13] Dunkels A, Gronvall B, Voigt T. Contiki—a lightweight and flexible operating system for tiny networked sensors. *Proceedings of the 29th Annual IEEE international Conference on Local Computer Networks (Lcn'04)*. LCN IEEE Computer Society: Washington, DC, 2004; pp. 455–462.

[14] Abrach H, Bhatti S, Carlson J, Dai H, Rose J, Sheth A, Shucker B, Deng J, Han R. MANTIS: system support for multimodal networks of in-situ sensors. *Proceedings 2nd ACM International Conference on Wireless Sensor Networks and Applications*. ACM Press: New York, 2003; pp. 50–59.

[15] Han C, Kumar R, Shea R, Kohler E, Srivastava M. SOS: a dynamic operating system for sensor nodes. *Proceedings of the 3rd international Conference on Mobile Systems, Applications, and Services (Seattle, Washington, June 06–08, 2005). MobiSys'05*. ACM Press: New York, 2005; pp. 163–176.

[16] Gay D, Levis P, von Behren R, Welsh M, Brewer E, Culler D. The nesC language: a holistic approach to networked embedded systems. *Proceedings of the ACM SIGPLAN 2003 Conference on Programming Language Design and Implementation*. San Diego, CA, 2003; pp. 1–11.

Chapter 70
Sensor Placement Optimization

Robert J. Barthorpe and Keith Worden
Department of Mechanical Engineering, University of Sheffield, Sheffield, UK

1 Introduction	1239
2 Overview of the Sensor Placement Optimization (SPO) Problem	1240
3 Literature Review	1240
4 Theory	1242
5 Case Study—Sensor Placement Optimization using an Ant Colony Metaphor	1244
6 Remarks	1249
References	1249

1 INTRODUCTION

The basic problem of fault detection is to deduce the existence of a defect in a structure from measurements taken at sensors distributed upon it. The quality of these measurements and thus the quality of structural health monitoring (SHM) achieved is, to a large extent, dependent upon where sensors are placed on the structure. Cost and practicality issues preclude the instrumentation of every point of interest on the

Encyclopedia of Structural Health Monitoring. Edited by Christian Boller, Fu-Kuo Chang and Yozo Fujino © 2009 John Wiley & Sons, Ltd. ISBN: 978-0-470-05822-0.

structure and lead to the selection of a smaller set of measurement locations. The purpose of this article is to state the problem of sensor placement optimization (SPO) and describe the approaches that have been investigated for its solution. The following discussion focuses on sensor placement techniques for the category of SHM methods based upon the analysis of structural dynamics, although many of the issues raised are of wider relevance.

Traditionally, successful sensor placement has been heavily reliant upon the knowledge and experience of those performing the testing. Practical methods, for example, choosing locations near the antinodes of low-frequency vibration modes, are combined to create *ad hoc* sensor distributions. Where resources allow, several combinations of possible sensor configurations may be experimentally tested with the one that performs best chosen as the final design. While this is certainly a significant improvement on arbitrary placement, recent research has attempted to formalize the location process by casting it as a problem of optimization.

This research effort has led to the development of a variety of approaches to SPO, which are suited to different applications. An overview of the problem is given in Section 2, a review of the technical literature in Section 3, and an outline theory for a selection of the most influential techniques is given in Section 4.

A case study to illustrate an application of SPO is presented in Section 5.

2 OVERVIEW OF THE SENSOR PLACEMENT OPTIMIZATION (SPO) PROBLEM

The aim of the sensor placement exercise can be stated as the need to select a subset of measurement locations from a large finite set of candidate locations, in order to represent the system as accurately as possible using the limited number of degrees of freedom (DoFs) available. This can be viewed as a three-step decision process:

1. Sensor quantity—How many sensors need to be installed on the structure to allow successful dynamic testing?
2. Sensor placement optimization—Where should these sensors be located to most accurately capture the required data?
3. Evaluation—How can the performance of different sensor configurations be measured?

In general, the first aspect would have been resolved during pretest planning. The minimum requirement for the system to be observable is that the number of sensors required cannot be less than the number of mode shapes to be identified, with an upper limit usually imposed either by the cost or availability of equipment. In practice, a greater number of sensors are likely to be required to allow the mode shapes to be visualized, and in cases where there are surplus sensors available, a decision must be made as to whether they are best used for improving visualization or as backup against sensor failure.

The second aspect is the area that has attracted the majority of research interest, and is the primary focus of this article. For the limited number of sensors available, the problem is the development of a suitable sensor placement performance measure to be optimized and the selection of an appropriate method with which it can be optimized. Some approaches require a single calculation to be performed, some are iterative, and many others take the form of an objective function to which an optimization technique must be applied. The majority of this article deals with presenting the available alternatives.

The third and final aspect includes several possibilities for assessing the performance of chosen sensor sets. While there is an ever-present temptation to leap into data collection at the earliest opportunity, time spent assessing the effectiveness of the chosen network is invariably well spent.

Throughout this article, the *candidate set* refers to the set of all DoFs that are available as sensor locations. The *measurement set* refers to the DoFs employed as sensor locations. The *full model* includes the measurements at all available points and the *reduced model* includes only the measurements available at the DoFs specified in the measurement set.

3 LITERATURE REVIEW

The problem of determining the optimum locations for sensor placement has been addressed from a variety of perspectives. It appears to have been first addressed by control engineers before finding broad application in the field of structural dynamics. In the recent years, there has been increasing interest in developing placement methods specific to the needs of SHM, typically the optimal identification of structural characteristics sensitive to damage. This development is reflected in the literature.

An influential class of techniques has emerged from the concept of assessing all the locations of the candidate sensor set against some objective function, and then iteratively deleting those sensors that perform least well until the required number of measurement locations remain. For the effective independence (EI) method introduced in [1], based upon earlier work in [2], the sensors are ranked according to their contribution in maintaining the determinant of the Fisher information matrix (FIM, described in Section 4). The location that contributes least at each iteration step is selected for deletion. The kinetic energy (KE) method, see for example [3], assumes that the sensors will have maximum observability of the modes of interest if the sensors are placed at points of maximum KE for that mode. It has often been noted that EI and KE methods can produce similar results, especially in structures with homogeneous mass distributions, and the inherent mathematical connection between the two methods is revealed in [4].

Further examples of the iterative approach include the eigenvalue vector product (EVP) [5], average driving-point residue (ADPR) [6], effective independence driving-point residue (EI-DPR) [7], strain energy distribution [8], and modal assurance criterion (MAC)-based [9] methods. The comparative performance of several of the iterative techniques is investigated in [10, 11].

It should be noted that the iterative process does not necessarily need to be a reduction from the candidate set. In [12], an alternative EI approach is presented whereby the sensor set is iteratively expanded to include those sensor locations that offer the greatest increase in the determinant of the FIM. The expansion approach reduces the computational expense incurred when the candidate set is large and allows for any desired sensor locations to be specified at the outset, with the remaining sensors placed optimally.

In [13], optimal sensor placement is formulated as a mixed variable programming (MVP) problem. The emphasis is on the development of a general framework using MVP optimization that could be applied to any number of objective functions. The MVP formulation allows variables to be categorical, taking their values from a predefined set or list. For the sensor placement problem, a categorical position variable is defined for each sensor.

A further class of methods takes advantage of developments in the field of combinatorial optimization. Perhaps the first use of genetic algorithms (GAs) for the sensor placement problem is presented in [14]. The authors propose the GA as an alternative to the EI method, with the determinant of the FIM chosen as the objective function (the *fitness* function in GA terminology). In [15], the fitness function is taken as a product of two terms: the first term measures the fitness from the point of view of observability; the second component is geometric and penalizes the clustering of sensors.

In one of the first sensor placement approaches specific to SHM [16], a structural damage localization approach based on eigenvector sensitivity is adopted and an SPO approach using the same method is developed. The EI approach is applied to the sensitivity matrix that is to be used for damage localization, with the DoFs providing the greatest amount of information for localization retained. In [17], the difficulties that can occur in solving the sensitivity matrix are highlighted. The problem is reformulated and solved using an improved GA.

In [18], the selection of optimal sensor locations for impact detection in composite plates is approached using a GA to optimize a fitness function based upon the concept of mutual information. The concept is used to eliminate redundancies in information between selected sensors and rank them based on their remaining information content. In [19], an equally spaced configuration of measurement points is assumed, and the sensor spacing is varied to minimize the average mutual information between measurement locations. In [20], a statistical method for optimally placing sensors for the purposes of updating structural models for subsequent damage detection and localization is presented. The optimization is performed by minimizing information entropy, a unique measure of the uncertainty in the model parameters. The uncertainties are calculated using Bayesian techniques, and the minimization is realized using a GA.

In one of the first studies of sensor placement for an SHM problem, a neural network (NN) for the location and classification of faults and a GA for the determination of an optimal (or near optimal) sensor configuration were used [21]. The NN is trained using mode-shape curvatures provided by an FE model of a cantilever plate. The probabilities of misclassification for the different damage conditions are obtained from the NN, and the inverse of this measure is employed as a fitness function for the GA. The use of simulated annealing (SA) for the optimization has also been investigated [22].

In [23], an NN/GA approach was used to place sensors for the location and quantification of impacts on a composite plate. An artificial NN is trained to locate the point of impact, and a second NN employed to estimate the impact force. A GA was used to select an optimal set of measurement locations from the candidate set, using the estimation of the impact force provided by the NN as the parameter to be maximized. The work is extended to cover fail-safe sensor placements [24]. In [25], a new damage location method is presented that combines classical triangulation procedures with experimental wave velocity analysis and GA optimization.

The NN problem studied in [23] was investigated using a different optimization technique reported in [26]. Here, an ant colony metaphor was used to place sensors based upon the fitness functions generated by the NN, and the results were compared with those from the GA and exhaustive search. This work is demonstrated in Section 5 (*see* **Chapter 32**).

4 THEORY

A large variety of performance indices have been developed for the problem of sensor placement, but it is only comparatively recently that the problem has been considered from an SHM perspective. Rather than attempting to cover this multiplicity of approaches in full, the intention of this section is to highlight some of the key considerations made in the selection and development of objective functions appropriate to the SHM practitioner. Several influential general approaches (notably the EI and KE methods) are covered, as are more recent techniques developed for the specific purpose of damage detection. The descriptions have been kept mathematically light intentionally; full descriptions of the algorithms may be found in the references.

4.1 Effective independence (EI)

The EI method makes use of the FIM, which offers a measure of the information that a sampled random variable contains about an unknown parameter; formally, Fisher information is the variance of the score with respect to the unknown parameter. Where there are multiple unknown parameters, it may be stated in matrix form with elements

$$(I(\boldsymbol{\theta}))_{ij} = E\left[\frac{\partial}{\partial \theta_i}\ln f(X;\boldsymbol{\theta})\frac{\partial}{\partial \theta_j}\ln f(X;\boldsymbol{\theta})\right] \quad (1)$$

where

- $\boldsymbol{\theta} = [\theta_1, \theta_2, \ldots, \theta_N]$ is the vector of unknown parameters
- $(I(\boldsymbol{\theta}))_{ij}$ is the Fisher information with respect to the unknown parameters θ_i and θ_j
- X is the sampled random variable
- $f(X;\boldsymbol{\theta}) = L(\boldsymbol{\theta})$ is the likelihood function of $\boldsymbol{\theta}$
- E denotes the expectation.

For the SPO problem, the target mode shapes may be regarded as the unknown, sought parameters, with the sampled data being that available from the given sensor distribution. Every DoF in the candidate set is ranked according to its contribution to the determinant of the FIM, and the lowest ranked DoF is eliminated. The new, reduced set is then re-ranked, and the process repeated in an iterative manner until the desired number of sensors remains. This is adopted as the optimal measurement set. Maintaining the determinant of the FIM leads to the selection of a set of sensor locations for which the mode shapes of interest are as linearly independent as possible, while retaining sufficient information about the target modal responses. The approach is based on the EI distribution vector $\mathbf{E}_D$, defined as the *diagonal of the prediction matrix*, $\mathbf{E}$:

$$\mathbf{E} = [\boldsymbol{\Phi}]\{[\boldsymbol{\Phi}]^T[\boldsymbol{\Phi}]\}^{-1}[\boldsymbol{\Phi}]^T \quad (2)$$

where $\boldsymbol{\Phi}$ is the matrix of FE target modes, in this case partitioned according to a given sensor distribution. Each diagonal element is the fractional contribution of each sensor location to the rank of E, which can only be full rank if the target mode partitions are linearly independent. The algorithm is iterative; at each step, terms in $\mathbf{E}_D$ are sorted to give the least important sensor, which is then deleted. The corresponding elements in $\boldsymbol{\Phi}$ are also deleted. The iteration concludes when the required number of sensors is obtained.

4.2 Average driving-point residue (ADPR)

A drawback of the EI approach is that the algorithm can select sensor locations that display low signal strength, making the system vulnerable to noisy conditions. The ADPR offers a measure of the contribution of any point to the overall modal response. If $j = 1\ldots N$ modes of interest are to be measured and ω_j is the eigenvalue of jth mode, the ADPR at the ith DoF can be calculated from FE data as

$$ADPR_i = \sum_{i=j}^{N}\frac{\Phi_{ij}^2}{\omega_j} \quad (3)$$

4.3 Effective independence driving-point residue (EI-DPR)

The values given by the EI algorithm are weighted by the ADPR values to give the EI-DPR vector. For the ith DoF

$$E_{D_i}^{\text{EI-DPR}} = E_{D_i}^{\text{EI}} ADPR_i \qquad (4)$$

This adaptation leads to a greater likelihood of sensors being placed in areas of high signal strength. In addition to improving signal-to-noise ratios, this tends to result in the selection of relatively uniformly spaced sensor locations.

4.4 Kinetic energy method (KE)

The KE method assumes that the sensors will have maximum observability of the modes of interest if the sensors are placed at points of maximum KE for that mode, and accordingly ranks sensor locations based on their dynamic contribution to the target mode shapes. It follows a similar procedure to that used in the EI method, the key difference being that a KE measure, rather than the determinant of the FIM, is maximized. It is alternatively known as the *modal kinetic energy* (*MKE*) *method* or the *kinetic energy method* (*KEM*) in the literature.

KE indices are calculated for all candidate sensor locations as follows:

$$KE_{ij} = \Phi_{ij} \sum_s M_{is} \Phi_{sj} \omega_j^2 \qquad (5)$$

where **M** is the mass matrix. The sensor locations that offer the highest KE indices are selected as the measurement locations. As the method selects those sensor locations with the largest available signal amplitudes, the signal-to-noise ratios tend to be high, making the method attractive for use in noisy conditions. However, in contrast to EI, the KE method does not consider the linear independence of the target modes, an important consideration for both modal identification and test–analysis correlation.

4.5 Eigenvalue vector product (EVP)

The EVP method computes the product of the eigenvector components for candidate sensor location for the range of modes to be measured N: a maximum for this product is a candidate measurement point. Some modification may be required if a point is a node of one of the modes. The EVP of the ith DoF is calculated as

$$EVP_i = \prod_{j=1}^{N} |\Phi_{ij}| \qquad (6)$$

4.6 Mutual information

Mutual information gives a measure of how much information one sensor location "learns" from another. If there are two sets of measurement locations, A and B, the amount of information learned by a_i about b_j is represented by the mutual information $I(a_i, b_j)$,

$$I(a_i, b_j) = \log_2 \left[\frac{P_{AB}(a_i, b_j)}{P_A(a_i) P_B(b_j)} \right] \qquad (7)$$

where

- a_i and b_j are the measurements from locations A and B, respectively
- $P_A(a_i)$ and $P_B(b_j)$ are the individual probability densities for A and B
- $P_{AB}(a_i, b_j)$ is the joint probability density for measurements A and B.

If the measurement of a_i is completely independent of the measurement of b_j, $I(a_i, b_j)$ becomes zero. The average mutual information between A and B is calculated by averaging over all the sensor locations, and the optimal sensor location determined by minimizing the mutual information between sensors.

4.7 Information entropy method

Optimal sensor placement is achieved by minimizing the change in the information entropy $H(D)$, given by

$$H(D) = E_\theta [-\ln p(\theta|D)]$$
$$= -\int p(\theta|D) \ln p(\theta|D) \, d\theta \qquad (8)$$

where

- θ is the uncertain parameter set (e.g., stiffness parameters, modal parameters, etc.)

- D is the dynamic test data
- E_θ is the mathematical expectation with respect to θ.

A rigorous mathematical description is given in [27].

4.8 Sensitivity-based methods

In the SHM-specific method proposed in [16], the prediction matrix **E** used in the EI method is adapted to use the sensitivity matrix developed for damage location. The modified matrix is given by

$$E = F(K)[F(K)^T F(K)]^{-1} F(K)^T \quad (9)$$

where

- $F(K)$ is the vector of sensitivity coefficients of the mode shape changes with respect to a damage vector.

As for the EI approach, the diagonal terms of E provide the fractional contribution of the corresponding measurement location to the rank of E. The location that contributes the least is removed, and the process is repeated iteratively until the required quantity of sensors remains.

5 CASE STUDY—SENSOR PLACEMENT OPTIMIZATION USING AN ANT COLONY METAPHOR

A comparatively recent addition to the canon of combinatorial optimization algorithms are those based on "ant colony metaphors". These are founded on the cooperative interaction of simple computational agents termed *ants*. The basic forms of the algorithms—"ant-density", "ant-quantity", and "ant-cycle"—were introduced in [28]. Of these basic forms, ant-cycle proved to be the most effective and this was renamed *ant-system* and discussed in more detail in [29]. The ant algorithms have proved to be a useful addition to the set of methods, combining aspects of greedy search with population-based cooperative search. The work put forward in [26] is presented here to illustrate an interesting approach to the problem of sensor placement.

5.1 The ant algorithm

Real ants are well known to be capable of finding the shortest path to a food source from their nest without using visual cues [30]. This is done by exploiting pheromone information. Ants deposit pheromone as they move and follow pheromone trails deposited by previous ants. If a trail has higher pheromone information, an ant will follow it in preference to other trails with less pheromone. In other words, an ant will follow a trail with higher pheromone, with higher probability. Consider Figure 1. Initially, the ants leave the nest and have no preference as to which path they will follow and they will choose a path with equal probability. Assuming that all ants travel with equal speed, the ants taking the lower path will reach the food before those taking the upper one. When returning to the nest, they will choose the lower path with higher probability because the upper path ants have not arrived yet to lay trail. This reinforces the pheromone path on the lower trail. When the lower path ants arrive back at the nest, new ants will initially see twice as much trail on the lower path and choose it with higher probability. It is not difficult to see that a positive feedback mechanism emerges, which reinforces the desirability of the lower trail. Eventually, all ants will follow the shorter path. This simple example illustrates the idea behind ant algorithms, which is that simple agents can communicate, using distributed memory about the problem, to cooperate and solve an optimization problem. Because the aim here and elsewhere in the engineering literature is to solve an engineering problem and not to model real ant colonies, various simplifications are assumed. Artificial ants differ from real ants in the following ways:

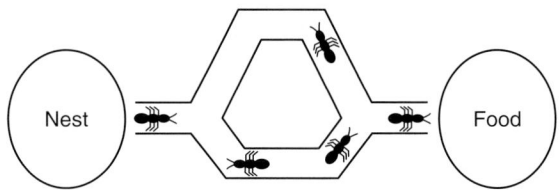

Figure 1. Ant foraging activity.

1. They are completely blind.
2. They have some memory.
3. They live in a discrete-time environment.

In order to explain the original ant-system algorithm, it is convenient to do so in the context of the classic traveling salesperson problem (TSP), which provided the original motivation for the algorithms. The basic TSP is formulated as follows: given N cities distributed randomly within the plane, find the shortest tour that visits each city only once and returns to the original city. The distance $d(i, j)$ between cities i and j with coordinates (x_i, y_i) and (x_j, y_j) is computed using the standard Euclidean norm. One can think of the problem in terms of a graph, where each pair of cities is potentially joined by an edge (i, j). The ants move around in this graph laying a pheromone trail with intensity $\tau(i, j)$ for each edge (i, j) they traverse. This trail can be updated locally as the ants move or can be updated globally at the end of an iteration when all ants have completed a tour. The global updating rule, which forms the basis of the ant-system algorithm, was shown to be superior to local rules in [29]. The ants are forced to complete a tour by maintaining a *tabu list* for each ant, which simply contains a list of the cities that the ant has already visited, and the ant is forbidden to revisit any city in the tabu list.

Given a state of the system at time t in a specific iteration, each ant k is at one of the cities. The probability that an ant will make the journey from city i to city j is assumed to be

$$P_k(i, j) = \frac{[\tau(i, j)]^\alpha [\eta(i, j)]^\beta}{\sum_{j \in J_k(t)} [\tau(i, j)]^\alpha [\eta(i, j)]^\beta} \quad (10)$$

where $J_k(t)$ are the allowed cities for ant k at time t, i.e., the complement of the tabu list. If j is not in $J_k(t)$, the transition is forbidden by setting $P_k(i, j) = 0$. In equation (10), $\tau(i, j)$ is the trail intensity associated with the edge (i, j). The *visibility* $\eta(i, j)$ is the inverse of the distance between cities i and j. This is included in order to allow a degree of greediness in the algorithm; the ants are more likely to move to a nearer city at a given step. This is not guaranteed to lead to a tour with overall minimum length as at step N; the ant may be very far from its starting city. The parameters α and β control the relative importance of trail and visibility. Note that the inclusion of visibility means that the ants are not totally blind in this variant of the algorithm. After N steps and the ants have completed a tour, the pheromone levels for each edge are updated using the rule

$$\tau(i, j) \longrightarrow \rho \tau(i, j) + \sum_{k=1}^{m} \tau_k(i, j) \quad (11)$$

where

$$\tau_k(i, j) = \frac{Q}{L_k} \quad (12)$$

if edge (i, j) is in the tour of the kth ant and $\tau_k(i, j) = 0$ otherwise. Q is a user-defined constant and L_k is the total tour length for ant k in that iteration. The total number of ants is m. The parameter ρ is also user defined. A number between 0 and 1 represents the trail persistence between iterations; $1 - \rho$ is a measure of trail *evaporation*.

This is how the ant-system algorithm is used to solve the TSP. In order to benchmark the code constructed to solve the sensor optimization problem here, the algorithm was applied to a known TSP problem—the Oliver30 problem. This was chosen as it was the problem used in the original ant-system paper [29]. Overall, 25 runs of the ant-system algorithm were carried out. The agreement between the runs was impressive. By 500 iterations, 16 of the runs had arrived at the best previously known solution, with a tour length of 423.74 (Figure 2).

Having established that the algorithm worked properly on the TSP problem, some modification was required before it could be applied to the SPO problem. The first important difference is that one is looking for an optimum subset of the candidate sensor locations and this means that a tour does not visit all locations. The second major difference is that the ants are truly blind. The sensor distributions can only be evaluated when all the members of the tour are known; there is no analog of visibility. The ant algorithm for SPO proceeds as follows. The ants are distributed randomly on the sensors with each edge initialized with the same trail intensity. At each step,

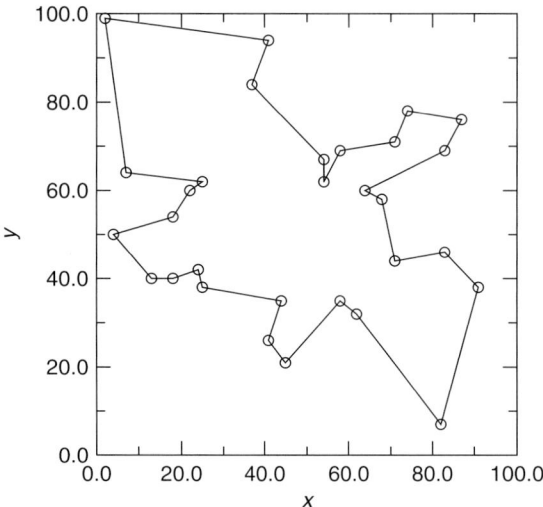

Figure 2. Best solution found to Oliver30 TSP problem.

each ant chooses a sensor to visit by means of a transition probability,

$$P_k(i,j) = \frac{\tau(i,j)}{\sum_{j \in J_k(t)} \tau(i,j)} \quad (13)$$

i.e., α is set arbitrarily to unity (and β is irrelevant). If the sensor has already been visited, transition is forbidden as before. The iteration ends when the required number of candidate locations has been visited. The (global) updating rule is very similar to the TSP variant,

$$\tau(i,j) \longrightarrow \rho\tau(i,j) + (1-\rho)\sum_{k=1}^{m} \tau_k(i,j) \quad (14)$$

where $\tau_k(i,j) = F_k$, if (i,j) is in the tour for the kth ant and zero otherwise. F_k is the fitness of the kth tour, i.e., the value of the objective function to be maximized for the sensor distribution.

A little trial and error yielded 0.2 as a good initial trail level and the number of ants was taken as 10 simply because that led to optimum solutions in the Oliver30 TSP problem. The rationale behind the value of *persistence* ρ used is given later.

5.2 Impact test of the composite panel

A simple impact experiment was performed to study the effectiveness of NNs for the impact identification problem and also to experimentally investigate the optimal sensor location problem.

The structure under examination consisted of a rectangular 530 mm × 300 mm composite plate made from laminated carbon fibre reinforced plastic (CFRP) and four aluminum channels. The structure is shown in Figure 3. The top flanges of the channels were attached to the plate by a line of rivets; the bottom flanges were fixed rigidly with screws to a pneumatic measuring table. This box structure was intended to simulate the skin panel of an aircraft. The composite plate was instrumented with 17 piezoceramics (*PZT Sonox P5*, 15 mm × 15 mm) (*see* **Chapter 156**) fixed on the lower surface of the plate, the impacts being performed on the upper surface. The piezoceramics were used as strain sensors. Figure 4 shows the total distribution of the sensors used.

Figure 3. Composite box structure.

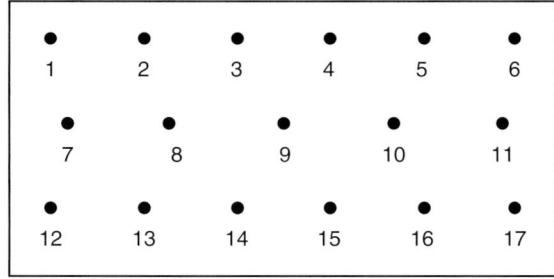

Figure 4. The candidate sensor locations.

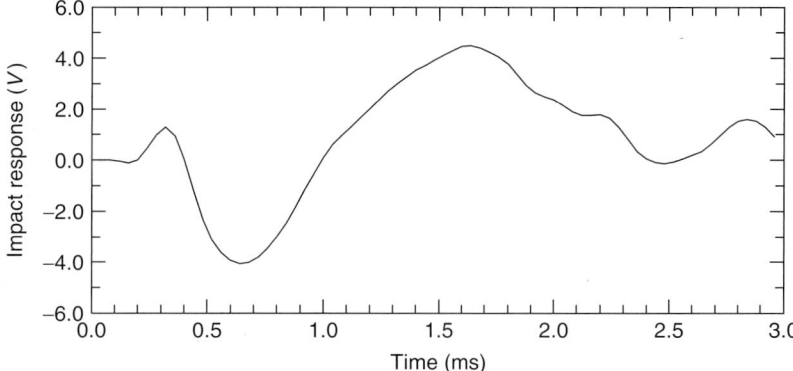

Figure 5. Example of a strain time-history.

The impacts were applied using a modally-tuned impact hammer. The levels of force applied were kept below 0.1 N in order not to damage the plate. As discussed below, two sets of measurements were made. The first comprised 80 impacts at random positions on the plate and was intended for use as network training data. The second set of measurements contained 95 impacts placed on a regular grid, and these were divided between the network validation and test sets: 48 in the former and 47 in the latter. In the validation and testing sets, each location was subjected to impacts of varying magnitudes.

The strain data were recorded using a DIFA SCADAS II 24 channel measuring system running the LMS 3.4.04 data-acquisition software. For each impact, 8192 samples were recorded at a frequency of 25 kHz. Figure 5 shows an example of the strain data recorded.

5.3 Impact identification strategy using neural networks

Before carrying out the subset optimization, a network training exercise was carried out using all 17 candidate sensors. This was in order to establish a lower bound for the error of the impact locator.

The NN paradigm used for this study was the standard multilayer perceptron (MLP) trained with the back-propagation learning rule. The particular implementation is described in more detail in [23].

The first problem in establishing the network analysis was to determine the appropriate features for the diagnosis, i.e., which data should be used to train the network. As described in [23], a preliminary study considered several different time and frequency domain features, namely, (i) time after impact of maximum response, (ii) magnitude of maximum response, (iii) peak-to-trough range of the response, and (iv) real and imaginary parts of the response *spectrum*, integrated over frequency. The features were investigated alone and in combination, and it was found that the best results were obtained using features A and B in tandem. This meant that each sensor contributed two features and the dimension of the pattern vectors for training was thus 34.

The next problem was to design the training strategy. A principled approach demanded the availability of three data sets. The first, the *training* set, is used to determine the network weights. The second, the *validation* set, is used to investigate the optimum structure for the network and the final *testing* set is used to assess the effectiveness of the optimized network. Part of the network training problem is to determine the best structure for the network, i.e., the number of layers and number of neurons per layer. The following approach was adopted as described in [31] and numerous structures were assessed. The number of neurons in the input and output layers of the network was fixed by the number of measurement features and diagnostic outputs, respectively. In this case, the network must have 34 inputs. As the network considered here was required to signal the location of damage, two outputs were required, namely, the x and y coordinates of the impact site. A single hidden layer was assumed, and the number of neurons in the layer was established by optimizing the network performance over the validation set; the objective function used was the product of the mean x and y errors, i.e., the mean "area error". This was normalized to give a percentage of the plate area. Because the data sets are comparatively small, noise was added during training as a form of regularization. The root-mean-square (RMS) level of this noise was also established by minimizing the validation error.

For the location problem, the optimization procedure produced a minimum error over the validation set when there were eight neurons in the hidden layer and the RMS noise level was 0.1. When the corresponding network was evaluated on the testing set, the mean (modulus) of the x error was 23.1 mm and the mean y error was 25.7 mm. This gives an area corresponding to 1.5% of the plate area.

In order to have a tractable optimization problem, it was assumed that eight hidden units would also be the best choice when only a subset of the candidate sensor data was used for training. A similar assumption was made for the noise level. Following trial and error experiments, a training run of 100 000 iterations was applied.

5.4 Results

The algorithm summarized by equations (13) and (14) was used for the SPO problem here. The fitness F_k was evaluated as follows. If an ant k produced a certain tour at a given iteration, i.e., (1, 3, 7) in the search for an optimum three-sensor distribution, the NN was trained only with data from the sensors 1, 3, and 7. If this yielded an area error δA, the fitness was returned as $F_k = 1/\delta A$. Because the network structure was fixed, δA was taken from the validation set.

The main parameter controlling the efficiency of the algorithm was found to be the persistence ρ. Several values were experimented with, using a three-sensor search and the corresponding results from exhaustive search from [23] for comparison. Values of 0.95, 0.9, 0.8, and 0.5 were considered. As it was found that only $\rho = 0.9$ leads to the optimum under 59 iterations, this was chosen as the value for the runs. As observed above, 10 ants were used in each run and these were distributed randomly among the cities at initiation. A maximum of 50 iterations were allowed; this bounded the number of function evaluations at 500.

The method was demonstrated for finding a three-sensor distribution. The reason for this is that the number of candidate three-sensor distributions is 684 and it was feasible to compare the results of the optimization algorithms with those from exhaustive search. Also, three sensors are the minimum number that would be needed to locate the impact event from time-of-flight data using triangulation.

As the NN returns a value for the fitness that depends on the starting conditions for training, six separate exhaustive searches were carried out with different initial conditions. The best distribution found over the six runs was found to be 3 : 10 : 12 with an error of 1.99% (Table 1). As observed in [23], this solution agreed with engineering intuition in spacing the sensors in such a way as to effectively triangulate over a large area of the plate (Figure 6). When the GA was applied using the six sets of network starting conditions used for the exhaustive searches, it returned the same distribution (3 : 10 : 12) as the exhaustive search in every instance. In most of the cases, the GA found the optimal result faster than the exhaustive search would have.

For comparison, in 10 runs, the ant algorithm found the above solution once out of 10 runs, but

Table 1. Best three-sensor distributions from six exhaustive searches [23]

Search no.	Distribution	Area error (%)
1	3 : 12 : 17	2.16
2	8 : 11 : 12	2.15
3	1 : 3 : 17	2.15
4	3 : 12 : 14	2.19
5	3 : 7 : 11	2.13
6	3 : 10 : 12	1.99

Reproduced from Ref. 23. © Blackwell, 2000.

```
○    ○    ●    ○    ○    ○
1    2    3    4    5    6

○    ○    ○    ●    ○
7    8    9    10   11

●    ○    ○    ○    ○    ○
12   13   14   15   16   17
```

Figure 6. The optimal three-sensor distribution found by exhaustive search and by the genetic algorithm.

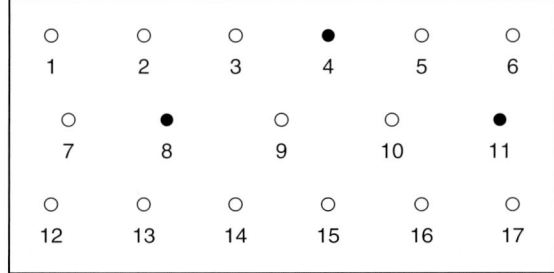

Figure 7. The optimal three-sensor distribution found by the ant colony algorithm.

also found a better solution with an error of 1.89% (distribution 4:8:11 as shown in Figure 7). The average area error over the 10 runs was 2.11%. The ant algorithm found the optimal distributions faster than the exhaustive search would have.

5.5 Conclusions

The main conclusion of this article is that ant colony metaphors provide an effective solution to the SPO problem. The algorithm is shown to find the optimum in a problem with three sensors, which is amenable to an exhaustive search. Further, the optimal solution is discovered faster than when a GA is used. This result was not entirely expected, as the effect of the parameters α and β was investigated in [29] and it was found that searches based on pheromone alone ($\beta = 0$) were suboptimal.

6 REMARKS

The need for SPO techniques specific to SHM has been recognized, and promising approaches are emerging. Future methods are likely to be specific to the SHM methodology employed (as in the case investigated in [16] and the various NN approaches) and should be developed alongside them. A great number of techniques developed for other structural dynamics applications are also of use to the SHM practitioner. Techniques based upon maximizing the FIM determinant are widely used and represent a significant improvement on the *ad hoc* approaches of the past, with approaches based upon combinatorial optimization appearing to offer further significant advantages. This said, engineering judgment will always play a vital part in successful sensor placement and is pervasive in the application of SPO techniques.

In terms of the development of new algorithms, it is desirable that they should, wherever feasible, be compared with the results of an exhaustive search. This is admittedly limited to cases with low numbers of sensors in the candidate and measurement sets. Where the computational demands of exhaustive search prove prohibitive, "optimal" solutions should be validated both against randomly generated configurations, and against the configurations suggested by the symmetry of the analyzed structure. The robustness of the placement algorithm should be investigated to assess performance when factors such as sensor quantity and analytical model fidelity are altered. There remains a shortage of work concentrating on comparative studies of different approaches and validation of results. The former will give some guidance in engineering applications; the latter will increase confidence in the SPO methodology.

Wherever possible, successful numerical simulation should be followed by experimental study. Where sensor placement is being optimized for the purpose of SHM, extending the experimental study to cover the results of an SHM study would be informative, especially where comparison can be made between placement algorithms and exhaustive search results.

REFERENCES

[1] Kammer DC. Sensor placement for on-orbit modal identification and correlation of large space structures. *Journal of Guidance, Control, and Dynamics* 1991 **14**(2):251–259.

[2] Shah PC, Udwadia FE. A methodology for optimal sensor locations for identification of dynamic systems. *Journal of Applied Mechanics-Transactions of the Asme* 1978 **45**(1):188–196.

[3] Heo G, Wang ML, Satpathi D. Optimal transducer placement for health monitoring of long span bridge. *Soil Dynamics and Earthquake Engineering* 1997 **16**(7–8):495–502.

[4] Li DS, Li HN, Fritzen CP. The connection between effective independence and modal kinetic energy methods for sensor placement. *Journal of Sound and Vibration* 2007 **305**(4–5):945–955.

[5] Jarvis B. Enhancements to modal testing using finite elements. *Sound and Vibration* 1991 **25**(8):28–30.

[6] Penny JET, Friswell MI, Garvey SD. Automatic choice of measurement locations for dynamic testing. *AIAA Journal* 1994 **32**(2):407–414.

[7] Imamovic N. *Validation of Large Structural Dynamics Models using Modal Test Data*, PhD thesis. Imperial College London, 1998.

[8] Hemez FM, Farhat C. An Energy based Optimum Sensor Placement Criterion and Its Application to Structural Damage Detection. *Proceedings of the 12th International Modal Analysis Conference.* Honolulu, HI, 1994; pp. 1568–1575.

[9] Breitfeld T. A method for identification of a set of optimal points for experimental modal analysis. *Proceedings of the 13th International Modal Analysis Conference.* Nashville, TN, 1995.

[10] Larson CB, Zimmerman DC, Marek EL. A comparison of modal test planning techniques: excitation and sensor placement using the NASA 8-bay truss. *Proceedings of the 12th International Modal Analysis Conference.* Honolulu, HI, 1994; pp. 205–211.

[11] Meo M, Zumpano G. On the optimal sensor placement techniques for a bridge structure. *Engineering Structures* 2005 **27**(10):1488–1497.

[12] Kammer DC. Sensor set expansion for modal vibration testing. *Mechanical Systems and Signal Processing* 2005 **19**(4):700–713.

[13] Beal JM, Shukla A, Brezhneva OA, Abramson MA. Optimal sensor placement for enhancing sensitivity to change in stiffness for structural health monitoring, *Optimization and Engineering* 2008 **9**(2):119–142.

[14] Yao L, Sethares WA, Kammer DC. Sensor placement for on-orbit modal identification via a genetic algorithm. *AIAA Journal* 1993 **31**(10):1922–1928.

[15] Frauchi CG, Gallieni D. Pre-test optimisation by genetic algorithm. *Proceedings of the 19th International Seminar on Modal Analysis.* Leuven, 1994.

[16] Shi ZY, Law SS, Zhang LM. Optimum sensor placement for structural damage detection. *Journal of Engineering Mechanics* 2000 **126**(11):1173–1179.

[17] Guo HY, Zhang L, Zhang LL, Zhou JX. Optimal placement of sensors for structural health monitoring using improved genetic algorithms. *Smart Materials and Structures* 2004 **13**(3):528–534.

[18] Said WM, Staszewski WJ. Optimal sensor location for damage detection using mutual information. *11th International Conference on Adaptive Structures and Technologies.* Nagoya, 2000; pp. 428–435.

[19] Trendafilova I, Heylen W, Van Brussel H. Measurement point selection in damage detection using the mutual information concept. *Smart Materials and Structures* 2001 **10**(3):528–533.

[20] Papadimitriou C, Beck JL, Au SK. Entropy-based optimal sensor location for structural model updating. *Journal of Vibration and Control* 2000 **6**(5):781–800.

[21] Worden K, Burrows AP, Tomlinson GR. A combined neural and genetic approach to sensor placement. *Proceedings of the 13th International Modal Analysis Conference.* Nashville, TN, 1995; pp. 1727–1736.

[22] Worden K, Burrows AP. Optimal sensor placement for fault detection. *Engineering Structures* 2001 **23**(8):885–901.

[23] Worden K, Staszewski WJ. Impact location and quantification on a composite panel using neural networks and a genetic algorithm. *Strain* 2000 **36**(2):61–70.

[24] Staszewski WJ, Worden K, Wardle R, Tomlinson GR. Fail-safe sensor distributions for impact detection in composite materials. *Smart Materials and Structures* 2000 **9**(3):298–303.

[25] Coverley PT, Staszewski WJ. Impact damage location in composite structures using optimized sensor triangulation procedure. *Smart Materials and Structures* 2003 **12**(5):795–803.

[26] Overton G, Worden K. Sensor optimisation using an ant colony metaphor. *Strain* 2004 **40**(2):59–65.

[27] Papadimitriou C. Optimal sensor placement methodology for parametric identification of structural systems. *Journal of Sound and Vibration* 2004 **278**(4–5):923–947.

[28] Dorigo M, Maniezzo V, Colorni A. *Positive Feedback as a Search Strategy*, Report No. 91-016, Politecnico di Milano, 1991.

[29] Dorigo M, Maniezzo V, Colorni A. The ant system: optimization by a colony of cooperating agents. *IEEE Transactions on Systems, Man, and Cybernetics. Part B* 1996 **26**(1):1–13.

[30] Hölldobler B, Wilson EO. *The Ants*, Springer-Verlag: Berlin, 1990.

[31] Tarassenko L. *A Guide to Neural Computing Applications*, Arnold: London, 1998.

Chapter 71
Sensor Network Paradigms

Charles R. Farrar, Gyuhae Park and Kevin M. Farinholt
Engineering Institute, Los Alamos National Laboratory, Los Alamos, NM, USA

1 Introduction	1251
2 Data Acquisition for SHM	1252
3 SHM Sensor System Design Considerations	1252
4 Current SHM Sensing Systems	1253
5 Sensor Network Paradigms	1256
6 Future Sensing Network Paradigms	1259
7 Practical Implementation Issues for SHM Sensing Networks	1262
8 Summary	1265
References	1265

1 INTRODUCTION

Structural health monitoring (SHM) is the process of detecting damage in structures. The goal of SHM is to improve the safety and reliability of aerospace, civil, and mechanical infrastructure by detecting damage before it reaches a critical state. To achieve this goal, technology is being developed to replace qualitative visual inspection and time-based maintenance procedures with more quantifiable and automated damage assessment processes. These processes are implemented using both hardware and software with the intent of achieving more cost-effective, condition-based maintenance. A more detailed general discussion of SHM can be found in [1, 2].

The authors believe that all approaches to SHM, as well as all traditional nondestructive evaluation procedures (e.g., ultrasonic inspection, acoustic emissions, and active thermography) can be cast in the context of a statistical pattern-recognition problem [2, 3]. Solutions to this problem require four steps: (i) operational evaluation, (ii) data acquisition, (iii) feature extraction, and (iv) statistical modeling for feature classification; these form the statistical pattern-recognition paradigm for SHM. A specific topic that has not been extensively addressed in the SHM literature [4, 5] is the development of mathematically and physically rigorous approaches to designing the SHM sensing system that is used to address the data-acquisition portion of the problem. To date, most SHM system designs are done in a somewhat *ad hoc* manner where the engineer picks a sensing system that is readily available and that he or she is familiar with, and then attempts to demonstrate that a specific type of damage can be detected with that system. If an appropriate level of damage-detection fidelity cannot be obtained, then the system is modified in some empirical manner with the hope that the fidelity improves. Alternatively, as new sensing systems are developed by engineers outside the SHM field, researchers in this

---

Encyclopedia of Structural Health Monitoring. Edited by Christian Boller, Fu-Kuo Chang and Yozo Fujino © 2009 John Wiley & Sons, Ltd. ISBN: 978-0-470-05822-0.

field apply these systems to their respective SHM studies in an effort to see if these systems provide an enhanced damage-detection capability. Through these approaches, several sensor network paradigms for SHM have emerged, and this article summarizes and compares these paradigms. When making such a comparison, it should be noted that the authors do not believe there is one sensor network paradigm that is optimal for all SHM problems. All these paradigms have relative advantages and disadvantages. Also, the paradigms described are not at the same level of maturity and, hence, some may require more development to obtain a field-deployable system while others are readily available with commercial off-the-shelf solutions.

This article first addresses the data-acquisition portion of the paradigm, where the various parameters of the system that must be considered in its design and subsequent field deployment are summarized. Several sensor systems that have been developed specifically for SHM are then discussed in terms of these parameters. These sensor systems suggest the definition of three general SHM sensor network paradigms that are then described along with a summary of their relative attributes and deficiencies. A fourth sensor network that is currently under development is proposed that provides an alternative approach to sensing for SHM. This article also summarizes the practical implementation issues for SHM sensor systems in an effort to suggest a more mathematically and physically rigorous approach to future SHM sensing system design. The article concludes by referring the reader to fundamental axioms of SHM that have been proposed [6] and more specifically the subset of these axioms that address sensing issues for SHM.

2 DATA ACQUISITION FOR SHM

The *data-acquisition* portion of the SHM process involves selecting the excitation methods, the sensor types, number and locations, and the data-acquisition/storage/processing/transmittal hardware. The actual implementation of this portion of the SHM process is application specific. A fundamental premise regarding data acquisition and sensing is that these systems do not measure damage. Rather, they measure the response of a system to its operational and environmental loading or the response to inputs from actuators embedded with the sensing system. Depending on the sensing technology deployed and the type of damage to be identified, the sensor readings may be more or less directly correlated to the presence and location of damage. Data interrogation procedures (feature extraction and statistical modeling for feature classification) are the necessary components of an SHM system that convert the sensor data into information about the structural condition. Furthermore, to achieve successful SHM, the data-acquisition system has to be developed in conjunction with these data interrogation procedures.

Inherent in the data acquisition, the feature extraction and statistical modeling portions of the SHM process are data normalization, cleansing, fusion, and compression. As it applies to SHM, data normalization is the process of separating changes in sensor reading caused by damage from those caused by varying operational and environmental conditions [7]. Data cleansing is the process of selectively choosing data to pass on to, or reject from, the feature selection process. Data fusion is the process of combining information from multiple sensors in an effort to enhance the fidelity of the damage-detection process. Data compression is the process of reducing the dimensionality of the data, or the feature extracted from the data, in an effort to facilitate efficient information storage and to enhance the statistical quantification of these parameters. These four activities can be implemented in either hardware or software and usually a combination of these two approaches is used.

3 SHM SENSOR SYSTEM DESIGN CONSIDERATIONS

All sensor systems are deployed for one of the following applications:

1. detection and tracking problems;
2. model development, validation, and uncertainty quantification;
3. control systems.

SHM is a detection and tracking problem. The goal of any SHM sensor system development is to make the sensor reading as directly correlated with, and as sensitive to, damage as possible (detection) and then have the sensor readings and associated damage-sensitive features extracted from these data change in a monotonic fashion with increasing damage levels (tracking). At the same time, one also strives to make the sensors as independent as possible from all other sources of environmental and operational variability. To best meet these goals for the SHM sensor and data-acquisition system, the following sensing system properties must be defined:

1. types of data to be acquired;
2. sensor types, number, and locations;
3. bandwidth, sensitivity, and dynamic range;
4. data acquisition/telemetry/storage system;
5. power requirements;
6. sampling intervals (continuous monitoring versus monitoring only after extreme events or at periodic intervals);
7. processor/memory requirements;
8. excitation source (active sensing).

Note that some sensor system properties are not independent. For example, increasing sensitivity usually is associated with decreasing dynamic range and increasing bandwidth is typically associated with decreasing frequency resolution. There can be even more issues that must be addressed when developing the sensing portion of the SHM process. Fundamentally, there are four issues related to a specific SHM application that control the selection of hardware to address these sensor system design parameters:

1. the length scales on which damage is to be detected;
2. the time scale on which damage evolves;
3. how varying and/or adverse operational and environmental conditions affect the sensing system;
4. cost.

In addition, the feature extraction, data normalization, and statistical modeling portions of the process can greatly influence the definition of the sensing system properties. Before such decisions can be made, two important questions must be addressed.

First, one must answer the question, "What is the damage to be detected?" The answer to this question must be provided in as quantifiable a manner as possible and address issues such as (i) type of damage (e.g., crack, loose connection, and corrosion), (ii) threshold damage size that must be detected, (iii) probable damage locations, and (iv) anticipated damage growth rates. The more specific and quantifiable this definition, the more likely it is that one will optimize one's sensor budget to produce a system that has the greatest possible fidelity for damage detection. Second, an answer must be provided to the question, "What are the environmental and operational variability that must be accounted for?" To answer this question, one will not only have to have some ideas about the sources of such variability, but one will also have to have thought about how to accomplish data normalization. Typically, data normalization is accomplished through some combination of sensing system hardware and data interrogation software. However, these hardware and software approaches are not optimal if they are not done in a coupled manner.

In summary, from the discussion in this section, it becomes clear that the ability to convert sensor data into structural health information is directly related to the coupling of the sensor system hardware development with the data interrogation procedures.

4 CURRENT SHM SENSING SYSTEMS

Sensing systems for SHM consist of some or all of the following components:

1. transducers that convert changes in the field variable of interest (e.g., acceleration, strain, and temperature) to changes in an electrical signal (e.g., voltage, impedance, and resistance);
2. actuators that can be used to apply a prescribed input to the system (e.g., a piezoelectric transducer bonded to the surface of a structure);
3. analog-to-digital (A/D) converters that transform the analog electrical signal into a digital signal that can subsequently be processed on digital hardware; for the case where actuators are used, a digital-to-analog (D/A) converter is also needed to change the prescribed digital signal to an analog voltage that can be used to control the actuator;
4. signal conditioning;
5. power;

6. telemetry;
7. processing;
8. memory for data storage.

The number of sensing systems available for SHM is enormous and these systems vary quite a bit depending upon the specific SHM activity. Two general types of SHM sensing systems are described below.

4.1 Wired systems

Here, wired SHM systems are defined as ones that telemeter data and transfer power to the sensor over a direct wired connection from the transducer to the central data analysis facility, as shown schematically in Figure 1. In some cases, the central data analysis facility is then connected to the internet such that the processed information can be monitored at a subsequent remote location. There are a wide variety of such systems. At one extreme is peak-strain or peak-acceleration sensing devices that notify the user when a certain threshold in the measured quantity has been exceeded. A more sophisticated system often used for condition monitoring of rotating machinery is a piezoelectric accelerometer with built-in charge amplifier connected directly to a hand-held, single-channel fast Fourier transform (FFT) analyzer. Here, the central data storage and analysis facility is the hand-held FFT analyzer. At the other extreme is custom-designed systems with hundreds of data channels containing numerous types of sensors that cost on the order of multiple millions of dollars such as the sensing system deployed on the Tsing Ma bridge in China [8].

There are a wide range of commercially available wired systems, some of which have been developed for general-purpose data acquisition and the others for SHM applications. Those designed for general-purpose data acquisition can typically interface with a wide variety of transducers and also have the capability to drive actuators. Most of these systems have integrated signal conditioning, data processing, and data storage capabilities and run off of AC power. Those designed to run off batteries typically have a limited number of channels and they are limited in their ability to operate for long periods of time.

One wired system that has been specifically designed for SHM applications consists of an array of piezoelectric lead zirconate titanate (PZT) patches embedded in a Mylar sheet that is bonded to a structure [9]. The PZT patches can be used as either an actuator or a sensor. Damage is detected, located, and, in some cases, quantified by examining the

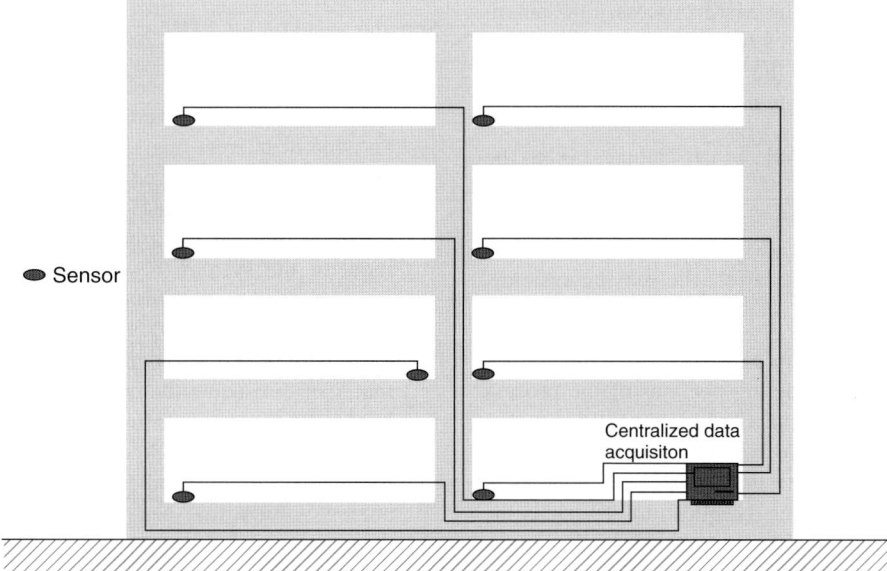

Figure 1. Paradigm I: a wired sensor network connected to a central data-acquisition system running off ac power.

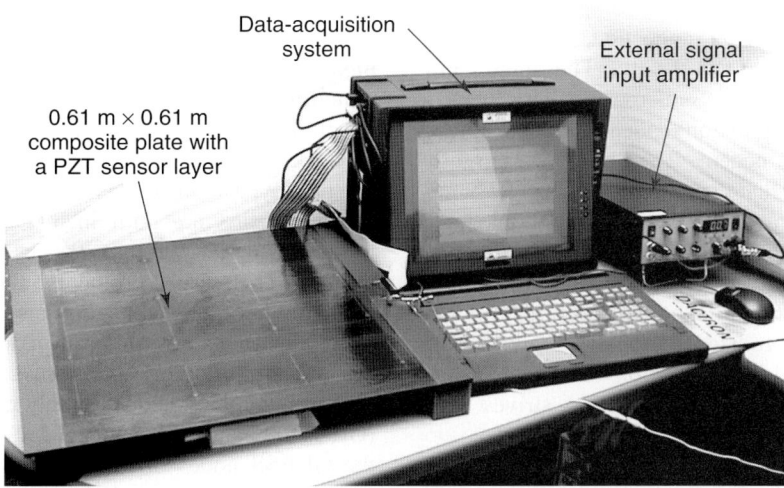

Figure 2. An example of a wired data-acquisition system designed specifically for SHM applications. This system consists of 16 piezoelectric patches in a Mylar sheet. The sensors are connected to a data-acquisition system through the ribbon wire.

attenuation of signals between different sensor–actuator pairs or by examining the characteristics of waves reflected from the damage. An accompanying computer is used for signal conditioning, A/D and D/A conversion, data analysis, and display of final results. The system, which runs on ac power, is shown in Figure 2. This system is described in more detail in **Chapter 78**.

4.2 Wireless transmission systems

More recently, researchers have been adapting general-purpose wireless embedded sensor nodes for SHM applications. Tanner *et al.* [10] modified an SHM algorithm to the limitations of commercial off-the-shelf wireless sensing and data processing hardware. A wireless sensing system of *motes* running TinyOS operating systems developed at the University of California, Berkeley, was chosen because of their commercial availability and their built-in wireless communication capabilities. A mote consists of modular circuit boards integrating a sensor, microprocessor, A/D converter, and wireless transmitter, all of which run off two AA batteries. A significant reduction in power consumption can be achieved by processing the data locally and only transmitting the results. The system was demonstrated using a small portal structure with damage induced by loss of preload in a bolted joint. The tested mote system is shown in Figure 3. However, the processor proved to be very limited, allowing only the most rudimentary data interrogation algorithms to be implemented. Another application of these sensor nodes to civil-engineering infrastructure can be found in **Chapter 69**.

Lynch *et al.* [11] presented hardware for a wireless peer-to-peer SHM system. Using off-the-shelf components, the authors couple sensing circuits and

Figure 3. A mote sensor node that includes a microprocessor, sensor, A/D converter, and radio. A penny has been placed on the node for scale.

wireless transmission with a computational core allowing a decentralized collection, analysis, and broadcast of a structure's health. The final hardware platform includes two microcontrollers for data collection and computation connected to a spread spectrum wireless modem. The software is tightly integrated with the hardware and includes the wireless transmission module, the sensing module, and application module. The application module implements the time-series-based SHM algorithm. This integrated data interrogation process requires communication with a centralized server to retrieve model coefficients. The close integration of hardware and software with the dual microcontrollers strives for a power efficient design.

Spencer [12] provides the state-of-the-art review of current "smart sensing" technologies that includes the compiled summaries of wireless work in the SHM field using small, integrated sensor, and processor systems. A smart sensor is defined as a sensing system with an embedded microprocessor and wireless communication. Many smart sensors covered in this article are still in the stage where they simply sense and transmit data. The mote platform is discussed as an impetus for development of the next generation of SHM systems and a new generation of mote is also outlined. The authors also raised the issues that current smart sensing approaches scale poorly to systems with densely instrumented arrays of sensors that will be required for future SHM systems.

To develop a truly integrated SHM system, the data interrogation processes must be transferred to embedded software and hardware that incorporate sensing, processing, and the ability to return a result either locally or remotely. Most off-the-shelf solutions currently available, or in development, have a deficit in processing power that limits the complexity of the software and SHM process that can be implemented. Also, many integrated systems are inflexible because of tight integration between the embedded software, the hardware, and sensing. More recently, researchers have implemented distributed data interrogation algorithms where processing is done across the sensor network to enhance the computational capabilities of these sensor systems [13, 14].

To implement computationally intensive SHM processes, Farrar et al. selected a single board computer as a compact form for increased processing power [15]. Also included in the integrated system is a digital signal-processing board with six A/D converters providing the interface to a variety of sensing modalities. Finally, a wireless network board is integrated to provide the ability for the system to relay structural information to a central host, across a network, or through local hardware. Figure 4 shows the prototype of this sensing system. Each of these hardware parts are built in a modular fashion and loosely coupled through the transmission control protocol or Internet protocols. By implementing a common interface, changing or replacing a single component does not require a redesign of the entire system. By allowing processes developed in the Graphical Linking and Assembly of Syntax Structure (GLASS) client to be downloaded and run directly in the GLASS node software, this system became the first SHM hardware solution where new processes can be created and loaded dynamically. This modular nature does not lead to the most power optimized design, but instead achieves a flexible development platform that is used to find the most effective combination of algorithms and hardware for a specific SHM problem. Optimization for power is of secondary concern and will be the focus of follow-on efforts [15].

5 SENSOR NETWORK PARADIGMS

The sensor systems discussed in the previous section have led to three types of sensor network paradigms that are either currently being used for SHM or are the focus of current research efforts in this field. These paradigms are described below. Note that the illustrations of these systems show them applied to a building structure. However, these paradigms can be applied to a wide variety of aerospace, civil, and mechanical systems, and the building structure is simply used for comparison purposes.

5.1 Sensor arrays directly connected to central processing hardware

Figure 1 shows a sensor network directly connected to the central processing hardware. Such a system is the most common one used for SHM studies. The advantage of this system is the wide variety of commercially available off-the-shelf systems that can

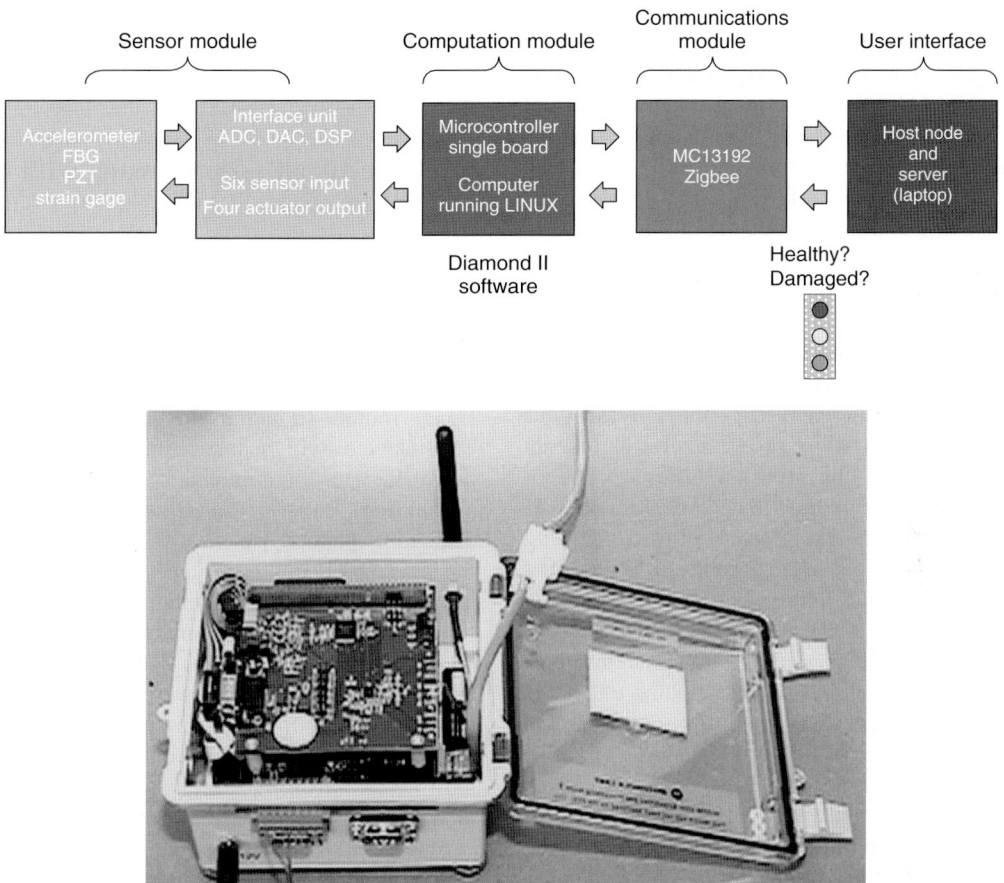

Figure 4. A sensor node incorporating a single board computer to increase processing power. Also included in the system is a digital signal-processing board with six A/D converters that interface to a variety of sensing modalities and a wireless network board that provides the ability to relay information to a central host, across a network, or through local hardware.

be used for this type of monitoring and the wide variety of transducers that can typically be interfaced with such a system. For SHM applications, these systems have been used in both a passive and active-sensing manner. Limitations of such systems are that they are difficult to deploy in a retrofit mode because they usually require ac power, which is not always available. Also, these systems are one-point failure sensitive as one wire can be as long as a few hundred meters. In addition, the deployment of such a system can be challenging with potentially over 75% of the installation time attributed to the installation of system wires and cables for large-scale structures such as those used for long-span bridges [16]. Furthermore, experience with field-deployed systems has shown that the wires can be costly to maintain because of general environmental degradation and damage caused by things such as rodents and vandals.

5.2 Decentralized processing with hopping connection

The integration of wireless communication technologies into SHM methods has been widely investigated to overcome the limitations of wired sensing networks. Wireless communication can remedy the cabling problem of the traditional monitoring system and significantly reduce the maintenance cost. The schematic of the decentralized wireless monitoring system, which is summarized in detail by Spencer et al. [12], is shown in Figure 5.

From the large-scale SHM practice, however, several very serious issues arise with the current design and deployment scheme of the decentralized wireless sensing networks [12, 17]. First, the current wireless sensing design usually adopts *ad hoc* networking and hopping that result in a problem referred to as *data collision*. Data collision is a phenomenon that results from a network device receiving several simultaneous requests to store or retrieve data from other devices on the network. With increasing numbers of sensors, a sensor node located close to the base station will experience tremendous data transmission, possibly resulting in a significant bottleneck. Because the workload of each sensor node cannot be evenly distributed, the chances of data collision increase with expansion of the sensing networks. In addition, this decentralized wireless sensing network scales very poorly in active-sensing system deployment. Because active sensors can serve as actuators as well as sensors, the time synchronization between multiple sensor/actuator units is a challenging task. Because of the processor scheduling or sharing, the use of multiple channels on one sensor node would reduce the sampling rate, which provides neither a practical nor equitable solution for active-sensing techniques that typically interrogate higher frequency ranges. Therefore, for *in situ* applications, the current design scheme can potentially be a very expensive operation.

5.3 Decentralized processing with hybrid connection

The hybrid connection network advantageously combines the previous two networks, as illustrated in Figure 6. At the first level, several sensors are connected to a relay-based piece of hardware, which can serve as both a multiplexer and general-purpose signal router, shown in Figure 6 as a black box. This device will manage the distributed sensing network, control the modes of sensing and actuation, and multiplex the measured signals. The device can also be expanded by means of daisy-chaining. At the next level, multiple pieces of this hardware are linked to a decentralized data control and processing station. This control station is equipped with data-acquisition boards, on-board computer processors, and wireless telemetry, which is similar to the architecture of current decentralized wireless sensors. This device will perform duties of a relay-based hardware control, data acquisition, local computing, and transmission of the necessary results of the computation to the central system. At the highest

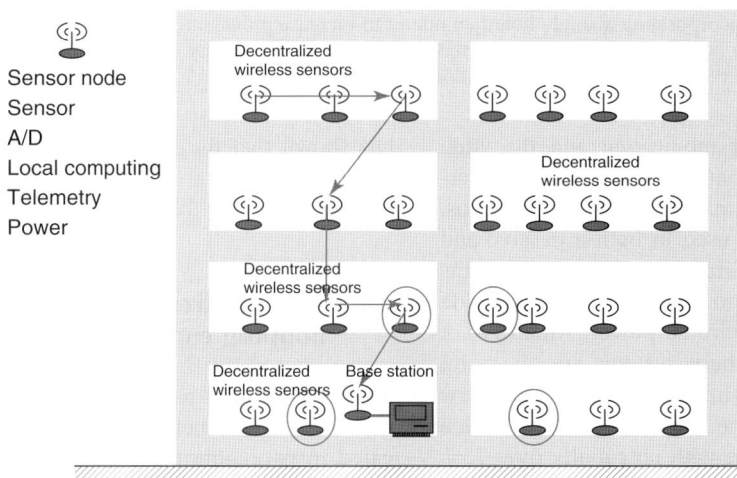

Figure 5. Paradigm II: decentralized processing with each sensor node running off battery power and utilizing a "hoping" telemetry protocol. The "mote" shown in Figure 3 is one such sensor node that can be deployed to form this type of sensor network.

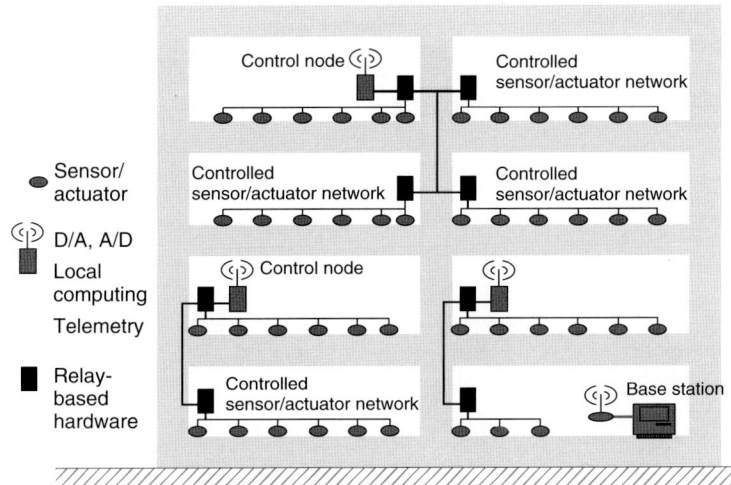

Figure 6. Paradigm III: a hybrid decentralized sensor system with local multiplexing and a hopping telemetry protocol.

level, multiple data processing stations are linked to a central monitoring station that delivers a damage report back to the user. Hierarchal in nature, this sensing network can efficiently interrogate large numbers of distributed sensors and active sensors while maintaining an excellent sensor–cost ratio because only a small number of data acquisition and telemetry units is necessary. This hierarchal sensing network is especially suitable for active-sensing SHM techniques, and is being investigated by Dove *et al.* [17]. In their study, the expandability or the sensing network was of most importance for significantly larger numbers of active sensors, as the number of channels on a decentralized wireless sensor is limited because of processor sharing and scheduling. The prototype of the relay-based hardware ("black box" shown in Figure 6) is illustrated in Figure 7.

6 FUTURE SENSING NETWORK PARADIGMS

The sensing network paradigms described in the previous section have one characteristic in common. The sensing system and associated power sources are installed at fixed locations of the structural system. As previously stated, the deployment of such sensing systems can be costly and the power source may not be always available. A new, energy-efficient future sensing network is currently being investigated collaboratively by Los Alamos National Laboratory and the University of California, San Diego, and is shown in Figure 8. This system couples energy-efficient embedded sensing technology and remote interrogation platforms based on either robots or unmanned aerial vehicles (UAV) to assess damage in structural systems [18]. This approach involves using an unmanned mobile host node (delivered via UAV or robot) to generate radio frequency (RF) signal near the receiving antennas connected to sensor nodes that have been embedded on the structure. Once a capacitor on the sensor node is charged by the RF energy emitted from the node on the UAV, the sensors measure the desired response (impedance, strain, etc.) at critical areas on the structure and transmit the signal back to a processor on the mobile host.

Figure 9 shows a sensor node that has been developed for this mode of remote powering and telemetry. This sensor node uses a low-power integrated circuit that can measure, control, and record an impedance measurement across a piezoelectric transducer. The sensor node integrates several components, including a microcontroller for local computing and sensor node control, a radio for wireless data transmission, multiplexers for managing up to seven piezoelectric transducers per node, energy storage mediums, and several triggering options including a wireless triggering into one package to realize a comprehensive, self-contained wireless active-sensor node for SHM applications. It was estimated that this sensor

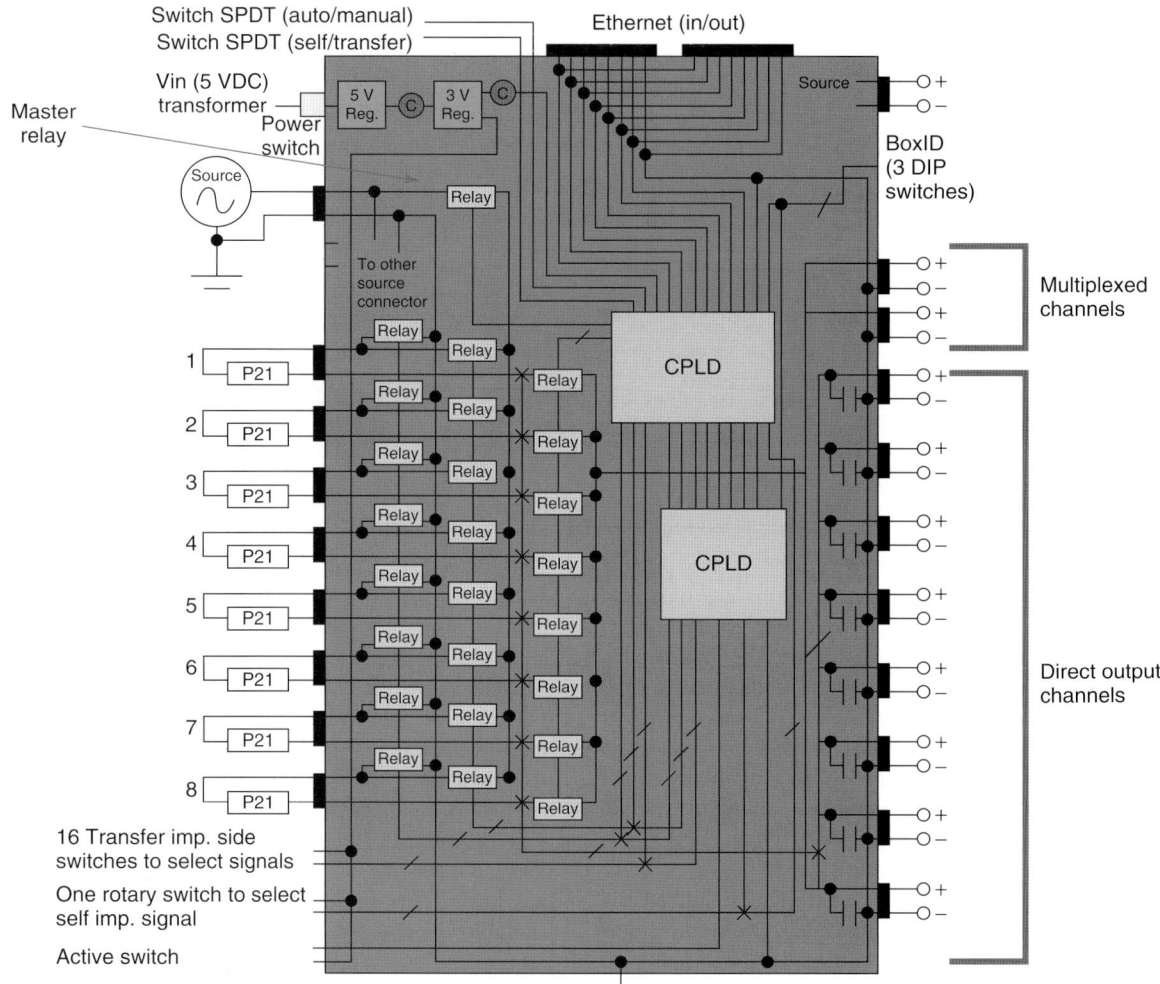

Figure 7. The relay-based hardware shown as a black box in Figure 6. This device manages the distributed sensing network, controls the modes of sensing and actuation, and multiplexes the measured signals.

node requires less than 60 mW of total power to operate, measure, compute, and transmit. Considering this amount of power consumption, the sensor node is within the range of the wireless energy transmission capabilities provided by the host node on the UAV as well as for energy harvesting devices such as small solar arrays.

One UAV with its power source, telemetry, and computing can be used to interrogate an entire sensor array placed on the structure and then can be used for other structures that have similar embedded sensor arrays. A recent field demonstration of this sensor network strategy is shown in Figure 10 where a remotely controlled helicopter was used to deliver power to sensor nodes mounted on a bridge structure. The next generation of this system will take traditional sensing networks to the next level of autonomy, as the mobile hosts (such as UAV), will fly to known critical infrastructure based upon a global positioning system (GPS) locator, deliver the required power, and then perform the SHM assessment without human intervention. This technology will be directly applicable to rapid structural condition assessment of buildings and bridges after an earthquake, where the sensor nodes may need to be deployed for decades during which conventional battery power will be

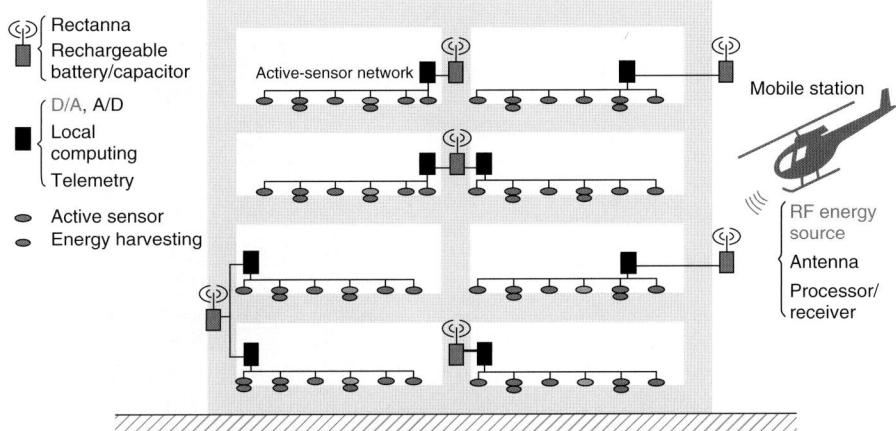

Figure 8. A new sensor network strategy where power and processing are brought to the sensor nodes on a robotic device. The sensor nodes are powered on demand by means of wireless energy transmission.

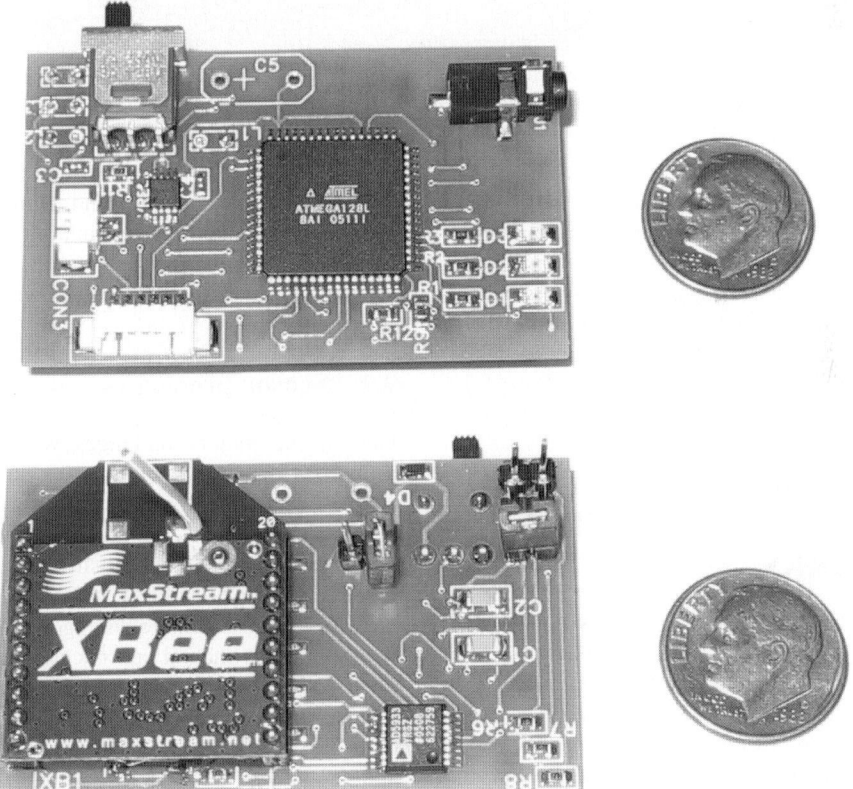

Figure 9. The sensor node that has been designed to receive power wirelessly from a remote host as depicted in Figure 8. This sensor node can measure impedance across up to seven piezoelectric sensors.

Figure 10. Field demonstration of wireless power delivery to a sensor node embedded on a bridge structure. The receiving antenna can be seen suspended from the bottom flange of the bridge girder.

depleted. Also, this technology may be adapted and applied to damage detection in a variety of other civilian and defense-related structures such as nuclear power plants where it is advantageous to minimize human exposure to hazardous environments during the inspection process. A review of other applications of robotic devices for SHM sensing can be found in [19].

7 PRACTICAL IMPLEMENTATION ISSUES FOR SHM SENSING NETWORKS

A major concern in the current sensing network development is the long-term reliability of the network and how to power the network. Other concerns are the abilities of the sensing systems to capture local and system level response, that is, the need to capture response on widely varying length and time scales, and to archive data in a consistent, retrievable manner for long-term analysis. These challenges are nontrivial because of the tendency for each technical discipline to work more or less in isolation. Therefore, an integrated systems engineering approach to the damage-detection process and regular, well-defined routes of information dissemination are essential. The subsequent portions of this article address specific sensing system issues associated with SHM.

7.1 Sensor properties

One of the major challenges with defining sensor properties is that these properties need to be defined *a priori* and typically cannot be changed easily once a sensor system is in place. These sensor properties include bandwidth, sensitivity (dynamic range), number, location, stability, reliability, cost, telemetry, etc. To address this challenge, a coupled analytical and experimental approach to the sensor system deployment should be used in contrast with the *ad hoc* procedures used for most current damage-detection studies. First, critical system failure modes should be well-defined in as quantifiable a manner as possible, using high-fidelity numerical simulations or from previous experiences, before the sensing system is designed. The high-fidelity numerical simulations/experiences can be used to define the required bandwidth, sensitivity, sensor location, and sensor number. Additional sensing requirements can also be ascertained if changing operational and environmental conditions are included in the models so as to determine how these conditions affect the damage-detection process. The outcome of such analyses may indicate that additional sensors are needed to quantify the effects of varying operational and environmental conditions on the damage-detection process. As an example, it has been shown that the dynamic properties of a bridge structure vary significantly with temperature [20]; however, a measure of ambient air temperature does not correlate with the change in dynamic properties. Instead, it was found that the change in dynamic properties was correlated with the temperature differential across the bridge, which implies the need for multiple temperature sensors in the sensing system.

Another potential level of integration between modeling and sensing resides in the integration of software and hardware components. Once the actuation and sensing capability has been selected, their location has been optimized for damage observability, and the specifications of the data-acquisition system have been met, it may be advantageous to integrate model output and sensing information as much as possible. For example, surrogate models can be programmed on local digital signal-processing chips and their predictions can be compared to sensor output in real time. One obvious benefit would be to minimize the amount of communication by integrating the analysis capability with real-time sensing. In an integrated approach, features can be extracted from sensing data and numerical simulation. Test-analysis comparison and parameter estimation can then be performed locally, which would greatly increase the efficiency of the damage-detection process.

7.2 Power consideration

A major consideration in using a dense sensor array is the problem of providing power to the sensors. This demand leads to the concept of "information as a form of energy". Deriving information costs energy. If the only way to provide power is by direct connections, then the need for wireless communications protocols is eliminated, as the cabled power link can also be used for the data transmission. However, if a wireless communication protocol is used, the development of micropower generators will provide significant advantages over battery power sources as the concept of autonomous embedded sensing cannot be realized if one has to periodically replace batteries. A possible solution to the problem of localized power generation is technologies that enable harvesting ambient energy to power the sensor nodes [21, 22] (*see* **Chapter 75**, **Chapter 76**; and **Chapter 77**). Forms of energy that may be harvested include thermal, vibration, acoustic, and solar. Because such energy harvesting is somewhat new technology, the overriding consideration of reliability still exists, as it does with other components of the monitoring system. With two-way communication capability, the local sensing and processing units can also turn themselves off or go into a "sleep" mode for energy conservation and they can be resuscitated when a "wake-up" signal is broadcast. This approach to power management is discussed in detail in [22].

7.3 Sensor calibration, stability, and ruggedness

When discussing calibration, stability, and ruggedness, it must be clear that these concepts are applied to the entire sensor system and not just the sensor itself. Calibration is the process of determining the relationship between the field variable to be measured and the electrical signal generated by the sensor. Stability refers to how the calibration varies with time. Because SHM involves a comparison of measured response before and after a damaging event, stability is the more critical property for SHM sensor systems. Well-defined sensor calibration procedures exist, but approaches for establishing sensor stability are less well-defined. Most sensors are calibrated at a specialized calibration facility with well-established protocols and standards. This type of calibration is expected to endure, but for embedded sensor systems it needs to be supplemented by incorporating a self-diagnosing and self-calibrating capability into the sensors. In some cases, measurements are acceptable with 20% error, as long as this error remains constant from one measurement to another one made at some future time. In other scenarios, absolute accuracies are necessary to ensure that the sensor has the fidelity to measure the changes in system response associated with the onset of damage.

Ruggedness of the sensors is a prime consideration for SHM. If part of the system is compromised, then the overall confidence in the system performance is undermined. For sensors implemented for SHM, several ruggedness considerations emerge:

1. The nontrivial problem of sensor selection for extreme environments (e.g. in-service turbine blades exposed to extreme temperatures, high-temperature components of an oil refinery and fluid systems of nuclear power plant that are exposed to radiation fields).
2. Sensors may be less reliable than the component they are monitoring—for example, reliable parts may have failure rates of 1 in 100 000 over several years time. Sensors are often small, complex assemblies with built-in microelectronics, so sensors subjected to the same operational and environmental loading conditions may fail more often than the component being monitored. Loss of sensor signal may then be falsely interpreted as component failure, not sensor failure.
3. Sensors may fail through outright sensor destruction while the component being monitored endures.

False indications of damage or damage precursors are extremely undesirable. If this occurs often, the sensor is either overtly or covertly ignored. Recently, several studies have focused on issues of sensor validation [23, 24]. However, in general, there is little data on the long-term stability and ruggedness of SHM sensor systems.

7.4 Multiscale sensing

Depending on the size and location of the structural damage and the loads applied to the system, the adverse effects of the damage can be either immediate or may take some time before it alters the system's performance. In terms of length scales, all damage begins at the material level and then under appropriate loading conditions progresses to component and system level damage at various rates. In terms of time scales, damage can accumulate incrementally over long periods of time such as that associated with fatigue or corrosion. Damage can also occur on much shorter time scales because of scheduled discrete events such as aircraft landings and from unscheduled discrete events such as enemy fire on a military vehicle. Therefore, the most fundamental issue that must be addressed when developing a sensing system for SHM is the need to capture the structural response on widely varying length and time scales.

The sensing systems that are able to capture the responses over varying length and time scales have not been substantially investigated by researchers, although it is quite possible to use the same piezoelectric patches in both an active, high-frequency mode and in a passive mode to capture the lower-frequency global response of the system. As an example, in the active mode, the piezoelectric sensors can be used to detect and locate damage on a local level using relatively higher frequency excitation and response measurements. This type of active sensing can be used to detect delamination in the composite skin on the wing of an unmanned aerial vehicle. In addition, the same sensors can be used in a passive mode to monitor the low-frequency global modal response of the wing when it is subjected to aerodynamic loading. This global response data can be used to assess the effect of the delamination on the flutter characteristics of that aircraft as determined by analysis of the coupling between the first bending and torsion mode of the wing.

7.5 Sensor–actuator optimization

Few researchers have addressed the issue of developing a systematic approach to the design of a sensor system for SHM. In very general terms, one approach is to consider the sensor system design as a constrained optimization problem. An example, of one such study that employed machine learning to optimize sensor number and location is given in [25]. In terms of a constrained optimization problem, the designer would like to maximize the "damage observability" subjected to a wide variety of possible constraints such as cost, weight, power (when active sensing is used), and allowable locations. A challenge to actually implementing this approach is coming up with accurate mathematical definitions for damage observability and its relation to the various sensor system properties. This challenge is confounded by the fact that quite a few sensor system parameters may influence observability and the interactions between these various parameters may not be well understood.

One approach to solve the optimization problem is to determine (or assume) that a particular sensor to be employed has a certain damage-detection resolution (i.e., can detect a 1-mm crack through the thickness of a plate within 15-cm radius of the sensor). Then assume that you have an infinite number of sensors, which in turn maximizes observability. Next, optimization procedures such as genetic algorithms or gradient descent methods are used to maximize the observability while retaining some fraction of the infinite sensor array. This process produces a sensor layout with a minimum number of sensors placed at locations that maximize damage observability. Note that this optimization problem will become much more complicated when "real-world" issues such as operational and environmental variability have to be addressed. Also, one must consider the trade-offs between an optimal sensing system and a redundant sensing system. If one sensor or sensor node fails in an optimal system, it is most likely no longer optimal.

With the advent of active-sensing approaches, there can be SHM applications where the excitation is selectable, and, this excitation should be chosen to maximize damage observability. As a simple example, consider a beam or column with a crack that is nominally closed because of a preload. If the provided excitation is not sufficient to open and close the crack, the detectability of the crack in the measured output will be severely limited. Thus, if possible, it is important to answer the question: "Given ever-present physical limits on the level of excitation, and limited outputs that can be measured, what excitation should be provided to a system to make damage most detectable?" When one considers that an excitation may be viewed as a time series with hundreds or thousands of free parameters, optimization in this high-dimensional space might be a daunting task. However, as is demonstrated in [26, 27], a gradient-based technique may be used in which the gradient can be calculated very efficiently. This method does require a model of the system and the accuracy of that model will influence the results.

8 SUMMARY

In this article, the current sensor system design research that is being done to address the data-acquisition portion of the SHM problem is summarized. Several sensor systems that have been developed specifically for SHM are discussed. These sensor systems lead to the definition of several general SHM sensor network paradigms. All of these paradigms have relative advantages and disadvantages. Also, the paradigms described are not at the same level of maturity and, hence, some may require more development to obtain a field-deployable system, while others are readily available with commercial off-the-shelf solutions. At this time, no formal and accepted design methodology exists for the development of an SHM sensing system. As such, this article has also summarized practical implementation issues associated with the SHM sensor system in an effort to suggest the need for a more mathematically and physically rigorous approach to future SHM sensing system design. Finally, it should be noted that recently fundament axioms for SHM have been proposed [6] on the basis of the information published in the extensive amount of literature on SHM over the last 20 years. Of the eight axioms proposed in this article, seven are closely related to sensing aspects of the SHM problem and, therefore, should be considered when designing any SHM sensor network.

REFERENCES

[1] Worden K, Dulieu-Barton JM. An overview of intelligent fault detection in systems and structures. *International Journal of Structural Health Monitoring* 2004 **3**(1):85–98.

[2] Farrar CR, Worden K. An introduction to structural health monitoring. *Philosophical Transactions of the Royal Society A* 2007 **365**:303–315.

[3] Farrar CR, Doebling SW, Nix DA. Vibration-based structural damage identification. *Philosophical Transactions of the Royal Society: Mathematical, Physical and Engineering Sciences* 2001 **359**(1778):131–149.

[4] Doebling SW, Farrar CR, Prime MB, Shevitz DW. *Damage Identification and Health Monitoring of*

Structural and Mechanical Systems from Changes in their Vibration Characteristics: A literature Review, Los Alamos National Laboratory report LA-13070-MS, 1996.

[5] Sohn H, Farrar CR, Hemez FM, Czarnecki JJ, Shunk DD, Stinemates DW, Nadler BR. *A Review of Structural Health Monitoring Literature from 1996–2001*, Los Alamos National Laboratory report LA-13976-MS, 2004.

[6] Worden K, Farrar CR, Manson G, Park G. The fundamental axioms of structural health monitoring. *Proceedings of the Royal Society A: Mathematical, Physical and Engineering Sciences* 2007 **463**(2082):1639–1664.

[7] Farrar CR, Sohn H, Worden K. Data normalization: a key to structural health monitoring *Proceedings of the Third International Structural Health Monitoring Workshop*. Stanford, CA, 2001.

[8] Ni YQ, Wang BS, Ko JM. Simulation studies of damage location in Tsing Ma Bridge deck. *Proceedings of Nondestructive Evaluation of Highways, Utilities, and Pipelines IV*. SPIE: Bellingham, Washington, 2001, pp. 312–323.

[9] Lin M, Qing X, Kumar A, Beard S. SMART layer and SMART suitcase for structural health monitoring applications. *Proceedings of SPIE on Smart Structures and Materials*. The International Society for Optical Engineering: Newport Beach, CA, March 2001; Vol. 4332, pp. 98–106.

[10] Tanner NA, Wait JR, Farrar CR, Sohn H. Structural health monitoring using modular wireless sensors. *Journal of Intelligent Material systems and Structures* 2003 **14**(1):43–56.

[11] Lynch JP, Law KH, Kiremidjian AS, Carryer E, Kenny TW, Partridge A, Sundararajan A. Validation of a wireless modular monitoring system for structures. *Procedings of SPIE 9th Annual International Symposium on Smart Structures and Materials*. San Diego, CA, 17–21 March 2002.

[12] Spencer BF, Ruiz-Sandoval ME, Kurata N. Smart Sensing Technology: Opportunities and Challenges. *Structural Control and Health Monitoring* 2004 **11**(4):349–368.

[13] Zimmerman AT, Shiraishi M, Swartz A, Lynch JP. Automated modal parameter estimation by parallel processing within wireless monitoring systems. *ASCE Journal of Infrastructure* 2008 **14**(1): 102–113.

[14] Swartz RA, Lynch JP. A multirate recursive arx algorithm for energy efficient wireless structural monitoring. *4th World Conference on Structural Control and Monitoring*. San Diego, CA, 2006.

[15] Farrar CR, Allen DW, Ball S, Masquelier MP, Park G. Coupling sensing hardware with data interrogation software for structural health monitoring. *Proceedings of 11th International Symposium on Dynamic Problems of Mechanics*. Ouro Preto, Brazil, March 2005.

[16] Lynch JP, Partridge A, Law KH, Kenny TW, Kiremidjian AS, Carryer E. Design of a Piezoresistive MEMS-based accelerometer for integration with a wireless sensing unit for structural monitoring. *ASCE Journal of Aerospace Engineering* 2003 **16**:108–114.

[17] Dove JR, Park G, Farrar CR. Hardware design of hierarchal active-sensing networks for structural health monitoring. *Smart Materials and Structures* 2006 **15**:139–146.

[18] Todd M, *et al.* A different approach to sensor networking for shm: remote powering and interrogation with unmanned aerial vehicles. *Structural Health Monitoring 2007 Quantification, Validation and Implementation*. DEStech Publication: Lancaster, PA, 2007, Vol. 1, pp. 29–43.

[19] Huston DR. Robotic surveillance approaches for SHM. *Structural Health Monitoring 2005 Advancement and Challenges for Implementation*. DEStech Publication: Lancaster, PA, 2005, Vol. 2, pp. 1586–1593.

[20] Farrar CR, Cornwell PJ, Doebling SW, Prime MB. *Structural Health Monitoring Studies of the Alamosa Canyon and I-40 Bridges*. Los Alamos National Laboratory report, LA-13635-MS, 2000.

[21] Sodano HA, Inman DJ, Park G. A review of power harvesting from vibration using piezoelectric materials. *The Shock and Vibration Digest* 2004 **36**(3):197–205.

[22] Park G, Farrar CR, Todd MD, Hodgkiss W, Rosing T. *Power Harvesting for Embedded Structural Health Monitoring Sensing Systems*. Los Alamos National Laboratory report, LA-14314-MS, 2007.

[23] Park G, Farrar CR, Rutherford CA, Robertson AN. Piezoelectric active sensor self-diagnostics using electrical admittance measurements. *ASME Journal of Vibrations and Acoustics* 2006 **128**:469–476.

[24] Kerschen G, Boe PD, Golinval J, Worden K. Sensor validation using principal component analysis. *Smart Materials and Structures* 2005 **14**(1):36–42.

[25] Worden K, Burrows AP, Tomlinson GR. A combined neural and genetic approach to sensor

placement. *Proceedings of 13th International Modal Analysis Conference*, Nashville, TN, USA, 1995; pp. 1727–1736.

[26] Bement MT, Bewley T. Optimal excitation design for damage detection using adjoint based optimization Part 1 theoretical development. *Mechanical Systems and Signal Processing*, submitted for publication.

[27] Bement MT, Bewley T. Optimal excitation design for damage detection using adjoint based optimization Part 2 experimental verification. *Mechanical Systems and Signal Processing*, submitted for publication.

Chapter 72
Nondestructive Evaluation of Cooperative Structures (NDECS)

Daniel L. Balageas
Structure and Damage Mechanics Department, ONERA (The French Aerospace Lab), Châtillon, France

1 Introduction	1269
2 Recent Evolution in NDE, SHM, and Maintenance Philosophy	1269
3 Between NDE and SHM Could We Imagine an Intermediate Way?	1270
4 Imaging Interactions of Lamb Waves with Damages using Lock-in Thermography	1271
5 Imaging Interactions of Lamb Waves with Damages using Stroboscopic Shearography	1274
6 Imaging Interactions of Lamb Waves with Damages using Scanned Laser Ultrasonics and Embedded PZT Receivers	1276
7 Conclusions	1278
References	1278

1 INTRODUCTION

The recent evolution of nondestructive evaluation (NDE) techniques used for maintenance of structures

Encyclopedia of Structural Health Monitoring. Edited by Christian Boller, Fu-Kuo Chang and Yozo Fujino © 2009 John Wiley & Sons, Ltd. ISBN: 978-0-470-05822-0.

is characterized by rapid and dramatic changes. These changes also concern the general philosophy of structure maintenance and monitoring.

Three major evolutions can be highlighted: (i) in the pure NDE field, the ever growing importance of full-field, real-time, noncontact imaging techniques (*see* **Chapter 65**); (ii) the birth and the impressive development of structural health monitoring (SHM), which could be superficially considered as an avatar of NDE, if only seen as a fully integrated NDE; and (iii) the importance of Lamb waves to elaborate SHM systems.

An analysis of these evolutions leads us to consider a possible third way, intermediate between NDE and SHM, called *nondestructive evaluation of cooperative structures* (NDECS). To illustrate this concept, three possible NDECS are presented. The first two techniques result from the combination of localized and embedded Lamb wave generation and noncontact optical detection; the last one combines laser ultrasonics and embedded piezoelectric detection.

2 RECENT EVOLUTION IN NDE, SHM, AND MAINTENANCE PHILOSOPHY

Presently, the maintenance of aerospace structures is based on the use of NDE techniques. One of the characteristic evolutions of NDE and of experimental mechanics techniques is the importance gained by the development of full-field, real-time, noncontact imaging techniques. Such techniques, thanks to charge-coupled device (CCD) cameras, are performing a parallel acquisition of information coming from a very large number of locations. This is achieved by cameras working in various spectral domains: ultraviolet, visible, near, or far infrared, associated with coherent light illumination (interferometric techniques such as electronic speckle pattern interferometry (ESPI) [1]) or incoherent sources (stimulated thermography [2], visible image correlation techniques [3], etc.). These techniques can image fields of displacements, deformation, temperature, etc. of structures submitted to various types of solicitations: loads [4], fatigue [5], vibrations [6, 7], radiant heat fluxes [8], eddy currents [9], electromagnetic fields [10], etc. A high-frequency periodical excitation often used in NDE consists in propagating ultrasounds in the structure. Nevertheless, it has been only very recently coupled to full-field imaging techniques, and more particularly with shearography and thermography.

SHM, which belongs to the domain of smart materials and structures, can be considered as a new form of NDE, characterized by the full integration of sensors, actuators, and intelligence inside the structure during its manufacturing process. It is the reason why the recent growth of SHM community is partially explained by the progressive aggregation of members of the NDE community, while SHM, at its beginning, was mainly done by the mechanical engineering community.

SHM techniques were introduced in the 1990s and show an impressive development [4]. If we look at the SHM literature, we see an increasing importance of Lamb wave-based work and, for this purpose, a prevalent use of piezoelectric patches (*see* **Chapter 14**; **Chapter 55**; **Chapter 57**). This is illustrated by a statistical analysis of the recent publications given in [4, 11].

These three evolutions (growing importance of full-field real-time imaging techniques, impressive development of SHM, and importance gained by the use of Lamb waves to elaborate SHM systems) are mainly driven by a cost-reduction strategy of maintenance operations (*short-term objective*). The second evolution—development of SHM—aims at the replacement of scheduled maintenance inspections by performance-based or condition-based inspections (*long-term objective*).

3 BETWEEN NDE AND SHM COULD WE IMAGINE AN INTERMEDIATE WAY?

If we consider the progress achieved in both NDE and SHM fields, a third approach seems possible now. This third way could consist in only embedding at well-chosen locations into the structure the stimulation function (actuators or emitters) and leaving the full-field noncontact detection (sensors or receivers) outside, or in adopting an opposite configuration with a full-field noncontact stimulation outside and a localized detection inside. To optimize such systems, Lamb waves could be used in conjunction with full-field noncontact optical techniques. So, using the more recent and promising techniques of both SHM and NDE domains, the proposed approach could be applied in a *short term*, with existing technologies, and would permit time and money saving for maintenance operations. A well-suited appellation for this type of technique could be NDECS.

Such an idea is not totally new. A similar approach has been made by Walsh [12] with a more limited field of application. Walsh proposed to replace conventional ultrasonic testing using surface contact probes by a semiembedded system in which the emitter is inside the structure and the detector is outside, which allows an improved resolution and a larger depth of penetration. Walsh called this concept as *nondestructive evaluation ready material* (*NDERM*) technology.

The NDECS concept proposed here is more efficient and more ambitious. Figure 1 presents a schematic view of what could be an NDECS system. Three possible systems are described using (i) lock-in thermography to monitor the thermal effects resulting from the interaction of the waves with possible damages like delaminations, cracks, or corrosions; (ii) stroboscopic shearography to monitor the full field of the surface displacements created by the Lamb waves; (iii) ultrasounds generated by a

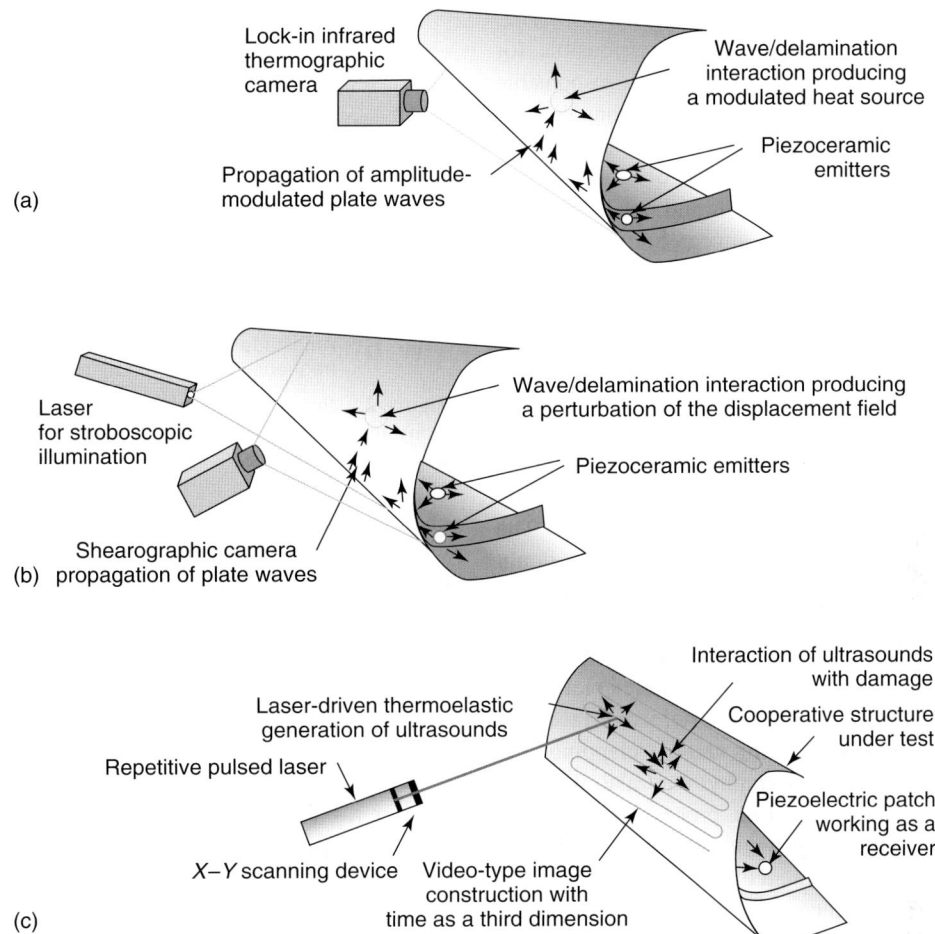

Figure 1. Three possible NDECS: (a) lock-in ultrasonic vibrothermography with Lamb waves generated by embedded piezoelectric emitters [Reproduced with permission from Ref. 11. © 2007.], (b) shearographic visualization of Lamb wave interaction with damage [11], and (c) piezoelectric detection of laser-generated ultrasounds. [Reproduced with permission from Ref. 13. © Takatsubo, 2006.]

pulse repetitive laser and detected by an embedded piezoelectric receiver. These three imaging techniques are chosen because they seem more promising and they are almost ready to be used, since their feasibility has been proved in laboratory. The two first systems are based on work done at Office National d'Etudes et de Recherches Aérospatiales (ONERA) and were presented as NDECS systems in [11]. The third system is studied by a group at National Institute of Advanced Industrial Science and Technology (AIST) (Japan) [13]. The following sections present the principle and some laboratory results already obtained with these techniques.

4 IMAGING INTERACTIONS OF LAMB WAVES WITH DAMAGES USING LOCK-IN THERMOGRAPHY

4.1 Principle of lock-in ultrasonic vibrothermography

Ultrasonic vibrothermography is an NDE technique based on the application of a modulated mechanical stress on the tested structure, while an IR camera maps the surface temperature [14]. Thermomechanical coupling (*see* **Chapter 19**) is responsible for heat production, which, in turn, is partly responsible for damping. In particular, damaged regions convert energy into heat through enhanced viscoelastic dissipation, collisions, and/or rubbing of internal free surfaces present in delaminations and cracks. Surface defects and internal defects appear hotter when the surface temperature is mapped, a fact that provides the basis for the use of joint mechanical excitation and thermography as a nondestructive technique.

Most of the experimenters have used high-power ultrasound emitters producing simultaneously several uncontrolled waves: mechanical shakers [15], a 500-W ultrasound cleaner [16], ultrasound generators such as sonotrodes (generally designed for plastic welding) with power up to 2 kW [17, 18], etc. The use of such powerful devices may cast some doubt on the nondestructive aspect of the control. The first danger is thus, in the case of polymer composite testing, of overheating the material in the contact area. The second danger is that small tolerable cracks may grow unexpectedly fast when the defect surfaces are submitted to energetic rubbing and/or "clapping".

Ultrasonic vibrothermography can also be performed through video lock-in processing [15] (*see* **Chapter 19**) if the high-frequency mechanical excitation is amplitude modulated at a low frequency. In this case, the heat generated by the *high-frequency vibrations* is modulated at this *low frequency* and the thermographic lock-in system is synchronized with this thermal wave frequency. At ONERA it has been proposed to evaluate whether lock-in ultrasonic vibrothermography could be applied by means of small, low-energy piezoelectric transducers embedded or surface bonded to the tested structure, and generating Lamb waves (Figure 2). These transducers are currently used in acousto-ultrasonic SHM systems. This has been successfully demonstrated in the case of C/epoxy plates [14]. This solution presents three main advantages: (i) low-power actuators (less than 1 W) can be used, inducing a low dissipation at the defect location, nevertheless detected, thanks to the lock-in procedure; (ii) embedding or bonding the piezoelectric patches guarantee a coupling constant in time and ascertainable by measuring the electromechanical impedance [19, 20]; and (iii) the use

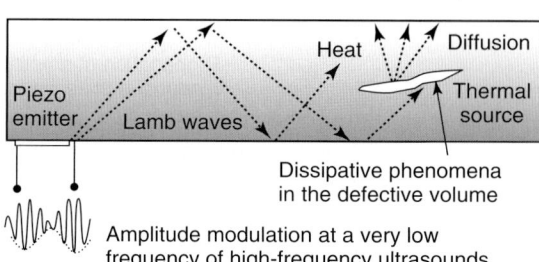

Figure 2. Principle of lock-in ultrasonic vibrothermography applied to the visualization of the interaction of Lamb waves with a damage.

of Lamb wave permits long-distance propagation with moderate attenuation, thus giving an almost uniform sensitivity to the defect detection in the full structure.

The choice of the frequency and the nature of the wave to excite the defects are very important. This has been demonstrated by several experimenters [21–23].

4.2 Low-energy detection of delaminations in composite panels

Following the principle presented in Figure 2, results have been obtained using Lamb waves and very low energies. Temperature mapping was performed using an IR focal plane array camera (Amber 4128, 128 × 128 pixels), with a noise equivalent temperature difference (NETD) of 7 mK. The lock-in technique was applied to demodulate the thermal signal. The ultrasound actuator was fed with a high-frequency electric modulation, between a few tens and a few hundreds of kilohertz, and the amplitude was modulated with a sinus function or a square function at a low frequency (33–300 mHz). By processing 500 images, a temperature modulation amplitude image with an NETD lower than about 0.5–0.6 mK was obtained, making the detection of defects such as delaminations, debonding, microcracking, and macrocracks possible.

The results presented here were obtained with a $[0_4/45_4/90_4/-45_4]_S$ C/epoxy plate of dimensions $70 \times 70 \, \text{cm}^2$ with a delamination produced by

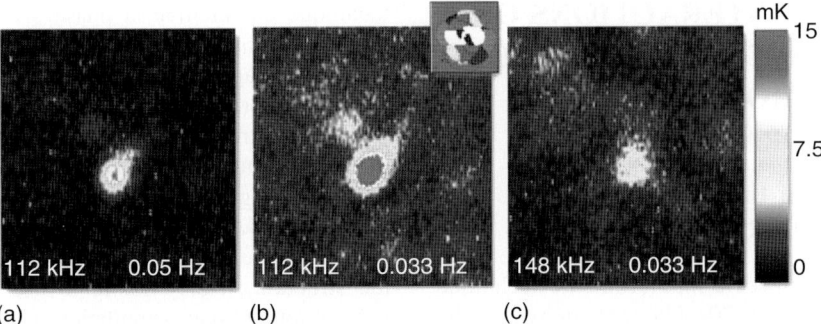

Figure 3. Temperature amplitude map on the front side of the impacted area of a C/epoxy coupon. Influence of the low-frequency modulation for a Lamb wave frequency of 112 kHz: (a) low amplitude modulation frequency of 50 mHz, (b) same as in (a) but with 33 mHz. Influence of the wave frequency (same Lamb mode) with amplitude modulation at 33 mHz: (b) 112 kHz, (c) 148 kHz.

a 5-J impact. The electric power injected in the piezoelectric emitter (disc-shaped, 30 mm in diameter and 200 micron thick) was 1 W. The amplitude of the Lamb waves is modulated at a low frequency (30–300 mHz) to cope with the thermal wave attenuation. In the present case, the delaminations can be detected from front face (impacted face) provided the modulation frequency is lower than 100 mHz. Figure 3(a–b) presents the thermal amplitude images obtained for two modulation frequencies: 50 and 33 mHz. Decreasing the modulation frequency has a beneficial effect on the sensitivity, but results in more blurring. These results highlight the well-known depth probing property of "thermal waves".

With the modulation frequency fixed at 33 mHz to get a good signal-to-noise ratio, several wave frequencies between 50 and 150 kHz were used. The temperature amplitude images obtained after synchronous demodulation are presented in Figure 3(b–c) for 112 and 148 kHz. The highest contrast, i.e. 22 mK, is observed for a frequency of 112 kHz. When the actuator is excited at 148 kHz, the defect is still detectable, but the contrast dropped to about 8 mK. Decreasing the wave frequency has the same detrimental effect. This shows how important the choice of the ultrasound frequency for a safe detection by thermography. For this optimum frequency, the apparent size of the defect is comparable to the image obtained by classical D-scan (see upper right part of Figure 3b).

These experiments show that defects can be detected whatever their depth be, provided that dissipation is high enough and that the modulation frequency is chosen sufficiently low (thick plates need higher mechanical energy and deep defects need lower lock-in modulation frequency to be revealed). The high selectivity of the ultrasound frequency regarding the thermomechanical coupling at the defect location has, however, to be taken into account. This requires some theoretical work to predict, for a given structure and a given defect geometry, which kind of ultrasound wave would lead to maximum dissipation at the defect locus.

4.3 Practical applications

The experimental results just presented concerned delaminations in composite structures. In fact, the field of application of the technique is much wider and ultrasonic vibrothermography has already been applied to real structures presenting realistic damages. These works have been essentially performed with sonotrodes or high-energy lead zirconate titanate (PZT) patches. The types of structures and damages detected until now in these conditions are varied. Let us mention, taken from [17, 22, 23] (i) delaminations in skin and stringer and stringer debondings in composite aircraft panels; (ii) cracks and debonds in metallic structures, hidden corrosions between skin and riveted stringers in aluminum structures; and (iii) heterogeneities in a multilayer C/C–SiC ceramic composite. We can suppose that by choosing tailored Lamb waves, similar results could be obtained with low or moderate energies.

5 IMAGING INTERACTIONS OF LAMB WAVES WITH DAMAGES USING STROBOSCOPIC SHEAROGRAPHY

5.1 Principle of the shearographic imaging of ultrasounds

Shearography is a speckle interferometric technique that appeared in the 1970s [24–26], in which the interfering photons are issued from two closed points of the structure thanks to an optical "shearing" device. This prevents the system from being sensitive to vibrations and solid motions of the structure. The measurement is performed for two successive states of deformation of the structure. By electronic data reduction, a subtraction of the two speckle images produces a resulting image containing information linked to the gradient of deformation between the two states in the direction of the shear. To make the technique quantitative, a phase stepping is required and, from four intensity images corresponding to four phase lags introduced thanks to a controllable mobile mirror, a phase image is obtained, which can be directly graded in gradient of deformation. This differential image is different from the images of deformation given by more classical interferometric techniques like ESPI [1].

To produce two different states of deformation of the structure, it is possible to use depressurization [27], photothermal stimulation [28], or vibrations. To define two states in the case of vibrations, it is necessary to produce a stroboscopic illumination with an acousto–optic modulator in the laser beam [29] as seen in Figure 4(a).

Gordon and Bard [30] were the first to propose and apply this technique to the visualization of ultrasounds. As shown in Figure 4(b), the two states of deformation of the structure correspond to two laser illuminations with a phase difference of 180° with respect to the ultrasounds. The phase stepping

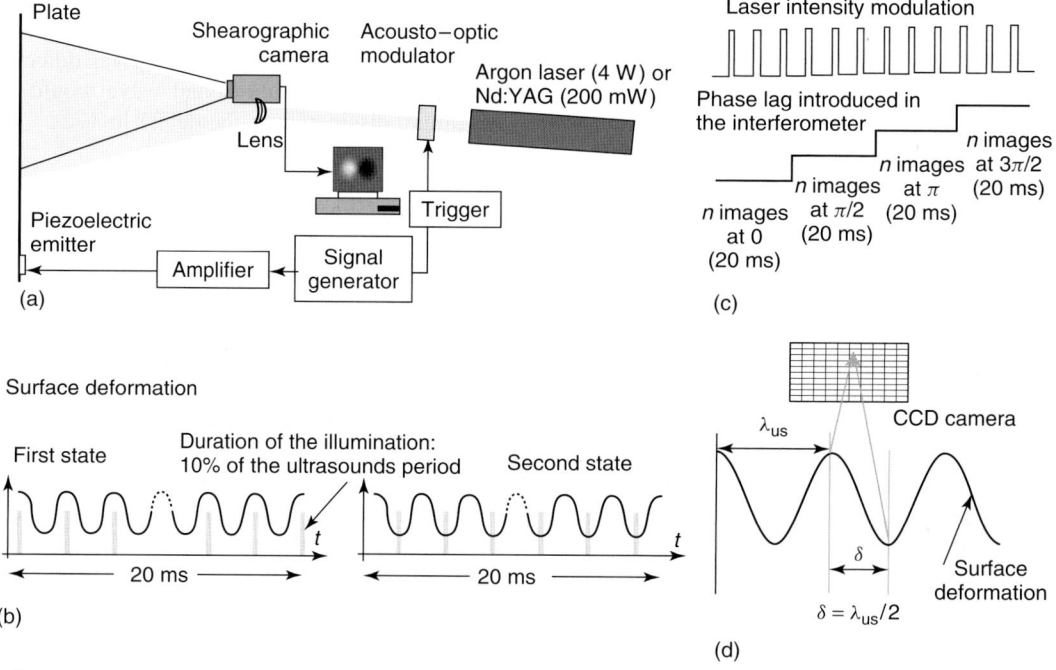

Figure 4. Visualization of Lamb waves by shearography: (a) experimental setup (the shearographic camera contains the three elements of the Michelson interferometer: shearing, phase stepping, and shear control devices). (b) synchronization between the signal feeding the emitter and the laser illumination, producing the stroboscopic effect; (c) synchronization between the laser illumination and the phase lag introduced in the arm of the interferometer (valid for the two states); and (d) shear adjustment (shear distance equal to half the wavelength).

technique is used (Figure 4c) and, to increase the signal-to-noise ratio, for every phase shift, image accumulation is performed for 20 ms. Finally, the better sensitivity is obtained with a shearing distance equal to half the ultrasound wavelength (Figure 4d), producing a difference of displacement between the two interfering points that is maximum. Under these conditions, the shearographic image level is strictly equal to four times the ultrasound amplitude, with a sensitivity of the order of 1 nm. If the shear is equal to the wavelength, then the wave is not seen, the deformation of the interfering points being identical, which can be interesting to visualize the waves diffracted by a damage [31, 32].

The use of continuous Lamb waves leads to results, which can be hard to interpret whether real structural parts are likely to induce reflections, mode conversions etc. Such elements can be fasteners, stringers, rivets, etc. or simply the plate edges. In particular, these effects are important in metallic structures for which the attenuation of Lamb waves is very low. To avoid this inconvenience, it is more convenient to generate short bursts and to follow their propagation [32]. Some modifications have to be introduced relative to the wave imaging mode: only one laser stroke is used per burst and its firing is delayed according to the propagation stage one wants to image. The second surface state is simply obtained by reversing the transducer input voltage.

In its simplest way of application, the shearographic camera views the structure perpendicularly to its surface, and then only the out-of-plane displacement is detected (present results). Nevertheless, it is possible to visualize the in-plane displacement, and this can be useful for Lamb wave imaging [33, 34].

5.2 Illustrative examples of shearographic imaging of Lamb waves

Figure 5 presents the visualization of the interaction of Lamb waves with a delamination in a C/epoxy plate [31, 32]. The coupon, equipped with a PZT emitter, is the one already used for the vibrothermographic experiment. The detection and localization are unambiguous and feasible on both front and rear faces of the plate. The interaction between the incident wave and the delamination is a diffraction phenomenon. It is possible to detect, although not so easily, this diffraction (emergent wave) around the damaged area (in particular, in Figure 5c).

Figure 6 presents an interaction in a more complex configuration: an artificial defect in a composite sandwich structure of a radome [35]. The defect is a lack of material, representative of a debond between the outer skin (glass/epoxy composite) and the core made of low-density foam. There is an interaction only in the outer skin, which allows us to delimit the extent of damage. No interaction is produced in the inner skin, which is thicker, because its dispersion curves are almost identical to that of the full sandwich.

The demonstration of the possibility of imaging the propagation of Lamb wave bursts has been achieved too [32]. Figure 7 presents the images of

Figure 5. Interaction of Lamb waves (S_0 mode, 68 kHz) propagating in the C/epoxy plate described in the section on vibrothermography. The dimensions of images are $17 \times 12 \, cm^2$ and their dynamic range is near to 50 nm. The electric power injected in piezoelectric emitter is 1 W. In figure (a) the shearographic image of the front face obtained with a shear distance equal to half the wavelength shows both incoming and diffracted waves. This is also the case for the image of the rear face given in figure (b). In figure (c) the image of the rear face obtained with a shear distance equal to the Lamb wavelength only shows the diffracted waves. [Reproduced with permission from Ref. 32. © EDP Sciences, 2006.]

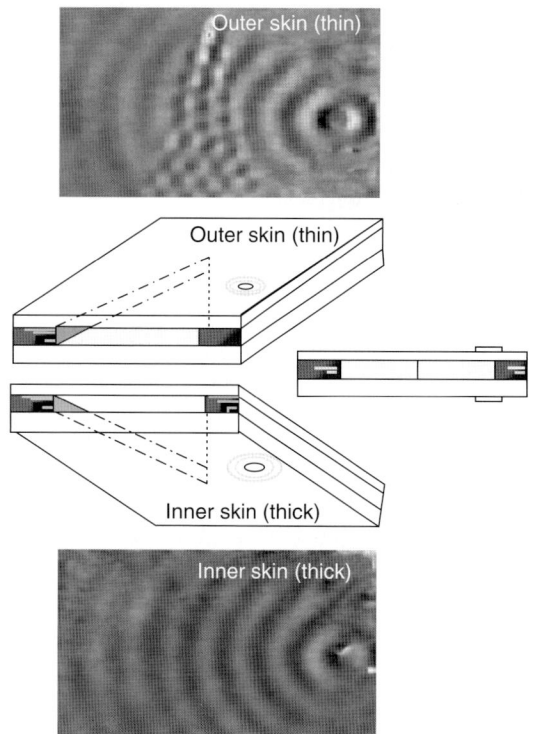

In the last image, a circular diffracted wave is seen around the defect. Five-period bursts have been used, with windowing, and this unavoidably induced higher frequency content for the mechanical excitation. Modes other than S_0 with different velocities and wavelengths could thus be present. Together with the fact that the bursts are repetitively emitted, this could explain why the defect already appears before the burst reaches it. Indeed small amplitude ripples can be seen in front of and behind the burst.

6 IMAGING INTERACTIONS OF LAMB WAVES WITH DAMAGES USING SCANNED LASER ULTRASONICS AND EMBEDDED PZT RECEIVERS

6.1 Principle

The concept of the third technique (Figure 1c) proposed by Takatsubo and colleagues from AIST (Japan) [13], is based on the fact that when ultrasonic waves propagate between two points using one emitter and one receiver having the same frequency characteristics, the same waveform would be detected if the emitter and the receiver replaced each other. The technique uses a pulsed laser generating ultrasounds by thermoelastic effect and a piezoelectric receiver. The waves detected by the piezoelectric sensor at a point B and generated by the laser at a point A would be almost the same as the waves detected at point A issued from the laser generation at point B. This visualization is based on the reversibility of the propagation of ultrasonic waves.

Figure 6. Shearographic visualization of the interaction of Lamb waves (A_0 mode, $f = 20$ kHz) with a defect in the core (lack of foam) in a composite sandwich of a radome.

the displacements recorded on the impacted sample previously monitored (Figure 5) for three different delays after the burst emission. In the first image, the Lamb wave front is still upstream from the impact defect and in the second image the burst has already passed the defect. Again a 30° wide wake with a 180° phase delay appears downstream.

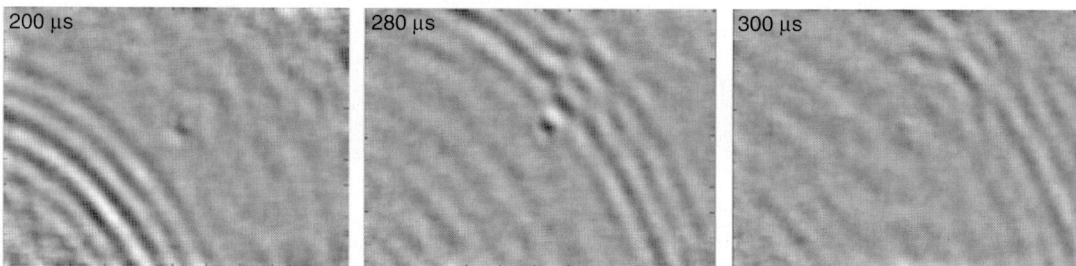

Figure 7. Shearograms of burst propagation in a delaminated C/epoxy plate (5-J impact) obtained for three delays after the burst. Transducer driven with repetitive five-period bursts (68 kHz). [Reproduced with permission from Ref. 32. © EDP Sciences, 2006.]

Let us consider now that the laser impact is scanned on the structure surface. Then, the train of waveforms detected by the fixed-position piezoelectric sensor during the laser scanning will be the same as those obtained by detecting the waves generated by directing the laser beam toward the position of the piezoelectric sensor while scanning the sensor on the structure surface. Therefore, images may be created by the waveforms detected during scanning of the laser. Displaying these images consecutively in the order of the measurement times produces an animation showing the propagation pattern that would be generated by a laser generation of ultrasounds at the location of the piezoelectric receiver. These authors have experimentally demonstrated that this principle is still valid if a defect is present in the propagation path [13].

The technique has several remarkable advantages: (i) it allows visualization of ultrasonic waves propagating in a complex-shaped 3D structure with curved surfaces, steps, and dents; (ii) there is no need to adjust the laser incidence angle and the focal distance; (iii) high detection sensitivity is obtained; and (iv) noncontact wide-field imaging of complex structures is made possible.

6.2 Practical applications

The technique has been applied to the detection of various types of structures and defects: holes and flaws in curved metallic pipes, slits in metallic plates, and debonding in a carbon fiber reinforced plastic (CFRP) structure. Figure 8 presents an illustration of the possibilities of the technique for detecting debonding between the skin and a hat stringer in a CFRP structure (airplane wing). The pulsed laser was a yttrium aluminum garnet (YAG) (1064 nm) delivering 5-mJ, 10-ns pulses with a repetition rate of 20 Hz. The laser is fixed on two rotation axis stages controlled by a PC. The image results from the impact of the 2-mm-diameter laser beam onto

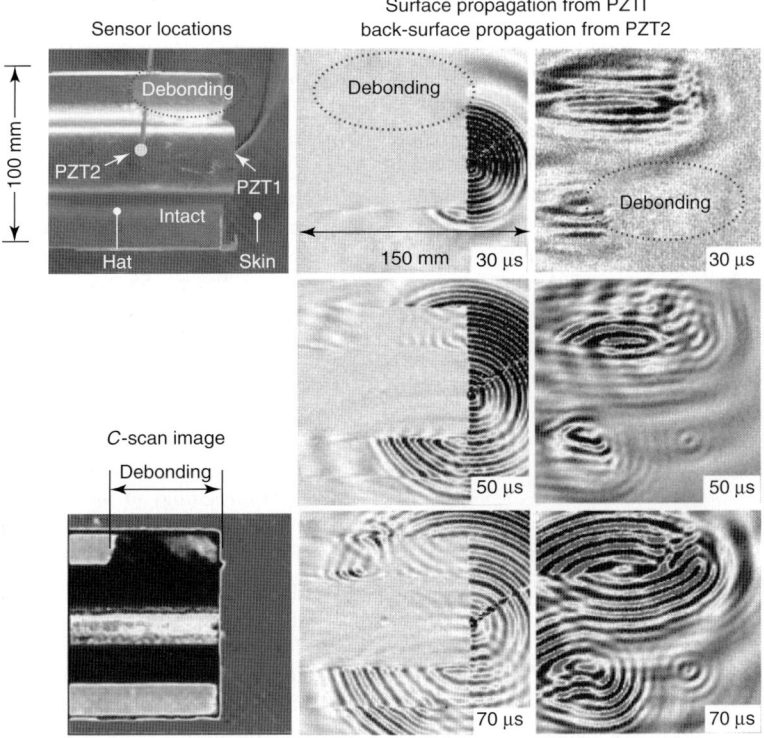

Figure 8. Detection of a debonding between skin and hat stringer of a CFRP structure. [Reproduced with permission from Ref. 13. © Takatsubo, 2006.]

100 × 100 points on the structure. For each impact, the PZT receiver signal is recorded. Thanks to the reversibility, the amplitudes of the recorded trains of waveforms at a given time correspond to the displacements of the ultrasounds at the irradiation points. Images can be created whose intensities are determined by these amplitudes. When displayed consecutively, the propagation at the surface of the structure of the ultrasonic waves can be observed, allowing the detection of damaged areas. Using Fourier analysis, it is possible to construct monofrequency images. The visualized signal in Figure 8 is a single-frequency (300 kHz) ultrasound signal, so there is very little velocity dispersion in the propagation image since such a dispersion depends on frequency. The debonding forms a thin air layer, and ultrasonic waves hardly propagate across it. The debonded part and the intact part can thus be clearly distinguished.

7 CONCLUSIONS

The concept of NDECS has been defined and proposed as a third possible way between the classical NDE and SHM. From the analysis of the present evolution of techniques in NDE and SHM fields, it has been deduced that the best NDECS solutions could be the combined use of Lamb waves and optical noncontact detection or emission systems. Three NDECS configurations have been described on the basis of the combined use of embedded PZT transducers with the following techniques: stroboscopic shearography, lock-in infrared thermography, and scanned laser ultrasonics. To demonstrate that these solutions are applicable in the short term, promising experimental results have been presented.

REFERENCES

[1] Rastogi PK. *Holographic Interferometry: Principles and Methods, Springer Series in Optical Sciences*. Springer-Verlag: Berlin, 1994; Vol. 68.

[2] Maldague X. *Theory and Practice of Infrared Technology for Nondestructive Testing, Wiley Series in Microwave and Optical Engineering*. John Wiley & Sons, 2001.

[3] Schmidt T, Tyson J. Full-field dynamic displacement and strain measurement using advanced 3D image correlation photogrammetry. *Experimental Techniques* 2003, Part I: **27**(3):47–50, Part II: **27**(4):44–47.

[4] Balageas D. Introduction to structural health monitoring. In *Structural Health Monitoring*, Balageas D, Fritzen CP, Güemes A (eds). ISTE: London, 2006, pp. 13–43.

[5] Arnould O, Hild F, Brémond P. Thermal evaluation of the mean fatigue limit of a complex structure. *Thermosense XXVII, SPIE Proceedings 5782*. SPIE: Bellingham, WA, 2005; pp. 255–263.

[6] Potet P, Bathias C, Degrigny B. Quantitative characterization of impact damage in composite materials: a comparison of computerized vibrothermography and X-ray tomography. *Materials Evaluation* 1988 **46**(8):1050–1054.

[7] Tenek LH, Henneke EGII, Gunzburger MD. Flaw dynamics and vibro-thermography thermoelastic NDE of advanced composite materials. *Thermosense XIII, SPIE Proc. 1467*. SPIE: Bellingham, WA, 1991; pp. 252–263.

[8] Balageas DL, Déom AA, Boscher DM. Characterization and nondestructive testing of carbon-epoxy composites by a pulsed photothermal method. *Materials Evaluation* 1987 **45**(4):461–465.

[9] Riegert G, Zweschper Th, Busse G. Lock-in thermography with eddy-current excitation. *QIRT Journal* 2004 **1**(1):21–31.

[10] Balageas DL, Levesque P, Nacitas M, Krapez JC, Gardette G, Lemistre M. Microwaves holography revealed by photothermal films and lock-in IR thermography: application to electromagnetic materials NDE. *Proceedings of SPIE 2944*. SPIE: Bellingham, WA, 1996; pp. 55–66.

[11] Balageas D. Non-destructive evaluation of cooperative structures (NDECS): a third way? *First Asia-Pacific Workshop on SHM*. Yokohama, December 2007.

[12] Walsh SM. Practical issues in the development and deployment of intelligent systems and structures. In *Proceedings of the 2nd International Workshop on SHM: SHM 2000*, Chang F-K (ed). Technomic Publishing: Lancaster-Basel, 1999, pp. 612–621.

[13] Takatsubo J, Yashiro S, Wang B, Tsuda H, Toyama N. Laser ultrasonics imaging technique for nondestructive inspection of defects in three-dimensional objects. *First Asia-Pacific Workshop on SHM*. Yokohama, December 2007 *see also, in Japanese* 高坪純治、王波、津田浩、遠山暢之、発振レーザ走査法による三次元任意形状物体を伝わる超音波の可視化、日本機械学会論文集 2006 72-718A〜: 945–950.

[14] Krapez JC, Taillade F, Balageas D. Ultrasound-lock-in thermography NDE of composite plates with low power actuators. Experimental investigation of the influence of the Lambwave frequency. *QIRT Journal* 2005 **2**(2):191–206.

[15] Rantala J, Wu D, Busse G. Amplitude modulated lockin vibrothermography for NDE of polymers and composites. *Research in Nondestructive Evaluation* 1996 **7**:215–218.

[16] Rantala J, Wu D, Busse G. NDT of polymer materials using lock-in thermography with water-coupled ultrasonic excitation. *NDT&E International* 1998 **31**(1):43–49.

[17] Busse G, Dillenz A, Zweschper T. Defect-selective imaging of aerospace structures with elastic-wave-activated thermography. *Thermosense XXIII, Proceedings of SPIE 4360*. SPIE: Bellingham, WA, 2001; pp. 580–586.

[18] Favro LD, Han X, Zhong O, Sun G, Sui H, Thomas RL. Infrared imaging of defects heated by a sonic pulse. *Review of Scientific Instruments* 2000 **71**(6):2418–2421.

[19] Giurgiutiu V, Zagrai A, Bao JJ. Piezoelectric wafer embedded active sensor for aging aircraft structural health monitoring. *Structural Health Monitoring—An International Journal* 2002 **1**(1):41–62.

[20] Pacou D, Pernice M, Dupont M, Osmont D. Study of the interaction between bonded piezo-electric devices and plates. In *Proceedings of First European Workshop on SHM: Structural Health Monitoring 2002*, Balageas DL. DEStech Publishing: Lancaster, PA, 2002, pp. 406–413.

[21] Han X. Frequency dependence of the thermosonic effect. *Review of Scientific Instruments* 2003 **74**(7):414–416.

[22] Dillenz A, Zweschper T, Riegert G, Busse G. Progress in phase angle thermography. *Review of Scientific Instruments* 2003 **74**(7):417–419.

[23] Zweschper T, Riegert G, Dillenz A, Busse G. Frequency modulated elastic wave thermography. *Thermosense XXV, Proceedings of SPIE 5073*. SPIE: Bellingham, WA, 2003; pp. 386–391.

[24] Leendertz J, Butters J. An image shearing speckle pattern interferometer for measuring bending moments. *Journal of Physics E: Scientific Instruments* 1973 **6**:1107–1110.

[25] Hung YY. A speckle-shearing interferometer: a tool for measuring derivatives of surface displacements. *Optics Communications* 1974 **11**(2):132–135.

[26] Hung YY. Shearography: a novel and practical approach for non-destructive inspection. *Journal of Nondestructive Evaluation* 1989 **8**(2):55–67.

[27] Clarady JF, Summers M. Electronic holography and shearography NDE for inspection of modern materials and structures. *Review of Progress in QNDE* 1993 **12A**:381–386.

[28] Paoletti D, Schirripa Spagnolo G, Zanetta P, Facchini M, Albrecht D. Manipulation of speckle fringes for non destructive testing of defects in composites. *Optics and Laser Technology* 1994 **26**(2):99–104.

[29] Hariharan P, Oreb B. Stroboscopic holographic interferometry: application of digital. *Optics Communications* 1986 **59**(2):83–86.

[30] Bard BA, Gordon GA, Wu S. Laser modulated phase-stepping digital shearography for quantitative full-field imaging of ultrasonic waves. *Journal of the Acoustical Society of America* 1998 **103**:3327.

[31] Krapez JC, Taillade F, Lamarque T, Balageas D. Shearography: a tool for imaging Lamb waves in composites and their interaction with delaminations. *Review of Progress in QNDE* 1999 **18A**:905–912.

[32] Taillade F, Krapez JC, Lepoutre F, Balageas D. Shearographic visualization of lamb waves in carbon epoxy plates: interaction with delaminations. *The European Physical Journal—Applied Physics* 2000 **AP9**:69–73.

[33] Rastogi PK. Measurement of in-plane strains using electronic speckle and electronic speckle-shearing pattern interferometry. *Journal of Modern Optics* 1996 **43**(8):1577–1581.

[34] Moulin E, Assaad J, Delebarre C, Kaczmarek H, Balageas D. Study of a piezoelectric transducer embedded in composite plate: application to Lamb waves generation. *Journal of Applied Physics* 1997 **82**(5):2049–2055.

[35] Devillers D, Taillade F, Osmont D, Krapez JC, Lemistre M, Lepoutre F. Shearographic imaging of the interaction of ultrasonic waves and defects in plates. *Proceedings of SPIE 3993*. SPIE: Bellingham, WA, 2000; pp. 142–149.

SECTION 2
Information Fusion and Data Management

Chapter 73
Web-based SHM

Vistasp M. Karbhari[1] and Hong Guan[2]
[1] University of Alabama in Huntsville, Huntsville, AL, USA
[2] HDR, Los Angeles, CA, USA

1 Introduction	1283
2 Components of a Web-based SHM System	1285
3 An Example for Web-based SHM of Bridges	1286
4 Summary	1296
References	1299

1 INTRODUCTION

Structural health monitoring (SHM) is increasingly being considered as a means of not only obtaining data related to the response of components and structural systems but also as a means of providing an assessment of the "health" of the system. While there are still substantial differences of opinion regarding the scope and applicability of SHM (i.e., does the system merely relate to the collection of data and provision of a rudimentary knowledge of response, or does it actually serve as an autonomous or semiautonomous means of interpreting structural

Encyclopedia of Structural Health Monitoring. Edited by Christian Boller, Fu-Kuo Chang and Yozo Fujino © 2009 John Wiley & Sons, Ltd. ISBN: 978-0-470-05822-0.

state in terms of characteristics such as capacity, remaining life, and necessity of repair), there is no doubt that the appropriate implementation of such a system can provide significant information of value to the owner, hitherto not available either in the same timescale or in quantity/depth, or both. In fact, recent developments in this field have led to the introduction of the term *civionics* to describe the "hardware and physical installation of the sensors, wires, conduits, termination, and control boxes" [1], which complements SHM in terms of a system that enables the acquisition and interpretation of data beyond that represented by traditional nondestructive evaluation (NDE).

Currently, civil infrastructure systems such as bridges, pipelines, waterways, and buildings are inspected at routine intervals till significant distress is noted, after which the period between inspections is decreased and the level of inspection is increased till such time that the distress has been corrected by replacement or repair. This time-based monitoring is inefficient not just in terms of resources but also since it does not minimize "downtime" of the structure. Since civil infrastructure forms a critical part of a nation's well-being, its deterioration has immense effect on the economies of the region, as it serves as the basis for the transfer of goods and services. The decrease in "downtime" due to maintenance, rehabilitation, or even rapid replacement is a critical aspect that needs to be addressed, and hence a move to a condition-based assessment of structures is advantageous. The implementation of condition-based assessment, however, requires that changes in structural response be monitored as a result of both normal operation and extreme events. This can be done through the implementation of a true SHM system, which could enable autonomous and continuous recording and assessment of predetermined response parameters associated with materials, components, and/or the entire system.

The development of an SHM system is essentially based on the ability to acquire data, transmit it, interrogate it, and then make decisions based on the cumulative sets of data stored in the database. Thus, in effect, the SHM system is essentially a decision system that is fronted by sensors and backed by a knowledge base, the critical elements of which are represented in Figure 1.

Tremendous advances in technologies related to sensor technologies, data compression and transmission, and interrogation make the deployment of an SHM system substantially easier today than in the past. For example, advances in image capture and analysis have made it possible to track vehicles in traffic for purposes of transportation planning such that vehicles can be tracked and classified over long periods of time to accumulate large volumes of tracking data, which can then be used to build models consisting of the traffic flow parameters such as density, flow, and speed [2]. In addition, tracking data can be used for event detection and definition of normal motion paths for detection of abnormal events, as shown schematically in Figure 2.

In addition, systems have already been developed and implemented for the rapid assessment of pavement condition in terms of the presence and extent of pavement cracks, potholes, and other damage, which can disrupt the smooth and efficient flow of traffic and overall safety. For example, the Hanshin Expressway Public Corporation (HEPC) in Japan has implemented an intelligent inspection system based on the use of Charge Coupled Device (CCD) cameras, laser emitters, and image-processing techniques—all on board vehicles that continuously traverse the road network to obtain data, which is then transmitted and used from a central location [3]. In this, as in other SHM systems, the main components are those of (i) data acquisition, (ii) data transmission, (iii) data processing and interrogation, (iv) data storage, (v) data retrieval, and (vi) diagnostics of long-term response. While a number of these aspects can be performed through direct assessment systems without need for comparison of historical data or excessive diagnostics, this hinders the full development of an SHM system that would enable true diagnosis and prognosis of a structure and the system that it is a part of (for example, a single bridge within a network

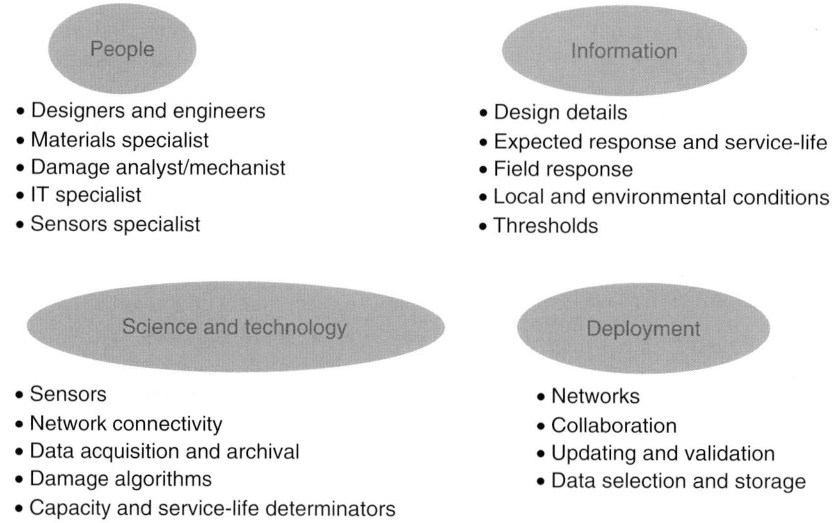

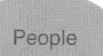

People
- Designers and engineers
- Materials specialist
- Damage analyst/mechanist
- IT specialist
- Sensors specialist

Information
- Design details
- Expected response and service-life
- Field response
- Local and environmental conditions
- Thresholds

Science and technology
- Sensors
- Network connectivity
- Data acquisition and archival
- Damage algorithms
- Capacity and service-life determinators

Deployment
- Networks
- Collaboration
- Updating and validation
- Data selection and storage

Figure 1. Critical elements of an SHM system.

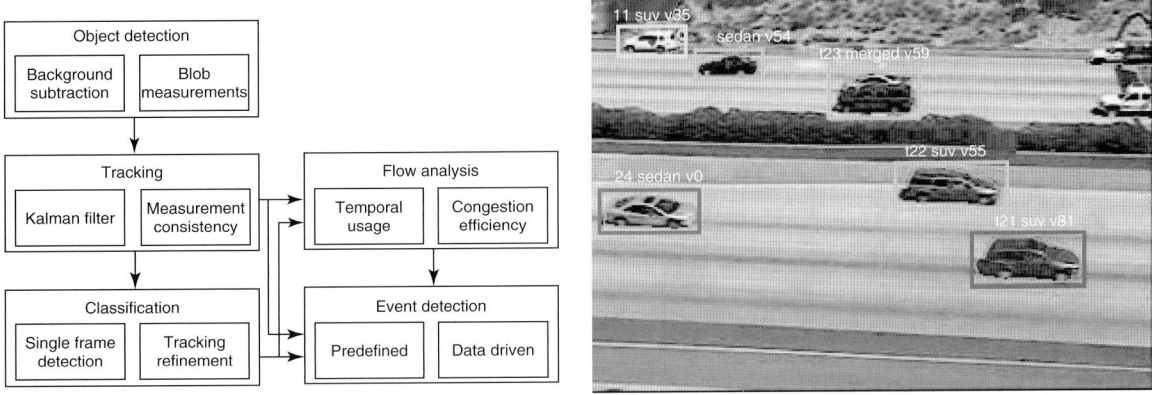

Figure 2. Schematic and basis for tracking of vehicles. [Courtesy of Mohan Trivedi, UCSD.]

of highways that have multiple bridges on them). For efficient use of such a system, it is incumbent that the SHM system be able to semiautonomously compare response at any point in time, not just with preset thresholds of performance but also with prior data to assess trends and develop self-learning models that can be used to predict future response. In addition, it is critical that the user be able to assess information rapidly about not just single structures but also about networks, so as to enable the assessment of the effect of deterioration in one element on the entire system and to thereby enable dynamic resource allocation. For example, in the case of earthquakes, it would be useful to be able to pull up, using regional maps, the location of critical structures, interrogate their health from a remote location, and reach decisions on which ones need immediate attention in terms of resources. All these aspects are best suited to be addressed through a web-based monitoring system.

2 COMPONENTS OF A WEB-BASED SHM SYSTEM

The essential components of a web-based SHM system are shown in Figure 3.

Data is collected from suites of sensors on individual structural components or systems and is brought back to a central location. While wireless systems are becoming ubiquitous, it must be remembered that vast amounts of data transmission, while possible, need large bandwidths. Depending on the location of the structures being monitored, it may be better to have packets of data sent at predetermined intervals, rather than on a continuous basis. Further, it may, in fact, be better to have the SHM itself be triggered to collect data only when specific thresholds are reached or exceeded (weight, deflection, acceleration, etc.). Once this data is brought into the system, it can be directly stored keeping in mind that storage can itself be a major issue. It is thus necessary to have a methodology in place by which data is assessed for storage through selection of "typical" sets for storage, events and novelties, and comparison with previously collected sets, and to store only the required values (such as maxima, minima, averages, etc.). The development and implementation of an appropriate selection and archival plan is essential, especially if the web-based SHM system is used for a large number of structures. A powerful back-end database to manage raw data and analysis is essential in these cases. Data itself can be processed using both model-free and model-based approaches, which when compared to structural response models and thresholds can provide important characteristics for decision making. For a true level IV SHM system, it is important that the analysis tools be linked to predictive and "what-if" capabilities to provide the results that are of use to the owner/engineer in making decisions. A prototype system for web-based monitoring is used to demonstrate the requirements in the next section.

The value of the web-based system is threefold. First, it allows for direct assessment of data sets, both streaming and stored, by the user, using a forum

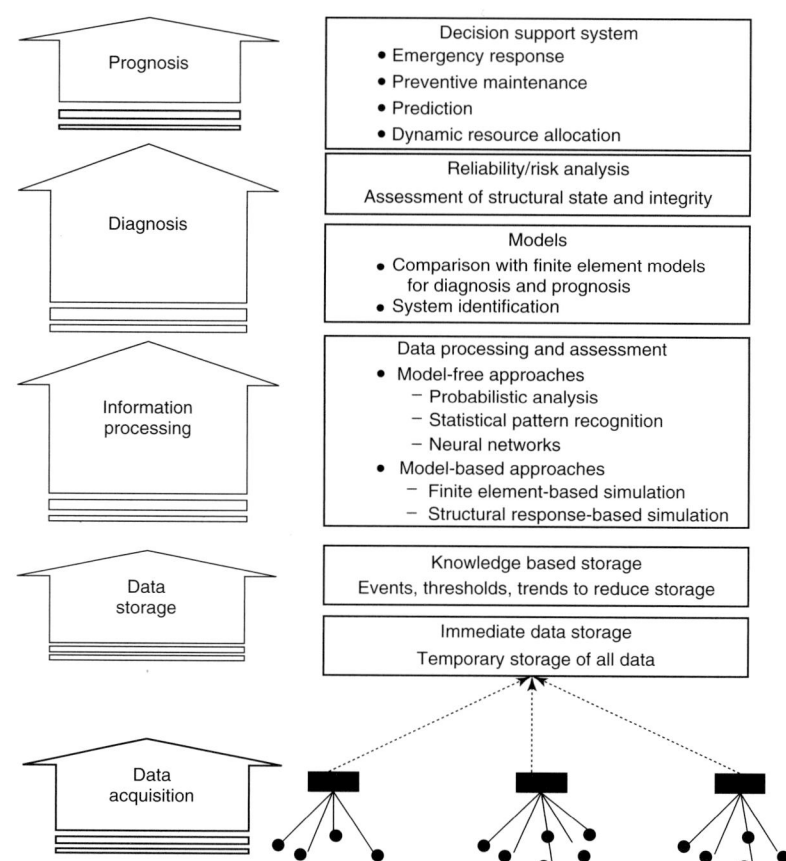

Figure 3. Components of a web-based SHM system.

that is now ubiquitous for rapid access of information. Secondly, it can be programmed to not only provide views of the data in sets chosen by the user (i.e., different users could simultaneously, from different sites, access different sets of sensors based on choices made from preset suites, or from the entire set) but to also interrogate the data, based on predetermined thresholds so as to enable rapid assessment of structural response. Thirdly, the use of a web-based schema allows for autonomous delivery of messages to personnel as needed. In addition, information from a set of bridges can be combined on a single system such that an entire network can be monitored remotely. It is essential, however, to recognize that the number of clients that can access the system at one time and the amount of data that can be accessed individually will depend intrinsically on bandwidth. In addition, the system has to be designed to both send and receive data, rather than the more commonly used schema of just receiving data, which is then accessed with a time delay. The use of a web-based interface also ensures that access is not restricted by location and that 24/7 communication and monitoring is available.

3 AN EXAMPLE FOR WEB-BASED SHM OF BRIDGES

3.1 Theoretical basis

Various health monitoring/NDE techniques have been proposed for continuous or time-based assessment of structural response, among which vibration-based damage detection (VDD) techniques have been widely accepted as a flexible choice for potential use in long-term, real-time, remote monitoring systems. VDD can be described as a class of techniques that are capable of detecting the stiffness/mass change of the structure based on its dynamic response due to forced or ambient excitation. The change of stiffness/mass of the structure is often related to the damage of structural components and the consequent loss of performance. The general procedure of VDD can be roughly divided into four steps: (i) measurement of structural dynamic response in terms of accelerations or displacements; (ii) extraction of modal parameters, such as natural frequencies and mode shapes from the measurements obtained; (iii) calculation of stiffness/mass change based on modal parameters, used in conjunction with the same set of parameters from the as-built or "baseline" structure; and (iv) damage localization and severity estimation based on the results of the third step. In the present example, the time domain decomposition (TDD) method is used, which allows the extraction of mode shapes as the first step, followed by the determination of the corresponding natural frequencies as the second step. Although the TDD needs frequency information, it does not require use of discrete Fourier transforms, which result in substantial computational effort in most other modal parameter extraction algorithms. Once the modal parameters have been identified through this procedure, the next step in the modal-based damage detection procedure is localization of damage and severity estimation. This is done following the damage index method initially proposed by Stubbs and Oseguela [4]. The use of this methodology not only enables the monitoring of structural health but also assessment of damage and its progression as a function of time. In the present setup for a linear, undamaged structure, the ith modal stiffness of a linear, undamaged structure can be described as

$$K_i = \vec{\phi}_i^T \mathbf{C} \vec{\phi}_i \qquad (1)$$

where $\vec{\phi}_i$ is the ith modal vector and $\mathbf{C}$ is the system stiffness matrix. The contribution of the jth member to the ith modal stiffness can then be given by

$$K_{ij} = \vec{\phi}_i^T C_j \vec{\phi}_i \qquad (2)$$

The fraction of modal energy of the ith mode contributed by the jth member, also called *modal sensitivity*, is defined as

$$F_{ij} = K_{ij}/K_i \qquad (3)$$

and correspondingly, using * to represent the damaged structure, the fraction of modal energy of a damaged structure can be defined as

$$F_{ij}^* = K_{ij}^*/K_i^* \qquad (4)$$

in which

$$K_{ij}^* = \vec{\phi}_i^{*T} C_j^* \vec{\phi}_i^* \qquad K_i^* = \vec{\phi}_i^{*T} C^* \vec{\phi}_i^* \qquad (5)$$

and

$$C_j = E_j C_{jo} \qquad C_j^* = E_j^* C_{jo} \qquad (6)$$

A fundamental aspect in this method is that the modal sensitivity for the ith mode and jth member remain unchanged for both the undamaged and damaged structures, i.e.,

$$F_{ij}/F_{ij}^* = (K_{ij}^* K_i)/(K_i^* K_{ij}) = 1 \qquad (7)$$

Therefore, a damage index β_j for the jth member, can be obtained as

$$\beta_j = \frac{E_j}{E_j^*} \quad \text{damage indicated when} \quad \beta_j > 1 \qquad (8)$$

$$\beta_j = \frac{\gamma_{ij}^* K_i}{\gamma_{ij} K_i^*} = \frac{\phi_i^{*T} C_{jo} \phi_i^* K_i}{\phi_i^T C_{jo} \phi_i K_i^*} \qquad (9)$$

and the severity of damage can be estimated by

$$E_j^* = E_j \left(1 + \frac{dE_j}{E_j}\right) = E_j(1+\alpha_j) \qquad (10)$$

where

$$\alpha_j = \frac{\gamma_{ij} K_i^*}{\gamma_{ij}^* K_i} - 1 \qquad (11)$$

in which

$$\gamma_{ij} = \vec{\phi}_i^T C_{jo} \vec{\phi}_i \qquad \gamma_{ij}^* = \vec{\phi}_i^{*T} C_{jo} \vec{\phi}_i^* \qquad (12)$$

1288 *Systems and System Design*

3.2 Description of sensors and protocol

The bridge discussed in this article is a two-span highway bridge with a total length of approximately 20.1 m and a width of approximately 13.0 m. It carries two north-bound lanes on State Highway 86 in Riverside County, California. The superstructure is of slab-on-girder type, with two equal

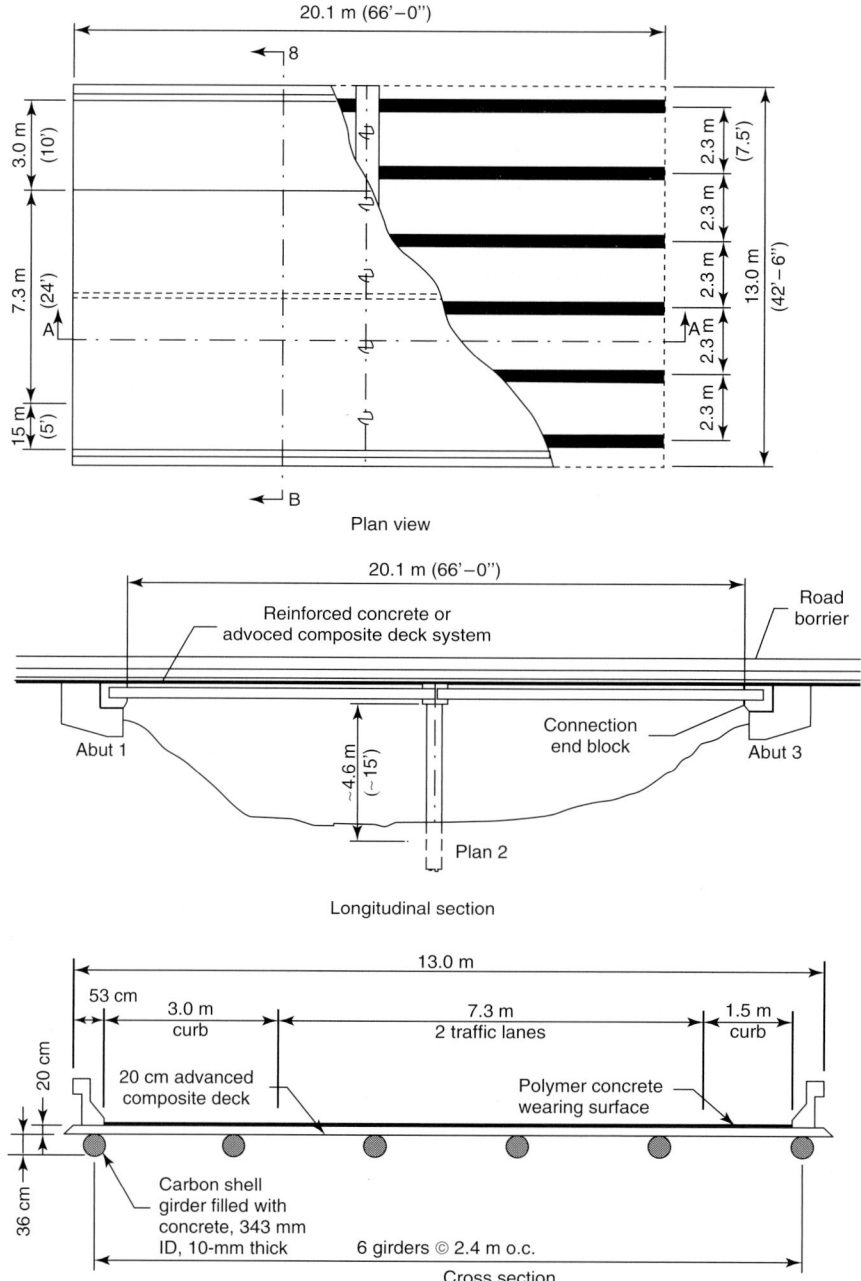

Figure 4. Schematic of the bridge.

spans of 10 m, and a cap beam connecting the two adjoining spans. The six main girders are composed of 10-mm-thick prefabricated filament-wound carbon/epoxy shells filled on-site with lightweight concrete, which support six modular E-glass fiber-reinforced polymer (GFRP) deck panels that serve as road surface, transfer vehicular loads to the girders, and also act as the transverse connections between girders. Schematics of the bridge are shown in Figure 4.

Details related to connections, proof testing at the component and systems levels, and design are reported in [5] and are hence not repeated herein. The bridge is open to heavy truck traffic and serves as a test bed for the implementation of SHM strategies, in addition to being one of the first fiber-reinforced polymer (FRP) vehicular bridges on a major route.

The response of the bridge is characterized through the placement of 63 accelerometers, 20 strain gauges, 4 linear potentiometers, 1 temperature sensor, and a pan-tilt-zoom (PTZ) camera. A total of 63 Model 3140 single axis accelerometers from IC Sensors, Inc. with a dynamic range of ±2g, a sensitivity of $1\,V\,g^{-1}$, and a frequency response of 0–200 Hz have been used. These provide an accurate measurement of the ambient-excitation-induced response of interest for this particular application. Forty-two accelerometers were placed at locations on the bottom of the composite deck, referred to as *nodes*, each protected by a sealed equipment housing. The nodes formed a 7×6 grid, with seven locations in the bridge longitudinal direction and six in the transverse direction (as shown in Figure 5), enabling the identification of several mode shapes required for damage detection, and for comparison with theoretical mode shapes already identified by finite-element analysis (FEA).

In addition to the 42 accelerometers measuring vertical acceleration, an additional horizontal accelerometer was installed in the same equipment housing at some of the nodes, to measure the horizontal accelerations that might be caused by an earthquake. A total of 20 bonded resistance gauges were also attached to critical locations on the bottom of the girder, the deck, and the middle section of the girder. Four linear potentiometers were used to measure the deflection of the composite girders at the midspan of two central girders, which experience the maximum deflection (as identified by previously conducted load tests).

A Campbell Scientific CR9000 modular high-speed data logger serves as the core of the data-acquisition system. The CR9000 is capable of making measurements at an aggregate sampling rate of 100 kHz for up to 200 channels. A Transmission Control Protocol/Internet Protocol (TCP/IP) interface enables communication and data collection via standard wired or wireless networks. The raw voltage data collected by the sensors are conditioned and digitized, and then stored into the cache memory of the CR9000 to be retrieved later. The High Performance Wireless Research and Education Network (HPWREN), funded by the National Science Foundation (NSF), which is operational in Southern California, is utilized as the wireless link for data transmission using a 900-MHz wireless antenna and a transmission rate of 45 Mbps. Figure 6 shows the bridge with the antenna used for data transmission.

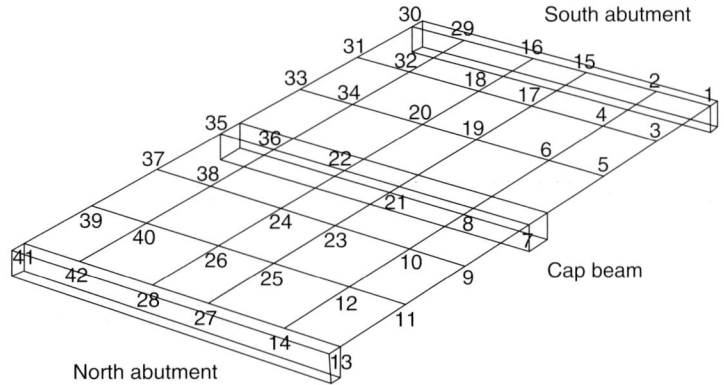

Figure 5. Location of accelerometers.

Figure 6. Bridge with antenna for data transmission.

The automated collection, digitization, and transmission of data are controlled by two collaborating programs. The first program, written in CRBASIC and running on a CR9000 DAQ system, serves the collection and digitization function. It enables data from all channels to be collected at preset intervals and when triggered either by an extreme event, such as an earthquake, or when a preset response threshold is exceeded, such as deflection or acceleration from extremely heavy traffic or permit loads. It also allows for data from a select number of channels, denoted as *streaming channels* to be collected and transmitted continuously in real time. The second program set is housed in the central server at the home node, and serves both data transmission and analysis functions. A schematic of flow is shown in Figure 7.

Continuous analyses performed on streaming data and results, such as peak displacement and strain, are then stored for retrieval on demand. Simultaneously, analyses are also performed on event-based data, as they become available, and the detailed results, such as estimates of damage localization and severity are also stored in the database. Event-based data are archived in an event data database. The server program also serves as a relay for streaming data to multiple end users.

Web-based SHM 1291

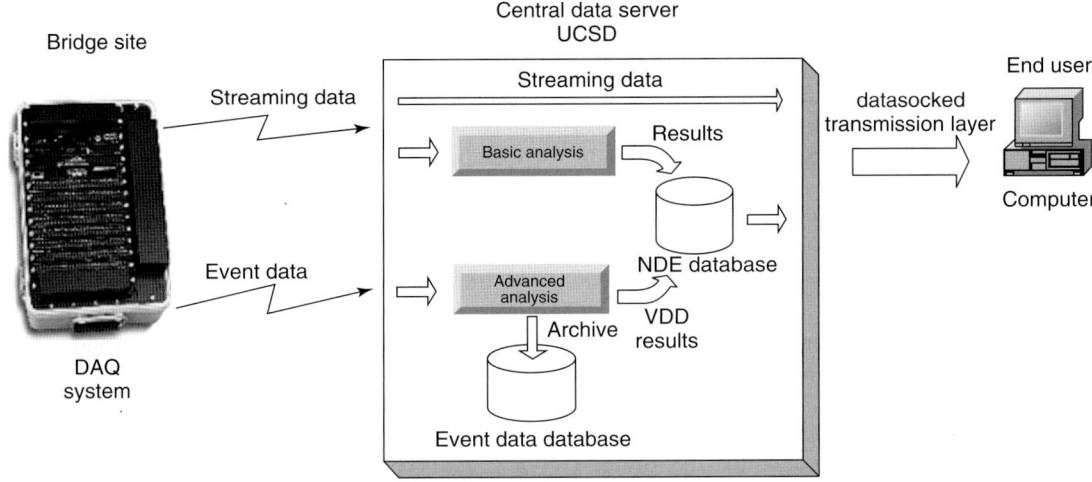

Figure 7. Schematic for information flow.

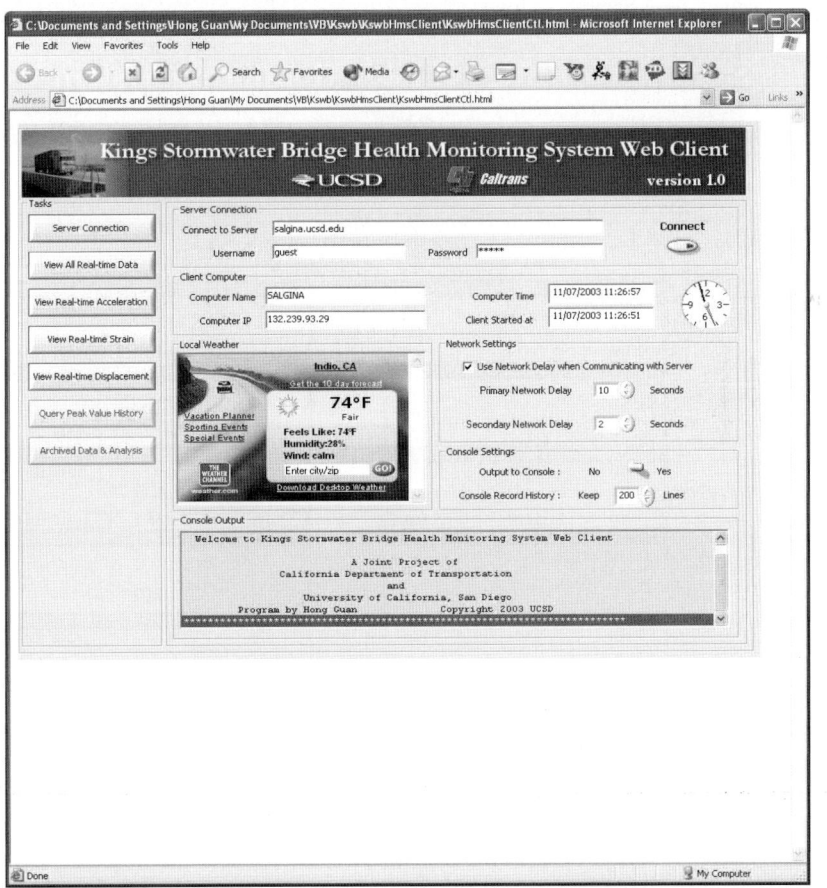

Figure 8. The web client page.

3.3 Web-based implementation

After the data is collected at the bridge site, controlled by a program running on the data logger, raw data is sent over the wireless network to a server located at the University of California, San Diego in La Jolla. A Matlab-based program then autonomously analyzes the raw data, generates results, and sends these to a web server. End users can then access these results using a standalone program or a specific web client, as shown in Figure 8.

The user can then select sensors for which data is desired using maps with sensor locations as in Figure 9.

The web-based user interface enables users with an appropriate connection to monitor the bridge's behavior and assess pertinent serviceability, reliability, and durability aspects related to the structure. By making use of the data stored on the central server, users can access historical data to make comparisons and/or to determine effects of specific events. For the purposes of the current investigation, an event is defined as a *significant excitation to the bridge*, usually caused by single or multiple vehicles crossing the bridge. In terms of measurement, an event is usually a time period over which the bridge is subject to excitation and measurements are taken. Both raw data and processed data are analyzed and stored with reference to the specific event through a date–time stamp. Users can either choose to plot sensor records pertaining to a single event, multiple events, or even records of

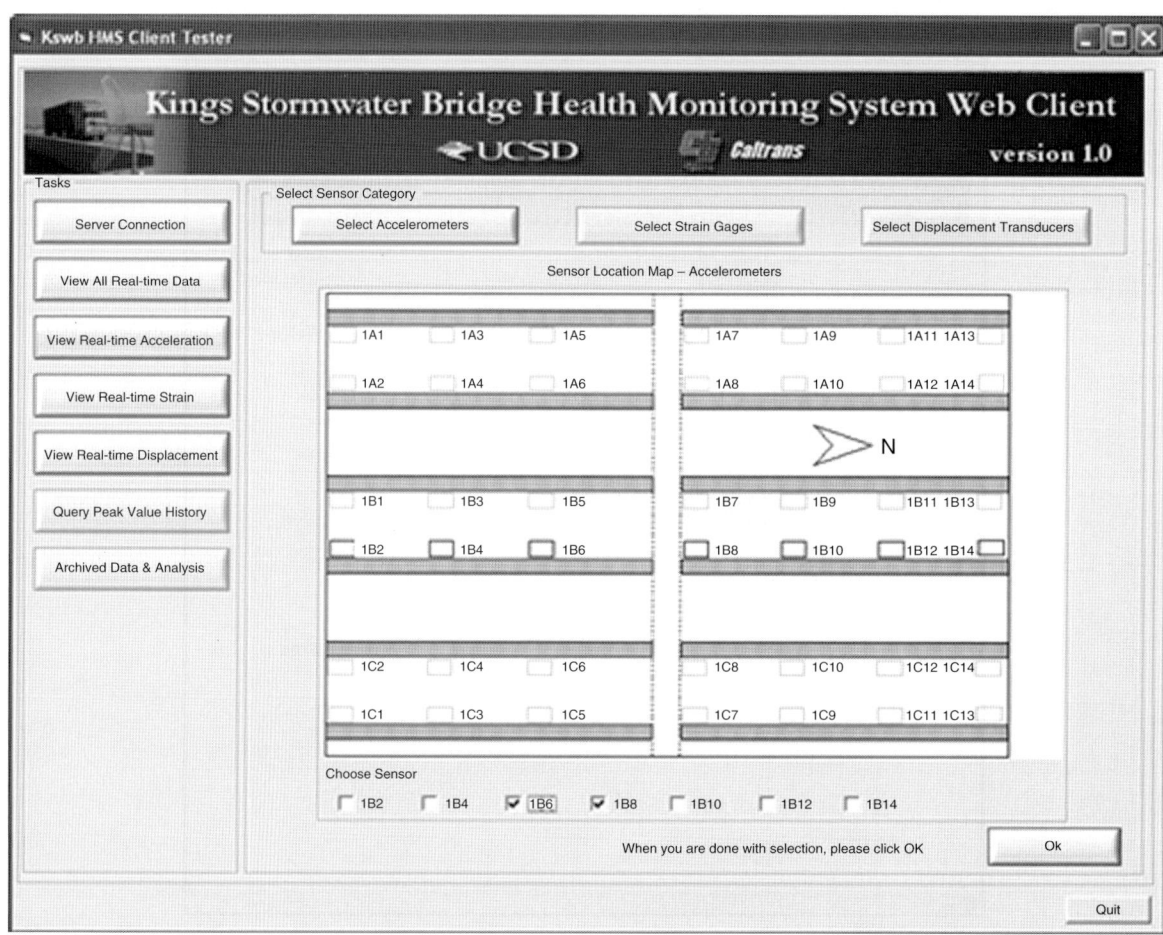

Figure 9. Example of map showing sensor locations.

multiple sensors selected by location. Results can then be viewed in real time (Figure 10a), with the capability of zooming in on details of any sensor (Figure 10b).

In addition, historical records can be assessed to compare peaks with current measurements, such as in Figure 11(a) for displacements, and in Figure 11(b) for strains.

3.4 Sample results of diagnostic analysis

At the current level of development, the vibration-based damage diagnostics information is supplemented by results from a finite-element model of the entire bridge (including deck panels, girders, deck-girder connections, cap beam, column/piers, and abutments), giving the user the capability of comparing structural response characteristics determined through the use of sensor measurements to FEA-based "ideal" or "updated" characteristics, as well as to predetermined response thresholds that signify bounds of acceptable structural behavior. The model was developed in ANSYS using design and construction drawings, and manufacturing specifications for the FRP composite components. The deck panels were modeled using elastic shell elements with equivalent properties, while the girders were modeled as beam elements. Connections between girders and deck panels were modeled as rigid constraints. The concrete barrier was also modeled using beam elements connected to the deck panels, while the connections to the abutments and the cap beam were modeled as rigid connections. The frequencies of the first five modes obtained from the initial FEA model are listed in Table 1 together with corresponding frequencies from a baseline forced

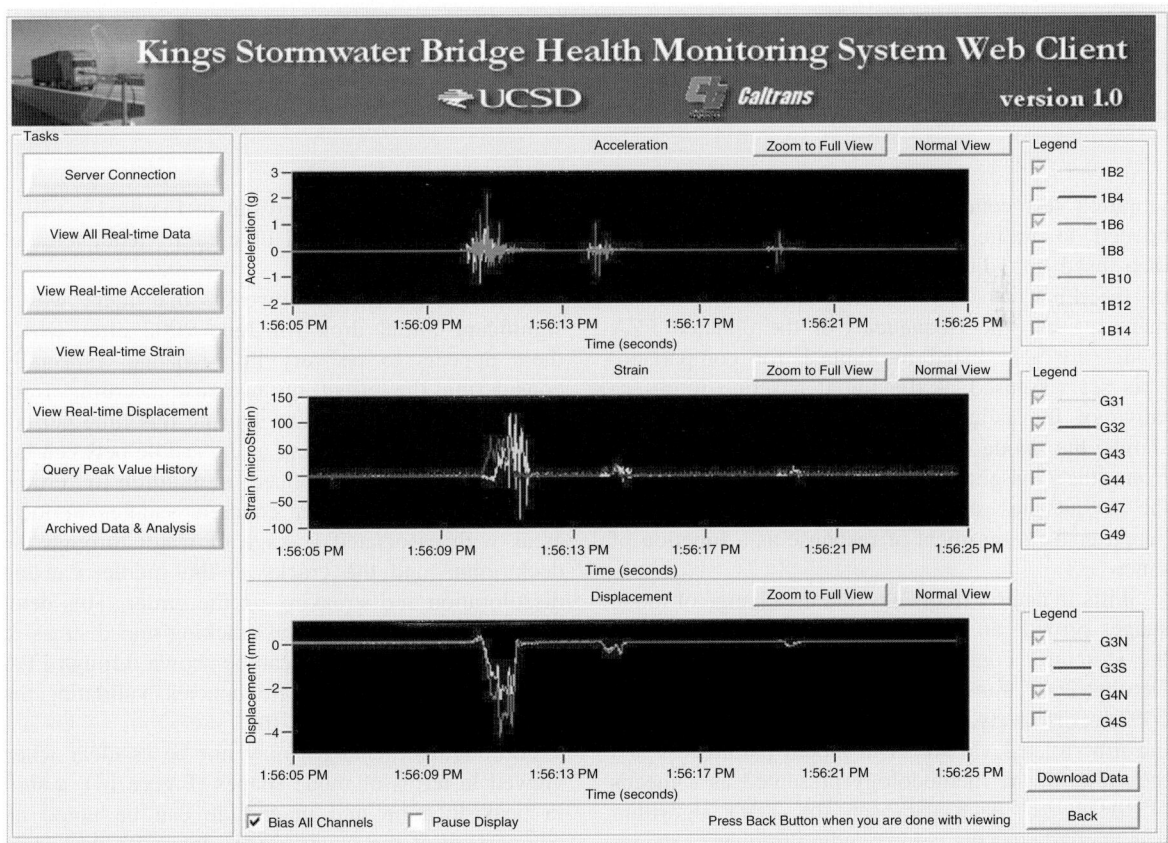

Figure 10a. (a) Real-time view of sensor-based response measurement. (b) "Zooming in" on data.

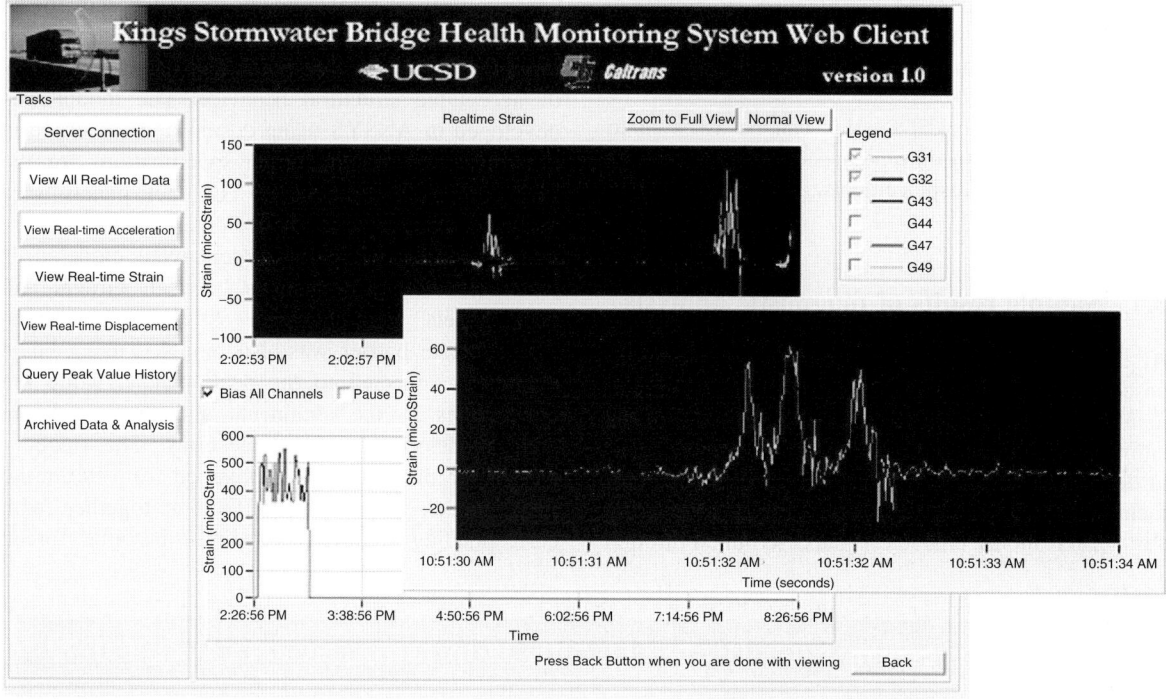

Figure 10b. *Continued.*

vibration test conducted just after the bridge was opened to traffic. A model updating process is then used to adjust the material properties used in the FEA model to better match the experiment results. The frequencies of the updated FEA model are also listed in Table 1 for comparison. This set was used as the initial baseline representative of the as-built nonaged structure. Figure 12 shows the mode shapes of the first two modes from the "baseline" finite-element model. Both the frequencies and the mode shapes are used hereafter for assessment of response as a function of time.

Shifts in natural frequency and the change of mode shapes provide an indication of stiffness changes in the structure. Four different data sets, collected in October 2001, June 2002, January 2003, and August 2003, respectively, are used as an example to show the time-related variations in modal parameters and the value of the methodology presented as a means of SHM. For each data set, the TDD algorithm was applied after the corresponding frequency ranges were determined. The change in natural frequencies is quite obvious when plotted out in Figure 13, which shows a decreasing trend of natural frequencies of both modes.

It should be noted, however, that there was a significant initial drop till June 2002 after which there appears to be very little change, especially considering the effect on response accruing from the changes in temperature. During this time period, some horizontal debonding at the saddle between the deck and girders was noticed, in addition to vertical cracking of the polymer concrete saddles in local areas with associated leakage of water through the deck joints and the cracks in the saddle, causing discoloration and streaking on the girder. This deterioration, in addition to the cracking and distress in the joints between decks, can be shown through FEA to lead to the change in frequencies, validating the experimental observations [6].

Although the changes may not be as clear when viewed in terms of mode shapes (Figure 14), a shift in peak location and magnitude can be noted on comparison of the August 2003 data with data from two other time periods.

Table 1. Comparison of frequencies of initial FE model and experiment

Mode number	Frequency (Hz)		
	Experiment	Initial FE model	Updated FE model
1	11.03	12.32	11.12
2	13.11	13.97	12.98
3	15.36	15.92	15.01
4	16.92	17.82	17.11
5	19.01	19.65	19.34

It is noteworthy that the analytically simulated mode shapes based on FEA models compare well with the experimental results, again showing the value of FEA use for both comparison at the threshold levels and for purposes of prediction. Translating the results into structural stiffness, this means the structure is becoming more flexible, i.e., decreasing structural stiffness with respect to time.

Use of the damage index method allows for determination of magnitude of damage indicators, which can provide assessment of damage in the structure. Changes in these indicators can be seen in Figure 15, and it is noted that the region with the highest damage indicator magnitude is concentrated around an area about 3.3 m (10 ft) on either side of the cap beam, which correlates with visual observation of damage through deterioration of the construction joint between deck panels. It is of interest to note that the increase in damage indicators in these

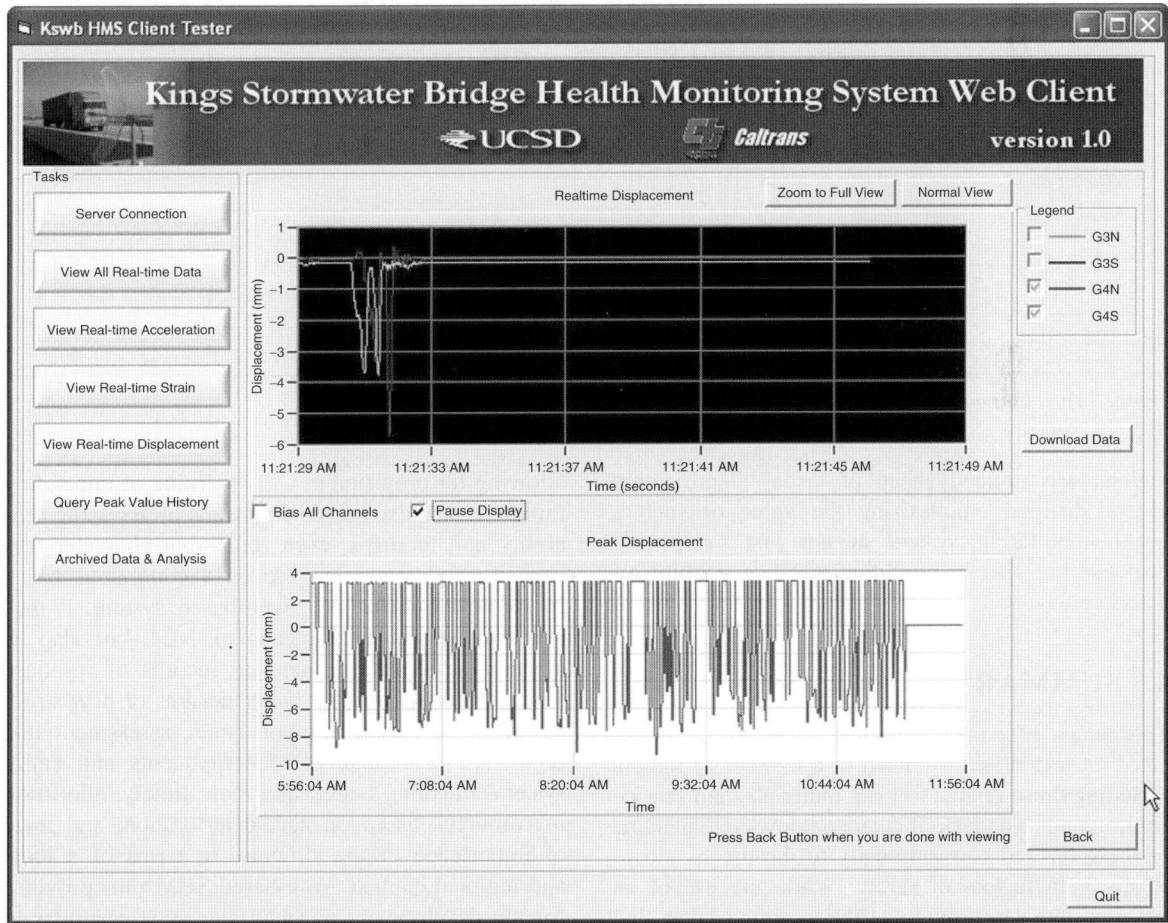

Figure 11a. (a) Access to peak displacements. (b) Access to peak strains from the overall data record.

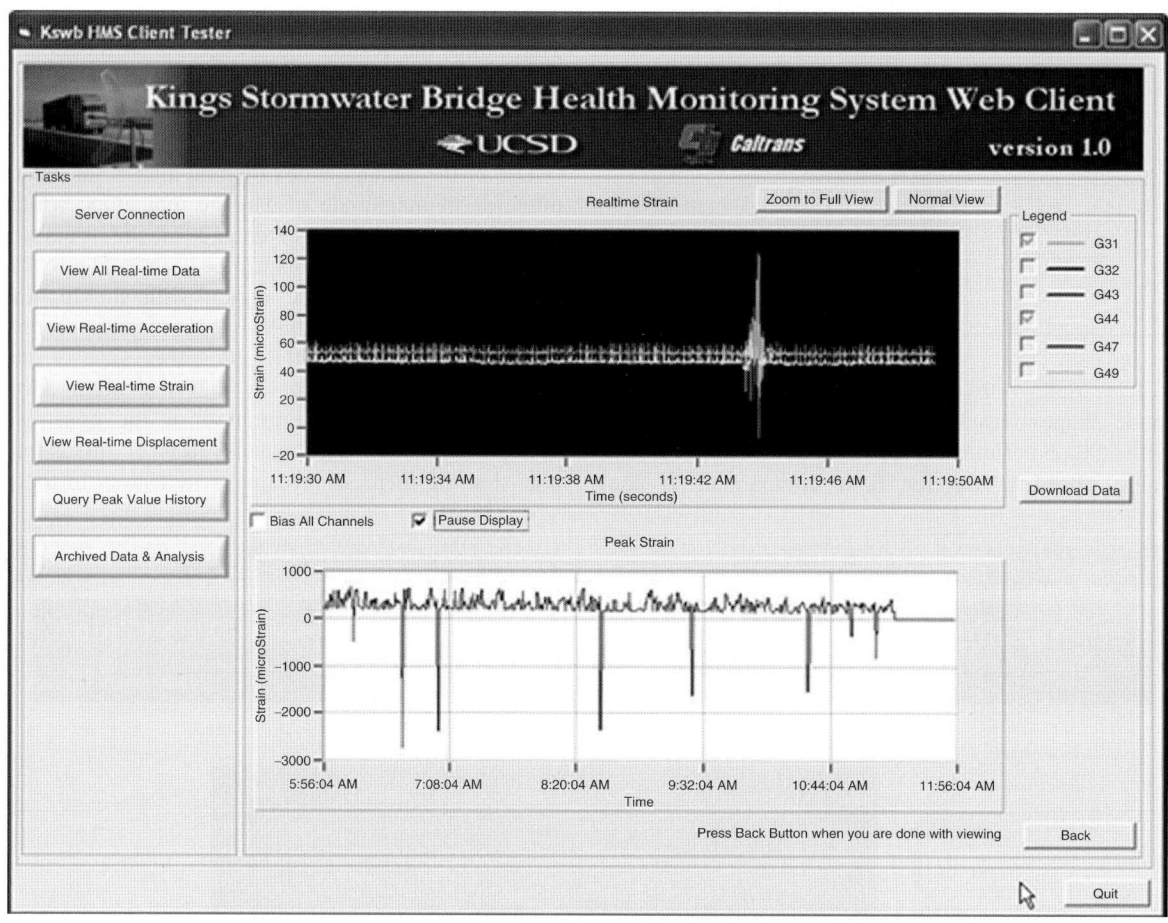

Figure 11b. *Continued.*

regions was noticed ahead of visual identification of the distress, emphasizing the value of such a health monitoring system in not just monitoring response but also in providing advance warning of deterioration.

4 SUMMARY

The design of components and structures is predicated on uncertainties related to a number of factors including those related to load, materials, and service environment. The lack of detailed knowledge in some of these has led conventionally to the use of intrinsically high factors of safety. While the use of an SHM system can provide, at the minimum, data that can be used to assess structural integrity in a manner allowing for provision of early warning of impending failure, it can also serve as a means of assessing reliability and, perhaps, even remaining characterizing capacity and remaining life on a continuous basis. In its development as a web-based system, it thus not only serves as a means of NDE but also as a means of collecting detailed records of response that will enable not just better designs of new structures but also the basis of better assessment and prediction of reliability of components and entire systems enabling better, and more timely, allocation of resources based on age, deterioration, and level of importance of the structure in terms of a local and regional context. Web-based SHM thus provides for the development of an *in situ* system similar to the system used by the federal aviation administration

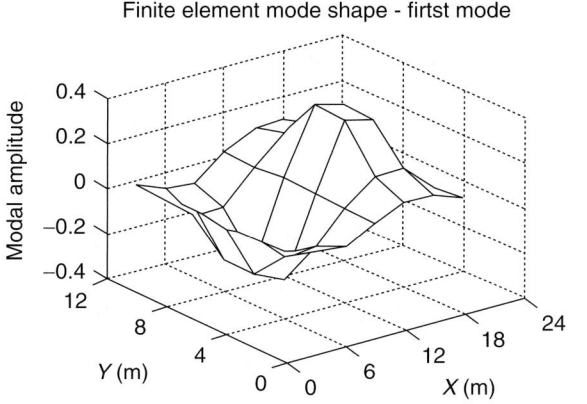

Figure 12. Mode shapes from the FEA model.

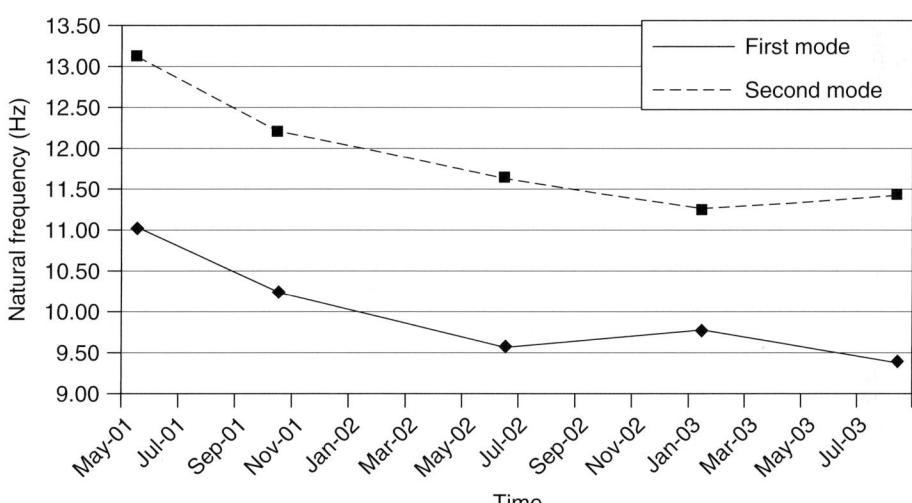

Figure 13. Change in frequency as a function of time.

(FAA) in tracking flights in the air, but based on fixed objects whose "health" varies and can be tracked in time.

The SHM system presented in this article provides an example of implementation of a system amenable to the use of autonomous monitoring. The system not only enables the collection and presentation of data from a suite of sensors but also provides for the assessment of the data in terms of structural characteristics, which can be compared to analytical results derived from an FEA model, which is calibrated to the initial response. This allows for the setting of response thresholds that would provide warnings to the user/owner when exceeded. In the current form of implementation, the web-based system has data transferred both in real time and in packages at preset intervals across a wireless network. Data is interrogated using a damage prognosis algorithm that provides indicators of damage severity and allows for comparison between response and damage state at various points in time. Thus subtle changes in bridge response, which may not be seen easily through just inspection of modal parameters, such as deterioration in

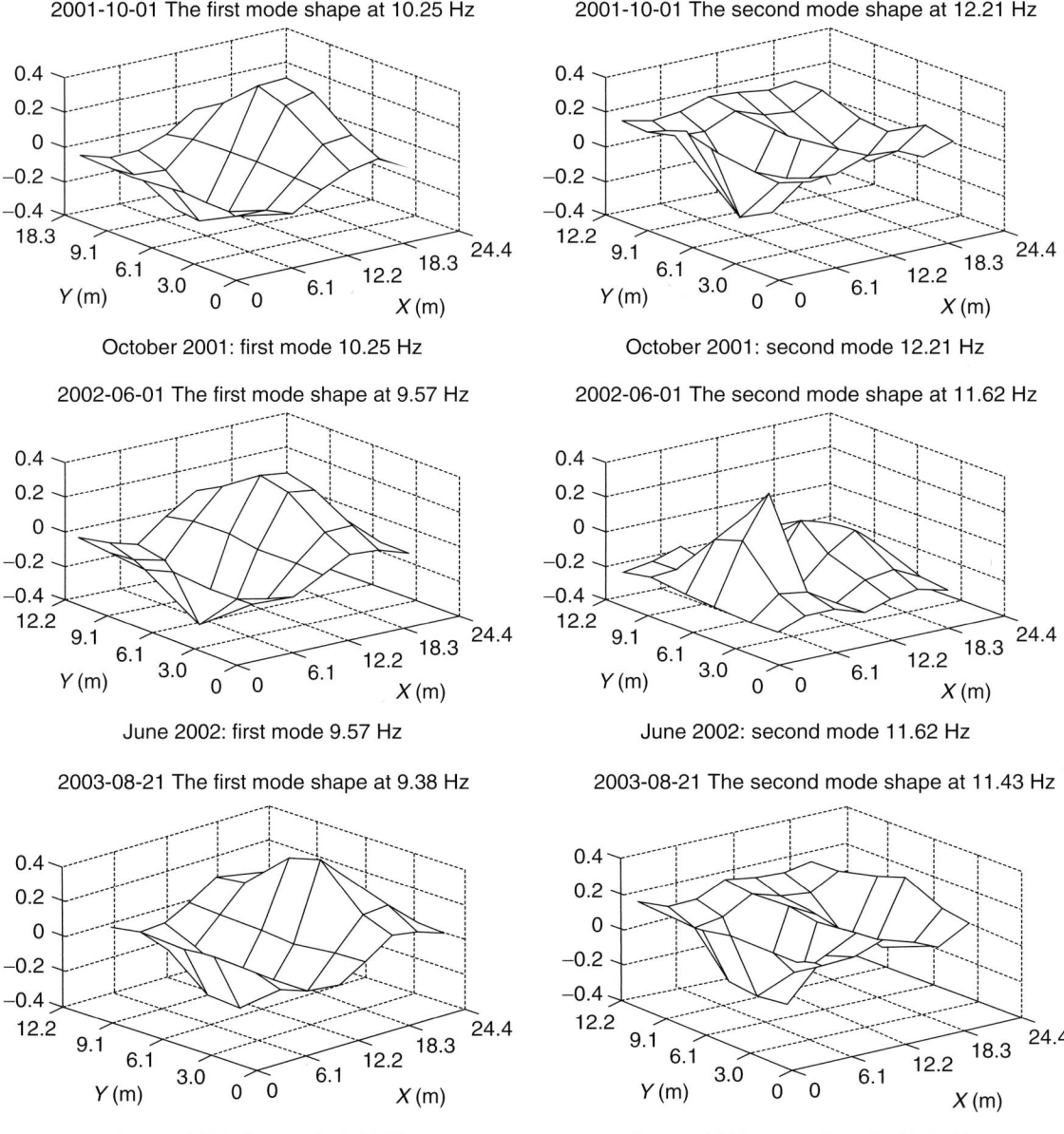

Figure 14. Changes in mode shapes over time.

expansion joints over time, can be easily determined. It is expected that the further development of such systems will lead to the establishment of a comprehensive methodology for autonomous health monitoring of structural systems to the point where true condition-based physical inspection and monitoring would become a reality. The integration of damage identification and finite-element-based tools further provides assistance to the engineer in assessing health immediately rather than having to resort to expensive closures while assessments are made off-line.

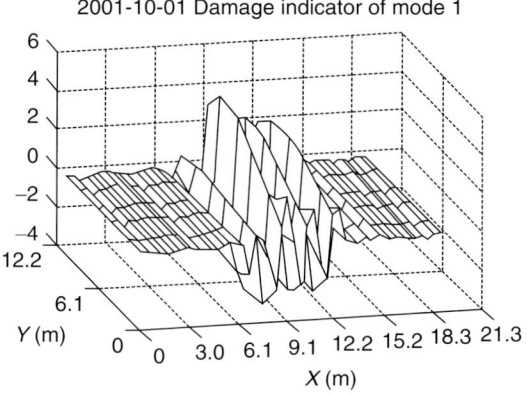
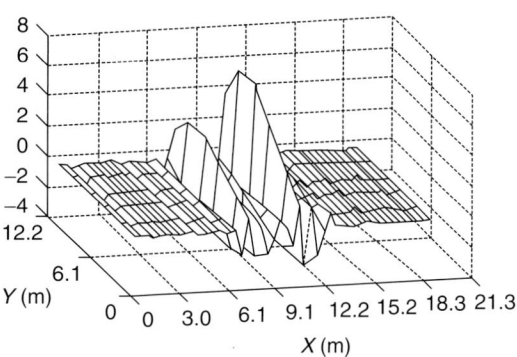

Figure 15. Examples of change in damage indicators.

REFERENCES

[1] Mufti AA, Bakht B, Tadros G, Horosko AT, Sparks G. Civionics—a new paradigm in design, evaluation, and risk analysis of civil structures. *Journal of Intelligent Material Systems and Structures* 2007 **18**:757–763.

[2] Morris B, Trivedi MM. Robust classification and tracking of vehicles in traffic video streams. *IEEE International Intelligent Transportation Systems Conference*, Toronto, Canada, September 2006.

[3] Adachi Y. Monitoring technologies for maintenance and management of urban highways in Japan. In *Sensing Issues in Civil Structural Health Monitoring*, Ansari F (ed). Springer: Dodrect, Netherlands, 2005, pp. 13–22.

[4] Stubbs N, Osegueda R. Global non-destructive damage evaluation in solids. *International Journal of Analytical and Experimental Modal Analysis* 1990 **5**(2):67–79.

[5] Karbhari VM, Seible F, Burgueño R, Davol A, Wernlli M, Zhao L. Structural characterization of fiber-reinforced composite short- and medium-span bridge systems. *Applied Composite Materials* 2000 **7**:151–182.

[6] Karbhari VM, Guan H, Sikorsky C. Web-based structural health monitoring of a FRP composite bridge. *Proceedings of the 1st International Conference on Structural Health Monitoring and Intelligent Infrastructure*, Tokyo, November 2003, pp. 217–226.

Chapter 74
Design of Active Sensor Network and Multilevel Data Fusion

Xiaoming Wang[1] and Zhongqing Su[2]

[1] *CSIRO Sustainable Ecosystems, Commonwealth Scientific and Industrial Research Organisation, Melbourne, VIC, Australia*
[2] *Department of Mechanical Engineering, Hong Kong Polytechnic University, Kowloon, Hong Kong, China*

1 Introduction	1301
2 Active Sensing	1304
3 Distributed Sensing System and Decision Fusion	1304
4 Distributed Sensor Network for Damage Detection of Composite Structures	1309
5 Summary	1312
Acknowledgments	1313
References	1313

1 INTRODUCTION

While traditional nondestruction evaluation (NDE), ranging from simple visual inspection to methods based on electrical properties, electromagnetic properties and acoustic or ultrasonic wave, has provided effective means in evaluating structural performance [1], it is understood that the application of structural health monitoring (SHM) has led to a paradigm shift in thinking from event-based structural evaluation to continuous on-line *in situ* monitoring. Such a philosophy may potentially reduce cost significantly without compromising performance and reliability of engineering structures such as airframes [2, 3] and bridges [4, 5]. It provides critical information of long-term structural performance for life cycle maintenance and management [6]. As one example, SHM has been introduced into a maintenance policy by the US Department of Defence, known as *Condition-Based Maintenance Plus* (*CBM+*), to reduce the cost in maintenance by a schedule-based approach [7].

SHM is a technique that uses sensors, on site or remotely, to collect information on structural, mechanical, or physical behavior, either continuously or periodically, for the diagnosis and prognosis of structural integrity and performance. It was comprehensively reviewed by many researchers [8–10]. Generally speaking, an SHM system may be specified

Encyclopedia of Structural Health Monitoring. Edited by Christian Boller, Fu-Kuo Chang and Yozo Fujino © 2009 John Wiley & Sons, Ltd. ISBN: 978-0-470-05822-0.

in terms of its functionality, sensing system design, and information acquisition and interpretation.

1.1 Functionality

An ideal SHM system should be capable of indicating structural health at four levels hieratically:

- level one: the occurrence of an adverse event to structure, affecting structural performance and functionality;
- level two: the location where an adverse event to structure occurs;
- level three: scale or severity of the adverse event to structure;
- level four: residual service life of structures or structural components to maintain required minimum performance and functionality with such an adverse event;

where an adverse event to structure can generally be considered as a consequence arising from an interaction between structures and their environment, which may include mechanical (loading, wind, impacts, and shock, etc.), chemical and biological (chloride, water, bacteria, etc.), thermal (radiation, heating, thermal shock, etc.), and so on. The consequence can be any form of damage or degradation in relation to performance and functionality of structures, such as crack, fatigue, buckling, delamination, dislocation, disconnection, etc.

1.2 Sensing system design

An appropriate sensing system is of vital importance to capture authentic signals to quantify the structural performance and identify any degradation in the performance. Much akin to the nerve cell of human beings, the sensor is a device for detecting the variation in physical, chemical, or biological properties, and, by proper transduction, transforming the measurands into interpretable information. Sensor technology is seen as a basic element in SHM. Basically, an ideal sensor system for SHM meets the following requirements: (i) veridical acquisition of changes in local or global structural responses; (ii) faithful delivery of captured changes; (iii) possibly less intrusion to host structure; (iv) endurance for general structural service conditions with its service life not less than that of the host structure; and (v) ease in handling, attachment, integration, and operation. In some demanding situations, e.g., for aerospace applications, the sensor should also feature small size, light mass, extremely high sensitivity and reliability, low cost and power consumption, little deterioration with aging, remote data transmission, robustness for noise, reduced wire, or even wireless, etc. While it is difficult to design a sensing system strictly satisfying all the requirements indicated above, an optimal design may be reached for a specific application by considering its most desired needs.

There are many sensing options including ultrasonic probe, acoustic emission sensor, magnetic sensor, eddy-current transducer, accelerometer, strain gauge, laser interferometer, optical fiber, electromagnetic acoustic transducer (EMAT), etc. In particular, piezoelectric elements have been generating a lot of interest for SHM, especially because of their potential of being used as either sensors or actuators [1–15].

By virtue of the utility of piezoelectric elements, an active sensing scheme can be designed on the basis of the assessment of electromechanical impedance spectrums (*see* **Chapter 5**). It is implemented by measuring the electric current in response to the voltage imposed on the piezoelectric element integrated with structures. The electromechanical impedance is directly related to structural dynamics [16], and therefore any structural change will, in principle, lead to variation in the impedance. Applications of such a scheme include detection for bridge joint deterioration [17, 18], disbond in composite repair patch [19], loose joints in pipelines [20], and cracks in a concrete beam [21].

The active sensing can also be achieved by using Lamb waves. Lamb waves have been widely explored [22] since 1961 when they were introduced as a means of damage detection [23]. Lamb waves, made up of a superposition of longitudinal and shear modes, are available in a thin plate or shell structures. The mode of a Lamb wave can be either symmetric or antisymmetric, and can be excited by a variety of means [10]. Details can also be found in Section 2 (*see* **Chapter 3**). Piezoelectric elements are capable of generating and acquiring Lamb waves, and are particularly suitable for integration into a host structure as an *in situ* actuator and sensor, as a result of their low mass, good mechanical strength, wide frequency band, low power consumption and acoustic impedance, as well as low cost. Theoretically, a sensing system can be designed on the basis of the concept to relate damage-modulated Lamb waves signal to damage location, size, or severity, offering an SHM system featured with functionality at all four levels.

1.3 Signal acquisition and interpretation

Signal acquisition is designed to capture any information related to structural health status, which can further be interpreted at the four levels of functionality required by an SHM system. The concept to apply piezoelectric elements for identifying structural damages by assessing Lamb waves can be schematically described by Figure 1. As seen, one piezoelectric element is used as an actuator to generate Lamb waves, and the other as a sensor to collect signals. With the occurrence of damage, the original signal is considerably modulated by the damage, as in the example described in Figure 2(a) and (b). In comparison with Figure 2a, without the effect of structural damage, Figure 2(b) shows that a structural damage such as delamination in a composite laminate induces a *shear horizontal* (SH) mode wave [24], and there is a time lag between the SH wave and the symmetric mode (S) wave.

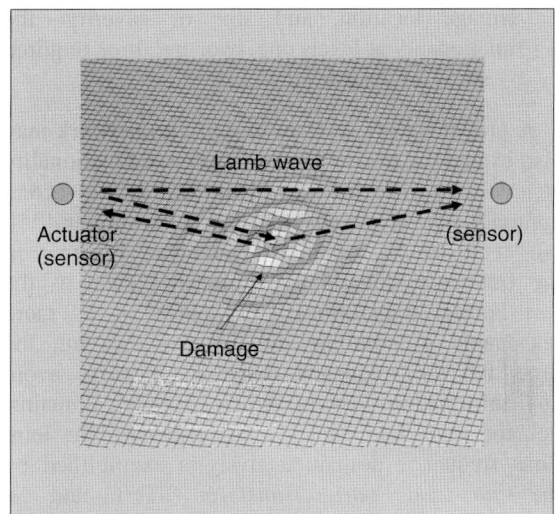

Figure 1. Active sensing scheme with Lamb waves.

Basically, there are two approaches to interpret the signals acquired by a sensor in an active sensing scheme based on Lamb waves:

- using *time of flight* (TOF) of damage-modulated Lamb waves between recipient wave and the damage-reflected wave, to identify a damage event and quantify the location of damage—the functionality at levels one and two required by a SHM system;
- correlating features of damage-modulated Lamb waves to identify a damage event, quantify

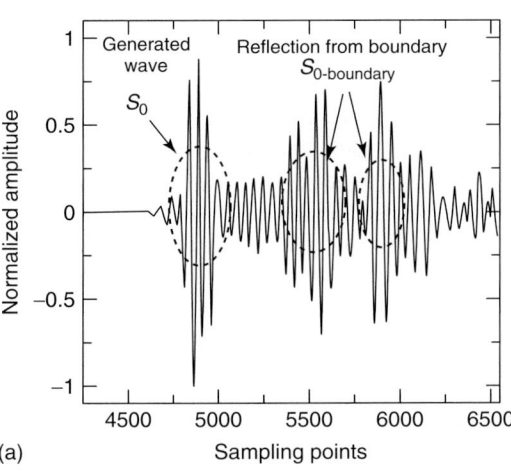
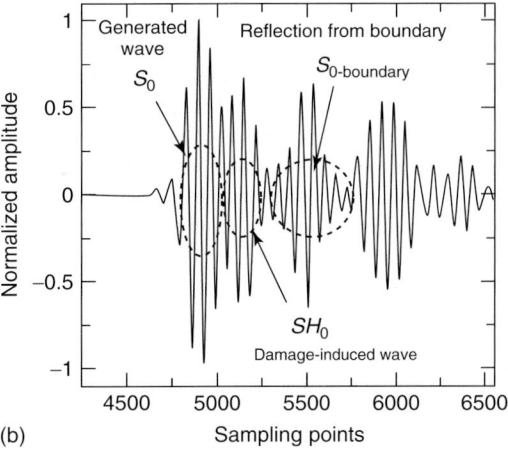

Figure 2. Acquired Lamb wave signals: (a) without damage modulation and (b) with damage modulation.

damage location, and size or severity—the functionality at levels one, two, and three required by an SHM system.

A proper sensing system design into network may also extend the first approach to achieve functionality at level four. The features mentioned in the second approach can be obtained in the time domain [25], but it is more common to analyze features in the frequency domain by approaches such as the fast Fourier transform (FFT) [26–28]. It is more reasonable to combine the analyses in both the time and the frequency domains so as to avoid any information loss in either of these domains, i.e., the time–frequency domain analysis. The joint time–frequency domain analysis is exemplified by the *short-term Fourier transform* (STFT) and the *Wigner–Ville distribution* (WVD) [29]. Recently, *wavelet transform* (WT) has become more popular in feature extraction for SHM [30–32], such as denoising a signal to facilitate feature extraction in the time domain [33].

2 ACTIVE SENSING

Active sensing design using piezoelectric elements has created great potential to advance the development of SHM techniques, especially in applications pertaining to plate and shell structures that largely exist in airframes, marine vehicles, containers or tanks, pipelines, and so on. An active piezoelectric sensing system is capable of both generating interrogative signals and capturing structural-modulated responses, which reflect the condition of structures, over *an area* instead of at *a point*. However, the identification of structural condition is normally shown as a typical inverse problem that represents the solution (structural condition) in terms of a *hidden* physical phenomenon to be obtained or estimated from *observed* signal from sensors. Mathematically, it often shows that the process is highly ill-conditioned or not well-posed that requires, (i) a solution to exist, (ii) the solution to be unique, and (iii) the solution to continuously depend on the observed data. Complication of structural damages, such as occurring at multiple locations, may impose even more difficulties to obtain the solution correctly. While a simple active piezoelectric sensing system may potentially monitor a broad structural area, it also creates uncertainties in signal interpretation caused by noises and signal scattering as a result of environment, structural boundary, and discontinuity interference. Distinguishing the damage-induced signal variation from noises, particularly in engineering practices, is very challenging.

Intrinsic problems in seeking an inverse solution as well as practical issues in view of system operations require a new path to increase the robustness and reliability of SHM. Artificial intelligence techniques can be adopted to reproduce human cognitive processes for the identification of structural abnormality and derive *beliefs* at the level of an individual sensor, which is essentially a part of a decentralized or distributed sensor network. An overall *decision* or consensus on structural condition becomes available by combining the *beliefs* within a group and then subsequently among groups of individual sensors in the network in multiple steps, known as *a multilevel fusion process*, which may considerably increase the robustness in obtaining inverse solutions and reduce uncertainties in damage identification for SHM. How to design a distributed active sensing system to suit the needs of a multilevel decision fusion process becomes important.

3 DISTRIBUTED SENSING SYSTEM AND DECISION FUSION

Neurological study [34] indicates that signals to different senses are initially segregated at the neural level and neurons do not interact with each other on what they sense until the signals are transmitted to the brain, where the sensory signals converge on the same target to supply perceptions and orient behavior. It is understood that there are three critical steps involved in such a process: (i) distributed sensing, (ii) information fusion, and (iii) establishing perception and orienting behavior. Extending the concept to SHM, sensors can be designed in a distributed way—not only in terms of a spatial network but also in terms of their functions—implying that they are able to sense different physical parameters at many locations in a specific pattern. Distributed sensors may firstly form their individual beliefs that represent their own interpretation of data or "the world" they sense.

Secondly, a fusion process then combines all their beliefs following conjunctive, disjunctive, or compromising rules to form a consensus about the view on "the world"-health status of the structures. The advantages of using a distributed sensor network include the combination of all kinds of information sources that can be the same type or completely different, the enhancement of the awareness of targets with superior system reliability and robustness [35], and the reduction in dependence of final perception/decision on the information from a single sensing source.

3.1 Distributed active sensing structure

Figure 3 gives a representative unit in a distributed active sensing structure, including the following four elements: (i) active sensor node, (ii) passive sensor node, (iii) sensor controller, and (iv) sensing path. The active sensor node plays an active role by generating an interrogative signal propagating through a sensing path and forming its belief about structural health on the basis of the response of the passive sensor node. The passive sensor node is used to receive a signal transmitted through structures from the active sensor and provide feedback to the active sensor node by the sensor controller. The sensor controller decides which sensor node takes an active role, relays and processes feedback to establish a belief for the active sensor node. The sensing path is the possible pathway of interrogative signal traveling from the active sensing node to the passive sensing nodes within a structure.

A distributed active sensing structure can be designed in a centralized format, as shown in Figure 4, where sensors may rotate to take an active role to generate interrogative signals and get feedback

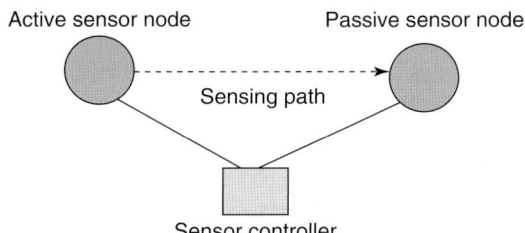

Figure 3. Sensor nodes, controller, and sensing path.

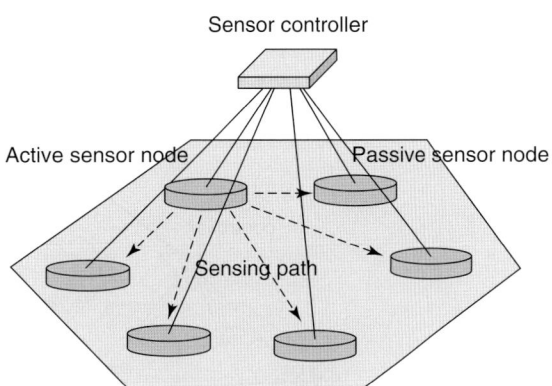

Figure 4. Centralized active sensing structure.

from all other (passive) sensors only through one sensor controller. All feedbacks then become inputs to form a belief of the active sensor group in regard to structural health status. It should be indicated that the entire process is controlled by one sensor controller, know as *a centralized structure*. Figure 5 shows a hierarchical structure, which has an extra sensor controller at a higher layer. As mentioned in the centralized structure, each sensor may have its own belief on the status of structural health. The sensor controller at the low layer may form a belief of a sensor group on the basis of the beliefs of individual sensors through the second level of the fusion process. In the hierarchical structure, the high-layer sensor control subsequently establishes a belief of entire sensor groups by fusing the beliefs formed at the low-layer sensor controllers.

Different from the hierarchical structure, the autonomous active sensing structure, as shown in Figure 6, does not possess any high-layer sensor controller. Sensor controllers are independent of each other, but maintain communication among themselves. Note that the belief of entire sensor groups is established at each sensor controller by acquiring information from all the others with their own rules, which are not necessarily the same.

More complex active distributed sensing structures can be established on the basis of the basic structures discussed above. One example is the hybrid structure, which is intended to integrate the features of centralized, hierarchical, and autonomous structures.

1306 *Systems and System Design*

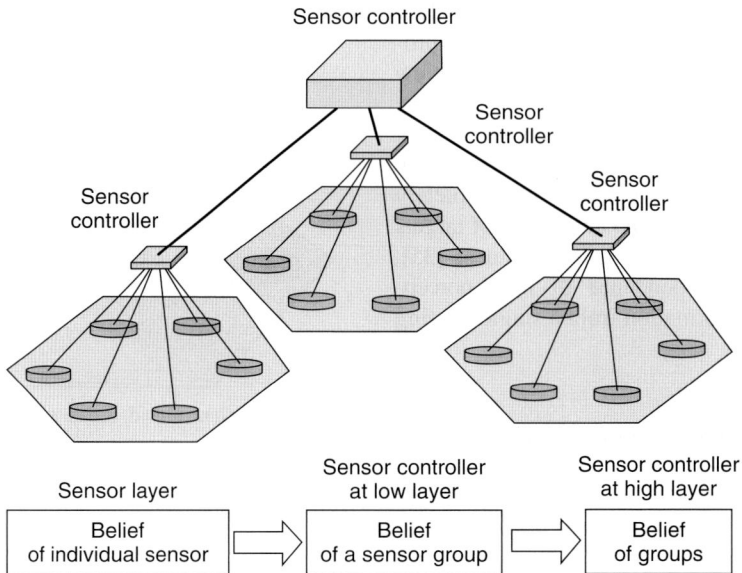

Figure 5. Hierarchical active sensing structure.

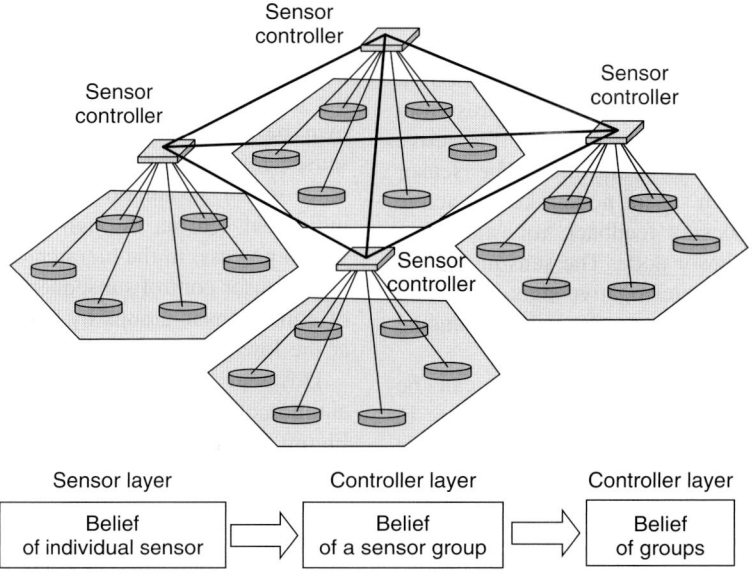

Figure 6. Autonomous active sensing structures.

3.2 Fusion in distributed active sensor network

As discussed previously, the establishment of a belief at each layer, i.e., sensor layer and sensor controller at a low or high layer, has to be implemented via fusion, which basically combines the beliefs at the lower layer to form a new belief or posterior belief at a higher layer. The fusion process is intended to make the posterior brief with less imprecision, uncertainty, and incomplete information, accurately characterizing the structural health status. More precisely, a brief is described by a degree that represents how certain or possible the perception of a sensor or a sensor controller may enhance such a view or consensus as the occurrence of a series of potential adverse structural health events. A fusion process is described by a projection function, $\Im$, which projects a vector of the prior belief, $\vec{B}_{\text{prior}}$, at a lower layer to a posterior belief, $B_{\text{posterior}}$, at a higher layer presented by

$$\Im : \vec{B}_{\text{prior}} \longrightarrow B_{\text{posterior}}, \text{ for } \vec{B}_{\text{prior}} \in I^n$$
$$\text{and } B_{\text{posterior}} \in I \qquad (1)$$

where I represents a measure set of the degree of the posterior belief as well as the degree of all subsets of the prior belief given in n dimension. It is normally defined as an interval, for example, [0, 1], where "0" indicates that the prior belief *cannot* enhance the posterior belief or provide information to strengthen the consensus, and "1" represents that the prior belief can enhance the posterior belief.

Character of fusion functions can be divided into three basics [36]. Assuming that there are two prior belief inputs with their degrees described by x and y, and a fusion function is given by $\Im$ with the posterior belief given by $\Im(x, y)$, then

- $\Im$ is conjunctive if $\Im(x, y) \leq \min(x, y)$;
- $\Im$ is disjunctive if $\Im(x, y) \geq \max(x, y)$;
- $\Im$ is a compromise if $\min(x, y) \leq \Im(x, y) \leq \max(x, y)$.

Conjunctive fusion gives the common part among the prior beliefs, reducing the less certain components and giving the smallest measure, as shown in Figure 7(a). Disjunctive fusion covers all possible beliefs, increasing the certainty and giving the greatest measure, as shown in Figure 7(b). Compromise fusion gives an intermediate measure, as in Figure 7(c).

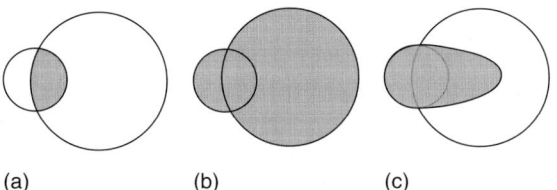

(a)   (b)   (c)

Figure 7. Fusion behavior: (a) conjunctive, (b) disjunctive, and (c) compromise.

In practice, a fusion function can have any one of the function character described above or their mixture, depending on the context of the fusion function. The selection of fusion schemes has to be considered in relation to the data sets and the fusion objectives. The following are a few examples of fusion functions [36]:

- conjunctive fusion functions

$$\Im(x, y) = \{\min(x, y); xy; \max(0, x + y - 1)\},$$
$$(x, y) \in [0, 1] \times [0, 1] \qquad (2)$$

- disjunctive fusion functions

$$\Im(x, y) = \{\max(x, y); x + y - xy; \min(1, x + y)\},$$
$$(x, y) \in [0, 1] \times [0, 1] \qquad (3)$$

- compromising fusion functions

$$\Im(x, y) = \left\{ \frac{2xy}{x+y}; \sqrt{xy}; \frac{x+y}{2}; \sqrt{\frac{x^2+y^2}{2}} \right\},$$
$$(x, y) \in [0, 1] \times [0, 1] \qquad (4)$$

More fusion techniques include the voting scheme, Bayesian inference [37], Dempster–Schafer theory [38, 39], and fuzzy logic [40].

3.2.1 Fusion based on voting scheme

In a fusion based on voting scheme, the posterior belief is formed according to the voting index [37]:

$$\Im(\vec{Y}_i) = \sum_{j=1}^{n_i} w_{ij} Y_{ij} \qquad (5)$$

where n_i is the total number of prior beliefs from all sensors used to assess the ith location ($i = 1, \ldots, N$), and Y_{ij} is the prior belief degree of the jth sensors on the structural condition at the ith location. Typically, it equals one when supporting a tested consensus and zero when against it. In equation (5), w_{ij} represents the voting weight of the jth sensor in relation to the ith location, and $\sum_j w_{ij} = 1$. The fusion based on voting scheme is compromising.

3.2.2 Fusion based on Bayesian theory

In Bayesian fusion, with multiple inputs, E_i, from a distributed sensor network, the posterior probability of the belief, $\Theta = \theta_i$, is given by [37]

$$P(\Theta = \theta_i | E_1, \ldots, E_j, \ldots, E_m)$$
$$\sim P(\Theta = \theta_i) \prod_{j=1}^{m} P(E_j | \Theta = \theta_i) \quad (6)$$

where $P(\Theta = \theta_i)$ is the prior probability of the belief of $\Theta = \theta_i$; $P(\Theta = \theta_i | E_i)$, $(i = 1, \ldots, m)$ is the posterior probability of the belief of $\Theta = \theta_i$, given inputs $E_i (i = 1, \ldots, m)$; and $P(E_i | \Theta = \theta_i|)$ is the likelihood of the occurrence of E_i at the assumption of $\Theta = \theta_i$. The fusion based on Bayesian theory is conjunctive.

3.2.3 Fusion based on Dempster–Shafer rule

Assuming mass functions $m_j(B_j)(j = 1, \ldots, m)$ that represent the degree of belief of the sources (e.g., sensors) on the proposition B_j, the Dempster–Shafer rule for the fusion of belief degree on the proposition Θ, where $\Theta = B_1 \cap \ldots B_j \cap \ldots B_m$, is then given by [38, 39]

$$m_1 \oplus \ldots m_j \oplus \ldots m_m(\Theta)$$
$$= \frac{\sum_{B_1 \cap \ldots B_j \cap \ldots B_m = \Theta} m_1(B_1) m_2(B_2) \ldots m_m(B_m)}{1 - k} \quad (7)$$

where

$$k = \sum_{B_1 \cap \ldots B_j \cap \ldots B_m = \emptyset} m_1(B_1) m_2(B_2) \ldots m_m(B_m) \quad (8)$$

where k is a measure of conflict between sources, reflecting the imprecision of information; and $B_1 \cap \ldots B_j \cap \ldots B_m \cap \ldots = \emptyset$ implies the scenario with null intersection among all $B_j (j = 1, \ldots, m)$. The fusion based on Dempster–Shafer rule is conjunctive.

3.2.4 Fusion based on fuzzy interference

Assuming membership functions, $\mu_{A_j}(x)(j = 1, \ldots, m)$ that define a fuzzy set of beliefs on the proposition, such as damage occurrence at a location, the fusion through fuzzy conjunction is given by [40]

$$\mu_{A_1 \cap A_2 \ldots \cap A_m}(x) = \min\{\mu_{A_1}(x), \mu_{A_2}(x), \ldots, \mu_{A_m}(x)\} \quad (9)$$

where x is the universe of discourse related to the degree of prior beliefs. To defuzzificate the combined fuzzy conclusion in equation (9), the centroid defuzzification method is applied by using the following equation:

$$\bar{x} = \frac{\sum_{j=1}^{m} \mu_{A_j}(x_j) x_j}{\sum_{j=1}^{m} \mu_{A_j}(x_j)} \quad (10)$$

where $\bar{x}$ is the posterior belief through fusing degree of prior beliefs, on the proposition of damage occurrence.

3.2.5 Fusion with neural network

A neural network consists of one input layer with α elements (im), two hidden neural processing layers respectively possessing λ and η computing elements (referred to as *neurons*), and one output layer containing β variables (ov). Mathematically, the ith output variable in the network is formularized as

$$ov_i = T_3\left(\left(\sum_{q=1}^{\eta} w^3_{q-i} \cdot T_2\left(\left(\sum_{r=1}^{\lambda} w^2_{r-q} \cdot T_1 \right.\right.\right.\right.$$
$$\left.\left.\left.\left. \times \left(\left(\sum_{p=1}^{\alpha} w^1_{p-r} \cdot im_p\right) + b^1_r\right)\right) + b^2_q\right)\right) + b^3_i\right) \quad (11)$$

where im_p denotes the pth input element. $w^r_{p-q} (r = 1, 2, 3)$, defined as *weight*, represents the linkage joining the pth input element/neuron in the rth layer with the qth neuron/output variable in the next layer. Similarly, b^i_q, called *bias*, is an offset constant for the qth element in ith layer, ($q = 1, 2, \ldots, \lambda/\eta/\beta$ for the first/second-processing/output layer). $T_s (s = 1, 2, 3)$ is the transfer function in the network [41]. In the fusion with a neural network, the inputs may be the features related to prior beliefs, while outputs are the parameters related to posterior beliefs.

4 DISTRIBUTED SENSOR NETWORK FOR DAMAGE DETECTION OF COMPOSITE STRUCTURES

The following gives a few examples to demonstrate the applications of distributed sensor network and one/multiple level of fusion in damage detection of composite structures, as shown in Figure 8. The simulation is concentrated on a representative active sensing unit (RASU), extracted from structures with a complex distributed sensor network.

In a distributed active sensor network, each sensor takes a role of Lamb wave generator (active sensor) or receiver (passive sensor). The information fed back to the active sensor from a passive sensor within a sensing path is represented using a set of digitalized signal features, termed digital damage fingerprint (DDF) in a study [42] pertaining to structural conditions. DDF is a kind of characteristic of a scattered Lamb waves signal that are extracted from the wavelet filter-processed signal and energy spectrum by gradually screening the noncharacteristic components and using principal signal components only, as described by a simple example shown in Figure 9.

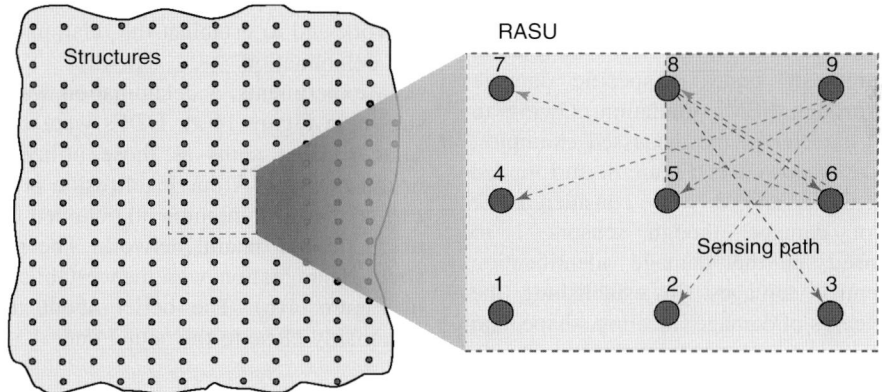

Figure 8. Representative active sensing unit (RASU) from structures.

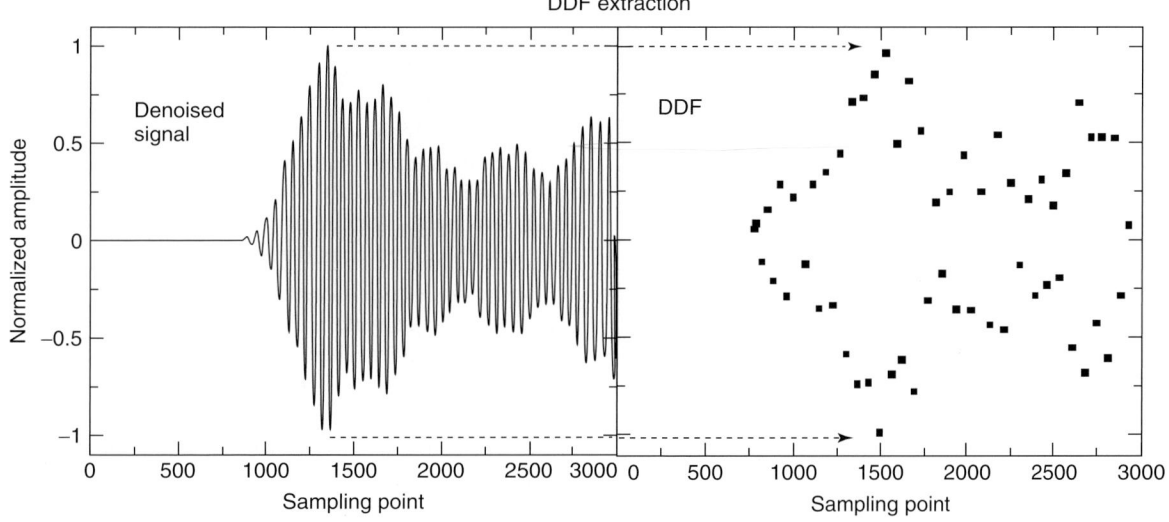

Figure 9. An example of DDF extracted from a denoised signal by a wavelet filter.

The one or multiple level fusion process is strongly related to DDF.

By way of illustration, a CF/EP (T650/F584) laminate was supported on all its four sides with a stacking sequence of $[45/-45/0/90]_s$ and measuring 500 mm × 500 mm × 1.275 mm [43]. A sensor network of nine piezoelectric elements was surface-bonded on the laminate, sensors being numbered Pi $(i = 1, 2, \ldots, 9)$, similar to the one in Figure 8. The laminate was pretreated with damage (hole or delamination) during fabrication. Without losing generality, the damage was presumed elliptic, defined with six parameters: presence (0 or 1), location (ξ, ζ), semimajor/minor axes (α, β) and orientation of θ, although it is very often that not all of those details are required. Knowledge about digital damage fingerprints DDFs in relation to each sensing path under a specific damage was obtained through FEM simulation. Owing to the symmetric character in the discussed examples, only the area as shadowed in Figure 8, surrounded by sensors 5, 6, 9, and 8, was considered for the location of damage, though sensors 1 to 9 were all used in the damage identification process to identify damages. In establishing the knowledge, the cases of damage (location, shape, and orientation of holes/delamination) assumed are shown in Figure 10.

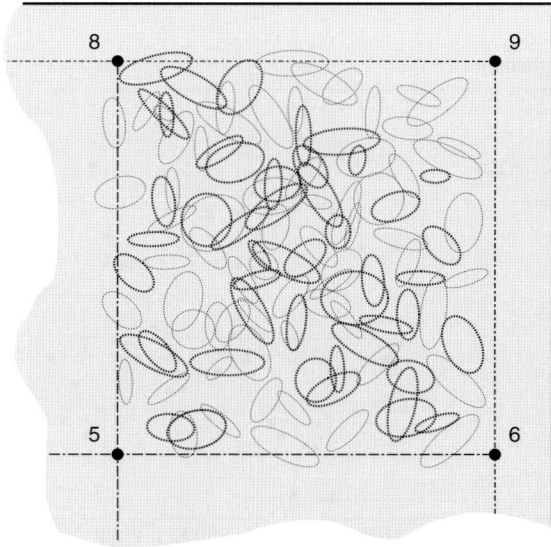

Figure 10. Samples of damages for acquiring knowledge of relevant DDFs.

4.1 Centralized sensing structure and one-level fusion process

In this example, the sensing structure was designed to be centralized, with one sensor controller to regulate one of sensors to generate Lamb waves and the others to collect signals, as shown in Figure 4. Feedback from all passive sensors became inputs to form a belief of structural condition status via a fusion process. Neural network was applied to implement the fusion process. The prior belief was described by the knowledge of delamination characteristics (location, shape, and orientation) in relation to DDFs obtained through FEM simulation. That knowledge was then employed to train a neural network and to establish the relation between DDFs and delamination.

In experiments, a delamination was pretreated as shown in Figure 11(a). DDFs were obtained on the same sensing paths as those utilized to train the neural network. A one-level fusion process was then carried out by combining all obtained DDFs as inputs of the trained neural network. The posterior belief about delamination was then established, as shown in Figure 11(b). The belief about the damage is reasonably close to the actual one.

4.2 Hierarchical distributed sensing structure and multilevel fusion

The belief of the sensors on structural damage status can also be quantified by correlating real-time acquired DDFs with preestablished knowledge about the DDFs, which are obtained from FEM simulation in association with specified damage patterns, illustrated in Figure 10. As a result, corresponding to each sensing path, there is one correlation coefficient C_{ij}^k, which can be considered as the degree of belief of an active sensor i within a sensing path $\vec{ij}$ on a specific damage pattern D_k defined in the preestablished knowledge. Theoretically, when there are N sensors, there exist $N - 1$ sensing paths for one active senor i, or $\vec{ij}(j = 1, \ldots, i-1, i+1, \ldots, N)$. The level-one fusion is required to consolidate the correlation coefficients $C_{ij}^k(j = 1, \ldots, i-1, i+1, \ldots, N)$ to form a belief, $Bel(k; i)$, of the active sensor i on the damage patterns $D_k(k = 1, \ldots, M)$, or

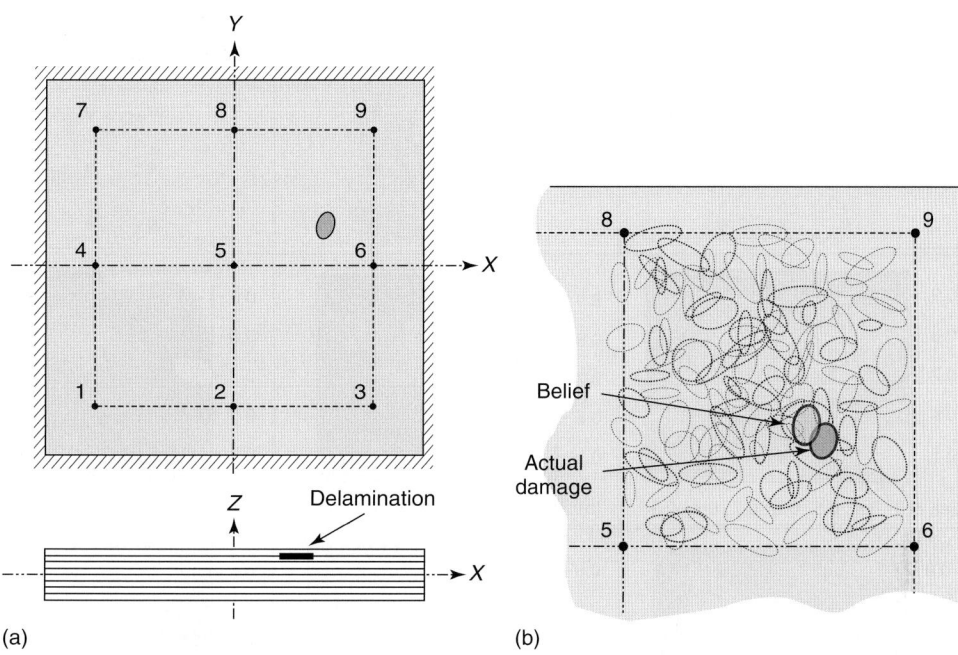

Figure 11. Experiment and identification: (a) A CF/EP laminate with a pretreated delamination and (b) comparison of the actual pretreated delamination with the belief on the damage formed on using an artificial neural network (ANN) [43].

described by

$$\Im_1(k) : C_{ij}^k \longrightarrow Bel(k; i), \text{ for } C_{ij}^k \in [0, 1]^{N-1}$$
$$(j = 1, \ldots, i-1, i+1, \ldots, N)$$
$$\text{and } Bel(k; i) \in [0, 1] \quad (12)$$

Meanwhile, all sensors may form their own beliefs, $Bel(k; i)(i = 1, \ldots, N)$, on each damage pattern. The level-two fusion is required to combine all beliefs of sensors to establish a belief of a group of sensors, which is normally at a sensor controller layer in the hierarchical active sensing structure as discussed previously. The fusion function for a sensor group p is described by

$$\Im_2(k) : Bel(k; i) \longrightarrow Bel(k; p), \text{ for } Bel(k; i)$$
$$\in [0, 1]^N (i = 1, \ldots, N)$$
$$\text{and } Bel(k; p) \in [0, 1] \quad (13)$$

When there are groups of sensors, or there are groups of fusion approaches for one sensor group, that are used to form a posterior belief, the level-three fusion, $\Im_3(k)$, has to be implemented at a higher layer in the hierarchical structure, but very much similar to the one described in equations (12) and (13).

With fusion processes based on Bayesian theory and voting scheme [44], the belief about damage can be identified through level-one fusion, as shown in Figure 12(b) and (e), and level-two fusion, as shown in Figure 12(c) and (f). The actual damage is highlighted by a dotted circle. It is shown that the belief about the damage from one sensing path, as shown in Figure 12(a) and (d), is very vague, and actually cannot identify the damage. The belief after level-one fusion increases the confidence of belief about the damage, while the accuracy in identifying damage improves significantly after level-two fusion.

It is interesting to find that the contrast of the result from the Bayesian theory is much better than the one from the voting scheme. As discussed previously, the fusion based on the Bayesian theory is conjunctive, which may reduce uncertain components in the fusion process and provide a smallest measure. The fusion based on the voting scheme is compromising, which take an intermediate measure, implying that it could not reduce uncertain data as good as the fusion

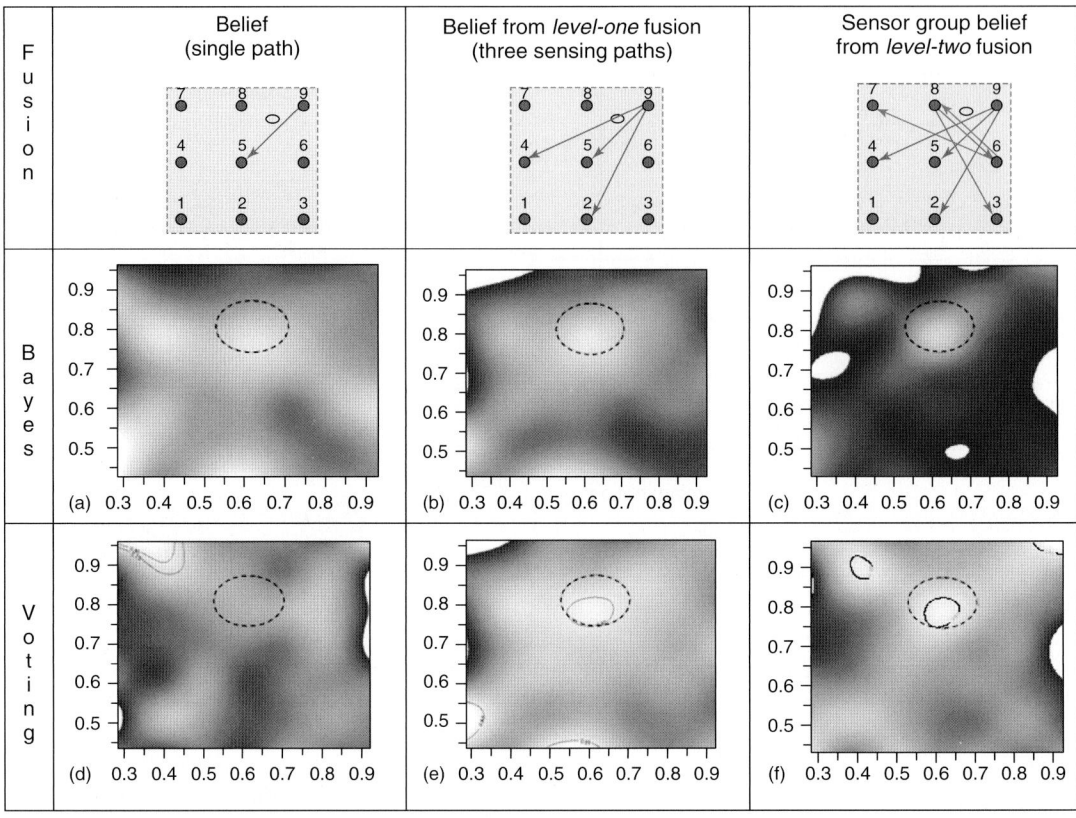

Figure 12. Damage identification by a fusion process, (a), (b), and (c) were obtained by fusion based on Bayesian theory, (d), (e), and (f) were obtained by fusion based on voting scheme (results of (c–f) extracted from [44]). The gray scale is equivalent to the probability of damage occurrence with the lighter corresponding to the higher possibility.

process based on the Bayesian theory. Therefore, the proper selection of fusion process is important.

5 SUMMARY

Design of a sensing network, and subsequently the interpretation of data acquired from sensors within the structure, are two critical issues in SHM for practical applications. It is understood that the process to identify structural health status is basically an inverse problem, which is often mathematically ill-posed or ill-conditioned creating considerable barriers to offer a robust and reliable identification algorithm, especially based on a simple sensor structure.

Active sensor network is designed to deal with the problems of health monitoring, capable of providing richer and sometime redundant information for structural condition identification by spatially distributed sensing structure. The sensing structure can be centralized, hierarchical, or even autonomous, in correspondence to the requirements of data interpretation.

The challenges faced by data interpretation for distributed active sensing network is how to consider information from all the sensors. The information from each sensor may be consistent or conflict with each other, and sometimes, meaningless. The combination of all information becomes crucial in the interpretation of data during the process of SHM. Data fusion can be utilized to implement the task. There have been a wide range of approaches in data fusion, including those algorithms based on voting scheme, Bayesian theory, Dempster–Schafer rules, fuzzy inference, and neural network. The behavior of data fusion operation can be conjunctive, disjunctive, compromising, or a mix of them. It determines the quality of fusion process, which is often involved with imprecision, uncertainties, and incompleteness. An example has been demonstrated to see the difference of results from a fusion process based Bayesian theory and voting scheme.

Fusion can be considered as a process to establish a posterior belief about a set of propositions, such as structural damage events, on the basis of a set of prior beliefs that are possessed by physical elements such as sensors. Basically, a data fusion process is utilized to increase robustness and reliability of an SHM algorithm by reducing imprecision, uncertainties, and incompleteness as much as possible, through rich information from distributed sensor network.

ACKNOWLEDGMENTS

Z. Su thanks Grant A-PA8G from the Hong Kong Polytechnic University and supports from Prof. Lin Ye of the University of Sydney.

REFERENCES

[1] Hellier C. *Handbook of Nondestructive Evaluation*. McGraw-Hill: New York, 2001.

[2] Talbot D. Boeing's flight for survival. *Technology Review* 2003: 35–44.

[3] Beral B, Speckmann H. Structural health monitoring (SHM) for aircraft structures: a challenge for system developers and aircraft manufacturers. In *Proceedings of the 4th International Workshop on SHM*, Stanford, CA, Chang FK (ed). DEStech Publications: Lancaster, PA, 2003.

[4] Chase SB. Smarter bridges, why and how? *Smart Materials Bulletin* 2001 **2**(10):9–13.

[5] Pines D, Aktan A. Status of SHM of long-span bridges in the United States. *Progress in Structural Engineering and Materials* 2002 **4**(4):372–380.

[6] Ko JM, Ni YQ. Technology developments in SHM of large-scale bridges. *Engineering Structures* 2005 **27**(12):1715–1725.

[7] Derriso MM, Pratt DM, Homan DB, Schroeder JB, Bortner RA. Integrated vehicle health management: the key to future aerospace systems. In *Proceedings of the 4th International Workshop on SHM*, Stanford, CA, Chang FK (ed). DEStech Publications: Lancaster, PA, 2003.

[8] Sohn S, Farrar CR, Hemez FM, Czarnecki JJ, Shunk DD, Stinemates DW, Nadler BR. *A Review of SHM Literature: 1996–2001*. Los Alamos National Laboratory: Los Alamos, NM, 2001.

[9] Giurgiutiu V, Cuc A. Embedded non-destructive evaluation for SHM, damage detection, and failure prevention. *The Shock and Vibration Digest* 2005 **37**(2):83–105.

[10] Su Z, Ye L, Lu Y. Guided Lamb waves for identification of damage in composite structures: a review. *Journal of Sound and Vibration* 2006 **295**(3–5):753–780.

[11] Crawley EF, Luis J. Use of piezoelectric actuators as elements of intelligent structures'. *AIAA Journal* 1987 **25**:1373–1385.

[12] Liang C, Sun FP, Rogers CA. Electro-mechanical impedance modeling of active material systems. *Proceedings of Mathematics and Control in Smart Structures, SPIE 2192*. SPIE, Bellingham, WA, 1994; pp. 232–253.

[13] Wang X, Ehlers C, Neitzel M. Electro-mechanical dynamic analysis of the piezo-electric stack. *Smart Materials and Structures* 1996 **5**(4):492–500.

[14] Wang X, Ehlers C, Neitzel M. Analytical investigation on static models of piezo-electric patches attached on beams and plates. *Smart Materials and Structures* 1997 **6**(2):204–213.

[15] Wang X, Shen YP. On the characterization of piezoelectric actuators attached to structures. *Smart Materials and Structures* 1998 **7**:389–395.

[16] Giurgiutiu V. Embedded self-sensing piezoelectric active sensors for on-line structural identification. *Journal of Sound and Vibration* 2002 **124**(1):116–125.

[17] Ayres JW, Lalande F, Chaudhry Z, Rogers A. Qualitative impedance-based health monitoring of civil infrastructures. *Smart Materials and Structures* 1998 **7**(5):599–605.

[18] Ritdumrongkul S, Abe M, Fujino Y, Miyashita T. Qualitative health monitoring of bolted joints using a piezoelectric actuator-sensor. *Smart Materials and Structures* 2004 **13**(1):20–29.

[19] Chiu WK, Galea S, Koss LL, Najic N. Damage detection in bonded repairs using piezoceramics. *Smart Materials and Structures* 2000 **9**(4):466–475.

[20] Park G, Cudney HH, Inman DJ. Feasibility of using impedance-based damage assessment for pipeline structures. *Earthquake Engineering and Structural Dynamics* 2001 **30**(10):1463–1474.

[21] Tseng KK, Wang L. Smart piezoelectric transducers for *in situ* health monitoring of concrete. *Smart Materials and Structures* 2004 **13**(5):1017–1024.

[22] Rose JL. A vision of ultrasonic guided wave inspection potential. *Proceedings of the Seventh ASME NDE Topical Conference*. San Antonio, TX, 2001; NDE-Vol. 20(1–5).

[23] Worlton DC. Experimental confirmation of Lamb waves at megacycle frequencies. *Journal of Applied Physics* 1961 **32**:967–971.

[24] Rose JL. *Ultrasonic Waves in Solid Media*. Cambridge University Press (UK): New York, 1999.

[25] Sohn H, Farrar CR. Damage diagnosis using time series analysis of vibration signals. *Smart Materials and Structures* 2001 **10**:1–6.

[26] Heller K, Jacobs LJ, Qu J. Characterization of adhesive bond properties using Lamb waves. *NDT and E International* 2000 **33**:555–563.

[27] Koh Y, Chiu WK, Rajic N. Effects of local stiffness changes and delamination on Lamb waves transmission using surface mounted piezoelectric transducers. *Composite Structures* 2002 **57**:437–443.

[28] Youbi FEI, Grondel S, Assaad J. Signal processing for damage detection using two different array transducers. *Ultrasonics* 2004 **42**:803–806.

[29] Niethammer M, Jacobs LJ, Qu J, Jarzynski J. Time-frequency representations of Lamb waves. *The Journal of the Acoustical Society of America* 2001 **109**(5):1841–1847.

[30] Wang Q, Deng X. Damage detection with spatial wavelets. *International Journal of Solids and Structures* 1999 **36**(23):3443–3468.

[31] Kim H, Melhem H. Damage detection of structures by wavelet analysis. *Engineering Structures* 2004 **26**(3):347–362.

[32] Rucka M, Wilde K. Application of continuous wavelet transform in vibration based damage detection method for beams and plates. *Journal of Sound and Vibration* 2006 **297**(3–5):536–550.

[33] Smith C, Akujuobi CM, Hamory P, Kloesel K. An approach to vibration analysis using wavelets in an application of aircraft health monitoring. *Mechanical Systems and Signal Processing* 2007 **21**(3):1255–1272.

[34] Murphy RR. Biological and cognitive foundations of intelligent sensor fusion. *IEEE Transactions of Systems, Man, and Cybernetics—Part A: Systems and Humans* 1996 **26**(1):42–51.

[35] Xiong N, Svensson P. Multi-sensor management for information fusion: issues and approaches. *Information Fusion* 2002 **3**:163–186.

[36] Bloch I. Information combination operators for data fusion: a comparative review with classification. *IEEE Transactions of Systems, Man, and Cybernetics—Part A: Systems and Humans* 1996 **26**(1):52–67.

[37] Byington CS, Garga AK. Data fusion for developing predictive diagnostics for electromechanical systems. In *Handbook of Data Fusion*, Hall D, Llinas J (eds). CRC Press: Boca Raton, FL, 2001.

[38] Bloch I. Some aspects of Dempster–Shafer evidence theory for classification of multi-modality medical images taking partial volume effect into account. *Pattern Recognition Letters* 1996 **17**:905–919.

[39] Schocken S, Hummel RA. On the use of the Dempster Shafer mode in information indexing and retrieval applications. *International Journal of Man-Machine Studies* 1993 **39**:843–879.

[40] Yen J, Langari R. *Fuzzy Logic: Intelligence, Control, and Information*. Prentice-Hall: Upper Saddle River, NJ, 1999.

[41] Haykin SS. *Neural Network: A Comprehensive Foundation*. Prentice-Hall, 1999.

[42] Su Z, Ye L. A damage identification technique for CF/EP composite laminates using distributed piezoelectric transducers. *Composite Structures* 2002 **57**:465–471.

[43] Su Z, Ye L. Digital damage fingerprints (DDF) and its application in quantitative damage identification. *Composite Structures* 2005 **67**:197–204.

[44] Wang X, Foliente G, Su Z, Ye L. Multilevel decision fusion in a distributed active sensor network for structural damage detection. *Structural Health Monitoring: An International Journal* 2006 **5**:45–58.

SECTION 3
Autonomous Sensing

Chapter 75

Energy Harvesting and Wireless Energy Transmission for SHM Sensor Nodes

Kevin M. Farinholt, Gyuhae Park and Charles R. Farrar
Engineering Institute, Los Alamos National Laboratory, Los Alamos, NM, USA

1 Introduction	1317
2 Energy Harvesting	1318
3 Radio Frequency Wireless Energy Transmission	1321
4 Conclusions	1325
References	1325

1 INTRODUCTION

The management of energy resources is an essential component in the success of any structural health monitoring (SHM) system. With applications that span aerospace, civil, and mechanical engineering infrastructure, it is necessary for sensors and sensor nodes to be both physically robust and energy efficient. In many applications, a sensor network must be installed in locations that are difficult to access,

This article is a US government work and is in the public domain in the United States of America. Copyright © 2009 John Wiley & Sons, Ltd in the rest of the world. ISBN: 978-0-470-05822-0.

and often these systems have a desired operation life span that exceeds the capabilities of conventional battery technologies. To augment or replace the need to manually recharge or replace batteries within sensors, it is desirable to utilize ambient energy sources that are present within or around the structure being monitored. This process of extracting energy from the environment or a surrounding system and converting it into usable electrical energy is known as *energy harvesting*. Recently, there has been a surge of research in this area, brought about by advances in wireless technology and low-power electronics such as microelectromechanical system (MEMS) devices.

In many SHM applications, there is a considerable amount of ambient energy present within the structure under analysis. This ambient energy is typically in the form of mechanical vibrations induced through environmental or operational conditions, or thermal gradients that develop throughout the day from solar heating or the operation of machinery. The purpose of this article is to provide an up-to-date assessment of available energy harvesting methods suitable for potential SHM sensing applications. This article is not intended to provide an exhaustive literature review, as this area is very broad and useful review articles are already available in the literature (*see* **Chapter 76**; **Chapter 77**). Instead, this article provides a concise introductory survey on the topic and outlines the current status of energy harvesting as applied to relevant themes in SHM. In the second part of the article, a recent alternative to energy harvesting based on vibrations and thermal gradients is presented with more details, using radio frequency (RF) signals to wirelessly deliver electrical energy to operate SHM sensing systems. In this approach, both energy and data interrogation commands are conveyed via a mobile host to each sensor node in order to perform the individual interrogation. Power does not have to reside at the sensor node, relaxing battery, or other such powering requirements. This article also discusses such a prototype system, which is used to interrogate piezoelectric impedance-based sensors on a full-scale bridge.

2 ENERGY HARVESTING

The source of ambient energy can take various forms, including sunlight, thermal gradients within a material, human motion and body heat, vibrations, and ambient RF energy. Several articles reviewing possible energy sources can be found in the literature [1–7]. Fry *et al.* [1] provide a survey of power supplies such as thermoelectric generators (TEGs), mechanical vibration devices using piezoelectric transducers, wind turbines, solar cells, and other exotic portable power sources that utilize ambient electromagnetic radiation, as well as traditional portable supplies such as batteries and fuel cells. While this report is concerned with sources for the US military special operations requirements, it provides insight on the future research trends in energy harvesting.

Roundy [2] compared the energy density of available and portable energy sources. He concludes that, for the device whose desired lifetime is in the range of 1 year or less, battery technology alone is sufficient to provide enough energy. However, if a device requires a longer service life, which is often the case, then the energy harvester can provide a better solution than the battery technologies. Paradiso and Starner [3] point out that the battery technology has evolved very slowly in mobile computing, with only a threefold increase in the battery energy density since 1990. At the same time, the disk storage density has been increased by 1300 times and the CPU speed by nearly 800 times. Park *et al.* [4] summarized several energy harvesting techniques that have been used in SHM applications. The summary also includes the issues associated with power-efficient hardware design and issues with energy harvesting system integration. Glynne-Jones and White [5], Qiwai *et al.* [6], and Mateu and Moll [7] also summarized the basic principles and components of energy harvesting techniques, including piezoelectric, electrostatic, magnetic induction, and thermal energy. A common suggestion listed in these articles is the combined use of several energy harvesting strategies in the same devices so that the harvesting capabilities in many different situations and applications can be increased. Furthermore, energy consumption can be minimized in an effort to close the gap between required and harvested energy [7].

2.1 Electrical energy from mechanical vibrations

One of the most effective methods of implementing an energy harvesting system is to use mechanical vibration to apply strain energy to the piezoelectric material or displace an electromagnetic coil. Energy generation from mechanical vibration usually uses ambient vibration around the energy harvesting device as an energy source, and then converts it into useful electrical energy. The research in this area has made use of mechanical vibration in order to quantify the efficiency and amount of energy capable of being generated and converted, as well as to power various electronic systems. Active materials, usually ceramic- or polymer-based, may be fabricated into a variety of shapes and sizes amenable to various applications. Some examples are shown in Figure 1.

The concept of utilizing piezoelectric material for energy generation has been studied by many researchers over the past few decades, which is well summarized in [8–10]. Piezoelectric materials form transducers that are able to interchange electrical energy and mechanical motion or force. These materials, therefore, can be used as mechanisms to transfer ambient vibration into electrical energy that

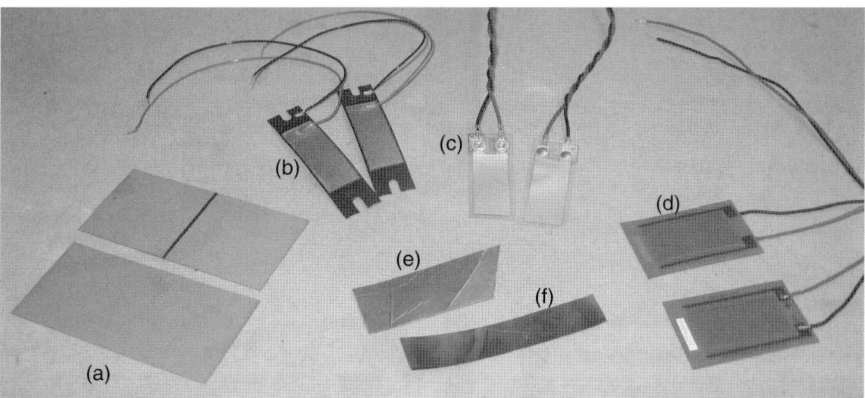

Figure 1. Some typical sensor/actuator materials: (a) PZT 5A4E from Piezo Systems; (b) thunder actuator from FACE International; (c) PVDF transducers from Measurement Specialties; (d) MFC actuators from Smart Material; (e) ionic polymer transducers from Discover Technologies; and (f) ionic polymer–metal composites from Environmental Robots.

may be stored and used to power other devices. A full description of the piezoelectric effect and the methods used to model the behavior of these materials is beyond the scope of this article. However, a significant number of journal articles and conference proceedings develop accurate models and discuss the fundamentals of these materials in great detail [11–13].

Basically, a variety of piezoceramic devices and piezo films have been examined in various settings to try and capture ambient vibration energy [14–16]. Most of them have been laboratory demonstrations and experiments. Some have focused on the electronics in an attempt to optimize the energy transferred from mechanical motion to electrical energy [17, 18]. Others have focused on using various different electrode patterns and packaging to optimize the electromechanical coupling [19]. Studies have also focused on using tuning to try and maximize the amount of energy transferred [20]; however, most of them have dealt directly with the random nature of mechanical energy. In all cases, the goal is to maximize the amount of energy flowing from the mechanical motion of vibration into usable electrical energy. The second feature addressed in the literature is, what should be done with the harvested energy. Some efforts examine storage through capacitors, immediate use, and storage through charging batteries [16, 21]. Other articles examine various applications of harvested power for powering the small electronics including SHM strain-sensing [22, 23] and environmental-sensing systems [24].

2.2 Electrical energy from thermal sources

A second method of obtaining energy from ambient sources is through the use of TEGs that capitalize on thermal gradients. TEGs use the Seebeck effect (Figure 2), which describes the current generated when the junction of two dissimilar metals experiences a temperature difference. Using this principle, numerous p-type and n-type junctions are arranged electrically in series and thermally in parallel to construct the TEG. Thus, when an electrical current is applied to the TEG, a thermal gradient is generated, allowing the device to function as a small solid-state heat pump. Inversely, if a thermal gradient is applied to the device, it will generate an electrical current that can be utilized to power other electronics.

TEGs have been used for capturing ambient energy in various applications. Lawrence and Snyder [25] suggest a potential method of retrieving electric energy from the temperature difference that exists between the soil and the air. To test their concept, a prototype was built without the TEG and the heat flow was measured to estimate the amount of power that could be obtained. The results showed that a maximum instantaneous power of approximately 0.4 mW could be generated by the thermoelectric device. Rowe *et al.* [26] investigate the ability to construct a large TEG capable of supplying 100 W of power from hot waste water. The system tested used numerous thermoelectric devices placed between two

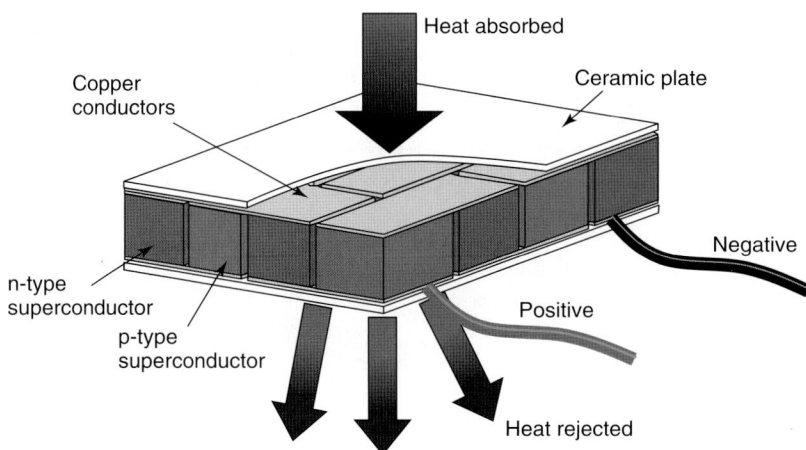

Figure 2. Schematic of the Seebeck effect.

chambers, one with flowing hot water and the other with cold water flowing in the opposite direction and thus maximizing the heat exchange. With a total of 36 modules, each with 31 thermocouples, 95 W of power could be generated.

As described, the idea of using thermoelectric devices to capture ambient energy from a system is not a new concept. However, in many cases, the research efforts utilize liquid heat exchangers or forced convection that significantly improve heat flow and power generation, but require complex cooling loops and systems. Therefore, Sodano *et al.* [27] investigated the use of TEGs as power harvesting devices that do not have an active heat exchanger, but function as a completely passive power scavenging system. Two potential applications are investigated, utilizing solar radiation and harvesting of waste heat. For each application, an experimental prototype was constructed and tested to determine the effectiveness in recharging a discharged nickel–metal hydride battery. The results showed that the TEG does produce significantly more power than a piezoelectric device and that the charge time needed to recharge a battery is significantly lower. This study is presented in more detail in **Chapter 77**.

The TEG is a mature technology and a reliable energy converter with no moving parts compared to vibration-based harvesters. The TEG has been actively studied for the last three decades and the literature in this area is extensive. One of the drawbacks of this technology is low efficiency (<5%) if there is low temperature gradient present. Further, the fabrication cost is high, and the volume and weight are still too large for microscale sensing systems. Therefore, with the recent advances made in nanotechnologies, the fabrication of MEMS-scale TEG devices has been actively studied [28–30]. For instance, $0.5\,cm^2$ of a new thermoelectric thin film developed by Applied Digital Solution produces $1.5\,\mu W$ of power with only a $5\,°C$ temperature gradient [31].

2.3 Current and future efforts in energy harvesting

Although extensive research work has been focused on energy harvesting, the amount of harvested energy appears to fall significantly behind that required by SHM sensing systems. The power requirement (not counting telemetry) for active-sensing SHM sensor nodes ranges in the order of tens to hundreds of milliwatts. Therefore, the major limitations facing researchers in the field of power harvesting revolve around the fact that the power generated by piezoelectric materials is far too small to power SHM devices. Therefore, methods of increasing the amount of energy generated by the power harvesting device or developing new and innovative methods of accumulating the energy are the key technologies that will allow energy harvesting to become a practical source of power for wireless SHM systems. Innovations in power storage must be developed before power harvesting technology will see widespread use. The energy harvesting materials have typically been used to determine the extent of power capable of being generated rather than investigating applications and uses of the harvested energy. However, some limited research has focused on field applications of energy harvesters in SHM systems, one example being the intelligent repair application discussed in **Chapter 76**. The practical applications for energy harvesting systems, such as wireless self-powered SHM sensing networks, must be clearly identified with emphasis on power management issues. Application-specific, design-oriented approaches are needed to help the practical use of this technology. Finally, the long-term reliability of energy harvesting devices under the field operating condition should be extensively studied and validated before the full-scale deployment can take place.

3 RADIO FREQUENCY WIRELESS ENERGY TRANSMISSION

Another potential solution for powering SHM sensor nodes and networks is the use of wireless energy transmission. This approach relies on the use of electromagnetic radiation to charge a capacitor or battery that is embedded within or near a sensor network. At short ranges (∼cm) this method is highly efficient owing to inductive or capacitive coupling through magnetic or electric fields [32]. At longer ranges, the efficiency decreases as the electromagnetic radiation is transmitted in the form of visible or near-infrared light, or microwave radiation in the gigahertz range [33, 34]. While the efficiency of short-range coupling mechanisms is ideal, the close proximity that is necessary between source and receiver makes this option difficult to implement in many SHM applications. At longer length scales, the use of visible or near-infrared light provides an effective method for transmitting energy from the source to the target because of the ability to focus light. Unfortunately, the inefficiencies in current photovoltaic cells are detrimental as the conversion capabilities are on the order of 12%. Some experimental triple-junction cells have been developed with efficiencies of 30–40%; however, they require a highly concentrated light source to operate effectively [35].

Another transmission mechanism that is being considered is microwave radiation. In this case, power is generated elsewhere and transmitted to a sensor node by some form of RF radiation. This concept can utilize two different RF energy sources, ambient or controlled RF sources. Previous studies showed that electronics can be used to efficiently capture the ambient radiation sources and convert them to useful electricity. Harrist [36] attempted to charge a cellular phone battery by capturing ambient 915 MHz RF energy. Although he was not able to fully charge the battery, he observed $4\,\text{mV s}^{-1}$ charging time from a typical cellular phone battery. Although there are several electronics that may derive their required power from ambient RF sources, the amount of captured energy is extremely low, typically in the range of a few microwatts. Therefore, the technology that has received the most attention is the microwave transmission with a controlled or so-called beamed RF sources, and has been significantly improved in the last several decades. A source antenna transmits microwaves across the atmosphere or space to a receiver, which can either be a typical antenna with rectifying circuitry to convert the microwaves to dc power, or a rectenna (rectifying antenna) that integrates the technology to receive and directly convert the microwaves into dc power.

A pair of excellent survey articles was written to discuss the history of microwave power [37, 38]. With the use of rectennas, efficiencies in the 50–80% range of dc-to-dc conversion have been achieved. Significant testing has also been done across long distances and with kilowatts power levels [39]. The study showed the feasibility of the wireless energy delivery systems for actuating large devices, including dc motors and piezoelectric actuators. Briles *et al.* [40] invented an RF wireless energy delivery system for underground gas or oil recovery pipes. The RF energy is generated on the surface, and travels through the conductive pipe acting as an antenna or a waveguide. The sensor module in the bottom of the pipe captures this energy and uses it to power the electrical equipment. With a 100-W transmitted power from the surface, it was estimated that around 48 mW of instant power can be captured after traveling 1.6 km along the pipe.

The fundamental limitation in this approach is the dispersion of microwaves, since they cannot be focused as readily as light. However, this loss from dispersion during transmission is made up for in the increased efficiency with which this RF energy can be converted into electrical energy. Relative to the photovoltaic cells, the conversion of RF energy is more efficient than the highest performing photovoltaics. To further improve the efficiency of the microwave transmission, larger arrays of receiving antennas can be assembled to harvest more of the transmitted energy. Current research efforts in RF wireless energy transmission focus on improving the conversion efficiency and attempt to maximize the output power by designing efficient antennas and rectennas. In particular, circular polarized antennas are being implemented in the rectenna design because it avoids the directionality of other antenna designs [41–43]. An array of rectennas is increasingly used to improve the output power [44] and several new rectenna design schemes are proposed [45, 46]. Different elements are also used for efficient rectification [47, 48] in an attempt to obtain optimum output power, and these research trends are similar to those typically pursued in the energy harvesting arena.

Originally considered for alleviating the wiring harness in space structures or microaerial vehicles or providing an extremely low power for those typically used in radio frequency identification (RFID) tags in the $1-100\,\mu\text{W}$ range, the application of an RF wireless energy transmission system for powering electronics typically used in distributed sensing networks has not been studied substantially in the past. In particular, the application of this technology for SHM sensor nodes in order to alleviate the challenges associated with power supply issues has never been addressed in the literature. Therefore, recently, a new and efficient SHM sensing network is proposed, whereby the electric power and interrogation commands are wirelessly provided by a mobile agent [49–51]. This approach involves using an unmanned mobile host node to generate an RF signal near sensors that have been embedded on the structure. The sensors measure the desired response at critical areas on the structure and transmit the signal back to the mobile host again via the RF communications. This "wireless" communication capability draws power from the RF energy transmitted between the host and sensor node and uses it to both power the sensing circuit and transmit the signal back to the host, which is schematically described in Figure 3. This research takes traditional sensing networks to the next level, as the mobile hosts (such as UAV) will fly to known critical infrastructure based upon a GPS locator, deliver required power, and then begin to perform an inspection without human intervention. The mobile hosts will search for the sensors on the structure and gather critical data needed to perform the structural health evaluation. This integrated technology will be directly applicable to rapid structural condition assessment of buildings and bridges after an earthquake. Also, this technology may be adapted and applied to damage detection in a variety of other civilian and defense-related structures such as pipelines, naval vessels, hazardous waste disposal

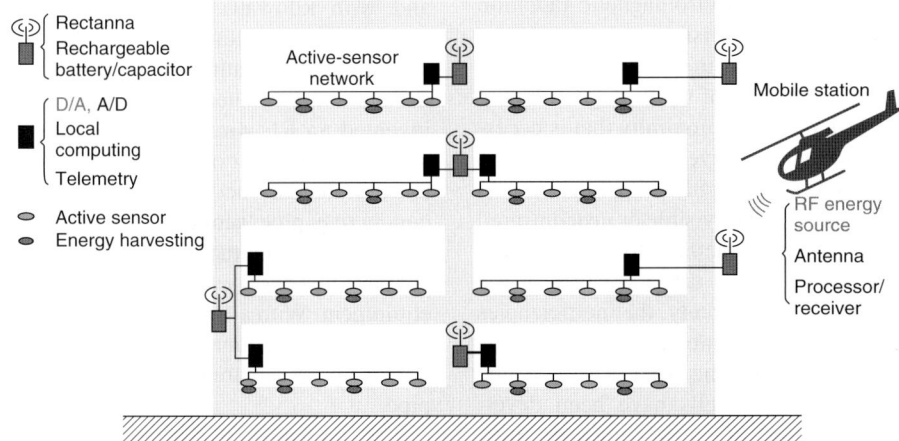

Figure 3. A new sensing network that includes wireless energy transmission and energy harvesting and is interrogated by an unmanned robotic vehicle.

containers, and commercial aircraft. It should be emphasized that this technology can be hybrid in that the sensor node is still equipped with energy harvesting devices and the mobile host would provide additional energy if the energy harvesting device is not able to provide enough power to operate the sensor nodes. Even if the energy harvesting device provides sufficient power, the mobile agent can wirelessly trigger the sensor nodes, collect the information, and/or provide computational resources, which significantly relax the power and computation demand at the sensor node level.

3.1 Patch rectenna design for SHM sensor node

There are two principal components in any wireless energy transmission system: the transmitting antenna and the receiving antenna. The transmitter is driven by an energy source such as a microwave generator and power amplifier, whereas the receiver is coupled with a rectifying circuit that converts the RF energy into usable electrical energy. A supercapacitor can serve as a storage device to collect the received energy and power the associated sensor node.

As stated previously, limitations in microwave transmission are associated with the dispersion of the wave as it travels through space. This dispersion can significantly reduce the amount of power received by the rectenna. This loss is seen to be a function of the square of the distance from the transmitting antenna, as seen in the Friis transmission equation:

$$P_R = \frac{G_T G_R \lambda^2}{(4\pi R)^2} P_T \quad (1)$$

where G_T and G_R are the transmitting and receiving gains, λ is the microwave wavelength in meters, R is the distance between antennas in meters, and P_T and P_R are the transmitted and received power in milliwatts. For a single microstrip patch antenna, such as the one shown in Figure 4, the power efficiency is calculated to be approximately 1.2% at 1 m spacing [52]. While this level of efficiency is relatively low, it can be increased considerably by assembling a larger array of these microstrip patch antennas as shown in Figure 4. This configuration greatly enhances performance as the array is capable of harvesting more of the incident wave from the source antenna. In this configuration, laboratory tests have shown that the antenna array is capable of charging a 0.1 F capacitor to 3.7 V in 28 s when located at 1.5 m from a source antenna that is emitting 900 mW of microwave power at 2.4 GHz.

3.2 SHM sensor node design for wireless energy transmission

This discussion of energy harvesting and wireless energy transmission is targeted toward one sensor

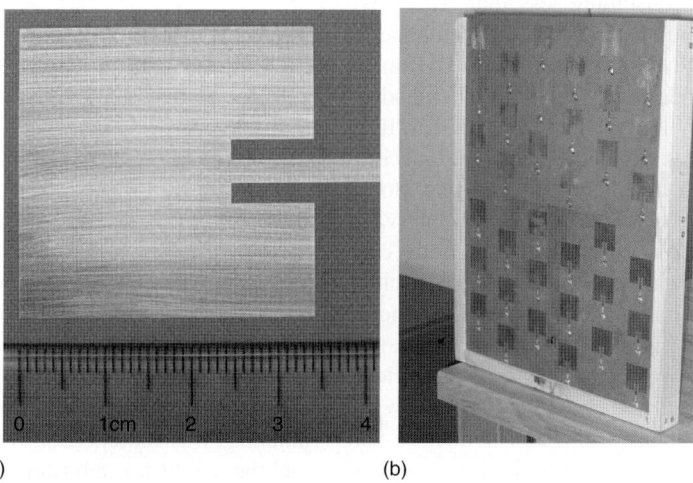

Figure 4. (a) A single microstrip patch antenna used to receive microwave energy and (b) a rectenna array of 36 of these patch antennas.

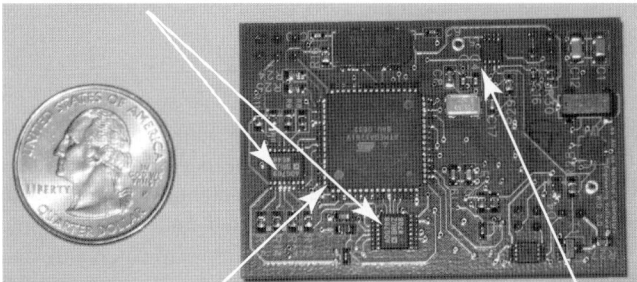

Figure 5. Wireless impedance device (WID2) developed by Los Alamos National Laboratory to measure, process, and transmit electrical impedance data for up to seven piezoelectric sensors.

node that has been developed by researchers at Los Alamos National Laboratory [53]. This system is designed as an active-sensor node that can interrogate the electrical impedance of seven piezoelectric patches, which utilizes impedance-based SHM [54], and is referred to as the *wireless impedance device* (WID2). The principal components of the design are shown in Figure 5, and include an ATmega1281V microcontroller, an AD5933 impedance chip with ADG708 multiplexers, an ATAK5278 wake-up chip, and an AT86RF230 telemetry module.

This sensor node is designed to provide onboard computing and data storage for a variety of SHM schemes. The standard operating condition for the WID2 is the idle mode in which the sensor node operates below 0.65 mA, consuming less than 1.82 mW of power. Proper configuration of the sleep modes should allow us to reduce current draw to 0.01 mA, providing a sensor node that is capable of extended operation at extremely low-power levels. The WID2 can be brought out of this idle mode at specific time intervals through an integrated real-time clock that is capable of waking the WID2 on intervals of 1 s to 1 year. Additionally, the sensor node can be activated through a low-frequency wakeup chip that monitors an integrated inductor for a magnetically coupled wakeup signal. Once the sensor node is activated, the microcontroller measures electrical impedance of each attached piezoelectric sensor, evaluates the status of the sensor, and either stores or transmits the result to a mobile host, which subsequently transmits the data back to the base station. Current and power consumption of the

Table 1. Current and power consumption for WID2 in different operational states

State	Current (mA)	Power (mW)
Measure	20	56
Transmit	22	61.6
Idle	0.65	1.82

WID2 are outlined in Table 1 for each operational state.

The measure and transmit operations take a combined time of 10 s to complete. When considering the power consumption, the power requirement of WID2 is within the range of the capabilities of several energy harvesting or wireless energy transmission methods. Under laboratory testing, it was found that a 0.1 F supercapacitor provides sufficient energy for more than 10 s of operation, when it is charged to 3.7 V, which combines energy harvesting and transmission techniques to provide decades of sustainable operation.

3.3 Field testing

In August 2007, this method was field tested on the Alamosa Canyon Bridge near Truth or Consequences, New Mexico. Experiments were designed to test the effectiveness of this wireless energy transmission system when mounted on a mobile host. The source antenna, was mounted on the side of an automobile and the height was adjusted to match the height of the receiving antenna array. The microwave source and power amplifier were powered by a power inverter

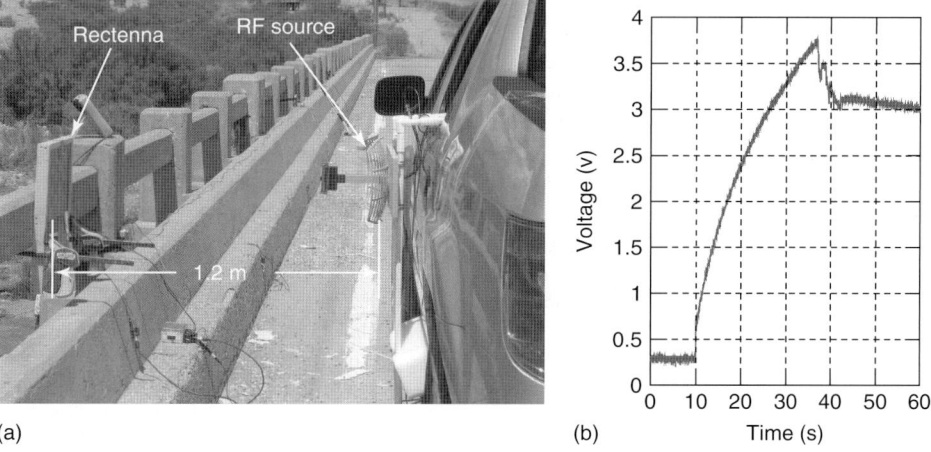

Figure 6. (a) Wireless energy delivery system field tested on Alamosa Canyon Bridge, NM. (b) The RF source was configured to emit 1 W of energy at 2.4 GHz, charging a 0.1-F capacitor to 3.7 V in 27 s.

that was plugged into the automobile's 12 V dc power supply. The receiving antenna was composed of 18 microstrip patch antennas clamped to the concrete side rail of the bridge as seen in Figure 6. This rectenna was used to charge a 0.1-F (P/N: PB-5R0V104) supercapacitor, which was connected to a WID2.0 sensor node and used to monitor the preload in two bolts on the bridge through piezoelectric sensors.

For this test, the automobile was stopped in parallel with the rectenna, as shown in Figure 6. The source antenna was activated and the voltage on the 0.1-F capacitor was visually monitored to prevent the capacitor from charging above 4 V (the WID2.0 operates on a nominal voltage between 2.7 and 3.3 V). During these experiments, the spacing between the transmitting and receiving antennas was maintained at 1.2 m, and the resulting charge time was measured to be 27 s to obtain a voltage of 3.7 V within the supercapacitor. Once charged, the sensor node was used to take measurements from two PZT active sensors, perform the local computing to determine bolt tightness, and then transmit 8 bytes of data back to the mobile host.

4 CONCLUSIONS

Energy harvesting is slowly coming into full view of the SHM and the more general sensing network communities. With continual advances in wireless sensor/actuator technology, improved signal-processing technique, and the continued development of power-efficient electronics, energy harvesting will continue to attract the attention of researchers and field engineers. However, it should be emphasized that energy harvesting still remains in its infancy and only a few successful examples are in practice. Still, a tremendous research effort is required to convert, optimize, and accumulate the necessary amount of energy to power such electronics.

One potential alternative that has shown successful performance in remotely powering sensor nodes is the use of wireless energy transmission. This technique builds upon an already existing technology that has been redirected to focus on low-power SHM systems. Laboratory and field tests have demonstrated that this method is highly effective in powering nodes that are designed to perform discrete measurements of a structure's health. Further development is needed to determine the optimal configuration of the receiving array and to couple this technique with charging electrochemical batteries to provide testing capabilities between charging cycles.

REFERENCES

[1] Fry D, Holcomb D, Munro J, Oakes L, Maston M. *Compact Portable Electric Power Sources*, ORNL/TM-13360. Oak Ridge National Laboratory Report, 1997.

[2] Roundy SJ. *Energy Scavenging for Wireless Sensor Nodes with a Focus on Vibration to Electricity Conversion*, Ph.D. Dissertation. Department of Mechanical Engineering, University of California: Berkeley, CA, 2003.

[3] Paradiso JA, Starner T. Energy scavenging for mobile and wireless electronics. *IEEE Pervasive Computing* 2005 **4**:18–27.

[4] Park G, Farrar CR, Todd MD, Hodgkiss W, Rosing T. *Energy Harvesting for Structural Health Monitoring Sensor Networks*, LA-14314-MS. Los Alamos National Laboratory Report, 2007.

[5] Glynne-Jones P, White N. Self-powered systems: a review of energy sources. *Sensor Review* 2001 **21**:91–97.

[6] Qiwai M, Thomas J, Kellogg J, Baucom J. Energy harvesting concepts for small electric unmanned systems. *Proceedings of the SPIE* 2004 **5387**:84–95.

[7] Mateu L, Moll F. Review of energy harvesting techniques and applications for microelectronics. *Proceedings of the SPIE* 2005 **5837**:359–373.

[8] Sodano H, Inman D, Park G. A review of power harvesting from vibration using piezoelectric materials. *The Shock and Vibration Digest* 2004 **35**:451–463.

[9] DuToit NE, Wardle BL, Kim SG. Design considerations for MEMS-scale piezoelectric mechanical vibration energy harvesters. *Integrated Ferroelectrics* 2005 **71**:121–160.

[10] Anton S, Sodano H. A review of power harvesting using piezoelectric materials (2003–2006). *Smart Materials and Structures* 2007 **16**:R1–R21.

[11] Niezrecki C, Brei D, Balakrishnam S, Moskalik A. Piezoelectric actuation technology: state of the art. *The Shock and Vibration Digest* 2001 **33**:269–280.

[12] Inman DJ, Cudney HH. *Structural and Machine Design Using Piezoceramic Materials: A Guide for Structural Design Engineers*, Final Report NASA Langley Grant NAG-1-1998, 2000.

[13] Sirohi J, Chopra I. Fundamental understanding of piezoelectric strain sensors. *Journal of Intelligent Material Systems and Structures* 2000 **11**:246–257.

[14] Goldfarb M, Jones LD. On the efficiency of electric power generation with piezoelectric ceramic. *ASME Journal of Dynamic Systems, Measurement, and Control* 1999 **121**:566–571.

[15] Clark W, Ramsay MJ. Smart material transducers as power sources for MEMS devices. *International Symposium on Smart Structures and Microsystems*. Hong Kong, 2000.

[16] Lesieutre GA, Hofmann HF, Ottman GK. Electric power generation from piezoelectric materials. *The 13th International Conference on Adaptive Structures and Technologies*. Potsdam, Berlin, 7–9 October 2002.

[17] Kasyap A, Lim J, Johnson D, Horowitz S, Nishida T, Ngo K, Sheplak M, Cattafesta L. Energy reclamation from a vibrating piezoceramic composite beam. *Proceedings of 9th International Congress on Sound and Vibration*, Paper No. 271. Orlando, FL, 2002.

[18] Han J, von Jouanne A, Le T, Mayaram K, Fiez TS. Novel power conditioning circuits for piezoelectric micro power generators. *IEEE Applied Power Electronics Conference and Exposition*, Anaheim, CA, USA, 2004; Vol. 3, pp. 1541–1546.

[19] Sodano HA, Inman DJ, Park G. Comparison of piezoelectric energy harvesting devices for recharging batteries. *Journal of Intelligent Material Systems and Structures* 2005 **16**:799–807.

[20] Cornwell PJ, Goethal J, Kowko J, Damianakis M. Enhancing power harvesting using a tuned auxiliary structure. *Journal of Intelligent Material Systems and Structures* 2005 **16**:825–834.

[21] Sodano HA, Inman DJ, Park G. Generation and storage of electricity from power harvesting devices. *Journal of Intelligent Material Systems and Structures* 2005 **16**:67–75.

[22] Elvin NG, Elvin AA, Spector M. A self-powered mechanical strain energy sensor. *Smart Materials and Structures* 2001 **10**:293–299.

[23] Inman DJ, Grisso BL. Towards autonomous sensing. *Proceedings of the SPIE* 2006 **6174**:T1740–T1749.

[24] Microstrain. http://www.microstrain.com, 2008.

[25] Lawrence EE, Snyder GJ. A study of heat sink performance in air and soil for use in a thermoelectric energy harvesting device. *Proceedings of the 21st International Conference on Thermoelectronics*. Portland, OR, 2002; pp. 446–449.

[26] Rowe MD, Min G, Williams SG, Aoune A, Matsuura K, Kuznetsov VL, Fu LW. Thermoelectric recovery of waste heat—case studies. *Proceedings of the 32nd Intersociety Energy Conversion Engineering Conference*. Honolulu, HI, July 27–August 1 1997; pp. 1075–1079.

[27] Sodano HA, Dereux R, Simmers GE, Inman DJ. Power harvesting using thermal gradients for recharging batteries. *Proceedings of the 15th International Conference on Adaptive Structures and Technologies*. Bar Harbor, ME, 25–27 October 2004.

[28] Bottner H. Thermoelectric micro devices: current state, recent developments and future aspects for technological progress and applications. *Proceedings of 21st International Conference on Thermoelectric*, La Grande Motte, France, 2003; pp. 511–518.

[29] Snyder GJ, Lim JR, Huang CK, Fleurial JP. Thermoelectric microdevice fabricated by a MEMS-like electrochemical process. *Nature Materials* 2003 **2**:528–531.

[30] Jovanovic V, Ghamaty S. Design, fabrication and testing of energy-harvesting thermoelectric generators. *Proceedings of the SPIE* 2006 **6173**:G–1–G–8.

[31] http://www.asdx.com, 2006.

[32] Jang J, Liu JF, Yue CP, Sohn H. Development of self-contained sensor skin for highway bridge monitoring. *Smart Structures and Materials, Proceedings of the SPIE*, San Diego, CA, USA, 2006; Vol. 6174, pp. 1291–1300.

[33] Blackwell T. Recent demonstrations of laser power beaming at DFRC and MSFC. *BEAMED ENERGY PROPULTION: Third International Symposium on Beamed Energy Propultion. AIP Conference Proceedings*. American Institute of Physics, 2005; Vol. 766, pp. 73–85.

[34] Choi H, Song K, Golembiewskii W, Chu S, King G. Microwave power of smart material actuators. *Smart Materials and Structures* 2004 **13**:38–48.

[35] Green M, Zhao J, Wang A, Wenham S. Progress and outlook for high-efficiency crystalline silicone solar cells. *Solar Energy Materials and Solar Cells* 2001 **65**:9–16.

[36] Harrist DW. *Wireless Battery Charging System Using Radio Frequency Energy Harvesting*, M.S. thesis. Department of Electrical Engineering, University of Pittsburgh: Pittsburgh, PA, 2004.

[37] Brown WC. The history of wireless power transmission. *Solar Energy* 1996 **56**:3–21.

[38] Maryniak GE. Status of international experimentation in wireless power transmission. *Solar Energy* 1996 **56**:87–91.

[39] Choi S, Song K, Golembiewskii W, Chu SH, King G. Microwave powers for smart material actuators. *Smart Materials and Structures* 2004 **13**:38–48.

[40] Briles SD, Neagley DL, Coates DM, Freund SM. *Remote Down-Hole Well Telemetry*, US Patent # 6,766,141, 2004.

[41] Strassner B, Chang K. 5.8 GHz circularly polarized dual-rhombic-loop traveling-wave rectifying antenna for low power-density wireless power transmission applications. *IEEE Transactions on Microwave Theory and Techniques* 2003 **51**:1548–1553.

[42] Ali M, Yang G, Dougal R. A new circularly polarized rectenna for wireless power transmission and data communication. *IEEE Antennas Wireless Propagation Letters* 2005 **4**:205–208.

[43] Ren YJ, Chang K. 5.8 GHz circularly polarized dual-diode rectenna and rectenna array for microwave power transmission. *IEEE Transactions on Microwave Theory and Technique* 2006 **54**:1495–1502.

[44] Kim J, Yang SY, Song DD, Jones S, Choi SH. Performance characterization of flexible dipole rectennas for smart actuator use. *Smart Materials and Structures* 2006 **15**:809–815.

[45] Park JY, Han SM, Itoh T. A rectenna design with harmonic-rejecting circular-sector antenna. *IEEE Antennas and Wireless Propagation Letters* 2004 **3**:52–54.

[46] Chin CH, Xue Q, Chan CH. Design of a 5.8 GHz rectenna incorporating a new patch antenna. *IEEE Antennas and Wireless Propagation Letters* 2005 **4**:175–178.

[47] Epp LW, Khan AR, Smith HK, Smith RP. A compact dual-polarized 8.51 GHz rectenna for high-voltage actuator applications. *IEEE Transactions on Microwave Theory and Technique* 2000 **48**:111–119.

[48] Zbitou J, Latrach M, Toutain S. Hybrid rectenna and monolithic integrated zero-bias microwave rectifier. *IEEE Transactions on Microwave Theory and Technique* 2006 **54**:147–152.

[49] Farrar CR, Park G, Allen DW, Todd MD. Sensor network paradigms for structural health monitoring. *Structural Control and Health Monitoring* 2006 **13**(1):210–225.

[50] Todd MD, *et al.* A different approach to sensor networking for SHM: remote powering and interrogation with unmanned aerial vehicles. *Proceedings of 6th International Workshop on Structural Health Monitoring*. Stanford, CA, 11–13 September 2007.

[51] Park G, Overly TG, Nathnagel M, Farrar CR, Mascarenas DL, Todd MD. A wireless active-sensor node for impedance-based structural health monitoring. *Proceedings of US-Korea Smart Structures Technology for Steel Structures*. Seoul, 16–18 November 2006.

[52] Nothnagel M, Park G, Farrar CR. Wireless energy transmission for structural health monitoring embedded sensor nodes. *Proceedings of the SPIE*, San Diego, CA, USA, 2007; Vol. 6532, pp. 653216.

[53] Overly T, Park G, Farrar CR. Development of impedance-based wireless active-sensor node for structural health monitoring. *The 6th International Workshop on Structural Health Monitoring*, Stanford, CA, USA, 2007.

[54] Park G, Sohn H, Farrar CR, Inman DJ. Overview of piezoelectric impedance-based health monitoring and path forward. *The Shock and Vibration Digest* 2003 **35**(6):451–463.

Chapter 76
On the Way to Autonomy: the Wireless-interrogated and Self-powered "Smart Patch" System

Stephen C. Galea, Stephen Van der Velden, Scott Moss and Ian Powlesland

Air Vehicles Division, Defence Science and Technology Organisation (DSTO), Melbourne, VIC, Australia

1 Introduction	1329
2 Background	1331
3 Detailed Design	1333
4 Safety-of-flight and Functional Testing	1344
5 System Implementation and Flight Test Results	1345
6 Discussions—Lessons Learned	1346
7 Conclusions	1348
Acknowledgments	1348
References	1348

1 INTRODUCTION

The application of bonded composite patches or doublers to repair or reinforce defective (secondary) metallic structures is becoming recognized as an effective and versatile repair procedure for many types of problems [1]. However, the application of bonded composite repairs to cracked aircraft primary structure is generally acceptable only on the basis that a margin on design limit-load (DLL) capability is retained in the event of loss (total absence) of the repair [2]. However, assuming that the static requirements and quality-assurance processes are satisfied, one approach to certify full credit for the patch in slowing crack growth could be justified by a *continuous safety-by-inspection approach.* This approach is based on the continuous self-assessment of the patch system integrity using a *smart patch* approach [2, 3], by incorporating *in situ* sensors to continuously monitor the structural condition of the patch system and associated remaining damage in the parent structure. The need to follow approved patch design, fabrication, and quality-assurance procedures is unchanged; this approach simply allows a relaxation of the probability of failure requirements, particularly in relation to environmental degradation. However, the viability of any *smart patch* or *in situ* structural health monitoring (SHM) approach now depends on establishing its reliability or probability of damage detection, which is similar to

Encyclopedia of Structural Health Monitoring. Edited by Christian Boller, Fu-Kuo Chang and Yozo Fujino © 2009 Commonwealth of Australia. ISBN: 978-0-470-05822-0.

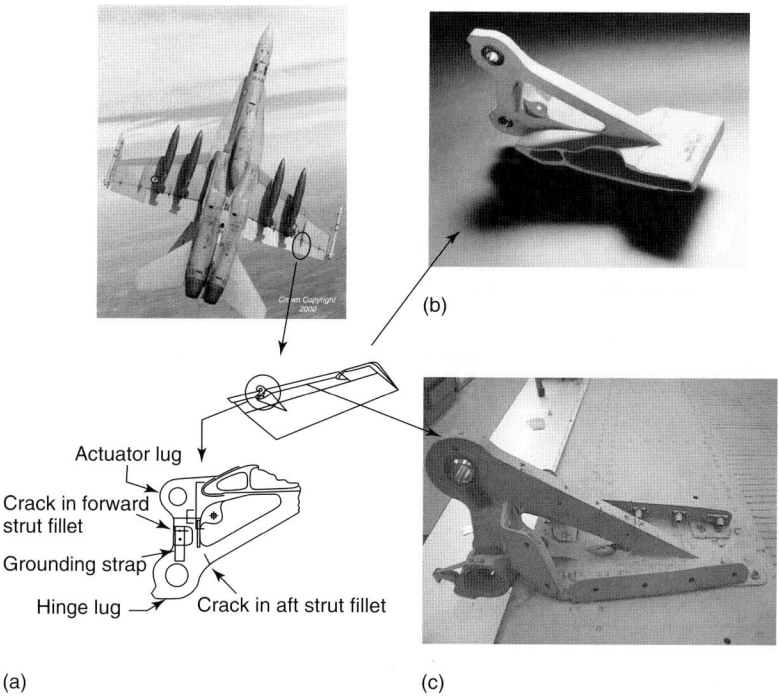

Figure 1. (a) Schematic of the aileron, hinge lug, and hinge aft strut with likely crack locations (taken from [4]). Photograph of F/A-18 (b) aluminum, and (c) titanium aileron hinges.

the problem of probability of detection in Non-Destructive Inspection (NDI), and should include system self-checking and redundancy to provide the required level of confidence.

To demonstrate and evaluate the feasibility of the *smart patch* concept, the Australian Defence Science and Technology Organisation (DSTO) has developed two SHM systems to interrogate the structural health of a boron/epoxy (b/ep) doubler (or reinforcement) on an F/A-18 inboard aileron hinge, as shown in Figure 1(a). One system, the battery-powered *smart patch* (BPSP), uses a lithium ion-based battery as the power source and measures load transfer from the structure to the patch, using conventional strain gauges, to monitor patch health. The patch health data is uploaded by the operator via an infrared (IR) link. The other concept, the self-powered (wireless) *smart patch* (SPSP), involves more technical risk and consists of a piezoelement-based self-powered sensing system powered by an array of piezotransducers, which convert structural dynamic strain to electrical energy, and monitors damage in the patch via piezoelectric film strain sensors. In this system, the patch health data is uploaded by the operator using a magnetic transceiver. The instrumented aileron was installed on a Royal Australian Air Force (RAAF), Aircraft Operation Support Group (AOSG) F/A-18 in early 2006 and was flown for 12 months.

This article describes some of the critical issues associated with the development, evaluation and implementation of the self-powered *smart patch* system, including system design, functionality and certification testing, and installation. The article gives a brief overview of the various available power-harvesting techniques and discusses several key issues when designing and developing the self-powered *smart patch* system, viz., (i) demand—power requirements, (ii) supply—energy generation from vibration or strain-based sources, and (iii) conversion—issues and efficiencies associated with energy conversion from mechanical to electrical energies. Flight data from the SHM system and lessons learned during the program are also presented.

2 BACKGROUND

2.1 General system specifications

The initial idea was to evaluate the *smart patch* concept on a bonded composite patch/reinforcement designed for an F/A-18 aluminum aileron hinge with a propensity to cracking [5] (Figure 1a). During the initial system design phase, the RAAF decided to refurbish the ailerons and, in so doing, replaced the aluminum hinges (Figure 1b) with redesigned titanium hinges (Figure 1c), thus eliminating the cracking problem. However, in consultation with senior RAAF engineering managers, it was decided to proceed with the *smart patch* demonstrator on a titanium aileron hinge.

The first phase of the project was to establish the operating envelope of the system, including temperature, strain, and vibration at the aileron hinge and, flying time and elapsed time between system installation and removal from the aircraft. All these issues are critical when designing and implementing such a system. These general specifications are listed below [6]:

- The expected operational temperature range was -40 to $+70\,^\circ$C.
- No strain time history data was available for the titanium aileron hinge; however, flight loads and design data for the aluminum aileron hinge indicated that the operational peak strains were of the order of $1500\,\mu\varepsilon$ at an excitation frequency of between 8 and 42 Hz.
- Flight times were expected to vary between 40 and 60 min per sortie.
- The system was expected to be operational for an elapsed time of one year, which would entail between 100 and 200 sorties.

2.2 Concept development and high-level system design

In general, the *smart patch* concept needs to continually inspect regions of the reinforcement likely to suffer disbond damage [2]. There are several methods of achieving this goal and providing an indication of structural health. One method involves subjecting the structure to a known excitation and then measuring a response at the "interrogation" time (*see* **Chapter 107**; **Chapter 111**). An alternative would be to use some form of "in-service" excitation (*see* **Chapter 111**). In this case, in-flight loading is used as the excitation mechanism and in-flight strain response as the damage indicator. A finite element (FE) analysis of the hinge is undertaken to determine the most likely regions of damage and to assist in developing the most appropriate damage indicator (see Section 3.1) [6].

An autonomous remotely accessed patch health monitoring (*smart patch*) system was the final objective. As discussed previously, the full *smart patch* demonstrator consisted of two, developed concurrent, wireless SHM systems, namely, the BPSP and the SPSP. However, even though the BPSP system is a lower risk solution it does suffer from the limitation of a finite power supply and would require occasional maintenance to replace the battery. The SPSP offers significant advantages over the BPSP since it is maintenance-free and therefore facilitates rapid acceptance by operators, maintainers, and certification authorities and would incur minimum additional through-life-support costs. This latter approach would enable minimum interruption to relatively rigid and well-established logistic and maintenance procedures associated with aging aircraft. In addition to monitoring the health of bonded patches, this approach would facilitate SHM of structural "hot spots" in "difficult-to-access" locations. Therefore, to encourage acceptance by operators and maintainers, a fully autonomous system design was chosen with no batteries. This self-powered system had the following major hardware components; (i) power-harvesting elements, (ii) electronics to manage power harvesting, interrogate the sensors, calculate/store patch health, and allow data download, (iii) sensing elements, (iv) system wireless interrogation, and (v) handheld data gathering unit. Issues associated with the selection and design of these components are discussed in Section 3.

2.2.1 Energy harvesting

DSTO is exploring structural power-harvesting techniques for use on defence platforms [7]. The main driver for this research is the development of powering systems for autonomous distributed sensor networks (DSN) applied to airframes for SHM. Using "smart" sensor concepts, damage and damage growth in the airframe, and other structural life-related problems, would be continuously monitored on board the aircraft to provide real-time damage assessment. This technology could potentially permit a safe reduction in inspection and regular maintenance costs with substantial impact on the through-life costs. Using a DSN for SHM has only recently become feasible because of advances in micropower electronics and microelectromechanical system (MEMS); however, there are still a number of issues to overcome.

One of the main hurdles to the use of a DSN is powering. It is unsatisfactory to use the conventional approach of running power through copper wires on a central power/data bus because of

1. certification issues;
2. the added mass of wiring;
3. the drain on the limited electrical power available on an aircraft;
4. the wiring occupying significantly more space on the aircraft than the sensors themselves;
5. the fact that the DSN is no longer unobtrusive on the airframe and would interfere with maintenance procedures;
6. installation of the DSN itself would be complicated, time consuming, and expensive;
7. the significant reliability and durability issues with excessive wiring.

An alternative to the central power/databus is to have a ubiquitous network of autonomous sensors with independent power supplies. One option is to use batteries as the primary power source; however, replacing batteries in a DSN would be a considerable maintenance overhead and therefore, alternative approaches for powering such systems are required. Because of the low-power nature of MEMS systems, the concept of self-powering a DSN via harvesting power from the local environment becomes an intriguing possibility. On an airframe, in particular, there can be significant accelerations and dynamic strains available that lend themselves to the power-harvesting concept. Power harvesting has the potential to

- reduce the logistical burden and costs incurred by replacing a number of battery-powered systems;
- increase system availability due to the longevity of a device that is "self-powered" (compared with a battery-powered device); and
- power large numbers of tiny sensors in a network that may be difficult (or impossible) to power any other way.

Power harvesting itself is the process of energy scavenging from the local environment. Depending on the operational environment, various energy sources may be available [8] (*see* **Chapter 75**), for example: vibration, kinetic, solar, thermal (*see* **Chapter 77**), chemical, and biological. On an aircraft, some or all of the first four sources could be available, depending on the location of the sensors. Within DSTO, the focus has been on the use of piezoelectric elements (both ceramic and polymer) for power-harvesting applications [9] using vibrational energy.

There is significant worldwide activity in the area of power harvesting, and a number of reviews of the literature are available [10]. Eggborn [11] discusses optimization of power harvesting from the piezoelectric/cantilever combination. Defense Advanced Research Projects Agency (DARPA) supported a "micro power generation" program that funded a number of energy harvesting investigations, for example, the "piezoelectric eel" project [12] and the "thermoelectric generator" [13]. Perhaps the best-known example of power harvesting is the Massachusetts Institute of Technology (MIT) piezoelectric shoe [14]. In 1996, Starner [15] discussed the possibility of using the everyday actions of the human body as a source of potentially harvestable energy. Starner calculated that one of the best prospects for power harvesting from the human body is walking—an average 68-kg person, walking with an average human gait of two steps per second, generates 67 W of "heel strike" power. Paradiso, at the MIT Media Lab [16], investigated the possibility of parasitic "shoe scavenging" and the result was a piezoelectric shoe-based power-harvesting system [17]. Both PVDF (polyvinylidene fluoride) piezoelectric films and PZT (lead zirconate titanate) piezoelectric ceramics and the associated harvesting electronics were placed in a running shoe. As the shoe wearer walked, the piezoelectric deformed, generating piezoelectric voltage that was then (through the power-harvesting electronics) used to charge a storage capacitor. The storage capacitor was then used to drive a radio frequency (RF) transmitter.

There are a number of other MEMS power harvesting examples available in the literature, for example, El-hami *et al.* [18] describe a micromachined magnet/coil power harvesting device. The so called smart dust being developed at the University of California, Berkeley is a significant MEMS-based project, which involves the development of a "self-powered" device [19]. More recently, Beeby unveiled a MEMS-based energy harvester with practical volume 0.15 cm^3 capable of producing 46 μW in a 4 kΩ load from vibration levels of 0.59 m s^{-2} at the device resonance frequency of 52 Hz [20].

However, for the self-powered system that was implemented on the F/A-18 aileron hinge, time and resource limitations led to the development of a relatively simple power-harvesting system based on the conversion of strain energy to electrical energy (charge) via two different electromechanical coupling devices, PVDF piezoelectric films, and PZT piezoelectric ceramics. The combination provided a degree of robustness in terms of mechanical and thermal durability. Also, since the operational strains in the titanium aileron hinge were not well understood, the combination of PVDF and PZT power harvesting elements ensured that enough electrical power would be generated to power the system. PVDF elements are durable, have an adequate operating temperature range, up to about 80 °C, and are chemically inert. The PZT wafers have better strain/electrical energy conversion efficiency than PVDF and have a better operating temperature range, up to ∼180 °C, but are not as mechanically durable as PVDF (i.e., low strain to failure of 0.1–0.2% strain).

2.2.2 *System operation and information flow*

The basic operation and information flow issues can be summarized as follows:

- When in flight, a combination of piezoelectric film (chosen for its durability) and piezoelectric ceramic (chosen for its higher output) elements are used to power up the system.

- A patch health measurement is taken by the piezoelectric film sensors when power is available and the temperature and strain are within a selected range. As damage growth is slow, the relatively sparse sampling is considered adequate.
- The reading is used to refine the current health estimate stored in a nonvolatile memory.
- A handheld interrogator, placed in close proximity to the hinge, powers the circuitry of the self-powered system via inductive coupling and allows readings to be downloaded via the same inductive link. Inductive coupling was chosen as it is able to carry power and information through a thin barrier (in this case, the plastic aileron hinge cowl, although transmission through thin aluminum is also possible).

3 DETAILED DESIGN

3.1 Damage detection scheme

The FE model of the hinge with a b/ep reinforcement, which is shown in Figure 2(a), was used to assess the proposed damage detection techniques [21]. Two load cases were considered in the FE analysis, viz., load case WO39 a tensile design ultimate load condition and WO42 a compressive design ultimate load condition [5]. Figure 2(b) shows the peel and shear stress distribution in the adhesive of the reinforcement, respectively, for load case WO39, and indicates that two critical regions exist, namely, the tapered region at the end of the patch on the hinge strut ($x \sim 200$ mm) and the region of high adhesive peel stresses in the concave portion of the reinforcement ($x \sim 70$ mm). The BPSP monitored damage in the former and the SPSP monitored damage in the latter. In both cases, damage was detected by monitoring the change in ratio of critical region strains to far-field strains (referred to here as the *patch health indicator*) [22]. The discussion below focuses on the analysis associated with the damage detection approach for the SPSP system.

In this case, the damage was simulated by a 10 mm-long disbond in the adhesive layer at the high peel stress region (at $x \sim 70$ mm) of the patch. The surface longitudinal strains (ε_x) plotted in Figure 2(c) for the tensile load case, WO39, show significant increases in surface strains when the damage is introduced,

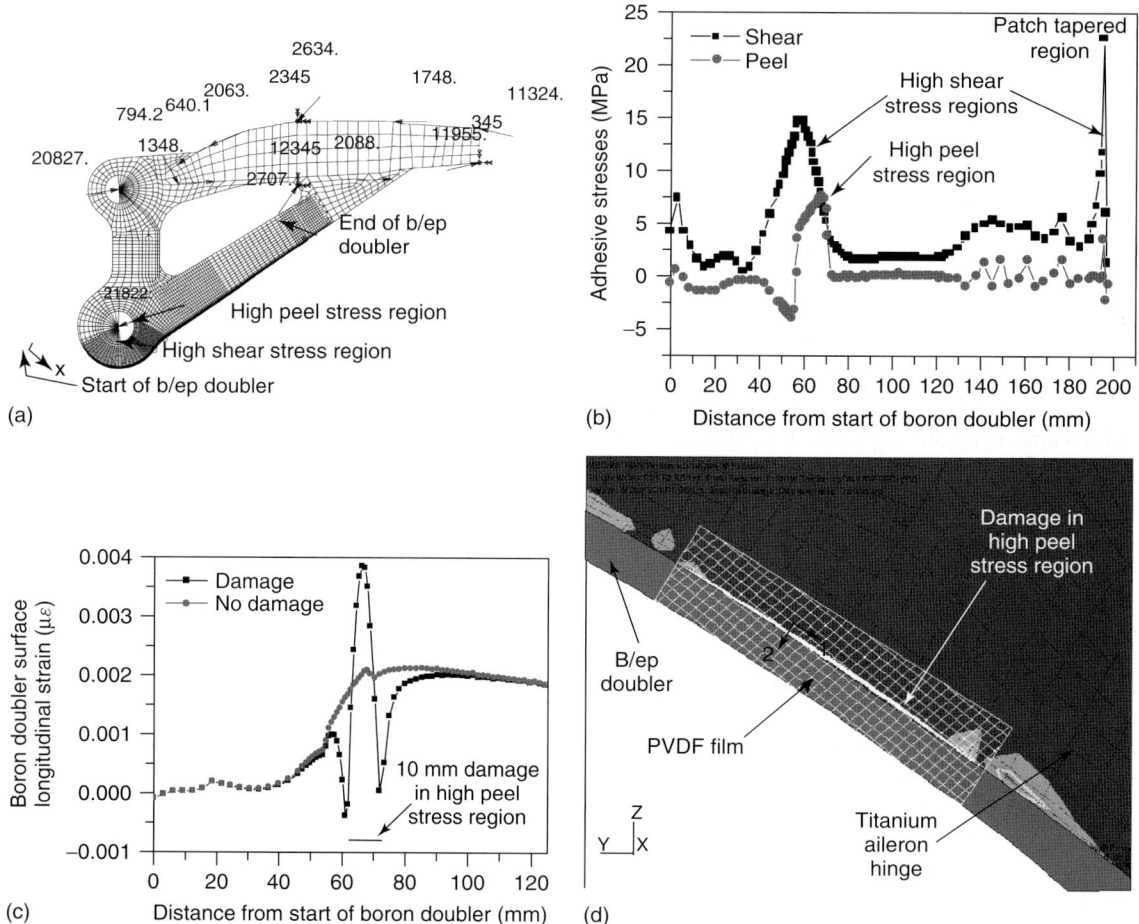

Figure 2. (a) Finite element model of titanium aileron hinge with b/ep doubler. (b) Predicted adhesive stresses in the reinforcement system with no damage present (load case WO39). (c) Longitudinal surface strain profile on b/ep doubler due to a 10 mm disbond in the adhesive high peel stress region. (d) FE displacement plot and damage detection scheme for damage in the high peel stress region.

indicating that damage detection is possible using surface strain sensors. However, the distinctive sharp peak/trough in the surface strains (over a 10 mm length) means that the probability of detection is very sensitive to the sensor size and proximity to the damage site. Thus sensors will only detect the damage if the sensing length is small, compared to the strain gradient length, and if the sensors are positioned over or very close to the damage or when the damage reaches a substantial size. Also, it is possible that the damage will initiate on one side of the strut before growing, through the width, to the other side of the strut. Therefore, to have a reasonable chance of detecting damage, one approach would be to use an array of short gauge length strain sensors on both sides of the b/ep reinforcement. However, this would require several dual-channel devices or one multichannel device to monitor a reasonable length of about 10–20 mm. A more simplistic (in terms of the electronics and number of sensors required) approach is to make use of the disbond opening under load (Figure 2d).

In this application, piezoelectric film sensors were proposed for measuring strain, or, more specifically, to measure strain produced in the film bridging the disbond as it opens. Piezoelectric film sensors

Table 1. Voltage outputs for two piezoelectric film sensors

Result case	$\int \varepsilon_x \, dS$ (mm²)	$\int \varepsilon_y \, dS$ (mm²)	$\int \delta u \, dy$ (mm²)	PVDF output (V)	Ratio $V_{\text{dam}}/V_{\text{nodam}}$
No damage case (4 mm × 15 mm sensor covering the anticipated damage)	0.0899	0.0345	0	2.7	2.5
Damage case (4 mm × 15 mm sensor covering the damage)	0.0899	0.0511	0.147	6.8	
No damage case (2 mm × 15 mm sensor covering the anticipated damage)	0.0470	0.0226	0	1.5	3.7
Damage case (2 mm × 15 mm sensor covering the damage)	0.0463	0.0394	0.147	5.5	
No damage case (5 mm of the 4 mm × 15 mm sensor covering the anticipated damage)	0.0836	0.0294	0	2.5	1.6
Damage case (5 mm of the 4 mm × 15 mm sensor covering the damage)	0.0835	0.0360	0.0506	3.9	
No damage case (5 mm of the 2 mm × 15 mm sensor covering the anticipated damage)	0.0449	0.0182	0	1.4	1.9
Damage case (5 mm of the 2 mm × 15 mm sensor covering the damage)	0.0443	0.0176	0.0506	2.7	

generate a charge when strained rather than conventional electrical-resistance foil strain gauges, which *require* electrical power to measure strain. Thus the use of such sensors also reduces the system power requirements. In this case, a piezoelectric film sensor covering the edge of the aileron hinge strut (over the edge of the patch, adhesive layer, and the titanium strut), as shown in Figure 2(d), was used to detect disbond opening. FE studies for two sensor configurations, one 2 mm × 15 mm long and the other 4 mm × 15 mm long, were undertaken [23]. The area integral of the longitudinal (ε_x) and transverse (ε_y) strains of the sensing area is tabulated in Table 1 for various sensing conditions, i.e., no damage, the entire sensor covering the damage, and only half the sensor covering the damage. Table 1 also includes the area integrals of the crack opening ($\int \delta u \, dy$) and the predicted electrical voltage calculated from the overall area integral of strain [24]. The results show that the 2 and 4 mm wide (by 15 mm long) sensors will have a 2–4 and 1.5–2.5 fold increase in response, respectively, due to a 10 mm disbond, when compared with the undamaged case. The piezoelectric film sensor (see Figure 3) was manufactured from 28 μm thick PVDF sheet where the silver ink electrodes were removed using a grit blasting process and a mask was placed over the sheet to protect the sensing regions during grit blasting.

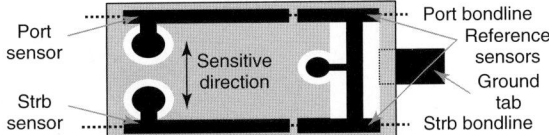

Figure 3. PVDF sensor configuration (hatched area indicates silver ink electrode on top surface and gray is the silver electrode on the lower surface).

3.2 Electronic design and power (demand) requirements

Geometric constraints and limitations in strain level and frequency response limit the amount of power available while using piezoelectric elements to fractions of a milliwatt. Since it is difficult to get a reasonable amount of signal processing done on this kind of power budget, the harvested power was stored in a capacitor. Consequently, a reading was taken only once the following conditions were met: (i) sufficient energy was available for a complete reading/computation/store cycle, (ii) the structure was experiencing significant strain (within a given window), and (iii) the temperature was also within a given window. Budget and time constraints necessitated a fairly large A-shaped printed circuit board (APCB) footprint with surface-mounted commercial off the shelf (COTS) electronic components, shown in Figure 4(a).

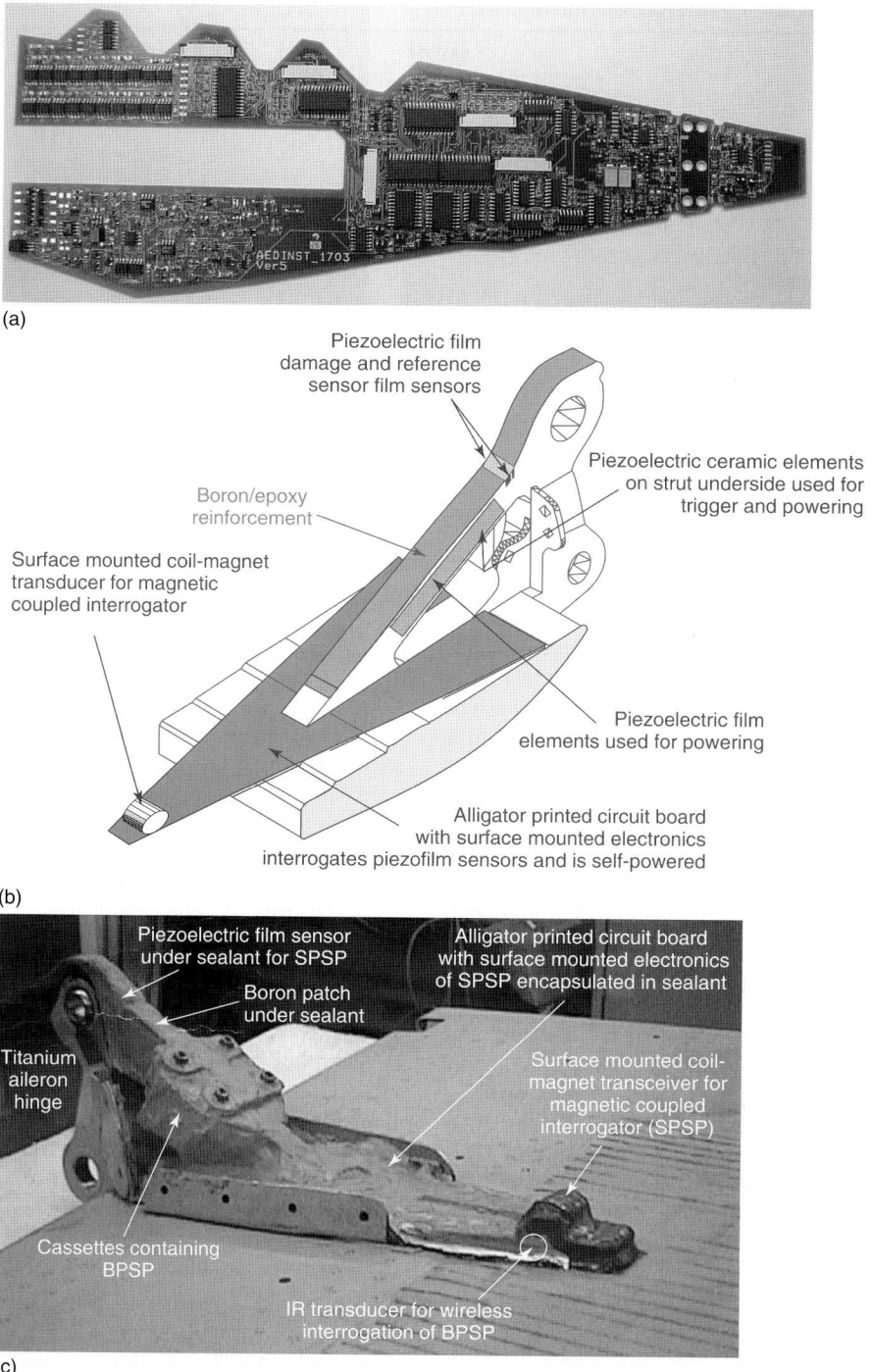

Figure 4. (a) Photo of the SPSP printed circuit board, (b) schematic of self-powered *smart patch* system, and (c) photograph of the installed *smart patch* system on an F/A-18 aileron hinge. (SPSP and BPSP are self-powered and battery-powered *smart patch* configurations, respectively.)

Figure 4(b) and (c) show the mounting arrangement of the *smart patch*, including the BPSP, and the location of the transducers on the aileron hinge. However, the design is such that it should be a relatively straightforward process to reduce the unit to very small proportions using standard complementary metal oxide semiconductor (CMOS) technology—this would have the added benefit of quite substantially reducing the energy requirements and overall weight.

To keep the energy requirements of the circuit as low as possible, it was necessary to minimize the number of components. This is illustrated in Figure 5 for the analog section and Figure 6 for the digital section. Figure 5 shows that the analog section consists of a small number of building blocks: (i) a power harvester (ii) a power controller (with temperature lockout), (iii) a trigger, (iv) a signal conditioner, (v) two logarithmic analog to digital converters, and (vi) an interrogation transceiver. It can be noted from the actual circuit, inserted under the block diagram, that there are, indeed, only a few components.

To minimize energy usage, the digital section, depicted in Figure 6, uses as few components as is practicable. The building blocks are (i) a flash to binary converter, (ii) a computation unit, (iii) data storage, (iv) a finite state machine, (v) a clock control, (vi) an interrogation control, (vii) a data output unit, and (viii) programming/testing ports. Items (i)–(iv) are all implemented in ferroelectric nonvolatile random access memory (FRAM) to reduce power requirements and to reprogram the circuit function; this proved extremely valuable in the test and verification stage of the system design.

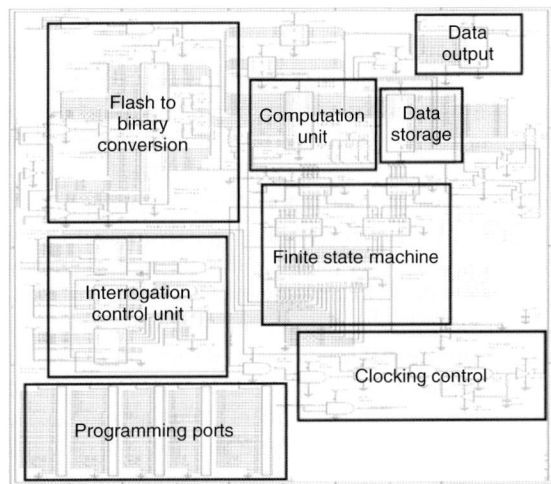

Figure 6. Electrical schematic block diagram—digital section.

As discussed previously, PVDF film was proposed as the sensing material because of the fact that it requires no excitation power. Other advantages with piezoelectric film sensors are their very good fatigue life and strain to failure properties, simple signal conditioning requirements, good high-frequency response; they are also light weight and conform easily to complex shapes. Some limitations with PVDF sensors that need to be considered are their limited operating temperature, poor thermal sensitivity, poor response in the low-frequency regime, and also the extreme care that is required to ensure the correct sensor polarity during installation. As there is a high probability of a substantial amount of strain being generated at low frequencies, it was considered important to ensure that the frequency roll-off characteristics were fairly accurately matched, especially in adverse environments. This in itself can be tedious, but as high impedances are involved, the greater challenge is ensuring that they remain matched in adverse environments. To assist with this problem, buffer amplifiers with low input leakage were incorporated, but the need for thorough protection against moisture ingress remained a critical issue. Considerable care was taken to reduce the risk of moisture ingress creating a shunting effect that would adversely affect this matching.

To minimize system power requirements, the patch health measurement was a ratio obtained using an

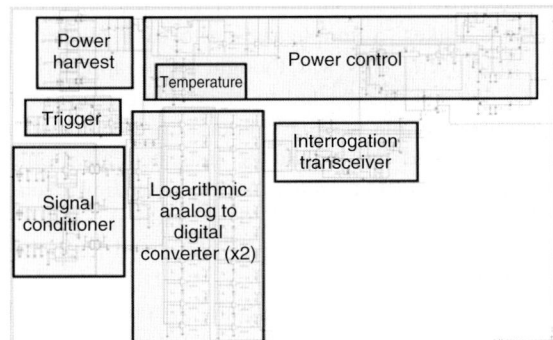

Figure 5. Electrical schematic block diagram—analog section.

analog technique, where the signal from the "reference" transducer was fed through a divider chain and compared with the signal from the active transducer (i.e., sensor over the damage region) via a chain of 10 comparators (with 10% resolution). Each digitizer/divider "reading" uses about 100 nJ (i.e., typically 2.5 µW per comparator, with a 1.2 ms power-up time for 10 comparators plus resistive chain). One advantage of this technique is that it has a frequency response in the kilohertz region, and minimal possibility of time skew problems between the two readings forming the ratio. A disadvantage is the relatively long (1.2 ms) "switch on" settling time of the comparators (and amplifiers). This necessitated a two-stage power supply, where the analog and digital circuits were powered for milliseconds and a few microseconds, respectively. Also, since the PVDF sensors could not be calibrated before installation, owing to time and resource constraints, it was likely that the sensor readings from the undamaged critical region might not exactly match those from the reference; consequently, perfect patch health readings might not result in a ratio value of unity.

The device was designed to measure two ratios, one on the starboard and the other on the port side of the strut, with respect to a common reference and to a 10% resolution. Each reading, taken when the conditions previously described were met, was recorded in FRAM, wrapping back to the initial position in storage after approximately 16 million readings. FRAM was chosen primarily because of its relatively low access energy requirement, which is in the region of 1 nJ/bit compared typically with a few microjoules per bit for electrically erasable programmable read-only memory (EEPROM) devices, and its tolerance of extended temperatures. The parallel variant was chosen, as the higher access speed helped in minimizing system energy requirements. Unfortunately, at the time of design, only the 5 V supply variant of these parts was available, thus to keep the design simple, most of the circuit operates at this relatively high voltage. The high-level choice of FRAM as the storage medium dictated that the device required some digital circuitry, and the choice of input transducers dictated analog input circuitry.

Besides the storage function, the other aspects of the electronic design that were chosen to be implemented in digital technology were sections of the computation and most of the interrogation functions.

A programmable logic device (PLD) was originally chosen on the grounds it provided low operating power with maximum flexibility. However, further investigation revealed a large and potentially problematic start-up energy requirement. As standard cell and custom integrated circuits (IC) were beyond our budget it was decided to use medium scale integration (MSI) CMOS as the most appropriate option. The information stored was two (4 bit) ratios and one (24 bit) integer indicating the number of reading cycles. A temperature-dependent lockout was fitted to the supply circuit to prevent any operations being attempted if the operating temperature was outside specified temperature limits. The limits set were quite tight, within a window range of 10–40 °C, to ensure that readings were not influenced by thermal effects. This was considered an acceptable option in this application as the data required was a very small number of samples and not considered to be particularly dependent on temperature.

In the most elementary case, the only value needed would be the last worst patch health indicator reading. However this was considered not to be sufficiently robust, since a single noise "spike" could cause a misleading reading. This led to the conclusion that filtering and some other confidence-building indicators were required. A time stamp was considered, but this would necessitate a permanent power source to keep the clock running. Also, since the aim was for "indefinite" operation, a reading counter was chosen as a compromise. Unfortunately, the only place to maintain the count is in the nonvolatile memory, thus necessitating a "measurement cycle" to consist of (i) reading the memory, (ii) taking the measurement, and (iii) writing the updated information back into memory. In an attempt to reduce the effects of spurious spikes, a simple slew rate filter algorithm was employed. This was thought to be appropriate as the damage is expected to grow relatively slowly with respect to the reading rate, and even if the patch suddenly failed this would be indicated after only 10 reading cycles at most. During a reading cycle, the system is synchronized to an internal ~1 MHz clock, the speed being basically chosen to minimize the energy required.

Measurements of the overall system indicated the prototype consumed about 210 nJ of energy to perform the digitizing and dividing function, 60 nJ to perform the storage and retrieval function, and 10 nJ for clocking. Hence, the total energy required of the system to make a reading, retrieve and store data, and undertake simple real-time data processing is approximately 280 nJ.

3.3 Power harvesting—energy harvesting and conversion issues

On the supply side, there were two PVDF stacks, each stack having three PVDF elements electrically in series, with poling directions perpendicular to the loading direction. The individual PVDF elements had a nominal thickness of 52 μm, and a manufacturer-rated capacitance of 5.7 nF [25] from which the PVDF relative permittivity $\varepsilon_R \sim 11.3$ was calculated. The PVDF elements were laminated prior to installation, bonded together into a stack using Hysol EA9309 adhesive, and found to have a fairly consistent measured series capacitance of $\sim$1.9 nF. Care was taken to ensure that PVDF electrodes of similar potential (while under cyclic loading) were colocated to minimize capacitive coupling. Insulation resistance was measured to be in excess of 50 GΩ, so both dc and low ac frequency dielectric loss appeared to be negligible for PVDF.

Using the typical published piezoelectric coefficient for PVDF (d_{31}) of 23 pC/N and assuming a simple one-dimensional piezoelectric case with no losses, the predicted peak voltage due to a peak strain of 300 με was $\sim$32.1 V across a stack of three PVDF elements in series. Each PVDF stack fed power through a separate bridge rectifier, which incorporated BAS70 [26] diodes to minimize electrical loss. Each diode had a typical forward voltage drop of 0.22 V (at relevant current levels) and 3 nA of leakage current. The PVDF stacks were directly bonded to the aircraft structure and had a calculated output of $\sim$0.1 V/με (open circuit). The PVDF stack capacitance of 1.9 nF meant that about 0.19 nC (0.1 V × 1.9 nF)/1 με of charge could potentially be generated.

The energy harvesting power supply was a simple diode bridge (Figure 7a) and capacitor, which can be thought of as having two states, diode conduction "ON" and "OFF" as shown in Figure 7(c). The circuit was designed for a nominal operating voltage of about 5.2 V. Thus the first 5.2 V/(0.1 V/1 με) = 52 με of mechanical load excursion did no work on the electrical load (assuming the previous excursion reached the conducting state) and, in fact, incurred a slight leakage current loss while it was occurring. At the nominal operating point, the storage capacitor holds an electrical potential energy of $E = 1/2\ CV^2 = 0.5 \times 1\,\mu F \times (5.2\,V)^2 \approx 14\,\mu J$.

Once the diodes turn "ON", the voltage can be thought of as being approximately constant (assuming the storage capacitance is much larger than the PVDF stack capacitance, 1 μF vs. 0.0019 μF in our case) and each 1 με would add energy of $1/2 \times 1.9\,nF \times [(5.2\,V + 0.1\,V)^2 - (5.2\,V)^2] \approx 1\,nJ$ (not allowing for losses).

With this type of system the primary "losses" associated with the diodes is because of leakage when the diode is reversed biased, and forward drop when the diode is conducting. The power "loss" associated with leakage can be approximated by multiplying the leakage current by the (mean) reverse voltage and by the fraction of time the device is reverse biased. The forward loss can be approximated by the ratio of forward voltage drop to load voltage.

As mentioned, the energy harvesting circuit implemented had a working load voltage of about 5.2 V and since the diodes had a forward voltage drop of about 0.22 V, the diodes of the full bridge (two diodes on at a time) would have dissipated about 100 × (2 × 0.22 V/5.2 V) $\sim$ 8% of the incoming power while in the conducting state.

If the input was truly sinusoidal, then it is expected that the diodes would be reverse biased for about half the time. Given that the leakage of the rectifying diodes was about 3 nA at the operating voltage of 5.2 V and with at least one pair being reverse biased at any point in time, then the expected drain would be about (2 × 3 nA) × 5.2V $\approx$ 31.2 nW while in operation. Should the excitation amplitude reduce for a period then this value is expected to double, since all four diodes would be reverse biased. Hence, a minimum of 31.2 nW/1 nW $\approx$ 31 με of "active" strain excursion would be required per second to replenish diode losses. This figure was kept low by the use of low-leakage diodes. If BAT54 diodes with 600 nA leakage [27] had been chosen, then an "active" strain excursion level of $\sim$6200 με s^{-1} would be required. Here the term *active* refers to the part of the strain excursion that occurs while the diodes are "ON".

The energy harvesting process was modeled via LTSpice [28] simulation, as shown in Figure 7(a),

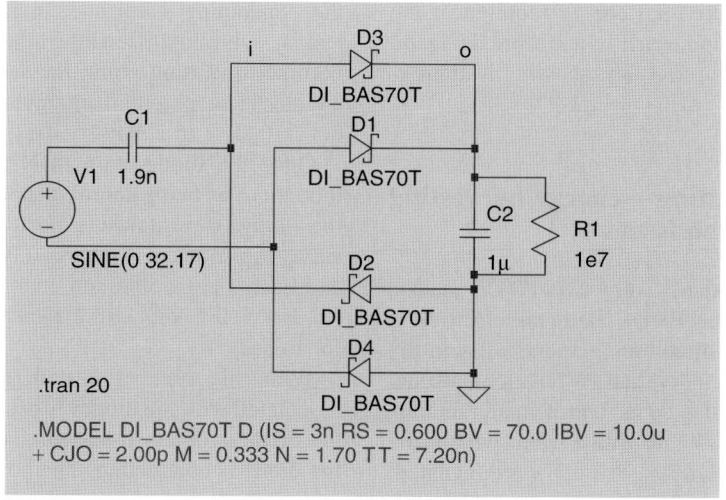
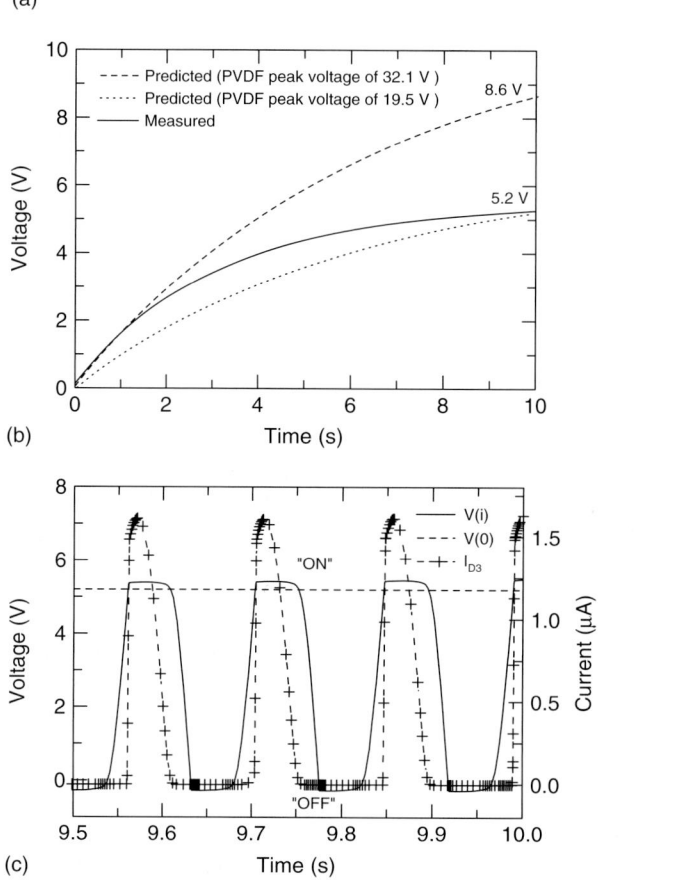

Figure 7. (a) Model of PVDF/diode rectifier energy harvesting circuit. (b) Measured versus predicted storage capacitor voltage. Predictions were made using two PVDF peak voltages viz., 32.1 and 19.5 V. The measured voltage was for a 1 μF storage capacitor being charged using a PVDF stack subject to a sinusoidal strain with a peak value of 300 με at a frequency of 7 Hz. Storage capacitor voltage was ∼5.2 V after 10 s, and ∼5.36 V after 14 s. (c) The current for diode D3 (I_{D3}) versus time for a PVDF peak voltage of 19.5 V and a storage capacitor voltage near the operation point of 5.2 V.

which incorporates the BAS70 full bridge diode rectifier, a 1.9 nF PVDF stack, and a 1 μF storage capacitor with a 10 MΩ parallel resistance to model a 10X oscilloscope probe (at low frequency, the probe capacitance of ∼13 pF was considered irrelevant). The model assumes that the energy harvester and aircraft structure are effectively decoupled, i.e., actions of the harvester have negligible effect on both the bulk and local strains in the aircraft structure. The PVDF is modeled as a 1.9 nF capacitor in series with an ac voltage source of 7 Hz with peak voltage 32.1 V. The manufacturer states that the BAS70 diode leakage is 3 nA (at 25 °C and 5 V reverse voltage); this was confirmed experimentally (the diode measured was from the same diode batch that was later flown) and hence the saturation current "I_s" in the LTSpice diode model was adjusted to 3 nA. LTSpice modeling showed that after 10 s of charging, from a sinusoidal strain excitation of 300 με peak strain at 7 Hz, the storage capacitor dc voltage should be ∼8.6 V assuming only diode losses.

Figure 7(b) also shows a comparison of the LTSpice simulation with the measured voltage of a 1 μF storage capacitor during the charge up phase, using a PVDF stack (viz., one stack of three 52 μm thick by 156 mm long by 19 mm wide PVDF elements) subjected to a sinusoidal strain with a peak value of 300 με at a frequency of 7 Hz. In this case, the PVDF stack is a similar configuration to those flown on the flight demonstrator. The storage capacitor voltage achieved was ∼5.2 V after 10 s, and after 14 s the final storage capacitor voltage was 5.36 V.

Given the stated mechanical loading conditions (7 Hz at a peak strain of 300 με), a comparison between the final simulated storage capacitor voltage (8.6 V) and the measured storage capacitor voltage (5.2 V), as shown in Figure 7(b), indicates that the inefficiencies (losses) totaled about 100 × (32.1 − 19.5/32.1) ∼ 39%. Three mechanisms have been postulated for this energy harvesting loses. In decreasing order of importance they are

1. the effect of Poisson's ratio, a tensile stress in the 1 direction will produce ∼1/3 compression in the 2 direction, which will reduce the piezoelectric output voltage by ∼1/3 (assuming $d_{31} \sim d_{32}$);
2. a "shear lag" effect that is the combination of bondline mechanical compliance issues and, more importantly, the reduction of strain in the PVDF elements that are located furthest from the metallic substructure due to the compliance of the piezoelectric film; since the PVDF elements are quite compliant, not all the strain is transferred from the metallic substructure through these elements to the ones above—the larger the stack the less strain will be transferred to the top elements;
3. the effect of capacitive coupling between the PVDF metallization layers.

It appears that Poisson's ratio effect alone cannot explain the 39% efficiency decrease measured. Using the measured storage capacitor voltage of 5.2 V (at $t \sim 10$ s) generated by a single PVDF stack, the energy stored in the 1 μF capacitor after 10 s was approximately,

$$E = \frac{1}{2} C V^2 = 0.5 \times 1 \ \mu F \times (5.2 \ V)^2 \sim 14 \ \mu J \quad (1)$$

This means 14 μJ of energy could be harvested every 10 s from each side of the strut, resulting in 1.4 μW, which was sufficient energy for the device to operate when attached to the aluminum hinge configuration.

The limited flight strain data available for the aluminum aileron hinge configuration, indicated that the strain loading scenario (i.e., 300 με peak at 7 Hz) should be easily achieved at the aileron aft hinge strut. Note also that the 7 Hz used in this experimental study is quite conservative since the in-flight strain and acceleration data indicate that structural resonances occurred at about 14, 30, and 55 Hz, as shown in Figure 8. These higher frequencies would produce more energy for harvesting within a given time period.

The replacement of the aluminum aileron hinge with the titanium hinge meant that there was a significant reduction in the operational strain levels on the hinge. FE studies indicated that the strains were reduced by about a factor of about 2. The maximum surface strains on the b/ep doubler, for load condition WO39, were reduced from 5000 με for the aluminum to 2100 με for the titanium hinge [21, 23]. Therefore, it was decided to add additional power-harvesting PZT elements on the lower side of the hinge strut, as shown in Figure 4(b). The PZT power-harvesting elements consisted of two 32 mm long by 12.5 mm

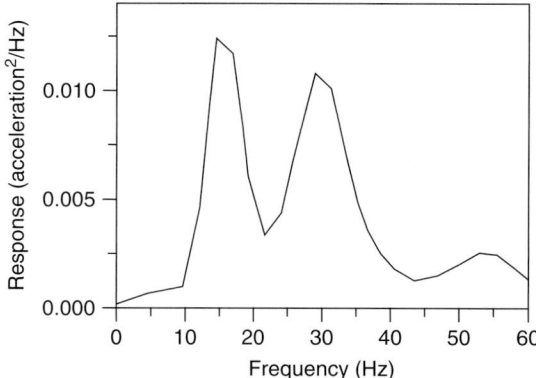

Figure 8. Typical acceleration frequency spectra measured on in the aileron hinge region.

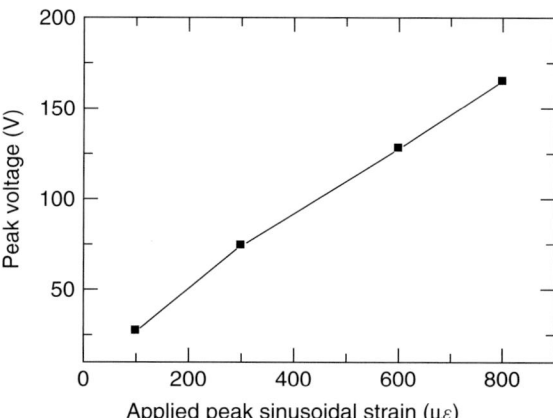

Figure 9. Peak strain versus open-circuit voltage for a T226-A4-303Y PZT-5A bimorph subjected to a sinusoidal peak strain (10–30 Hz).

wide Piezo Systems T226-A4-303Y bimorph transducers, consisting of a 0.12 mm thick brass shim sandwiched between dual parallel poled 0.27 mm thick PZT wafers [29]. The calculated capacitance of the PZT-5A bimorph was 12.4 nF compared with the measured capacitance of approximately 9.2 nF.

When a load equaling 300 $\mu\varepsilon$ was applied to the PZT-5A bimorph the calculated open-circuit voltage was $\sim$115 V, assuming no losses or inefficiencies. To experimentally confirm that the voltage produced by the PZT-5A bimorph was as expected, a specimen was manufactured from 4340 steel plate with dimensions 260 mm × 75 mm × 4 mm. The PZT-5A bimorph element was bonded to the plate using silver-loaded epoxy with a resultant bondline thickness of less than 200 μm. The top surface of the bimorph was also coated with silver-loaded epoxy to approximate the bonding conditions that the bimorph would experience in service.

As shown in Figure 9, when the PZT-5A bimorph was subjected to a sinusoidal 300 $\mu\varepsilon$ peak strain (at excitation frequencies between 10 and 30 Hz), the recorded peak voltage produced by the bimorph was about $\sim$75 V across a 10 MΩ oscilloscope probe, which is an 100 × (115 V − 75 V/115 V) $\approx$ 35% efficiency decrease. The difference between the calculated and measured PZT-5A bimorph voltages appears to be almost entirely explainable by the Poisson's ratio effect.

For the purpose of energy harvesting, the output from each of the PZT bimorphs was fed into a separate bridge rectifier that was effectively working in a half-wave manner. The main benefit of the half-bridge approach was that it allowed one electrode of each of the PZT-5A bimorphs to be bonded to, and hence, electrically grounded with, the aircraft structure.

To summarize, assuming no mechanical or electrical losses, the individual PVDF elements appeared to be rated at $\sim$2.5 nJ/(m^2·$\mu\varepsilon$) and the PZT-5A bimorphs at $\sim$4.4 μJ/(m^2·$\mu\varepsilon$). Maintaining the assumption of no mechanical or electrical losses, and also assuming 100% strain transfer efficiency and with no maximum placed on voltages, with a constant sinusoidal mechanical loading of 7 Hz at 300 $\mu\varepsilon$ peak strain, it was calculated that

1. $\sim$0.96 μW of power could be harvested from a single PVDF element, resulting in a total of $\sim$5.8 μW harvestable from the two PVDF stacks;
2. $\sim$2.2 mW from a single PZT bimorph, resulted in a total of 4.4 mW from the two PZT-5A bimorphs attached to the aileron strut; and
3. the overall minimum harvestable power from the PVDF stacks and the PZT bimorphs was $\sim$4.406 mW.

As can be seen, the two PZT-5A bimorphs were capable of producing significantly more power than the two PVDF stacks; however, there was a risk that the ceramic PZT-5A bimorphs might not have been robust enough to cope with the stresses induced by operational flight loads.

Further expanding upon the energy harvesting efficiency and loss factors discussed above, we arrive at the following results:

1. Dielectric loss: this appears to be sufficiently small as to be considered negligible for both for both PVDF and PZT-5A bimorphs. For example, a PVDF charged to 5 V, with 50 GΩ insulation resistance, has a continuous dc power loss of $\sim (5\,\text{V})^2/50 \times 10^{12}\,\Omega = 0.5$ pW, and assuming low-frequency structural vibration, ac power loss to the dielectric will also be negligible.
2. Geometrical strain compliance between the underlying structure and the piezoactive material: this is an efficiency loss due to the single sided mechanical coupling that is normally used to attach a piezoelectric element to the structure.
3. Diode loss: the diode loss in the bridge rectifier is caused by both the forward voltage drop across the diode and the diode leakage current, both of which produce real energy losses via heat production. Generally speaking, the peak currents produced by the energy harvesting process are small, and assuming no losses, the peak currents will be $<3\,\mu$A for the PVDF stack and $<50\,\mu$A for the PZT-5A bimorph. The Schottky diodes used in the rectifiers must be carefully chosen; however, it is not a trivial task to determine the best diode because the LTSpice diode models provided by manufacturers are generally quite inaccurate when currents are small (i.e., 100 μA or less), so accurate diode losses cannot be simulated. For energy harvesting from low-frequency structural excitations the authors consider diode leakage to be the main diode loss component. A laboratory testing program appears to be the most efficient method of determining the best rectification diodes for the particular energy harvesting task.

At this point it should be noted that the simple energy harvesting circuits implemented on the *smart patch* may not have been optimal since the voltage across the piezoelements, and hence the electrical component of the force was rather low. On the other hand, having very small forces feeding back from the energy harvester in the aircraft structure is conservative from a structural dynamics point of view. In any case, circuit board space and design time limitations prevented us from using a more elaborate circuit.

It is difficult to assess the exact mechanical and electrical losses in the harvesting system during any particular flight because of the

1. lack of low current fidelity in the diode spice models provided by manufacturers;
2. action of the diode bridge rectifiers during the energy harvesting process being nonlinear; and
3. mechanical loading of the aileron hinge during flight being unknown.

However, for a specific sinusoidal loading condition of 300 $\mu\varepsilon$ peak strain at 7 Hz, measurements have shown (Figure 7b) that a PVDF stack can charge up a 1 μF capacitor to the operational load voltage of 5.2 V in $\sim$10 s. Under the same loading conditions, taking into account the inefficiencies and loss mechanisms discussed above, LTSpice simulation calculated that a PZT-5A bimorph will charge up a 1 μF capacitor to the operating load voltage of 5.2 V in $\sim$0.57 s. Given that the time to charge the 1 μF storage capacitor to the operating load voltage 5.2 V is known, the expected power-harvesting level for the combination of the two PVDF stacks ($2 \times [0.5 \times 1\,\mu\text{F} \times (5.2\,\text{V})^2/10\,\text{s}] \approx 2.8\,\mu$W) and the two PZT-5A bimorphs ($2 \times [0.5 \times 1\,\mu\text{F} \times (5.2\,\text{V})^2/0.57\,\text{s}] \approx 90\,\mu$W) was estimated to be $\sim$93 μW, meaning that $\sim$0.55 s of time was required to generate enough energy to power the electronics outlined in Section 3.2.

3.4 Interrogation

During the information uploading operation, the circuit was both powered and clocked by an alternating magnetic field provided by the interrogator. This field was modulated with a serial representation of the reading information, which was continually repeated until the interrogator received several identical samples at which time it accepted the data.

The antenna consists of a coil of copper wire wound on a ferrite core. The device is energized by the current produced in the coil in response to a low-frequency magnetic field produced by the interrogator. The data is modulated on this carrier by changing the reflected impedance. The data rate is derived as a submultiple (1/16th) of the carrier frequency.

4 SAFETY-OF-FLIGHT AND FUNCTIONAL TESTING

A number of testing activities were undertaken, during the development of the *smart patch* system, to cover two main aspects:

1. safety-of-flight issues, which required testing on a prototype *smart patch* system to characterize the system response from specified mechanical vibration/shock, electromagnetic and pressure/temperature environments;
2. system functional testing consisting of both bench-top testing of the electronic components (including interrogator) and a prototype *smart patch* system on an "original" aluminum aileron hinge component (removed from an F/A-18 aileron).

4.1 Safety-of-flight testing

The vibration test on a system mock up consisted of a resonance search test, and a low-frequency swept sine, a low-frequency sine dwell and a high-frequency random vibration test, and finally a shock test in each of the three principal axes. Testing requirements are based on specifications given in [30, 31]. Visual inspection of the test article after testing did not reveal any deformation, delamination, fragmentation, or breakage [32]. Results of the tests showed that the system survived all vibration and shock testing without failure and that the equipment performed normally after the test.

Temperature and altitude testing included temperature cycles between -40 and $+80\,°C$ with concurrent altitude pressure cycles from sea level to 9000 m. No adverse structural or chemical problems were encountered; however, significant variations in the coefficient of thermal expansion between the potting compound and the wiring caused solder joints to fail—these issues meant that a more compliant potting compound needed to be used for the in-flight demonstrator.

The electromagnetic test consisted of a radiated emissions test in accordance with MIL-STD-461E [33] and Boeing requirements [34] for RF susceptibility from 1 MHz to 18 GHz. No emissions were detected from the *smart patch* when it was optically isolated from the interrogator and monitor computer [35].

4.2 Functional testing

Electronic components (such as the APCB and handheld interrogator) were bench tested using simulated voltage inputs for the piezoelectric film sensor and power-harvesting elements. A trial installation was also performed on a subcomponent "original" aluminum aileron hinge taken from an F/A-18 aileron, similar to the component shown in Figure 1(b) and then subjected to functional and environmental testing. That was undertaken to "fine tune" and validate the installation procedure and evaluate the performance of the *smart patch* after adverse environmental (i.e., hot/wet/cold) loading. The environmental testing involved placing the subcomponent in an environmental chamber and subjecting it to several moderate and severe thermal cycles of -30 to $40\,°C$ and -40 to $70\,°C$, respectively. All thermal cycling was undertaken at the maximum relative humidity of the chamber (which was about 100%), and each temperature extreme was maintained for at least 30 min, to ensure condensation would occur on the subcomponent during the environmental cycling; the aim was to ensure that the environmental cycling would comprehensively test the protective coating process. The specimen was then installed in the testing machine to perform a number of functional tests. This testing program was extremely useful and outlined several electronic and installation deficiencies, involving the environmental protection [6]. The main issues were problems associated with the FRAM units, as well as issues with the protective coating process associated with the electromagnetic shielding layer bonding to the protective polysulfide PR1750 sealant. Polysulfide PR1750 was used for the environmental protection for the electrical components and piezoelectric transducers, as well as to bond the APCB to the aileron hinge surface. Electromagnetic shielding was achieved by wrapping aluminum foil around the APCB, after applying the PR1750, and wiring. Care needed to be taken to ensure that air pockets were removed between the foil and the

PR1750. A coating of conductive paint provided the final electromagnetic protective layer.

5 SYSTEM IMPLEMENTATION AND FLIGHT TEST RESULTS

A comprehensive installation procedure was written and approved by the appropriate airworthiness authorities [6]. The system was then installed on an Aircraft Research and Development Unit (ARDU), Flight Test Squadron (AFTS) aileron, over about a two-week period in November 2005. The aileron with the installed *smart patch* is shown in Figure 10.

The instrumented aileron hinge was installed on Hornet A21-101 in February 2006. The patch health information was to be downloaded by ground personnel from the instrumented aileron using a handheld unit, as shown in Figure 11, at the end of every week. No downloads were attempted if the aircraft did not fly during the week or if operational requirements meant that the aircraft was located at another base. After each download, the handheld unit was connected to a nearby docking station, which ensured that the handheld device was fully charged and allowed the patch health data, in the handheld unit, to be downloaded via a GSM mobile (cell) phone link for analysis. A plot of this data is shown in Figure 12.

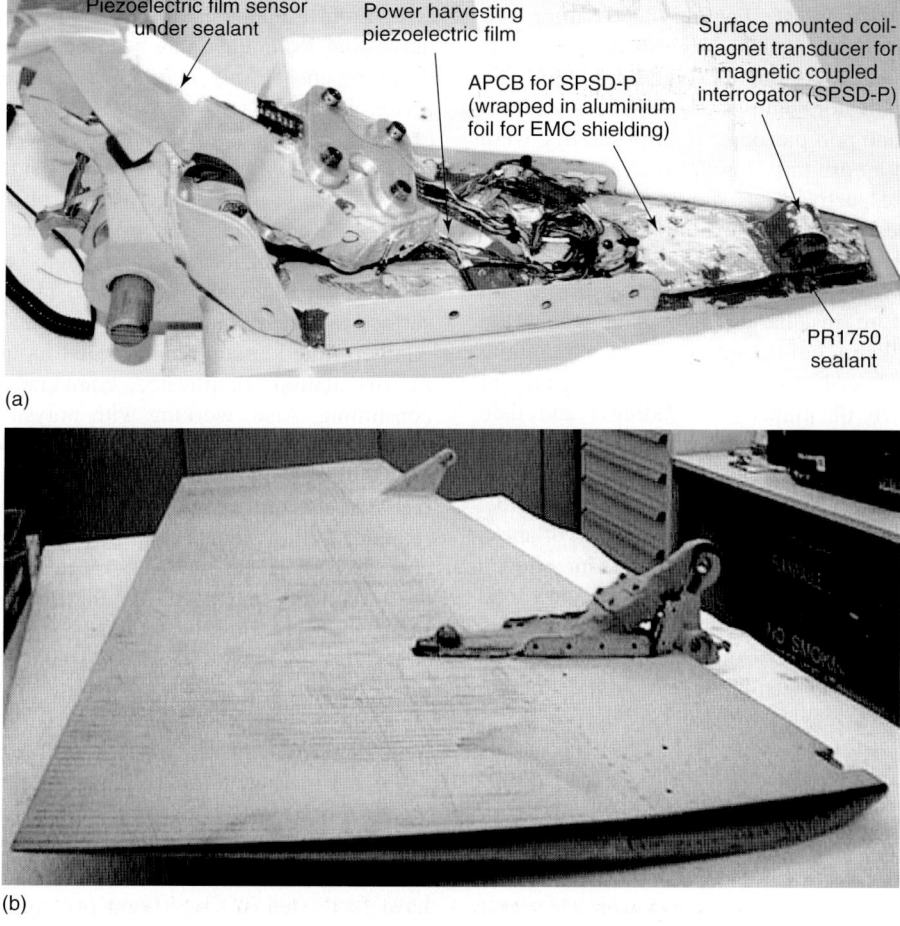

Figure 10. (a) *Smart patch*, without protective coating and electromagnetic shielding, installed on aileron hinge; (b) photograph of *smart patch* on the F/A-18 aileron hinge.

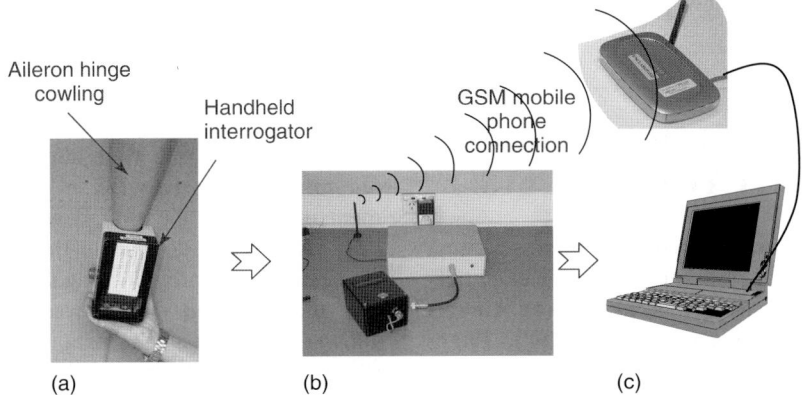

Figure 11. Patch health data download from *smart patch* (a) using handheld interrogator, (b) connecting handheld interrogator to docking station, and (c) download data to base station via GSM mobile phone link.

The trial ended in late February 2007 after 110 flights had been completed. During this time, the self-powered unit had registered 178 valid readings. The patch health indicator readings stored in FRAM, from the starboard and port piezoelectric film sensors, were initially set to an arbitrary low health value of about 0.8 and 0.1, respectively, as shown in Figure 12(a). Both patch health readings exhibited similar trends toward their respective perfect patch health ratio. As discussed previously, the PVDF sensors were not calibrated before installation and, as a result, the port and starboard ratios may settle to a nonunity ratio. The starboard sensor converged to a ratio of about 1.0 after 14 flights (37 readings) and then maintained a consistent value until the end of the trial. Therefore, any divergence from this value would indicate damage in the patch on the starboard side. The port sensor converged to a different ratio of about 0.63 after 97 flights (134 readings). As mentioned above, 178 valid readings were recorded after a total of 110 flights, as shown in Figure 12(b). The figure shows a steady increase in valid readings, of between one and two readings per flight, which was another indication that the device was working correctly.

6 DISCUSSIONS—LESSONS LEARNED

Considerable experience in low-power electronic design, design and implementation of packaging, as well as understanding issues associated with the design and implementation of a strain-based power-harvesting techniques has been achieved during the development of this demonstrator. One significant lesson learnt was to be cautious in the choice of electronic components that have only recently been released. This is because the project suffered considerable delays in the early part of the program due to the early adoption of immature COTS electronics, in particular, early versions of the Ti430 and a nonoperational FRAM chip. The use of COTS surface-mounted electronic components necessitated a large printed circuit board (PCB) footprint, making the installation complicated, cumbersome, and time consuming. Also, working with polysulfide PR1750 sealant to provide environmental protection, provided an additional level of complexity mainly due to the fact that the sealant was difficult to work with and required 24 h to cure, besides posing health and safety concerns. Hence, it would be desirable to eliminate the use of this sealant during any future *smart patch* installations.

Overall, there is a need to simplify the installation process into as few steps as possible. If the electronics were miniaturized (on a single chip), then this would greatly assist with packaging and installation, as well as reducing energy requirements, improving robustness, and reducing weight. A vibration-based energy harvesting approach would have also been feasible in this situation and would have facilitated in simplifying packaging and installation. This is because the system could be installed as a self-contained device thus eliminating the need

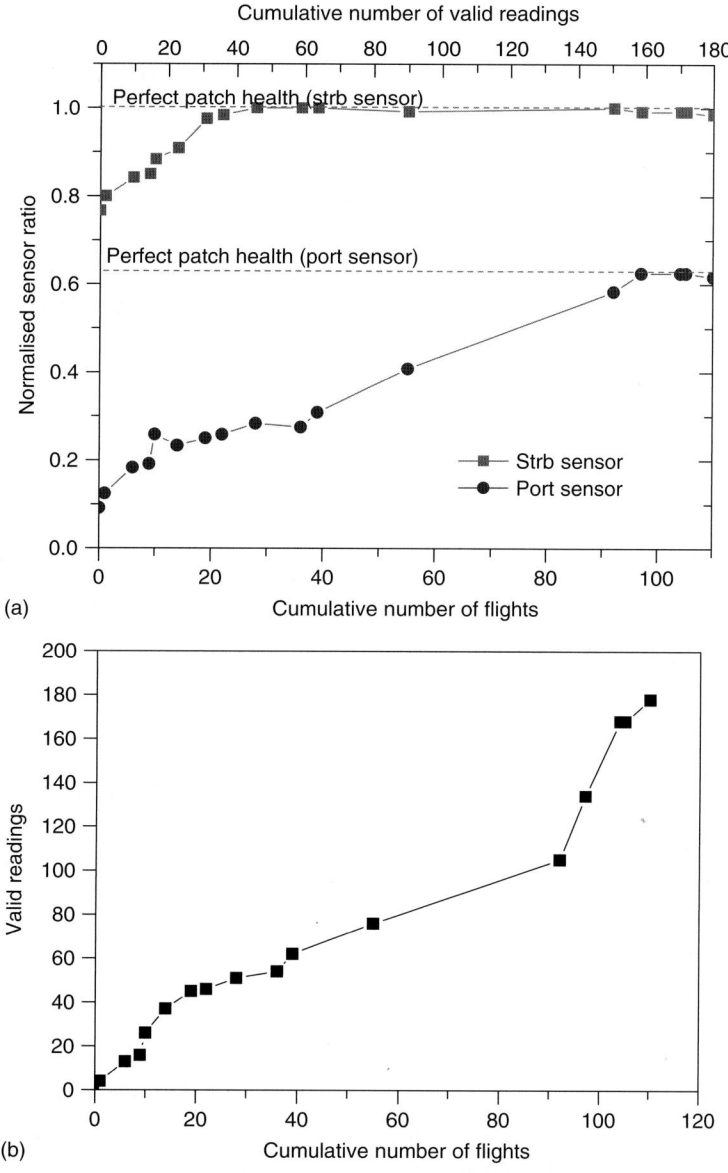

Figure 12. (a) Variation of *smart patch* normalized sensor ratios (patch health indicator readings) with increasing number of flights (from Mar 2006 to Feb 2007). (b) Plot shows the number of valid readings for the self-powered system with increasing number of flights.

to install separate power-harvesting piezoelectric elements, which required a significant footprint, to the structure.

After the instrumented aileron was removed from the aircraft, the *smart patch*, including the b/ep reinforcement, was removed from the aileron. Before removing the b/ep reinforcement from the aileron hinge, a standard tap test was undertaken, indicating no disbonds were present in the patch. Removal of the sensors, power-harvesting elements, PCB, and electromagnetic shielding indicated no significant adverse environmental effects on the various components.

However, it was observed that, in some regions, the PR1750 had not adhered to the shielding, suggesting that care needs to be taken when handling the shielding. Also, some of the PVDF layers in the power-harvesting stacks had disbonded. This observation reinforced the fact that more durability and robustness testing of the various sensing and power-harvesting elements would be required if the system were to be installed for a longer period.

7 CONCLUSIONS

The *smart patch* approach, which, in this case, is based on a self-powered *in situ* SHM system, would assist in alleviating certification concerns associated with implementing bonded composite repairs to primary aircraft structures. This approach relies on the ability to autonomously detect disbonding in the patch and is basically a continuous safety-by-inspection approach for the bonded repair. The main aspects of a self-powered *in situ* patch health monitoring system, applied to a composite bonded reinforcement on an operational F/A-18 aircraft, were outlined in this article. The system incorporates piezoelectric film PVDF sensors and is designed to operate using the electrical power generated by an array of PVDF and PZT power-harvesting piezoelements, which convert structural dynamic strain to electrical energy. The system was successfully installed on an operational RAAF F/A-18 aircraft. Patch health information downloaded from the *smart patch* system after 12 months of operational flying has indicated that the device is performing as expected and that the patch has "good" health. This was confirmed by applying a standard tap test to the composite bonded patch.

The system was designed with the focus on minimizing power requirements. This meant that issues such as the type of sensor, the amount and type of patch health information stored, electronic design/components including memory type and the interrogation technique were chosen from the standpoint of minimizing overall energy consumption. Owing to time constraints, the self-powered approach was strain-based, rather then vibration-based, since this required minimal development time even though it significantly complicated the installation process. To further lower risk, two types of power-harvesting piezoelectric elements, polymer and ceramic, were used in the system. This ensured adequate durability while ensuring enough power was generated at the reduced strain levels anticipated in the refurbished titanium hinges. The *smart patch* approach also demonstrated to the operators and maintainers of the aircraft that the interrogation of such systems required minimal interruption to their current maintenance routine.

ACKNOWLEDGMENTS

This program was undertaken under a RAAF sponsored task with support from Aircraft Structural Integrity-Directorate General Technical Airworthiness (ASI-DGTA) especially Officer-In-Charge ASI WGCDR Jason Agius and Dr Madabhushi Janardhana. Also, significant support was provided by AOSG, Aerospace Systems Engineering Squadron, especially Chief Engineer WGCDR Greg Young, Mr Des Cass, SGT Gavin Jones and SGT Andrew Schutz. The authors would like to also acknowledge Andrew Rider for development of the surface treatment procedure for the titanium aileron hinge and Ivan Stoyanovski for the application of the composite patch to the hinge; David Rowlands for assistance in installation of the *smart patch* and fabrication of the handheld interrogation unit; Keith Muller, Rodney Gray, Carlos Rey and Bruce Crosbie for assistance in mechanical testing; Peter Virtue and Howard Morton for undertaking the NDI of the aileron; Anthony Walley for assistance in developing the installation procedure and testing the PZT elements; Mike Konak, Quang Nguyen, Peter Smith and Sami Weinberg for assistance in circuit design and software development; and Richard Callinan for undertaking FE analysis. Support from Richard Chester and Alan Baker during the program is also gratefully acknowledged. Significant technical support from Aerostructures Technologies Pty Ltd is also gratefully acknowledged—Bryan Stade and Karmal Gill for designing the attachment and mounting arrangements; Mathew Goldstraw, Greg Rowlinson, Simon Maan and Ron Westcott for undertaking the FE analyses.

REFERENCES

[1] Baker AA. Introduction and overview. In *Advances in the Bonded Composite Repair of Metallic*

Airframe Structures, ISBN 0080429939, Baker AA, Rose LRF, Jones R (eds). Elsevier, 2002; Chapter 1, pp. 1–17.

[2] Baker AA. Certification issues for critical repairs. In *Advances in the Bonded Composite Repair of Metallic Airframe Structures*, ISBN 0080429939, Baker AA, Rose LRF, Jones R (eds). Elsevier, 2002; Chapter 22, pp. 643–656.

[3] Galea SC. Smart patch systems. In *Advances in the Bonded Composite Repair of Metallic Airframe Structures*, ISBN 0080429939, Baker AA, Rose LRF, Jones R (eds). Elsevier, 2002; Chapter 20, pp. 571–612.

[4] *F/A-18 A/B/C/D Trailing Edge Flap and Aileron Hot Spot Analysis—Final Report*, McDonnell Douglas Aerospace Report MDA 96A0138, Revision A, March 1997.

[5] Chester R (ed). *Life Extension of F/A-18 Inboard Aileron Hinges by Shape Optimization and Composite Reinforcement*, Defence Science and Technology Organisation (DSTO), Technical Report DSTO-TR-0699, January 1999.

[6] Galea S, et al. *Smart Patch Flight Demonstrator—System Implementation and Lessons Learned*, Defence Science and Technology Organisation (DSTO), Technical Report DSTO-RR in draft, 2008.

[7] Galea SC, Moss SD, Powlesland IG, Baker AA. Application of a smart patch on an F/A-18 aileron hinge. *Proceedings of ACUN-4 Composite Systems: Macrocomposites, Microcomposites, Nanocomposites*. University of New South Wales, Sydney, 2002.

[8] Roundy S, Wright PK, Rabaey JM. *Energy Scavenging for Wireless Sensor Networks with a Special Focus Vibrations*, ISBN: 1-4020-7663-0. Kluwer Academic Publishers, 2004.

[9] Konak MJ, Powlesland IG, van der Velden SP, Galea SC. A self powered discrete time piezoelectric vibration damper. *Proceedings of the Far East and Pacific Rim Symposium on Smart Materials Structures and MEMS*. SPIE, 1997; Vol. 3241, pp. 270–279.

[10] Sodano G, Inman DJ. A review of power harvesting using piezoelectric materials. *The Shock and Vibration Digest* 2004 **36**(3):197–205.

[11] Eggborn T. *Analytical Models to Predict Power Harvesting with Piezoelectric Materials*, Master's thesis. Virginia Polytechnic Institute and State University, May 2003.

[12] Taylor GW, Burns JR, Kammann SM, Powers WB, Welsh TR. The energy harvesting eel: a small subsurface ocean/river power generator. *IEEE Journal of Oceanic Engineering* 2001 **26**(4):539–547.

[13] Fleurial JP. Thermoelectric energy conversion: future directions and technology development needs. *Indo-US Workshop on Emerging Trends in Energy Technology*. New Delhi, 11–16 March 2007.

[14] Shenck NS, Paradiso JA. Energy scavenging with shoe-mounted piezoelectrics. *IEEE Micro* 2001 **21**(3):30–42.

[15] Starner T. Human powered wearable computing. *IBM Systems Journal* 1996 **35**(3–4):618–629.

[16] Starner T, Paradiso JA. Human-generated power for mobile electronics. In *Low Power Electronics Design*, ISBN-10: 0849319412, Piguet C (ed). CRC Press, 2004, pp. 45-1–45.26.

[17] Shenck NS. *Generation from Piezoceramics in a Shoe*, Master's thesis. MIT, May 1999.

[18] El-hami M, Glynne-Jones P, White NM, Hill M, Beeby S, James E, Brown AD, Ross JN. Design and fabrication of a new vibration based electromechanical power generator. *Sensors and Actuators, A* 2001 **92**:335–342.

[19] Warneke B, Last M, Liebowitz B, Pister KSJ. Smart dust: communicating with a cubic-millimeter computer. *Computer* 2001 **34**(1):44–51.

[20] Beeby SP, Torah RN, Tudor MJ, Glynne-Jones P, O'Donnell T, Saha CR, Roy S. A micro electromagnetic generator for vibration energy harvesting. *Journal of Micromechanics and Microengineering* 2007 **17**:1257–1265.

[21] Armitage RP. *FE Analysis of a Boron Reinforcement on an F/A–18 Aileron Hinge*, Aerostructures Letter Report, Ref. 4-13-12-6. PM1058, May 2002.

[22] Moss SD, Galea SC, Powlesland IG, Konak M, Baker AA. In-situ health monitoring of a bonded composite patch using the strain ratio technique. *SPIE's 2000 Symposium on Smart Materials and MEMS, Smart Structures and Devices Conference*, Paper 4235-41. SPIE: Melbourne, 2000; Vol. 4325.

[23] Rowlinson GR. *FE Analysis of F/A–18 Aileron Hinge (Blueprint Profile) for Active Sensor Design*, Aerostructures Letter Report, Ref. 4-13-12-3. PM661, March 2001.

[24] Zhang H, Galea SC, Chiu WK, Lam YC. An investigation of thin PVDF films as fluctuating strain measuring and damage monitoring devices. *Smart Materials and Structures* 1993 **2**:208–216.

[25] Measurement Specialities (MEAS), *Piezo Film Product Guide and Price List*, 2007, http://www.meas-spec.com.

[26] Diodes Incorporated, *BAS70 Diode Specification Sheet*, 2007, http://www.diodes.com/datasheets/ds11007.pdf.

[27] Diodes Incorporated, *BAT54 Diode Specification Sheet*, 2007, http://www.diodes.com/datasheets/ds11005.pdf.

[28] Engelhardt M. *LTspice/SwitcherCAD III*. Linear Technology Corporation, 2007, http://www.linear.com.

[29] Piezo Systems, *Catalog 7B*, 2007, http://www.piezo.com.

[30] MIL-STD-810E, *Environmental Test Methods and Engineering Guidelines*, July 1989.

[31] *F-18 Vibration, Shock, and Acoustic Noise Design Requirements and Test Procedures for Aircraft Equipment*, McDonald-Douglas Report MDC A2276, March 1978.

[32] Accredited Test Services, *Vibration Testing oF F/A18 Smart Patch Aileron Hinge*, Report TS1167, April 2002.

[33] MIL-STD-461E, *Requirements for the Control of Electromagnetic Interference Characteristics of Subsystems and Equipment*, 1999.

[34] O'Byrne MA. *Electromagnetic Interference Control Requirements*, Boeing Document Number D6-16050-4, July 1991.

[35] Accredited Test Services, *EMC Testing on F/A 18 Smart Patch Aileron Hinge*, Report TS1150, March 2002.

Chapter 77
Energy Harvesting using Thermoelectric Materials

Daniel J. Inman[1] and Henry A. Sodano[2]

[1] *Center for Intelligent Material Systems and Structures, Virginia Polytechnic Institute and State University, Blacksburg, VA, USA*
[2] *Department of Mechanical and Aerospace Engineering, Arizona State University, Tempe, AZ, USA*

1 Introduction	1351
2 Basic Theory	1353
3 Examples	1354
4 Experimental Testing	1355
5 Results	1356
6 Comparison of Harvesting Technologies	1358
7 Conclusions	1359
References	1359
Further Reading	1360

1 INTRODUCTION

The concept of developing completely self-power electronics has received significant interest over the past decade partially fueled by the recent advances in wireless and structural health monitoring (SHM) technology [1]. As described in earlier articles, SHM is an integrated process of data acquisition, signal processing, and statistical inference used to track and assure the safety and performance of a structure. SHM has immediate benefits and market potential. However, because of the need for a widely dispersed sensor network to effectively monitor large civil structures and the need to dispense with increasing wiring harnesses in vehicle applications, it has been realized that wireless SHM sensors should be used to reduce the system complexity (*see* **Chapter 69**). However, when using wireless sensors it is necessary that a portable power supply be used, such as batteries. Because batteries must be periodically replaced, the use of batteries to power wireless sensors greatly restricts sensor placement and location. For instance, it is often necessary to embed the sensor in the structure to achieve the desired response or to be close to areas of high damage probability. By implementing methods of obtaining ambient energy from the sensors surroundings, a wireless device could be designed to function indefinitely. Additionally, one major use of SHM systems would be to quickly inspect civil structures or military vehicles after catastrophic events, which typically cause widespread power outages thus making the use of wired power sources impossible.

Encyclopedia of Structural Health Monitoring. Edited by Christian Boller, Fu-Kuo Chang and Yozo Fujino © 2009 John Wiley & Sons, Ltd. ISBN: 978-0-470-05822-0.

The recent advances in low power electronics and wireless technology have made the hardware necessary to perform the required tasks available, but an effective power source has yet to be identified. Therefore, the development of wireless SHM systems revolves around the ability to capture ambient energy surrounding the device and convert it into usable electrical power. In many studies, piezoelectric materials have been utilized to capture the ambient vibrations around a system and convert them into electrical power [2] (*see* **Chapter 76**). However, the energy generated by piezoelectric materials is typically far too small to directly power most electronic devices. The issue of too little energy can be compensated for by using energy storage methods to accumulate sufficient energy to power the electronics in short bursts. A second and more developed method of obtaining energy from ambient sources is through the use of thermoelectric generators (TEGs), which capitalize on thermal gradients. TEGs use the Seebeck effect [3], which describes the current generated when the junction of two dissimilar metals/semiconductors experiences a temperature difference. Using this idea, numerous p-type and n-type junctions are arranged electrically in series and thermally in parallel to construct the TEG. Thus, if a thermal gradient is applied to the device, it will generate an electrical current that can be utilized to power other electronics. By implementing power harvesting devices, autonomous portable systems can be developed that do not depend on traditional methods for providing power, such as the battery, which has a limited operating life.

TEGs are an established technology and have been used for capturing ambient energy in various applications. Lawrence and Snyder [4] suggest a potential method of retrieving electric energy from the temperature difference that exists between the soil and the air. To test their concept, a prototype was built without the TEG and the heat flow was measured to estimate the amount of power that could be obtained. The results showed that a maximum instantaneous power of approximately 0.4 mW could be generated by the thermoelectric device. Rowe *et al.* [5] investigate the ability to construct a large TEG capable of supplying 100 W of power from hot waste water. The system tested used numerous thermoelectric devices placed between two cambers, one with flowing hot water and the other with cold water flowing in the opposite direction thus maximizing the heat exchange. With a total of 36 modules, each with 31 thermocouples, 95 W of power could be generated. Fleming *et al.* [6] investigated the use of TEG to provide electrical power in micro air vehicles. A TEG was mounted on the exhaust system of an OS max 61 internal combustion engine and was shown to generate 380 mW of power.

Several authors have studied the use of TEGs for obtaining waste energy from the exhaust of automobiles. Birckolz *et al.* [7] worked with Porsche to develop a TEG unit that would fit around the exhaust pipe of the 944 engine. The unit was experimentally tested and found to generate an open circuit voltage of 22 V and a total power of 58 W. Similarly, Matsubara [8] constructed an exhaust system using ten TEG modules and a liquid heat exchanger to maximize the thermal gradient. The system was tested on a 2000-cc class automobile and shown to produce 266 W of power. Bass *et al.* [9] investigated the placement of a TEG in the vertical muffler of a class 8 diesel truck. The system generated 1 kW of power, thus allowing it to be employed as a substitute for the truck's alternator. By removing the alternator from the engine, the power delivered to the driveshaft was increased by 3–5 hp, providing an increase in fuel efficiency and a reduction in emissions. For more information on TEG applications in automobiles, see Vázquez *et al.* [10]. A review of previous research in TEGs shows that there are several key parameters that dictate the amount of energy generated; the most important of these are the surface area in contact with the thermal source, temperature gradient across the device, and the thermal conductivity between the TEG and the source.

The idea to use thermoelectric devices to capture ambient energy from a system is not new. However, TEGs have typically been used simply to determine the extent of power capable of being generated rather than investigating applications and uses of the energy. Furthermore, most previous research efforts have utilized liquid heat exchangers or forced convection to significantly improve heat flow and power generation, but require complex cooling loops and systems. These previous studies also commonly do not consider the amount of energy applied to the cooling system and therefore only report gross levels of power. In this article, the use of TEGs as power harvesting devices that do not have an active heat exchanger, but function as a completely passive power scavenging system with SHM applications, is reviewed. The motivation for investigating a passive power generation device stems from the need to identify effective power sources for the development of self-powered wireless SHM systems. These systems could be placed in a desired location without regular replacement of batteries or maintenance, as most wireless devices currently do. Because of the remote placement of the TEG, it becomes impossible to incorporate active heat exchangers into the system.

When deploying a wireless SHM network, the sensors are placed in a variety of locations, some of which have ambient vibration and others that have ambient thermal gradients. Because the ambient source of energy may vary around the structure, it is important to have a variety of methods to capture the energy. In the case of vibrating structures such as a bridge or tall building, piezoelectric materials may form the ideal energy harvesting device, but if ambient vibration is not present, then alternative power generation devices must be studied. Here two potential locations for a wireless SHM system that does not have significant vibration are examined: first, those that are exposed to the sun but do not vibrate, such as a dam; and second, structures that are subjected to high temperatures, such as a boiler, rocket, or combustion engine. When performing SHM, it is typically not necessary to continuously check its integrity, but rather to perform an analysis once a day or perhaps a week. Therefore, the energy harvesting device should be able to store the energy until it is needed by the electronics. The ability to store the generated energy from piezoelectric materials in a rechargeable battery has been shown by Sodano *et al.* [11, 12]. Here two TEG systems are discussed in the context of using them to recharge a completely discharged nickel–metal hydride battery and the relative charge time will be compared with that of piezoelectric-based power harvesting systems.

2 BASIC THEORY

TEGs use the Seebeck effect, which describes the current generated when the junction of two dissimilar metals experiences a temperature difference. This effect is also responsible for the current generated by a thermocouple. Using this idea, numerous *p*-type and *n*-type semiconductor junctions are arranged electrically in series and thermally in parallel to construct the TEG. Thus, when an electrical current is applied to the TEG a thermal gradient is generated, allowing the device to function as a small solid state heat pump. Inversely, if a thermal gradient is applied to the device, it will generate an electrical current that can be utilized to power other electronics. The simplest model of TEG is

$$V = \alpha \Delta T \tag{1}$$

Here V is the voltage generated, ΔT is the temperature difference (gradient) across the TEG and the constant of proportionality α is called the *Seebeck coefficient*. Thus the basic TEG model produces a dc voltage proportional to the temperature difference across the device. If the TEG is connected to a load, then one simple model is to consider the TEG electrically as a voltage source with some internal resistance applied to some resistive load. This is illustrated in Figure 1. In the figure V_{TEG} is the voltage generated according to equation (1), R_{TEG} is the internal electrical resistance provided by the TEG material and R_{load} is the electrical resistance provided by an external load to the TEG. The load resistance is that of the electrical resistance associated with a direct use device or the resistance offered by a storage device (either a capacitor or battery). For the case when R_{load} is a battery or a capacitor being charged, a diode must be included in this circuit between the generator and the load to prevent a backflow of energy from the battery to the TEG when ambient energy is not being harvested. As a general rule, the power delivered is maximized when the load resistance or impedance is equal to the internal resistance of the TEG.

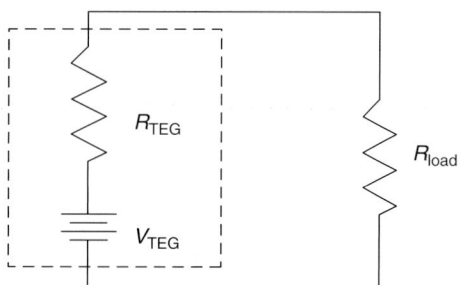

Figure 1. Simple electric circuit model of a TEG.

Practical TEG devices consist of many *pn* junctions connected electrically in series and thermally in parallel (so the same thermal gradient appears across each *pn* pair) as illustrated in Figure 2. If *n* is the number of such junctions, then the voltage can be modeled simply as

$$V = n\alpha \Delta T \qquad (2)$$

Therefore, the voltage or power output is increased by increasing the number of junctions. Traditional thermoelectrics are built by physically assembling discrete blocks of *n*-doped or *p*-doped thermoelectric elements onto electrical circuits as illustrated in Figure 2. The circuits are typically on ceramic plates that have metal lines to route the electrical current. The thermoelectric elements are placed onto the ceramic plates using pick-and-place assembly equipment and attached to the metal lines on the ceramic plates using solder. On the basis of this method of manufacturing, the ability to scale the size of the thermoelectrics to smaller scales has been limited.

Scaling thermoelectric devices to a smaller size is desirable because the power density (watts per square centimeter that it can generate) is inversely proportional to the length (height in Figure 2) of the thermoelectric element. By scaling to smaller geometries, higher power densities can be achieved, leading to a more efficient device. This advantage has spawned a rapid growth of research into microelectromechanical system (MEMS) thermoelectric devices constructed using thin-film technology. Thin-film thermoelectric materials can be grown by conventional semiconductor deposition methods, and devices can be fabricated using conventional semiconductor microfabrication techniques. The resulting TEG devices can be produced in large quantities with small dimensions, allowing them to be easily integrated into the SHM sensing platform.

3 EXAMPLES

Two thermoelectric systems will be presented here to determine their ability to convert solar energy and waste heat into electrical energy and its storage in a conventional rechargeable battery. The results will be compared with energy harvesting using photovoltaic materials (solar cells). The first system developed used solar radiation to heat one side of the TEG, while the cold side was bonded to a metallic structure that functioned as a heat sink with a large thermal mass or capacity. Because direct solar energy was not sufficient to heat the hot side of the TEG, the idea of a "greenhouse" was combined with a solar concentrator to elevate the temperature. A greenhouse functions by allowing visible light emitted by the sun to pass through the transparent surface, thus heating the objects inside and converting the visible light into thermal infrared waves that cannot penetrate the surrounding medium to exit, causing the heat to build up. Because dark objects do not reflect much visible light, but rather convert almost all of it into thermal energy, a blackbody heat sink was placed inside the container to increase the thermal energy stored. To further increase the amount of thermal energy applied to the hot side of the heat sink a solar concentrator was used. A solar concentrator focuses a large area of sunlight onto a smaller area, thus increasing the thermal energy stored in the greenhouse. A schematic describing the layout of the system is shown in Figure 3. This system does not preclude the use of a solar cell, but could be applied with one to capitalize on both the incident light as well as the thermal energy released.

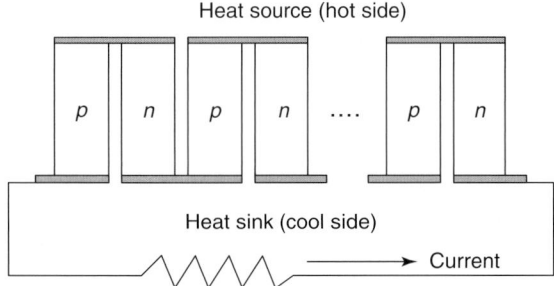

Figure 2. Arrangement of a thermal electric generator device.

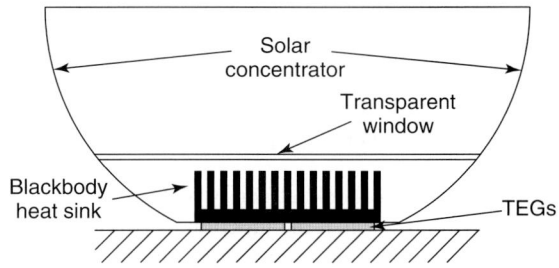

Figure 3. Schematic of solar harvesting device.

The second method of capturing ambient energy was from waste heat that would be available from combustion engines, boilers, or furnaces. To simulate the energy available in these locations, a hot plate was used. A heat sink was attached to the hot plate to allow more energy to be removed from the cold side of the TEG and facilitate a larger power output. To investigate completely passive power harvesting methods, the systems did not include a means of providing forced convection, which would greatly increase the energy produced by the TEG, but would require additional energy for the active cooling systems.

4 EXPERIMENTAL TESTING

A prototype of each power harvesting system was constructed and experimentally tested to identify their ability to capture thermal energy, convert it to electrical power and then use the electrical output to charge a discharged nickel–metal hydride battery. Both systems used eight Melcor thermoelectric coolers (model HT-4-30) to generate the electrical signal. The physical properties of this device are shown in Table 1. The first system used solar energy to develop a thermal gradient over the TEG. To increase the amount of solar energy applied to the power harvesting device, an aluminum parabolic solar concentrator was used to reflect the visible light onto a black body aluminum heat sink (7.62 × 7.62 × 3.175 cm). The heat sink was painted flat black to allow more of the visible light to be converted into thermal energy. Above the heat sink was a Plexiglas window that trapped the thermal energy, thus allowing it to accumulate and increase the thermal gradient over the TEG. The solar concentrator, heat sink, and TEGs were bonded to a steel plate, which

Table 1. Dimensions and electrical properties of the TEG

Property	Value
Dimensions ($w \times l \times h$)	30 mm × 34 mm × 3.2 mm
Maximum temperature difference	77 °C
Number of thermocouple junctions	127
Device resistance	3.78 Ω
Resistivity	1.37 $\Omega\,cm^{-1}$

Figure 4. Experimental setup of the solar harvesting system.

acted as the host structure; this setup is shown in Figure 4.

The second source of ambient thermal energy tested here is simply a structure with an elevated temperature available from aircraft engines, Heating Ventilation and Air Conditioning (HVAC) systems, boilers, or furnaces. To simulate the energy available in these locations, a hot plate was used. Because the purpose of this study is to investigate completely passive power harvesting methods, the systems did not include a means of providing forced convection, which would greatly increase the energy produced by the TEG, but would require additional energy for the active cooling systems. A prototype of the power harvesting system was constructed and experimentally tested to identify its ability to capture thermal energy, convert it to electrical power, and then use the electrical output to charge a discharged nickel–metal hydride battery. The setup used eight Melcor thermoelectric coolers (model HT-4-30) to generate the electrical signal. To simulate this environment, the TEGs were fixed between a heat sink and a thin aluminum plate, which was then attached to a hot plate using thermal grease. The experimental setup is shown in Figure 5. To monitor the temperature of the hot and cold sides of the TEG, Omega CO-1 thermocouples were used. Because the thermocouple was only 0.13-mm thick, it could be bonded to the hot and cold sides of the TEG.

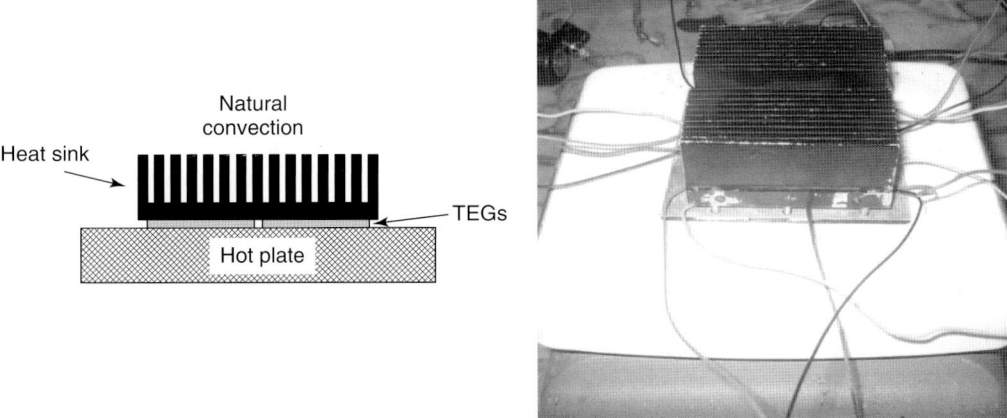

Figure 5. Experimental setup of the energy harvesting system.

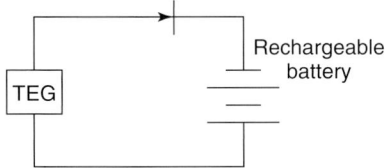

Figure 6. Diagram of circuit used to recharge batteries.

When an SHM system is applied, it is often only necessary that the integrity of that structure is not monitored continuously, but rather only periodically—every day, week, or even month. Continuous testing is not necessary because in many cases, damage will occur over an extended period of time from aging or fatigue. If the SHM system operates only for a small percentage of the time, the energy generated by the power harvesting system must be stored until needed by the electronics. The storage of the electricity is also necessary to ensure that when the electronics are switched on, sufficient power is available. Typically, if energy is to be stored for an extended period of time, it is more effective to use a rechargeable battery. Therefore, a simple circuit can be been constructed to take the electrical energy generated by the two power harvesting devices and store it in a nickel–metal hydride battery. The circuit used in this study is shown in Figure 6. The diode is a necessary piece of this circuit because it forces current to only flow in one direction. If the diode is not present, the TEG would draw power form the battery during times when the voltage generated was less than the voltage of the battery or if the thermal gradient is reversed. Because the output of the TEG is a dc signal, it does not require a means of rectifying, which is a source of energy loss in piezoelectric power harvesting.

5 RESULTS

To experimentally illustrate the power generated by the TEG, it is placed on a hot plate, thus allowing the applied temperature to be accurately monitored. For the system of eight TEGs electrically in series used in this power harvesting device, the Seebeck coefficient is not defined and therefore was fit using experimental data and equation (2). To determine this coefficient, the voltage output of the TEG was measured as the temperature of the hot plate was varied for two different configurations of the TEG modules. A schematic of the different TEG configurations is provided in Figure 7. The resulting voltage output for each configuration and the linear fits for each case are shown in Figure 8. From this figure, it can be see that the Seebeck coefficient for the two different configurations varies. The change in the Seebeck coefficient occurs because the stacking of the TEG modules causes variation in the thermal gradient over each module and in the heat sink's ability to remove energy from the cold side due to a change in surface area in contact with the thermal source. The effect of the configuration on the heat sink's performance can be

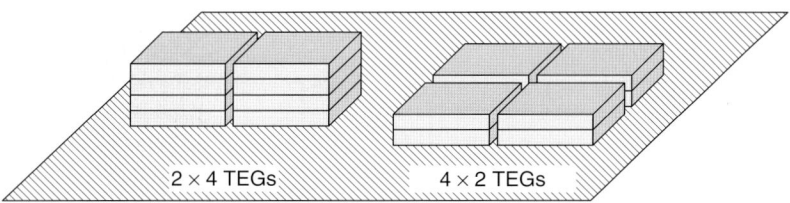

Figure 7. Three different configurations of the TEGs with the hot plate, the footprint is 20.4 cm² with two TEGs and 40.8 cm² with four TEGs.

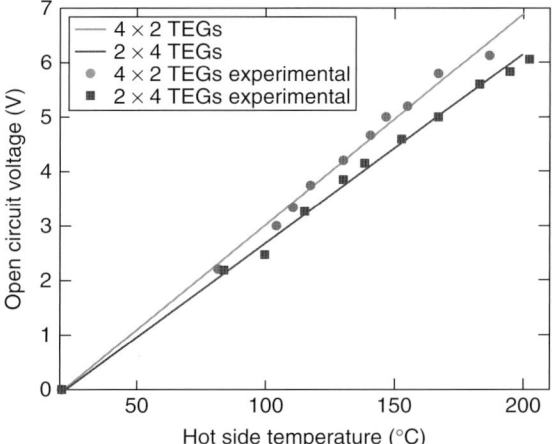

Figure 8. Output voltage of the thermoelectric generator as a function of the hot side temperature.

illustrated by the air temperature around the heat sink. When the heat sink is close to the hot plate, the temperature of the air surrounding it is higher and thus reduces the total thermal difference between the hot and cold sides. However, if too many TEGs are stacked, the gradient between the hot plate and heat sink is divided over each TEG and a large portion of energy is allowed to be transmitted from the sides of the stack, thus reducing the overall power output. The resulting Seebeck coefficient of the configuration using two layers of four TEGs and four layers of two TEGs are 2.96×10^{-4} and 2.68×10^{-4} V K^{-1} respectively.

The current output for each of these systems can now be found using the following relationship [12]:

$$I = \frac{G \alpha D_T}{2\rho} \qquad (3)$$

where I is the current output, G is the area per unit length of the thermoelectric element, and ρ is the resistivity. Now the power generated by each system can easily be determined by equating the power output of the energy harvesting system to the product of the output voltage and current. The results show that the power harvesting module does indeed vary linearly and the thermal gradient or number of modules necessary to generate a particular amount of power for an electronic device can be easily determined. Furthermore, the amount of power that can be generated from TEGs when placed on a 200 °C surface can be as high as 40 mW without any form of convective heat transfer, as shown in Figure 9. As a comparison, when a piezoelectric device subjected to the level of vibration typically found in an automobile engine, it can only generate a maximum of 2 mW [10]. This point illustrates the substantial difference in the power available from TEGs and the idea that more powerful electronics can be powered when using them.

After the amount of power generated by the TEG for various hot side temperatures had been quantified, the ability of each device to recharge a nickel–metal

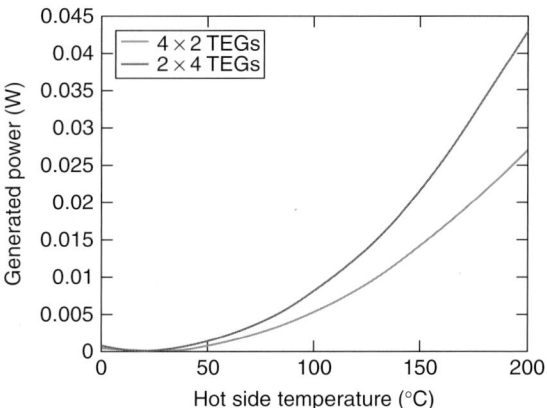

Figure 9. Power generated by the thermal harvesting system as a function of the hot side temperature.

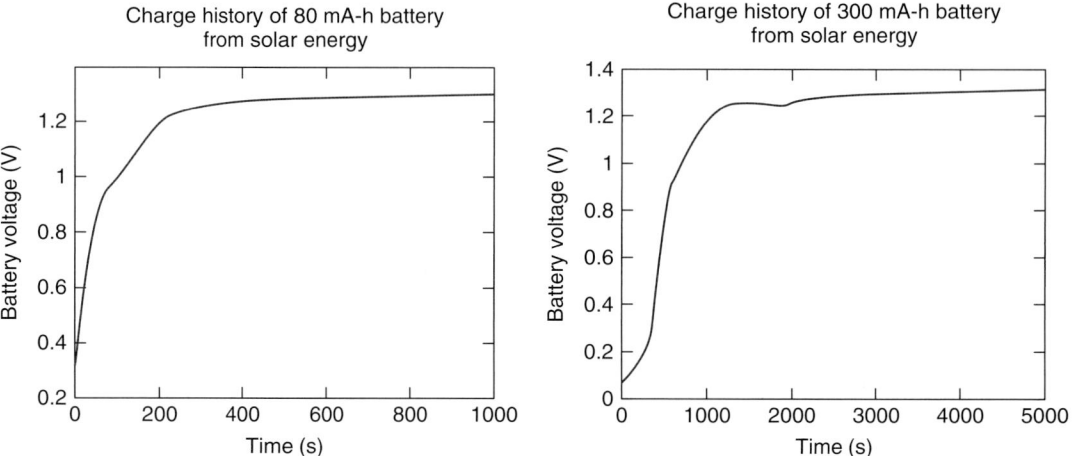

Figure 10. Charge histories of rechargeable batteries from solar harvesting system for an 80 mA·h battery and a 300 mA·h battery.

hydride battery was determined. The circuit layout shown in Figure 6 was used to collect the electrical output of the TEG and store it in the rechargeable battery. For this study, an 80 and a 300 mA·h battery were tested with the solar harvesting system and only a 300 mA·h battery was tested for the waste heat device. (The unit milliampere-hour "mA·h" indicates the capacity of a battery: an 80 mA·h capacity implies the battery will last for 1 h if subjected to 80-mA discharge current.) An 80 mA·h battery is roughly the size of a watch battery and a 300 mA·h battery is slightly less than half the size of a typical AAA battery.

First, the solar harvesting device was investigated by placing the system in direct sunlight with an ambient temperature of 29.4 °C. Both batteries were charged on the same day with relatively short time between each test, therefore limiting the variation between tests. The hot side temperature of the solar harvesting device was measured to be approximately 52.8 °C. The resulting charge times for both batteries are shown in Figure 10. From these figures it is apparent that the TEG system developed in this study can effectively harvest solar radiation and use that energy to quickly recharge a discharged battery. Next, the waste heat system was tested and the resulting charge time is shown in Figure 11. Once again, it is clear that the TEG is very capable of storing converted energy in a battery.

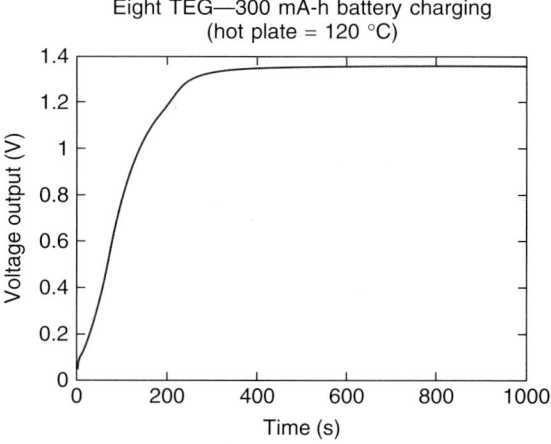

Figure 11. Charge history of a 300 mA·h nickel–metal hydride battery from waste heat harvesting system.

6 COMPARISON OF HARVESTING TECHNOLOGIES

After identifying the ability to use a TEG for the purpose of recharging small batteries, the results were compared with those found using piezoelectric materials by Sodano *et al.* [11]. The time required for each system to charge the battery to a cell voltage of 1.2 V was the measure in each case and the results are provided in Table 2. The charge time listed for the TEG devices are from Figures 10 and 11,

Table 2. Time needed to achieve the battery's cell voltage by a TEG with solar energy and waste heat, and for a piezoelectric material experiencing random vibration [11]

Capacity of the battery (mA·h)	Charge time from solar energy (min)	Charge time from waste heat (min)	Charge time from vibration using piezoelectric materials (h)
80	3.3	NA	2
300	17.3	3.5	5.8

NA, Not Available

while the charge time for the piezoelectric system are for a PZT patch excited at an amplitude and random frequency measured from a typical automobile [11]. The results from charging a battery using piezoelectric materials are only provided to give an idea of the results from previous studies performed under realistic conditions and do not represent that a TEG will always outperform piezoelectric materials. (Note: this comparison is not intended to imply that piezoelectric materials are less effective or efficient than a TEG and these tests were performed on two very different systems.) The charge times in Table 2 may seem to be lower than possible; the time listed is simply to achieve the battery's cell voltage and not a full charge. The time needed to take the battery up to capacity would be longer, however, the times listed for both the piezoelectric and the TEGs both represent the time required to reach the cell voltage from a fully discharged state. To determine the time needed to provide a complete charge to the battery, a charge controller would be needed. The TEG is capable of quick recharge times because of its large current output, whereas the piezoelectric material supplies a very high voltage at a low current. To give an idea of the difference in these two devices, the impedance of one TEG is approximately $3\,\Omega$ while that of the piezoelectric device is approximately $10\,000\,\Omega$. Owing to the lower impedance of the TEG, eight modules had to be connected electrically in series to boost the output voltage to the required $1.2\,V$ of the battery; however, this lower impedance also makes the TEG far more suited for use with rechargeable batteries, which charge faster with larger currents.

7 CONCLUSIONS

The usefulness of wireless SHM systems is established in many articles in this encyclopedia. Energy harvesting is an enabling technology for SHM. The long term goal of many SHM systems is to develop a completely self-powered electronic module that may be placed in a remote location without the concern of replacing the power supply. Generating energy from ambient sources using piezoelectric materials to harvest the vibration energy surrounding a system is one approach. For those cases where the structure in question does not experience sufficient vibration to generate the power levels needed to sustain the operation of the electronics, thermoelectric harvesting may be appropriate. This article has focused on using TEGs to generate an electrical signal from ambient thermal gradients. Two potential examples of this technology are presented: harvesting of solar radiation and harvesting of waste heat. For each application, an experimental prototype was constructed and tested to determine the effectiveness in recharging a discharged nickel–metal hydride battery.

Each application of the thermal generator was implemented such that only conductive heat transfer was present. Most of the previous studies have utilized an active heat exchanger to increase the energy output. However, the use of active convection causes the net power output to be reduced due to the energy applied to the heat exchanger. Therefore, this article has focused on demonstrating that with fairly small thermal gradients and only conductive heat transfer the thermal electric generator can form an effective power harvesting source. The results indicate that TEG does produce significantly more power than a piezoelectric device and that the time needed to recharge a battery is significantly lower. Thus in applications where thermal gradients are present, TEGs make a suitable choice for powering SHM applications [12].

REFERENCES

[1] Spencer BF, Ruiz-Sandoval ME, Kurata N. Smart sensing technology: opportunities and challenges.

Journal of Structural Control and Health Monitoring 2004 **11**(4):349–368.

[2] Sodano HA, Park G, Inman DJ. A review of power harvesting using piezoelectric materials. *The Shock and Vibration Digest* 2003 **36**(3):197–206.

[3] Goldsmith HE. *Applications of Thermo-Electricity, Methuen's Monographs on Physical Science.* John Wiley & Sons: 1960.

[4] Lawrence EE, Snyder GJ. A study of heat sink performance in air and soil for use in a thermoelectric energy harvesting device. *Proceedings of the 21st International Conference on Thermoelectronics.* Portland, OR, 25–29 August 2002; pp. 446–449.

[5] Rowe MD, Min G, Williams SG, Aoune A, Matsuura K, Kuznetsov VL, Fu LW. Thermoelectric recovery of waste heat—case studies. *Proceedings of the 32nd Intersociety Energy Conversion Engineering Conference.* Honolulu, HI, 27 July–1 August 1997; pp. 1075–1079.

[6] Fleming J, Ng W, Ghamaty S. Thermoelectric-based power system for unmanned-air-vehicle/microair-vehicle applications. *Journal of Aircraft* 2004 **41**(3):674–676.

[7] Birckolz U, Grob E, Stohrer U, Voss K. Conversion of waste exhaust heat in automobile using FeSi2 thermoelements. *Proceeding of the 7th International Conference on Thermoelectric Energy Conversion.* Arlington, VA, 1988; pp. 124–128.

[8] Matsubara K. Development of a high efficient thermoelectric stack for a waste exhaust heat recovery of vehicles. *Proceedings of the 21st International Conference on Thermoelectronics.* Portland, OR, 25–29 August 2002; pp. 418–423.

[9] Bass JC, Elsner, NB, Leavitt FA. Performance 1 kW thermoelectric generator for diesel engines. *Proceedings of AIP Conference* 1994 **316**(1):295–298.

[10] Vázquez J, Sanz-Bobi MA, Palacios R, Arenas A. State of the art of thermoelectric generators based on heat recovered from the exhaust gases of automobiles. *Proceedings of the 7th European Workshop on Thermoelectrics.* Pamplona, Spain, 2002, Paper No. 17.

[11] Sodano HA, Park G, Inman DJ. Generation and storage of electricity from power harvesting devices. *Journal of Intelligent Material Systems and Structures* 2005 **16**(1):67–75.

[12] Sodano HA, Park G, Inman DJ. Comparison of piezoelectric energy harvesting devices for recharging batteries. *Journal of Intelligent Material Systems and Structures* 2005 **16**(10):799–807.

FURTHER READING

Grisso BL, Kim J, Farmer JR, Ha DS, Inman DJ. Autonomous impedance-based SHM utilizing harvested energy. In *Structural Health Monitoring 2007*, Chang F-K (ed). DEStech Publications, September 11–13, 2007; pp. 1373–1380.

SECTION 4
Examples of Systems

Chapter 78

Stanford Multiactuator–Receiver Transduction (SMART) Layer Technology and Its Applications

Xinlin P. Qing[1], Shawn J. Beard[1], Amrita Kumar[1], Irene Li[1], Mark Lin[2] and Fu-Kuo Chang[2]

[1] *Acellent Technologies, Inc., Sunnyvale, CA, USA*
[2] *Department of Aeronautics and Astronautics, Stanford University, Stanford, CA, USA*

1 Introduction	1363
2 Smart Layer	1364
3 Integration of Smart Layer with a Host Structure	1367
4 Survivability and Reliability of Built-in Sensor Network	1372
5 Effect of Smart Layer on Structural Integrity	1378
6 Sensor Network-based Structural Health Monitoring Systems	1382
7 Damage Detection in Composite Structures	1383
8 Conclusion	1385
Acknowledgments	1385
Related Articles	1385
References	1385

Encyclopedia of Structural Health Monitoring. Edited by Christian Boller, Fu-Kuo Chang and Yozo Fujino © 2009 John Wiley & Sons, Ltd. ISBN: 978-0-470-05822-0.

1 INTRODUCTION

A typical structural health monitoring (SHM) system consists of three basic components: (i) actuators and/or sensors, (ii) diagnostic software, and (iii) integrated hardware to monitor the "health state" of in-service structures [1–4]. Many research activities have been done to evaluate new sensor technologies for health monitoring of structures, such as optical fibers [5], piezoelectric materials [6], magnetostrictive materials [7, 8], and microelectromechanical system (MEMS) [9]. Among various types of transducers, piezoelectric materials are being widely used for SHM because they can be used as either actuators or sensors due to their piezoelectric effect and vice versa (*see* **Chapter 55**; **Chapter 57**). In the active-sensing mode, both wave propagation and impedance-based SHM methods have been developed [10–15]. In the passive sensing mode, the piezoelectric sensors are used as sensors that continuously monitor external impact events [16]. In the wave propagation-based SHM, guided waves, such as Lamb and Rayleigh waves, are most widely used for damage detection in metallic and composite structures. Guided waves used for damage detection are introduced into a structure at one point by a piezoelectric actuator and sensed by another piezoelectric sensor at a different position. A key advantage of using piezoelectric elements is that a larger area of the structure can be monitored with fewer transducers, which is vitally important for the monitoring of large-scale structures. Other sensors, like optical fiber-based types, can only scrutinize smaller, specific areas, thus leaving larger areas of a structure unmonitored.

However, it is well understood in the field of SHM that the network of transducers plays a key role in the performance of the SHM system. The ability of sensors and actuators in the network to communicate with each other establishes the intelligence of the system. The type, location, and number of sensors and actuators critically affect the sensitivity of the SHM system.

As the number of sensors increases, the integration of such a network with a structure can be very challenging or become impractical. Obviously, for SHM, the network sensors must be able to

1. integrate/adapt easily within/onto the structure;
2. accommodate any structural configuration;
3. carry a minimal weight; and
4. operate under variable environments.

Stanford multiactuator–receiver transduction (SMART) Layer technology, originally developed by Lin and Chang [17], can overcome these major challenges listed above. In this article, the progress of the SMART Layer technology is summarized in terms of its adaptability for practical applications.

2 SMART LAYER

2.1 SMART Layer design

An important part of the SHM system is the proper integration/adaptation of the sensors with the structure. Sensors permanently mounted onto structures provide the capability to monitor the condition of these structures throughout their service life. The SMART Layer is a unique and cost-effective method for integrating a network of piezoelectric elements with a structure [17, 18]. The layer is made of a thin dielectric film with an embedded network of distributed piezoelectric elements that can be used as either actuators or sensors. The novelty of the SMART Layer lies in its networking capabilities with any type of sensor that enhances its monitoring capabilities and eliminates the need to place each type of sensor individually on the structure. The major features of the SMART Layer include

- actuating and sensing capabilities;
- built-in sensor network for area sensing;
- signal consistency and sensor reliability;
- multiple wires from every transducer to improve reliability of the circuit;
- shielded layer to reduce electromagnetic (EM) noise;
- "hardwired" sensors for direct hardware diagnostic;
- ease of installation.

As shown in Figure 1, the SMART Layer utilizes a layered construction: a circuit layer, an insulation layer, and a sensor layer. In the sensor layer, PZT (lead zirconate titanate) discs with a diameter of 6.35 mm and a thickness of 0.25 mm are most often used, while other PZTs of different shapes and sizes can also be included in the layer as needed. Typical properties of the SMART Layer are given in Table 1.

2.2 Double/multiwire SMART Layer

The SMART Layer can be designed and manufactured in different configurations based on the requirements of the application. To increase the reliability

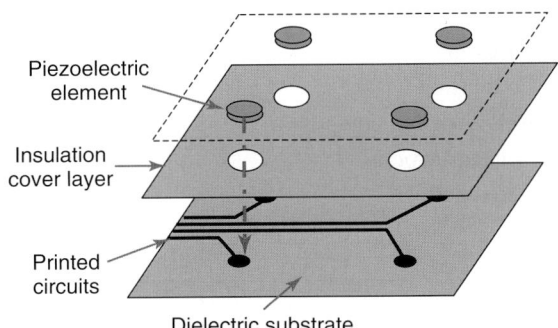

Figure 1. Overview of SMART Layer.

Table 1. Typical properties of SMART Layer

Network sensors	Dielectric material	Thickness of the layer	Temperature range	Environment
PZT-5A disc $D = 6.35$ mm $T = 0.25$ mm	Polyimide	~0.15 mm thick at the area without PZT	−54 to 121 °C for continuous operation	Coated SMART Layer passed salt fog and humidity tests

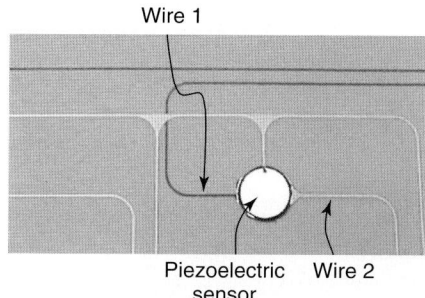

Figure 2. A double/multiwire version of SMART Layer.

of the layer, multiple printed circuitry has been developed for the layer design. With a single-wired layer, a small cut caused by impact or damage in the circuit can break the communication between the sensor and the diagnostic hardware system, rendering it worthless. The method of placing multiple circuits on the layer has been developed so that if one circuit breaks, the other could still be used. Figure 2 shows a picture of the sample of a layer with double wires for each sensor that has been manufactured in this way. Each sensor/actuator is connected via double wires to the final terminals.

This wire redundancy design in the layer eliminates the "fear of breakage of the wires" and makes the layer more reliable. This concept can be extended to any number of wires emanating from a single piezoelectric sensor. The circuit is designed so that the wires emanating from the same piezoelectric sensor are as far away from each other as possible. This design aspect ensures that any damage to one circuit does not damage the other circuit as well.

2.3 Layer for 3D structures

The SMART Layers can be fabricated in different shapes for integrating with different contours of structures. They vary in complexity ranging from a simple flat strip with one or two sensors to a complex 3-D "shell" with more than 50 sensors [18]. For structures with multiple curvatures and complex geometry, SMART Layers can be custom designed with special shapes and cutouts to provide a perfect fit. Two innovative manufacturing methods are described below.

One method of fabricating three-dimensional complex-shaped SMART Layers is to use mechanical locks at preselected locations and then shape the layer based on the required geometry. Upon curing, the layer will hold its shape. A schematic of this concept is presented in Figure 3.

A two-dimensional SMART Layer can also be fabricated through a compressed molding process into a layer with a three-dimensional configuration. A typical fabrication process for a three-dimensional SMART Layer was developed [19] and is briefly outlined as follows (Figure 4):

1. A two-dimensional SMART Layer is first fabricated based on a sensor location map from the 3-D structure.
2. Upon completion of the manufacturing process of the 2-D layer, appropriate cutting, trimming, and scissoring may be applied to conform the layer into the shape of the targeted structure.
3. The 2-D layer is then placed between the molds and subjected to a required processing temperature and pressure.
4. The pressure is maintained during the cooling cycle. After cooling down to room temperature, the layer can be removed from the molds and maintains its 3-D shape.

Figure 5 shows a 3-D layer that was fabricated on the basis of the procedures from items 1 to 4 outlined above, to conform to a composite mold. The 3-D layer was then embedded into a composite frame through

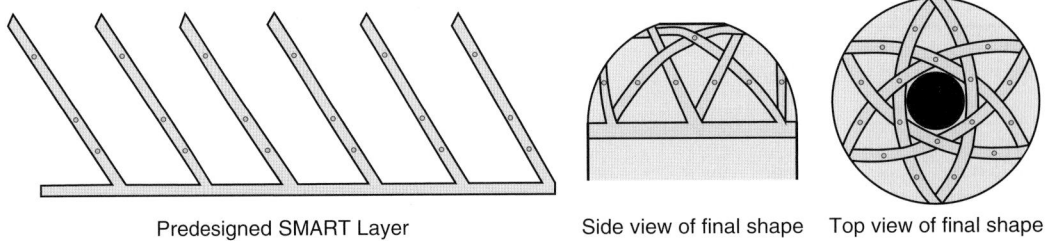

Figure 3. Fabrication method for 3-D SMART Layer.

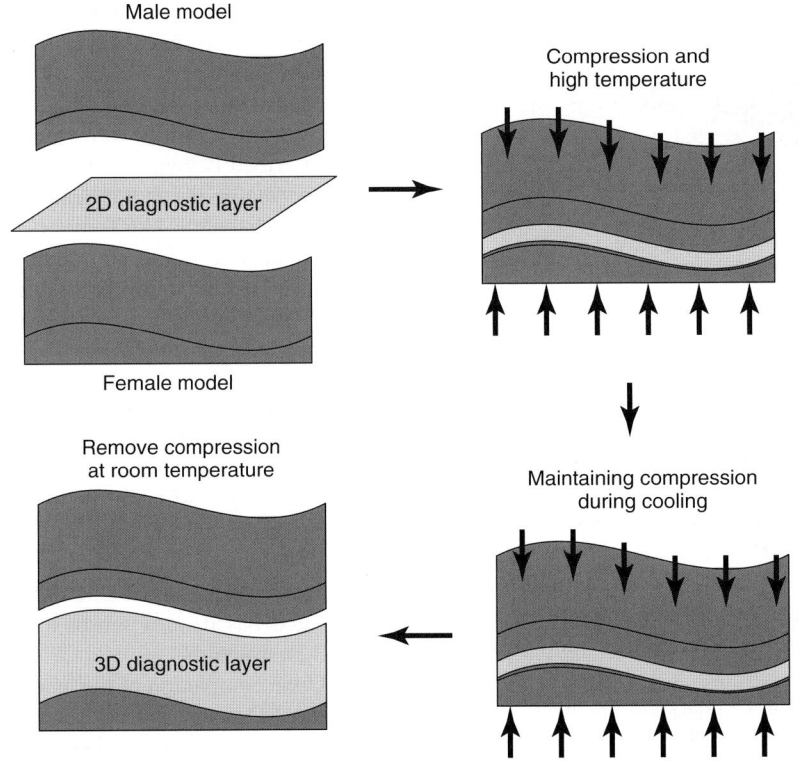

Figure 4. Conversion process of 2-D SMART Layer to 3-D.

a resin transfer molding process, which is described in the next section.

2.4 Hybrid SMART Layer

The SMART Layer can also accommodate other types of transducers along with piezoelectric elements. For instance, techniques have been developed to incorporate optical fibers into the layer. As shown in Figure 6(a), fiber Bragg grating (FBG) sensors are incorporated in the same layer with piezoelectric elements to create a hybrid piezoelectric/fiber-optic SMART Layer. Typical procedures for manufacturing such a diagnostic layer and the physical principles of the hybrid system can be found in the literature [20]. There is another article in the same section of this Encyclopedia that describes the hybrid SHM system (*see* **Chapter 79**). Similarly, other types of sensors, such as strain gauges, MEMS, and temperature sensors can also be incorporated in

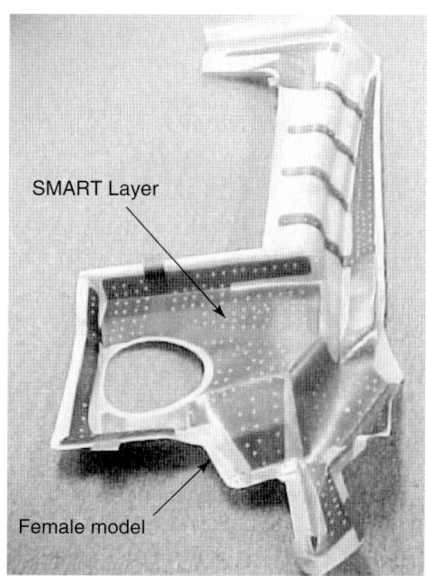

Figure 5. 3-D SMART Layer with 50 piezoelectric elements embedded.

As shown in Figure 7, the SMART Layer can either be surface-mounted onto an existing structure or embedded into a composite structure during fabrication.

3.1 Surface Mounting the SMART Layer

As an example of a surface mounting the SMART Layer, the design and installation of the layers on bonded repairs on an F-16 airplane [21] (*see* **Chapter 99**) is briefly described here. The center keel area of the F-16 fuselage station 341 bulkhead is susceptible to fatigue crack growth due to a maintenance-induced initiation site. The bulkhead is a large single-piece machined structure, and replacement of a damaged bulkhead is time consuming and costly. A system of adhesively bonded repairs was developed by Southwest Research Institute and applied to the keel structure as a potential alternative to replacing the bulkhead. To assess the health of the bonded structure as the airplane continues to be used, a health monitoring system was developed to inspect the bonded repairs.

Structure in the keel area of the F-16 station 341 bulkhead is complex, with material interfaces, fastened joints, and system penetrations. Figure 8 shows two views of the area in the main landing gear wheel well of the airplane that will be repaired and monitored. To ensure that the sensors would fit on the airplane and still provide adequate coverage of the repair area, a full-scale paper mock-up of the sensor layer, including positioning the transducers, routing

the SMART Layer. Figure 6(b) shows a PZT/strain gauge hybrid layer mounted on a composite panel.

3 INTEGRATION OF SMART LAYER WITH A HOST STRUCTURE

In an SHM system, a sensor network needs to be permanently mounted onto the host structures.

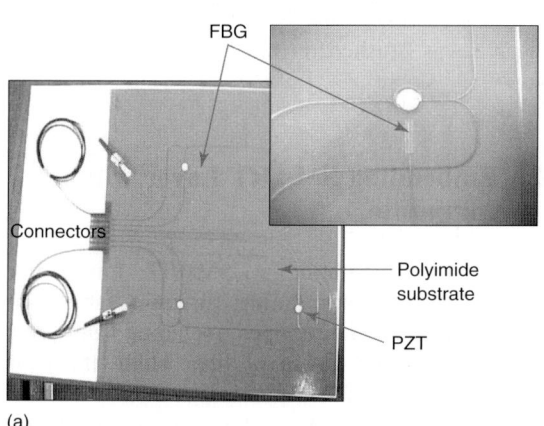
(a)

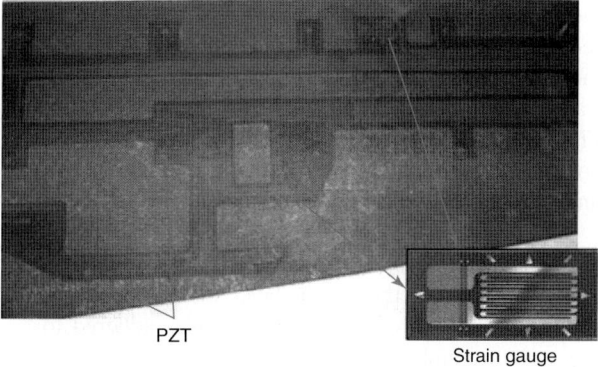
(b)

Figure 6. Hybrid SMART Layer.

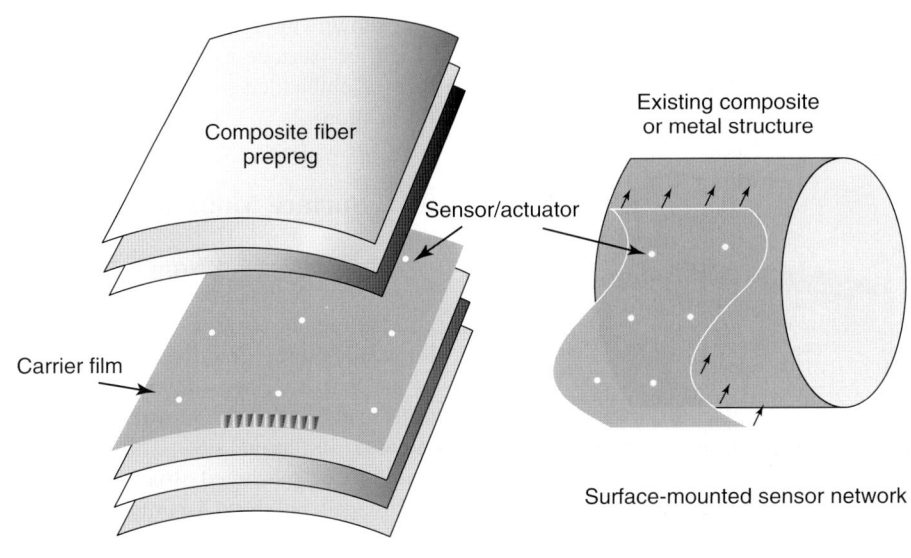

Figure 7. Diagnostic layer integrated with structure.

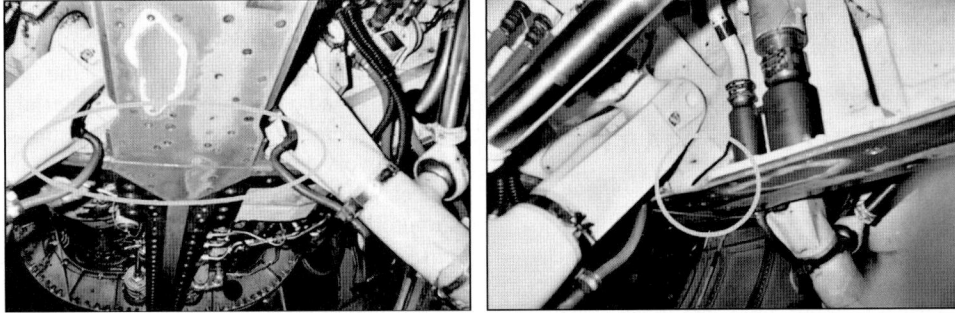

Figure 8. The main landing gear wheel well of an F-16 contains complex structures and multiple system installations [21].

the wiring, and finding an accessible but protected location for the connector, was created by using the F-16 at the US Air Force Museum. A SMART Layer was fabricated to match the paper mock-up. Figure 9 shows the layer prior to installation. The completed system consists of three separate layers and an integral connector.

The installation was performed at Hill Air Force Base in Ogden, Utah. Hysol EA 9394 epoxy adhesive was used to bond the sensor layer on the structure. As shown in Figure 10(b), the layer was adhered to the structure and fixed temporarily with clamps and Kapton tape. After room temperature curing, the tape and clamps were removed, and an overcoat of epoxy adhesive was brushed on the exterior of the sensor layer.

3.2 Embedding SMART Layer within composite

Methods for integrating a SMART Layer into a composite structure during different fabrication processes have been developed, including hand layup of prepregs or wet layup of fiber cloth, the resin transfer molding (RTM) process, and the filament-winding process. For hand layup of prepregs, the layer is simply treated as an additional ply laid

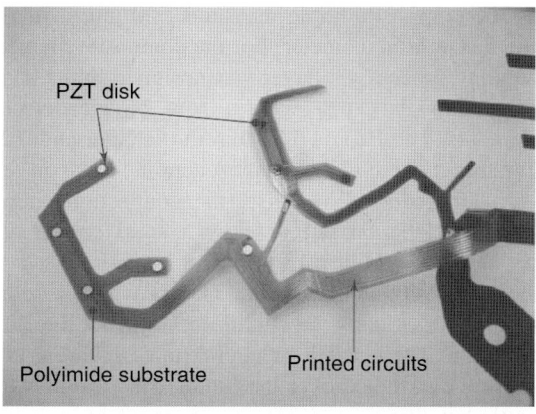 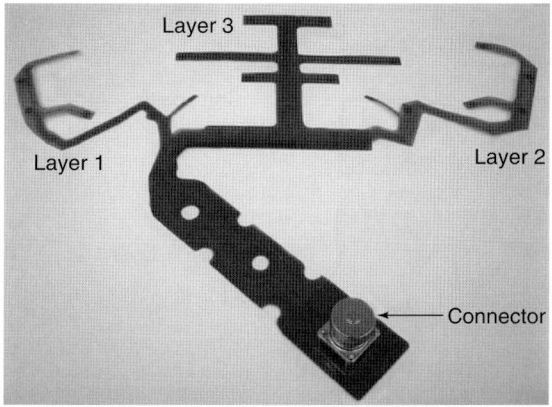

Figure 9. SMART Layer designed for monitoring the bonded repairs of F-16 [21].

Figure 10. The SMART Layers were mounted on the three repairs installed in the keel area of the station 341 bulkhead. (a) In-service F-16, (b) installation, and (c) mounted SMART Layer with surface coating [21].

down during layup. When inserting the layer inside composites, the sensor side preferably should face the thicker side of the laminate to account for the extra thickness of the sensors. For most of the cases, no other adhesive is needed when cocured with composite. Examples for integrating SMART Layers inside composites with RTM process and filament-winding process are given below.

3.2.1 Diagnostic layer integrated into composite during the RTM process

In this section, the study for integrating a sensor network into a three-dimensional composite foam core structure during the RTM process is presented [19]. RTM, which is a popular composite manufacturing process, is a low-pressure, closed molding process, where a mixed resin and catalyst are injected into a closed mold containing a fiber pack or preform. After the resin has cured, the mold can be opened and the finished component can be removed. The inclusion of a SMART Layer inside composite structure during RTM process is shown in Figure 11. The layer is inserted in the mold as an extra preform.

As an example, the 3-D layer shown in Figure 5 was inserted into the interface of the composite and foam core during the RTM manufacturing process. The open sides of piezoceramic disks on the diagnostic layer were oriented toward the composite. The cured composite sandwich structure with sensor network integrated is shown in Figure 12.

3.2.2 Sensor network integrated into filament-wound structure

Filament winding is the process of winding resin-impregnated fiber or tape on a mandrel surface in a precise geometric pattern. This is accomplished by rotating the mandrel while a delivery head precisely positions fibers on the mandrel surface. It is used for the manufacture of parts with high fiber volume

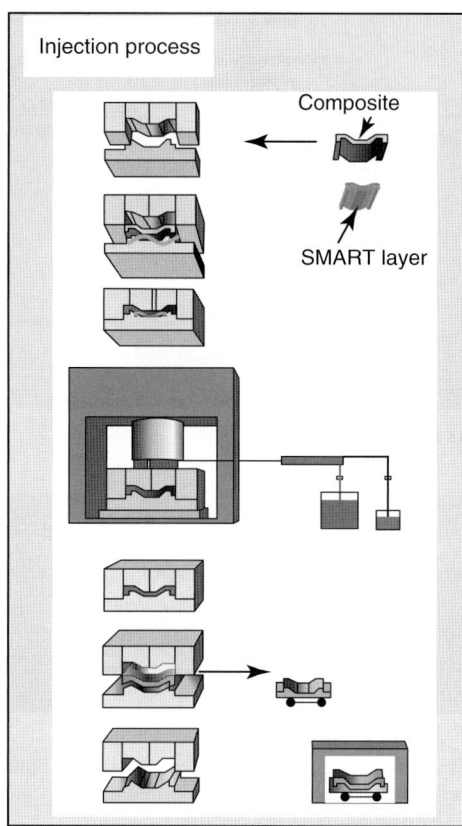

Figure 11. Composite RTM process and inclusion of SMART Layer.

fractions and controlled fiber orientation. To monitor the integrity of filament-wound composite structures, such as solid rocket motors and liquid fuel bottles, a way to embed a sensor network into a filament-wound composite structure was investigated [19, 22].

The composite bottle measures 500 mm in length and 381 mm in diameter. For the composite filament-winding process, the layer was designed in the shape of a thin flat strip with a row of piezoelectric elements. The PZT is 6.35 mm in diameter and 0.25-mm thick. There were five piezoceramic disks on each strip. To reduce EM noise, the connection circuits were shielded on one side by a solid copper foil. As shown in Figure 13, eight strips were used for the composite-wound bottle. This provided a sensor spacing of slightly less than 153 mm in the hoop direction. The result is an approximate square grid of sensors distributed over the bottle. The design of the bottle has five hoop composite layers and two helical composite layers in the cylindrical section, but only two helical layers in the dome sections.

Figure 14 shows some pictures of how the strips of SMART Layer were incorporated into the composite bottle during the filament-winding process. The main process involves winding composite prepreg tows onto an aluminum liner mandrel in either a hoop direction or a helical pattern. First, a hoop layer was wound onto the cylindrical section of the bottle. Four sensor strips were then placed on the top of the hoop layer and held in place by Teflon tape affixed on the dome part. Then two hoop layers were wound over the cylindrical section again. Next, the tapes used to fix the location of the four strips were removed before winding two helical layers over the entire bottle. Then, another hoop layer was wound over the cylindrical section. Next, the four outer sensor strips were placed on the bottle, held in place by Teflon tape. Finally, a hoop layer was wound on the top of the cylindrical section, leaving the outer strips exposed on the dome part. All the open

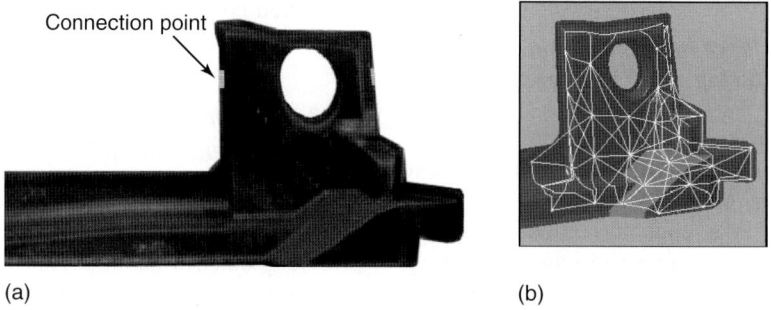

Figure 12. SMART Layer embedded inside a composite sandwich structure in RTM process: (a) photo of the structure and (b) diagram of the sensor network paths.

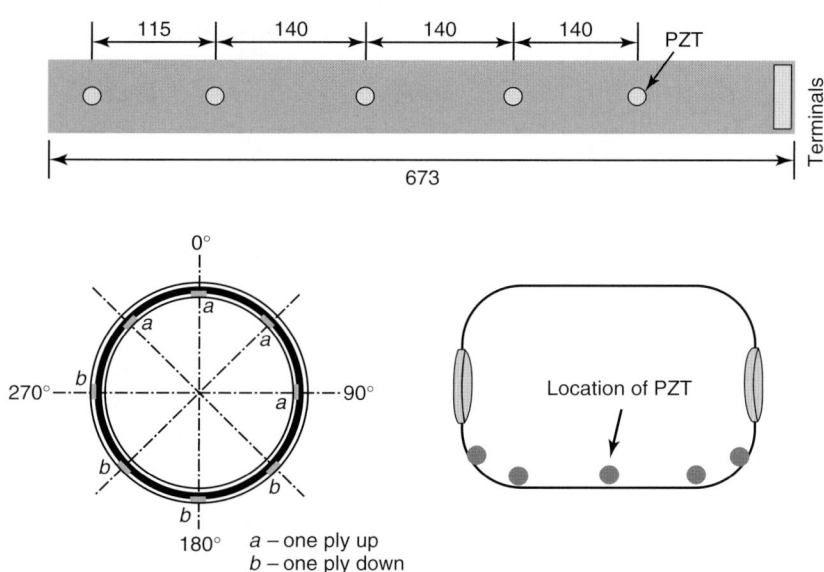

Figure 13. Piezoelectric sensors and their locations in filament-wound bottle.

Figure 14. Embedding the sensor network in the filament-wound bottle. (a) The liner, (b) SMART Layers applied after winding of the hoop section, (c) helical winding, (d) preparation of bottle for curing, (e) bottle ready to be cured in the oven, and (f) complete filament-wound bottle with embedded sensor network.

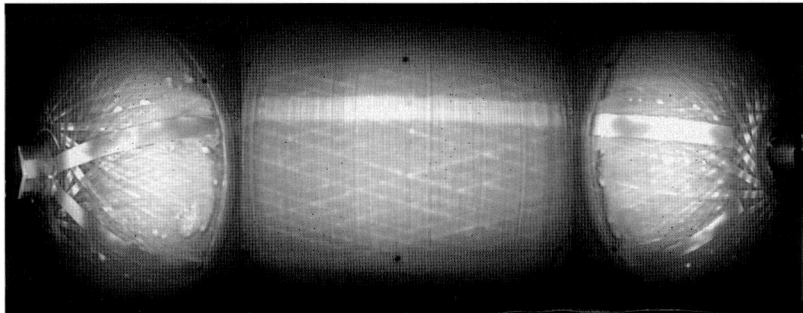

Figure 15. Finished bottle and thermographic image showing embedded SMART Layer.

sides of piezoelectric elements on the strips were oriented toward the thicker side of the composite: the piezoelectric elements on the inner strips face outward and the piezoelectric elements on the outer strips face inward. Since the composite material used to fabricate the bottle is a prepreg, i.e., has both fiber and resin in it, no epoxy resin was transferred into the bottle prior to cure. Since a vacuum bag was not used during curing, Teflon tape was used to push the part of four sensor strips on the outer surface of the dome to stick with the dome surface. The composite bottle was then cured at 177 °C (350 °F) for 2 h in the oven. During curing, the bottle was rotated continuously to obtain an even flow of resin and to eliminate any resin pools.

Figure 15 shows the finished bottle and the corresponding image from a thermographic scan (for thermal imaging techniques, *see* **Chapter 19**). The imager used was an Indigo Merlin running under Thermal Wave Imaging software. The heating source was a TWI Flash System. Prior to scanning, the outer surface of the bottle was painted flat black [23]. From the thermographic image, the strips looked well bonded to the composite. This clearly demonstrated the compatibility of the SMART Layer with the filament-winding process typically used for the manufacturing of composite rocket motor cases.

4 SURVIVABILITY AND RELIABILITY OF BUILT-IN SENSOR NETWORK

The reliability of the sensor network for SHM must be entirely guaranteed through the life of structures. But the environments, such as mechanical loading, temperature, and chemicals, can significantly degrade the performances of sensors due to the nature of sensor materials [24] (*see* **Chapter 156**). A series of tests have been conducted to determine the survivability and functionality of the SMART Layer and its sensors under the different environment conditions. These tests include

- static and fatigue tests;
- temperature tests;
- vibration tests;
- salt fog and humidity tests.

4.1 Characterization of loading effect on the performance of PZT

4.1.1 Mechanical tests on coupon specimens

Tensile tests were conducted on three types of specimens: (i) aluminum specimens with SMART Layer mounted on the surface; (ii) composite specimens with SMART Layer mounted on the surface; and (iii) composite specimens with the layer embedded in the plies [25]. There were two PZT disks on each SMART Layer. For the composite specimens, the layer was cocured one ply down or on the surface of the specimen at 180 °C in an autoclave, while for the aluminum specimen the layer was bonded with an epoxy adhesive film Hysol EA 9696 and cured at 120 °C. During the test, the actuating signal was a five-cycle sinusoidal tone burst with an amplitude of 50 V and a frequency of 50 kHz. The performance of

PZT was studied by the difference between the amplitudes of sensing signals taken at different load histories. The piezoelectric property of PZT was evaluated at intervals of certain strain with and without loading.

Figure 16 shows the results of monotonic tensile tests. As shown in Figure 16(b) and (c), for the PZTs bonded on the composite specimens, the amplitude, η/η_o (η_o is the amplitude of the signal before loading), was almost constant up to the failure strain of PZT, $\varepsilon = 0.1\%$. The difference between the surface-mounted and embedded sensors was negligible except for the region of high strain. However, as shown in Figure 16(a), the amplitude, η/η_o, was almost constant up to about $\varepsilon = 0.3\%$, which is much larger than the failure strain of PZT.

The PZTs mounted on the aluminum specimen could survive much higher tensile strain because of the compressive prestress applied to the PZT during the adhesive mounting process of PZTs [25]. It is clear that the performance of PZT remains unchanged when the applied strain does not exceed the static failure strain of PZT, whereas the degradation of PZT occurs when the applied strain exceeds the static failure strain of PZT.

In addition, it can be concluded that the performance of PZT depends not on the materials of the host structure but on the thermal residual strain induced during the cure cycle. It is important to note that the PZT embedded in the structure can be used without any degradation even if the applied strain to the host structure exceeds the failure strain of PZT by introducing compressive prestress via a specific cure cycle.

The fatigue loading tests were carried out by controlling the strain applied to the specimen on the servo-hydraulic testing machine. The frequency of loading was 10 Hz and the ratio of minimum strain to maximum strain was 0.1. Figure 17 shows the results of fatigue loading tests. For the aluminum specimens, the amplitude, η/η_o, remained unchanged after 10^6 cycles when the maximum strain did not exceed 0.3%. The critical point of the onset of degradation in the fatigue loading tests of aluminum specimens, $\varepsilon = 0.3\%$, agreed with that in the monotonic tensile

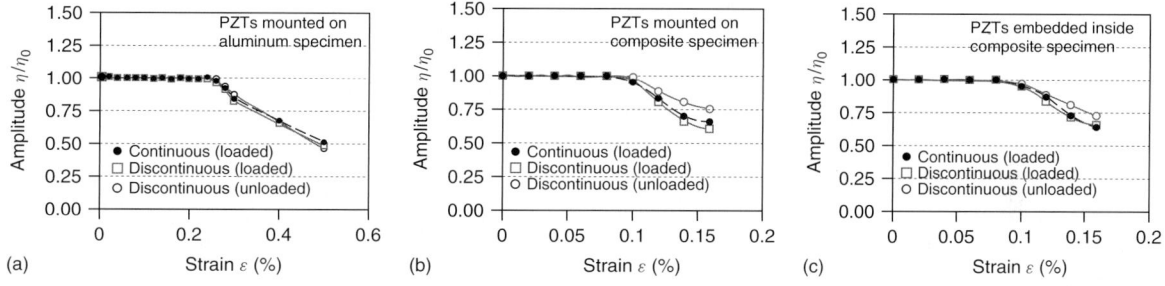

Figure 16. Results of tensile test for (a) aluminum specimen, (b) composite specimen with surface-mounted layer, and (c) composite with embedded layer.

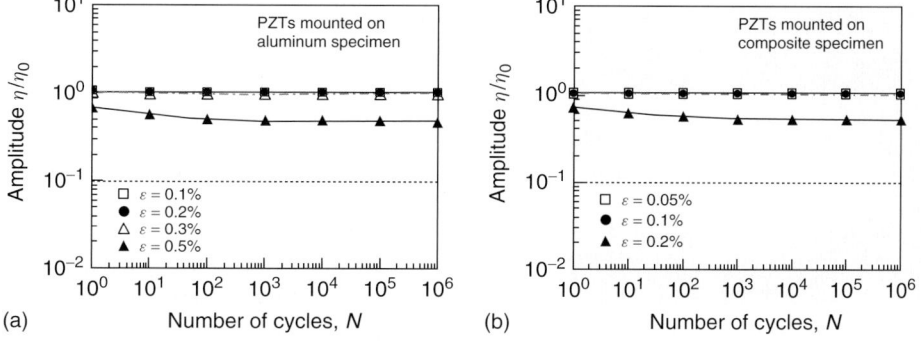

Figure 17. Effect of fatigue loading on the performance of PZT: (a) aluminum and (b) composite.

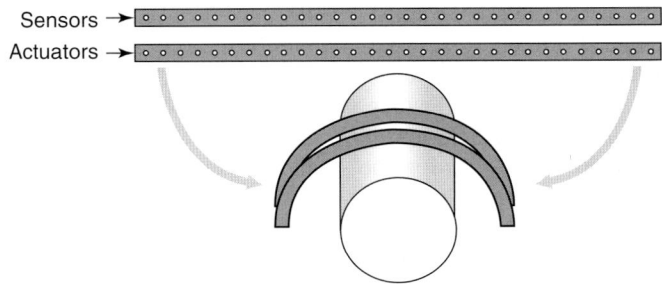

Figure 18. Thirty actuators and 30 sensors on two SMART Layers.

tests. For composite specimens, the amplitude, η/η_o, was almost constant for maximum $\varepsilon < 0.1\%$ when increasing the number of cycles up to 10^6. From the results, it can be concluded that the performance of PZT is not degraded under fatigue loading when the applied strain does not exceed the static failure strain of PZT, whereas the degradation of PZT is considerably stable after the first or several cycles of loading when the applied strain exceeds the static failure strain of PZT.

4.1.2 Fatigue test on a Thunder Horse steel pipe sample

The purpose of the test was to demonstrate the ability of SMART technology to detect and monitor the growth of a fatigue crack at the girth weld of a steel pipe used in offshore industry, and also to investigate the reliability and survivability of PZTs under fatigue load for real-world applications.

Test article
A Thunder Horse steel pipe sample was used in the test. The pipe, consisted of two 3-m-long segments joined together with a girth weld, has a 324-mm outer diameter with a wall thickness of 40 mm. Two 30-sensor SMART Layer strips were mounted on either side of the girth weld as shown in Figures 18 and 19. The sensors used in the strips were PZTs, each having a diameter of 6.35 mm and a thickness of 0.76 mm.

Test procedures and results
The testing was conducted with 2.8 MPa (400 psi) internal pressure and the fatigue cycling ran at ±206 MPa (30 ksi) at 27 Hz. For this test, the 30 piezos on one strip were used as actuators while the 30 piezos on the other strip were used as sensors. Each actuator, in turn, was excited with a five-peak modulated sine wave burst. The excitation generates stress waves that propagate through the structure and these are recorded by the sensor directly across from the actuator. Data was collected at 25 different intervals during the test. The test was stopped at about 12 300 000 cycles because a through-wall crack was observed in the pipe. All sensors were functioning when the tests were stopped.

The effect of cycling electric loading on the performance of PZTs was also investigated. The maximum voltage, which a 0.25-mm PZT can survive, is 70 VAC. Both surface-mounted PZTs and embedded PZTs inside the composite were tested. In each cycle, five-peak modulated sine wave bursts with maximum amplitude of 50 V were input to the PZT actuators at 10 different frequencies (50, 100, 150, 200, 250, 300, 350, 400, 450, and 500 kHz), respectively. The amplitude of sensor response to the excitation generated by the actuator is employed to evaluate the performance of the actuator. Test results that showed sensor signals for their actuator–sensor paths remained unchanged after 10^7 cycles for all PZTs on both specimens.

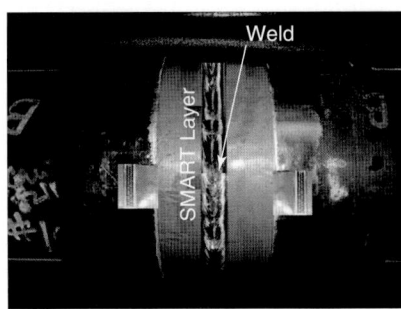

Figure 19. SMART Layers were mounted right up against the cap of the girth weld.

4.2 Effect of temperature and vibration

4.2.1 Temperature effect

Owing to the temperature dependence of material properties of the PZT, adhesive, and host structure, the sensor signals responding to the same voltage input to the actuator are temperature dependent. Figure 20 shows the temperature dependence of piezoelectric coefficients of PZT. It is expected that the sensor signal will change when the temperature changes. Tests on both metal and composite structures were conducted to investigate the temperature effect on the performance of SMART Layer. The SMART Layers were mounted on the structures with Hysol EA 9396. Hysol EA 9396 is a low-viscosity room temperature curing adhesive system with excellent strength properties at temperatures from -55 to $177\,^\circ\text{C}$ (-67 to $350\,^\circ\text{F}$). Before mounting the SMART Layers on the structures, the surfaces of the structures were well prepared. The adhesive was cured at room temperature, followed by a post-curing of 1 h at $93\,^\circ\text{C}$ ($200\,^\circ\text{F}$).

Sensor signals from the SMART Layers on both metal and composite structures were recorded at different temperatures and cycles within the temperature range from -51 to $93\,^\circ\text{C}$ (-60 to $200\,^\circ\text{F}$). Typical sensor signals from the SMART Layers on an Alloy 718 duct are shown in Figure 21. On the basis of the results, it is clear that the sensor signals for all frequencies tested here are repeatable at the same temperature after several temperature cycles. However, both amplitude and phase of the signals are different at different temperatures. According to the signals recorded from the SMART Layers on the metal structures, it can be seen that the amplitude of the signals slightly changed when the temperature changed. Also the phase of the signals shifted. However, the amplitude of signals from the layer on the composite panels significantly reduced when the temperature increased from -51 to $93\,^\circ\text{C}$ (-60 to $200\,^\circ\text{F}$), depending on the resin used in the

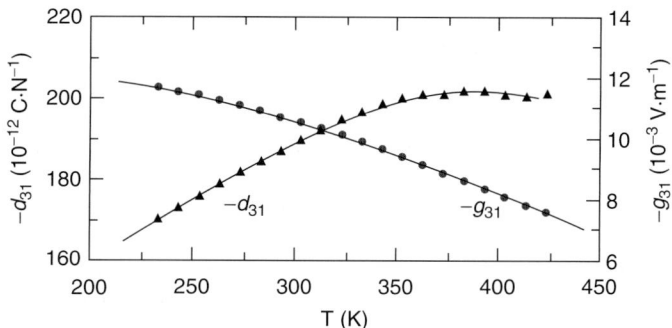

Figure 20. Temperature dependence of piezoelectric coefficients (d_{31}, g_{31}) of PZT (APC 850).

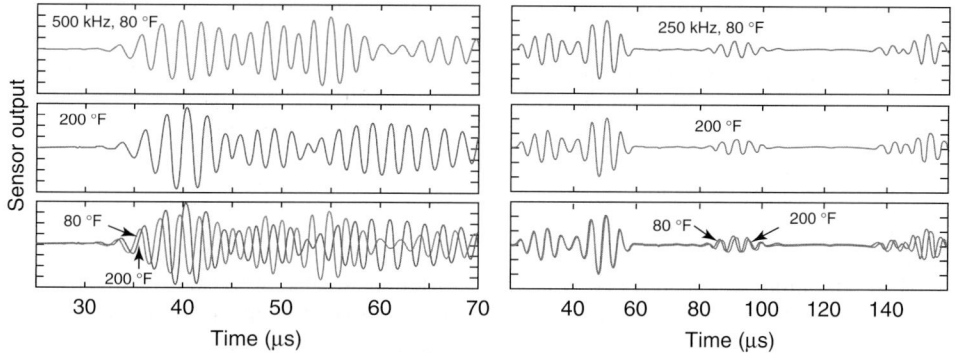

Figure 21. Typical sensor signals recorded at different temperatures.

composites. It can be concluded that the SMART Layer works well within the temperature range from −51 to 93 °C if the host structure can survive the environment [26].

For some specific applications, such as monitoring crack growth in a liquid rocket engine pipe, the sensor network must be able to survive a much harsher environment. Acellent has conducted the tests to determine the ability of the current SMART Layer to survive the extreme cold temperatures and high pressures of rocket engines. The test procedures and results are described in the following section [27]. From the successful application of composite manufacturing process monitoring, it demonstrated that the SMART Layer indeed survive high temperatures up to 200 °C [18].

4.3 Performance of SMART Layer under combined cryogenic temperature and vibration environment

This section describes how the physical robustness of the SMART Layer as well as operational survivability and functionality were verified with a duct simulator, conditioned to flight vibration and shock environments on a simulated large booster LOX-H_2 engine [27].

4.3.1 Test specimen and setup

Two layers (L1 and L3) with four piezoelectric single crystals on each were fabricated and mounted on the surface of an Alloy 718 duct with a diameter of 152 mm and length of 406 mm, as shown in Figure 22. The other two layers (L2 and L4) with PZTs were also manufactured and mounted on the duct to compare the performance of different piezoelectric materials. The low-temperature adhesive EP29LPSP was used to bond the layers on the Alloy 718 duct.

As shown in Figure 23, the duct with flanges was bolted between two aluminum "bookend" fixtures on the shaker machine. Liquid nitrogen (LN_2) flowed into one flange, through the duct, and out of the other flange. The vibration was controlled via an arithmetic average between accelerometers on each bookend. The test environments were derived from operational measurements on large booster LOX-H_2

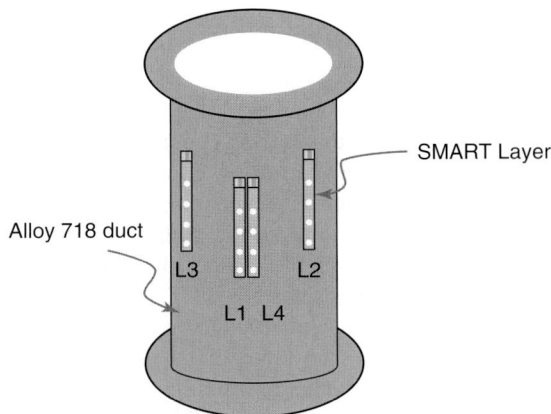

Figure 22. Schematic of the test specimen with SMART Tapes mounted.

engines during engine hot-fire testing. The defined random and sine environments represent an average of the data from the large propellant duct with the highest overall vibration level.

4.3.2 Test procedure

The duct assembly was installed in the test setup and a functional checkout of the sensors was conducted to verify their pretest operational condition. Test instrumentation was installed, including the triaxial response accelerometer and two temperature-monitoring type-K thermocouples. After completion of the flat random characterization run and instrumentation checks, the duct assembly was chilled down and attempt made to perform the same characterization at cryogenic temperatures. Following the completion of 30 min at 0 dB for the cryogenic test, four shock pulses were applied to the test setup at 0 dB.

Pitch–catch signals for both types of piezoelectric sensors were taken at ambient and cryogenic temperatures prior to applying the dynamic environments, during full-level vibration testing, and after completion of full-level testing.

4.3.3 Measurement results

Figure 24 shows some typical sensor signals at different stages. The sensor signals for piezoelectric sensors after the full-level test are the same as the signals before the test. It demonstrated that the piezoelectric sensors could withstand the operational

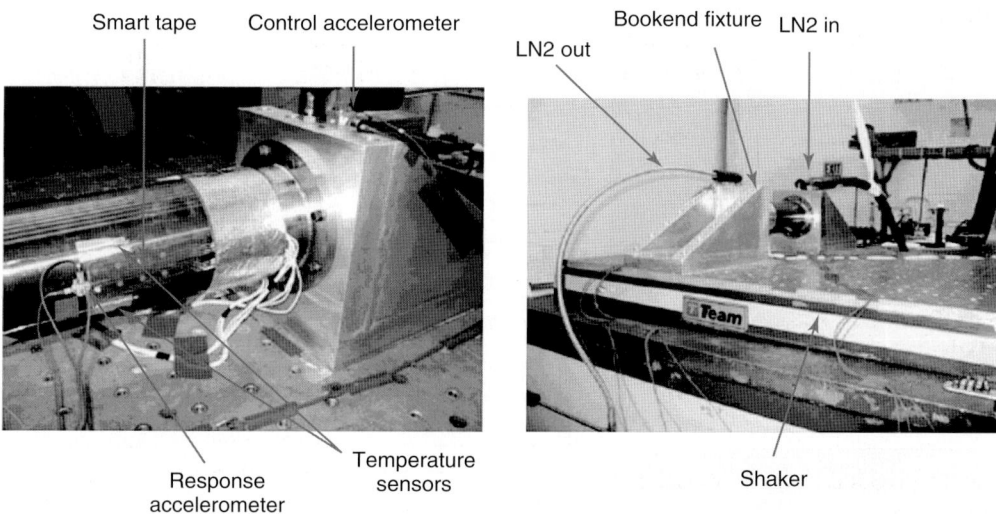

Figure 23. Setup for the combined cryogenic temperature and vibration test.

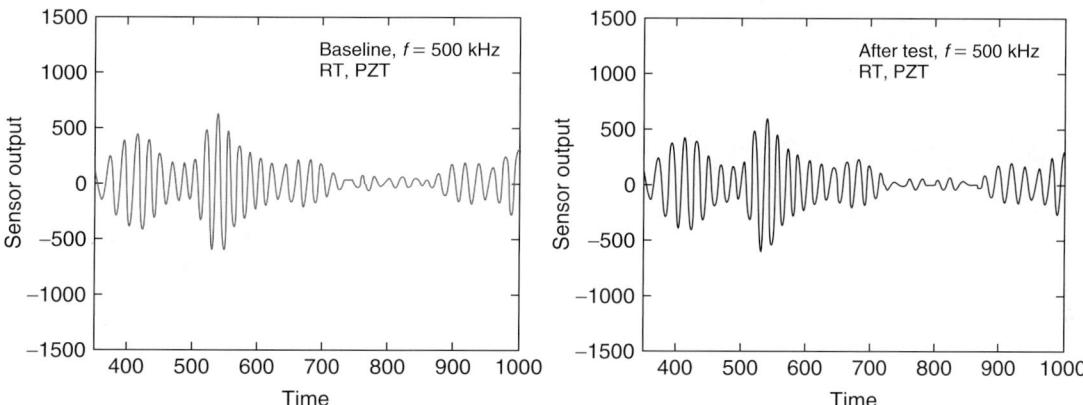

Figure 24. Comparison of 500-kHz signals from PZTs at room temperature before and after the full-level test.

levels of vibratory and shock energy on a representative rocket engine duct assembly within a restricted frequency band in laboratory testing.

4.4 Effect of moisture and aggressive chemical environment

Generally, the piezoelectric materials and polyimide used in the SMART Layer do not have a very good durability in high humidity or aggressive chemical environments. Protective coatings or encapsulation is recommended if the layer is used in high humidity or aggressive chemical environments. To evaluate the effectiveness and quality of protective coatings for the SMART Layer, both humidity testing and salt fog testing were conducted in a humidity chamber and a salt fog chamber per MIL-STD-810F, respectively. The coating tested is an epoxy primer per MIL-PRF-23377, Class C, which is then top coated per MIL-C-46168, Class H. The thickness of the epoxy primer is 15–25 µm, while the thickness of the top coating is 45–75 µm.

The comparisons of signals for the specimens before and after the salt fog test are shown in Figure 25. Test results showed that the SMART

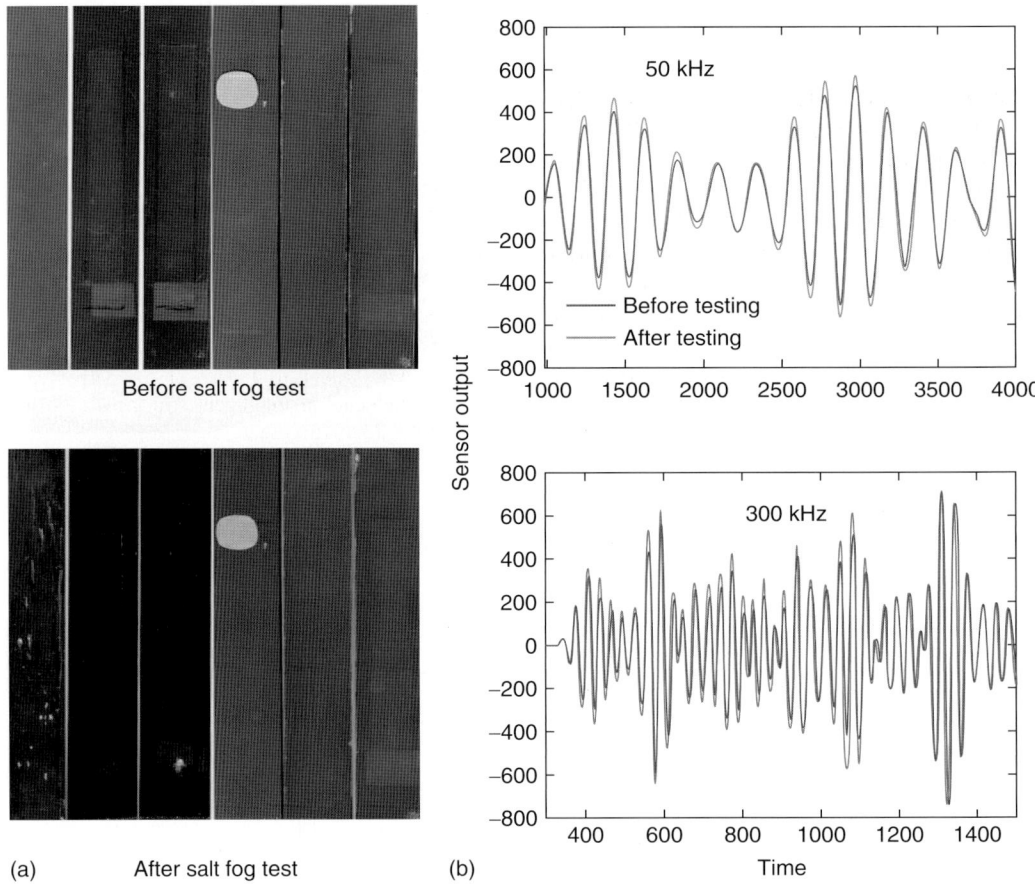

Figure 25. Salt fog test results: (a) aluminum specimens with protective coatings and (b) sensor signals.

Layers can survive humidity and salt fog exposure with proper surface treatment (sanding or chemical treatment), bonding the layers on the surface with Hysol EA 9396, and use proper coatings (epoxy primer per MIL-PRF-23377, class C, which is then top coated per MIL-C-46168, class H). Note that there is no protective coating necessary if the layer is embedded inside a composite.

5 EFFECT OF SMART LAYER ON STRUCTURAL INTEGRITY

When the SMART Layer is embedded inside a composite structure, the effect of the layer on the structural integrity is a concern. It is important to investigate the performance of the composite structure with embedded SMART Layer. Mechanical tests on composite coupon specimens with and without embedded polyimide layer were conducted to assess the change in structural integrity due to inclusion of the SMART Layer.

5.1 Quasi-static impact test

To determine the effects of embedding a SMART Layer into a composite, measurements of the damage tolerance of composites to transverse impact are needed. Quasi-static impact tests were used to simulate a low velocity impact and create a delamination. The tests were performed on both woven graphite/epoxy composites and toughened prepreg carbon–fiber/epoxy composites.

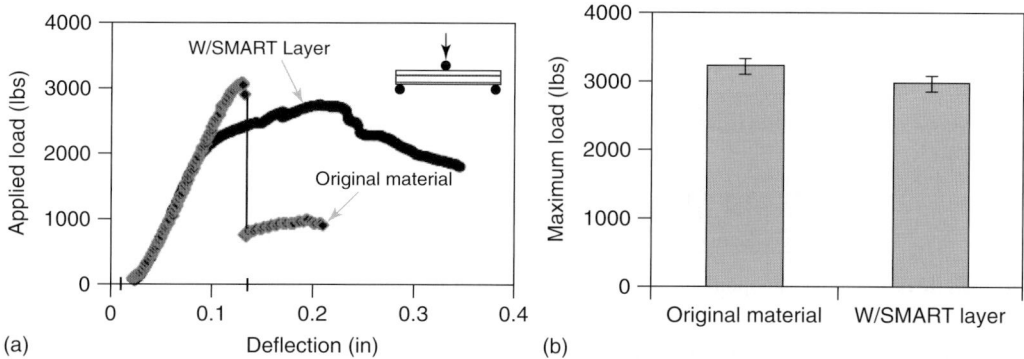

Figure 26. Mechanical test results of specimens with and without embedded SMART Layer (1 lb = 0.45 kg, 1 in. = 25.4 mm). (a) Load–deflection curve and (b) strength comparison.

5.1.1 Woven graphite/epoxy composite

The woven graphite/epoxy coupon specimens tested have a $[0_4/90_4/0_4]$ stacking sequence, specifically designed to promote delamination at the two ply-group interfaces. The specimens measured $140 \times 76 \times 5.2\,\text{mm}^3$. A SMART Layer was placed at the lower 0/90 interface. The thickness of the layer was 0.15 mm, while the thickness of the coupon specimens was 5.2 mm. For testing, the specimens were simply supported on two sides by steel rods spaced 90.0 mm apart and loaded in the center by a 12.7-mm diameter spherical indenter. The specimens were loaded at a displacement rate of $0.127\,\text{mm min}^{-1}$ ($0.05\,\text{in. min}^{-1}$). Test results on these specimens are presented in Figure 26. The test results indicate that the presence of the SMART Layer does not noticeably affect the strength of the host composite structure.

An examination of the cross section of the specimens corroborates these findings. Magnified views of the cross sections of the specimens are shown in Figure 27. It is clear that delamination in the specimens without the embedded SMART Layer occurs at the lower 0/90 interface, as expected, because of the high interfacial shear stress at the ply-group interface. However, in the specimens with an embedded SMART Layer at the lower 0/90 interface, there is no delamination. The actual delamination occurs one or two plies away from the interface, indicating that the SMART Layer does not promote delamination.

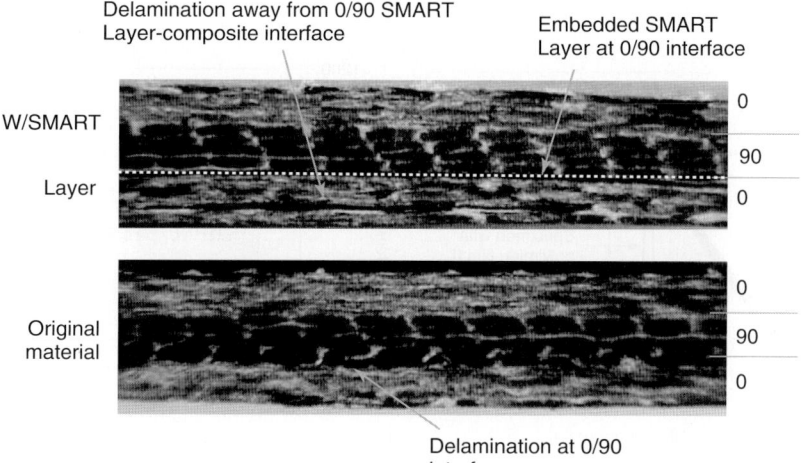

Figure 27. Cross-sectional view of test specimens.

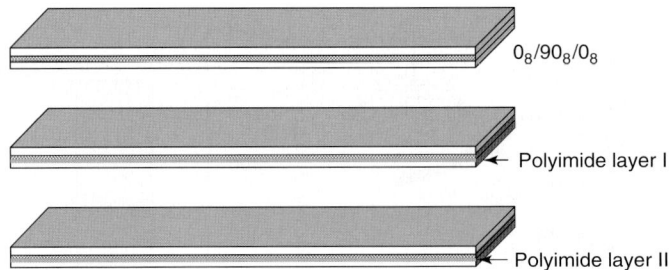

Figure 28. EX-1522 composite coupon specimens with or without polyimide layer inserted at the 0/90 interface.

5.1.2 Toughened carbon–fiber/epoxy composite

The toughened carbon–fiber/epoxy (EX-1522) composite coupon specimens used for quasi-static impact test have a layup $[0_8/90_8/0_8]$. For the embedded specimens, a 0.076-mm-thick polyimide layer I or a 0.152-mm-thick polyimide layer II is inserted at the interface to determine the effect of the thickness of SMART Layer on delamination growth, as shown in Figure 28. The specimens measured $130.0 \times 24 \times 4.0 \, \text{mm}^3$. Similar to the woven graphite/epoxy composite, the specimens were supported on two sides by steel rods spaced 75.0 mm apart and loaded in the center by a 12.7-mm diameter spherical indenter. The specimens were loaded at a displacement rate of $01.27 \, \text{mm} \, \text{min}^{-1}$. Test results on these specimens are presented in Figure 29.

The effect of embedding a SMART Layer on the damage tolerance of T300 and T800 composites to transverse impact was also studied by Lin and Chang [17]. Typical X-ray photographs for the tested specimens with $[0_8/90_8/0_8]$ stacking sequence were shown in Figure 30. As indicated by the X-ray photographs, the delamination area is actually smaller for the specimens with an embedded polyimide bondply; except in the case of the Toray T800H/3900–2 material, which is a resin system with a very high toughness, that it showed a slightly larger delamination. The improvement in impact resistance for most materials is attributed to the fact that the ductile polyimide layer toughens the interface and reduces matrix cracking, thus suppresses delamination.

5.2 Short beam shear test

Short beam shear tests on toughened carbon–fiber/epoxy (EX-1522) composite coupon specimens with and without polyimide layer were conducted. As shown in Figure 31, three types of specimens with unidirectional layup were tested: regular specimen

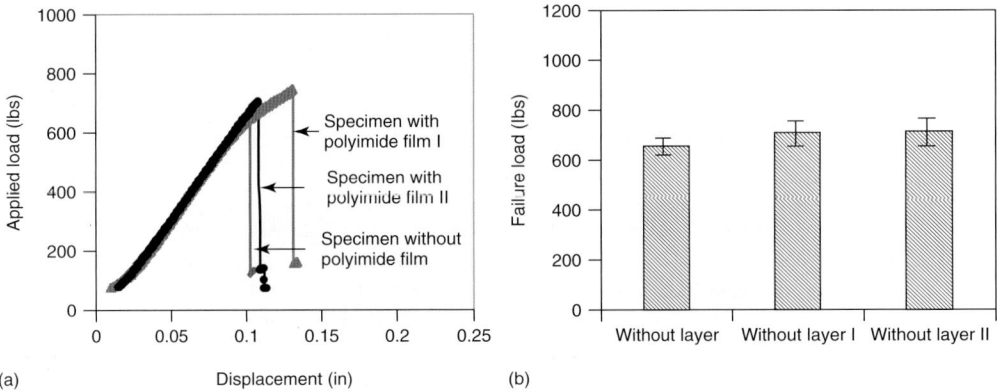

Figure 29. Three-point bending test results of layup $[0_8/90_8/0_8]$ with and without polyimide layer embedded at 0/90 interface. (a) Typical load–deflection curve and (b) strength comparison.

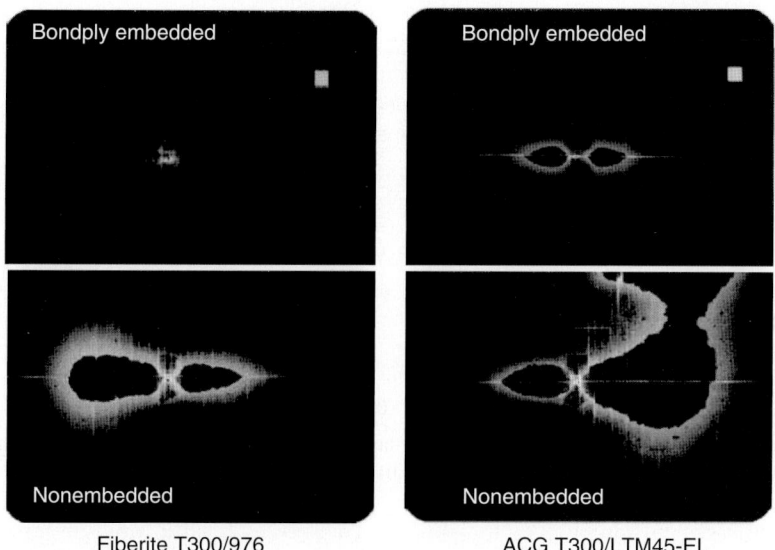

Figure 30. X-ray photographs of delaminations created by the impact test.

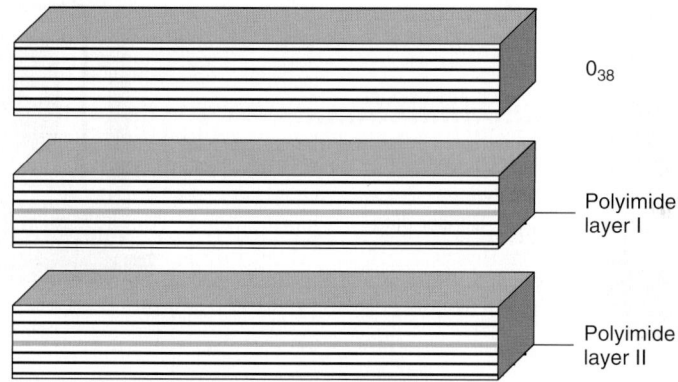

Figure 31. Short beam shear test specimen configurations. Polyimide layer was inserted in the middle plane of some specimens.

without polyimide embedded (Beam 1), specimen with 0.076-mm-thick polyimide layer I embedded (Beam 2), and specimen with 0.152-mm-thick polyimide layer II embedded in the middle plane (Beam 3). The specimens with a layup [0_{38}] measured 40 × 6.35 × 6.35 mm^3. The specimens were loaded under three-point bending with a 25.4-mm span. A loading rate of 0.127 mm min^{-1} was applied. Test results on these specimens are presented in Figure 32.

As shown in Figure 32, there is not much difference between the short beam shear strength of composite with and without polyimide layer. All types of specimens failed in a similar fashion—multiple simultaneous shear cracks between plies, i.e., delamination.

Once again, the test results indicate that the presence of the SMART Layer does neither noticeably affect the strength of the host composite structure nor promote delamination. Besides the studies mentioned above, more investigations on the change in structural integrity due to inclusion of the layer can be found in the literature [17, 28–30]. Results showed that the embedded layer would not decrease either the stiffness or strength of a composite structure.

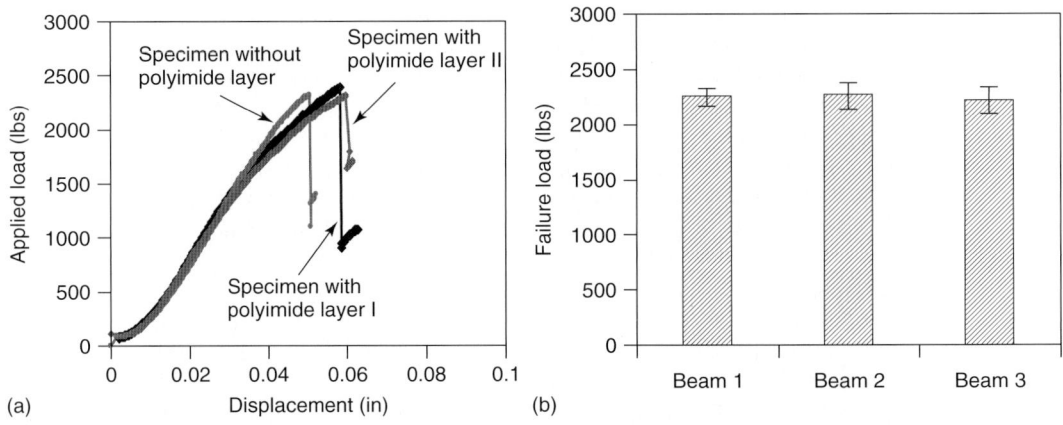

Figure 32. Short beam shear test results of unidirectional layup specimens with and without polyimide layer embedded in the middle plane. (a) Typical load–deflection curve and (b) strength comparison.

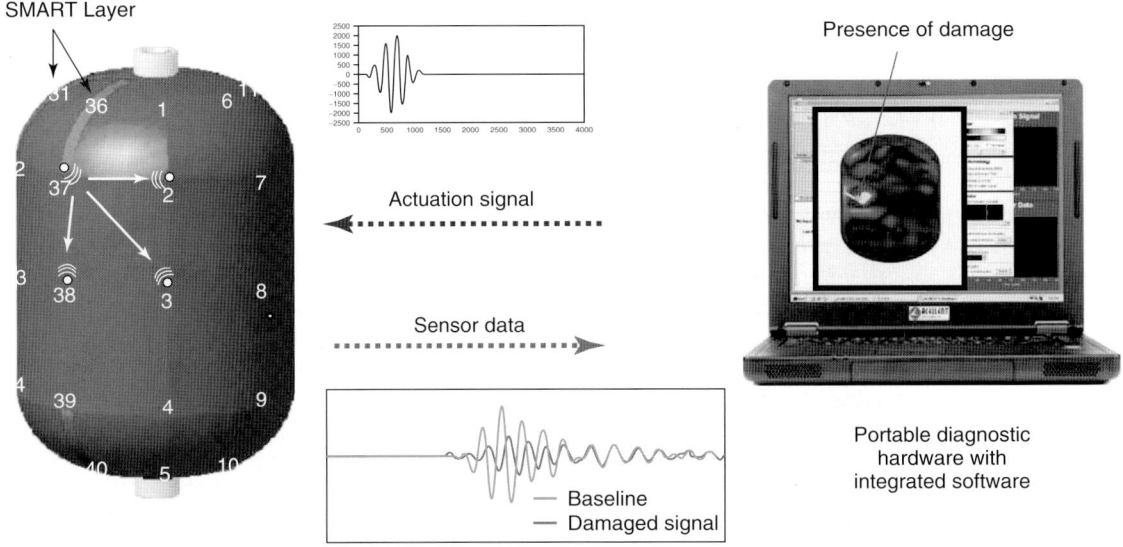

Figure 33. Diagram of Acellent's active-sensing system.

6 SENSOR NETWORK-BASED STRUCTURAL HEALTH MONITORING SYSTEMS

The SMART Layer integrated with a structure can be used for both active and passive sensing. The functioning of the active-sensing system is analogous to that of a built-in acousto-ultrasonic nondestructive evaluation (NDE) with a network of miniaturized piezoelectric transducers. The signal generator of the diagnostic hardware produces a diagnostic signal capable of propagating in structures with minimum distortion while being sensitive to any induced damage. Figure 33 shows the principles of Acellent's portable SHM system. The signal wave generated by the diagnostic hardware drives the actuator to produce a stress wave that propagates across the structure and is measured by the sensors.

When a propagating stress wave encounters a discontinuity in the geometry or material property of the structure, the wave is reflected or scattered.

The methodology used in the diagnostic process is based on the comparison of the current sensor responses with previously recorded sensor responses (baselines) from the undamaged structure. The differences between the two sets of signals are what contain the information about any existing damage or other anomalies. Some applications of the active sensing system can be found in the literature [31–34].

The passive sensing can be used in real time to detect impact events including both impact location and energy. The passive sensing system is set up to continuously listen for impacts at all times. When a sufficiently large force strikes the structure, the piezoceramic sensors that are bonded onto the structure pick up the stress waves traveling through the surface of the structure. A trigger mechanism enables software control to determine the trigger condition, which could be related to a specific set of features of the sensor measurements. As an example, to monitor the safety of a thermal protection system (TPS), a built-in passive sensing system is being developed for detecting the impact location and impact force in real time on TPS panels supplied by Lockheed Martin Space Systems [35].

7 DAMAGE DETECTION IN COMPOSITE STRUCTURES

A SmartComposite system based on SMART Layer technology has been developed to handle some practical issues for the application of SHM on composite structures in real world [36]. Regardless of the application, there are a number of elements that are essential to the practical usage and implementation of any SHM system. These elements can be grouped into three main categories: (i) the system must be easy to use, (ii) it must provide a well-defined resolution, and (iii) it must be very reliable. The essential elements for each category are listed below:

- Easy to use
  - the sensor installation should be straightforward with minimal training;
  - the system should have simple calibration procedures—preferably automated;
  - all data analysis shall be automated;
  - results should be output in standard formats.

- Well-defined resolution
  - the system must provide a quantifiable probability of detection (POD);
  - when damage is detected, the system must indicate size/severity along with the measure of uncertainty.

- High reliability
  - the system must be able to compensate for environmental changes;
  - the system must have built-in test for hardware and sensor self-diagnostics;
  - the sensors must survive extreme environmental conditions;
  - in the event of damage to a sensor, the system must be repairable.

Each element listed above has, by itself, been studied by many researchers and developers for a variety of applications. But there has been little work conducted to combine all the essential elements into a common framework. On the basis of the SMART Layer technology, Acellent has addressed this issue by focusing on all of the above elements, customizing each one for composite structures, and then unifying all of them into an integrated system called *the SmartComposite System*.

The key features of the system include sensor self-diagnostics and an adaptive algorithm to automatically compensate for damaged sensors, reliable damage detection under different environmental conditions, and generation of POD curves. In addition, state-of-the-art techniques to optimize sensor placement, automated calibration, and automated damage detection with no user interpretation of data make it an efficient and user-friendly system. A few of the features are discussed in the following sections.

7.1 Self-diagnostics

If one or more sensors are degraded, damaged, or missing, the SHM system may not function properly and can give false indications of structural damage. Measuring the impedance of each channel can be used to find an open or short circuit. This can indicate a missing sensor or damaged connection/wiring.

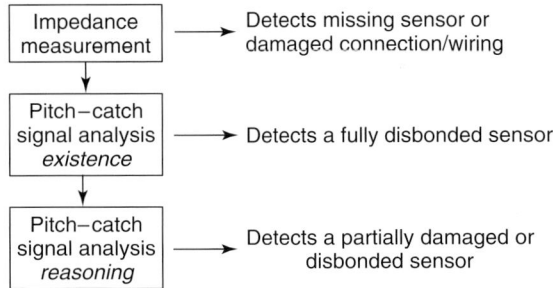

Figure 34. Integrated three-step method to automatically detect faulty sensors.

But a degraded or damaged sensor may go undetected using the impedance method. To resolve this, a reasoning process using the active sensor signals has been developed and implemented to detect degraded or damaged sensors that the impedance method may miss.

An integrated three-step method shown in Figure 34 has been developed to automatically detect

- faulty sensors caused by a missing sensor or damaged connection/wiring;
- a sensor that is still connected to the electronics, but is disbonded from the structure; and
- a partially damaged or disbonded sensor.

The impedance measurement of each channel can be used to find an open or short circuit. If a sensor is flagged as degraded, damaged, or missing, all signal data from the faulty sensor are removed from the analysis routines.

7.2 Temperature compensation

Current state-of-the-art damage detection methodologies rely on the use of baseline data collected from the structure in the undamaged state. The methodologies are based on comparing the current sensor responses to the previously recorded baseline sensor responses, and using the differences to glean information about structural damage. However, it is known that environmental effects, such as temperature differences, will also cause changes in the sensor signals, and will thus interfere with most damage detection schemes.

Acellent has developed a calibration technique utilizing multiple baselines that can be employed to mitigate the effects of environmental changes. The technique has been tested and verified for both global and local changes in temperature and can be used to compensate for global temperature changes in the structure as well as temperature gradients [36].

With this method, data is collected from the healthy structure at various temperatures and stored in a so-called baseline space. The baseline signals from each temperature can be used to create a baseline surface for each actuator–sensor path. At a later time, when a sensor scan is performed to search for damage, the newly recorded signals are compared with the

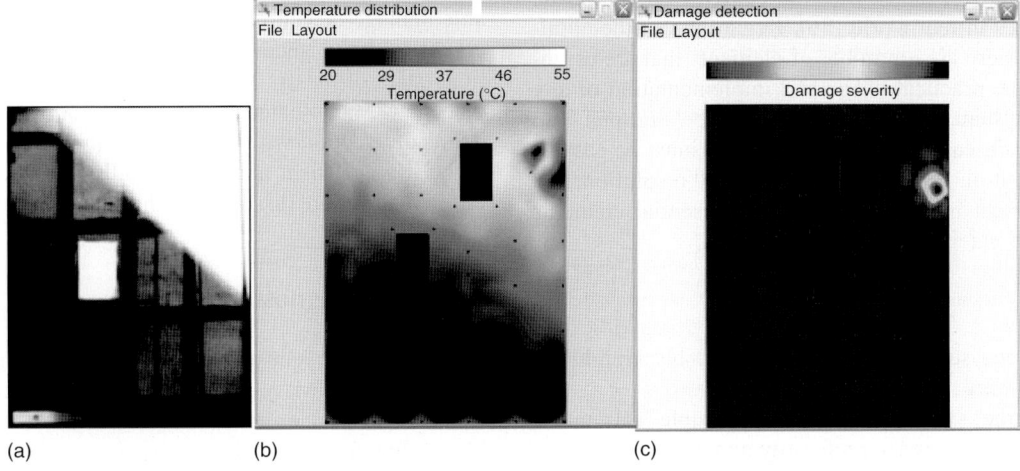

Figure 35. Environment compensation to eliminate temperature effect on a composite panel. (a) Structure in sun, (b) temperature distribution, and (c) detected damage.

corresponding baseline surfaces to determine a best fit along the temperature axis. An example of the temperature compensation is shown in Figure 35. The specimen shown in Figure 35 is a 1.5 m × 1.5 m fiberglass stiffener panel with surface-mounted SMART Layers containing a total of 50 actuators/sensors.

8 CONCLUSION

SHM offers the promise of a paradigm shift from schedule-driven maintenance to condition-based maintenance of assets. SMART Layer technology is a viable and cost-effective means of monitoring the structural condition and detecting damage while structures are in service. The techniques for fabricating the diagnostic layers and the methods for integrating the layer into structures are developed. The performance of the piezoelectric elements embedded in the SMART Layer under different loading and environment conditions was investigated. Other practical issues, such as the self-diagnostics of the embedded network and environmental compensation, are also discussed. On the basis of the study, the following remarks can be made:

1. Using SMART Layer, large sensor (and actuator) networks can be easily integrated with metal/composite structures. The characteristic features of the diagnostic layer include (i) ease of installation; (ii) adaptation to any structure with complex geometry; (iii) use of an area sensing network; (iv) actuation and sensing capabilities; (v) signal consistency and sensor reliability; and (vi) shielding to reduce EM noise.
2. The performance of PZT remains unchanged when the applied strain does not exceed the failure strain of PZT and the electric cycling input is below the maximum voltage.
3. The SMART Layer is functional within a wide temperature range. Proper protective coatings can effectively isolate the layer from the moisture and aggressive chemical environment when the layer is mounted on the surface of a structure.
4. Test results showed that the presence of the SMART Layer does neither noticeably affect the strength of the host composite structure nor promote delamination.
5. A SmartComposite system based on SMART Layer technology has been developed to handle some practical issues for the application of SHM in real world. The major features of the system include sensor network self-diagnostics, environment compensation, and POD curves with quantified damage size.

ACKNOWLEDGMENTS

The authors would like to acknowledge the financial support of the National Institute of Standards and Technology (NIST), Army, Air Force, Missile Defense Agency, Defense Advanced Research Projects Agency (DARPA), Federal Aviation Administration (FAA), and National Aeronautics and Space Administration (NASA) of the United States for sponsoring some of the developments presented in this article. The authors also would like to thank Dr. Roy Ikegami, Dr. David Zhang, and other colleagues at Acellent Technologies Inc. for the technical consultation and assistance in the development of the technology.

RELATED ARTICLES

Chapter 4: Acoustic Emission

REFERENCES

[1] Boller C. Identification of life cycle cost reductions in structures with self-diagnostic devices. *Proceedings of the NATO RTO Symposium on Design Issues*. Ottawa, 1999; pp. 1–8.

[2] Chang F-K. Ultra reliable and super safe structure for the new century. In *Proceedings of the First European Workshop on SHM, Structural Health Monitoring 2002*, Balageas D (ed). Cachan, July 2002. DEStech Publications: Lancaster, PA, 2002, pp. 3–12.

[3] Beral B, Speckmann H. Structural health monitoring (SHM) for aircraft structures: a challenge for system developers and aircraft manufactures. In *Proceedings of the 4th International Workshop on SHM, SHM 2005: From Diagnostics and Prognostics to Structural Health Management*, Chang F-K (ed). Stanford University, September 2003. DEStech Publications: Lancaster, PA, 2003, pp. 12–29.

[4] Trego A, Akdeniz A, Haugse E. Proceedings of the Second European Workshop on SHM, Structural Health Monitoring 2004. In *Structural Health Management Technology on Commercial Airplanes*, Boller C, Staszewski W (eds). DEStech Publications: Munich, Lancaster, PA, 2004, pp. 317–323.

[5] Ansari F. Fiber optic health monitoring of civil structures using long gage and acoustic sensors. *Smart Materials and Structures* 2005 **14**:S1–S7.

[6] Giurgiutiu V, Zagrai A. Characterization of piezoelectric wafer active sensors. *Journal of Intelligent Material Systems and Structures* 2000 **11**:959–975.

[7] Kwun H, Kim S-Y, Light GM. Magnetostrictive sensor guided-wave probes for structural health monitoring of pipelines and pressure vessels. In *Proceedings of the 5th International Workshop on SHM, SHM 2005: Advancements and Challenges for Implementation*, Chang F-K (ed). Stanford University, September 2005. DEStech Publications: Lancaster, PA, 2005, pp. 694–701.

[8] Calkin FT, Flatau AB, Dapino MJ. Overview of magnetostrictive sensor technology. *Journal of Intelligent Material Systems and Structures* 2007 **18**:1057–1066.

[9] Varadan VK, Varadan VV. Microsensors, microelectromechanical systems (MEMS), and electronics for smart structures and systems. *Smart Materials and Structures* 2000 **9**:953–972.

[10] Giurgiutiu V, Zagral A, Bao JJ. Piezoelectric wafer embedded active sensors for aging aircraft structural health monitoring. *Structural Health Monitoring* 2002 **1**(1):41–61.

[11] Lee BC, Staszewski WJ. Modeling of Lamb waves for damage detection in metallic structures: part II. Wave interactions with damage. *Smart Materials and Structures* 2003 **12**:815–824.

[12] Kessler S, Spearing S, Soutis C. Damage detection in composite materials using Lamb wave methods. *Smart Materials and Structures* 2002 **11**:269–278.

[13] Qing X, Chan H, Beard S, Kumar A. An active diagnostic system for structural health monitoring of rocket engines. *Journal of Intelligent Material Systems and Structures* 2006 **17**:619–628.

[14] Ihn J, Chang F-K. Detection and monitoring of hidden fatigue crack growth using a built-in piezoelectric sensor/actuator network: I. Diagnostics. *Smart Materials and Structures* 2004 **13**(3):609–620.

[15] Park G, Sohn H, Farrar CR, Inman D. Overview of piezoelectric impedance-based health monitoring and path forward. *The Shock and Vibration Digest* 2003 **35**(6):451–463.

[16] Seydel R, Chang F-K. Impact identification of stiffened composite panel: I. System development. *Smart Materials and Structures* 2001 **10**:354–369.

[17] Lin M, Chang F-K. The manufacture of composite structures with a built-in network of piezoceramics. *Composite Science and Technology* 2002 **62**:919–939.

[18] Lin M, Qing X, Kumar A, Beard S. SMART layer and SMART suitcase for structural health monitoring applications. In *Proceedings of SPIE on Smart Structures and Materials 2001: Industrial and Commercial Applications of Smart Structures Technologies*, McGowan A-MR (ed). Newport Beach, CA, March 2001. SPIE, 2001; Vol. 4332, pp. 98–106.

[19] Qing X, Beard B, Kumar A, Ooi T, Chang F-K. Built-in sensor network for structural health monitoring of composite structure. *Journal of Intelligent Material Systems and Structures* 2007 **18**:39–49.

[20] Qing X, Kumar A, Zhang C, Gonzalez IF, Guo G, Chang F-K. Hybrid piezoelectric/fiber optic diagnostic system for structural health monitoring. *Smart Materials and Structures* 2005 **14**(3):S98–S103.

[21] Malkin M, Qing X, Leonard M, Derriso M. Flight demonstration: health monitoring for bonded structural repairs. In *Proceedings of the Third European Workshop on SHM, Structural Health Monitoring 2006*, Güemes A (ed). Granada, July 2006. DEStech Publications: Lancaster, PA, 2006, pp. 167–175.

[22] Qing X, Beard S, Kumar A, Chan H, Ikegami R. Advances in the development of built-in diagnostic system for filament wound composite structures. *Composite Science and Technology* 2006 **66**:1694–1702.

[23] Russell S, Walker J, Workman G. Efficient nondestructive evaluation of prototype carbon fiber reinforced structures. *Proceedings of the 10th US-Japan Conference on Composite Materials*. Stanford University, Stanford, CA, 16–18 September 2002.

[24] Blackshire JL, Jata KV. *Integrated sensor durability and reliability*. Air Force Research Laboratory, Wright-Patterson Air Force Base: Ohio, 2008.

[25] Kusaka T, Qing X. Characterization of loading effect on the performance of SMART layer embedded or surface mounted on structures. In *Proceedings of the 4th International Workshop on SHM, SHM 2005:*

[26] Qing XP, Kumar A, Beard S, Yu P, Zhang D, Liu C, Hannum R. Advanced self-sufficient structural health monitoring system. In *Proceedings of the Third European Workshop on SHM, Structural Health Monitoring 2006*, Güemes A (ed). Granada, July 2006. DEStech Publications: Lancaster, PA, 2006, pp. 807–814.

[27] Qing XP, Beard SJ, Kumar A, Sullivan K, Aguilar R, Merchant M, Taniguchi M. Performance of piezoelectric sensors based SHM system under combined cryogenic temperature and vibration environment. *Smart Materials and Structures* 2008 **17**(5):055010.

[28] Yang SM, Hung CC, Chen KH. Design and fabrication of a smart layer module in composite laminated structures. *Smart Materials and Structures* 2005 **14**(2):315–320.

[29] Tang S, Xiong K, Liang D, Li D. The development of SMART layer used in structural health monitoring. *Journal of Experimental Mechanics (in Chinese)* 2005 **20**(2):226–234.

[30] Qi B, Bannister M. Mechanical performance of carbon/epoxy composites with embedded polymeric films. *Key Engineering Materials* 2007 **334–335**:469–472.

[31] Qing X, Beard S, Kumar A, Hannum R. A real-time active smart patch system for monitoring the integrity of bonded repair on an aircraft structure. *Smart Materials and Structures* 2006 **15**:N66–N73.

From Diagnostics and Prognostics to Structural Health Management, Chang F-K (ed). Stanford University, September 2003. DEStech Publications: Lancaster, PA, 2003, pp. 1539–1546.

[32] Qing X, Wu Z, Chang F-K, Ghosh K, Karbhari V, Sikorsky C. Monitoring the disbond of externally bonded CFRP composite strips for rehabilitation of bridges. In *Proceedings of the Third European Workshop on SHM, Structural Health Monitoring 2006*, Güemes A (ed). Granada, July 2006. DEStech Publications: Lancaster, PA, 2006, pp. 463–470. Also in *FRP International* 2006 **3**(3):11–14.

[33] Qing X, Kumar A. Integrated active-passive "SMART layer" system monitoring structural defects. *Technology Advances, MRS Bulletin* 2005 **30**(7):506.

[34] Qing X, Beard S, Kumar A, Yu P, Chan HL, Zhang D, Ooi T, Marotta SA. Practical requirements for implementation and usage of SHM systems on aerospace structures. In *Proceedings of the 5th International Workshop on SHM, SHM 2005: Advancements and Challenges for Implementation*, Chang F-K (ed). Stanford University, September 2005. DEStech Publications: Lancaster, PA, 2005, pp. 1502–1509.

[35] Yu P. Real time impact detection system for thermal protection system. In *Proceedings of the 7th International Workshop on SHM, SHM 2007: Quantification, Validation, and Implementation*, Chang F-K (ed). Stanford University, September 2007. DEStech Publications: Lancaster, PA, 2007, pp. 153–159.

[36] Beard S, Liu B, Qing P, Zhang D. Challenges in implementation of SHM. In *Proceedings of the 7th International Workshop on SHM, SHM 2007: Quantification, Validation, and Implementation*, Chang F-K (ed). Stanford University, September 2007. DEStech Publications: Lancaster, PA, 2007, pp. 65–84.

Chapter 79
Hybrid PZT/FBG Sensor System

Zhanjun Wu[1], Xinlin P. Qing[2] and Fu-Kuo Chang[1]
[1] Department of Aeronautics and Astronautics, Stanford University, Stanford, CA, USA
[2] Acellent Technologies, Inc., Sunnyvale, CA, USA

1 Introduction	1389
2 FBG as Ultrasonic Stress Wave Sensors	1390
3 Damage Detection	1392
4 Active Hybrid Piezoelectric/Fiber-optic SHM System	1392
5 Concluding Remarks	1399
Acknowledgments	1399
References	1399
Further Reading	1401

Encyclopedia of Structural Health Monitoring. Edited by Christian Boller, Fu-Kuo Chang and Yozo Fujino © 2009 John Wiley & Sons, Ltd. ISBN: 978-0-470-05822-0.

1 INTRODUCTION

Damage detection techniques have been studied intensively in almost every aspect of engineering. In recent years, tremendous progress has been made. This is apparently a response to the high demand for safer structures and lower cost. Technological progress on novel smart actuators and sensors, such as piezoelectric transducers (lead–zirconate–titanate—PZTs) and fiber-optic sensors (fiber Bragg grating—FBG), has paved the way for the development of structural health monitoring (SHM) technology [1, 2]. Owing to their advantages, piezoelectric ceramics have been widely employed in noise and vibration control and active and passive sensing [3–6]. However, when the piezoelectric sensors are used in the active-sensing mode, such as using Lamb wave in pitch–catch manner to detect structural damage with a highly integrated diagnostic system, crosstalk between PZT actuation signals and sensor signals is always a problem. Further, electronic signals from PZT sensors to the data storage or processing unit may attenuate significantly during long-distance transmission, for example, from a bridge deck to its monitoring and control center. FBG sensor systems have offered an attractive solution to the problem of making strain measurements, as they have the advantages of small size, high resolution, high multiplexibility, electromagnetic immunity, and potentially high-density quasi-distributed measurements. FBGs serving as strain sensors have recently been used corresponding to $\lambda_S/L \gg 1$, S_λ approaches the static value $S_m \approx 0.78$.

3 DAMAGE DETECTION

3.1 Lamb wave source location

Betz and Lamb [18] suggested a fit function to predict the directivity dependence of FBG sensors.

$$\hat{A}(\alpha) = a_2 \sin^2\left(\pi \frac{\alpha - a_1}{180}\right) \quad (4)$$

where $\hat{A}$ is the normalized amplitude of the strain amplitude, α is the angle between the direction of FBG sensors and the longitudinal wave propagation, and a_1 and a_2 are fit parameters that can be determined by test. He designed a test scheme to obtain those parameters and then used the inverse of equation (4) to calculate the direction of the Lamb wave, and by two sets of FBG sensor rosettes the location of the Lamb wave source can be found.

3.2 Acoustic emission detection

Udd [14] and his coworkers studied the acoustic emission detection ability of FBG sensors. However, a further study was carried out by Baldwin and Vizzini [23]. Baldwin and Vizzini successfully demonstrated the ability of an FBG sensor to detect a pencil lead break event, metal to metal impact events, and AE events from composite specimens with loosely mounted and embedded FBG sensors using a similar interrogation technique as Udd adopted.

3.3 Composite delamination detection

Ogisu et al. [15] reported damage detection with hybrid FBG sensor/PZT actuators using small diameter FBG sensors. They proposed a novel FBG sensor signal interrogation system capable of detecting frequencies of up to 1 MHz. The system employs an arrayed wave guide grating (AWG)-type filter to obtain a high-sensitivity filter characteristic for detecting small displacements in the grating of the FBG sensors. Furthermore, the AWG-type systems are also capable of interrogating multiple sensors at the same time. The sensor length in their study was less than one-seventh of the wavelength of the Lamb wave to ensure sensitivity. They also carried out double-lap joint-type composite coupon tests with the proposed method to demonstrate the damage detection capability.

4 ACTIVE HYBRID PIEZOELECTRIC/FIBER-OPTIC SHM SYSTEM

To take the advantages of both PZT and fiber-optic sensor, an active hybrid piezoelectric/fiber-optic SHM system has been developed recently [13]. With this system, the emerging concept of SHM can become a commercially viable option in structural engineering, allowing a new generation of safer, more reliable, and lower maintenance structures. As shown in Figure 2, the developed structural diagnostic system can permit quantitative characterization and event determination pertaining to aerospace and civil structures in hostile service environments. More specifically, the hybrid system can potentially be used to perform:

- *in situ* material property characterization;
- detect material and structural defects;
- detect damage including delaminations and corrosion; and
- characterize load environments (fatigue, overload).

4.1 Principle of the PZT/FBG hybrid active structural health monitoring system

The hybrid diagnostic system uses piezoelectric actuators to input a controlled excitation to the structure and fiber-optic sensors to capture the corresponding structural response. The system consists of three major parts: a diagnostic layer with a network of piezoelectric elements and fiber gratings to offer a simple and efficient way to integrate a large network of transducers onto a structure; diagnostic hardware consisting of an arbitrary waveform generator and a high-speed fiber grating demodulation unit together with a high-speed data-acquisition card to provide actuation input, data collection, and

Chapter 79
Hybrid PZT/FBG Sensor System

Zhanjun Wu[1], Xinlin P. Qing[2] and Fu-Kuo Chang[1]

[1] Department of Aeronautics and Astronautics, Stanford University, Stanford, CA, USA
[2] Acellent Technologies, Inc., Sunnyvale, CA, USA

1 Introduction 1389
2 FBG as Ultrasonic Stress Wave Sensors 1390
3 Damage Detection 1392
4 Active Hybrid Piezoelectric/Fiber-optic SHM System 1392
5 Concluding Remarks 1399
Acknowledgments 1399
References 1399
Further Reading 1401

1 INTRODUCTION

Damage detection techniques have been studied intensively in almost every aspect of engineering. In recent years, tremendous progress has been made. This is apparently a response to the high demand for safer structures and lower cost. Technological progress on novel smart actuators and sensors, such as piezoelectric transducers (lead–zirconate–titanate—PZTs) and fiber-optic sensors (fiber Bragg grating—FBG), has paved the way for the development of structural health monitoring (SHM) technology [1, 2]. Owing to their advantages, piezoelectric ceramics have been widely employed in noise and vibration control and active and passive sensing [3–6]. However, when the piezoelectric sensors are used in the active-sensing mode, such as using Lamb wave in pitch–catch manner to detect structural damage with a highly integrated diagnostic system, crosstalk between PZT actuation signals and sensor signals is always a problem. Further, electronic signals from PZT sensors to the data storage or processing unit may attenuate significantly during long-distance transmission, for example, from a bridge deck to its monitoring and control center. FBG sensor systems have offered an attractive solution to the problem of making strain measurements, as they have the advantages of small size, high resolution, high multiplexibility, electromagnetic immunity, and potentially high-density quasi-distributed measurements. FBGs serving as strain sensors have recently been used

Encyclopedia of Structural Health Monitoring. Edited by Christian Boller, Fu-Kuo Chang and Yozo Fujino © 2009 John Wiley & Sons, Ltd. ISBN: 978-0-470-05822-0.

in numerous harsh field environments, such as civil structures [7–9], aerospace vehicles [10], and ships [11] (*see* **Chapter 61**).

A hybrid piezoelectric/fiber-optic sensor system offers the best decoupling of actuator and sensor signals (minimum interference), because the two devices use different mechanisms for signal transmission: the piezoelectric actuators use electrical channels while the FBG sensors use optical means [12, 13]. However, building a hybrid system is not simply a matter of putting the devices together. One of the most important issues arises from their different characteristics. PZTs are often used to actuate structures or detect dynamic responses, while FBG sensors are mainly used for quasi-static measuring or relatively low-frequency responses. This does not mean that FBG sensors cannot measure dynamic responses. But there are some issues relating to such sensors needed to be investigated, such as the relationship between sensitivity and grating length, the part of the optical fiber on which gratings are engraved, strain resolution, frequency range, and also the signal-to-noise ratio. Udd [14] demonstrated the capability of FBG sensors to detect acoustic emission and ultrasonic stress waves. Ogisu *et al.* [15] reported damage detection with the FBG sensor/PZT actuator hybrid system. Minardoa *et al.* [16] gave the full characterization of the ultrasonic waves by wavelength shift detection, indicating that the measurement can be achieved only if the grating length is smaller than the ultrasonic wavelength. Pierce *et al.* [17] successfully demonstrated the technique of damage inspection for CFRP (carbon fiber–reinforced plastic) plates with both fiber-optic Michelson interferometer and fiber-optic Mach–Zehnder interferometer using ultrasonic Lamb waves. However, this technique is not capable of multiplexing. Betz and his colleagues [18–20] have done a thorough study on both theoretical analysis and experimental validation of using PZT/FBG hybrid system to sense acoustic stress wave for damage detection.

In this article, a general review of the hybrid piezoelectric/fiber-optic SHM systems is presented, which includes the ultrasonic stress wave detectability study based on both experimental demonstration and theoretical analysis, the integrated hybrid PZT/FBG active diagnostic system and the applications, as well as the demonstrations of damage detection in both metallic and composite structures.

2 FBG AS ULTRASONIC STRESS WAVE SENSORS

2.1 Experimental demonstration

FBG is a selective reflector that reflects optical signals of a certain wavelength called the *Bragg wavelength* λ_b, according to the physical state of the gratings. An FBG consists of a series of periodically located cross sections with a higher refractive index (RI) at the core of a length of optical fiber. At each of the RI steps, a small fraction of the optical signal is reflected. With variations of temperature and strain, the fiber's physical properties change. This results in a change of the RI (n) and the period of the Bragg grating (Δ_0) and causes a Bragg wavelength shift (λ_b). This shift in the Bragg wavelength can be measured by an optical spectrum analyser (*see* **Chapter 4**; **Chapter 55**; **Chapter 78**).

$$\lambda_b = 2n_{\text{eff}_0}\Delta_0 \qquad (1)$$

There are several methods for interrogating the FBG signals to effect quasi-static or low-frequency multisensor measurements [21]. Micron Optics fiber Fabry–Perot tunable filter (FFP-TF), developed by Micron Optics, is commonly used. It utilizes a Fabry–Perot etalon that passes wavelengths that are equal to integer fractions of the cavity (etalon) length; all other wavelengths are attenuated according to the Airy function [22]. However, for high-frequency measurements up to $\sim 10^5$ Hz or higher, another interrogation scheme, developed by Udd [14], is well suited. As shown in Figure 1, a matched fiber grating filter is employed instead of FFP-TF.

Udd and his coworkers presented a systematic study on acoustic emission detection using FBG sensors. In the first test, they attached an FBG sensor onto a PZT actuator and designed an interrogation scheme shown in Figure 1, in which a tunable matching FBG filter was employed to interrogate the

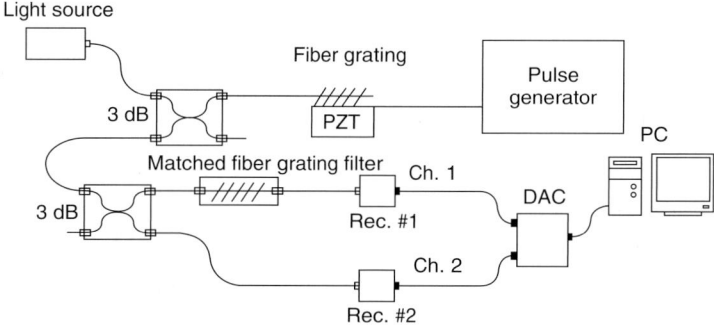

Figure 1. Schematic diagram of the FBG system with matched fiber grating filter. [Reproduced with permission from Ref. 14. © SPIE, 2001.]

optical response signal. Data were taken in the form of photoreceiver voltage response to the light signal. The ratio of signal from channel 1 to channel 2 can be correlated to the corresponding strain under measurement. From their experimental investigation, a lower detection limit has been established for acoustic emission (AE) detection using FBGs. They also carried out tests to study the capability of Lamb wave detection by FBGs on an aluminum plate. The detection results showed that the sensitivity of the FBG sensor varied according to the distance between the FBG sensor and the ultrasonic transducer. Furthermore, the directional dependency of the detection sensitivity was also studied.

2.2 Theoretical analysis

Minardoa et al. [16] investigated the response of an FBG to the dynamic strain induced by a symmetric mode Lamb wave theoretically and presented numerical simulation results. They studied the interaction between a uniform FBG of length L, written into the core of a standard single-mode fiber, and an ultrasonic wave. The FBG that is not subjected to any external stress is described by a modulation of the effective RI of the fundamental guided mode along the fiber axis z.

$$n_{\text{eff}}(z) = n_{\text{eff}0} - \Delta n \sin^2\left(\frac{\pi}{\Lambda_0} z\right) \quad (2)$$

where Δn is the maximum index change and Λ_0 is the grating pitch. The Bragg wavelength, as determined by applying the Bragg condition, is given by equation (1).

They assumed an ultrasonic plane wave, with the acoustic wavefront normal to the optical fiber. Strain field is modeled by a longitudinal strain wave propagating along the fiber axis. In addition, the time dependence is assumed to be sinusoidal and can be expressed as

$$\varepsilon(z, t) = \varepsilon_m \cos(k_s z - \omega_s t) \quad (3)$$

Here, ε_m denotes the strain wave amplitude, ω_s its angular frequency, and k_s its wave number related to its wavelength by $k_s = 2\pi/\lambda_s$.

The strain wave influences the Bragg grating response by modulating its geometrical and physical properties, which is the same as quasi-static measurement using FBG sensors. Then the new effective RI of the Bragg grating under the ultrasonic wave action can be evaluated as the sum of two contributions. The first one is a mechanical contribution, due to the modulation of the grating pitch under the strain wave, which can be determined by the deformation along the z axis of the grating; the second one is an optical contribution due to the change in RI via the elasto-optic effect. Through the study the sensitivity of FBG response has been shown to decrease when reducing the ratio between the ultrasound (US) wavelength and the grating length. Moreover, a significant spectrum shape distortion has been shown to occur for high-power ultrasonic waves. It also showed there are essentially three main operating regions that can be distinguished. The first region, when $\lambda_S/L \ll 1$, the wavelength shift sensitivity S_λ approaches zero, and so no apparent Bragg wavelength modulation occurs. In the second region, corresponding to $\lambda_S/L \approx 1$, S_λ increases with the ratio λ_S/L, and in the third region, corresponding to $\lambda_S/L \gg 1$, S_λ approaches the static value $S_m \approx 0.78$.

3 DAMAGE DETECTION

3.1 Lamb wave source location

Betz and Lamb [18] suggested a fit function to predict the directivity dependence of FBG sensors.

$$\hat{A}(\alpha) = a_2 \sin^2\left(\pi \frac{\alpha - a_1}{180}\right) \quad (4)$$

where $\hat{A}$ is the normalized amplitude of the strain amplitude, α is the angle between the direction of FBG sensors and the longitudinal wave propagation, and a_1 and a_2 are fit parameters that can be determined by test. He designed a test scheme to obtain those parameters and then used the inverse of equation (4) to calculate the direction of the Lamb wave, and by two sets of FBG sensor rosettes the location of the Lamb wave source can be found.

3.2 Acoustic emission detection

Udd [14] and his coworkers studied the acoustic emission detection ability of FBG sensors. However, a further study was carried out by Baldwin and Vizzini [23]. Baldwin and Vizzini successfully demonstrated the ability of an FBG sensor to detect a pencil lead break event, metal to metal impact events, and AE events from composite specimens with loosely mounted and embedded FBG sensors using a similar interrogation technique as Udd adopted.

3.3 Composite delamination detection

Ogisu *et al.* [15] reported damage detection with hybrid FBG sensor/PZT actuators using small diameter FBG sensors. They proposed a novel FBG sensor signal interrogation system capable of detecting frequencies of up to 1 MHz. The system employs an arrayed wave guide grating (AWG)-type filter to obtain a high-sensitivity filter characteristic for detecting small displacements in the grating of the FBG sensors. Furthermore, the AWG-type systems are also capable of interrogating multiple sensors at the same time. The sensor length in their study was less than one-seventh of the wavelength of the Lamb wave to ensure sensitivity. They also carried out double-lap joint-type composite coupon tests with the proposed method to demonstrate the damage detection capability.

4 ACTIVE HYBRID PIEZOELECTRIC/FIBER-OPTIC SHM SYSTEM

To take the advantages of both PZT and fiber-optic sensor, an active hybrid piezoelectric/fiber-optic SHM system has been developed recently [13]. With this system, the emerging concept of SHM can become a commercially viable option in structural engineering, allowing a new generation of safer, more reliable, and lower maintenance structures. As shown in Figure 2, the developed structural diagnostic system can permit quantitative characterization and event determination pertaining to aerospace and civil structures in hostile service environments. More specifically, the hybrid system can potentially be used to perform:

- *in situ* material property characterization;
- detect material and structural defects;
- detect damage including delaminations and corrosion; and
- characterize load environments (fatigue, overload).

4.1 Principle of the PZT/FBG hybrid active structural health monitoring system

The hybrid diagnostic system uses piezoelectric actuators to input a controlled excitation to the structure and fiber-optic sensors to capture the corresponding structural response. The system consists of three major parts: a diagnostic layer with a network of piezoelectric elements and fiber gratings to offer a simple and efficient way to integrate a large network of transducers onto a structure; diagnostic hardware consisting of an arbitrary waveform generator and a high-speed fiber grating demodulation unit together with a high-speed data-acquisition card to provide actuation input, data collection, and

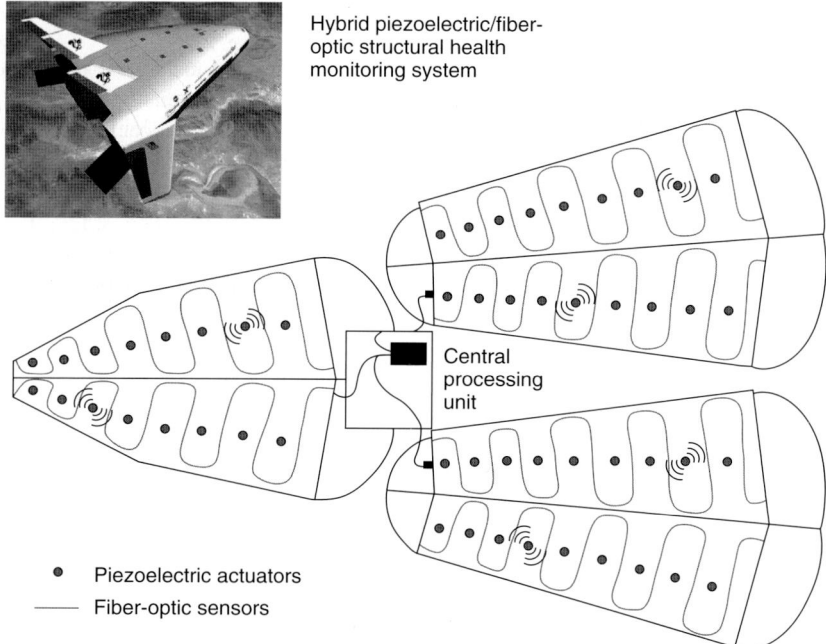

Figure 2. Schematic of a hybrid piezoelectric/fiber-optic structural health monitoring system for aerospace vehicles. [Reproduced from Ref. 13. © Institute of Physics Publishing, 2005.]

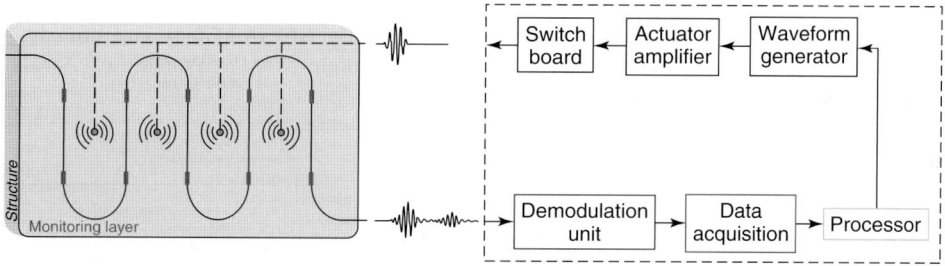

Figure 3. Piezoelectric/fiber-optic hybrid sensing system scheme. [Reproduced from Ref. 13. © Institute of Physics Publishing, 2005.]

information processing; and diagnostic software to determine the condition of the structure. Figure 3 shows a schematic diagram of this system.

One major advantage of the hybrid diagnostic system is that it offers the best actuator/sensor decoupling (minimum interference between actuation input signal and sensor output signal) because the transducers use different mechanisms for signal transmission: the piezoelectric actuators use electrical channels while the fiber-optic sensors use optical means. Since they use two separate mechanisms for transmitting signals, there is virtually no interference between them.

4.2 Hybrid piezoelectric/fiber-optic sensor sheets

Hybrid piezoelectric/fiber-optic (HyPFO) sensor sheets have been developed [13, 24]. The concept of a HyPFO sensor sheet is a generalization of the concept of a SMART layer [25], which is a device that comprises a thin dielectric film containing an embedded network of distributed piezoelectric actuator/sensors. Such a device can be mounted on the surface of a metallic structure or embedded inside a composite material structure during fabrication of the structure. Besides the piezoelectric and fiber-optic sensors, other types of sensors, such as strain gauges, MEMS (microelectromechanical systems) and TRD (time-rate-of-decay) temperature sensors, can also be integrated in the sensor layer. The advantages of a hybrid sensor layer include the following:

- It is not necessary to install each sensor individually on a structure. Sensors are embedded in thin, flexible films that can easily be mounted on structures in minimal amounts of installation time.
- Multiple measurements can be performed. For example, fiber-optic sensors can be used to measure temperatures, PZTs can be used to measure concentrations of hydrogen, and sensors of both types can be used to monitor acoustic emissions.

4.3 Demonstration of damage detectability of the system

In order to demonstrate the damage detection capability of the active hybrid diagnostic system, tests were conducted on both aluminum and fiber-reinforced composite plates.

4.3.1 Damage detection on an aluminum plate

Damage detection tests were performed on an aluminum plate with a dimension of $500 \times 500 \times 1.5 \, mm^3$. Four PZT actuators, each having a diameter of 6.35 mm and thickness of 0.25 mm, and a single grating sensor were mounted on the top surface of the aluminum plate. The single FBG is written at the center wavelength of 1550 nm and has a gauge length of 10 mm. The layout of the PZT actuators and the FBG sensors is shown in Figure 4. In the tests, five-peak burst waveforms were used to excite the structure [26].

To demonstrate the ability of the hybrid PZT/FBG system to detect damage, a $50 \times 8 \times 2 \, mm^3$ small stick-on patch was attached to the path between actuators and fiber grating sensors to simulate damage. Using the same five-peak burst waveforms as above, a series of tests with different frequencies of actuation signal were carried out. The sensor output was recorded and compared with the baseline taken before a stick-on patch was put on the aluminum plate. The typical test results are shown in Figure 5.

There are many modes of Lamb waves that can be actuated in a plate with a PZT. In our test, according to the product of frequency and thickness, only two kinds of modes were activated, i.e., A_0 and S_0. When

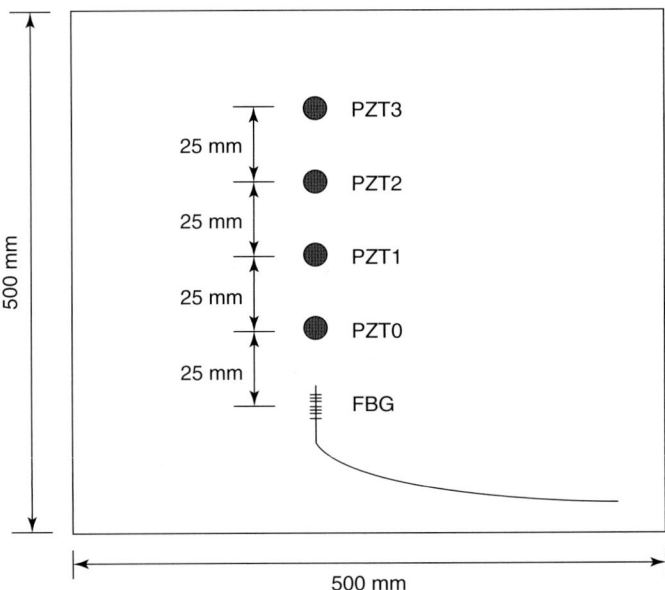

Figure 4. Layout of PZT actuators and the FBG sensor on the aluminum plate.

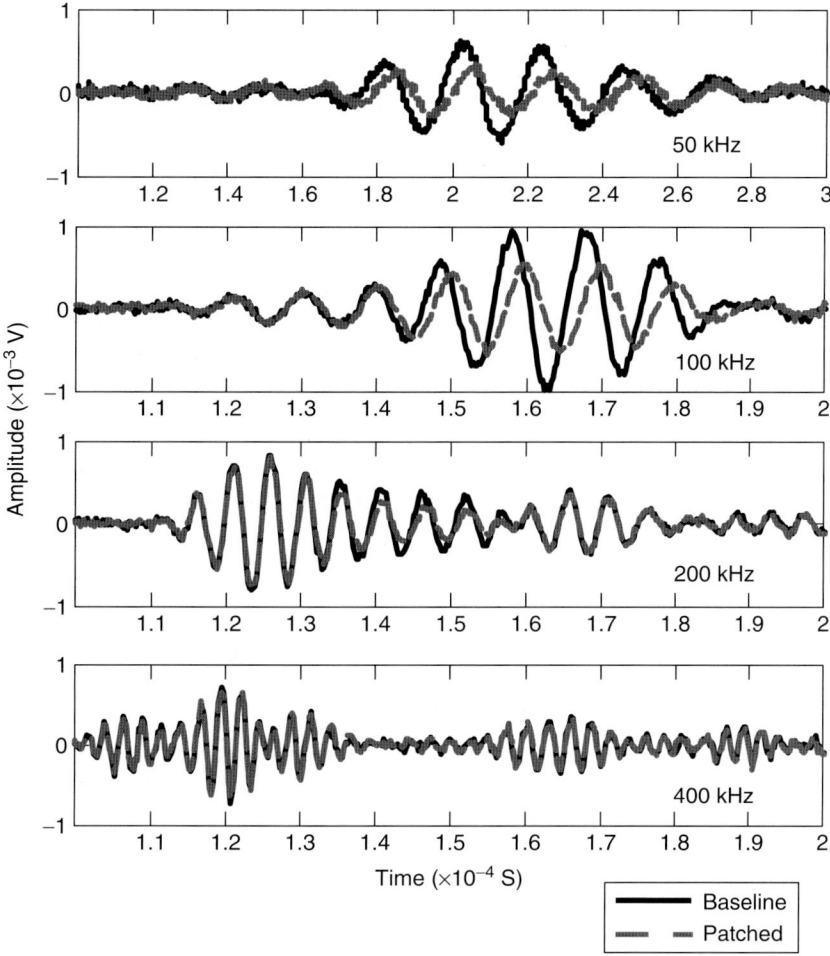

Figure 5. Damage detection test results from the aluminum plate.

the wavelengths are large in relation to the thickness of the plate, the fundamental symmetric mode and the fundamental antisymmetric mode are equivalent to the extensional and flexural waves, respectively.

The data presented in Figure 5 were signals actuated by PZT1 and collected by the FBG, the direction of which is parallel to the path of signal transmission. It was demonstrated that the signal amplitude is related to the distance and the cross angle between the signal transmission path and the FBG's direction [25]. From the signals we obtained, it can be seen that both A_0 and S_0 mode waves are detected. S_0 mode waves travel faster in this case, so they arrive first. When the center frequencies of the actuation signal are 50 and 100 kHz, respectively, S_0 mode waves carry less energy and are not significant. When the center frequency of the actuation signal is 200 kHz, S_0 mode waves and A_0 mode waves are both significant. We cannot see apparent changes in S_0 mode waves owing to the stick-on patch. However, we can see significant changes in both amplitude and phase in A_0 mode waves, which arrived later than S_0 mode waves. The results are consistent with established theories. The A_0 mode flexural waves show high sensitivity to weight added on their transmission path. The simulated damage was thus clearly identified. When the center frequency of actuation signal is 400 kHz, there are no A_0 mode waves showing up in the sensor signal, which is caused by the limitation of FBG sensors. When FBG sensors are utilized for stress wave detection, they cannot pick up a clear signal when the wavelength of the stress wave in question is smaller than the gauge length of the FBG sensor. In the case of 400 kHz, the wavelength of the A_0 mode decreases to about 6 mm, which is less than the 10-mm gauge length of the FBG. As Minardo [16] has demonstrated, no A_0 mode stress wave could be detected. There are only S_0 mode signals detected in this case, which are extensional waves and not sensitive to weight added on its traveling path, and hence the signal after damage and baseline signal are almost identical.

4.3.2 A quasi-distributed damage-detection scheme for composite plates

A damage detection scheme for a composite plate was implemented using the hybrid PTZ/FBG active-sensing system. The configuration of the specimen tested is given in Figure 6. Two PZT actuators and three FBG sensors (engraved on the same optical fiber) were embedded in a Hybrid SMART layer [13], which was then bonded onto the composite plate.

A delamination damage was introduced by repeated impacts on the center of the zone covered by the sensor network. A pitch–catch test was then carried out with a five-peak wave actuation signal of 100 kHz. Typical signals are shown in Figure 7. It can be seen that there are significant changes in both the amplitude and phase of the detected signals due to the delamination damage. To locate and evaluate the extent of damage, a damage index can be employed. When multiple paths are affected by the damage area, which results in big damage indices for these paths, their effects add up to show a heightened intensity of colors. This display technique can be used as a fast imaging method to help visualize the approximate location and extent of damage [28], as shown in Figure 8.

The extent of delamination damage in the composite plate was examined by an ultrasonic scan. The results clearly showed that severe damage was inflicted on the center area of the zone. The damage detected with the hybrid system is consistent with the result of the ultrasonic scan and X-ray image of the impact damage. In summary, it has been demonstrated that debond in composite structures can be conveniently identified by the proposed hybrid PZT/FBG active-sensing scheme [26].

4.4 Potential application for monitoring a large area

As described in [29], there is a major challenge in the networking of a multitude of piezoelectric sensors applied to physically large structures because of a large number of connection wires and big signal noise from long-distance communication. The hybrid system could be a potential solution for the applications of SHM on large structures, such as the health monitoring of bridge and long pipeline structures [27].

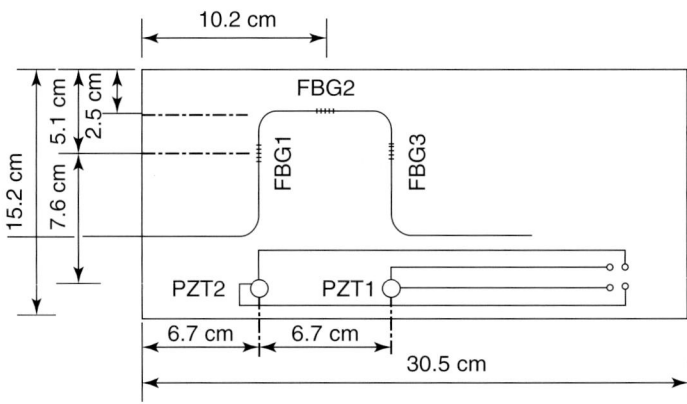

Figure 6. Configuration of the composite specimen. [Reproduced with permission from Ref. 27. © 2006, Qing *et al.*]

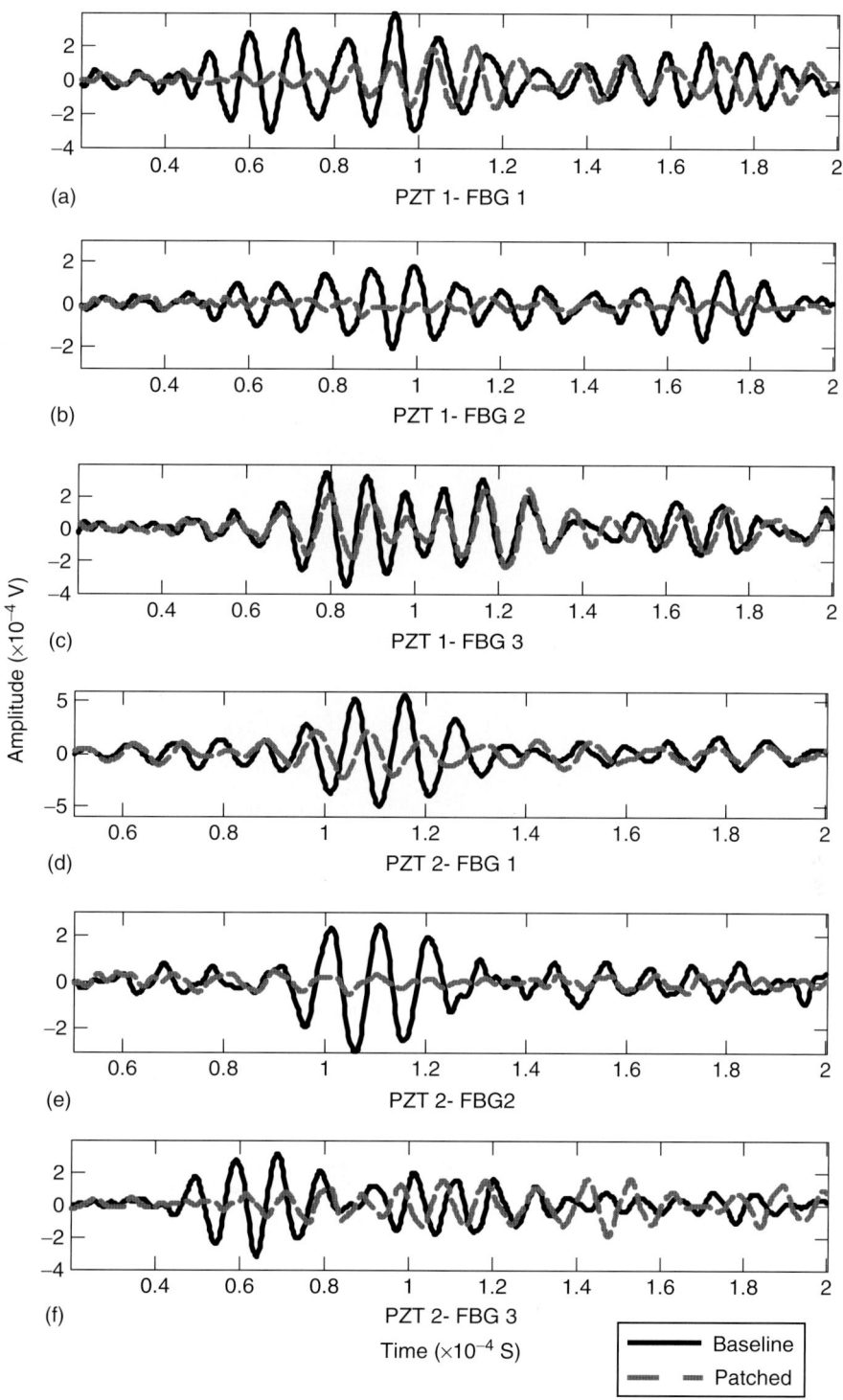

Figure 7. Damage detection results with the hybrid active-sensing system.

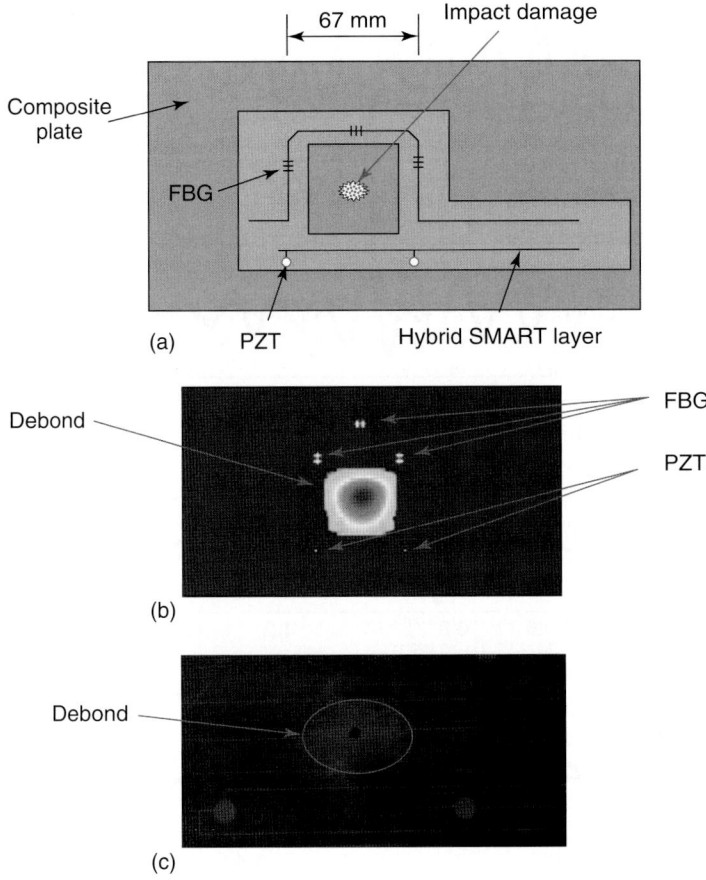

Figure 8. Hybrid sensor network used for damage detection. (a) Hybrid layer mounted on the surface of a composite plate, (b) diagnostic image of impact damage on the composite plate, and (c) X-ray image of the damage on the composite plate.

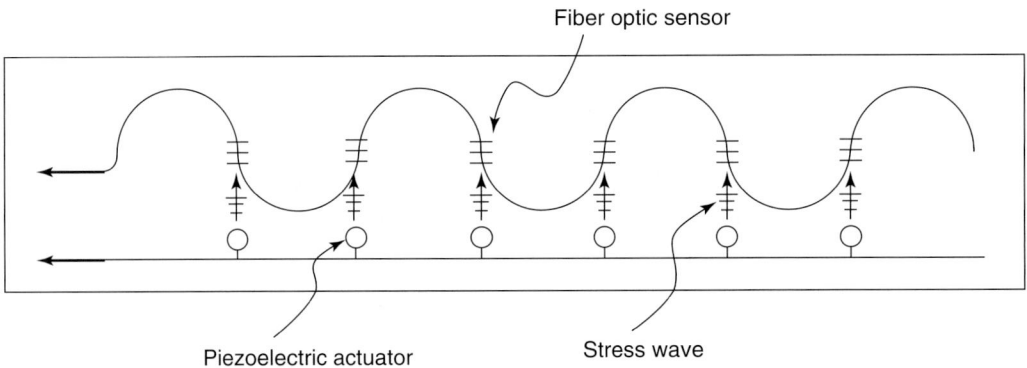

Figure 9. A typical design of hybrid piezoelectric/fiber-optic sensor network. [Reproduced with permission from Ref. 27. © 2006, Qing *et al.*]

Figure 9 shows a typical design of hybrid piezoelectric/fiber-optic sensor network used in the SHM system. In the network, all piezoelectric actuators are connected to a diagnostic instrument through a shielded cable. The distributed FBGs are used as sensors. The piezoelectric actuators can be simply connected together in series or in parallel, and then they generate stress waves in the structure simultaneously. By using a set of control wires, each piezoelectric actuator can also be used to generate stress wave in the structure individually. The smart cable with piezoelectric actuators and the optic fiber with distributed gratings can be placed on a thin carrier film, and then will be permanently bonded on the composite or metal structure to be monitored. The changes caused from the degradation of material properties (or corrosion) around all diagnostic paths will be identified. The design is particularly good for global damage monitoring of large composite or metal structures, such as fuel storage tanks for space ships, bridges with composite repairs, and metal pipelines. The advantages of this design include the following: (i) simpler wiring; (ii) long-distance sensor signal transmission; and (iii) capability to detect ultrasonic stress waves, quasi-static strain, and impacts. With all those features, an ideal health management solution based on hybrid FBG/PZT network can be offered by monitoring multiple physical parameters using the same system. At first, strain, temperature, and impact can be monitored continuously using FBG sensors, which can provide environment loading information and is critical to structural heath assessment. Then, when there is any abnormal sign, such as high stress or impact at certain locations, the active system can be triggered to send out diagnostic signals using the PZT actuators to detect whether any damage occurred.

5 CONCLUDING REMARKS

In order to resolve some issues in the piezoelectric sensor–based SHM system, such as coupling between sensors and actuators and long-distance signal transmission for large structure applications, extensive research on the development of hybrid piezoelectric/fiber-optic diagnostic systems has been conducted. A general review of the efforts made is presented in the article. An integrated hybrid active system and its application approach are also introduced, and damage detection capability with the system is demonstrated. One of the key issues for the hybrid PZT/FBG systems is the interrogation technique for FBG sensors, which dominates the sensitivity and dynamic range of the system. Besides, the fabrication technology of FBG sensors also needs to be improved to achieve high sensitivity for ultrasonic stress wave, which requires smaller grating length compared to quasi-static strain measurement.

ACKNOWLEDGMENTS

The authors would like to acknowledge the financial support from NASA to Acellent and the National Science Foundation (Grant No. CMS-0200399) to Stanford University for this development.

REFERENCES

[1] Choi K, Chang FK. Identification of impact force and location using distributed sensors. *Journal of American Institute of Aeronautics and Astronautics* 1996 **34**(1):136–142.

[2] Kersey AD, Davis MA, Patrick HJ, LeBlanc M, Koo KP, Atkins CG, Putnam MA, Friebele EJ. Fiber grating sensors. *Journal of Lightwave Technology* 1997 **15**(8):1442–1463.

[3] Han JH, Rew KH, Lee I. An experimental study of active vibration control of composite structures with a piezo-ceramic actuator and a piezo-film sensor. *Smart Materials and Structures* 1997 **6**(5):549–558.

[4] Qing X, Beard S, Kumar A, Chan H, Ikegami R. Advances in the development of built-in diagnostic system for filament wound composite structures. *Composite Science and Technology* 2006 **66**:1694–1702.

[5] Rose JL, Rajana K, Hansch MKT. Ultrasonic guided waves for NDE of adhesively bonded structures. *Journal of Adhesion* 1995 **50**:71–82.

[6] Rose JL, Ditri J. Pulse-echo and through transmission lamb wave techniques for adhesive bond inspection. *British Journal Of Non-Destructive Testing* 1992 **34**(12):591–594.

[7] Merzbachery CI, Kersey AD, Friebele EJ. Fiber optic sensors in concrete structures: a review. *Smart Materials and Structures* 1996 **5**(2):196–208.

[8] Idrissy RL, Kodindoumay MB, Kersey AD, Davis MA. Multiplexed Bragg grating optical fiber sensors for damage evaluation in highway bridges. *Smart Materials and Structures* 1998 **7**(2):209–216.

[9] Todd M, Johnson GA, Vohra S, Chen-Chang C, Danver B, Malsawma L. Civil infrastructure monitoring with fiber Bragg gratings sensor arrays. In *Proceedings of Structural Health Monitoring 2000*, Chang FK (ed). Technomic: Lancaster, PA, 1999, pp. 359–368.

[10] Fox JJ, Glass BJ. Impact of integrated vehicle health management (IVHM) technologies on ground operations for reusable launch vehicles (RLVs) and spacecraft. *IEEE Aerospace Conference Proceedings*, 2000; Vol. 2, pp. 179–186.

[11] Wang G, Pran K. Ship hull structure monitoring using fiber optic sensors. *Proceedings of European COST F3 Conference on System Identification and Structure Health Monitoring*, Universidad Politécnica de Madrid: Spain, 2000; Vol. 1, pp. 15–17.

[12] Lin M, Powers WT, Qing X, Kumar A, Beard SJ. Hybrid piezoelectric/fiber optic SMART layers for structural health monitoring. *Proceeding of 1st European Workshop on Structural Health Monitoring*, France, 2002; pp. 641–648.

[13] Qing X, Kumar A, Zhang C, Gonzalez IF, Guo G, Chang FK. A hybrid piezoelectric/fiber optic diagnostic system for structural health monitoring. *Smart Materials and Structures* 2005 **14**(5):98–103.

[14] Perez I, Cui HL, Udd E. Acoustic emission detection using fiber Bragg gratings. *Proceeding of SPIE* 2001 **4328**:209–215.

[15] Ogisu T, Shimanuki M, Kiyoshima S, Okabe Y, Takeda N. Damage growth detection of composite laminate structure using embedded FBG sensor/PZT actuator hybrid system. *Proceedings of SPIE* 2005 **5758**:93–104.

[16] Minardo A, Cusano A, Bemini R, Zeni L, Giordano M. Fiber Bragg gratings as ultrasonic waves sensors. *Proceedings of SPIE* 2004 **5502**:84–87.

[17] Pierce SG, Philp WR, Culshaw B, Gachagan A, McNab A, Hayward G, Lecuyer F. Surface-bonded optical fiber sensors for the inspection of CFRP plates using ultrasonic lamb waves. *Smart Materials and Structures* 1996 **5**(6):776–787.

[18] Betz DC. Lamb wave detection and source location using fiber Bragg grating rosettes. *Proceedings of SPIE* 2003 **5050**:117–127.

[19] Betz DC, Thursby G, Culshaw B, Staszewski WJ. Identification of structural damage using multifunctional Bragg grating sensors: I. Theory and implementation. *Smart Materials and Structures* 2006 **15**(5):1305–1312.

[20] Betz DC, Staszewski WJ, Thursby G, Culshaw B. Structural damage identification using multifunctional Bragg grating sensors: II. Damage detection results and analysis. *Smart Materials and Structures* 2006 **15**(5):1313–1322.

[21] Kersey AD, Davis MA, Patrick HJ, LeBlanc M, Koo KP, Askins CG, Putnam MA, Friebele EJ. Fiber grating sensors. *Journal of Lightwave Technology* 1997 **15**(8):1442–1463.

[22] Kersey AD, Berkoff TA, Morey WW. Multiplexed fiber Bragg grating strain-sensor system with a fiber fabry–perot wavelength filter. *Optics Letters* 1993 **18**(16):1370–1372.

[23] Baldwin CS, Vizzini AJ. Acoustic emission crack detection with FBG. *Proceedings of SPIE* 2003 **5050**:133–143.

[24] Lin Mark, Qing X, *Hybrid Piezoelectric/Fiber Optic Sensor Sheets—Multiple Sensors of Different Types Could be Installed On Or In Structures*, MFS-31846-1, *NASA Tech Briefs*, July, 2004.

[25] Lin M, Qing X, Kumar A, Beard S. SMART Layer and SMART suitcase for structural health monitoring applications. *Proceedings of SPIE on Smart Structures and Material Systems* 2001 **3329**:98–106.

[26] Wu Z, Qing X, Chang FK. Debond detection for composite laminate plates with a distributed hybrid PZT/FBG sensor network. *Journal of Intelligent Material Systems and Structures (Submitted)*.

[27] Qing X, Wu Z, Chang FK, Ghosh, K, Karbhari V, Sikorsk C. Monitoring the disbond of externally bonded CFRP composite strips for rehabilitation of bridges. *Proceeding of the Third European Workshop on Structural Health Monitoring*, Granada, Spain, 2006; pp. 463–470.

[28] Beard S, Qing PX, Hamilton M, Zhang DC. Multifunctional software suite for structural health monitoring using smart technology. *Proceedings of the 2nd European Workshop on Structural Health Monitoring*, Munich, 2004; pp. 101–108.

[29] Qing X, Beard S, Kumar A, Yu P, Chan HL, Zhang D, Ooi T, Marotta SA. Practical requirements for implementation and usage of SHM systems on aerospace structures. *Proceedings of the 5th International Workshop on Structural Health Monitoring*, Stanford University, 2005; pp. 1502–1509.

FURTHER READING

González IF, Wu ZJ, Chang FK. Health monitoring by Means of hybrid diagnostic System. In *Proceeding of Structural Health Monitoring*, Chang FK (ed). DEStech Publications: Stanford, CA, 2005, pp. 732–740.

Chapter 80
The HELP-Layer® System

Michel B. Lemistre
Laboratoire SATIE/CNRS, Ecole Normale Supérieure de Cachan, Cachan, France

1 Introduction 1403
2 The HELP-Layer® System 1403
3 Numerical Simulation 1405
4 Conclusion 1411
References 1411

Encyclopedia of Structural Health Monitoring. Edited by Christian Boller, Fu-Kuo Chang and Yozo Fujino © 2009 John Wiley & Sons, Ltd. ISBN: 978-0-470-05822-0.

1 INTRODUCTION

The major risk for composite materials is the creation of delaminations resulting from impacts being often associated by fiber breaking. Besides, other types of damages can be caused by thermal aggressions, inducing pyrolysis of the polymeric matrix and by liquid ingress linked to aggressive environments. The most current structural health monitoring (SHM) system (based on fiber-optic sensors and acousto-ultrasonics), well suited to damages of mechanical origin, are unfortunately poorly sensitive to these last two types of damages. Assuming that all damages can affect the main electrical properties of structures such as electrical conductivity and dielectric permittivity, a measurement of these two parameters may allow detecting all kinds of damages. This is the reason why a new SHM system has been developed. This system, based on the interaction between a low-frequency electromagnetic field and the composite structure, allows the electrical characterization of materials. As shown here, this method detects and characterizes all kind of damages with a good sensitivity.

2 THE HELP-Layer® SYSTEM

2.1 Short recall of the principle

The HELP-Layer® system [1, 2] is a low-frequency electromagnetic technique (*see* **Chapter 63**) applied to an SHM concept for composite structures. Its principle is based on a simple concept: a carbon fiber reinforced plastic (CFRP) structure is a double medium made, on the one hand, of a conductive medium (carbon fibers of conductivity $\sigma \approx 10^4 \, \text{S} \cdot \text{m}^{-1}$) and, on the other hand, of a perfectly dielectric medium (the polymeric matrix). The electromagnetic behavior of the first medium is sensitive to conductivity variations and the second medium to the orientational polarization phenomenon. So, one can define a polarization vector **P** linked with the local electric field E_1 by the following relation:

$$\mathbf{P} = \chi_e \varepsilon_0 E_1 \qquad (1)$$

where ε_0 is the dielectric permittivity of the free space (8.84×10^{-12} F·m^{-1}) and χ_e is the electric susceptibility, itself linked with the relative dielectric permittivity ε_r by the following expression:

$$\varepsilon_r = 1 + \chi_e \qquad (2)$$

After induction of eddy currents in the structure, the measurement of the reflected electric field E_r gives access to both electrical parameters σ and ε_r, by the following relation:

$$E_r = \frac{J}{\sigma} + E_1(\varepsilon_r - 1) \qquad (3)$$

with J being the eddy current density. This method allows for detection of a possible local variation of σ and/or ε_r. Therefore, all damages inducing a local variation of one or both these parameters will be detected.

In the case of a glass fiber reinforced plastic (GFRP) structure, the induction of eddy currents being impossible, one excites the structure with an electric field. So, only one parameter is measured, the relative dielectric permittivity ε_r.

2.2 Technology used

The HELP-Layer® system is a complete system having two parts. The first part is the sensitive layer itself. The second part is the associated electronics having in charge the excitation for inducing eddy currents (or electric field) in the structure, and the data reduction process to extract the relevant information and to build an image of the structure where the damages clearly appear.

The sensitive layer is made of a printed circuit on a 200-µm-thick dielectric substrate including a double network of crossed wires; this layer is bonded or embedded in the structure under test. Figures 1 and 2 respectively present the principle of the layer and a photo of a carbon/epoxy plate equipped with such a layer. By scanning these networks, one can perform local measurements of the electric field, with each analyzed zone having a dimension of 20 mm × 20 mm corresponding to the distance between two successive wires of each network (Figure 1). The first network, called the *inductive network*, is short-circuited at one end (or in open circuit in the case of a GFRP structure);

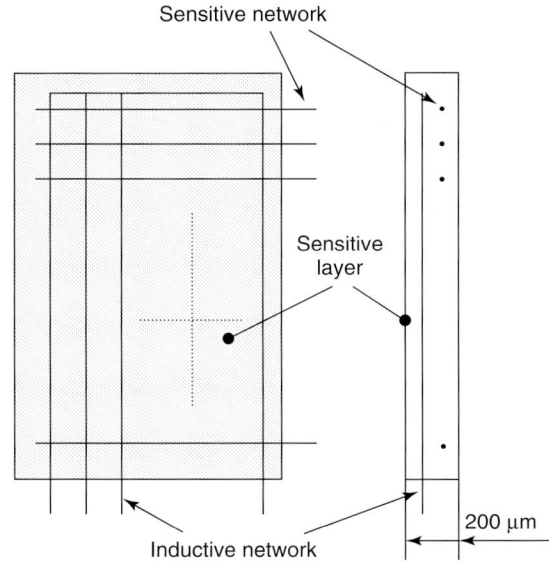

Figure 1. Geometry of the sensitive layer.

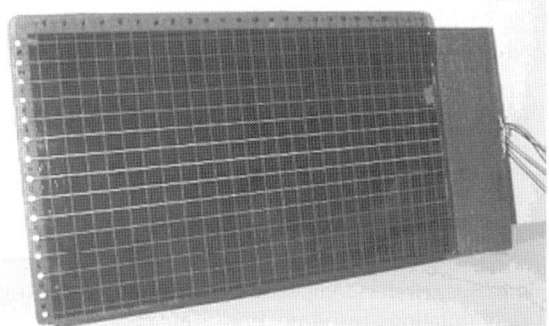

Figure 2. Instrumented structure.

each pair of wires represents a loop of induction (or a capacitance in the case of a GFRP structure). The second network, called the *sensitive network*, is not short-circuited, and each pair of successive wires can be considered as a capacitance allowing the in-plane component of the electric field perpendicular to the wires to be measured. Furthermore, each wire of this network can be considered as an elementary antenna, measuring the in-plane component of the electric field parallel to the wires (Figure 3). There are two possible exploitation methods of this technique. The first one consists of using only the modulus of the in-plane electric field to build an image of the damaged structure. The second one exploits separately the two in-plane components of the electric

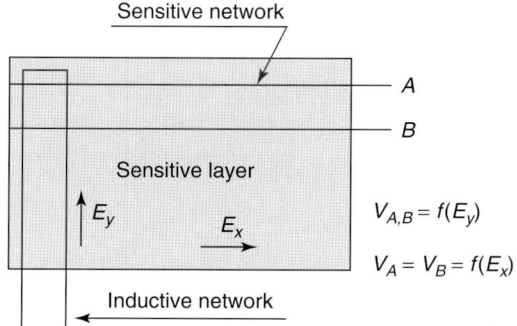

Figure 3. Measurement of the electric field components.

field and allows computation of the new values of the electric parameters to fully characterize the damages and to estimate their severity. This second method of exploitation is explained in the following subsections.

2.3 Example of application

An example of results obtained with the first method of exploitation is shown in Figure 5. Figures 2, 4, and 5 respectively present a photograph of an instrumented structure (a 16-ply orthotropic carbon/epoxy plate $[0_2, 90_2]_{2s}$ of dimensions: 610 mm × 305 mm × 2 mm), the types and locations of the damages in the structure (as seen from the opposite face of the plate), and the electromagnetic image obtained. The plate has been damaged by six different defects: a 4 J impact (I2) inducing a severe delamination with fiber breakage, a 2 J impact (I1) inducing a light delamination, and four local burns produced by "high energy sparks" (30 V, 5 A) of various duration, delivering energies of 40 J (B2), 80 J (B3), 120 J (B1), and 400 J (B4) (Figure 4).

In Figure 5, one can see that all damages are perfectly detected except the lower 2 J impact, probably because of the fact that there is no fiber breakage, so there is no variation of electric properties inside the structure under test. The inductive network has been excited by a continuous signal having a frequency of 700 kHz. The data reduction process is based on a multiresolution processing, using wavelet transform [3–5].

3 NUMERICAL SIMULATION

3.1 Basic principle

The numerical simulation method used is a concept developed at the Ecole Normale Supérieure of Cachan/France, called *distributed point source method* (DPSM) [6–8]. The main originality of this method is that, contrary to a classical finite

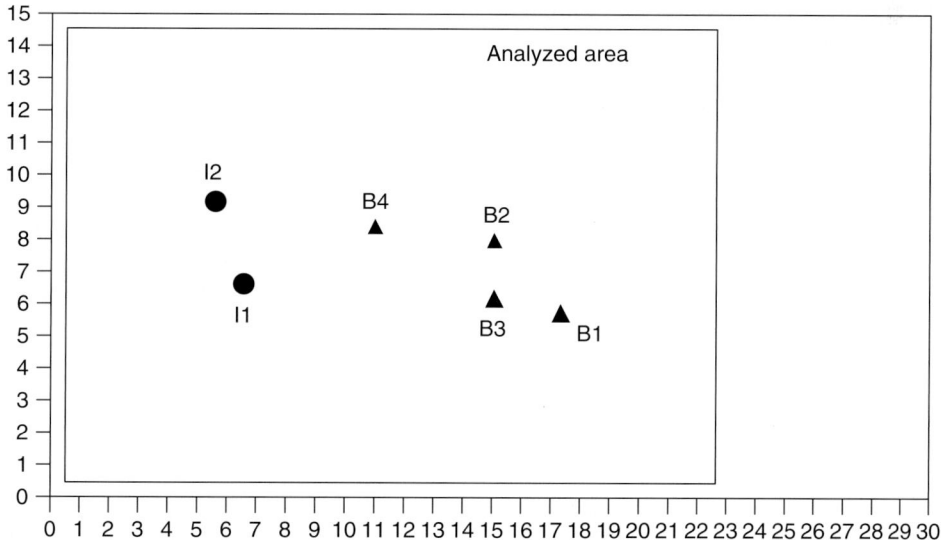

Figure 4. Damaged structure.

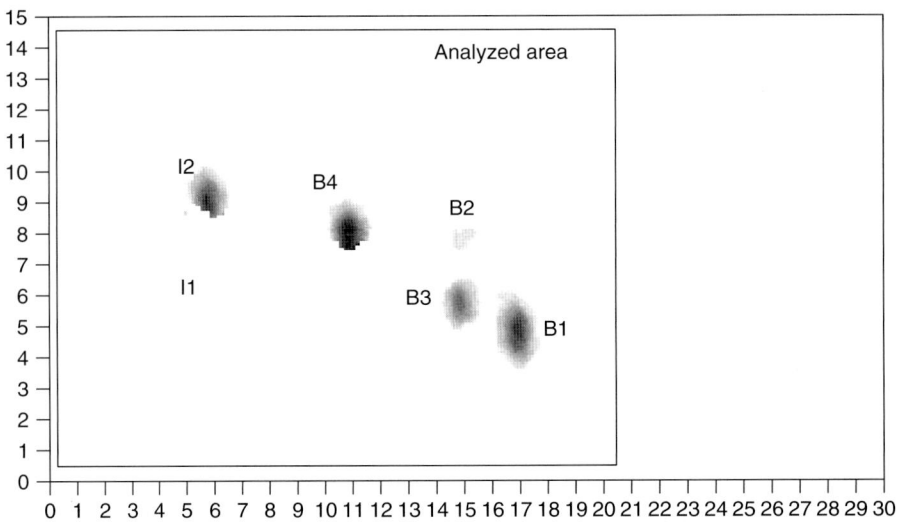

Figure 5. Electromagnetic image of the damaged structure shown in Figure 4.

elements method, it is not necessary to mesh the totality of the computation volume, but only the surface of interest. The implementation of the model simply requires discretization of the active surface of the transducer or the interfaces to obtain an array of point sources, so that the initial complexity is changed into a superposition of elementary problems. The active surfaces like transducers, emitters, or interfaces reflecting a part of an incident field, are discretized into a finite number of elementary surfaces, a point source being placed at the centroid of every elementary surface. It is interesting to note that the energy (or the power) radiated by such a system is the product of a scalar quantity by the flux of a vector (or the time derivative of the flux, for power). Let us call the scalar quantity θ_k the scalar potential, and ϕ_k the flux emitted by the source k, the vector being the field V. For magnetic systems, θ_k and ϕ_k represent the magnetic potential and the flux of magnetic induction for the N elementary sources (Figure 6). For instance, one can calculate the potential θ at point M by superposition of elementary charges as follows:

$$\theta(M) = k \left(\sum_{n=1}^{N} \frac{q_n}{R_n} \right) \quad (4)$$

where R_n represents the distance between the source k and the point M, and q_n is the elementary charge at the point n (i.e., the charge generating the source k).

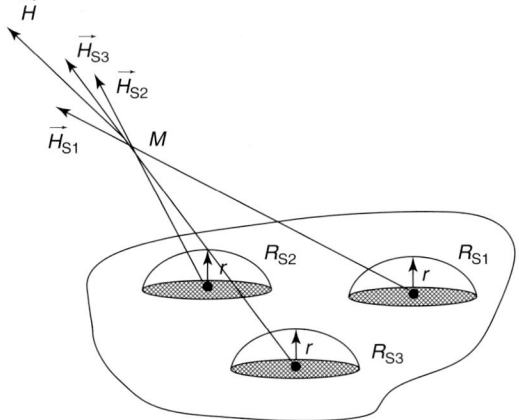

Figure 6. Discretization of a sensor surface into a finite number of hemispherical surfaces.

However, it is necessary to calculate the potential θ at the top point of each hemispherical surface (contribution of each elementary source toward others) by the following equation:

$$\begin{bmatrix} \theta_1 \\ \theta_2 \\ \vdots \\ \theta_n \end{bmatrix} = k \begin{bmatrix} F_{1,1} & F_{1,2} & :: & F_{1,n} \\ F_{2,1} & F_{2,2} & :: & F_{2,n} \\ :: & :: & :: & :: \\ F_{n,2} & :: & :: & F_{n,n} \end{bmatrix} \cdot \begin{bmatrix} q_1 \\ q_2 \\ \vdots \\ q_n \end{bmatrix} \quad (5)$$

where the function F represents the inverse of the distance R between the charge q_n and the calculation point. From matrix $[F]$, which is a regular matrix, it is possible to calculate the elementary charges by the following relation:

$$[q] = \frac{1}{k}[G] \cdot [\theta] \quad (6)$$

with

$$[G] = [F]^{-1} \quad (7)$$

One can thus calculate the three quantities θ, V, and ϕ in each point of the space by using the knowledge of the elementary sources. If we place a target at point M, one can calculate the same quantities on its surface by using the matrix of reflection coefficients. A major issue of this method is that the surface of interest is meshed uniquely and the thickness of the target is taken into account by the matrix of reflection coefficients. This method allows the behavior of the HELP-Layer® system to be simulated.

3.2 Simulation conditions

As shown before, the HELP-Layer® system includes two crossed conductive networks. One of these networks is in charge of inducing eddy currents inside the carbon structure and appears in the form of parallel lines, with a spacing of 20 mm, and is short-circuited at one end. To induce significant eddy currents inside the structure (i.e., a significant electric field), it is necessary to sequentially inject a current of 1 A into each line of the inductive network. The frequency of excitation is chosen taking into account the depth of the structure (i.e., the skin effect). To simplify the problem, only one element of the HELP-Layer® of 60 mm × 60 mm is modeled, including only one inductive line. The structure is set at 0.1 mm above the HELP-Layer®. The current source is an inductive line of 60-mm length located in the xy plane at $x = 30$ mm and $y = 0-60$ mm, meshed by 60 current elements Idl (Figures 7 and 8). This current source is called *DPSM primary sources JAp*. The formulation used is the DPSM/Green's formulation [9].

The first computation consists of calculating the electromagnetic field on the surface of the structure ($\vec{E}$ and $\vec{H}$) or, more precisely, the magnetic vector potential $\vec{A}_1$ in medium 1 (free space) computed by superimposing the effect of current sources JAp and JA1 (DPSM virtual sources; see Figure 8. The magnetic vector potential in medium 2 (the structure) $\vec{A}_2$ is only defined by the current sources JAs and called *DPSM secondary sources*, the boundary

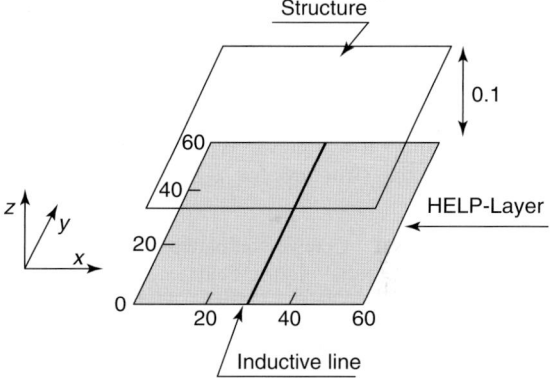

Figure 7. Geometry of the simulation (dimensions are given in millimeters).

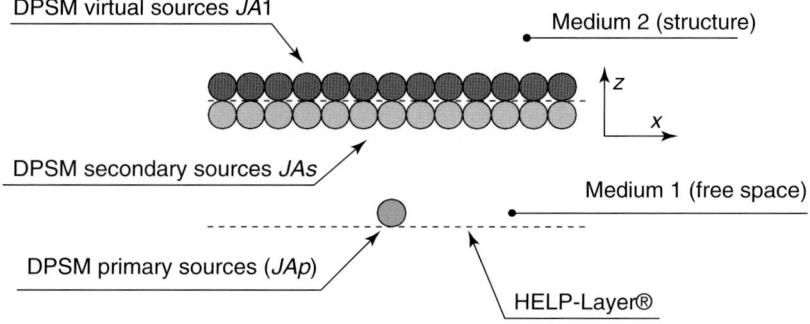

Figure 8. DPSM current sources.

conditions being the continuity on the vector potential and its first derivative along the z axis:

$$\begin{cases} \vec{A}_1 = \vec{A}_2 \\ \dfrac{1}{\mu_1} \dfrac{\partial \vec{A}_1}{\partial z} = \dfrac{1}{\mu_2} \dfrac{\partial \vec{A}_2}{\partial z} \end{cases} \quad (8)$$

In the case of harmonic excitation, one can compute the electric field by the following relation:

$$\vec{E} = -j\omega \vec{A} \quad (9)$$

3.3 Validation with experimental data

A computation of the in-plane component of the electric field E_y for three kinds of materials allows the results obtained for the same values measured to be compared by the HELP-Layer® system. Table 1 shows this comparison.

3.4 Modeling a damage inside the structure

Assuming that the defect is located at 1-mm depth inside the structure for $x \in (30, 40)$ mm and

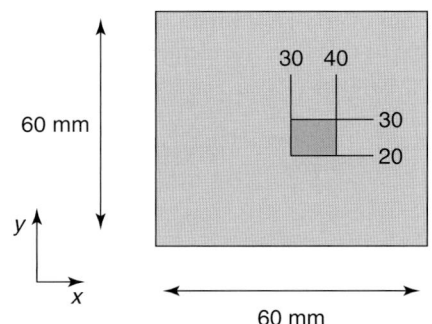

Figure 9. Location of the virtual damage.

$y \in (20, 30)$ mm (Figure 9), generating a variation of one electric parameter of the structure (i.e., σ or ε). The first step is to construct the secondary sources JAs inside the structure. After that one performs the same process as presented previously, with new virtual sources $JA2$ and new secondary sources $JAs2$, medium 2 being the structure and medium 3 being the damaged zone of the structure (Figure 10). Note that we do not solve the global problem but a new elementary problem with new DPSM sources $JAs2$.

Two kinds of damages (damage 1 and damage 2) have been simulated, by variation of the local electrical conductivity σ and by variation of the local relative dielectric permittivity ε_r, respectively. Table 2

Table 1. Comparison between computed values and experimental values of the E_y component of the tangential electric field

Type of structures	E_y component (V·m^{-1}): computed values	E_y component (V·m^{-1}): experimental values	ΔE_y (%)
$[0_2, 45_2, 90_2, -45_2]_S$	1.59	1.65	+4
$[0_2, 45_2, 90_2, -45_2]_{2S}$	1.40	1.40	−3.5
$[0_2, 90_2]_{2S}$	1.60	1.65	+4

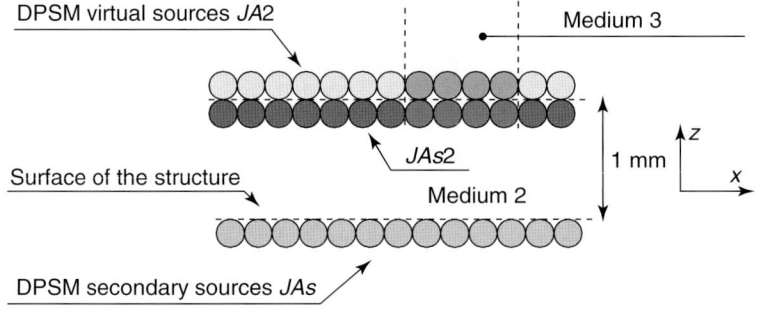

Figure 10. New DPSM sources.

Table 2. Numerical values of σ and ε_r for each kind of damage

Kind of damage	Sound area		Damaged area	
	σ (S·m^{-1})	ε_r	σ (S·m^{-1})	ε_r
1	10^4	4	5×10^3	4
2	10^4	4	10^4	2

shows the numerical values of these two parameters for each kind of damage.

Figure 11(a) and (b) present the modulus of the resulting electric field $|E_y|$ and $|E_x|$, respectively for the damaged structure 1. Figure 12 (a) and (b) present the same parameters for the damaged structure 2. These electric fields expressed in volts per millimeter represent for each one of the damaged structures (i.e., 1 or 2) the difference between the electric field obtained from a perfectly sound structure and the electric field obtained from the damaged structure.

One can see that in the case of a damage generating a variation of the electric conductivity σ, the y component of the electric field is dominating. In Figure 11(a) and (b), showing $|\vec{E}_y|$ and $|\vec{E}_x|$ respectively, the maximum value of the y component is 10^{-3} V·mm^{-1}, whereas the maximum value of the x component is only 1.3×10^{-7} V·mm^{-1}. On the contrary, in the case of a damage generating a variation of the dielectric permittivity ε_r, the major contribution on the electric field is given by the x component (Figure 12a and b): i.e.,

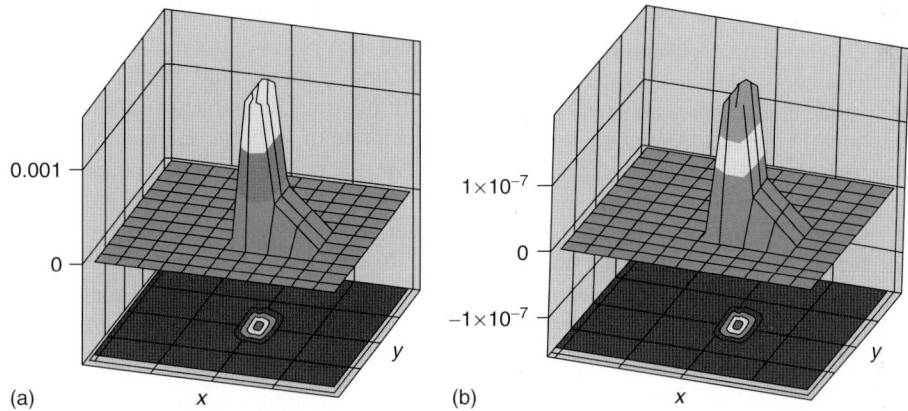

Figure 11. Electric fields resulting from a damaged structure 1. (a) Modulus of the E_y component and (b) modulus of the E_x component.

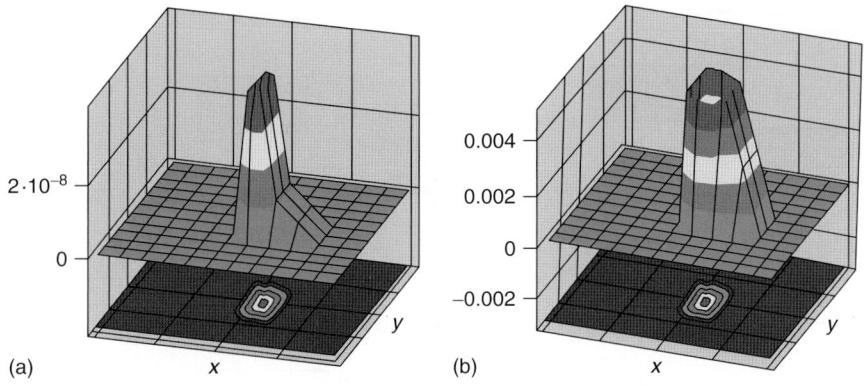

Figure 12. Electric fields resulting from a damaged structure 2. (a) Modulus of the E_y component and (b) modulus of the E_x component.

the maximum value of the y component is only 2.5×10^{-8} V·mm^{-1}, while the x component is 3×10^{-3} V·mm^{-1}. This is a very important result because it allows a kind of damage (i.e., the origin of the damage, *see* **Chapter 63**) to be determined.

3.5 Solving the inverse problem

To solve the inverse problem (i.e., computation of the new values of σ and ε_r, and determination of the kind of damage), the preceding remarks allow an algorithm to be developed (Figure 13). The first step helps to determine if the modulus of the x component of the electric field measured by the HELP-Layer® system E_{x_m} is significant (i.e., whether $>10^{-3}$ V·m^{-1}). If the E_{x_m} component is not significant, the damage has a mechanical origin (delamination, fiber breaking, crack); so, one computes the modulus of the E_y component (resulting from the difference: sound structure—damaged structure) with variation of the σ value and compares it to the experimental E_{y_m} component. When the computed E_y value is equal to the experimental E_{y_m} value, the last value of σ is the local conductivity of the structure due to the damage. If the E_{x_m} component is significant, the damage has no mechanical origin, but may be due to burning, liquid ingress, etc. One compares the computed E_x value with the experimental E_{x_m} value and determines the local permittivity ε_r due to the damage, by the same process. To determine the possible local variation of σ one can then perform the same process with the E_y components.

The first evaluation concerns a quasi-isotropic plate of 2-mm thickness $[45_2, 0_2, -45_2, 90_2]_S$, including various delaminations generated by calibrated impacts (impact energies of 0.75, 2, 2.5, 3, and 4 J). The HELP-Layer® system measures the two components of the electric field (i.e., E_{x_m} and E_{y_m}), while the simulation program computes the values ε_r and σ on the damaged area corresponding to the same electric field component variations. For a sound structure, the electrical parameters are $\varepsilon_r = 4$ and $\sigma = 10^4$ S·m^{-1}. Table 3 shows the results obtained. One can see that a delamination resulting from an impact of energy lower than 2.5 J does not cause a fiber breakage and consequently cannot induce a variation of σ that would detect the fiber breakage. The threshold of detection for this kind of defect is a delamination resulting from an impact of about 2.5 J for a 2-mm-thick plate of such a composite.

The second evaluation concerns a 2-mm-thick orthotropic plate $[0_2, 90_2]_S$, including various burning generated by electric sparks, for electric energies of 10, 40, 80, and 120 J. The field variations measured are given in Table 4. Assuming the same sound material properties as in the first case, the simulation

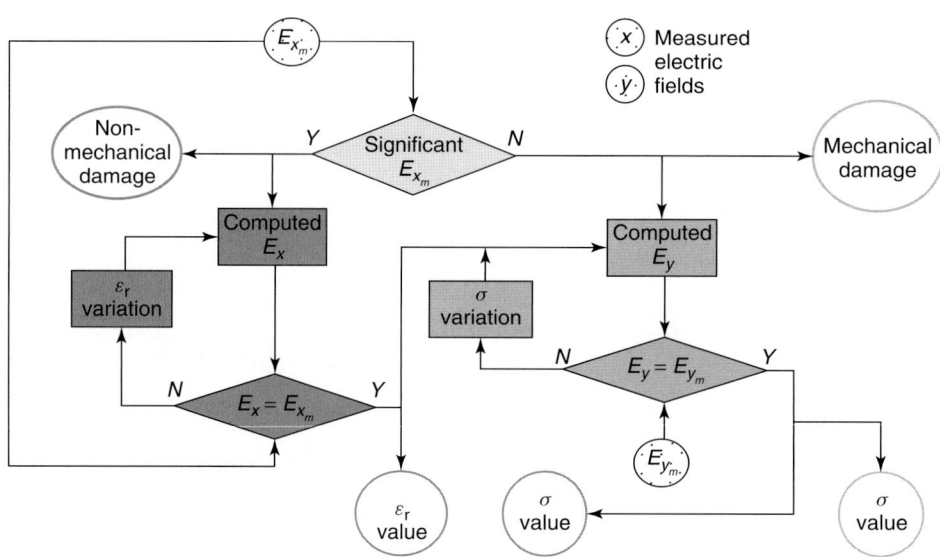

Figure 13. Algorithm allowing to solve the inverse problem.

Table 3. Equivalent σ and ε_r for impact delamination

Impact energies (J)	Measured fields		Deduced properties	
	ΔE_x (V·m^{-1})	ΔE_y (V·m^{-1})	σ (S·m^{-1})	ε_r
0.75	NS[a]	NS	X[b]	X
2	NS	NS	X	X
2.5	NS	1.1×10^{-2}	8×10^3	X
3	NS	1.6×10^{-2}	7×10^3	X
4	NS	5.0×10^{-2}	4.7×10^3	X

[a] Nonsignificant values.
[b] Same values as a sound structure.

Table 4. Equivalent σ and ε_r for electric burning

Electric energies (J)	Measured fields		Deduced properties	
	ΔE_x (V·m^{-1})	ΔE_y (V·m^{-1})	σ (S·m^{-1})	ε_r
10	0.07	1.8×10^{-3}	NV[a]	3.93
40	0.26	6×10^{-3}	NV	3.80
80	0.53	1.3×10^{-3}	NV	3.56
120	0.80	2×10^{-3}	NV	3.39

[a] Nonsignificant variation (see subsection 3.5).

leads to values of ε_r and σ as given in this table. Here an extra hypothesis was necessary. It was assumed that in the case of light burning, the low variation of the conductivity of the resin is masked by the relatively high conductivity of the carbon and that in the case of a fiber breakage phenomenon induced by hard burning, the increase of σ in the resin because of pyrolysis and the decrease of σ in the carbon fiber bundles would compensate each other. On the basis of these considerations, it has been assumed that only the permittivity ε_r was affected. In fact, it would have been more accurate if the two parameters, ε_r and σ, could have been varied. However, that would require the establishment of a new experimental procedure allowing the identification of both parameters. This improvement of the method is an objective for future development.

4 CONCLUSION

This article has been dedicated to an original system of structural health monitoring called the *HELP-Layer® system*, which is based on the interaction of an electromagnetic field and materials such as dielectric or conductive composite structures. The system shows that it is perfectly possible to detect, localize, and characterize various damages in a composite structure by using electromagnetic methods. Simulations allow validation of this electromagnetic method for use in a SHM system such as the HELP-Layer®, based on the analysis of electrical properties of materials. They also permit to distinguish between two types of damage (e.g., mechanical and thermal damage). The main interest of this method lies in its high sensitivity to "nonmechanical" damages such as resin pyrolysis and liquid ingress.

Work in progress consists in determining the effect of electrically qualified damage on the residual mechanical behavior of the composite material linking directly the HELP-Layer® system measurements to the degradation of structural performance, which, in fact, is the most interesting point.

REFERENCES

[1] Lemistre M. Low frequency electromagnetic techniques. *Structural Health Monitoring*. ISTE: London, UK, 2006; Chapter 6, pp. 412–461.

[2] Lemistre MB, Placko D. *HELP-Layer System*, French Patent FR0403310, 2004, US Patent US2005/0228 208A1, 2005.

[3] Lemistre MB, Balageas DL. A new concept for structural health monitoring applied to composite materials. In *Structural Health Monitoring*, Balageas DL (ed). DEStech Publications, 2002; pp. 493–507.

[4] Lemistre MB, Balageas DL. Structural health monitoring system based on diffracted Lamb waves analysis by multiresolution processing. *Smart Materials*. IOP Publishing, 2001; Vol. 10, No. 3, pp. 504–511.

[5] Mallat S. *A Wavelet Tour of Signal Processing*. Academic Press: New York, 1998.

[6] Lemistre MB, Placko D, Liebeaux N. Simulation of an health monitoring concept for composite materials, comparison with experimental data. *Proceedings of SPIE* 2003 **5047**:130–139.

[7] Placko D, Liebeaux N, Kundu T. *Modélisation par Sources Réparties*, Patent in progress no. 0214108.

[8] Dufour I, Placko D. An original approach of eddy current problems through a complex electrical image concept. *IEEE Transaction on Magnetics* 1996 **32**(2):348–365.

[9] Lemistre M. In *DPSM for Modeling Engineering Problems*, Placko D, Kundu T (eds). John Wiley & Sons, 2007; Chapter 10, pp. 333–347.

Chapter 81
Microelectromechanical Systems (MEMS)

Jonas Meyer, Reinhard Bischoff and Glauco Feltrin
Structural Engineering Research Laboratory, Empa, Swiss Federal Laboratories for Materials Testing and Research, Dübendorf, Switzerland

1 Introduction	1413
2 Mechanical MEMS Sensors	1414
3 MEMS-Based Structural Health Monitoring System	1415
4 Field Tests with the MEMS-based SHM System	1417
5 Results	1418
6 Conclusions	1421
Related Articles	1421
References	1422

1 INTRODUCTION

Microelectromechanical systems (MEMS) are integrated devices or systems of devices that combine electrical and mechanical components and that have a size, which ranges from the submicrometer to the millimeter level. Miniature mechanical elements such as beams, diaphragms, and springs are fabricated by micromachining techniques from silicon wafers and are combined with microelectronic components

Encyclopedia of Structural Health Monitoring. Edited by Christian Boller, Fu-Kuo Chang and Yozo Fujino © 2009 John Wiley & Sons, Ltd. ISBN: 978-0-470-05822-0.

and circuits to form microsensors, microactuators, and microengines. The microfabrication technology allows for fabrication of large systems of MEMS devices, which individually perform simple tasks, but in combination can accomplish complicated functions. MEMS allow sensing, controlling, and activating physical and chemical processes on the micro and, by a suitable combination of clusters of MEMS devices, also on the macro scale.

Typical examples of MEMS devices are accelerometers, gyroscopes, strain gauges, pressure and flow sensors, miniature robots, fluid pumps, microvalves, and micromirrors. The term *MEMS* was coined in 1987, long after the first micromachined devices, in particular, microsensors, were commercially available. Equivalent terms for MEMS are microsystems and micromachines.

The term micromachining designates the fabrication of micromechanical parts (such as diaphragms or beams). These parts were fabricated by etching away selected areas of the silicon substrate to obtain the desired micromechanical components. Since the early 1960s, various etching techniques were developed to improve the fabrication of micromechanical components and these techniques form the basis of the so-called bulk micromachining processing techniques. The need for higher design flexibility and better performance, however, gave rise to surface micromachining techniques, in which the so-called sacrificial layers are deposited between structural layers for mechanical separation and isolation. These sacrificial layers are then removed by etching to free the structural layers and to enable mechanical components to move relative to the substrate. Surface micromachining enables the fabrication of complex multicomponent integrated micromechanical structures that would not be possible with traditional bulk micromachining. Details on MEMS and their fabrication technologies can be found in [1–3].

2 MECHANICAL MEMS SENSORS

MEMS devices for sensing mechanical quantities are the most important class of microsensors. The first fabrication of silicon-based MEMS devices started in the late 1950s with the development of pressure microsensors. In 1974, National Semiconductor launched the first high-volume pressure sensor in the market. Silicon pressure sensors are at present the commercially most important microsensor type with a billion-dollar market and large-scale technical applications in different industries like the automotive and aeronautical industry.

The discovery of the piezoresistive effect in silicon and germanium in 1954 enabled the development of silicon-based micromachined strain gauges with a gauge factor 10–20 times greater than those based on metal films. Micromachined piezoresistive strain gauges are now a standard component in accelerometers and pressure sensors.

2.1 Accelerometers

Accelerometers are the commercially second most important type of mechanical microsensors. The design principle of MEMS accelerometers is the same as traditional accelerometers: an inertial mass suspended by a linear elastic mechanical component (micromachined cantilever beam, bridge, or membrane). Accelerations cause inertial forces, which deflect the suspended mass from its zero position. This deflection is converted by a pickup to an electrical signal, which, after a suitable signal conditioning by an internal integrated circuit, appears at the sensor output. The two most prevalent pickup types of MEMS accelerometers are: capacitive and piezoresistive pickup of the seismic mass movement, where capacitive polysilicon surface-micromachined and single-crystal micromachined devices are the most important types.

The amplitude range of capacitive MEMS accelerometers varies between a few g up to $50g$ for applications in air bag systems. Currently, several low-g MEMS accelerometers are commercially available. Table 1 displays a selection of these accelerometers, which have the characteristics to be applicable in structural health monitoring (SHM).

2.2 Application aspects

MEMS sensors have several advantages compared to conventional sensors. They are small, generally low power, highly integrated, and, usually, cheap. These

Table 1. Selection of commercially available MEMS accelerometers

Product	ST microelectronics LIS2L06AL	Analog devices ADXL204	Colybris MS8002.C	Colybris SI-Flex SF1500S	PCB 3711D1FB3G
Number of axes	2	2	1	1	1
Amplitude range (g)	±2.0 (±6.0)	±1.7	±2.0	±3.0	±3.0
Bandwidth (Hz)	0–2000	0–2500	0–200	0–1500	0–100
Sensitivity (mV g^{-1})	660 (220)	595	1000	1200	700
Noise µg/$\sqrt{Hz}$	30	170	18	0.5	110
Temperature range (°C)	−40 to +85	−40 to +125	−55 to +125	−40 to +125	−54 to +121
Input voltage (V dc)	2.4–5.5	3–6	2.5–5.5	6–15	5–30
Power consumption (mW)	2.8	1.7	<2	>60	>50
Package size (mm)	5 × 5 × 1.5	5 × 5 × 2	14.2 × 14.2 × 3.8	24.4 × 24.4 × 16.6	21.6 × 21.6 × 11.4[a]
Weight (g)	0.08	<1	1.64	—	77.8[a]

[a] Rugged titanium housing.

qualities enable the deployment of SHM systems with a large number of small sensors, partly integrated into the structure, at affordable costs. When deploying a large number of sensors, the cabling of all sensors with data logging units becomes so labor and cost intensive that it cancels all the advantages of applying MEMS sensors. Therefore, to overcome the limitations of cabling, MEMS sensors are often used in combination with wireless communication technologies. In this area, wireless sensor networks (WSNs) are an emerging technology that heavily bases its sensing capability on small, low-power, and cheap sensors. A WSN is a network of many small intercommunicating computers that are equipped with one or several sensors (*see* **Chapter 69**). Since WSNs rely completely on batteries, the application of low-power sensors is a key requirement, and minimizing power consumption is fundamental for extending the operation lifetime of the network. WSNs are being investigated for use in a variety of military, environmental, home, health, and SHM applications. SHM systems based on MEMS devices have been developed and tested in aerospace [4] and automotive [5] engineering. A review of WSNs for SHM is found in [6] and in-depth information about architectures and protocols for WSNs in [7].

3 MEMS-BASED STRUCTURAL HEALTH MONITORING SYSTEM

An application that illustrates very well the potentiality of an MEMS-based WSN is cable tension force monitoring of stay cable bridges. Cable stay forces can be monitored by means of natural frequency estimations based on vibration measurements. By using an appropriate cable model, the relation between the natural frequencies and the tensile cable force can be described.

3.1 Sensor node hardware

A typical network node is composed of one or more sensors, a signal conditioning unit, an analog to digital converter, a data processing unit with memory, a radio transceiver, and a power supply. These components are integrated into an enclosure, which protects the hardware components from mechanical, chemical, and environmental impacts. Figure 1 shows the open rugged and waterproof enclosure, which is designed for applications in harsh conditions. The enclosure contains the hardware components and the power supply, which consists of two 1.5-V batteries with 16 500 mA-h each. External status light emitting diodes (LEDs), switches and connectors allow supervising, interacting, and connecting external sensors, which have to be mounted directly to the structure (e.g., strain gauges).

Many hardware platforms are commercially available that are optimized in terms of power consumption. The prototype network presented in this article is based on the Tmote Sky platform [8]. It features a 6-channel ADC with a resolution of 12 bit, a 16-bit processor with 10-kB RAM and 48-kB program flash memory, and a radio transceiver operating in the 2.4-GHz ISM band with a raw data rate of 250 kbps.

The signal conditioning unit enables to interface various sensing elements. A Sensirion humidity sensor SHT11 [9] is mounted into an opening in the enclosure and allows for temperature and humidity measurements outside of the box. For vibration measurements, the MEMS accelerometer LSI2L06 of ST Microelectronics [10] has been applied because of its good noise performance, low-power consumption, and low costs. The accelerometer and the signal conditioning circuitry, consisting of an amplifier and a low-pass filter, are mounted on a dedicated board (Figure 2).

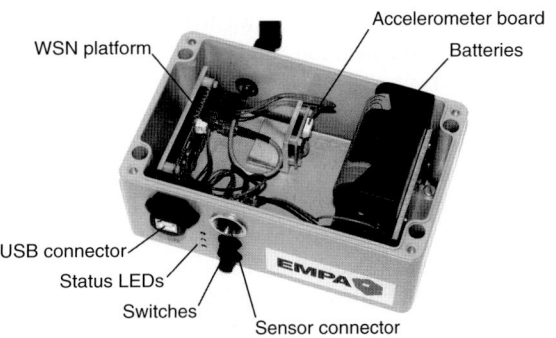

Figure 1. View of a physical sensor node with rugged and waterproof enclosure.

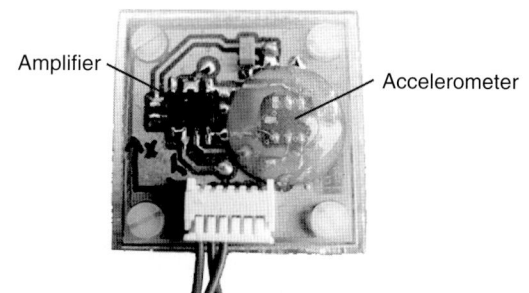

Figure 2. MEMS accelerometer board with amplification and filtering circuitry.

3.2 Sensor node software

The software running on each sensor node sets up the communication links between the sensor nodes, organizes the network topology, synchronizes the nodes, acquires measurements, and performs the data processing. The software is implemented as TinyOS components [11]. TinyOS is a component-based software framework designed for sensor networks and tailored to fit the memory constraints of the sensor nodes. It provides a concurrency model and mechanisms for structuring, naming, and linking software components to form a robust network embedded system.

The basic network functionality is provided by low-level network management components that operate independently from the actual monitoring applications. They are responsible for establishing the wireless links between adjacent nodes, for building the routing tree and for network-wide time synchronization. The monitoring application, which is built on top of these modules, uses this functionality to send and receive data and to have access to global time information. This allows for flexible exchange of the communication and time synchronization components.

A scheduler component forms the core of the actual monitoring application. It manages the data acquisition performed by the sensor node. Its clock is synchronized to the global time. The scheduler configures the measurement and data processing parameters like sampling rates, filter coefficients, thresholds, etc.; it triggers the data acquisition at the scheduled time. In addition to temperature, humidity, and acceleration measurements, information about the internal state of each sensor node (battery voltage) as well as communication parameters of the sensor network (e.g., routing tree) are monitored.

3.3 Data processing

The limited energy resources on each sensor node present the most restricting factor in designing and implementing WSN-based SHM systems. In terms of power consumption, wireless data transmission is much more expensive than data processing. In order to extend the system lifetime, it is therefore preferable to process the raw sensor readings in each sensor node with the aim to significantly reduce the data items that need to be transmitted to the data sink. This strategy is particularly recommended when monitoring vibration-based processes, which produce large amounts of raw data. There are several methods to reduce the size of raw data:

- Data compression encodes the data in a new representation that uses fewer bits than the original, not encoded data. This data reduction is done by using specific encoding schemes, which can either be lossless or lossy [12, 13].
- Data transformation transforms the raw data into a new kind of information that requires less space in terms of bits. Examples of simple data reductions can be maxima, minima, mean values, rms (root mean square), or statistical probability distributions of a physical quantity.
- Data analysis on the sensor node level reduces the amount of data transferred to the network. It differs from the methods described above because the raw data is subjected to an evaluation. The data is analyzed according to given criteria and a decision is taken if the data is relevant or not. Irrelevant data can already be discarded at the sensor node level.

Hence, long-term monitoring with WSN implies decentralized data processing and analysis. However, this is by far not possible for every analysis method. The limited energy resources restrict the complexity of the computational hardware of the sensor node, basically the memory size and computing speed of the hardware, and consequently affect the achievable analysis complexity.

In conventional monitoring systems, natural frequencies can be determined by identifying the peaks in a frequency spectrum that are computed via averaged spectrogram based on fast Fourier transform (FFT). An efficient in-place FFT computation of 1024 data samples of 16-bit length requires approximately 2 kB. This is a quite large amount of memory usage for low-power microcontrollers, with typical memory sizes of 2–10 kB, since the programs for data processing, task scheduling, time synchronization, and networking must be stored in the very same memory. Therefore, a much less memory demanding method for computing natural frequencies is needed.

Parametric methods of spectral analysis fit this requirement. The natural frequencies are estimated by computing the poles of a spectral model based on a rational function. If the natural frequencies are well separated in the frequency spectrum, a requirement that cable stays usually fulfill quite well, the vibration components associated to a vibration mode can be isolated by filtering the recorded data with a band pass filter. The use of a very simple two-parameter discrete time autoregressive (AR) model is then sufficient to estimate the natural frequency. An algorithm that enables a data reduction by a factor of 500 performs the following steps:

1. The analog signals of the accelerometer are digitalized and stored in a buffer.
2. The offset in the recorded data produced by the earth's gravity is removed by subtracting the average of the recorded data.
3. The recorded data is filtered with a band pass filter to isolate the frequency components close to one of the natural frequencies.
4. A data block is extracted from the filtered data. The size of the data block should contain at least one period of the natural frequency that will be estimated.
5. Using the data block, the parameters of the AR model are fitted. With these parameters, the natural frequency and the damping ratio of the AR model are estimated.
6. The quality of the natural frequency is tested using the estimated damping ratio, since a low damping ratio correlates with a nearly pure harmonic. If the damping ratio is greater than a given threshold, the estimated natural frequency is rejected and step 7 is skipped.
7. The natural frequency estimations that passed the quality test are stored in an array.
8. A new data block is extracted from the filtered data and the steps 5, 6, and 7 are repeated until all data blocks have been processed.
9. The mean value of the natural frequency estimations stored in the array is computed. This represents the estimated natural frequency that is transmitted to the network.

A detailed description of the implemented algorithm can be found in Feltrin *et al.* [14].

4 FIELD TESTS WITH THE MEMS-BASED SHM SYSTEM

The field tests were performed on the Stork Bridge, a two-span cable stayed road bridge with a total length of 124 m. The monitoring was performed on 6 of the 24 cables. Before deploying the monitoring system, a preliminary investigation with standard data-acquisition equipment was performed with the goal to determine the natural frequencies of the six cables and identify the vibration modes with the highest vibration level. This information was used to select the natural frequencies to be tracked by the WSN and to design the band pass filters.

4.1 Overall SHM system

The logical structure of the WSN monitoring system that has been deployed on the bridge is displayed in Figure 3. It is composed of three subsystems. The first subsystem is the WSN that consists of seven sensor nodes: six sensor nodes mounted on the cables, labeled as C21–C26, and the root node, labeled as C0, which is situated under the bridge deck at the abutment (Figure 4). The root node is connected via USB to the base station, which was placed inside the abutment. The base station is powered via the mains supply. The second subsystem is the remote control center that collects all data generated by the WSN and is responsible for the long-term storage of the data. It implements the data visualization and representation tools. Furthermore, this subsystem provides an interface to the operator to observe, control, and configure the WSN remotely. This subsystem was located at the EMPA site in Duebendorf, at a distance of 16 km from the Stork Bridge. The third subsystem forms

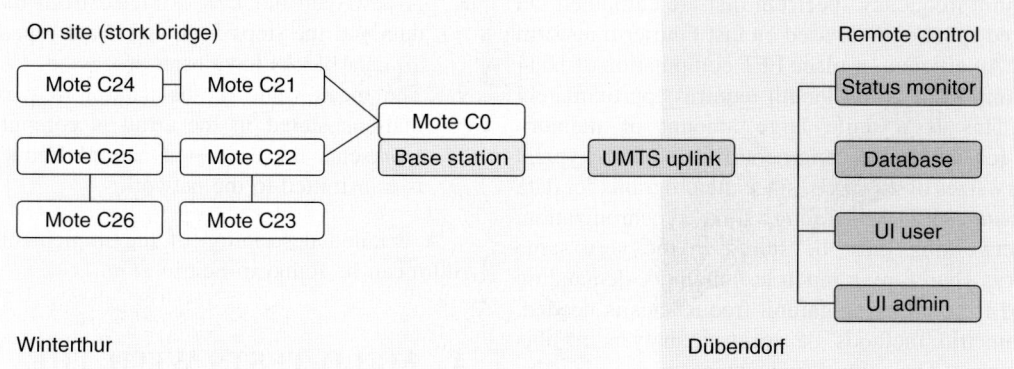

Figure 3. Logical structure of the structural health monitoring system.

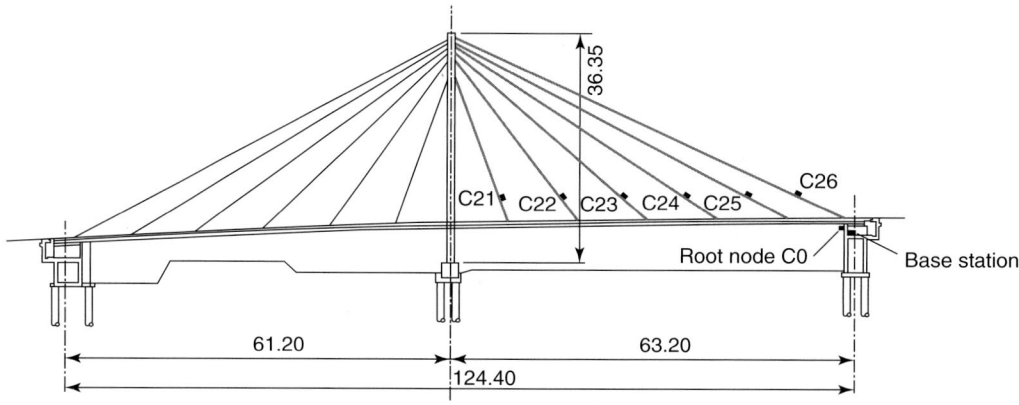

Figure 4. Elevation of the Stork Bridge with structural health monitoring setup.

the communication link between the WSN deployed on the bridge and the remote control center. This link is established via the base station using a standard wireless Universal Mobile Telecommunications System (UMTS) connection. A view to a network node mounted on a cable and to the root node below the bridge is shown in Figure 5.

The most challenging issue was to achieve overall system stability. A major source of instability was data processing. The computation of natural frequencies in one shot occupied the CPU for a long period and spoiled the execution of processes that guarantees the basic network functionality producing frequent system break downs. To overcome this problem, the algorithm was split up into tiny threaded code sections that required limited CPU time and that permitted an execution of basic network processes between two sequential threaded code sections. Furthermore, the integration of data acquisition, data processing, time synchronization, process scheduling, etc. into one software system turned out to be very sensitive to many tiny details regulating their interrelations. A change of duty cycle, for example, could destabilize the system producing system breakdowns within a short time. The modest CPU and RAM resources of the Tmote Sky platform significantly accentuated these problems.

5 RESULTS

Figure 6 displays a typical time history of the accelerations that were captured on the longest cable (C26) with the MEMS accelerometer LSI2L06 at a sampling rate of 100 Hz. The magnitude of the ambient cable vibration is very low. With respect to the 12-bit AD

Figure 5. Views of a sensor node mounted on a cable and the root node in the bridge abutment below the bridge deck.

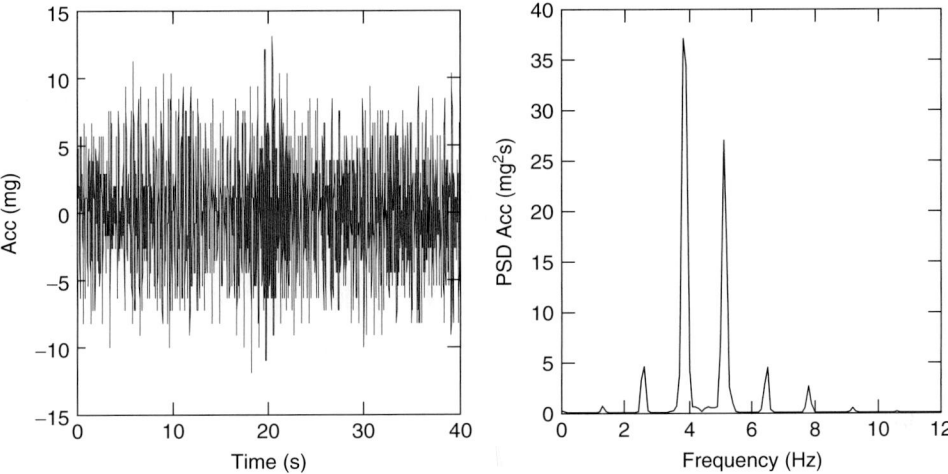

Figure 6. Time history of the accelerations sensed with the MEMS accelerometer LSI2L06 and its power density spectrum.

converter, which maps 1 mg approximately to 1 bit, the maximum of 13 mg is equivalent to 4 bits. Nevertheless, the natural frequencies can still be extracted from the time history as is demonstrated by the power density spectrum shown in Figure 6.

Figure 7 displays the natural frequencies of the cables C24, C25, and C26 of the Stork Bridge during a period of 60 days. The natural frequencies were estimated every minute from ambient vibration data using the algorithm described in this article. The typical rms magnitude of the ambient vibration data was 4–20 mg. The computation of the natural frequency lasts approximately 8 s. The algorithm was implemented in a series of threaded code sections to enable concurrent processes (e.g., time synchronization) to access the CPU. The three bands displayed in Figure 7; demonstrate that the algorithm generates estimations with a significant scattering. The accuracy of individual frequency estimations is within 5–10%, which is a direct consequence of the low level of accelerations and the short data blocks used for estimating the natural frequency (blocks of 50 samples, which correspond to 2–2.5 cycles).

A more accurate estimation of the natural frequency is obtained by using a moving average filter with a span of 200 samples (black curves inside the bands). Relatively small variations of natural frequencies are still detectable. This data processing step was done at the off-site control center with data retrieved from the data base. For monitoring

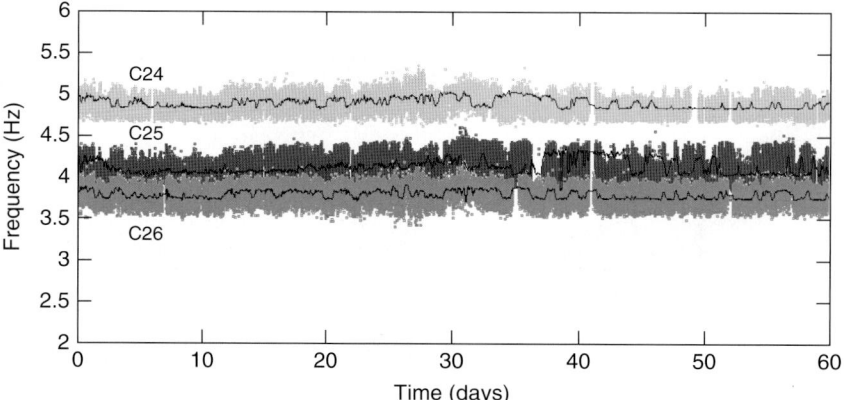

Figure 7. Time history of the natural frequencies of the cables C24, C25, and C26.

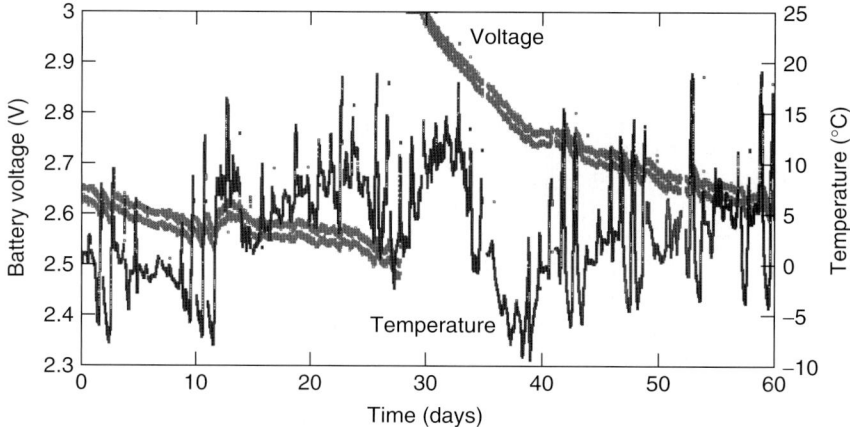

Figure 8. Time history of the battery voltage of the sensor node and the temperature measured on the cable C26.

of cable tension, the accuracy is good enough, since only significant changes are of concern for ensuring structural safety of a bridge.

Figure 8 shows the decay of battery voltage and the temperature on sensor node C26 over a 60-day period. It clearly depicts the dependency of battery capacity on temperature. The voltage graph consists of two lines. This effect is due to the fact that the battery voltage drops about 100 mV when the radio chip is turned on. Since voltage measurements are not synchronized to this switching, some measurements are taken when the radio is on and some when it is off. The voltage drop within 30 days is approximately 0.2 V. The theoretical lifetime of the WSN is approximately three months. This lifetime can be easily extended by a factor of 2 or 3 by extending the time between natural frequency estimations or by decreasing the duty cycle (the ratio of the system on time in a given period of time to the period of time), which was 40% during this test period. The voltage jump at day 29 is due to the replacement of batteries.

The graphs shown in Figures 7 and 8 reveal data losses during some periods of time. The causes of these losses are manifold: data from the sensor nodes is lost during the transport to the base station, stability issues in the communication software on the sensor nodes which lead to communication link breakdown, bugs in the software that render the base station irresponsive and block the reception of the data packets from the sensor nodes, and UMTS link breakdown caused by the telecommunication provider. The tests

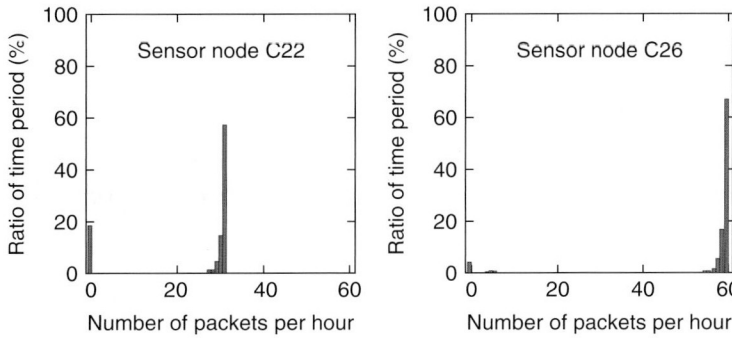

Figure 9. Frequency distribution of the number of valid data packets acquired within an hour of the cables C26 and C22.

demonstrate that data loss is intrinsically linked to the WSN since a lossless communication protocol would be too energy-consuming for field applications. Figure 9 displays the frequency distribution of the number of valid data packets within an hour of the cables C22 and C26 that were recorded at the remote control center during the 60-day period. For cable C26, which is closest to the root node, most of the time (67%), 60 packets out of 60 arrived at the control center. In 5.1% of the time no data was recorded. The same figures of the cable C22, which is farther away from the root node, are 57% with 30 packets out of 30 and 18.7% of the time with no data records. For the nodes C22, the natural frequency estimations occurred every 2 min. The mean data packet arrival rate over the 60 days was 24 out of 30 for C22 and 55 out of 60 for C26. As expected, the data loss increases with increasing distance to the base station. Moreover, the frequency distributions demonstrate that the wireless network is either up with a high reception rate (greater than 90%) or totally down. The case of an intermediate reception rate occurs very infrequently.

6 CONCLUSIONS

The field test on the Stork Bridge demonstrates that long-term monitoring with a low-power WSN is feasible. The feasibility is mainly based on the application of low-cost and low-power MEMS sensing technology and on a significant reduction of raw data that is achieved by decentralized data processing. The test demonstrates that the current technology does not provide the same reliability of mature wired monitoring systems in terms of stability, accuracy, and data loss. This outcome is not surprising since research and application of WSN technology in monitoring is still at a very early stage. The most critical issue is the handling of data processing and basic network functionality on a single CPU with very limited computational and memory resources. However, in the near future, these problems will be less critical since WSN platforms will be available that significantly provide more resources. Furthermore, to avoid conflicts between data processing and basic network functionality, the two tasks can be allocated to two separate low-power CPUs. Energy efficiency of hardware and software components will continue to play a central role. However, the progress in low-power microelectronics, which is driven by the huge market of portable electronic devices, and the application of energy harvesting technologies will mitigate the current limitations regarding power consumption. Nevertheless, the experience on the Stork Bridge demonstrates that with a well-balanced resource distribution between data processing and basic network functionality, a stable system can be achieved with existing low-power WSN platforms that provides useful information with an accuracy that is compliant to the monitoring objectives.

RELATED ARTICLES

Chapter 76: On the Way to Autonomy: the Wireless-interrogated and Self-powered "Smart Patch" System

Chapter 77: Energy Harvesting using Thermoelectric Materials

REFERENCES

[1] Gardner JW, Varadan VK, Awadelkarim OO. *Microsensors, Mems, and Smart Devices*. John Wiley & Sons: Chichester, 2001.

[2] Franssila S. *Introduction to Microfabrication*. John Wiley & Sons: Chichester, 2004.

[3] Varadan VK, Vinoy KJ, Gopalakrishnan S. *Smart Material Systems and Mems: Design and Development Methodologies*. John Wiley & Sons: Chichester, 2006.

[4] Osiander R, Darrin MAG, Champion JL (eds). *MEMS and Microstructures in Aerospace Applications*. Taylor & Francis: Boca Raton, FL, 2006.

[5] Valldorf J, Gessner W (eds). Advanced microsystems automotive applications. In *International Forum on Advanced Microsystems for Automotive Applications (AMAA)*. Springer: Berlin, 2007.

[6] Lynch JP, Loh K. A summary review of wireless sensors and sensor networks for structural health monitoring. *Shock and Vibration Digest* 2005 **38**(2):91–128.

[7] Karl H, Willig A. *Protocols and Architectures for Wireless Sensor Networks*. John Wiley & Sons: Chichester, 2005.

[8] Polastre J, Szewczyk R, Culler D. Telos: enabling ultra-low power research. *Proceedings of the Information Processing in Sensor Networks/SPOTS*. Berkeley, CA, April 2005.

[9] SHT1x/SHT7x Humidity & Temperature Sensor, http://www.sensirion.com/en/pdf/product_information/Data_Sheet_humidity_sensor_SHT1x_SHT7x_E.pdf, 2007.

[10] ST LIS2L06AL, MEMS inertial sensor, http://www.st.com/stonline/products/literature/ds/11665/lis2l06al.pdf, 2006.

[11] Levis P, *et al*. Tinyos: an operating system for wireless sensor networks. In *Ambient Intelligence*, Weber W, Rabaey JM, Aarts E (eds). Springer: New York, 2005, pp. 115–148.

[12] Lynch JP, Sundararajan A, Law KH, Kiremidjian AS, Carryer E. Power-efficient data management for a wireless structural monitoring system. *Proceedings of the 4th International Workshop on Structural Health Monitoring*. Stanford, CA, 15–17 September 2003; pp. 1177–1184.

[13] Caffrey J, *et al*. Networked sensing for structural health monitoring. *Proceedings of the 4th International Workshop on Structural Control*. New York, 10–11 June 2004; pp. 57–66.

[14] Feltrin G, Meyer J, Bischoff R, Saukh O. A wireless sensor network for force monitoring of cable stays. *Proceedings of the 3rd International Conference on Bridge Maintenance, Safety and Management, IABMAS 06*. Porto, 16–19 July 2006, on CD.